Finite Mathematics with Applications

In the Management, Natural, and Social Sciences

Ninth Edition

Finite Mathematics with Applications

In the Management, Natural, and Social Sciences

Ninth Edition

Margaret L. Lial
American River College

Thomas W. Hungerford
Saint Louis University

John P. Holcomb, Jr.
Cleveland State University

PEARSON

Addison
Wesley

Boston   San Francisco   New York
London   Toronto   Sydney   Tokyo   Singapore   Madrid
Mexico City   Munich   Paris   Cape Town   Hong Kong   Montreal

Publisher: Greg Tobin
Sponsoring Editor: Carter Fenton
Project Editor: Katie Nopper
Editorial Assistant: Rachel Monaghan
Senior Production Supervisor: Jeffrey Holcomb
Senior Marketing Manager: Becky Anderson
Marketing Assistant: Maureen McLaughlin
Media Producers: Michelle Murray, Christine Stavrou
Software Editor/MathXL: Janet McHugh
Software Editor/TestGen: Mary Durnwald
Senior Manufacturing Buyer: Evelyn Beaton
Senior Designer: Barbara T. Atkinson
Text Design: WestWords, Inc.
Cover Design: Suzanne Heiser/Night & Day Design
Production Coordination and Composition: WestWords, Inc.
Illustrations: Techsetters, Inc.
Cover Photo: Modern Architecture © PNC/Getty Images

Library of Congress Cataloging-in-Publication Data
Lial, Margaret L.
    Finite mathematics with applications: in the management, natural, and social sciences.—9th ed.
    Margaret L. Lial, Thomas W. Hungerford, John P. Holcomb.
        p. cm.
    ISBN 0-321-38672-8 (alk. paper)—ISBN 0-321-38962-X (alk. paper)
    1. Mathematics—Textbooks. 2. Social sciences—Mathematics—Textbooks. I.
Hungerford, Thomas W. II. Holcomb, John P. III. Title.

QA37.3.L55 2006
510—dc22   2005050720

3 4 5 6 7 8 9 10—RRDW—0807

Contents

Preface

Like its predecessors, this edition of *Finite Mathematics with Applications* is an applications-oriented text for students in business, management, social sciences, and health sciences. The only prerequisite is a course in algebra. Chapter 1 provides a thorough review of basic algebra for those students who need it.

The book is written at a level appropriate for its intended audience. We have done our best to present sound mathematics in an understandable manner, proceeding from the familiar to new material and from concrete examples to general rules and formulas. There is an ongoing focus on real-world problem solving, and almost every section includes relevant contemporary applications.

Key Organizational and Content Changes

Although the table of contents for this edition is substantially the same as that of the previous edition, there are some significant changes:

- Each chapter begins with a brief overview of applications of the mathematics treated in that chapter, together with references to appropriate exercises.
- The sequence of topics in Chapter 9 has been reorganized so that random variables, probability distributions, and expected value are introduced right away and used throughout the chapter.
- There has been a significant increase in real-data examples and exercises in Chapters 8, 9, and 10.
- More than 30% of the exercises in the book are new or revised.

Continuing Pedagogical Features

- *Balanced Approach* Multiple representations of a topic (symbolic, numerical, graphical, verbal) are given when appropriate. However, we do not believe that all representations are useful for all topics, so we are pedagogically selective in presenting alternative methods. Effective alternatives are discussed only when they are likely to increase student understanding.
- *Real-Data Examples and Explanations* Real-data exercises have long been a popular feature of this book. For this edition, most of these exercises have been updated and some new ones added. A significant number of real-data examples have also been introduced into the text.
- *Margin Exercises,* keyed to the discussion in the text, provide students with an opportunity to practice immediately after learning a new skill.

■ *Cautions* highlight common student difficulties or warn against frequently made mistakes.

■ *Technology Tips* appear within the text as appropriate to inform students of various features of their graphing calculators or spreadsheet programs and to guide them in using these tools.

■ In addition to the usual drill, conceptual, application-based exercises, and questions from past CPA exams, there are some specially marked exercises:

Writing Exercises ✎

Connection Exercises ⟳ that relate current topics to earlier sections; and

exercises that require technology ⟍.

■ *End-of-Chapter* materials include a summary of key terms and symbols and key concepts.

■ *Case Studies* appear at the end of each chapter and offer contemporary real-world applications of some of the mathematics presented in the chapter, providing at least a partial answer to the question, "What's this stuff good for?"

■ The pedagogical use of color makes this book's art program more effective in enhancing the mathematical exposition and clarifying many graphical presentations.

Technology

It is assumed that all students have a calculator that will handle exponential and logarithmic functions. Beyond that, however,

the use of technology in this text is optional.

Examples and exercises that definitely require some sort of technology (graphing calculators, spreadsheets, or other computer programs) are marked with the icon ⟍, so instructors who want to omit these discussions and exercises can easily do so. They should note, however, that some examples that do *not* require technology may include some mention of it or a graphing calculator screen (for instance, one showing the matrix associated with a system of linear equations).

Instructors who routinely use technology in their courses—and expect their students to do the same—will find more than enough material here to satisfy their needs. Here are some of the features they may want to incorporate into their courses:

■ *Examples and Exercises marked with* ⟍ A number of examples show students how various features of graphing calculators and spreadsheets can be applied to the topics in this book. These topics include, but are not limited to, graphing functions, solving equations, row-reducing matrices, finding linear regression models for data, statistical analysis, and approximations of definite integrals. There are also many exercises marked for technology use.

■ *Technology Tips* These are placed at appropriate points in the text to inform students of various features of their graphing calculator, spreadsheet, or other computer programs and to guide them in using those tools. Please note that

Technology Tips for:	Also apply to:
TI-84+	TI-83+ and usually TI-83
HP-39+	HP-39 and usually HP-38
Casio 9850	Casio 9750 and 9860

■ *Graphing Calculator Appendix* Part 1 is a brief introduction to the features of graphing calculators that are relevant to topics in the text. Part 2 contains programs for carrying out various procedures discussed in the text (including financial functions, simplex and two-stage methods for linear programming, and a RREF program for Casio).

An outline of the Graphing Calculator Appendix (including the complete list of programs) is on page 623. In order to facilitate easy updating of this material, the full appendix is available on www.aw-bc.com/MWA9. Printed copies of the appendix can be made available; contact your local Addison-Wesley sales representative.

Course Flexibility

This book can be used for a variety of courses, including the following:

Finite Mathematics (one semester or two quarters). Use as much of Chapters 1–4 as needed, and then go into Chapters 5–10 as time permits and local needs require.

College Algebra with Applications (one semester or quarter). Use Chapters 1–8, with Chapters 7 and 8 being optional.

Chapter interdependence is as follows:

	Chapter	*Prerequisite*
1	Algebra and Equations	None
2	Graphs, Lines, and Inequalities	Chapter 1
3	Functions and Graphs	Chapters 1 and 2
4	Exponential and Logarithmic Functions	Chapter 3
5	Mathematics of Finance	Chapter 4
6	Systems of Linear Equations and Matrices	Chapters 1 and 2
7	Linear Programming	Chapters 3 and 6
8	Sets and Probability	None
9	Counting, Probability Distributions, and Further Topics in Probability	Chapter 8
10	Introduction to Statistics	Chapter 8

Contact your local Addison-Wesley sales representative to order a customized version of this text.

Supplements

For the Instructor

Instructor's Edition (ISBN 0-321-38962-X)
This book contains answers to all exercises in the text.

Instructor.'s Solutions Manual (ISBN 0-321-33597-X)
This manual contains detailed solutions to all text problems, suggested course outlines, and a chapter interdependence chart.

Printed TestBank (ISBN 0-321-39668-5)
This test bank includes four tests per chapter, paralleling the text's Chapter Tests.

TestGen with QuizMaster (ISBN 0-321-33598-8)
TestGen enables instructors to build, edit, print, and administer tests by using a computerized bank of questions developed to cover all the objectives of the text. TestGen is algorithmically based, allowing instructors to create multiple, but equivalent, versions of the same question or test with the click of a button. Instructors can also modify test bank questions or add new questions by using the built-in question editor, which allows users to create graphs, import graphics, and insert math notation, variable numbers, or text. Tests can be printed or administered online via the Internet or another network. TestGen comes packaged with QuizMaster, which allows students to take tests on a local area network. The software is available on a dual-platform Windows/Macintosh CD-ROM.

MyMathLab
MyMathLab is a series of text-specific, easily customizable online courses for Addison-Wesley textbooks in mathematics and statistics. MyMathLab is powered by CourseCompass™—Pearson Education's online teaching and learning environment—and by MathXL®—our online homework, tutorial, and assessment system. MyMathLab gives you the tools you need to deliver all or a portion of your course online, whether your students are in a lab setting or working from home. MyMathLab provides a rich and flexible set of course materials, featuring free-response exercises that are algorithmically generated for unlimited practice and mastery. Students can independently use online tools, such as video lectures, animations, and a multimedia textbook, to improve their understanding and performance. Instructors can use MyMathLab's homework and test managers to select and assign online exercises correlated directly with the textbook, and they can also create and assign their own online exercises and import TestGen tests for added flexibility. MyMathLab's online gradebook—designed specifically for mathematics and statistics—automatically tracks students' homework and test results and gives the instructor control over how to calculate final grades. Instructors can also add offline (paper-and-pencil) grades to the gradebook. MyMathLab is available to qualified adopters. For more information, visit our website at www.mymathlab.com or contact your Addison-Wesley sales representative.

For the Student

Student's Solutions Manual (ISBN 0-321-33595-3)
This manual contains detailed, carefully worked-out solutions to all odd-numbered section exercises and all Chapter Review and Case Study exercises.

Digital Video Tutor (ISBN 0-321-40963-9)
The video lectures for this text are also available on CD-ROM, making it easy and convenient for students to watch the videos from a computer at home or on campus. The complete digitized video set, affordable and portable for students, is ideal for distance learning or supplemental instruction.

InterAct Math Tutorial Website: www.interactmath.com
Get practice and tutorial help online! This interactive tutorial website provides algorithmically generated practice exercises that correlate directly with the exercises in the textbook. Students can retry an exercise as many times as they like with new values each time for unlimited practice and mastery. Every exercise is accompanied by an interactive guided solution that provides helpful feedback for incorrect answers, and student can also view a worked-out sample problem that steps them through an exercise similar to the one they're working on.

MathXL®
MathXL® is a powerful online homework, tutorial, and assessment system that accompanies your Addison-Wesley textbook in mathematics or statistics. With MathXL, instructors can create, edit, and assign online homework and tests using algorithmically generated exercises correlated with your textbook at the objective level. All student work is tracked in MathXL's online gradebook. Students can take chapter tests in MathXL and receive personalized study plans based on their test results. The study plan diagnoses weaknesses and links students directly to tutorial exercises for the objectives they need to study and retest. For more information, visit our website at www.mathxl.com.

Addison-Wesley Math Tutor Center
The Addison-Wesley Math Tutor Center is staffed by qualified mathematics instructors who provide students with tutoring on examples and exercises answered at the back of the textbook. Tutoring is available via toll-free telephone, fax, e-mail, or whiteboard technology—which allows tutors and students to actually see the problems worked while they "talk" in real time over the Internet. This service is available five days a week, seven hours a day. For more information, go to www.aw.com/tutorcenter. An access card is required.

Acknowledgments

The authors wish to thank the following reviewers for their helpful comments and suggestions for this and previous editions of the text. (Reviewers of the ninth edition are noted with an asterisk.)

Erol Barbut, University of Idaho

Bob Beul, St. Louis University–Metropolitan College

Richard Bieberich, Ball State University

Chris Boldt, Eastfield College

Michael J. Bradley, Merrimack College

James F. Brown, Midland College

James E. Carpenter, Iona College

Jesus Carreon, Mesa Community College

Faith Y. Chao, Golden Gate University

Jan S. Collins, Embry-Riddle University

Jerry Currence, University of South Carolina–Lancaster*

Juli D'Ann Ratheal, Western Texas A & M University*

Frederick Davidson, Old Dominion University

Jean Davis, Southwest Texas State University

Duane E. Deal, Ball State University

Richard D. Derderian, Providence College

Carol E. DeVille, Louisiana Tech University

Wayne Ehler, Anne Arundel Community College

George A. Emerson, National University

Garret Etgen, University of Houston

George Evanovich, Iona College

Richard Fast, Mesa Community College

Gordon Feathers, Iona College

J. Franklin Fitzgerald, Boston University

Leland J. Fry, Kirkwood Community College

Dauhrice K. Gibson, Gulf Coast Community College

Robert E. Goad, Sam Houston State University

Richard E. Goodrick, University of Washington

Kim Gregor, Delaware Technical and Community College

Kay Gura, Ramapo College of New Jersey

Joseph A. Guthrie, University of Texas at El Paso

Patricia Hirschy, Delaware Technical and Community College

Arthur M. Hobbs, Texas A & M University

Irene Hollman, Southwestern College*

Miles Hubbard, St. Cloud State University

Katherine J. Huppler, St. Cloud State University

Carol M. Hurwitz, Manhattan College

Donald R. Ignatz, Lorain County Community College

Alec Ingraham, New Hampshire College

Robert H. Johnston, Virginia Commonwealth University

June Jones, Macon College

Paul Kaczur, Phoenix College

Michael J. Kallaher, Washington State University

Akihiro Kanamori, Boston University

Terence J. Keegan, Providence College

Hubert C. Kennedy, Providence College

Surinder Sehgal, Ohio State University

Gordon Shilling, University of Texas at Arlington

Calvin Shipley, Henderson State University

Pradeep Shukla, Suffolk University

James L. Southam, San Francsico State University

John Spellman, Southwest Texas State University

Joan M. Spetich, Baldwin-Wallace College

William D. Stark, Navarro College

Jo Steig, Phoenix College

David Stoneman, University of Wisconsin–Whitewater

David P. Sumner, University of South Carolina

Daniel F. Symancyk, Anne Arundel Community College

Giovanni Viglino, Ramapo College of New Jersey

Deborah A. Vrooman, Coastal Carolina College

H. J. Wellenzohn, Niagara University

Stephen H. West, Coastal Carolina College

Thelma West, University of Louisiana at Lafayette

Richard J. Wilders, North Central College

John L. Wisthoff, Anne Arundel Community College

Hing-Sing Yu, University of Texas at San Antonio

Cathy Zucco-Teveloff, Trinity College

We also wish to thank our accuracy checkers, who did an excellent job of checking both text and exercise answers: Jerry Currence, Douglas Ewert, Raja Khoury, Shannon Michaux, Ian C. Walters, Jr., Jon Booze, and John Morin. Thanks also to the supplements authors: Charles Odion, David Bridge, Victoria Smith, and Stela Pudar-Hozo. Special thanks to Lucie Haskins, who created an accurate and complete index for us; and to Becky Troutman, who carefully compiled the Index of Applications.

We want to thank the staff of Addison-Wesley for their assistance with, and contributions to, this book, particularly Greg Tobin, Carter Fenton, Katie Nopper, Rachel Monaghan, Jeff Holcomb, Becky Anderson, Maureen McLaughlin, Christine Stavrou, and Michelle Murray. Finally, we wish to express our appreciation to Pat McCutcheon of WestWords, Inc., who was a pleasure to work with.

Margaret L. Lial
Thomas W. Hungerford
John P. Holcomb, Jr.

Raja Khoury, Collin County Community College*
Clint Kolaski, University of Texas at San Antonio
Archille J. Laferriere, Boston College
Steve Laroe, University of Alaska–Fairbanks
Jeffrey Lee, Texas Tech University
Arthur M. Lieberman, Cleveland State University
Norman Lindquist, Western Washington University
Laurence P. Maher, Jr., North Texas State University
Norman R. Martin, Northern Arizona University
Donald Mason, Elmhurst College
James Mazzarella, Holy Family College
Walter S. McVoy, Illinois State University
C. G. Mendez, Metropolitan State College of Denver
Shannon Michaux, University of Colorado at Color
W. W. Mitchell, Jr., Phoenix College
Robert A. Moreland, Texas Tech University
Ruth M. Murray, College of DuPage
Kandasamy Muthuvel, University of Misconsin–Os
Carol Nessmith, Georgia Southern College
Peter Nicholls, Northern Illinois University
Ann O'Connell, Providence College
Kathy O'Dell, University of Alabama at Huntsville
Charles Odion, Houston Community College
Thomas J. Ordoyne, University of South Carolina a
Marian Paysinger, University of Texas at Arlington
Julienne K. Pendleton, Brookhaven College
Sandra Peskin, Queensborough Community Colleg
S. Pierce, West Coast University
John M. Plachy, Metropolitan State College
Elizabeth Polenzani, Pasadena City College
Donald G. Poulson, Mesa Community College
Wayne B. Powell, Oklahoma State University
Michael I. Ratliff, Northern Arizona University
Clark P. Rhoades, Loyola University at New Orlear
Sarah Sabinson, Queensborough Community Colle
Leon Sagan, Anne Arundel Community College
Subhash C. Saxena, University of South Carolina,
Harold Schachter, Queensborough Community Col
Steven A. Schonefeld, Tri-State University
Robert Seaver, Lorain County Community College

To the Student

Several features of the text are designed to assist you in understanding the concepts presented and learning the mathematical procedures involved. To help you learn new concepts and reinforce your understanding, there are numerous *side problems* in the margin. They are referred to in the text by numbers in colored diamonds, such as ◇2◇. When you see that symbol, you should work the indicated problem in the margin before going on.

Technology (such as graphing calculators, spreadsheets, or other computer programs) is not required to use this book. However, there are a number of optional examples and exercises that use graphing calculators; they are marked with the icon ⟋. There is also a *Graphing Calculator Appendix* that covers the basics of calculator use and provides a number of helpful programs for dealing with some of the topics in this book. The appendix can be found on www.aw-bc.com/MWA9. In addition, there are *Technology Tips* throughout the text that describe the proper menu or keys to be used in order to carry out a particular procedure on a graphing calculator. Please note that

Technology Tips for:	Also apply to:
TI-84+	TI-83+ and usually TI-83
HP-39+	HP-39 and usually HP-38
Casio 9850	Casio 9750 and 9860

The key to succeeding in this course is to remember that

mathematics is not a spectator sport.

You can't expect to learn mathematics without *doing* mathematics any more than you could learn to swim without getting wet. You have to take an active role, making use of all the resources at your disposal: your instructor, your fellow students, and this book.

There is no way that your instructor can cover every aspect of a topic during class time. You simply won't develop the level of understanding you need to succeed unless you read the text carefully. In particular, you should read the text *before* starting the exercises. However, you can't read a math book the way you read a novel. You should have pencil, paper, and calculator handy to do the side problems, work out the statements you don't understand, and make notes on things to ask your fellow students and/or your instructor.

Finally, remember the words of the great Hillel: "The bashful do not learn." There is no such thing as a "dumb question" (assuming, of course, that you have read the book and your class notes and attempted the homework). Your instructor will welcome questions that arise from a serious effort on your part. So get your money's worth: Ask questions.

Finite Mathematics with Applications

In the Management, Natural, and Social Sciences

Ninth Edition

Chapter **1**

Algebra and Equations

Mathematics is widely used in business, finance, and the biological, social, and physical sciences, from developing efficient production schedules for a factory to mapping the human genome. Mathematics also plays a role in determining interest on a loan from a bank and the cancer risk from a pollutant, as well as in the study of atmospheric pressure. See Exercises 61–63 on page 55 and Exercise 60 on page 65.

Because algebra and equations are the basic mathematical tools for many applications, we begin by reviewing the fundamental ideas of algebra. Your success in this course will depend on knowing the algebraic skills presented here.

1.1 The Real Numbers

Only real numbers will be used in this book.* The names of the most common types of real numbers are as follows.

The Real Numbers

Natural (counting) numbers   1, 2, 3, 4, . . .

Whole numbers   0, 1, 2, 3, 4, . . .

Integers   . . . , $-3, -2, -1, 0, 1, 2, 3,$ . . .

Rational numbers   All numbers that can be written in the form p/q, where p and q are integers and $q \neq 0$

Irrational numbers   Real numbers that are not rational

The relationships among these types of numbers are shown in Figure 1.1, in which each set of numbers is contained in the set to its right. So, for example, integers are also rational numbers and real numbers, but not irrational numbers.

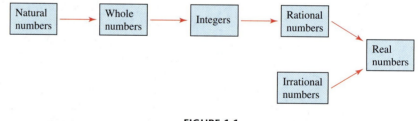

FIGURE 1.1

One example of an irrational number is π, the ratio of the circumference of a circle to its diameter. The number π can be approximated as $\pi \approx 3.14159$ ($\approx$ means "is approximately equal to"), but there is no rational number that is exactly equal to π.

▼ **EXAMPLE 1**   What kind of number is each of the following?

(a) 6

   Solution The number 6 is a natural number, a whole number, an integer, a rational number, and a real number.

(b) 3/4

   Solution This number is rational and real.

---

*Not all numbers are real numbers. An example of a number that is not a real number is $\sqrt{-1}$.

(c) 3π

Solution Because π is not a rational number, 3π is irrational and real. ▲ *

All real numbers can be written in decimal form. A rational number, when written in decimal form, is either a terminating decimal, such as .5 or .128, or a repeating decimal, in which some block of digits eventually repeats forever, such as 1.3333 . . . or 4.7234234234 . . . .† Irrational numbers are decimals that neither terminate nor repeat.

The only real numbers that can be entered exactly into a calculator are rational numbers that are terminating decimals of no more than 10 or 12 digits (depending on the calculator). Similarly, the answers produced by a calculator are often 10–12-digit decimal *approximations*—accurate enough for most applications. As a general rule, *you should not round off any numbers during a long calculator computation,* so that your final answer will be as accurate as possible. For convenience, however, we usually round off the final answer.

The important basic properties of the real numbers are as follows.

1

Name all the types of numbers that apply to the following.

(a) -2

(b) $-5/8$

(c) $\pi/5$

Answers:

(a) Integer, rational, real

(b) Rational, real

(c) Irrational, real

Properties of the Real Numbers

For all real numbers, a, b, and c, the following properties hold true:

Commutative properties $\quad a + b = b + a \qquad\qquad ab = ba$

Associative properties $\quad (a + b) + c = a + (b + c) \quad (ab)c = a(bc)$

Identity properties $\quad$ There exists a unique real number 0, called the **additive identity,** such that

$$a + 0 = a \quad \text{and} \quad 0 + a = a.$$

There exists a unique real number 1, called the **multiplicative identity,** such that

$$a \cdot 1 = a \quad \text{and} \quad 1 \cdot a = a.$$

Inverse properties $\quad$ For each real number a, there exists a unique real number $-a$, called the **additive inverse** of a, such that

$$a + (-a) = 0 \quad \text{and} \quad (-a) + a = 0.$$

If $a \neq 0$, there exists a unique real number $1/a$, called the **multiplicative inverse** of a, such that

$$a \cdot \frac{1}{a} = 1 \quad \text{and} \quad \frac{1}{a} \cdot a = 1.$$

Distributive property $\quad a(b + c) = ab + ac$

---

*The use of margin problems is explained in the "To the Student" section preceding this chapter.

†Some graphing calculators have a FRAC key that automatically converts some repeating decimals to fraction form. FRAC programs for other graphing calculators are in the Program Appendix.

⟨2⟩

Name the property illustrated in each of the following examples.

(a) $(2 + 3) + 9$
   $= (3 + 2) + 9$

(b) $(2 + 3) + 9$
   $= 2 + (3 + 9)$

(c) $(2 + 3) + 9$
   $= 9 + (2 + 3)$

(d) $(4 \cdot 6)p = (6 \cdot 4)p$

(e) $4(6p) = (4 \cdot 6)p$

Answers:

(a) Commutative property

(b) Associative property

(c) Commutative property

(d) Commutative property

(e) Associative property

The next five examples illustrate the properties listed in the preceding box.

▼ **EXAMPLE 2**   The commutative property says that the order in which you add or multiply two quantities doesn't matter.

(a) $(6 + x) + 9 = 9 + (x + 6)$ **(b)** $5 \cdot (9 \cdot 8) = (9 \cdot 8) \cdot 5$ ▲

▼ **EXAMPLE 3**   When the associative property is used, the order of the numbers does not change, but the placement of parentheses does.

(a) $4 + (9 + 8) = (4 + 9) + 8$ **(b)** $3(9x) = (3 \cdot 9)x$ ▲ ⟨2⟩

▼ **EXAMPLE 4**   By the identity properties,

(a) $-8 + 0 = -8$ **(b)** $(-9) \cdot 1 = -9.$ ▲

> **TECHNOLOGY TIP**   To enter -8 on a calculator, use the negation key (labeled $(-)$ or $+/-$), *not* the subtraction key. On most one-line scientific calculators, key in $8 +/-$. On graphing calculators or two-line scientific calculators, key in either $(-) 8$ or $+/- 8$.

▼ **EXAMPLE 5**   By the inverse properties, the statements in parts (a) through (d) are true.

(a) $9 + (-9) = 0$ **(b)** $-15 + 15 = 0$

(c) $-8 \cdot \left(\dfrac{1}{-8} \right) = 1$ **(d)** $\dfrac{1}{\sqrt{5}} \cdot \sqrt{5} = 1$ ▲

NOTE There is no real number x such that $0 \cdot x = 1$, so 0 has no multiplicative inverse. ⟨3⟩

⟨3⟩

Name the property illustrated in each of the following examples.

(a) $2 + 0 = 2$

(b) $-\dfrac{1}{4} \cdot (-4) = 1$

(c) $-\dfrac{1}{4} + \dfrac{1}{4} = 0$

(d) $1 \cdot \dfrac{2}{3} = \dfrac{2}{3}$

Answers:

(a) Additive identity property

(b) Multiplicative inverse property

(c) Additive inverse property

(d) Multiplicative identity property

▼ **EXAMPLE 6**   By the distributive property,

(a) $9(6 + 4) = 9 \cdot 6 + 9 \cdot 4$

(b) $3(x + y) = 3x + 3y$

(c) $-8(m + 2) = (-8)(m) + (-8)(2) = -8m - 16$

(d) $(5 + x)y = 5y + xy.$ ▲

NOTE As shown in Example 6(d), by the commutative property, the distributive property can also be written as $(a + b)c = ac + bc.$ ⟨4⟩

Order of Operations

We avoid ambiguity when working problems with real numbers by using the **order of operations** shown next, which has been agreed on as the most useful one. This order of operations is used by computers and calculators.

4

Use the distributive property to complete each of the following.

(a) $4(-2 + 5)$

(b) $2(a + b)$

(c) $-3(p + 1)$

(d) $(8 - k)m$

(e) $5x + 3x$

Answers:

(a) $4(-2) + 4(5) = 12$

(b) $2a + 2b$

(c) $-3p - 3$

(d) $8m - km$

(e) $(5 + 3)x = 8x$

5

Evaluate the following if $m = -5$ and $n = 8$.

(a) $-2mn - 2m^2$

(b) $\dfrac{4(n - 5)^2 - m}{m + n}$

Answers:

(a) 30

(b) $\dfrac{41}{3}$

6

Use a calculator to evaluate the following.

(a) $4^2 \div 8 + 3^2 \div 3$

(b) $[-7 + (-9)] \cdot (-4) - 8(3)$

(c) $\dfrac{-11 - (-12) - 4 \cdot 5}{4(-2) - (-6)(-5)}$

(d) $\dfrac{36 \div 4 \cdot 3 \div 9 + 1}{9 \div (-6) \cdot 8 - 4}$

Answers:

(a) 5

(b) 40

(c) $\dfrac{19}{38} = \dfrac{1}{2} = .5$

(d) $-\dfrac{1}{4} = -.25$

Order of Operations

If parentheses or square brackets are present,

1. Work separately above and below any fraction bar.
2. Use the rules below within each set of parentheses or square brackets, starting with the innermost set and working outward.

If no parentheses are present,

1. Find all powers and roots, working from left to right.
2. Do any multiplications or divisions in the order in which they occur, working from left to right.
3. Do any additions or subtractions in the order in which they occur, working from left to right.

▼ **EXAMPLE 7** Use the order of operations to evaluate each expression if $x = -2$, $y = 5$, and $z = -3$.

(a) $-4x^2 - 7y + 4z$

Solution Use parentheses when replacing letters with numbers.

$$-4x^2 - 7y + 4z = -4(-2)^2 - 7(5) + 4(-3)$$
$$= -4(4) - 7(5) + 4(-3) = -16 - 35 - 12 = -63$$

(b) $\dfrac{2(x - 5)^2 + 4y}{z + 4} = \dfrac{2(-2 - 5)^2 + 4(5)}{-3 + 4}$

$$= \dfrac{2(-7)^2 + 20}{1}$$

$$= 2(49) + 20 = 118 \quad ▲ \; ⟨5⟩$$

▼ **EXAMPLE 8** Use a calculator to evaluate

$$\dfrac{-9(-3) + (-5)}{3(-4) - 5(2)}.$$

Solution Use extra parentheses (shown here in blue) around the numerator and denominator when you enter the number in your calculator, and be careful to distinguish the negation key from the subtraction key.

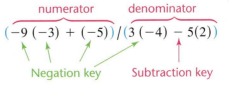

If you don't get -1 as the answer, then you are entering something incorrectly.

▲ ⟨6⟩

Square Roots

There are two numbers whose square is 16, namely, 4 and -4. The positive one, 4, is called the **square root** of 16. Similarly, the square root of a nonnegative number d is defined to be the *nonnegative* number whose square is d; this number is denoted $\sqrt{d}$. For instance,

$$\sqrt{36} = 6 \text{ because } 6^2 = 36, \quad \sqrt{0} = 0 \text{ because } 0^2 = 0 \text{ and}$$
$$\sqrt{1.44} = 1.2 \text{ because } (1.2)^2 = 1.44.$$

No negative number has a square root that is a real number. For instance, there is no real number whose square is -4, so -4 has no square root.

Every nonnegative real number has a square root. Unless an integer is a perfect square (such as $64 = 8^2$), its square root is an irrational number. A calculator can be used to obtain a rational approximation of these square roots.

TECHNOLOGY TIP   On one-line scientific calculators, $\sqrt{40}$ is entered as 40 $\sqrt{\ }$. On graphing calculators and two-line scientific calculators, key in
   $\sqrt{\ }$ 40 ENTER (or = or EXE).

▼ **EXAMPLE 9**   Estimate each of the following quantities. Verify your estimates with a calculator.

(a) $\sqrt{40}$

   Solution Since $6^2 = 36$ and $7^2 = 49$, $\sqrt{40}$ must be a number between 6 and 7. A typical calculator shows that $\sqrt{40} \approx 6.32455532$.

(b) $5\sqrt{7}$

   Solution $\sqrt{7}$ is between 2 and 3 because $2^2 = 4$ and $3^2 = 9$, so $5\sqrt{7}$ must be a number between $5 \cdot 2 = 10$ and $5 \cdot 3 = 15$. A calculator shows that $5\sqrt{7} \approx 13.22875656$. ▲ ⑦

CAUTION If c and d are positive real numbers, then $\sqrt{c + d}$ is *not* equal to $\sqrt{c} + \sqrt{d}$. For example, $\sqrt{9 + 16} = \sqrt{25} = 5$, but $\sqrt{9} + \sqrt{16} = 3 + 4 = 7$.

The Number Line

The real numbers can be illustrated geometrically with a diagram called a **number line.** Each real number corresponds to exactly one point on the line and vice versa. A number line with several sample numbers located (or **graphed**) on it is shown in Figure 1.2. ⑧

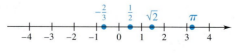

FIGURE 1.2

When comparing the sizes of two real numbers, the following symbols are used.

---

⑦

Estimate each of the following.

(a) $\sqrt{73}$

(b) $\sqrt{22} + 3$

(c) Confirm your estimates in parts (a) and (b) with a calculator.

Answers:

(a) Between 8 and 9

(b) Between 7 and 8

(c) 8.5440; 7.6904

---

⑧

Draw a number line, and graph the numbers $-4, -1, 0, 1, 2.5$, and $13/4$ on it.

Answer:

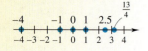

Symbol	Read	Meaning
$a < b$	a is less than b.	a lies to the *left* of b on the number line.
$b > a$	b is greater than a.	b lies to the *right* of a on the number line.

Note that $a < b$ means the same thing as $b > a$. The inequality symbols are sometimes joined with the equals sign, as follows.

Symbol	Read	Meaning
$a \leq b$	a is less than or equal to b.	either $a < b$ or $a = b$
$b \geq a$	b is greater than or equal to a.	either $b > a$ or $b = a$

TECHNOLOGY TIP   If your graphing calculator has inequality symbols (usually located on the TEST menu), you can key in statements such as "$5 < 12$" or "$-2 \geq 3$." When you press ENTER, the calculator will display 1 if the statement is true and 0 if it is false.

Only one part of an "either . . . or" statement needs to be true for the entire statement to be considered true. So the statement $3 \leq 7$ is true because $3 < 7$, and the statement $3 \leq 3$ is true because $3 = 3$.

⑨

Write *true* or *false* for the following.

(a) $-9 \leq -2$

(b) $8 > -3$

(c) $-14 \leq -20$

Answers:

(a) True

(b) True

(c) False

▼ **EXAMPLE 10**   Write *true* or *false* for each of the following.

(a) $8 < 12$

Solution This statement says that 8 is less than 12, which is true.

(b) $-6 > -3$

Solution The graph in Figure 1.3 shows that -6 is to the *left* of -3. Thus, $-6 < -3$, and the given statement is false.

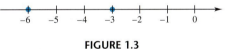

FIGURE 1.3

(c) $-2 \leq -2$

Solution Because $-2 = -2$, this statement is true. ▲ ⑨

A number line can be used to draw the graph of a set of numbers, as shown in the next few examples.

⑩

Graph all real numbers x such that

(a) $-5 < x < 1$

(b) $4 < x < 7$.

Answers:

(a)

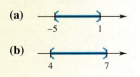

(b)

▼ **EXAMPLE 11**   Graph all real numbers x such that $1 < x < 5$.

Solution This graph includes all the real numbers between 1 and 5 and not just the integers. Graph these numbers by drawing a heavy line from 1 to 5 on the number line, as in Figure 1.4. Parentheses at 1 and 5 show that neither of these points belongs to the graph. ▲ ⑩

FIGURE 1.4

A set that consists of all the real numbers between two points, such as $1 < x < 5$ in Example 11, is called an **interval.** A special notation called **interval notation** is used to indicate an interval on the number line. For example, the interval including all numbers x such that $-2 < x < 3$ is written as $(-2, 3)$. The parentheses indicate that the numbers -2 and 3 are *not* included. If -2 and 3 are to be included in the interval, square brackets are used, as in $[-2, 3]$. The following chart shows several typical intervals, where $a < b$.

Intervals

Inequality	Interval Notation	Explanation
$a \leq x \leq b$	$[a, b]$	Both a and b are included.
$a \leq x < b$	$[a, b)$	a is included; b is not.
$a < x \leq b$	$(a, b]$	b is included; a is not.
$a < x < b$	(a, b)	Neither a nor b is included.

Interval notation is also used to describe sets such as the set of all numbers x, with $x \geq -2$. This interval is written $[-2, \infty)$. The set of all real numbers is written $(-\infty, \infty)$ in interval notation.

▼ **EXAMPLE 12**   Graph the interval $[-2, \infty)$.

Solution   Start at -2 and draw a heavy line to the right, as in Figure 1.5. Use a square bracket at -2 to show that -2 itself is part of the graph. The symbol ∞, read "infinity, " *does not* represent a number. This notation simply indicates that *all* numbers greater than -2 are in the interval. Similarly, the notation $(-\infty, 2)$ indicates the set of all numbers x such that $x < 2$. ▲

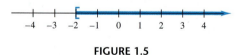

FIGURE 1.5

Absolute Value

Distance is always a nonnegative number. For example, the distance from 0 to -2 on a number line is 2, the same as the distance from 0 to 2. The **absolute value** of a number a is the distance on the number line from a to 0. Thus, the absolute value of both 2 and -2 is 2. We write the absolute value of the real number a as $|a|$. For example, the distance on the number line from 9 to 0 is 9, as is the distance from -9 to 0. (See Figure 1.6.) By definition, $|9| = 9$ and $|-9| = 9$.

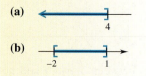

⟨11⟩

Graph all real numbers x in the given interval.

(a) $(-\infty, 4]$

(b) $[-2, 1]$

Answers:

(a)

(b)

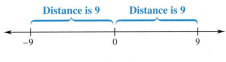

FIGURE 1.6

The facts that $|9| = 9$ and that $|-9| = 9 = -(-9)$ suggest the following algebraic definition of absolute value.

Absolute Value

For any real number a,

$$|a| = a \qquad \text{if } a \geq 0$$
$$|a| = -a \qquad \text{if } a < 0.$$

Note the second part of the definition: for a negative number, say, -5, the negative of -5 is the positive number $-(-5) = 5$. Similarly, if a is any negative number, then $-a$ is a *positive* number. Thus, *for every real number a, $|a|$ is nonnegative.*

▼ **EXAMPLE 13**   Evaluate $|8 - 9|$.

Solution First, simplify the expression within the absolute value bars:
$$|8 - 9| = |-1| = 1.$$

Similarly, $-|-5 - 8| = -|-13| = -13.$ ▲ ⟨12⟩

⟨12⟩

Find the following.

(a) $|-6|$

(b) $-|7|$

(c) $-|-2|$

(d) $|-3-4|$

(e) $|2-7|$

Answers:

(a) 6

(b) -7

(c) -2

(d) 7

(e) 5

◆ ## 1.1 Exercises

Label each of the following as true or false.

1. Every integer is a rational number.

2. Every real number is an irrational number.

3. The decimal expansion of the irrational number π begins 3.141592653589793 . . . . Use your calculator to determine which of the following rational numbers is the best approximation for the irrational number π:

$$\frac{22}{7}, \quad \frac{355}{113}, \quad \frac{103,993}{33,102}, \quad \frac{2,508,429,787}{798,458,000}.$$

Your calculator may tell you that some of these numbers are equal to π, but that just indicates that the number agrees with π for as many decimal places as your calculator can handle (usually 10–14). No rational number is exactly equal to π.

Identify the properties that are illustrated in each of the following. (See Examples 2–6.)

4. $-5 + 0 = -5$

5. $6(t + 4) = 6t + 6 \cdot 4$

6. $3 + (-3) = (-3) + 3$

7. $0 + (-7) = -7 + 0$

8. $8 + (12 + 6) = (8 + 12) + 6$

9. How is the additive inverse property related to the additive identity property? the multiplicative inverse property to the multiplicative identity property?

10. Explain the distinction between the commutative and associative properties.

Evaluate each of the following if $p = -2$, $q = 4$, and $r = -5$. (See Examples 7 and 8.)

11. $-3(p + 5q)$

12. $2(q - r)$

13. $\dfrac{q + r}{q + p}$ **14.** $\dfrac{3q}{3p - 2r}$

Business   *Lenders are required to state the annual percentage rate (APR) for every loan, using the formula APR = 12r, where r is the monthly interest rate. Find the APR when*

15. $r = 1.5$ **16.** $r = 1.67$

Find the monthly interest rate when

17. $APR = 9$ **18.** $APR = 19.5$

Evaluate each expression, using the order of operations given in the text. (See Examples 7 and 8.)

19. $3 - 4 \cdot 5 + 5$ **20.** $(4 - 5) \cdot 6 + 6$

21. $8 - 4^2 - (-12)$ **22.** $8 - (-4)^2 - (-12)$

23. $-(3 - 5) - [2 - (3^2 - 13)]$

24. $\dfrac{2(3 - 7) + 4(8)}{4(-3) + (-3)(-2)}$

25. $\dfrac{2(-3) + 3/(-2) - 2/(-\sqrt{16})}{\sqrt{64} - 1}$

26. $\dfrac{6^2 - 3\sqrt{25}}{\sqrt{6^2 + 13}}$

Use a calculator and list the given numbers in order from smallest to largest.

27. $\dfrac{189}{37}, \quad \dfrac{4587}{691}, \quad \sqrt{47}, \quad 6.735, \quad \sqrt{27}, \quad \dfrac{2040}{523}$

28. $\dfrac{385}{177}, \quad \sqrt{10}, \quad \dfrac{187}{63}, \quad \pi, \quad \sqrt{\sqrt{85}}, \quad 2.9884$

Express each of the following statements in symbols, using $<$, $>$, $\leq$, or $\geq$.

29. 12 is less than 18.5.

30. -2 is greater than -20.

31. x is greater than or equal to 5.7.

32. y is less than or equal to -5.

33. z is at most 7.5.

34. w is negative.

Graph the following intervals on a number line. (See Examples 11 and 12.)

35. $(-8, -1)$ **36.** $[-1, 10]$ **37.** $[-2, 2)$

38. $(3, 7]$ **39.** $(-2, 3]$ **40.** $(-2, \infty)$

41. $(3, \infty)$ **42.** $(-\infty, 5)$ **43.** $(-\infty, -2]$

44. $[-4, \infty)$

Business   *The Consumer Price Index (CPI) tracks the cost of a typical sample of consumer goods. The table shows the percentage increase for each year in a ten-year period*.*

Year	1995	1996	1997	1998	1999
% change in CPI	2.8	3.0	2.3	1.6	2.2

Year	2000	2001	2002	2003	2004
% change in CPI	3.4	2.8	1.6	2.3	2.7

Let r denote the yearly percentage increase in the CPI. For each of the following inequalities, find the number of years during the given period that r satisfied the inequality.

45. $r > 2.7$ **46.** $r < 2.7$ **47.** $r \leq 2.7$

48. $r \geq 3$ **49.** $r > 4$ **50.** $r \leq 1.7$

Health   *The body mass index (BMI), a number B that measures the relationship between a person's height H (in inches) and weight W (in pounds), is given by the formula*

$$B = \dfrac{.455W}{(.0254H)^2}.^*$$

Federal guidelines suggest that the desirable range for B is $19 \leq B \leq 25$.
(a) *Find the BMI of the following athletes.*
(b) *Determine whether or not each athlete's BMI falls in the desirable range.*

51. Steffi Graf (119 pounds; 5 ft, 9 in)

52. Jackie Joyner-Kersee (153 pounds; 5 ft, 10 in)

53. Tiger Woods (180 pounds; 6 ft, 2 in)

54. Shaquille O'Neal (300 pounds; 7 ft, 1 in)

*U.S. Bureau of Labor Statistics.

*Washington Post.

Physical Science   *The windchill factor is a measure of the cooling effect that the wind has on a person's skin. It calculates the equivalent cooling temperature if there were no wind.**

	Wind (mph)								
	Calm	**5**	**10**	**15**	**20**	**25**	**30**	**35**	**40**
	40	36	34	32	30	29	28	28	27
	30	25	21	19	17	16	15	14	13
	20	13	9	6	4	3	1	0	−1
	10	1	−4	−7	−9	−11	−12	−14	−15
Temperature (°F)	0	−11	−16	−19	−22	−24	−26	−27	−29
	−10	−22	−28	−32	−35	−37	−39	−41	−43
	−20	−34	−41	−45	−48	−51	−53	−55	−57
	−30	−46	−53	−58	−61	−64	−67	−69	−71
	−40	−57	−66	−71	−74	−78	−80	−82	−84

Suppose that we wish to determine the difference between two of these entries and we are interested only in the magnitude, or absolute value, of this difference. Then we subtract the two entries and find the absolute value. For example, the difference in windchill factors for wind at 20 miles per hour with a 20° air temperature and wind at 30 miles per hour with a 10° air temperature is $|-12° - 4°| = 16°$, or equivalently, $|4° - (-12°)| = 16°$.

Find the absolute value of the difference of the two indicated windchill factors.

55. Wind at 20 miles per hour with a 30° air temperature and wind at 10 miles per hour with a −10° air temperature

56. Wind at 20 miles per hour with a −20° air temperature and wind at 5 miles per hour with a 30° air temperature

57. Wind at 35 miles per hour with a −30° air temperature and wind at 15 miles per hour with a −20° air temperature

58. Wind at 40 miles per hour with a 40° air temperature and wind at 25 miles per hour with a −30° air temperature

Evaluate each of the following. (See Example 13.)

59. $|8| - |-4|$

60. $|-9| - |-12|$

61. $-|-4| - |-1 - 14|$

62. $-|6| - |-12 - 4|$

In each of the following problems, fill in the blank with either =, <, or >, so that the resulting statement is true.

63. $|5|$ _____ $|-5|$

64. $-|-4|$ _____ $|4|$

65. $|10 - 3|$ _____ $|3 - 10|$

66. $|6 - (-4)|$ _____ $|-4 - 6|$

67. $|-2 + 8|$ _____ $|2 - 8|$

68. $|3| \cdot |-5|$ _____ $|3(-5)|$

69. $|3 - 5|$ _____ $|3| - |5|$

70. $|-5 + 1|$ _____ $|-5| + |1|$

Write the expression without using absolute value.

71. $|a - 7|$ if $a < 7$

72. $|b - c|$ if $b \geq c$

73. If a and b are any real numbers, is it always true that $|a + b| = |a| + |b|$? Explain your answer.

74. If a and b are any two real numbers, is it always true that $|a - b| = |b - a|$? Explain your answer.

75. For which real numbers b does $|2 - b| = |2 + b|$? Explain your answer.

76. According to data from the Center for Science in the Public Interest, the healthy weight range for a person depends on the person's height. For example,

Height	Healthy Weight Range (lb)
5 ft 8 in.	143 ± 21
6 ft 0 in.	163 ± 26

Express each of these ranges as an absolute value inequality in which x is the weight of the preson.

77. The use of digital devices (cell phones, DVD players, PCs, etc.) continues to grow. The following graph shows the approximate number of digital devices (in billions) in use worldwide from 2000 to 2004:*

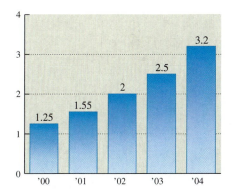

Consider the statement

$$|x - 2{,}000{,}000{,}000| \geq 500{,}000{,}000,$$

where x is the number of digital devices in use in that year. In what years was this statement true?

---

*Table from the Joint Action Group for Temperature Indices, 2001.

*Data, estimates, and projections from IDC.

Use inequality symbols to rewrite each of the following statements, which are based on an article in the St Louis Post-Dispatch.* Using x as the variable, describe what x represents in each exercise and then write an inequality. Example: For the statement "At least 6600 international students attend the University of Southern California (USC)," let x be the number of international students attending USC. Then x ≥ 6600.

_____

*"Foreign student enrollment declines at nation's universities," by Susan C. Thomson, *St. Louis Post-Dispatch,* November 13, 2004.

78. There are at most 600,000 international students enrolled at U.S. colleges and universities.

79. More than 13% of the international students are from India.

80. Less than 14% of the international students are from China.

81. When schools are ranked by the number of international students enrolled, the University of North Carolina (UNC) at Chapel Hill does not rank in the top 25.

82. When schools are ranked by the number of international students enrolled, the University of Illinois (UI) at Champaign-Urbana ranks in the top ten.

1.2 Polynomials

Polynomials are the fundamental tools of algebra and will play a central role in this course. In order to do polynomial arithmetic, you must first understand exponents. So we begin with them. You are familiar with the usual notation for squares and cubes:

$$5^2 = 5 \cdot 5 \text{ and } 6^3 = 6 \cdot 6 \cdot 6.$$

We now extend this convenient notation to other cases.

> If n is a natural number and a is any real number, then
>
> $$a^n \qquad \text{denotes the product} \qquad a \cdot a \cdot a \cdots a \ (n \text{ factors}).$$
>
> The number a is the **base** and n is the **exponent.**

▼ **EXAMPLE 1** 4^6, which is read "four to the sixth," or "four to the sixth power," is the number

$$4 \cdot 4 \cdot 4 \cdot 4 \cdot 4 \cdot 4 = 4096.$$

Similarly, $(-5)^3 = (-5)(-5)(-5) = -125$ and

$$\left(\frac{3}{2}\right)^4 = \frac{3}{2} \cdot \frac{3}{2} \cdot \frac{3}{2} \cdot \frac{3}{2} = \frac{81}{16}. \ ▲$$

① Evaluate the following.

(a) 6^3

(b) 5^{12}

(c) 1^9

(d) $\left(\dfrac{7}{5}\right)^8$

Answers:

(a) 216

(b) 244,140,625

(c) 1

(d) 14.75789056

▼ **EXAMPLE 2** Use a calculator to approximate the following.

(a) $(1.2)^8$

Solution Key in 1.2 and then use the x^y key (labeled $\wedge$ on some calculators); finally, key in the exponent 8. The calculator displays the (exact) answer 4.29981696.

(b) $\left(\dfrac{12}{7}\right)^{23}$

Solution Don't compute 12/7 separately. Use parentheses and key in $(12/7)$, followed by the x^y key and the exponent 23 to obtain the approximate answer 242,054.822. ▲ ①

CAUTION A common error in using exponents occurs with expressions such as $4 \cdot 3^2$. The exponent of 2 applies only to the base 3, so that

$$4 \cdot 3^2 = 4 \cdot 3 \cdot 3 = 36.$$

On the other hand,

$$(4 \cdot 3)^2 = (4 \cdot 3)(4 \cdot 3) = 12 \cdot 12 = 144,$$

so

$$4 \cdot 3^2 \neq (4 \cdot 3)^2.$$

Be careful to distinguish between expressions like -2^4 and $(-2)^4$.

$$-2^4 = -(2^4) = -(2 \cdot 2 \cdot 2 \cdot 2) = -16$$

$$(-2)^4 = (-2)(-2)(-2)(-2) = 16,$$

so

$$-2^4 \neq (-2)^4. \; \langle 2 \rangle$$

By the definition of an exponent,

$$3^4 \cdot 3^2 = (3 \cdot 3 \cdot 3 \cdot 3)(3 \cdot 3) = 3^6.$$

Since $6 = 4 + 2$, we can write the preceding equation as $3^4 \cdot 3^2 = 3^{4+2}$. This suggests the following fact, which applies to any real number a and natural numbers m and n.

Multiplication with Exponents

To multiply a^m by a^n, *add* the exponents:

$$a^m \cdot a^n = a^{m+n}.$$

▼ **EXAMPLE 3** Verify each of the following simplifications.

(a) $7^4 \cdot 7^6 = 7^{4+6} = 7^{10}$

(b) $(-2)^3 \cdot (-2)^5 = (-2)^{3+5} = (-2)^8$

(c) $(3k)^2 \cdot (3k)^3 = (3k)^5$

(d) $(m+n)^2 \cdot (m+n)^5 = (m+n)^7$ ▲ $\langle 3 \rangle$

The multiplication property of exponents has a convenient consequence. By definition,

$$(5^2)^3 = 5^2 \cdot 5^2 \cdot 5^2 = 5^{2+2+2} = 5^6.$$

Note that $2 + 2 + 2$ is $3 \cdot 2 = 6$, which corresponds to the fact that there are 3 copies of 5^2. This is an example of a more general fact about any real number a and natural numbers m and n.

$\langle 2 \rangle$

Evaluate the following.

(a) $3 \cdot 6^2$

(b) $5 \cdot 4^3$

(c) -3^6

(d) $(-3)^6$

(e) $-2 \cdot (-3)^5$

Answers:

(a) 108

(b) 320

(c) -729

(d) 729

(e) 486

$\langle 3 \rangle$

Simplify the following.

(a) $5^3 \cdot 5^6$

(b) $(-3)^4 \cdot (-3)^{10}$

(c) $(5p)^2 \cdot (5p)^8$

Answers:

(a) 5^9

(b) $(-3)^{14}$

(c) $(5p)^{10}$

Power of a Power

To find a power of a power, $(a^m)^n$, *multiply* the exponents:

$$(a^m)^n = a^{mn}.$$

▼ **EXAMPLE 4**   Compute the following.

(a) $(x^3)^4 = x^{3 \cdot 4} = x^{12}$.

(b) $[(-3)^5]^3 = (-3)^{5 \cdot 3} = (-3)^{15}$.

(c) $[(6z)^4]^4 = (6z)^{4 \cdot 4} = (6z)^{16}$. ▲ ④

It will be convenient to give a zero exponent a meaning. If the multiplication property of exponents is to remain valid, we should have, for example, $3^5 \cdot 3^0 = 3^{5+0} = 3^5$. But this will be true only when $3^0 = 1$. So we make this definition.

④

Compute the following.

(a) $(6^3)^7$

(b) $[(4k)^5]^6$

Answers:

(a) 6^{21}

(b) $(4k)^{30}$

Zero Exponent

If a is any nonzero real number, then

$$a^0 = 1.$$

For example, $6^0 = 1$ and $(-9)^0 = 1$. Note that 0^0 is *not* defined. ⑤

⑤

Evaluate the following.

(a) 17^0

(b) 30^0

(c) $(-10)^0$

(d) $-(12)^0$

Answers:

(a) 1

(b) 1

(c) 1

(d) -1

Polynomials

A **polynomial** is an algebraic expression such as

$$5x^4 + 2x^3 + 6x, \quad 8m^3 + 9m^2 + \frac{3}{2}m + 3, \quad -10p, \quad \text{or} \quad 8.$$

The letter used is called a **variable,** and a polynomial is a sum of **terms** of the form

$$(\text{constant}) \times (\text{nonnegative integer power of the variable}).$$

We assume that $x^0 = 1$, $m^0 = 1$, etc., so terms such as 3 or 8 may be thought of as $3x^0$ and $8x^0$, respectively. The constants that appear in each term of a polynomial are called the **coefficients** of the polynomial. The coefficient of x^0 is called the **constant term.**

▼ **EXAMPLE 5**   Identify the coefficients and the constant term of the polynomial.

(a) $5x^2 - x + 12$

     Solution   The coefficients are 5, -1, and 12, and the constant term is 12.

(b) $7x^3 + 2x - 4$

Solution The coefficients are 7,0, 2, and -4, because the polynomial can be written $7x^3 + 0x^2 + 2x - 4$. The constant term is -4. ▲

A polynomial that consists only of a constant term, such as 15, is called a **constant polynomial.** The **zero polynomial** is the constant polynomial 0. The **degree** of a polynomial is the *exponent* of the highest power of x that appears with a *nonzero* coefficient, and the nonzero coefficient of this highest power of x is the **leading coefficient** of the polynomial. For example,

Polynomial	Degree	Leading Coefficient	Constant Term
$6x^7 + 4x^3 + 5x^2 - 7x + 10$	7	6	10
$-x^4 + 2x^3 + \frac{1}{2}$	4	-1	$\frac{1}{2}$
x^3	3	1	0
12	0	12	12

The degree of the zero polynomial is *not defined,* since no exponent of x occurs with nonzero coefficient. First-degree polynomials are often called **linear polynomials.** Second- and third-degree polynomials are called **quadratics** and **cubics,** respectively. ⟨6⟩

⟨6⟩

Find the degree of each polynomial.

(a) $x^4 - x^2 + x + 5$

(b) $7x^5 + 6x^3 - 3x^8 + 2$

(c) 17

(d) 0

Answers:

(a) 4

(b) 8

(c) 0

(d) Not defined

Addition and Subtraction

Two terms having the same variable with the same exponent are called **like terms;** other terms are called **unlike terms.** Polynomials can be added or subtracted by using the distributive property to combine like terms. Only like terms can be combined. For example,

$$12y^4 + 6y^4 = (12 + 6)y^4 = 18y^4$$

and

$$-2m^2 + 8m^2 = (-2 + 8)m^2 = 6m^2.$$

The polynomial $8y^4 + 2y^5$ has unlike terms, so it cannot be further simplified. The next example shows how to add and subtract polynomials by combining terms.

▼ **EXAMPLE 6**   Add or subtract as indicated.

(a) $(8x^3 - 4x^2 + 6x) + (3x^3 + 5x^2 - 9x + 8)$

Solution Combine like terms.

$$(8x^3 - 4x^2 + 6x) + (3x^3 + 5x^2 - 9x + 8)$$
$$= (8x^3 + 3x^3) + (-4x^2 + 5x^2)$$ Commutative property
$$\qquad\qquad + (6x - 9x) + 8$$ and associative property
$$= 11x^3 + x^2 - 3x + 8$$ Distributive property

(b) $(-4x^4 + 6x^3 - 9x^2 - 12) + (-3x^3 + 8x^2 - 11x + 7)$
$$= -4x^4 + 3x^3 - x^2 - 11x - 5$$

(c) $(2x^2 - 11x + 8) - (7x^2 - 6x + 2)$

Solution Use the definition of subtraction: $a - b = a + (-b)$. Here, a and b are polynomials; a is $2x^2 - 11x + 8$ and $-b$ is

$$-(7x^2 - 6x + 2) = -7x^2 + 6x - 2.$$

Now perform the subtraction.

$$(2x^2 - 11x + 8) - (7x^2 - 6x + 2)$$
$$= (2x^2 - 11x + 8) + (-7x^2 + 6x - 2)$$
$$= -5x^2 - 5x + 6 \ \blacktriangle \ ⟨7⟩$$

Multiplication

The distributive property is also used to multiply polynomials. For example, the product of $8x$ and $6x - 4$ is found as follows.

$$8x(6x - 4) = 8x(6x) - 8x(4) \qquad \text{Distributive property}$$
$$= 48x^2 - 32x \qquad\qquad x \cdot x = x^2$$

▼ **EXAMPLE 7** Find each product.

(a) $2p^3(3p^2 - 2p + 5) = 2p^3(3p^2) + 2p^3(-2p) + 2p^3(5)$
$$= 6p^5 - 4p^4 + 10p^3$$

(b) $(3k - 2)(k^2 + 5k - 4) = 3k(k^2 + 5k - 4) - 2(k^2 + 5k - 4)$
$$= 3k^3 + 15k^2 - 12k - 2k^2 - 10k + 8$$
$$= 3k^3 + 13k^2 - 22k + 8 \quad \blacktriangle ⟨8⟩$$

▼ **EXAMPLE 8** The product $(2x - 5)(3x + 4)$ can be found by using the distributive property twice.

$$(2x - 5)(3x + 4) = 2x(3x + 4) - 5(3x + 4)$$
$$= 2x \cdot 3x + 2x \cdot 4 + (-5) \cdot 3x + (-5) \cdot 4$$
$$= 6x^2 + 8x - 15x - 20$$
$$= 6x^2 - \quad 7x \quad - 20 \ \blacktriangle$$

Observe the pattern in the second line of Example 8 and its relationship to the terms being multiplied.

$$(2x - 5)(3x + 4) = 2x \cdot 3x + 2x \cdot 4 + (-5) \cdot 3x + (-5) \cdot 4$$

First terms

$(2x - 5)(3x + 4)$

Outside terms

$(2x - 5)(3x + 4)$

Inside terms

$(2x - 5)(3x + 4)$

Last terms

⟨7⟩

Add or subtract.

(a) $(-2x^2 + 7x + 9)$
$+ (3x^2 + 2x - 7)$

(b) $(4x + 6) - (13x - 9)$

(c) $(9x^3 - 8x^2 + 2x)$
$- (9x^3 - 2x^2 - 10)$

Answers:

(a) $x^2 + 9x + 2$

(b) $-9x + 15$

(c) $-6x^2 + 2x + 10$

⟨8⟩

Find the following products.

(a) $-6r(2r - 5)$

(b) $(8m + 3) \cdot$
$(m^4 - 2m^2 + 6m)$

Answers:

(a) $-12r^2 + 30r$

(b) $8m^5 + 3m^4 - 16m^3$
$+ 42m^2 + 18m$

This pattern is easy to remember by using the acronym **FOIL** (**F**irst, **O**utside, **I**nside, **L**ast). The FOIL method makes it easy to find products such as this one mentally, without the necessity of writing out the intermediate steps.

▼ **EXAMPLE 9**   Use FOIL to find the product of these polynomials.

(a) $(3x + 2)(x + 5) = 3x^2 + 15x + 2x + 10 = 3x^2 + 17x + 10$

First   Outside   Inside   Last

(b) $(x + 3)^2 = (x + 3)(x + 3) = x^2 + 3x + 3x + 9 = x^2 + 6x + 9$

(c) $(2x + 1)(2x - 1) = 4x^2 - 2x + 2x - 1 = 4x^2 - 1$ ▲ ⟨9⟩

In business, the *revenue* from the sales of an item is given by

Revenue = (price per item) × (number of items sold).

The *cost* to manufacture and sell these items is given by

Cost = Fixed Costs + Variable Costs,

where the fixed costs include such things as buildings and machinery (which don't depend on how many items are made) and variable costs include such things as labor and materials (which vary, depending on how many items are made). Then

Profit = Revenue − Cost.

▼ **EXAMPLE 10**   Hot Rocks Music sells CDs for $8 each (wholesale) and can produce a maximum of 200,000 CDs. The variable cost of producing x thousand CDs is $3550x - 8x^2$ dollars, and the fixed costs for the manufacturing operation are $215,000. If x thousand CDs are manufactured and sold, find expressions for the revenue, cost, and profit.

Solution If x thousand CDs are sold at $8 each, then

Revenue = (price per item) × (number of items sold)
$$R = 8 \times 1000x = 8000x,$$

where $x \le 200$ (because only 200,000 CDs can be made). The variable cost of making x thousand CDs is $3550x - 8x^2$, so that

Cost = Fixed Costs + Variable Costs
$$C = 215{,}000 + (3550x - 8x^2) \quad (x \le 200)$$

Therefore, the profit is given by

$$P = R - C = 8000x - (215{,}000 + 3550x - 8x^2)$$
$$= 8000x - 215{,}000 - 3550x + 8x^2$$
$$P = 8x^2 + 4450x - 215{,}000 \quad (x \le 200). ▲ ⟨10⟩$$

⟨9⟩
Use FOIL to find these products.

(a) $(5k - 1)(2k + 3)$

(b) $(7z - 3)(2z + 5)$

Answers:

(a) $10k^2 + 13k - 3$

(b) $14z^2 + 29z - 15$

⟨10⟩
Suppose revenue is given by $7x^2 - 3x$, fixed costs are $500, and variable costs are given by $3x^2 + 5x - 25$. Write an expression for

(a) Cost

(b) Profit.

Answers:

(a) $C = 3x^2 + 5x + 475$

(b) $P = 4x^2 - 8x - 475$

1.2 Exercises

Use a calculator to approximate these numbers. (See Examples 1 and 2.)

1. 11.2^6

2. $(-6.54)^{11}$

3. $(-18/7)^6$

4. $(5/9)^7$

5. Explain how the value of -3^2 differs from $(-3)^2$. Do -3^3 and $(-3)^3$ differ in the same way? Why or why not?

6. Describe the steps used to multiply 4^3 and 4^5. Is the product of 4^3 and 3^4 found in the same way? Explain.

Simplify each of the following. Leave answers with exponents. (See Examples 3 and 4.)

7. $4^2 \cdot 4^3$

8. $(-4)^4 \cdot (-4)^6$

9. $(-6)^2 \cdot (-6)^5$

10. $(2z)^5 \cdot (2z)^6$

11. $[(5u)^4]^7$

12. $(6y)^3 \cdot [(6y)^5]^4$

List the degree of the polynomial, its coefficients, and its constant term. (See Example 5.)

13. $6.2x^4 - 5x^3 + 4x^2 - 3x + 3.7$

14. $6x^7 + 4x^6 - x^3 + x$

State the degree of the polynomial.

15. $1 + x + 2x^2 + 3x^3$

16. $5x^4 - 4x^5 - 6x^3 + 7x^4 - 2x + 8$

Add or subtract as indicated. (See Example 6.)

17. $(3x^3 + 2x^2 - 5x) + (-4x^3 - x^2 - 8x)$

18. $(-2p^3 - 5p + 7) + (-4p^2 + 8p + 2)$

19. $(-4y^2 - 3y + 8) - (2y^2 - 6y + 2)$

20. $(7b^2 + 2b - 5) - (3b^2 + 2b - 6)$

21. $(2x^3 + 2x^2 + 4x - 3) - (2x^3 + 8x^2 + 1)$

22. $(3y^3 + 9y^2 - 11y + 8) - (-4y^2 + 10y - 6)$

Find each of the following. (See Examples 7–9.)

23. $-9m(2m^2 + 6m - 1)$

24. $2a(4a^2 - 6a + 8)$

25. $(3z + 5)(4z^2 - 2z + 1)$

26. $(2k + 3)(4k^3 - 3k^2 + k)$

27. $(6k - 1)(2k + 3)$

28. $(8r + 3)(r - 1)$

29. $(3y + 5)(2y + 1)$

30. $(5r - 3s)(5r - 4s)$

31. $(9k + q)(2k - q)$

32. $(.012x - .17)(.3x + .54)$

33. $(6.2m - 3.4)(.7m + 1.3)$

34. $2p - 3[4p - (8p + 1)]$

35. $5k - [k + (-3 + 5k)]$

36. $(3x - 1)(x + 2) - (2x + 5)^2$

Business *Find expressions for the revenue, cost, and profit from selling x thousand items. (See Example 10.)*

Item Price	Fixed Costs	Variable Costs
37. $5.00	$150,000	$2250x$
38. $7.95	$220,000	$4300x$
39. $7.50	$250,000	$-3x^2 + 3480x - 325$
40. $13.25	$450,000	$-4x^2 + 2880x - 295$

Business *The bar graph that follows shows the number of Internet users (in millions), as projected by the U.S. Department of Commerce and Jupiter Media Metrix. The polynomial*

$$-.47x^2 + 19.12x + 101.55$$

gives a good approximation of the number of Internet users in year x, where x = 0 corresponds to 1999, x = 1 to 2000, and so on. For each year,

(a) *use the bar graph to determine the number of Internet users;*

(b) *use the polynomial to determine the number of Internet users.*

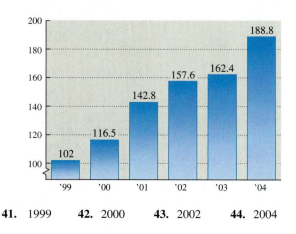

41. 1999   **42.** 2000   **43.** 2002   **44.** 2004

Assuming that the polynomial approximation remains accurate, estimate the number of Internet users in each of the following years.

45. 2005   **46.** 2006   **47.** 2008   **48.** 2010

Health *According to data from a leading insurance company, if a person is 65 years old, the probability that he or she will live for another x years is approximated by the polynomial*

$$1 - .0058x - .00076x^2.*$$

Find the probability that a 65-year-old person will live to the following ages.

49. 75 (that is, 10 years past 65)   **50.** 80

---

*Provided by Ralph DeMarr, University of New Mexico.

51. 87   **52.** 95

53. Physical Science One of the most amazing formulas in all of ancient mathematics is the formula discovered by the Egyptians to find the volume of the frustum of a square pyramid, as shown in the following figure:

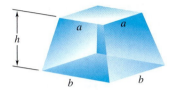

The volume of this pyramid is given by

$$(1/3)h \cdot (a^2 + ab + b^2),$$

where b is the length of the base, a is the length of the top, and h is the height.*

(a) When the Great Pyramid in Egypt was partially completed to a height h of 200 feet, b was 756 feet and a was 314 feet. Calculate its volume at this stage of construction.

(b) Try to visualize the figure if $a = b$. What is the resulting shape? Find its volume.

(c) Let $a = b$ in the Egyptian formula and simplify. Are the results the same?

54. Physical Science Refer to the formula and the discussion in Exercise 53.

(a) Use the expression $(1/3)h(a^2 + ab + b^2)$ to determine a formula for the volume of a pyramid with a square base b and height h by letting $a = 0$.

_____
*H. A. Freebury, *A History of Mathematics*. (New York: MacMillan Company, 1968).

(b) The Great Pyramid in Egypt had a square base of 756 feet and a height of 481 feet. Find the volume of the Great Pyramid. Compare it with the volume of the 273-foot-tall Louisiana Superdome, which has an approximate volume of 125 million cubic feet.*

(c) The Superdome covers an area of 13 acres. How many acres does the Great Pyramid cover? (*Hint:* 1 acre = 43, 560 ft².)

55. Suppose one polynomial has degree 3 and another also has degree 3. Find all possible values for the degree of their

(a) sum   (b) difference   (c) product.

Business *The following exercise is suitable for group work. A graphing calculator is needed. Use the table feature to make a table of values for the profit function in Example 10, with x = 0, 5, 10, . . . , 225. Use the table to answer the following questions.*

56. (a) What is the profit or loss (= negative profit) when 20,000 CDs are sold? When 40,000 are sold? Explain these answers.

(b) Approximately how many CDs must be sold in order for the company to make a profit?

(c) What is the profit from selling 100,000 CDs? From 150,000 CDs? From 200,000 CDs?

(d) Explain why the profit amounts shown in the table for 205,000 CDs and beyond are meaningless.

_____
*Louisiana Superdome (www.superdome.com).

1.3 Factoring

The number 18 can be written as a product in several ways: $9 \cdot 2$, $(-3)(-6)$, $1 \cdot 18$, etc. The numbers in each product $(9, 2, -3, \text{etc.})$ are called **factors,** and the process of writing 18 as a product of factors is called **factoring.** Thus, factoring is the reverse of multiplication.

Factoring of polynomials is a means of simplifying many expressions and of solving certain types of equations. As is the usual custom, factoring of polynomials in this book will be restricted to finding factors with *integer* coefficients (otherwise there may be an infinite number of possible factors).

Greatest Common Factor

The algebraic expression $15m + 45$ is made up of two terms: $15m$ and 45. Each of these terms has 15 as a factor. In fact, $15m = 15 \cdot m$ and $45 = 15 \cdot 3$. By the distributive property,

$$15m + 45 = 15 \cdot m + 15 \cdot 3 = 15(m + 3).$$

Both 15 and $m + 3$ are factors of $15m + 45$. Since 15 divides evenly into all terms of $15m + 45$ and is the largest number that will do so, it is called the **greatest common factor** for the polynomial $15m + 45$. The process of writing $15m + 45$ as $15(m + 3)$ is called **factoring out** the greatest common factor.

▼ **EXAMPLE 1**   Factor out the greatest common factor.

(a) $12p - 18q$

Solution Both $12p$ and $18q$ are divisible by 6, and
$$12p - 18q = \mathbf{6} \cdot 2p - \mathbf{6} \cdot 3q$$
$$= \mathbf{6}(2p - 3q).$$

(b) $8x^3 - 9x^2 + 15x$

Solution Each of these terms is divisible by x.
$$8x^3 - 9x^2 + 15x = (8x^2) \cdot x - (9x) \cdot x + 15 \cdot x$$
$$= x(8x^2 - 9x + 15)$$

(c) $5(4x - 3)^3 + 2(4x - 3)^2$

Solution The quantity $(4x - 3)^2$ is a common factor. Factoring it out gives
$$5(4x - 3)^3 + 2(4x - 3)^2 = (4x - 3)^2[5(4x - 3) + 2]$$
$$= (4x - 3)^2(20x - 15 + 2)$$
$$= (4x - 3)^2(20x - 13). \; \blacktriangle \; \langle 1 \rangle$$

⟨1⟩

Factor out the greatest common factor.

(a) $12r + 9k$

(b) $75m^2 + 100n^2$

(c) $6m^4 - 9m^3 + 12m^2$

(d) $3(2k + 1)^3 + 4(2k + 1)^4$

Answers:

(a) $3(4r + 3k)$

(b) $25(3m^2 + 4n^2)$

(c) $3m^2(2m^2 - 3m + 4)$

(d) $(2k + 1)^3(7 + 8k)$

Factoring Quadratics

If we multiply two first-degree polynomials, the result is a quadratic. For instance, using FOIL, we see that $(x + 1)(x - 2) = x^2 - x - 2$. Since factoring is the reverse of multiplication, factoring quadratics requires using FOIL backward.

▼ **EXAMPLE 2**   Factor $x^2 + 9x + 18$.

Solution We must find integers b and d such that
$$x^2 + 9x + 18 = (x + b)(x + d)$$
$$= x^2 + dx + bx + bd$$
$$x^2 + 9x + 18 = x^2 + (b + d)x + bd.$$

Since the constant coefficients on each side of the equation must be equal, we must have $bd = 18$; that is, b and d are factors of 18. Similarly, the coefficients of x must be the same, so that $b + d = 9$. The possibilities are summarized in this table:

Factors b, d of 18	Sum $b + d$
$18 \cdot 1$	$18 + 1 = 19$
$9 \cdot 2$	$9 + 2 = 11$
$6 \cdot 3$	$6 + 3 = 9$

There's no need to list negative factors, such as $(-3)(-6)$, because their sum is negative. The table suggests that 6 and 3 will work. Verify that

$$(x + 6)(x + 3) = x^2 + 9x + 18. \ \blacktriangle$$

2

Factor the following.

(a) $r^2 + 7r + 10$

(b) $x^2 + 4x + 3$

(c) $y^2 + 6y + 8$

Answers:

(a) $(r + 2)(r + 5)$

(b) $(x + 3)(x + 1)$

(c) $(y + 2)(y + 4)$

▼ **EXAMPLE 3** Factor $4y^2 - 11y + 6$.

Solution We must find integers a, b, c, and d such that

$$4y^2 - 11y + 6 = (ay + b)(cy + d)$$
$$= acy^2 + ady + bcy + bd$$
$$4y^2 - 11y + 6 = acy^2 + (ad + bc)y + bd.$$

Since the coefficients of y^2 must be the same on both sides, we see that $ac = 4$. Similarly, the constant terms show that $bd = 6$. The positive factors of 4 are 4 and 1 or 2 and 2. Since the middle term is negative, we consider only negative factors of 6. The possibilities are -2 and -3 or -1 and -6. Now we try various arrangements of these factors until we find one that gives the correct coefficient of y.

$$(2y - 1)(2y - 6) = 4y^2 - \mathbf{14y} + 6 \quad \text{Incorrect}$$
$$(2y - 2)(2y - 3) = 4y^2 - \mathbf{10y} + 6 \quad \text{Incorrect}$$
$$(y - 2)(4y - 3) = 4y^2 - \mathbf{11y} + 6 \quad \text{Correct}$$

The last trial gives the correct factorization. ▲

3

Factor the following.

(a) $x^2 - 4x + 3$

(b) $2y^2 - 5y + 2$

(c) $6z^2 - 13z + 6$

Answers:

(a) $(x - 3)(x - 1)$

(b) $(2y - 1)(y - 2)$

(c) $(3z - 2)(2z - 3)$

▼ **EXAMPLE 4** Factor $6p^2 - 7pq - 5q^2$.

Solution Again, we try various possibilities. The positive factors of 6 could be 2 and 3 or 1 and 6. As factors of -5, we have only -1 and 5 or -5 and 1. Try different combinations of these factors until the correct one is found.

$$(2p - 5q)(3p + q) = 6p^2 - \mathbf{13pq} - 5q^2 \quad \text{Incorrect}$$
$$(3p - 5q)(2p + q) = 6p^2 - \mathbf{7pq} - 5q^2 \quad \text{Correct}$$

Finally, $6p^2 - 7pq - 5q^2$ factors as $(3p - 5q)(2p + q)$. ▲

NOTE In Examples 2–4, we chose positive factors of the positive first term. Of course, we could have used two negative factors, but the work is easier if positive factors are used.

4

Factor the following.

(a) $r^2 - 5r - 14$

(b) $3m^2 + 5m - 2$

(c) $6p^2 + 13pq - 5q^2$

Answers:

(a) $(r - 7)(r + 2)$

(b) $(3m - 1)(m + 2)$

(c) $(2p + 5q)(3p - q)$

▼ **EXAMPLE 5** Factor $x^2 + x + 3$.

Solution There are only two ways to factor 3, namely, $3 = 1 \cdot 3$ and $3 = (-1)(-3)$. They lead to these products:

$$(x + 1)(x + 3) = x^2 + 4x + 3 \quad \text{Incorrect}$$
$$(x - 1)(x - 3) = x^2 - 4x + 3. \quad \text{Incorrect}$$

Therefore, this polynomial cannot be factored. ▲

Factoring Patterns

In some cases, you can factor a polynomial with a minimum of guesswork by recognizing common patterns. The easiest pattern to recognize is the *difference of squares*.

$$x^2 - y^2 = (x + y)(x - y). \qquad \text{Difference of squares}$$

To verify the accuracy of the preceding equation, multiply out the right side.

▼ **EXAMPLE 6**   Factor each of the following.

(a) $4m^2 - 9$

Solution Notice that $4m^2 - 9$ is the difference of two squares, since $4m^2 = (2m)^2$ and $9 = 3^2$. Use the pattern for the difference of two squares, letting $2m$ replace x and 3 replace y. Then the pattern $x^2 - y^2 = (x + y)(x - y)$ becomes

$$4m^2 - 9 = (2m)^2 - 3^2$$
$$= (2m + 3)(2m - 3).$$

(b) $128p^2 - 98q^2$

Solution First factor out the common factor of 2.

$$128p^2 - 98q^2 = 2(64p^2 - 49q^2)$$
$$= 2[(8p)^2 - (7q)^2]$$
$$= 2(8p + 7q)(8p - 7q)$$

(c) $x^2 + 36$

Solution The *sum* of two squares cannot be factored. To convince yourself of this, check some possibilities.

$$(x + 6)(x + 6) = (x + 6)^2 = x^2 + 12x + 36$$
$$(x + 4)(x + 9) = x^2 + 13x + 36$$

(d) $(x - 2)^2 - 49$

Solution Since $49 = 7^2$, this is a difference of two squares. So it factors as follows:

$$(x - 2)^2 - 49 = (x - 2)^2 - 7^2$$
$$= [(x - 2) + 7][(x - 2) - 7]$$
$$= (x + 5)(x - 9). \; ▲$$

⟨5⟩

Factor the following.

(a) $9p^2 - 49$

(b) $y^2 + 100$

(c) $(x + 3)^2 - 64$

Answers:

(a) $(3p + 7)(3p - 7)$

(b) Cannot be factored

(c) $(x + 11)(x - 5)$

Another common pattern is the *perfect square*. Verify each of the following factorizations by multiplying out the right side.

$$x^2 + 2xy + y^2 = (x + y)^2$$
$$x^2 - 2xy + y^2 = (x - y)^2 \qquad \text{Perfect Squares}$$

▼ **EXAMPLE 7**   Factor each polynomial.

(a) $16p^2 - 40pq + 25q^2$

Solution Because $16p^2 = (4p)^2$ and $25q^2 = (5q)^2$, use the second pattern shown in the preceding box, with $4p$ replacing x and $5q$ replacing y, to get

$$16p^2 - 40pq + 25q^2 = (4p)^2 - 2(4p)(5q) + (5q)^2$$
$$= (4p - 5q)^2.$$

Make sure that the middle term of the polynomial being factored, $-40pq$ here, is twice the product of the two terms in $4p - 5q^2$.

$$-40pq = 2(4p)(-5q)$$

(b) $169x^2 + 104xy^2 + 16y^4 = (13x + 4y^2)^2$, since $2(13x)(4y^2) = 104xy^2$. ▲ ⟨6⟩

▼ **EXAMPLE 8**   Factor each of the following.

(a) $12x^2 - 26x - 10$

Solution Look first for a greatest common factor. Here, the greatest common factor is 2: $12x^2 - 26x - 10 = 2(6x^2 - 13x - 5)$. Now try to factor $6x^2 - 13x - 5$. Possible factors of 6 are 3 and 2 or 6 and 1. The only factors of -5 are -5 and 1 or 5 and -1. Try various combinations. You should find the trinomial factors as $(3x + 1)(2x - 5)$. Thus,

$$12x^2 - 26x - 10 = 2(3x + 1)(2x - 5).$$

(b) $4z^2 + 12z + 9 - w^2$

Solution There is no common factor here, but notice that the first three terms can be factored as a perfect square.

$$4z^2 + 12z + 9 - w^2 = (2z + 3)^2 - w^2$$

Written in this form, the expression is the difference of squares, which can be factored as follows:

$$(2z + 3)^2 - w^2 = [(2z + 3) + w][(2z + 3) - w]$$
$$= (2z + 3 + w)(2z + 3 - w).$$

(c) $16a^2 - 100 - 48ac + 36c^2$

Solution Factor out the greatest common factor of 4 first.

$$16a^2 - 100 - 48ac + 36c^2 = 4[4a^2 - 25 - 12ac + 9c^2]$$
$$= 4[(4a^2 - 12ac + 9c^2) - 25] \quad \text{Rearrange terms and group.}$$
$$= 4[(2a - 3c)^2 - 25] \quad \text{Factor.}$$
$$= 4(2a - 3c + 5)(2a - 3c - 5) \quad \text{Factor the difference of squares.}$$
▲ ⟨7⟩

CAUTION Remember always to look first for a greatest common factor.

⟨6⟩ Factor.
(a) $4m^2 + 4m + 1$
(b) $25z^2 - 80zt + 64t^2$
Answers:
(a) $(2m + 1)^2$
(b) $(5z - 8t)^2$

⟨7⟩ Factor.
(a) $6x^2 - 27x - 15$
(b) $9r^2 + 12r + 4 - t^2$
(c) $18 - 8xy - 2y^2 - 8x^2$
Answers:
(a) $3(2x + 1)(x - 5)$
(b) $(3r + 2 + t) \cdot (3r + 2 - t)$
(c) $2(3 - 2x - y) \cdot (3 + 2x + y)$

Higher Degree Polynomials

Polynomials of degree greater than 2 are often difficult to factor. However, factoring is relatively easy in two cases: *the difference and the sum of cubes*. By multiplying out the right side, you can readily verify each of the following factorizations.

$$x^3 - y^3 = (x - y)(x^2 + xy + y^2) \qquad \text{Difference of cubes}$$
$$x^3 + y^3 = (x + y)(x^2 - xy + y^2) \qquad \text{Sum of cubes}$$

▼ **EXAMPLE 9** Factor each of the following.

(a) $k^3 - 8$

Solution Since $8 = 2^3$, use the pattern for the difference of two cubes to obtain

$$k^3 - 8 = k^3 - 2^3 = (k - 2)(k^2 + 2k + 4).$$

(b) $m^3 + 125 = m^3 + 5^3 = (m + 5)(m^2 - 5m + 25)$

(c) $8k^3 - 27z^3 = (2k)^3 - (3z)^3 = (2k - 3z)(4k^2 + 6kz + 9z^2)$ ▲ ⟨8⟩

Substitution and appropriate factoring patterns can sometimes be used to factor higher degree expressions.

▼ **EXAMPLE 10** Factor the following.

(a) $x^8 + 4x^4 + 3$

Solution The idea is to make a substitution that reduces the polynomial to a quadratic or cubic that we can deal with. Note that $x^8 = (x^4)^2$. Let $u = x^4$. Then

$$
\begin{aligned}
x^8 + 4x^4 + 3 &= (x^4)^2 + 4x^4 + 3 &&\text{Power of a power}\\
&= u^2 + 4u + 3 &&\text{Substitute } x^4 = u.\\
&= (u + 3)(u + 1) &&\text{Factor}\\
&= (x^4 + 3)(x^4 + 1) &&\text{Substitute } u = x^4.
\end{aligned}
$$

(b) $x^4 - y^4$.

Solution Note that $x^4 = (x^2)^2$ and similarly for the y term. Let $u = x^2$ and $v = y^2$. Then

$$
\begin{aligned}
x^4 - y^4 &= (x^2)^2 - (y^2)^2 &&\text{Power of a power}\\
&= u^2 - v^2 &&\text{Substitute } x^2 = u \text{ and } y^2 = v.\\
&= (u + v)(u - v) &&\text{Difference of squares}\\
&= (x^2 + y^2)(x^2 - y^2) &&\text{Substitute } u = x^2 \text{ and } v = y^2.\\
&= (x^2 + y^2)(x + y)(x - y) &&\text{Difference of squares} \quad ▲ ⟨9⟩
\end{aligned}
$$

Once you understand Example 10, you can often factor without explicit substitutions.

⟨8⟩

Factor the following.

(a) $a^3 + 1000$

(b) $z^3 - 64$

(c) $1000m^3 - 27z^3$

Answers:

(a) $(a + 10) \cdot$
$\qquad (a^2 - 10a + 100)$

(b) $(z - 4)(z^2 + 4z + 16)$

(c) $(10m - 3z) \cdot$
$\qquad (100m^2 + 30mz + 9z^2)$

⟨9⟩

Factor each of these.

(a) $2x^4 + 5x^2 + 2$

(b) $3x^4 - x^2 - 2$

Answers:

(a) $(2x^2 + 1)(x^2 + 2)$

(b) $(3x^2 + 2)(x + 1) \cdot$
$\qquad (x - 1)$

⟨10⟩

Factor $81x^4 - 16y^4$.

Answer:

$(9x^2 + 4y^2)(3x + 2y) \cdot$
$(3x - 2y)$

▼ **EXAMPLE 11**   Factor $256k^4 - 625m^4$

Solution Use the difference of squares twice, as follows:

$$256k^4 - 625m^4 = (16k^2)^2 - (25m^2)^2$$
$$= (16k^2 + 25m^2)(16k^2 - 25m^2)$$
$$= (16k^2 + 25m^2)(4k + 5m)(4k - 5m). \; ▲ \; ⟨10⟩$$

◆ 1.3 Exercises

Factor out the greatest common factor in each of the following. (See Example 1.)

1. $12x^2 - 24x$
2. $5y - 65xy$
3. $r^3 - 5r^2 + r$
4. $t^3 + 3t^2 + 8t$
5. $6z^3 - 12z^2 + 18z$
6. $5x^3 + 55x^2 + 10x$
7. $3(2y - 1)^2 + 7(2y - 1)^3$
8. $(3x + 7)^5 - 4(3x + 7)^3$
9. $3(x + 5)^4 + (x + 5)^6$
10. $3(x + 6)^2 + 6(x + 6)^4$

Factor the polynomial. (See Example 2.)

11. $x^2 + 5x + 4$
12. $u^2 + 7u + 6$
13. $x^2 + 7x + 12$
14. $y^2 + 8y + 12$
15. $z^2 + 10z + 24$
16. $r^2 + 16r + 60$

Factor the polynomial. (See Examples 3–5.)

17. $2x^2 - 9x + 4$
18. $3w^2 - 8w + 4$
19. $15p^2 - 23p + 4$
20. $8x^2 - 14x + 3$
21. $4z^2 - 16z + 15$
22. $12y^2 - 29y + 15$
23. $6x^2 - 5x - 4$
24. $12z^2 + z - 1$
25. $10y^2 + 21y - 10$
26. $15u^2 + 4u - 4$
27. $6x^2 + 5x - 4$
28. $12y^2 + 7y - 10$

Factor each polynomial completely. Factor out the greatest common factor as necessary. (See Examples 2–8.)

29. $3a^2 + 2a - 5$
30. $6a^2 - 48a - 120$
31. $x^2 - 81$
32. $x^2 + 17xy + 72y^2$
33. $9p^2 - 12p + 4$
34. $3r^2 - r - 2$
35. $r^2 + 3rt - 10t^2$
36. $2a^2 + ab - 6b^2$
37. $m^2 - 8mn + 16n^2$
38. $8k^2 - 16k - 10$

39. $4p^2 - 9$
40. $8r^2 + r + 6$
41. $3x^2 - 24xz + 48z^2$
42. $9m^2 - 25$
43. $a^2 + 4ab + 5b^2$
44. $6y^2 - 11y - 7$
45. $-x^2 + 7x - 12$
46. $4y^2 + y - 3$
47. $3a^2 - 13a - 30$
48. $3k^2 + 2k - 8$
49. $21m^2 + 13mn + 2n^2$
50. $81y^2 - 100$
51. $20y^2 + 39yx - 11x^2$
52. $12s^2 + 11st - 5t^2$
53. $y^2 - 4yz - 21z^2$
54. $49a^2 + 9$
55. $121x^2 - 64$
56. $4z^2 + 56zy + 196y^2$
57. $5m^3(m^3 - 1)^2 - 3m^5(m^3 - 1)^3$
58. $9(x - 4)^5 - (x - 4)^3$

Factor each of these. (See Example 9.)

59. $a^3 - 216$
60. $b^3 + 125$
61. $8r^3 - 27s^3$
62. $1000p^3 + 27q^3$
63. $64m^3 + 125$
64. $216y^3 - 343$
65. $1000y^3 - z^3$
66. $125p^3 + 8q^3$

Factor each of these. (See Examples 10 and 11.)

67. $x^4 + 5x^2 + 6$
68. $y^4 + 7y^2 + 10$
69. $b^4 - b^2$
70. $z^4 - 3z^2 - 4$
71. $x^4 - x^2 - 12$
72. $4x^4 + 27x^2 - 81$
73. $16a^4 - 81b^4$
74. $x^6 - y^6$
75. $x^8 + 8x^2$
76. $x^9 - 64x^3$

77. When asked to factor $6x^4 - 3x^2 - 3$ completely, a student gave the following result:

$$6x^4 - 3x^2 - 3 = (2x^2 + 1)(3x^2 - 3).$$

Is this answer correct? Explain why.

78. When can the sum of two squares be factored? Give examples.

79. Explain why $(x + 2)^3$ is not the correct factorization of $x^3 + 8$, and give the correct factorization.

80. Describe how factoring and multiplication are related. Give examples.

1.4 Rational Expressions

A **rational expression** is an expression that can be written as the quotient of two polynomials, such as

$$\frac{8}{x - 1}, \qquad \frac{3x^2 + 4x}{5x - 6}, \qquad \text{and} \qquad \frac{2y + 1}{y^4 + 8}.$$

It is sometimes important to know the values of the variable that make the denominator 0 (in which case the quotient is not defined). For example, 1 cannot be used as a replacement for x in the first expression above and $6/5$ cannot be used in the second one, since these values make the respective denominators equal 0. *Throughout this section, we assume that all denominators are nonzero*, which means that some replacement values for the variables may have to be excluded. ◇①

Simplifying Rational Expressions

A key tool for simplification is the following fact.

For all expressions P, Q, R, and S, with $Q \neq 0$ and $S \neq 0$

$$\frac{PS}{QS} = \frac{P}{Q}. \qquad \text{Cancellation Property}$$

▼ **EXAMPLE 1** Write each of the following rational expressions in lowest terms (so that the numerator and denominator have no common factor with integer coefficients except 1 or -1).

(a) $\dfrac{12m}{-18}$

Solution Both $12m$ and -18 are divisible by 6. By the cancellation property,

$$\frac{12m}{-18} = \frac{2m \cdot 6}{-3 \cdot 6}$$

$$= \frac{2m}{-3}$$

$$= -\frac{2m}{3}.$$

(b) $\dfrac{8x + 16}{4} = \dfrac{8(x + 2)}{4} = \dfrac{4 \cdot 2(x + 2)}{4} = \dfrac{2(x + 2)}{1} = 2(x + 2)$

Solution The numerator, $8x + 16$, was factored so that the common factor could be identified. The answer could also be written as $2x + 4$ if desired.

(c) $\dfrac{k^2 + 7k + 12}{k^2 + 2k - 3} = \dfrac{(k + 4)(k + 3)}{(k - 1)(k + 3)} = \dfrac{k + 4}{k - 1}$ ▲ ②

①
What values of the variable make each denominator equal 0?

(a) $\dfrac{5}{x - 3}$

(b) $\dfrac{2x - 3}{4x - 1}$

(c) $\dfrac{x + 2}{x}$

(d) Why do we need to determine these values?

Answers:

(a) 3

(b) 1/4

(c) 0

(d) Because division by 0 is undefined

②
Write each of the following in lowest terms.

(a) $\dfrac{12k + 36}{18}$

(b) $\dfrac{15m + 30m^2}{5m}$

(c) $\dfrac{2p^2 + 3p + 1}{p^2 + 3p + 2}$

Answers:

(a) $\dfrac{2(k + 3)}{3}$ or $\dfrac{2k + 6}{3}$

(b) $3(1 + 2m)$ or $3 + 6m$

(c) $\dfrac{2p + 1}{p + 2}$

Multiplication and Division

The rules for multiplying and dividing rational expressions are the same fraction rules you learned in arithmetic.

For all expressions P, Q, R, and S, with $Q \neq 0$ and $S \neq 0$,

$$\frac{P}{Q} \cdot \frac{R}{S} = \frac{PR}{QS}.$$
Multiplication Rule

$$\frac{P}{Q} \div \frac{R}{S} = \frac{P}{Q} \cdot \frac{S}{R} \qquad (R \neq 0)$$
Division Rule

▼ **EXAMPLE 2**

(a) Multiply $\dfrac{2}{3} \cdot \dfrac{y}{5}$.

Solution Use the multiplication rule. Multiply the numerators and then the denominators.

$$\frac{2}{3} \cdot \frac{y}{5} = \frac{2 \cdot y}{3 \cdot 5} = \frac{2y}{15}$$

The result, $2y/15$, is in lowest terms.

(b) $\dfrac{3y + 9}{6} \cdot \dfrac{18}{5y + 15}$

Solution Factor where possible.

$$\frac{3y + 9}{6} \cdot \frac{18}{5y + 15} = \frac{3(y + 3)}{6} \cdot \frac{18}{5(y + 3)}$$

$$= \frac{3 \cdot 18(y + 3)}{6 \cdot 5(y + 3)} \qquad \text{Multiply numerators and denominators.}$$

$$= \frac{3 \cdot 6 \cdot 3(y + 3)}{6 \cdot 5(y + 3)} \qquad 18 = 6 \cdot 3$$

$$= \frac{3 \cdot 3}{5} \qquad \text{Write in lowest terms.}$$

$$= \frac{9}{5}$$

(c) $\dfrac{m^2 + 5m + 6}{m + 3} \cdot \dfrac{m^2 + m - 6}{m^2 + 3m + 2}$

Solution

$$= \frac{(m + 2)(m + 3)}{m + 3} \cdot \frac{(m - 2)(m + 3)}{(m + 2)(m + 1)} \qquad \text{Factor.}$$

$$= \frac{(m + 2)(m + 3)(m - 2)(m + 3)}{(m + 3)(m + 2)(m + 1)} \qquad \text{Multiply.}$$

$$= \frac{(m-2)(m+3)}{m+1} \qquad \text{Lowest terms}$$

$$= \frac{m^2 + m - 6}{m+1} \quad \blacktriangle \; \langle 3 \rangle$$

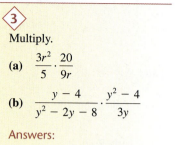

Multiply.

(a) $\dfrac{3r^2}{5} \cdot \dfrac{20}{9r}$

(b) $\dfrac{y-4}{y^2 - 2y - 8} \cdot \dfrac{y^2 - 4}{3y}$

Answers:

(a) $\dfrac{4r}{3}$

(b) $\dfrac{y-2}{3y}$

▼ **EXAMPLE 3**

(a) Divide $\dfrac{8x}{5} \div \dfrac{11x^2}{20}$.

    Solution   Invert the second expression and multiply [division rule].

$$\frac{8x}{5} \div \frac{11x^2}{20} = \frac{8x}{5} \cdot \frac{20}{11x^2} \qquad \text{Invert and multiply.}$$

$$= \frac{8x \cdot 20}{5 \cdot 11x^2} \qquad \text{Multiply.}$$

$$= \frac{32}{11x} \qquad \text{Lowest terms}$$

(b) $\dfrac{9p - 36}{12} \div \dfrac{5(p-4)}{18}$

    Solution

$$= \frac{9p - 36}{12} \cdot \frac{18}{5(p-4)} \qquad \text{Invert and multiply.}$$

$$= \frac{9(p-4)}{12} \cdot \frac{18}{5(p-4)} \qquad \text{Factor.}$$

$$= \frac{27}{10} \qquad \text{Cancel, multiply and write in lowest terms.} \quad \blacktriangle \; \langle 4 \rangle$$

4

Divide.

(a) $\dfrac{5m}{16} \div \dfrac{m^2}{10}$

(b) $\dfrac{2y - 8}{6} \div \dfrac{5y - 20}{3}$

(c) $\dfrac{m^2 - 2m - 3}{m(m+1)} \div \dfrac{m+4}{5m}$

Answers:

(a) $\dfrac{25}{8m}$

(b) $\dfrac{1}{5}$

(c) $\dfrac{5(m-3)}{m+4}$

Addition and Subtraction

As you know, when two numerical fractions have the same denominator, they can be added or subtracted. The same rules apply to rational expressions.

For all expressions P, Q, R, with $Q \neq 0$,

$$\frac{P}{Q} + \frac{R}{Q} = \frac{P + R}{Q}. \qquad \text{Addition Rule}$$

$$\frac{P}{Q} - \frac{R}{Q} = \frac{P - R}{Q}. \qquad \text{Subtraction Rule}$$

▼ **EXAMPLE 4**   Add or subtract as indicated.

(a) $\dfrac{4}{5k} + \dfrac{11}{5k}$

Solution Since the denominators are the same, we add the numerators.

$$\frac{4}{5k} + \frac{11}{5k} = \frac{4+11}{5k} = \frac{15}{5k} \qquad \text{Addition rule}$$

$$= \frac{3}{k} \qquad \text{Lowest terms}$$

(b) $\dfrac{2x^2 + 3x + 1}{x^5 + 1} - \dfrac{x^2 - 7x}{x^5 + 1}$

Solution The denominators are the same, so we subtract numerators, paying careful attention to parentheses.

$$\frac{2x^2 + 3x + 1}{x^5 + 1} - \frac{x^2 - 7x}{x^5 + 1} = \frac{(2x^2 + 3x + 1) - (x^2 - 7x)}{x^5 + 1} \qquad \begin{array}{l}\text{Subtraction}\\ \text{rule}\end{array}$$

$$= \frac{2x^2 + 3x + 1 - x^2 - (-7x)}{x^5 + 1} \qquad \begin{array}{l}\text{Subtract}\\ \text{numerators.}\end{array}$$

$$= \frac{2x^2 + 3x + 1 - x^2 + 7x}{x^5 + 1}$$

$$= \frac{x^2 + 10x + 1}{x^5 + 1} \qquad \begin{array}{l}\text{Simplify the}\\ \text{numerator.}\end{array}$$

▲

When fractions do not have the same denominator, you must first find a common denominator before you can add or subtract. A common denominator is a denominator that has each fraction's denominator as a factor.

▼ **EXAMPLE 5**   Add or subtract as indicated.

(a) $\dfrac{7}{p^2} + \dfrac{9}{2p} + \dfrac{1}{3p^2}$

Solution These three denominators are different, so we must find a common denominator that has each of p^2, $2p$, and $3p^2$ as factors. Observe that $6p^2$ satisfies these requirements. Use the cancellation property to rewrite each fraction as one that has $6p^2$ as denominator, then add them.

$$\frac{7}{p^2} + \frac{9}{2p} + \frac{1}{3p^2} = \frac{6 \cdot 7}{6 \cdot p^2} + \frac{3p \cdot 9}{3p \cdot 2p} + \frac{2 \cdot 1}{2 \cdot 3p^2} \qquad \text{Cancellation property}$$

$$= \frac{42}{6p^2} + \frac{27p}{6p^2} + \frac{2}{6p^2}$$

$$= \frac{42 + 27p + 2}{6p^2} \qquad \text{Addition rule}$$

$$= \frac{27p + 44}{6p^2} \qquad \text{Simplify.}$$

(b) $\dfrac{k^2}{k^2 - 1} - \dfrac{2k^2 - k - 3}{k^2 + 3k + 2}$

Solution Factor the denominators to find a common denominator.

$$\frac{k^2}{k^2 - 1} - \frac{2k^2 - k - 3}{k^2 + 3k + 2} = \frac{k^2}{(k+1)(k-1)} - \frac{2k^2 - k - 3}{(k+1)(k+2)}$$

A common denominator here is $(k+1)(k-1)(k+2)$ because each of the preceding denominators is a factor of this common denominator. Write each fraction with the common denominator.

$$\frac{k^2}{(k+1)(k-1)} - \frac{2k^2 - k - 3}{(k+1)(k+2)}$$

$$= \frac{k^2(k+2)}{(k+1)(k-1)(k+2)} - \frac{(2k^2 - k - 3)(k-1)}{(k+1)(k-1)(k+2)}$$

$$= \frac{k^3 + 2k^2 - (2k^2 - k - 3)(k-1)}{(k+1)(k-1)(k+2)} \qquad \text{Subtract fractions.}$$

$$= \frac{k^3 + 2k^2 - (2k^3 - 3k^2 - 2k + 3)}{(k+1)(k-1)(k+2)} \qquad \begin{array}{l}\text{Multiply}\\(2k^2 - k - 3)(k-1).\end{array}$$

$$= \frac{k^3 + 2k^2 - 2k^3 + 3k^2 + 2k - 3}{(k+1)(k-1)(k+2)} \qquad \text{Polynomial subtraction}$$

$$= \frac{-k^3 + 5k^2 + 2k - 3}{(k+1)(k-1)(k+2)} \qquad \text{Combine terms.} \ \blacktriangle \ ⑤$$

Complex Fractions

Any quotient of rational expressions is called a **complex fraction.** Complex fractions are simplified as follows.

▼ **EXAMPLE 6** Simplify the complex fraction.

$$\frac{6 - \dfrac{5}{k}}{1 + \dfrac{5}{k}}$$

Solution Multiply both numerator and denominator by the common denominator k.

$$\frac{6 - \dfrac{5}{k}}{1 + \dfrac{5}{k}} = \frac{k\left(6 - \dfrac{5}{k}\right)}{k\left(1 + \dfrac{5}{k}\right)} \qquad \text{Multiply by } \dfrac{k}{k}.$$

$$= \frac{6k - k\left(\dfrac{5}{k}\right)}{k + k\left(\dfrac{5}{k}\right)} \qquad \text{Distributive property}$$

$$= \frac{6k - 5}{k + 5} \qquad \text{Simplify.} \ \blacktriangle$$

⑤

Add or subtract.

(a) $\dfrac{3}{4r} + \dfrac{8}{3r}$

(b) $\dfrac{1}{m-2} - \dfrac{3}{2(m-2)}$

(c) $\dfrac{p+1}{p^2 - p} - \dfrac{p^2 - 1}{p^2 + p - 2}$

Answers:

(a) $\dfrac{41}{12r}$

(b) $\dfrac{-1}{2(m-2)}$

(c) $\dfrac{-p^3 + p^2 + 4p + 2}{p(p-1)(p+2)}$

◆ 1.4 Exercises

Write each of the following in lowest terms. Factor as necessary. (See Example 1.)

1. $\dfrac{8x^2}{56x}$

2. $\dfrac{27m}{81m^3}$

3. $\dfrac{25p^2}{35p^3}$

4. $\dfrac{18y^4}{24y^2}$

5. $\dfrac{5m+15}{4m+12}$

6. $\dfrac{10z+5}{20z+10}$

7. $\dfrac{4(w-3)}{(w-3)(w+6)}$

8. $\dfrac{-6(x+2)}{(x+4)(x+2)}$

9. $\dfrac{3y^2-12y}{9y^3}$

10. $\dfrac{15k^2+45k}{9k^2}$

11. $\dfrac{m^2-4m+4}{m^2+m-6}$

12. $\dfrac{r^2-r-6}{r^2+r-12}$

13. $\dfrac{x^2+4x-5}{x^2-1}$

14. $\dfrac{z^2-5z+6}{z^2-4}$

Multiply or divide as indicated in each of the following. Write all answers in lowest terms. (See Examples 2 and 3.)

15. $\dfrac{4p^3}{49} \cdot \dfrac{7}{2p^2}$

16. $\dfrac{24n^4}{6n^2} \cdot \dfrac{18n^2}{9n}$

17. $\dfrac{21a^5}{14a^3} \div \dfrac{8a}{12a^2}$

18. $\dfrac{2x^3}{6x^2} \div \dfrac{10x^2}{18x}$

19. $\dfrac{2a+b}{3c} \cdot \dfrac{15}{4(2a+b)}$

20. $\dfrac{4(x+2)}{w} \cdot \dfrac{3w^2}{8(x+2)}$

21. $\dfrac{15p-3}{6} \div \dfrac{10p-2}{3}$

22. $\dfrac{2k+8}{6} \div \dfrac{3k+12}{3}$

23. $\dfrac{9y-18}{6y+12} \cdot \dfrac{3y+6}{15y-30}$

24. $\dfrac{12r+24}{36r-36} \div \dfrac{6r+12}{8r-8}$

25. $\dfrac{4a+12}{2a-10} \div \dfrac{a^2-9}{a^2-a-20}$

26. $\dfrac{6r-18}{9r^2+6r-24} \cdot \dfrac{12r-16}{4r-12}$

27. $\dfrac{k^2-k-6}{k^2+k-12} \cdot \dfrac{k^2+3k-4}{k^2+2k-3}$

28. $\dfrac{n^2-n-6}{n^2-2n-8} \div \dfrac{n^2-9}{n^2+7n+12}$

29. In your own words, explain how to find the least common denominator of two fractions.

30. Describe the steps required to add three rational expressions. You may use an example to illustrate.

Add or subtract as indicated in each of the following. Write all answers in lowest terms. (See Example 4.)

31. $\dfrac{3}{5z} - \dfrac{1}{3z}$

32. $\dfrac{7}{4z} - \dfrac{5}{3z}$

33. $\dfrac{r+2}{3} - \dfrac{r-2}{3}$

34. $\dfrac{3y-1}{8} - \dfrac{3y+1}{8}$

35. $\dfrac{4}{x} + \dfrac{1}{5}$

36. $\dfrac{6}{r} - \dfrac{3}{4}$

37. $\dfrac{1}{m-1} + \dfrac{2}{m}$

38. $\dfrac{8}{y+2} - \dfrac{3}{y}$

39. $\dfrac{8}{3(a-1)} + \dfrac{3}{a-1}$

40. $\dfrac{5}{2(k+3)} + \dfrac{2}{k+3}$

41. $\dfrac{2}{5(k-2)} + \dfrac{5}{4(k-2)}$

42. $\dfrac{11}{3(p+4)} - \dfrac{5}{6(p+4)}$

43. $\dfrac{2}{x^2-4x+3} + \dfrac{5}{x^2-x-6}$

44. $\dfrac{3}{m^2-3m-10} + \dfrac{7}{m^2-m-20}$

45. $\dfrac{2y}{y^2+7y+12} - \dfrac{y}{y^2+5y+6}$

46. $\dfrac{-r}{r^2-10r+16} - \dfrac{3r}{r^2+2r-8}$

47. $\dfrac{3k}{2k^2+3k-2} - \dfrac{2k}{2k^2-7k+3}$

48. $\dfrac{4m}{3m^2+7m-6} - \dfrac{m}{3m^2-14m+8}$

In each of the following exercises, simplify the complex fraction. (See Example 5.)

49. $\dfrac{1+\dfrac{1}{x}}{1-\dfrac{1}{x}}$

50. $\dfrac{2-\dfrac{2}{y}}{2+\dfrac{2}{y}}$

51. $\dfrac{\dfrac{1}{x+h}-\dfrac{1}{x}}{h}$

52. $\dfrac{\dfrac{1}{(x+h)^2}-\dfrac{1}{x^2}}{h}$

53. Social Sciences   When ten cars per minute, on average, are arriving at the entrance to an amusement park and the average number being admitted per minute is x (with $x > 10$), then the average waiting time in minutes for each car is given by

$$\frac{x^2 - 10x + 25}{x^3 - 15x^2 + 50x}.^*$$

(a) Reduce this fraction to lowest terms.
(b) Use the reduced fraction to determine the average waiting time *in seconds* when $x = 11$, 15, and 20.

54. Business   In Example 10 of Section 1.2, we saw that the cost C of producing x thousand CDs is given by

$$C = -8x^2 + 3550x + 215{,}000 \quad (x \le 200).$$

(a) Write a rational expression in lowest terms that gives the average cost per CD when x thousand are produced. (*Hint:* The average cost is the total cost C divided by the number of CDs produced.)
(b) Find the average cost per CD for each of these production levels: 20,000, 50,000, and 150,000.

55. Business   The cost (in millions of dollars) for a 30-second ad during the TV broadcast of the Superbowl can be approximated by

$$\frac{.072x^2 + .744x + 1.2}{x + 2},$$

where $x = 0$ corresponds to 1980.*
(a) How much did an ad cost in 2005?
(b) If this trend continues, will the cost of an ad reach $4 million by 2020?

56. Health   The average company cost per hour of an employee's health insurance in year x is approximated by

$$\frac{.07x^2 + 1.15x + 1.08}{x + 1},$$

where $x = 0$ corresponds to 1998.†
(a) What was the hourly health insurance cost in 2004?
(b) Assuming that this model remains accurate and that an employee works 2100 hours per year, what is the annual company cost of her health insurance in 2008? Will annual costs reach $4200 by 2011?

---

*L. Haefner, *Introduction to Transportation Systems,* New York: Holt, Rinehart and Winston, 1986.

*Based on data from *Ad Age* and the *St. Louis Post–Dispatch.*

†Based on data from the U.S. Department of Labor.

1.5 Exponents and Radicals

Exponents were introduced in Section 1.2. In this section, the definition of exponents will be extended to include negative exponents and rational number exponents such as 1/2 and 7/3.

Integer Exponents

Positive integer and zero exponents were defined in Section 1.2, where we noted that

$$a^m \cdot a^n = a^{m+n}$$

for nonegative integers m and n. Now we develop an analogous property for quotients. By definition,

$$\frac{6^5}{6^2} = \frac{6 \cdot 6 \cdot 6 \cdot 6 \cdot 6}{6 \cdot 6} = 6 \cdot 6 \cdot 6 = 6^3.$$

Because there are 5 factors of 6 in the numerator and 2 factors of 6 in the denominator, the quotient has $5 - 2 = 3$ factors of 6. In general, we can make the following statement, which applies to any real number a and nonnegative integers m and n with $m > n$.

Division with Exponents

To divide a^m by a^n, *subtract* the exponents:

$$\frac{a^m}{a^n} = a^{m-n}.$$

▼ **EXAMPLE 1**   Compute each of the following.

(a) $\dfrac{5^7}{5^4} = 5^{7-4} = 5^3$

(b) $\dfrac{(-8)^{10}}{(-8)^5} = (-8)^{10-5} = (-8)^5$

(c) $\dfrac{(3c)^9}{(3c)^3} = (3c)^{9-3} = (3c)^6$ ▲ ◇1◇

When an exponent applies to the product of two numbers, such as $(7 \cdot 19)^3$, use the definitions carefully. For instance,

$$(7 \cdot 19)^3 = (7 \cdot 19)(7 \cdot 19)(7 \cdot 19) = 7 \cdot 7 \cdot 7 \cdot 19 \cdot 19 \cdot 19 = 7^3 \cdot 19^3.$$

In other words, $(7 \cdot 19)^3 = 7^3 \cdot 19^3$. This is an example of the following fact, which applies to any real numbers a, b and any nonnegative integer exponent n.

◇1◇

Evaluate each of these.

(a) $\dfrac{2^{14}}{2^5}$

(b) $\dfrac{(-5)^9}{(-5)^5}$

(c) $\dfrac{(xy)^{17}}{(xy)^{12}}$

Answers:

(a) 2^9

(b) $(-5)^4$

(c) $(xy)^5$

Product to a Power

To find $(ab)^n$, apply the exponent to *every* term inside the parentheses:

$$(ab)^n = a^n b^n.$$

CAUTION A common mistake is to write an expression such as $(2x)^5$ as $2x^5$, rather than the correct answer $(2x)^5 = 2^5 x^5 = 32x^5$

Analogous conclusions are valid for quotients (where a, b are any real numbers with $b \neq 0$ and n is a nonnegative integer exponent).

Quotient to a Power

To find $\left(\dfrac{a}{b}\right)^n$, apply the exponent to both numerator and denominator:

$$\left(\frac{a}{b}\right)^n = \frac{a^n}{b^n}.$$

▼ **EXAMPLE 2**   Compute each of the following.

(a) $(5y)^3 = 5^3y^3 = 125y^3$ Product to a power

(b) $(c^2d^3)^4 = (c^2)^4(d^3)^4$ Product to a power

$= c^8d^{12}$ Power of a power

(c) $\left(\dfrac{x}{2}\right)^6 = \dfrac{x^6}{2^6} = \dfrac{x^6}{64}$ Quotient to a power

(d) $\left(\dfrac{a^4}{b^3}\right)^3 = \dfrac{(a^4)^3}{(b^3)^3}$ Quotient to a power

$= \dfrac{a^{12}}{b^9}$ Power of a power

(e) $\left(\dfrac{(rs)^3}{r^4}\right)^2$

Solution Use several of the preceding properties in succession.

$$\left(\frac{(rs)^3}{r^4}\right)^2 = \left(\frac{r^3s^3}{r^4}\right)^2$$ Product to a power in numerator

$$= \left(\frac{s^3}{r}\right)^2$$ Cancel

$$= \frac{(s^3)^2}{r^2}$$ Quotient to a power

$$= \frac{s^6}{r^2}$$ Power of a power in numerator

As is often the case, there is another way to reach the last expression. You should be able to supply the reasons for each of the following steps:

$$\left(\frac{(rs)^3}{r^4}\right)^2 = \frac{[(rs)^3]^2}{(r^4)^2} = \frac{(rs)^6}{r^8} = \frac{r^6s^6}{r^8} = \frac{s^6}{r^2}. \quad ▲ \; ⧫②$$

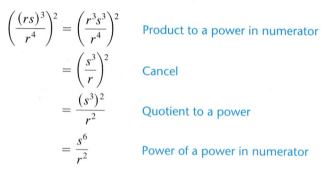

②
Compute each of these.

(a) $(3x)^4$

(b) $(r^2s^5)^6$

(c) $\left(\dfrac{2}{z}\right)^5$

(d) $\left(\dfrac{3a^5}{(ab)^3}\right)^2$

Answers:

(a) $81x^4$

(b) $r^{12}s^{30}$

(c) $\dfrac{32}{z^5}$

(d) $\dfrac{9a^4}{b^6}$

Negative Exponents

The next step is to define negative integer exponents. If they are to be defined in such a way that the quotient rule for exponents remains valid, then we must have, for example,

$$\frac{3^2}{3^4} = 3^{2-4} = 3^{-2}.$$

However,

$$\frac{3^2}{3^4} = \frac{3 \cdot 3}{3 \cdot 3 \cdot 3 \cdot 3} = \frac{1}{3^2},$$

which suggests that 3^{-2} should be defined to be $1/3^2$. Thus, we have the following definition of a negative exponent.

Negative Exponent

If n is a natural number, and if $a \neq 0$, then

$$a^{-n} = \frac{1}{a^n}.$$

▼ **EXAMPLE 3**   Evaluate the following.

(a) $3^{-2} = \dfrac{1}{3^2} = \dfrac{1}{9}$

(b) $5^{-4} = \dfrac{1}{5^4} = \dfrac{1}{625}$

(c) $x^{-1} = \dfrac{1}{x^1} = \dfrac{1}{x}$

(d) $-4^{-2} = -\dfrac{1}{4^2} = -\dfrac{1}{16}$

(e) $\left(\dfrac{3}{4}\right)^{-1} = \dfrac{1}{\left(\dfrac{3}{4}\right)^1} = \dfrac{1}{\dfrac{3}{4}} = \dfrac{4}{3}$ ▲ ◈3

There is a useful property that makes it easy to raise a fraction to a negative exponent. Consider, for example,

$$\left(\frac{2}{3}\right)^{-4} = \frac{1}{\left(\frac{2}{3}\right)^4} = \frac{1}{\left(\frac{2^4}{3^4}\right)} = 1 \cdot \frac{3^4}{2^4} = \left(\frac{3}{2}\right)^4.$$

This example is easily generalized to prove the following fact (in which a/b is a nonzero fraction and n a positive integer).

Inversion Property

$$\left(\frac{a}{b}\right)^{-n} = \left(\frac{b}{a}\right)^n$$

▼ **EXAMPLE 4**   Use the inversion property to compute each of the following.

(a) $\left(\dfrac{2}{5}\right)^{-3} = \left(\dfrac{5}{2}\right)^3 = \dfrac{5^3}{2^3} = \dfrac{125}{8}$

(b) $\left(\dfrac{3}{x}\right)^{-5} = \left(\dfrac{x}{3}\right)^5 = \dfrac{x^5}{3^5} = \dfrac{x^5}{243}$ ▲ ◈4

When keying in negative exponents on a calculator, be sure to use the negation key (labeled $(-)$ or $+/-$), not the subtraction key. Calculators normally display

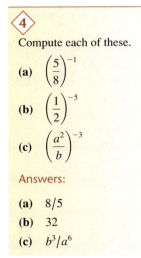

◇3

Evaluate the following.

(a) 6^{-2}

(b) -6^{-3}

(c) -3^{-4}

(d) $\left(\dfrac{5}{8}\right)^{-1}$

Answers:

(a) $1/36$

(b) $-1/216$

(c) $-1/81$

(d) $8/5$

◇4

Compute each of these.

(a) $\left(\dfrac{5}{8}\right)^{-1}$

(b) $\left(\dfrac{1}{2}\right)^{-5}$

(c) $\left(\dfrac{a^2}{b}\right)^{-3}$

Answers:

(a) $8/5$

(b) 32

(c) b^3/a^6

TECHNOLOGY TIP   The FRAC key is in the MATH menu of TI graphing calculators. A FRAC program for other graphing calculators is in the Program Appendix. Fractions can be displayed on some graphing calculators by changing the number display format (in the MODES menu) to "fraction" or "exact."

answers as decimals, as shown in Figure 1.7. Some graphing calculators have a FRAC key that converts these decimals to fractions, as shown in Figure 1.8.

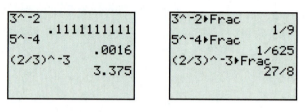

FIGURE 1.7   FIGURE 1.8

Roots and Rational Exponents

There are two numbers whose square is 16: 4 and −4. As we saw in Section 1.1, the positive one, 4, is called the *square root* (or second root) of 16. Similarly, there are two numbers whose fourth power is 16: 2 and −2. We call 2 the **fourth root** of 16. This suggests the following generalization.

> If n is even, the **nth root of a** is the positive real number whose nth power is a.

All nonnegative numbers have nth roots for every natural number n, but *no negative number has a real, even nth root*. For example, there is no real number whose square is −16, so −16 has no square root.

We say that the **cube root** (or third root) of 8 is 2 because $2^3 = 8$. Similarly, since $(-2)^3 = -8$, we say that −2 is the cube root of −8. Again, we can generalize.

> If n is odd, the **nth root of a** is the real number whose nth power is a.

Every real number has an nth root for every *odd* natural number n.

We can now define rational exponents. If they are to have the same properties as integer exponents, we want $a^{1/2}$ to be a number such that

$$(a^{1/2})^2 = a^{1/2} \cdot a^{1/2} = a^{1/2+1/2} = a^1 = a.$$

Thus, $a^{1/2}$ should be a number whose square is a, and it is reasonable to *define* $a^{1/2}$ to be the square root of a (if it exists). Similarly, $a^{1/3}$ is defined to be the cube root of a, and we have the following definition.

> If a is a real number and n is a positive integer, then
>
> $a^{1/n}$ is defined to be the nth root of a (if it exists).

▼ **EXAMPLE 5**   Examine the reasoning used to evaluate the following roots.

(a) $36^{1/2} = 6$ because $6^2 = 36$.

(b) $100^{1/2} = 10$ because $10^2 = 100$.

(c) $-(225^{1/2}) = -15$ because $15^2 = 225$.

(d) $625^{1/4} = 5$ because $5^4 = 625$.

(e) $(-1296)^{1/4}$ is not a real number, although $-1296^{1/4} = -6$ because $6^4 = 1296$.

(f) $(-27)^{1/3} = -3$ because $(-3)^3 = -27$.

(g) $-32^{1/5} = -2$ because $2^5 = 32$. ▲

A calculator can be used to evaluate expressions with fractional exponents. Whenever it's easy to do so, enter the fractional exponents in their equivalent decimal form. For instance, to find $625^{1/4}$, enter $625^{.25}$ into the calculator. When the decimal equivalent of a fraction is an infinitely repeating decimal, however, it is best to enter the fractional exponent directly, using parentheses, as in Figure 1.9. If you omit the parentheses or use a shortened decimal approximation (such as .333 for 1/3), you will not get the correct answers. Compare the answers in Figure 1.10 with the correct ones in Figure 1.9.

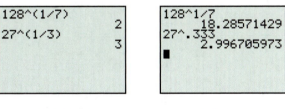

FIGURE 1.9   FIGURE 1.10

For other rational exponents, the symbol $a^{m/n}$ should be defined so that the properties for exponents still hold. For example, by the product property, we want

$$(a^{1/3})^2 = a^{1/3} \cdot a^{1/3} = a^{1/3+1/3} = a^{2/3}$$

This suggests the following definition.

For all integers m and all positive integers n, and for all real numbers a for which $a^{1/n}$ is a real number,

$$a^{m/n} = (a^{1/n})^m.$$

▼ **EXAMPLE 6**   Verify each of the following.

(a) $27^{2/3} = (27^{1/3})^2 = 3^2 = 9$

(b) $32^{2/5} = (32^{1/5})^2 = 2^2 = 4$

(c) $64^{4/3} = (64^{1/3})^4 = 4^4 = 256$

(d) $25^{3/2} = (25^{1/2})^3 = 5^3 = 125$ ▲

5

Evaluate the following.

(a) $16^{1/2}$

(b) $16^{1/4}$

(c) $-256^{1/2}$

(d) $(-256)^{1/2}$

(e) $-8^{1/3}$

(f) $243^{1/5}$

Answers:

(a) 4

(b) 2

(c) −16

(d) Not a real number

(e) −2

(f) 3

6

Evaluate the following.

(a) $16^{3/4}$

(b) $25^{5/2}$

(c) $32^{7/5}$

(d) $100^{3/2}$

Answers:

(a) 8

(b) 3125

(c) 128

(d) 1000

CAUTION When the base is negative, as in $(-8)^{2/3}$, some calculators produce an error message. On such calculators, you should first compute $(-8)^{1/3}$ and then square the result; that is, compute $[(-8)^{1/3}]^2$.

Rational exponents were defined so that one of the familiar properties of exponents remains valid. In fact, it can be proved that *all* of the rules developed earlier for integer exponents are valid for rational exponents. The following box summarizes these rules, which are illustrated in Examples 7–9.

Properties of Exponents

For any rational numbers m and n, and for any real numbers a and b for which the following exist,

(a) $a^m \cdot a^n = a^{m+n}$ Product property

(b) $\dfrac{a^m}{a^n} = a^{m-n}$ Quotient property

(c) $(a^m)^n = a^{mn}$ Power of a power

(d) $(ab)^m = a^m \cdot b^m$ Product to a power

(e) $\left(\dfrac{a}{b}\right)^m = \dfrac{a^m}{b^m}$ Quotient to a power

(f) $a^0 = 1$ Zero exponent

(g) $a^{-n} = \dfrac{1}{a^n}$ Negative exponent

(h) $\left(\dfrac{a}{b}\right)^{-n} = \left(\dfrac{b}{a}\right)^n$.   Inversion property

The power of a power property provides another way to compute $a^{m/n}$ (when it exists):

(1) $$a^{m/n} = a^{m(1/n)} = (a^m)^{1/n}.$$

For example, we can now find $4^{3/2}$ in two ways:

$$4^{3/2} = (4^{1/2})^3 = 2^3 = 8 \quad \text{or} \quad 4^{3/2} = (4^3)^{1/2} = 64^{1/2} = 8.$$

Definition of $a^{m/n}$ Statement (1)

▼ **EXAMPLE 7** Simplify each of the following.

(a) $7^{-4} \cdot 7^6 = 7^{-4+6} = 7^2 = 49$ Product property

(b) $5x^{2/3} \cdot 2x^{1/4} = 10x^{2/3}x^{1/4}$

$\qquad\qquad\qquad = 10x^{2/3+1/4}$ Product property

$\qquad\qquad\qquad = 10x^{11/12}$ $\dfrac{2}{3} + \dfrac{1}{4} = \dfrac{8}{12} + \dfrac{3}{12} = \dfrac{11}{12}$

(c) $\dfrac{9^{14}}{9^{-6}} = 9^{14-(-6)} = 9^{20}$ Quotient property

(d)
$$\frac{c^5}{2c^{4/3}} = \frac{1}{2} \cdot \frac{c^5}{c^{4/3}}$$

$$= \frac{1}{2}c^{5-4/3} \qquad \text{Quotient property}$$

$$= \frac{1}{2}c^{11/3} = \frac{c^{11/3}}{2} \qquad 5 - \frac{4}{3} = \frac{15}{3} - \frac{4}{3} = \frac{11}{3}$$

(e)
$$\frac{27^{1/3} \cdot 27^{5/3}}{27^3} = \frac{27^{1/3+5/3}}{27^3} \qquad \text{Product property}$$

$$= \frac{27^2}{27^3} = 27^{2-3} \qquad \text{Quotient property}$$

$$= 27^{-1} = \frac{1}{27} \qquad \text{Definition of negative exponent} \ \blacktriangle \ \langle 7 \rangle$$

You can use a calculator to check numerical computations, such as those in Example 7, by computing the left and right sides separately and confirming that the answers are the same in each case. Figure 1.11 shows this for part (e) of Example 7.

```
(27^(1/3)*27^(5/
3))/27^3
              .037037037
1/27
              .037037037
■
```

FIGURE 1.11

▼ **EXAMPLE 8** Perform the indicated operations.

(a) $(2^{-3})^{-4/7} = 2^{(-3)(-4/7)} = 2^{12/7}$ Power of a power

(b)
$$\left(\frac{3m^{5/6}}{y^{3/4}}\right)^2 = \frac{(3m^{5/6})^2}{(y^{3/4})^2} \qquad \text{Quotient to a power}$$

$$= \frac{3^2(m^{5/6})^2}{(y^{3/4})^2} \qquad \text{Product to a power}$$

$$= \frac{9m^{(5/6)2}}{y^{(3/4)2}} \qquad \text{Power of a power}$$

$$= \frac{9m^{5/3}}{y^{3/2}} \qquad \frac{5}{6} \cdot 2 = \frac{10}{6} = \frac{5}{3} \ \text{ and } \ \frac{3}{4} \cdot 2 = \frac{3}{2}$$

(c) $m^{2/3}(m^{7/3} + 2m^{1/3}) = m^{2/3}m^{7/3} + m^{2/3}2m^{1/3}$ Distributive property

$$= m^{2/3+7/3} + 2m^{2/3+1/3} = m^3 + 2m \qquad \text{Product rule} \ \blacktriangle \ \langle 8 \rangle$$

▼ **EXAMPLE 9** Simplify each expression. Give answers with only positive exponents.

(a) $\dfrac{(m^3)^{-2}}{m^4} = \dfrac{m^{-6}}{m^4} = m^{-6-4} = m^{-10} = \dfrac{1}{m^{10}}$

(b) $6y^{2/3} \cdot 2y^{-1/2} = 12y^{2/3-1/2} = 12y^{1/6}$

⟨7⟩

Simplify each of these.

(a) $9^7 \cdot 9^{-5}$

(b) $3x^{1/4} \cdot 5x^{5/4}$

(c) $\dfrac{8^7}{8^{-3}}$

(d) $\dfrac{5^{2/3} \cdot 5^{-4/3}}{5^2}$

Answers:

(a) 81

(b) $15x^{3/2}$

(c) 8^{10}

(d) $5^{-8/3}$ or $1/5^{8/3}$

⟨8⟩

Simplify the following.

(a) $(7^{-4})^{-2} \cdot (7^4)^{-2}$

(b) $\dfrac{c^4 c^{-1/2}}{c^{3/2}d^{1/2}}$

(c) $a^{5/8}(2a^{3/8} + a^{-1/8})$

Answers:

(a) 1

(b) $c^2/d^{1/2}$

(c) $2a + a^{1/2}$

(c) $\dfrac{x^{1/2}(x-2)^{-3}}{5(x-2)} = \dfrac{x^{1/2}}{5}\cdot\dfrac{(x-2)^{-3}}{x-2} = \dfrac{x^{1/2}}{5}\cdot(x-2)^{-3-1}$

$= \dfrac{x^{1/2}}{5}\cdot\dfrac{1}{(x-2)^4} = \dfrac{x^{1/2}}{5(x-2)^4}$

(d) Write $a^{-1}+b^{-1}$ as a single quotient.

Solution Be careful here. $a^{-1}+b^{-1}$ does *not* equal $(a+b)^{-1}$; the exponent properties deal only with products and quotients, not with sums. However, using the definition of negative exponents and addition of fractions, we have

$$a^{-1}+b^{-1} = \frac{1}{a}+\frac{1}{b} = \frac{b+a}{ab}. \; \blacktriangle$$

Radicals

Earlier, we denoted the nth root of a as $a^{1/n}$. An alternative notation for nth roots uses the radical symbol $\sqrt[n]{\ }$.

If n is an even natural number and $a \geq 0$, or if n is an odd natural number,

$$\sqrt[n]{a} = a^{1/n}.$$

In the radical expression $\sqrt[n]{a}$, a is called the radicand and n is called the index. When $n=2$, the familiar square-root symbol $\sqrt{a}$ is used instead of $\sqrt[2]{a}$.

▼ **EXAMPLE 10**   Simplify the following.

(a) $\sqrt[4]{16} = 16^{1/4} = 2$

(b) $\sqrt[5]{-32} = -2$

(c) $\sqrt[3]{1000} = 10$

(d) $\sqrt[6]{\dfrac{64}{729}} = \left(\dfrac{64}{729}\right)^{1/6} = \dfrac{64^{1/6}}{729^{1/6}} = \dfrac{2}{3} \; \blacktriangle$

Recall that $a^{m/n} = (a^{1/n})^m$ by definition and $a^{m/n} = (a^m)^{1/n}$ by statement **(1)** on page 38 (provided that all terms are defined). We translate these facts into radical notation as follows.

For all rational numbers m/n and all real numbers a for which $\sqrt[n]{a}$ exists,

$$a^{m/n} = (\sqrt[n]{a})^m \quad\text{or}\quad a^{m/n} = \sqrt[n]{a^m}.$$

9 Simplify the following. Give answers with only positive exponents.

(a) $(3x^{2/3})(2x^{-1})(y^{-1/3})^2$

(b) $\dfrac{(t^{-1})^2}{t^{-5}}$

(c) $\left(\dfrac{2k^{1/3}}{p^{5/4}}\right)^2\cdot\left(\dfrac{4k^{-2}}{p^5}\right)^{3/2}$

(d) $x^{-1}-y^{-2}$

Answers:

(a) $\dfrac{6}{x^{1/3}y^{2/3}}$

(b) t^3

(c) $32/(p^{10}k^{7/3})$

(d) $\dfrac{y^2-x}{xy^2}$

10 Simplify.

(a) $\sqrt[3]{27}$

(b) $\sqrt[4]{625}$

(c) $\sqrt[6]{64}$

(d) $\sqrt[3]{\dfrac{64}{125}}$

Answers:

(a) 3

(b) 5

(c) 2

(d) 4/5

Notice that $\sqrt[n]{x^n}$ cannot be written simply as x when n is even. For example, if $x = -5$, then

$$\sqrt{x^2} = \sqrt{(-5)^2} = \sqrt{25} = 5 \neq x.$$

However, $|-5| = 5$, so that $\sqrt{x^2} = |x|$ when x is -5. This is true in general.

For any real number a and any natural number n,

$$\sqrt[n]{a^n} = |a| \quad \text{if } n \text{ is even}$$

and

$$\sqrt[n]{a^n} = a \quad \text{if } n \text{ is odd.}$$

To avoid this difficulty that $\sqrt[n]{a^n}$ is not necessarily equal to a, we shall assume that all variables in radicands represent only nonnegative numbers, as they usually do in applications.

The properties of exponents can be written with radicals as follows.

For all real numbers a and b, and for positive integers n for which all indicated roots exist,

(a) $\quad \sqrt[n]{a} \cdot \sqrt[n]{b} = \sqrt[n]{ab}$ 　　　　　(b) $\quad \dfrac{\sqrt[n]{a}}{\sqrt[n]{b}} = \sqrt[n]{\dfrac{a}{b}} \quad (b \neq 0).$

▼ **EXAMPLE 11**　Simplify the following.

(a) $\sqrt{6} \cdot \sqrt{54} = \sqrt{6 \cdot 54} = \sqrt{324} = 18$

Alternatively, simplify $\sqrt{54}$ first.
$$\sqrt{6} \cdot \sqrt{54} = \sqrt{6} \cdot \sqrt{9 \cdot 6}$$
$$= \sqrt{6} \cdot 3\sqrt{6} = 3 \cdot 6 = 18$$

(b) $\sqrt{\dfrac{7}{64}} = \dfrac{\sqrt{7}}{\sqrt{64}} = \dfrac{\sqrt{7}}{8}$

(c) $\sqrt{75} - \sqrt{12}$

Solution　Note that $12 = 4 \cdot 3$ and that 4 is a perfect square. Similarly, $75 = 25 \cdot 3$ and 25 is a perfect square. Consequently,

$$\sqrt{75} - \sqrt{12} = \sqrt{25 \cdot 3} - \sqrt{4 \cdot 3} \qquad \textcolor{orange}{\text{Factor.}}$$
$$= \sqrt{25}\sqrt{3} - \sqrt{4}\sqrt{3} \qquad \textcolor{orange}{\text{Property (a)}}$$
$$= 5\sqrt{3} - 2\sqrt{3} = 3\sqrt{3} \qquad \textcolor{orange}{\text{Simplify.}} \quad ▲ \; ⑪$$

⑪
Simplify.

(a) $\sqrt{3} \cdot \sqrt{27}$

(b) $\sqrt{\dfrac{3}{49}}$

(c) $\sqrt{50} + \sqrt{72}$

Answers:

(a) 9

(b) $\dfrac{\sqrt{3}}{7}$

(c) $11\sqrt{2}$

Caution When a and b are nonzero real numbers,

$$\sqrt[n]{a + b} \text{ is } \textbf{NOT} \text{ equal to } \sqrt[n]{a} + \sqrt[n]{b}.$$

For example,

$$\sqrt{9 + 16} = \sqrt{25} = 5, \text{ but } \sqrt{9} + \sqrt{16} = 3 + 4 = 7,$$

so $\sqrt{9 + 16} \neq \sqrt{9} + \sqrt{16}$.

Multiplying radical expressions is much like multiplying polynomials.

▼ **EXAMPLE 12**   Multiply the following.

(a) $(\sqrt{2} + 3)(\sqrt{8} - 5) = \sqrt{2}(\sqrt{8}) - \sqrt{2}(5) + 3\sqrt{8} - 3(5)$ FOIL
$$= \sqrt{16} - 5\sqrt{2} + 3(2\sqrt{2}) - 15$$
$$= 4 - 5\sqrt{2} + 6\sqrt{2} - 15$$
$$= -11 + \sqrt{2}$$

(b) $(\sqrt{7} - \sqrt{10})(\sqrt{7} + \sqrt{10}) = (\sqrt{7})^2 - (\sqrt{10})^2$
$$= 7 - 10 = -3 \qquad ▲ ⟨12⟩$$

⟨12⟩

Multiply.

(a) $(\sqrt{5} - \sqrt{2})(3 + \sqrt{2})$

(b) $(\sqrt{3} + \sqrt{7})(\sqrt{3} - \sqrt{7})$

Answers:

(a) $3\sqrt{5} + \sqrt{10} - 3\sqrt{2} - 2$

(b) -4

Rationalizing Denominators and Numerators

Before the invention of calculators, it was customary to **rationalize the denominators** of fractions (that is, write equivalent fractions with no radicals in the denominator), because this made many computations easier. Although there is no longer a computational reason, rationalizing denominators (and sometimes numerators) is still used today to simplify expressions and to derive useful formulas.

▼ **EXAMPLE 13**   Rationalize each denominator.

(a) $\dfrac{4}{\sqrt{3}}$

Solution The key is to multiply by 1, with 1 written as a radical fraction:

$$\frac{4}{\sqrt{3}} = \frac{4}{\sqrt{3}} \cdot \mathbf{1} = \frac{4}{\sqrt{3}} \cdot \frac{\sqrt{3}}{\sqrt{3}} = \frac{4\sqrt{3}}{3}.$$

⟨13⟩

Rationalize the denominator.

(a) $\dfrac{2}{\sqrt{5}}$

(b) $\dfrac{1}{2 + \sqrt{3}}$

Answers:

(a) $\dfrac{2\sqrt{5}}{5}$

(b) $2 - \sqrt{3}$

(b) $\dfrac{1}{3 - \sqrt{2}}$

Solution The same technique works here, using $1 = \dfrac{3 + \sqrt{2}}{3 + \sqrt{2}}$:

$$\frac{1}{3 - \sqrt{2}} = \frac{1}{3 - \sqrt{2}} \cdot \mathbf{1} = \frac{1}{3 - \sqrt{2}} \cdot \frac{3 + \sqrt{2}}{3 + \sqrt{2}} = \frac{3 + \sqrt{2}}{(3 - \sqrt{2})(3 + \sqrt{2})}$$
$$= \frac{3 + \sqrt{2}}{9 - 2} = \frac{3 + \sqrt{2}}{7}. \qquad ▲ ⟨13⟩$$

▼ **EXAMPLE 14**   Rationalize the numerator of $\dfrac{2 + \sqrt{5}}{1 + \sqrt{3}}$.

Solution   As in part (b), we must write 1 as a suitable fraction. Since we want to rationalize the numerator here, we multiply by the fraction $1 = \dfrac{2 - \sqrt{5}}{2 - \sqrt{5}}$:

$$\frac{2 + \sqrt{5}}{1 + \sqrt{3}} = \frac{2 + \sqrt{5}}{1 + \sqrt{3}} \cdot \frac{2 - \sqrt{5}}{2 - \sqrt{5}} = \frac{4 - 5}{2 - \sqrt{5} + 2\sqrt{3} - \sqrt{3}\sqrt{5}}$$

$$= \frac{-1}{2 - \sqrt{5} + 2\sqrt{3} - \sqrt{15}}. \quad \blacktriangle$$

◆ 1.5 Exercises

Perform the indicated operations and simplify your answer (See Examples 1 and 2.)

1. $\dfrac{7^5}{7^3}$

2. $\dfrac{(-6)^{14}}{(-6)^6}$

3. $(4c)^2$

4. $(-3x)^3$

5. $\left(\dfrac{2}{x}\right)^5$

6. $\left(\dfrac{5}{xy}\right)^3$

7. $(3u^2)^3(2u^3)^2$

8. $\dfrac{(5v^2)^3}{(2v)^4}$

Perform the indicated operations and simplify your answer, which should not have any negative exponents. (See Examples 3 and 4.)

9. 6^{-1}

10. 10^{-3}

11. 2^{-5}

12. -5^{-3}

13. -7^{-4}

14. $(-x)^{-4}$

15. $(-y)^{-3}$

16. $\left(\dfrac{1}{6}\right)^{-2}$

17. $\left(\dfrac{1}{7}\right)^{-3}$

18. $\left(\dfrac{2}{5}\right)^{-4}$

19. $\left(\dfrac{4}{3}\right)^{-2}$

20. $\left(\dfrac{x}{y^2}\right)^{-2}$

21. $\left(\dfrac{a}{b^3}\right)^{-1}$

22. Explain why $-2^{-4} = -1/16$, but $(-2)^{-4} = 1/16$.

Evaluate each expression. Write all answers without exponents. Round decimal answers to two places. (See Examples 5 and 6.)

23. $49^{1/2}$

24. $8^{1/3}$

25. $(5.71)^{1/4}$

26. $(93.68)^{1/5}$

27. $27^{2/3}$

28. $24^{3/2}$

29. $-64^{2/3}$

30. $-64^{3/2}$

31. $(8/27)^{-4/3}$

32. $(27/64)^{-1/3}$

Simplify each expression. Write all answers using only positive exponents. (See Example 7.)

33. $\dfrac{4^{-2}}{4^3}$

34. $\dfrac{7^{-4}}{7^{-3}}$

35. $4^{-3} \cdot 4^6$

36. $9^{-9} \cdot 9^{10}$

37. $8^{2/3} \cdot 8^{-1/3}$

38. $12^{-3/4} \cdot 12^{1/4}$

39. $\dfrac{8^9 \cdot 8^{-7}}{8^{-3}}$

40. $\dfrac{5^{-4} \cdot 5^6}{5^{-1}}$

41. $\dfrac{9^{-5/3}}{9^{2/3} \cdot 9^{-1/5}}$

42. $\dfrac{3^{5/3} \cdot 3^{-3/4}}{3^{-1/4}}$

Simplify each expression. Assume all variables represent positive real numbers. Write answers with only positive exponents. (See Example 8 and 9.)

43. $\dfrac{z^6 \cdot z^2}{z^5}$

44. $\dfrac{k^6 \cdot k^9}{k^{12}}$

45. $\dfrac{2^{-1}(p^{-1})^3}{2p^{-4}}$

46. $\dfrac{(5x^3)^{-2}}{x^4}$

47. $(q^{-5}r^3)^{-1}$

48. $(2y^2z^{-2})^{-3}$

49. $(2p^{-1})^3 \cdot (5p^2)^{-2}$

50. $(5^{-1}m^2)^{-3} \cdot (3m^{-2})^4$

51. $(2p)^{1/2} \cdot (2p^3)^{1/3}$

52. $(5k^2)^{3/2} \cdot (5k^{1/3})^{3/4}$

53. $p^{2/3}(2p^{1/3} + 5p)$

54. $2z^{1/2}(3z^{-1/2} + z^{1/2})$

55. $\dfrac{(x^2)^{1/3}(y^2)^{2/3}}{3x^{2/3}y^2}$

56. $\dfrac{(c^{1/2})^3(d^3)^{1/2}}{(c^3)^{1/4}(d^{1/4})^3}$

57. $\dfrac{(7a)^2(5b)^{3/2}}{(5a)^{3/2}(7b)^4}$

58. $\dfrac{(6a)^{1/2}\sqrt{ab}}{a^2b^{3/2}}$

59. $x^{1/2}(x^{2/3} - x^{4/3})$

60. $x^{1/2}(3x^{3/2} + 2x^{-1/2})$

61. $(x^{1/2} + y^{1/2})(x^{1/2} - y^{1/2})$

62. $(x^{1/3} + y^{1/2})(2x^{1/3} - y^{3/2})$

Match the rational exponent expression in Column I with the equivalent radical expression in Column II. Assume that x is not zero.

I	II
63. $(-3x)^{1/3}$	(a) $\dfrac{3}{\sqrt[3]{x}}$
64. $-3x^{1/3}$	(b) $-3\sqrt[3]{x}$
65. $(-3x)^{-1/3}$	(c) $\dfrac{1}{\sqrt[3]{3x}}$
66. $-3x^{-1/3}$	(d) $\dfrac{-3}{\sqrt[3]{x}}$
67. $(3x)^{1/3}$	(e) $3\sqrt[3]{x}$
68. $3x^{-1/3}$	(f) $\sqrt[3]{-3x}$
69. $(3x)^{-1/3}$	(g) $\sqrt[3]{3x}$
70. $3x^{1/3}$	(h) $\dfrac{1}{\sqrt[3]{-3x}}$

Simplify each of the following. (See Examples 10–12.)

71. $\sqrt[3]{125}$ **72.** $\sqrt[6]{64}$

73. $\sqrt[4]{625}$ **74.** $\sqrt[5]{-243}$

75. $\sqrt[7]{-128}$ **76.** $\sqrt{44}\cdot\sqrt{11}$

77. $\sqrt[3]{81}\cdot\sqrt[3]{9}$ **78.** $\sqrt{49-16}$

79. $\sqrt{81}-4$ **80.** $\sqrt{49}-\sqrt{16}$

81. $\sqrt{81}-\sqrt{4}$ **82.** $\sqrt{6}\sqrt{12}$

83. $\sqrt{8}\sqrt{96}$ **84.** $\sqrt{50}-\sqrt{72}$

85. $\sqrt{75}+\sqrt{192}$ **86.** $5\sqrt{20}-\sqrt{45}+2\sqrt{80}$

87. $(\sqrt{2}+3)(\sqrt{2}-3)$

88. $(\sqrt{5}+\sqrt{2})(\sqrt{5}-\sqrt{2})$

89. $(\sqrt{3}+4)(\sqrt{5}-4)$

90. What is wrong with the statement $\sqrt[3]{4}\cdot\sqrt[3]{4}=4$?

Rationalize the denominator of each of the following. (See Example 13.)

91. $\dfrac{3}{1-\sqrt{2}}$ **92.** $\dfrac{2}{1+\sqrt{5}}$

93. $\dfrac{4-\sqrt{2}}{2-\sqrt{2}}$ **94.** $\dfrac{\sqrt{3}-1}{\sqrt{3}-2}$

Rationalize the numerator of each of the following. (See Example 14.)

95. $\dfrac{2-\sqrt{3}}{2+\sqrt{3}}$

96. $\dfrac{1+\sqrt{7}}{2-\sqrt{3}}$

The following exercises are applications of exponentiation and radicals.

97. Business The theory of economic lot size shows that, under certain conditions, the number of units to order to minimize total cost is

$$x=\sqrt{\frac{kM}{f}},$$

where k is the cost to store one unit for one year, f is the (constant) setup cost to manufacture the product, and M is the total number of units produced annually. (See Section 12.3.) Find x for the following values of f, k, and M.

  (a) $k=\$1$, $f=\$500$, $M=100{,}000$

  (b) $k=\$3$, $f=\$7$, $M=16{,}700$

  (c) $k=\$1$, $f=\$5$, $M=16{,}800$

98. Health The threshold weight T for a person is the weight above which the risk of death increases greatly. One researcher found that the threshold weight in pounds for men aged 40–49 is related to height in inches by the equation $h=12.3T^{1/3}$. What height corresponds to a threshold of 216 pounds for a man in this age group?

Business *The Consumer Price Index (CPI) in U.S. cities stood at 100 in 1982. For recent years, it can be approximated by*

$$111.371x^{.19234}\quad(x\geq4),$$

where x = 4 corresponds to 1994.* Approximate the CPI in the following years.

99. First half of 2002 $(x=12)$

100. Last half of 2003 $(x=13.5)$

101. Last half of 2005

102. First half of 2006

Health *The number of kidney transplants in year x can be approximated by*

$$2260.323x^{.59743}\quad(x\geq10),$$

where x = 10 corresponds to 1990.† Find the approximate number of kidney transplants in each year.

103. 1995

104. 2000

105. 2002

106. 2003

107. 2004

108. In early 2005, there were approximately 61,000 people waiting for a kidney transplant.‡ About how many of these people did *not* get a kidney transplant?

---

*Based on data from the U.S. Bureau of Labor Statistics.

†Based on data from the United Network for Organ Sharing.

‡Based on data from the United Network for Organ Sharing.

109. Business   On the basis of data from *Ward's Communications,* the number of SUVs sold each year (in millions) can be approximated by

$$\frac{12.8}{1 + 15.3817x^{-.844}} \quad (1 \le x \le 16),$$

where $x = 1$ corresponds to 1991.

(a) Approximately how many SUVs were sold in 2000? In 2003? In 2005?

(b) Assume that the formula in part (a) remains valid over the next 40 years (which is probably unlikely). Use the table feature of a graphing calculator to determine when SUV sales would reach 6,000,000. According to this approximation, would sales ever reach 9,000,000 in the next 40 years?

1.6   First-Degree Equations

An **equation** is a statement that two mathematical expressions are equal; for example,

$$5x - 3 = 13, \qquad 8y = 4, \quad \text{and} \quad -3p + 5 = 4p - 8$$

are equations.

The letter in each equation is called the variable. This section concentrates on **first-degree equations,** which are equations that involve only constants and the first power of the variable. All of the equations displayed above are first-degree equations, but neither of the following equations is of first degree: $2x^2 = 5x + 6$ (the variable has an exponent greater than 1) and $\sqrt{x} + 2 = 4$ (the variable is under the radical).

A **solution** of an equation is a number that can be substituted for the variable in the equation to produce a true statement. For example, substituting the number 9 for x in the equation $2x + 1 = 19$ gives

$$2x + 1 = 19$$
$$2(9) + 1 \stackrel{?}{=} 19 \qquad \text{Let } x = 9.$$
$$18 + 1 = 19. \qquad \text{True}$$

This true statement indicates that 9 is a solution of $2x + 1 = 19$. ◇①

The following properties are used to solve equations.

①

Is -4 a solution of the following equations?

(a) $3x + 5 = -7$

(b) $2x - 3 = 5$

(c)   Is there more than one solution of the equation in part (a)?

Answers:

(a)   Yes

(b)   No

(c)   No

Properties of Equality

1. The same number may be added to or subtracted from both sides of an equation:

$$\text{If } a = b, \text{ then } a + c = b + c \text{ and } a - c = b - c.$$

continued

2. Both sides of an equation may be multiplied or divided by the same nonzero number:

$$\text{If } a = b \text{ and } c \neq 0, \text{ then } \quad ac = bc \quad \text{ and } \quad \frac{a}{c} = \frac{b}{c}.$$

▼ **EXAMPLE 1**   Solve the equation $5x - 3 = 12$.

Solution Using the first property of equality, add 3 to both sides. This isolates the term containing the variable on one side of the equation.

$$5x - 3 = 12$$
$$5x - 3 + \mathbf{3} = 12 + \mathbf{3} \qquad \text{Add 3 to both sides.}$$
$$5x = 15$$

Now arrange for the coefficient of x to be 1 by using the second property of equality.

$$5x = 15$$
$$\frac{5x}{5} = \frac{15}{5} \qquad \text{Divide both sides by 5.}$$
$$x = 3$$

The solution of the original equation, $5x - 3 = 12$, is 3. Check the solution by substituting 3 for x in the original equation. ▲ ⟨2⟩

▼ **EXAMPLE 2** Solve $2k + 3(k - 4) = 2(k - 3)$.

Solution First, simplify the equation by using the distributive property on the left-side term $3(k - 4)$ and right-side term $2(k - 3)$:

$$2k + 3(k - 4) = 2(k - 3)$$
$$2k + 3k - 12 = 2(k - 3) \qquad \text{Distributive property}$$
$$2k + 3k - 12 = 2k - 6 \qquad \text{Distributive property}$$
$$5k - 12 = 2k - 6 \qquad \text{Collect like terms on left side.}$$

One way to proceed is to add $-2k$ to both sides.

$$5k - 12 + (\mathbf{-2k}) = 2k - 6 + (\mathbf{-2k}) \qquad \text{Add} -2k \text{ to both sides.}$$
$$3k - 12 = -6$$
$$3k - 12 + \mathbf{12} = -6 + \mathbf{12} \qquad \text{Add 12 to both sides.}$$
$$3k = 6$$
$$\frac{1}{3}(3k) = \frac{1}{3}(6) \qquad \text{Multiply both sides by } \frac{1}{3}.$$
$$k = 2$$

The solution is 2. Check this result by substituting 2 for k in the original equation. ▲ ⟨3⟩

⟨2⟩ Solve the following.
(a) $3p - 5 = 19$
(b) $4y + 3 = -5$
(c) $-2k + 6 = 2$

Answers:
(a) 8
(b) -2
(c) 2

⟨3⟩ Solve the following.
(a) $3(m - 6) + 2(m + 4) = 4m - 2$
(b) $-2(y + 3) + 4y = 3(y + 1) - 6$

Answers:
(a) 8
(b) -3

▼ **EXAMPLE 3**   The percentage y of workers who are covered by a "defined contribution" retirement plan (such as a 401(k)) in year x is approximated by the equation

$$4.0215(x - 1979) = 2.1y - 34.503.\text{*}$$

Use a calculator to determine when 59% of workers were covered by such a plan.

Solution Let $y = 59$ in the equation and solve for x. To avoid any round-off errors in the intermediate steps, it is often a good idea to do all the algebra first, without using the calculator.

$$4.0215(x - 1979) = 2.1y - 34.503$$

$$4.0215(x - 1979) = 2.1 \cdot 59 - 34.503 \qquad \text{Substitute } y = 59.$$

$$4.0215x - 4.0215 \cdot 1979 = 2.1 \cdot 59 - 34.503 \qquad \text{Distributive property}$$

$$4.0215x = 2.1 \cdot 59 - 34.503 + 4.0215 \cdot 1979 \qquad \begin{array}{l}\text{Add} \\ 4.0215 \cdot 2001 \\ \text{to both sides.}\end{array}$$

$$x = \frac{2.1 \cdot 59 - 34.503 + 4.0215 \cdot 1979}{4.0215} \qquad \begin{array}{l}\text{Divide both} \\ \text{sides by } 4.0215.\end{array}$$

Now use the calculator (inserting parentheses around the numerator) to determine that $x \approx 2001.23$, as shown in Figure 1.12. So, in early 2001, 59% of workers were covered by a defined contribution plan. ▲ ⟨4⟩

⟨**4**⟩

In Example 3, when were 50% of workers covered by a defined contribution plan?

Answer:

About mid-1996
($x \approx 1996.53$)

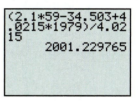

FIGURE 1.12

The next three examples show how to simplify the solution of first-degree equations involving fractions. We solve these equations by multiplying both sides of the equation by a *common denominator*. This step will eliminate the fractions.

▼ **EXAMPLE 4**   Solve $\dfrac{r}{10} - \dfrac{2}{15} = \dfrac{3r}{20} - \dfrac{1}{5}$.

Solution Here, the denominators are 10, 15, 20, and 5. Each of these numbers can be divided into 60; therefore, 60 is a common denominator. Multiply both sides of the equation by 60.

$$60\left(\frac{r}{10} - \frac{2}{15}\right) = 60\left(\frac{3r}{20} - \frac{1}{5}\right)$$

$$60\left(\frac{r}{10}\right) - 60\left(\frac{2}{15}\right) = 60\left(\frac{3r}{20}\right) - 60\left(\frac{1}{5}\right) \qquad \begin{array}{l}\text{Distributive} \\ \text{property}\end{array}$$

*Based on data from the Center for Retirement Research of Boston College.

$$6r - 8 = 9r - 12$$

$$6r - 8 + (-6r) + 12 = 9r - 12 + (-6r) + 12 \qquad \text{Add } -6r \text{ and 12 to both sides.}$$

$$4 = 3r$$

$$r = \frac{4}{3} \qquad\qquad \text{Multiply both sides by } 1/3.$$

Check this solution in the original equation. ▲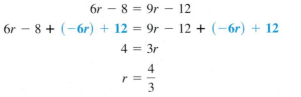

CAUTION Multiplying *both* sides of the *equation* by a number to eliminate fractions is valid. But multiplying a single fraction to simplify it is not valid. For instance, multiplying $\dfrac{3x}{8}$ by 8 *changes* it to $3x$, which is *not equal* to $\dfrac{3x}{8}$.

The second property of equality (page 46) applies only to *nonzero* quantities. Multiplying or dividing both sides of an equation by a quantity involving the variable (which might be zero for some values) may lead to an **extraneous solution**—that is, a number that does not satisfy the original equation. To avoid errors in such situations, always *check your solutions in the original equation.*

▼ **EXAMPLE 5**   Solve $\dfrac{4}{3(k + 2)} - \dfrac{k}{3(k + 2)} = \dfrac{5}{3}$.

Solution Multiply both sides of the equation by the common denominator $3(k + 2)$. Here, $k \neq -2$, since $k = -2$ would give a 0 denominator, making the fraction undefined.

$$3(k + 2) \cdot \frac{4}{3(k + 2)} - 3(k + 2) \cdot \frac{k}{3(k + 2)} = 3(k + 2) \cdot \frac{5}{3}$$

Simplify each side and solve for k.

$$4 - k = 5(k + 2)$$

$$4 - k = 5k + 10 \qquad\qquad \text{Distributive property}$$

$$4 - k + k = 5k + 10 + k \qquad\qquad \text{Add } k \text{ to both sides.}$$

$$4 = 6k + 10$$

$$4 + (-10) = 6k + 10 + (-10) \qquad \text{Add } -10 \text{ to both sides.}$$

$$-6 = 6k$$

$$-1 = k \qquad\qquad \text{Multiply both sides by } \tfrac{1}{6}.$$

The solution is -1. Substitute -1 for k as a check.

$$\frac{4}{3(-1 + 2)} - \frac{-1}{3(-1 + 2)} \overset{?}{=} \frac{5}{3}$$

$$\frac{4}{3} - \frac{-1}{3} \overset{?}{=} \frac{5}{3}$$

$$\frac{5}{3} = \frac{5}{3}$$

The check shows that -1 is the solution. ▲ ⟨6⟩

⟨5⟩

Solve the following.

(a) $\dfrac{x}{2} - \dfrac{x}{4} = 6$

(b) $\dfrac{2x}{3} + \dfrac{1}{2} = \dfrac{x}{4} - \dfrac{9}{2}$

Answers:

(a) 24

(b) -12

⟨6⟩

Solve the equation

$$\frac{5p + 1}{3(p + 1)} = \frac{3p - 3}{3(p + 1)} + \frac{9p - 3}{3(p + 1)}.$$

Answer:

1

▼ **EXAMPLE 6** Solve $\dfrac{3x-4}{x-2}=\dfrac{x}{x-2}$.

Solution Multiplying both sides by $x-2$ produces

$$3x-4=x$$
$$2x-4=0 \qquad \text{Subtract } x \text{ from both sides.}$$
$$2x=4 \qquad \text{Add 4 to both sides.}$$
$$x=2 \qquad \text{Divide both sides by 2.}$$

Substituting 2 for x in the original equation produces fractions with 0 denominators. Since division by 0 is not defined, $x=2$ is an extraneous solution. So the original equation has no solution. ▲ ⟨7⟩

Sometimes an equation with several variables must be solved for one of the variables. This process is called **solving for a specified variable.**

▼ **EXAMPLE 7** Solve for x: $3(ax-5a)+4b=4x-2$.

Solution Use the distributive property to get

$$3ax-15a+4b=4x-2.$$

Treat x as the variable, the other letters as constants. Get all terms with x on one side of the equation and all terms without x on the other side.

$$3ax-4x=15a-4b-2 \qquad \text{Isolate terms with } x \text{ on the left.}$$
$$(3a-4)x=15a-4b-2 \qquad \text{Distributive property}$$
$$x=\frac{15a-4b-2}{3a-4} \qquad \text{Multiply both sides by } \frac{1}{3a-4}.$$

The final equation is solved for x, as required. ▲ ⟨8⟩

Absolute Value Equations

Recall from Section 1.1 that the absolute value of a number a is either a or $-a$, whichever one is positive. For instance, $|4|=4$ and $|-7|=-(-7)=7$.

▼ **EXAMPLE 8** Solve $|x|=3$.

Solution Since $|x|$ is either x or $-x$, the equation says that

$$x=3 \quad \text{or} \quad -x=3$$
$$x=-3.$$

The solutions of $|x|=3$ are 3 and -3. ▲

▼ **EXAMPLE 9** Solve $|p-4|=2$.

Solution Since $|p-4|$ is either $p-4$ or $-(p-4)$, we have

$$p-4=2 \qquad \text{or} \qquad -(p-4)=2$$
$$p=6 \qquad\qquad -p+4=2$$
$$-p=-2$$
$$p=2,$$

⟨7⟩ Solve each equation.

(a) $\dfrac{3p}{p+1}=1-\dfrac{3}{p+1}$

(b) $\dfrac{8y}{y-4}=\dfrac{32}{y-4}-3.$

Answer:

Neither equation has a solution.

⟨8⟩ Solve for x.

(a) $2x-7y=3xk$

(b) $8(4-x)+6p=-5k-11yx$

Answers:

(a) $x=\dfrac{7y}{2-3k}$

(b) $x=\dfrac{5k+32+6p}{8-11y}$

so that 6 and 2 are possible solutions. Checking them in the original equation shows that both are solutions. ▲ ⟨9⟩

▼ **EXAMPLE 10**   Solve $|4m - 3| = |m + 6|$.

Solution To satisfy the equation, the quantities in absolute value bars must either be equal or be negatives of one another. That is,

$$4m - 3 = m + 6 \qquad \text{or} \qquad 4m - 3 = -(m + 6)$$
$$3m = 9 \qquad\qquad\qquad 4m - 3 = -m - 6$$
$$m = 3 \qquad\qquad\qquad 5m = -3$$
$$m = -\frac{3}{5}.$$

Check that the solutions for the original equation are 3 and $-3/5$. ▲ ⟨10⟩

Applications

One of the main reasons for learning mathematics is to be able to use it to solve practical problems. There are no hard-and-fast rules for dealing with real-world applications, except perhaps to use common sense. However, you will find it much easier to deal with such problems if you don't try to do everything at once. After reading the problem carefully, attack it in stages, as suggested in the following guidelines.

Solving Applied Problems

Step 1   Decide on the unknown. Name it with some variable that you *write down*. Many students try to skip this step. They are eager to get on with the writing of the equation. But this is an important step. If you don't know what the variable represents, how can you write a meaningful equation or interpret a result?

Step 2   Decide on a variable expression to represent any other unknowns in the problem. For example, if x represents the width of a rectangle, and you know that the length is one more than twice the width, then *write down* the fact that the length is $1 + 2x$.

Step 3   Draw a sketch or make a chart, if appropriate, showing the information given in the problem.

Step 4   Using the results of Steps 1–3, write an equation that expresses a condition that must be satisfied.

Step 5   Solve the equation.

Step 6   Check the solution in the words of the *original problem*, not just in the equation you have written.

The following examples illustrate this approach.

▼ **EXAMPLE 11**   A financial manager has $14,000 to invest for her company. She plans to invest part of the money in tax-free bonds at 6% interest and the remainder in taxable bonds at 9%. She wants to earn $1005 per year in interest from the investments. Find the amount she should invest at each rate.

Solution

Step 1   Let x represent the amount to be invested at 6%.

Step 2   After x dollars are invested, the remaining amount is $14,000 - x$ dollars, which is to be invested at 9%.

Step 3   Interest for one year is given by rate × amount invested. For instance, 6% of x dollars is $.06x$. The given information is summarized in the following chart.

Investment	Amount Invested	Interest Rate	Interest Earned in 1 Year
Tax-free bonds	x	6% = .06	$.06x$
Taxable bonds	$14,000 - x$	9% = .09	$.09(14,000 - x)$
Totals	14,000		1005

Step 4   Because the total interest is to be $1005, the last column of the tables shows that

$$.06x + .09(14,000 - x) = 1005.$$

Step 5   Solve the preceding equation.

$$.06x + .09(14,000 - x) = 1005$$
$$.06x + .09(14,000) - .09x = 1005$$
$$.06x + 1260 - .09x = 1005$$
$$-.03x = -255$$
$$x = 8500.$$

The manager should invest $8500 at 6% and $14,000 - $8500 = $5500 at 9%.

Step 6   Check these results in the original problem. If $8500 is invested at 6%, the interest is $.06(8500) = 510. If $5500 is invested at 9%, the interest is $.09(5500) = 495. So the total interest is $510 + $495 = $1005, as required. ▲ ⑪

⟨11⟩

An investor owns two pieces of property. One, worth twice as much as the other, returns 6% in annual interest, while the other returns 4%. Find the value of each piece of property if the total annual interest earned is $8000.

Answers:

6% return: $100,000; 4% return: $50,000

▼ **EXAMPLE 12**   Chuck travels 80 kilometers in the same time that Mary travels 180 kilometers. Mary travels 50 kilometers per hour faster than Chuck. Find the speed of each person.

Solution

Step 1 Use x to represent Chuck's speed and $x + 50$ to represent Mary's speed, which is 50 kilometers per hour faster than Chuck's.

Steps 2 and 3 Constant-rate problems of this kind require the distance formula

$$d = rt,$$

where d is the distance traveled in t hours at a constant rate of speed r. The distance traveled by each person is given, along with the fact that the time traveled by each person is the same. Solve the formula $d = rt$ for t.

$$d = rt$$

$$\frac{1}{r} \cdot d = \frac{1}{r} \cdot rt$$

$$\frac{d}{r} = t$$

For Chuck, $d = 80$ and $r = x$, giving $t = 80/x$. For Mary, $d = 180$, $r = x + 50$, and $t = 180/(x + 50)$. Use these facts to complete a chart, which organizes the information given in the problem.

	d	r	t
Chuck	80	x	$\dfrac{80}{x}$
Mary	180	$x + 50$	$\dfrac{180}{x + 50}$

Step 4 Because both people traveled for the *same time*, the equation is

$$\frac{80}{x} = \frac{180}{x + 50}.$$

Step 5 Multiply both sides of the equation by $x(x + 50)$.

$$x(x + 50)\frac{80}{x} = x(x + 50)\frac{180}{x + 50}$$

$$80(x + 50) = 180x$$

$$80x + 4000 = 180x$$

$$4000 = 100x$$

$$40 = x$$

Step 6 Since x represents Chuck's speed, Chuck went 40 kilometers per hour. Mary's speed is $x + 50$, or $40 + 50 = 90$ kilometers per hour. Check these results in the words of the original problem. ▲ ⟨12⟩

⟨12⟩

(a) Tom and Tyrone enter a run for charity. Tom runs at 7 mph and Tyrone runs at 5 mph. If they start at the same time, how long will it be until they are $1/2$ mile apart?

(b) In part (a), suppose the run has a staggered start. If Tyrone starts first, and Tom starts 10 minutes later, how long will it be until they are neck and neck?

Answers:

(a) 15 minutes (1/4 hour)

(b) After Tom has run 25 minutes.

▼ **EXAMPLE 13** An oil company distributor needs to fill orders for 89-octane gas, but has only 87- and 93-octane gas on hand. How much of each type should be mixed together to produce 54,000 gallons of 89-octane gas?

Solution The octane rating of a gasoline indicates that it has the same antiknock qualities as a standard fuel made of heptane and isooctane. The octane rating is the percentage of isooctane in the standard fuel.* So we assume that standard fuels are being mixed. Let x be the amount of 87-octane gas. Then $54,000 - x$ is the amount of 93-octane gas. We can summarize the relevant information in a chart:

Type of Gas	Quantity	% Isooctane	Amount of Isooctane
87 octane	x	87%	$.87x$
93 octane	$54,000 - x$	93%	$.93(54,000 - x)$
Mixture	$54,000$	89%	$.89(54,000)$

The amount of isooctane satisfies this equation:

$$(\text{Amount in 87-octane}) + (\text{Amount in 93-octane}) = \text{Amount in mixture}$$
$$.87x + .93(54,000 - x) = .89(54,000).$$

Use the distributive property on the left side and multiply on the right side:

$$.87x + 50,220 - .93x = 48,060$$
$$-.06x = -2160$$
$$x = \frac{-2160}{-.06} = 36,000.$$

So the distributor should mix 36,000 gallons of 87-octane gas with $54,000 - 36,000 = 18,000$ gallons of 93-octane gas to obtain 54,000 gallons of 89-octane gas ▲ ⟨13⟩

⟨13⟩

How much 89-octane gas and how much 94-octane gas are needed to produce 1000 gallons of 91-octane gas?

Answer:

600 gallons of 89-octane gas; 400 gallons of 94-octane gas

*This is one of several possible ways of determining octane ratings.

◆ **1.6 Exercises**

Solve each equation. (See Examples 1–6.)

1. $3x + 8 = 20$

2. $4 - 5y = 19$

3. $.6k - .3 = .5k + .4$

4. $2.5 + 5.04m = 8.5 - .06m$

5. $2a - 1 = 4(a + 1) + 7a + 5$

6. $3(k - 2) - 6 = 4k - (3k - 1)$

7. $2[x - (3 + 2x) + 9] = 3x - 8$

8. $-2[4(k + 2) - 3(k + 1)] = 14 + 2k$

9. $\dfrac{3x}{5} - \dfrac{4}{5}(x + 1) = 2 - \dfrac{3}{10}(3x - 4)$

10. $\dfrac{4}{3}(x - 2) - \dfrac{1}{2} = 2\left(\dfrac{3}{4}x - 1\right)$

11. $\dfrac{5y}{6} - 8 = 5 - \dfrac{2y}{3}$

12. $\dfrac{x}{2} - 3 = \dfrac{3x}{5} + 1$

13. $\dfrac{m}{2} - \dfrac{1}{m} = \dfrac{6m + 5}{12}$

14. $-\dfrac{3k}{2} + \dfrac{9k - 5}{6} = \dfrac{11k + 8}{k}$

15. $\dfrac{4}{x - 3} - \dfrac{8}{2x + 5} + \dfrac{3}{x - 3} = 0$

16. $\dfrac{5}{2p + 3} - \dfrac{3}{p - 2} = \dfrac{4}{2p + 3}$

17. $\dfrac{3}{2m + 4} = \dfrac{1}{m + 2} - 2$

18. $\dfrac{8}{3k - 9} - \dfrac{5}{k - 3} = 4$

Use a calculator to solve each equation. Round your answer to the nearest hundredth. (See Example 3.)

19. $9.06x + 3.59(8x - 5) = 12.07x + .5612$

20. $-5.74(3.1 - 2.7p) = 1.09p + 5.2588$

21. $\dfrac{2.63r - 8.99}{1.25} - \dfrac{3.90r - 1.77}{2.45} = r$

22. $\dfrac{8.19m + 2.55}{4.34} - \dfrac{8.17m - 9.94}{1.04} = 4m$

Solve each equation for x. (See Example 7.)

23. $4(a + x) = b - a + 2x$

24. $(3a - b) - bx = a(x - 2) \quad (a \neq -b)$

25. $5(b - x) = 2b + ax \quad (a \neq -5)$

26. $bx - 2b = 2a - ax$

Solve each equation for the specified variable. Assume that all denominators are nonzero. (See Example 7.)

27. $PV = k$ for V

28. $i = prt$ for p

29. $V = V_0 + gt$ for g

30. $S = S_0 + gt^2 + k$ for g

31. $A = \dfrac{1}{2}(B + b)h$ for B

32. $C = \dfrac{5}{9}(F - 32)$ for F

Solve each equation. (See Examples 8–10.)

33. $|2h - 1| = 5$ **34.** $|4m - 3| = 12$

35. $|6 + 2p| = 10$ **36.** $|-5x + 7| = 15$

37. $\left|\dfrac{5}{r - 3}\right| = 10$ **38.** $\left|\dfrac{3}{2h - 1}\right| = 4$

39. $|3y - 2| = |4y + 5|$ **40.** $|1 - 3z| = |z + 2|$

Solve the following applied problems.

Natural Science *The equation that relates Fahrenheit temperature F to Celsius temperature C is*

$$C = \frac{5}{9}(F - 32).$$

Find the Fahrenheit temperature corresponding to these Celsius temperatures.

41. -10 **42.** -22

43. 18 **44.** 32

Finance *The gross federal debt y (in trillions of dollars) in year x is approximated by*

$$y = .79x + 3.93 \quad (x \geq 3),$$

where x is the number of years after 2000.* In what year will the federal debt be

45. $8.67 trillion   **46.** $11.04 trillion

47. $12.62 trillion   **48.** $14.2 trillion

Health *The total health care expenditures E in the United States (in billions of dollars) can be approximated by*

$$E = 73.04x + 625.6,$$

where x is the number of years since 1990.† Determine the year in which health care expenditures are at the given level.

49. $991 billion   **50.** $1356 billion

51. $1794.25 billion   **52.** $1940.3 billion

Business *The percentage of workers covered by defined contribution retirement plans was discussed in Example 3. The percentage of workers y who are covered by a traditional pension in year x is given by the equation*

$$-2.1977(x - 2001) = y - 12.94.‡$$

Find the year in which the given percentage of workers covered by a traditional pension plan was:

53. 50%   **54.** 25%   **55.** 15%

56. Find the year in which the percentage of workers covered by a traditional pension was the same as the percentage covered by a defined benefits plan. What was that percentage? (*Hint:* Solve both the defined benefits equation in Example 3 and the preceding traditional pension equation for y. Set the results equal to each other and solve for x. Use this value of x to determine y.)

Social Science *The population P of the United States (in millions) is approximated by*

$$P = 2.831x + 224.361,$$

where x is the number of years since 1980. Determine the year in which the United States had or will have the following population figures.

57. 280,980,000

58. 289,474,000

59. 300,798,000

60. 309,291,000

Business *When a loan is paid off early, a portion of the finance charge must be returned to the borrower. By one method of calculating the finance charge (called the rule of 78), the amount of unearned interest (finance charge to be returned) is given by*

$$u = f \cdot \frac{n(n + 1)}{q(q + 1)},$$

---

*Based on data and projections from the Congressional Budget Office in 2005.

†Based on data from the U.S. Centers for Medicare and Medicaid Services.

‡Based on data from the Center for Retirement Research of Boston College.

where u represents unearned interest, f is the original finance charge, n is the number of payments remaining when the loan is paid off, and q is the original number of payments. Find the amount of the unearned interest in each of the following.

61. Original finance charge = $800, loan scheduled to run 36 months, paid off with 18 payments remaining

62. Original finance charge = $1400, loan scheduled to run 48 months, paid off with 12 payments remaining

63. **Natural Science**   The excess lifetime cancer risk R is a measure of the likelihood that an individual will develop cancer from a particular pollutant. For example, if $R = .01$, then a person has a 1% increased chance of developing cancer during a lifetime. The value of R for formaldehyde can be calculated from the equation $R = kd$, where k is a constant and d is the daily dose in parts per million. The constant k for formaldehyde can be calculated from the formula $k = .132\left(\frac{B}{W}\right)$, where B is the total number of cubic meters of air a person breathes in one day and W is a person's weight in kilograms.*
  (a) Find k for a person who breathes in 20 cubic meters of air per day and weighs 75 kg.
  (b) Mobile homes in Minnesota were found to have a mean daily dose d of .42 parts per million.† Calculate R.
  (c) For every 5000 people, how many cases of cancer could be expected each year from the preceding levels of formaldehyde? Assume an average life expectancy of 72 years.

Business   *Solve the following investment problems. (See Example 11.)*

64. Joe Gonzalez received $52,000 profit from the sale of some land. He invested part at 5% interest and the rest at 4% interest. He earned a total of $2290 interest per year. How much did he invest at 5%?

65. Weijen Luan invests $20,000 received from an insurance settlement in two ways: some at 6% and some at 4%. Altogether, she makes $1040 per year interest. How much is invested at 4%?

66. Maria Martinelli bought two plots of land for a total of $120,000. On the first plot, she made a profit of 15%. On the second, she lost 10%. Her total profit was $5500. How much did she pay for each piece of land?

67. Suppose $20,000 is invested at 5%. How much additional money must be invested at 4% to produce a yield of 4.8% on the entire amount invested?

Solve the following applied problems. (See Example 12.)

68. A plane flies nonstop from New York to London, cities that are about 3500 miles apart. After 1 hour and 6 minutes in the air, the plane passes over Halifax, Nova Scotia, which is 600 miles from New York. Estimate the flying time from New York to London.

69. On vacation, Le Hong averaged 50 mph traveling from Denver to Minneapolis. Returning by a different route that covered the same number of miles, he averaged 55 mph. What is the distance between the two cities if his total traveling time was 32 hours?

70. Russ and Janet are running in the Apple Hill Fun Run. Russ runs at 7 mph, Janet at 5 mph. If they start at the same time, how long will it be before they are 2/3 mile apart?

Natural Science   *Using the same assumptions about octane ratings as in Example 13, solve the following problems.*

71. How many liters of 94-octane gasoline should be mixed with 200 liters of 99-octane gasoline to get a mixture that is 97-octane gasoline?

72. A service station has 92-octane and 98-octane gasoline. How many liters of each gasoline should be mixed to provide 12 liters of 96-octane gasoline for a chemistry experiment?

Solve the following applied problems.

73. **Business**   A major car rental firm charges $45.56 a day with unlimited mileage. A discount firm offers a similar car for $20 a day plus 18 cents per mile. How far must you drive in a day in order for the cost to be the same at both firms?

74. **Natural Science**   A car radiator contains 8 quarts of fluid, 40% of which is antifreeze. How much fluid should be drained and replaced with pure (100%) antifreeze in order that the new mixture be 60% antifreeze?

75. **Transportation**   In Massachusetts, speeding fines are determined by the equation

$$y = 10(x - 65) + 50, \qquad x \geq 65,$$

where y is the amount of the fine (in dollars) for driving x miles per hour. If Paul was fined $100 for speeding, how fast was he driving?

76. **Finance**   Jack borrowed his father's luxury car and promised to return it with a full tank of premium gas, which costs $2.50 per gallon. From experience, he knows that he needs 15.5 gallons. But he has only $38 (and no credit card), which isn't enough. He decides to get as much premium as possible and fill the remainder of the tank with regular gas, which costs $2.30 per gallon. To the nearest thousandth of a gallon, how much of each type of gas should he get?

---

*A. Hines, T. Ghosh, S. Layalka, and R.Warder, *Indoor Air Quality and Control* Prentice Hall, 1993 (TD 883.1.I476 1993).

†I. Ritchie, and R. Lehnen, "An Analysis of Formaldehyde Concentration in Mobile and Conventional Homes" *Journal of Environmental Health* 47:300–305.

Geometry *Solve each of these geometry problems. (Hint: In each case, draw an appropriate figure and label its parts.)*

77. The length of a rectangular label is 3 centimeters less than twice the width. The perimeter is 54 centimeters. Find the width.

78. A puzzle piece in the shape of a triangle has a perimeter of 30 centimeters. Two sides of the triangle are each twice as long as the shortest side. Find the length of the shortest side.

79. A triangle has a perimeter of 27 centimeters. One side is twice as long as the shortest side. The third side is 7 centimeters longer than the shortest side. Find the length of the shortest side.

80. A closed recycling bin is in the shape of a rectangular box. Find the height of the bin if its length is 18 feet, its width is 8 feet, and its surface area is 496 square feet.

1.7 Quadratic Equations

An equation that can be written in the form

$$ax^2 + bx + c = 0,$$

where a, b, and c are real numbers with $a \neq 0$, is called a **quadratic equation.** For example, each of

$$2x^2 + 3x + 4 = 0, \qquad x^2 = 6x - 9, \qquad 3x^2 + x = 6, \qquad \text{and} \qquad x^2 = 5$$

is a quadratic equation. A solution of an equation that is a real number is said to be a **real solution** of the equation.

One method of solving quadratic equations is based on the following property of real numbers.

Zero-Factor Property

If a and b are real numbers, with $ab = 0$, then $a = 0$ or $b = 0$ or both.

▼ EXAMPLE 1 Solve the equation $(x - 4)(3x + 7) = 0$.

Solution By the zero-factor property, the product $(x - 4)(3x + 7)$ can equal 0 only if at least one of the factors equals 0. That is, the product equals zero only if $x - 4 = 0$ or $3x + 7 = 0$. Solving each of these equations separately will give the solutions of the original equation.

$$x - 4 = 0 \quad \text{or} \quad 3x + 7 = 0$$
$$x = 4 \quad \text{or} \quad 3x = -7$$
$$x = -\frac{7}{3}$$

The solutions of the equation $(x - 4)(3x + 7) = 0$ are 4 and $-7/3$. Check these solutions by substituting into the original equation. ▲ ◇①

---

①

Solve the following equations.

(a) $(y - 6)(y + 2) = 0$

(b) $(5k - 3)(k + 5) = 0$

(c) $(2r - 9)(3r + 5) \cdot$
$\qquad (r + 3) = 0$

Answers:

(a) $6, -2$

(b) $3/5, -5$

(c) $9/2, -5/3, -3$

▼ **EXAMPLE 2**   Solve $6r^2 + 7r = 3$.

Solution   Rewrite the equation as

$$6r^2 + 7r - 3 = 0.$$

Now factor $6r^2 + 7r - 3$ to get

$$(3r - 1)(2r + 3) = 0.$$

By the zero-factor property, the product $(3r - 1)(2r + 3)$ can equal 0 only if

$$3r - 1 = 0 \quad \text{or} \quad 2r + 3 = 0.$$

Solving each of these equations separately gives the solutions of the original equation.

$$3r = 1 \quad \text{or} \quad 2r = -3$$
$$r = \frac{1}{3} \qquad r = -\frac{3}{2}$$

Verify that both $1/3$ and $-3/2$ are solutions by substituting them into the original equation. ▲ ②

An equation such as $x^2 = 5$ has two solutions: $\sqrt{5}$ and $-\sqrt{5}$. The same idea is true in general.

②

Solve each equation by factoring.

(a) $y^2 + 3y = 10$

(b) $2r^2 + 9r = 5$

(c) $4k^2 = 9k$

Answers:

(a) $2, -5$

(b) $1/2, -5$

(c) $9/4, 0$

Square-Root Property

If $b > 0$, then the solutions of $x^2 = b$ are $\sqrt{b}$ and $-\sqrt{b}$.

The two solutions are sometimes abbreviated $\pm\sqrt{b}$.

▼ **EXAMPLE 3**   Solve each equation.

(a) $m^2 = 17$

Solution   By the square-root property, the solutions are $\sqrt{17}$ and $-\sqrt{17}$, abbreviated $\pm\sqrt{17}$.

(b) $(y - 4)^2 = 11$

Solution   Use a generalization of the square-root property, working as follows:

$$(y - 4)^2 = 11$$
$$y - 4 = \sqrt{11} \quad \text{or} \quad y - 4 = -\sqrt{11}$$
$$y = 4 + \sqrt{11} \qquad y = 4 - \sqrt{11}.$$

Abbreviate the solutions as $4 \pm \sqrt{11}$. ▲ ③

③

Solve each equation by using the square-root property.

(a) $p^2 = 21$

(b) $(m + 7)^2 = 15$

(c) $(2k - 3)^2 = 5$

Answers:

(a) $\pm\sqrt{21}$

(b) $-7 \pm \sqrt{15}$

(c) $(3 \pm \sqrt{5})/2$

As suggested by Example 3(b), it is easy to solve any quadratic equation that can be written in the form $(x + d)^2 = k$. Fortunately, there is a simple algebraic technique that enables us to write any quadratic equation in this form, as in the next example.

▼ **EXAMPLE 4**   Solve $4x^2 - 24x + 19 = 0$.

Solution First, rearrange the equation so that the x terms are on the left and the constants on the right:

$$4x^2 - 24x = -19.$$

Now divide both sides by 4 so that 1 is the coefficient of x^2:

$$x^2 - 6x = -\frac{19}{4}.$$

Now take half the coefficient of x (namely, $-6/2 = -3$) and square it, obtaining 9. (The reason for this step is explained below.) Add 9 to both sides and factor the left side:

$$x^2 - 6x + 9 = -\frac{19}{4} + 9$$

$$(x - 3)^2 = \frac{17}{4}.$$

Use the square-root property to complete the solution.

$$x - 3 = \pm\sqrt{\frac{17}{4}} = \pm\frac{\sqrt{17}}{2}$$

$$x = 3 \pm \frac{\sqrt{17}}{2} = \frac{6 \pm \sqrt{17}}{2}$$

The two solutions are $\dfrac{6 + \sqrt{17}}{2}$ and $\dfrac{6 - \sqrt{17}}{2}$. ▲

To see why the technique used in Example 4 always works, look at this multiplication:

$$\left(x + \frac{b}{2}\right)^2 = x^2 + 2\left(\frac{b}{2}\right)x + \left(\frac{b}{2}\right)^2 = x^2 + bx + \left(\frac{b}{2}\right)^2.$$

If you read it from right to left, it shows that the expression $x^2 + bx$ can be made into a perfect square by adding the square of half of the coefficient of x, namely, $\left(\dfrac{b}{2}\right)^2$. That's exactly what was done in Example 4 (where $b = -6$). This technique is called **completing the square.**

The method of completing the square can be used on the general quadratic equation,

$$ax^2 + bx + c = 0 \qquad (a \neq 0),$$

to convert it to one whose solutions can be found by the square-root property. This will give a general formula for solving any quadratic equation. Going through the necessary algebra produces the following important result, which you should memorize.

Quadratic Formula

The solutions of the quadratic equation $ax^2 + bx + c = 0$, where $a \neq 0$, are given by

$$x = \frac{-b \pm \sqrt{b^2 - 4ac}}{2a}.$$

CAUTION When using the quadratic formula, remember that the equation must be in the form $ax^2 + bx + c = 0$. Also, notice that the fraction bar in the quadratic formula extends under *both* terms in the numerator. Be sure to add $-b$ to $\pm\sqrt{b^2 - 4ac}$ *before* dividing by $2a$.

▼ **EXAMPLE 5**   Solve $x^2 + 1 = 4x$.

Solution First add $-4x$ to both sides to get 0 alone on the right side.

$$x^2 - 4x + 1 = 0$$

Now identify the letters a, b, and c. Here, $a = 1$, $b = -4$, and $c = 1$. Substitute these numbers into the quadratic formula to obtain

$$x = \frac{-(-4) \pm \sqrt{(-4)^2 - 4(1)(1)}}{2(1)}$$

$$= \frac{4 \pm \sqrt{16 - 4}}{2}$$

$$= \frac{4 \pm \sqrt{12}}{2}$$

$$= \frac{4 \pm 2\sqrt{3}}{2} \qquad \sqrt{12} = \sqrt{4 \cdot 3} = \sqrt{4} \cdot \sqrt{3} = 2\sqrt{3}$$

$$= \frac{2(2 \pm \sqrt{3})}{2} \qquad \text{Factor } 4 \pm 2\sqrt{3}.$$

$$x = 2 \pm \sqrt{3}. \qquad \text{Cancel 2.}$$

The $\pm$ sign represents the two solutions of the equation. First use $+$ and then use $-$ to find each of the solutions: $2 + \sqrt{3}$ and $2 - \sqrt{3}$. ▲

Example 5 shows that the quadratic formula produces exact solutions. In many real-world applications, however, you must use a calculator to find decimal approximations of the solutions. The approximate solutions in Example 5 are

$$x = 2 + \sqrt{3} \approx 3.732050808 \quad \text{and} \quad x = 2 - \sqrt{3} \approx .2679491924,$$

as shown in Figure 1.13.

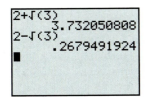

4
Use the quadratic formula to solve each equation.

(a) $x^2 - 2x = 2$

(b) $u^2 - 6u + 4 = 0$

Answers:

(a) $x = 1 + \sqrt{3}$ or $1 - \sqrt{3}$

(b) $u = 3 + \sqrt{5}$ or $3 - \sqrt{5}$

FIGURE 1.13

▼ **EXAMPLE 6**   The number of identity theft complaints (in thousands) in year x is approximated by

$$-5.5x^2 + 77.1x + 26.3 \qquad (0 \le x \le 9),$$

where x is the number of years since 2000.* Use the quadratic formula and a calculator to find the year in which there were 247,000 complaints.

To find the year x, we must solve the equation

$$-5.5x^2 + 77.1x + 26.3 = 247 \qquad \text{Complaints are measured in thousands.}$$

$$-5.5x^2 + 77.1x - 220.7 = 0 \qquad \text{Subtract 247 from both sides}$$

First compute the radical in the quadratic formula:

$$\sqrt{b^2 - 4ac} = \sqrt{77.1^2 - 4(-5.5)(-220.7)} \approx 33.00015151.$$

Then store the result (which we denote by M) in the calculator memory. (Check your instruction manual for how to store and recall numbers.) By the quadratic formula, the exact solutions are

$$x = \frac{-b \pm \sqrt{b^2 - 4ac}}{2a} = \frac{-b \pm M}{2a} = \frac{-77.1 \pm M}{2(-5.5)}.$$

Figure 1.14 shows that the approximate solutions are

$$x = \frac{-77.1 + M}{2(-5.5)} \approx 4.009 \qquad \text{and} \qquad x = \frac{-77.1 - M}{2(-5.5)} \approx 10.009.$$

Since we are given that $0 \le x \le 9$, the only applicable solution here is $x \approx 4.009$, which corresponds to early 2004. ▲ ⑤

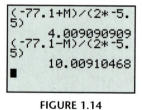

FIGURE 1.14

▼ **EXAMPLE 7**   Solve $9x^2 - 30x + 25 = 0$.

Solution Applying the quadratic formula with $a = 9$, $b = -30$, and $c = 25$, we have

$$x = \frac{-(-30) \pm \sqrt{(-30)^2 - 4(9)(25)}}{2(9)}$$

$$= \frac{30 \pm \sqrt{900 - 900}}{18} = \frac{30 \pm 0}{18} = \frac{30}{18} = \frac{5}{3}.$$

Therefore, the given equation has only one real solution. The fact that the solution is a rational number indicates that this equation could have been solved by factoring. ▲

⑤

Use the method in Example 6 to find approximate solutions for $5.1x^2 - 3.3x - 240.624 = 0$.

Answer:

$x \approx 7.2$ or $x \approx -6.5529$

TECHNOLOGY TIP   You can approximate the solutions of quadratic equations on a graphing calculator by using a quadratic formula program (see the Program Appendix) or using a built-in quadratic equation solver if your calculator has one. Then you need only enter the coefficients a, b, c to obtain the approximate solutions. (See your instruction manual for details.)

*Based on data from the Identity Theft Clearinghouse of the Federal Trade Commission.

▷ **6**

Solve each equation.

(a) $9k^2 - 6k + 1 = 0$

(b) $4m^2 + 28m + 49 = 0$

(c) $2x^2 - 5x + 5 = 0$

Answers:

(a) $1/3$

(b) $-7/2$

(c) No real solutions

▷ **7**

Use the discriminant to determine the number of real solutions of each equation.

(a) $x^2 + 8x + 3 = 0$

(b) $2x^2 + x + 3 = 0$

(c) $x^2 - 194x + 9409 = 0$

Answers:

(a) 2

(b) 0

(c) 1

▼ **EXAMPLE 8** Solve $x^2 - 6x + 10 = 0$.

Solution Apply the quadratic formula with $a = 1$, $b = -6$, and $c = 10$.

$$x = \frac{-(-6) \pm \sqrt{(-6)^2 - 4(1)(10)}}{2(1)}$$

$$= \frac{6 \pm \sqrt{36 - 40}}{2}$$

$$= \frac{6 \pm \sqrt{-4}}{2}$$

Since no negative number has a square root in the real number system, $\sqrt{-4}$ is not a real number. Hence, the equation has no real solutions. ▲ ▷ **6**

Examples 5–8 show that the number of solutions of the quadratic equation $ax^2 + bx + c = 0$ is determined by $b^2 - 4ac$, the quantity under the radical, which is called the **discriminant** of the equation. ▷ **7**

The Discriminant

The equation $ax^2 + bx + c = 0$ has either two, or one, or no real solutions.

If $b^2 - 4ac > 0$, there are two real solutions. (*Examples 5 and 6*)

If $b^2 - 4ac = 0$, there is one real solution. (*Example 7*)

If $b^2 - 4ac < 0$, there are no real solutions. (*Example 8*)

Applications

Quadratic equations arise in a variety of settings, as illustrated in the examples below. Example 9 depends on the following useful fact from geometry.

The Pythagorean Theorem

In a right triangle with legs of lengths a and b and hypotenuse of length c,

$$a^2 + b^2 = c^2.$$

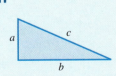

▼ **EXAMPLE 9** The size of a computer monitor is the diagonal measurement of its screen. The height of the screen is approximately $3/4$ of its width. Kathy claims that John's 14-inch monitor has less than half the viewing area of her 21-inch monitor. John says, "No way! 14/21 is 2/3, so my monitor has 2/3 of the viewing area of yours." Who is right?

Solution First, find the area of Kathy's screen. Let x be its width. Then its height is $3/4$ of x (that is, $.75x$, as shown in Figure 1.15).

FIGURE 1.15

By the Pythagorean Theorem,

$$x^2 + (.75x)^2 = 21^2$$
$$x^2 + .5625x^2 = 441 \qquad \text{Expand } (.75x)^2$$
$$(1 + .5625)x^2 = 441 \qquad \text{Distributive property}$$
$$1.5625x^2 = 441$$
$$x^2 = 282.24 \qquad \text{Divide both sides by 1.5625}$$
$$x = \pm\sqrt{282.24} \qquad \text{Square-root property}$$
$$x = \pm 16.8$$

We can ignore the negative solution, since x is a width. Thus, the width is 16.8 inches and the height is $.75x = .75(16.8) = 12.6$ inches, so the area is

$$\text{Area} = \text{width} \times \text{height} = 16.8 \times 12.6 = 211.68 \text{ square inches.}$$

Next, find the area of John's screen by working Problem ⟨**8**⟩ at the side, which shows that John's screen has an area of 94.08 square inches. So who is right? ▲ ⟨**8**⟩

⟨**8**⟩

Let x be the width of John's screen in Example 9.

(a) Write an equation that expresses the relationship between the width, height, and diagonal of the screen.

(b) Find the dimensions of the screen.

(c) Find its area.

Answer:

(a) $x^2 + (.75x)^2 = 14^2$

(b) 11.2 by 8.4 inches

(c) 94.08 square inches

▼ **EXAMPLE 10**   A landscape architect wants to make an exposed gravel border of uniform width around a small shed behind a company plant. The shed is 10 feet by 6 feet. He has enough gravel to cover 36 square feet. How wide should the border be?

  Solution A sketch of the shed with border is given in Figure 1.16. Let x represent the width of the border. Then the width of the large rectangle is $6 + 2x$, and its length is $10 + 2x$.

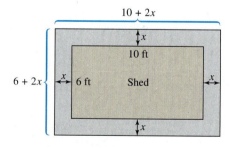

FIGURE 1.16

⟨**9**⟩

The length of a picture is 2 inches more than the width. It is mounted on a mat that extends 2 inches beyond the picture on all sides. What are the dimensions of the picture if the area of the mat is 99 square inches?

Answer:

  5 inches by 7 inches

We must write an equation relating the given areas and dimensions. The area of the large rectangle is $(6 + 2x)(10 + 2x)$. The area occupied by the shed is $6 \cdot 10 = 60$. The area of the border is found by subtracting the area of the shed from the area of the large rectangle. This difference should be 36 square feet, giving the equation

$$(6 + 2x)(10 + 2x) - 60 = 36.$$

Solve this equation with the following sequence of steps:

$$60 + 32x + 4x^2 - 60 = 36 \qquad \text{Multiply out left side.}$$
$$4x^2 + 32x - 36 = 0 \qquad \text{Simplify.}$$
$$x^2 + 8x - 9 = 0 \qquad \text{Divide both sides by 4.}$$
$$(x + 9)(x - 1) = 0 \qquad \text{Factor.}$$

10
Solve each of the following equations for the indicated variable. Assume that all variables are positive.

(a) $k = mp^2 - bp$ for p

(b) $r = \dfrac{APk^2}{3}$ for k

Answers:

(a) $p = \dfrac{b \pm \sqrt{b^2 + 4mk}}{2m}$

(b) $k =$

$\pm\sqrt{\dfrac{3r}{AP}}$ or $\dfrac{\pm\sqrt{3rAP}}{AP}$

$x + 9 = 0$ or $x - 1 = 0$ Zero-factor property

$x = -9$ or $x = 1$

The number -9 cannot be the width of the border, so the solution is to make the border 1 foot wide. ▲ ⑨

In some applications, it may be necessary to solve an equation in several variables for a specific variable.

▼ **EXAMPLE 11** Solve $v = mx^2 + x$ for x. (Assume that m and v are positive.)

Solution The equation is quadratic in x because of the x^2 term. Use the quadratic formula, first writing the equation in standard form.

$$v = mx^2 + x$$
$$0 = mx^2 + x - v$$

Let $a = m$, $b = 1$, and $c = -v$. Then the quadratic formula gives

$$x = \frac{-1 \pm \sqrt{1^2 - 4(m)(-v)}}{2m}$$

$$x = \frac{-1 \pm \sqrt{1 + 4mv}}{2m}. \; ▲ \; ⑩$$

◆ ## 1.7 Exercises

Use factoring to solve each equation. (See Examples 1 and 2.)

1. $(x + 4)(x - 14) = 0$
2. $(p - 16)(p - 5) = 0$
3. $x(x + 6) = 0$
4. $x^2 - 2x = 0$
5. $3z^2 = 9z$
6. $x^2 - 64 = 0$
7. $y^2 + 15y + 56 = 0$
8. $k^2 - 4k - 5 = 0$
9. $2x^2 = 7x - 3$
10. $2 = 12z^2 + 5z$
11. $6r^2 + r = 1$
12. $3y^2 = 16y - 5$
13. $2m^2 + 20 = 13m$
14. $10a^2 + 17a + 3 = 0$
15. $m(m + 7) = -10$
16. $z(2z + 7) = 4$
17. $9x^2 - 16 = 0$
18. $25y^2 - 64 = 0$
19. $16x^2 - 16x = 0$
20. $12y^2 - 48y = 0$

Solve each equation by using the square-root property. (See Example 3.)

21. $(r - 2)^2 = 7$
22. $(b + 5)^2 = 8$
23. $(4x - 1)^2 = 20$
24. $(3t + 5)^2 = 11$

Use the quadratic formula to solve each equation. If the solutions involve square roots, give both the exact and approximate solutions. (See Examples 5–8.)

25. $2x^2 + 7x + 1 = 0$
26. $3x^2 - x - 7 = 0$
27. $4k^2 + 2k = 1$
28. $r^2 = 3r + 5$
29. $5y^2 + 5y = 2$
30. $2z^2 + 3 = 8z$

31. $6x^2 + 6x + 4 = 0$
32. $3a^2 - 2a + 2 = 0$
33. $2r^2 + 3r - 5 = 0$
34. $8x^2 = 8x - 3$
35. $2x^2 - 7x + 30 = 0$
36. $3k^2 + k = 6$
37. $1 + \dfrac{7}{2a} = \dfrac{15}{2a^2}$
38. $5 - \dfrac{4}{k} - \dfrac{1}{k^2} = 0$

Use the discriminant to determine the number of real solutions of each equation. You need not solve the equations.

39. $25t^2 + 49 = 70t$
40. $9z^2 - 12z = 1$
41. $13x^2 + 24x - 5 = 0$
42. $20x^2 + 19x + 5 = 0$

Use a calculator and the quadratic formula to find approximate solutions of the equation. (See Example 6.)

43. $4.42x^2 - 10.14x + 3.79 = 0$
44. $3x^2 - 82.74x + 570.4923 = 0$
45. $7.63x^2 + 2.79x = 5.32$
46. $8.06x^2 + 25.8726x = 25.047256$

Solve the following problems. (See Example 3.)

47. **Physical Science** When wind resistance is ignored, a falling object travels a distance of d feet in t seconds, where $d = 16t^2$.
    (a) How long does it take for the object to fall 36 feet?
    (b) How long does it take for the object to fall five times farther than in part (a)?

48. Transportation According to the Federal Aviation Administration, the maximum recommended taxiing speed x (in miles per hour) for a plane on a curved runway exit is given by $R = .5x^2$, where R is the radius of the curve (in feet). Find the maximum taxiing speed for planes on such exits when the radius of the exit is
(a) 450 ft
(b) 615 ft
(c) 970 ft.

Solve the following problems. (See Example 6.)

49. Social Science The number of drivers who die in automobile accidents is related to the ages of the drivers, with teenagers and the elderly having the worst records. According to data from the National Highway Traffic Safety Administration, the driver fatality rate D per 1000 licensed drivers every 100 million miles can be approximated by the equation $D = .0031x^2 - .291x + 7.1$, where x is the age of the driver.
(a) For what ages is the driver fatality rate about 1 death per 1000?
(b) For what ages is the rate three times greater than in part (a)?

50. Finance The total resources T (in billions of dollars) of the Pension Benefit Guaranty Corporation, the government agency that insures pensions, can be approximated by the equation $T = -.26x^2 + 3.62x + 30.18$, where x is the number of years after 2000.* Determine when the total resources will be at the given level.
(a) $42.5 billion
(b) $30 billion
(c) When will the Corporation be out of money $(T = 0)$?

51. Business The number N of Walgreens drugstores in year x can be approximated by $N = 6.82x^2 - 1.55x + 666.8$, where $x = 0$ corresponds to 1980.† Determine when the number of stores was or will be
(a) 4240
(b) 5600
(c) 7000

52. Health The number N of AIDS cases diagnosed to date (in thousands) is approximated by $N = -.37x^2 + 59.5x + 247.26$, where x is the number of years since 1990.‡ Assuming that this equation remains valid through 2011, determine when the number of diagnosed cases of AIDS was or will be
(a) 825,000
(b) 1.2 million

---

*Based on data from the Center on Federal Financial Institutions.

†Based on data from the Walgreen Company.

‡Based on data from the U.S. Department of Health and Human Services.

Solve the following problems. (See Examples 9 and 10.)

53. Transportation Two trains leave the same city at the same time, one going north and the other east. The eastbound train travels 20 mph faster than the northbound one. After 5 hours, the trains are 300 miles apart. Determine the speed of each train, using the following steps.
(a) Let x denote the speed of the northbound train. Express the speed of the eastbound train in terms of x.
(b) Write expressions that give the distance traveled by each train after 5 hours.
(c) Use part (b) and the fact that the trains are 300 miles apart after 5 hours to write an equation. (A diagram of the situation may help.)
(d) Solve the equation and determine the speeds of the trains.

54. Chris and Josh have received walkie-talkies for Christmas. If they leave from the same point at the same time, Chris walking north at 2.5 mph and Josh walking east at 3 mph, how long will they be able to talk to each other if the range of the walkie-talkies is 4 miles? Round your answer to the nearest minute.

55. An ecology center wants to set up an experimental garden. It has 300 meters of fencing to enclose a rectangular area of 5000 square meters. Find the length and width of the rectangle as follows.
(a) Let $x =$ the length and write an expression for the width.
(b) Write an equation relating the length, width, and area, using the result of part (a).
(c) Solve the problem.

56. A landscape architect has included a rectangular flower bed measuring 9 feet by 5 feet in her plans for a new building. She wants to use two colors of flowers in the bed, one in the center and the other for a border of the same width on all four sides. If she can get just enough plants to cover 24 square feet for the border, how wide can the border be?

57. Joan wants to buy a rug for a room that is 12 feet by 15 feet. She wants to leave a uniform strip of floor around the rug. She can afford 108 square feet of carpeting. What dimensions should the rug have?

58. In 1991 Rick Mears won the (500 mile) Indianapolis 500 race. His speed (rate) was 102 mph (to the nearest mph) faster than that of the 1911 winner, Ray Harroun. Mears completed the race in 3.87 hours less time than Harroun. Find Mears's rate to the nearest whole number.

59. Physical Science If a ball is thrown upward with an initial velocity of 64 feet per second, then its height after t seconds is $h = 64t - 16t^2$. In how many seconds will the ball reach
(a) 64 ft?
(b) 28 ft?
(c) Why are two answers possible?

60. Physical Science The atmosphere pressure a (in pounds per square foot) at height h thousand feet above sea level is approximately $a = .8315h^2 - 73.93h + 2116.1$.
   (a) Find the atmospheric pressure at sea level and at the top of Mount Everest, the tallest mountain in the world (29,035 feet). Remember that h is measured in thousands.
   (b) The atmospheric pressure at the top of Mount Rainier is 1223.43 pounds per square foot. How high is Mount Rainier?

Solve each of the following equations for the indicated variable. Assume that all denominators are nonzero and that all variables represent positive real numbers. (See Example 11.)

61. $S = \dfrac{1}{2}gt^2$ for t
62. $a = \pi r^2$ for r

63. $L = \dfrac{d^4 k}{h^2}$ for h
64. $F = \dfrac{kMv^2}{r}$ for v

65. $P = \dfrac{E^2 R}{(r + R)^2}$ for R
66. $S = 2\pi rh + 2\pi r^2$ for r

67. Solve the equation $z^4 - 2z^2 = 15$ as follows.
   (a) Let $x = z^2$ and write the equation in terms of x.
   (b) Solve the new equation for x.
   (c) Set z^2 equal to each positive answer in part (b) and solve the resulting equation.

Solve each of the following equations. (See Exercise 67.)

68. $6p^4 = p^2 + 2$
69. $2q^4 + 3q^2 - 9 = 0$
70. $4a^4 = 2 - 7a^2$
71. $z^4 - 3z^2 - 1 = 0$
72. $2r^4 - r^2 - 5 = 0$

Chapter 1 Summary

Key Terms and Symbols

1.1 $\approx$ is approximately equal to
π pi
$|a|$ absolute value of a
real number
natural (counting) number
whole number
integer
rational number
irrational number
properties of real numbers
additive inverse
multiplicative inverse
order of operations
square roots
number line
interval
interval notation
absolute value

1.2 a^n a to the power n
exponent or power
multiplication with exponents
power of a power rule
zero exponent
base
polynomial
variable

coefficient
term
constant term
degree of a polynomial
zero polynomial
leading coefficient
quadratics
cubics
like terms
FOIL
revenue
fixed cost
variable cost
profit

1.3 factor
factoring
greatest common factor
difference of squares
perfect squares
sum and difference of cubes

1.4 rational expression
cancellation property
operations with rational
   expressions
complex fraction

1.5 $a^{1/n}$ nth root of a
$\sqrt{a}$ square root of a
$\sqrt[n]{a}$ nth root of a
properties of exponents
radical
radicand
index
rationalizing the denominator
rationalizing the numerator

1.6 first-degree equation
solution of an equation
properties of equality
extraneous solution
solving for a specified variable
absolute value equations
solving applied problems

1.7 quadratic equation
real solution
zero-factor property
square-root property
completing the square
quadratic formula
discriminant
Pythagorean Theorem

Key Concepts

Factoring

$$x^2 + 2xy + y^2 = (x + y)^2 \qquad x^3 - y^3 = (x - y)(x^2 + xy + y^2)$$
$$x^2 - 2xy + y^2 = (x - y)^2 \qquad x^3 + y^3 = (x + y)(x^2 - xy + y^2)$$
$$x^2 - y^2 = (x + y)(x - y)$$

Properties of Radicals

Let a and b be real numbers, n be a positive integer, and m be any integer for which the following exist. Then

$$a^{m/n} = \sqrt[n]{a^m} = (\sqrt[n]{a})^m \qquad \sqrt[n]{a^n} = |a| \text{ if } n \text{ is even} \qquad \sqrt[n]{a^n} = a \text{ if } n \text{ is odd}$$

$$\sqrt[n]{a} \cdot \sqrt[n]{b} = \sqrt[n]{ab} \qquad \frac{\sqrt[n]{a}}{\sqrt[n]{b}} = \sqrt[n]{\frac{a}{b}} \quad (b \neq 0)$$

Properties of Exponents

Let a, b, r, and s be any real numbers for which the following exist. Then

$$a^{-r} = \frac{1}{a^r} \qquad a^0 = 1 \qquad \left(\frac{a}{b}\right)^r = \frac{a^r}{b^r}$$

$$a^r \cdot a^s = a^{r+s} \qquad (a^r)^s = a^{rs} \qquad a^{1/r} = \sqrt[r]{a}$$

$$\frac{a^r}{a^s} = a^{r-s} \qquad (ab)^r = a^r b^r \qquad \left(\frac{a}{b}\right)^{-r} = \left(\frac{b}{a}\right)^r$$

Absolute Value

Assume that a and b are real numbers with $b > 0$.

The solutions of $|a| = b$ or $|a| = |b|$ are $a = b$ or $a = -b$.

Quadratic Equations

Facts needed to solve quadratic equations (in which a, b, and c are real numbers):

Factoring: If $ab = 0$, then $a = 0$ or $b = 0$ or both.

Square-Root Property: If $b > 0$, then the solutions of $x^2 = b$ are $\sqrt{b}$ and $-\sqrt{b}$.

Quadratic Formula: The solutions of $ax^2 + bx + c = 0$ (with $a \neq 0$) are

$$x = \frac{-b \pm \sqrt{b^2 - 4ac}}{2a}.$$

Discriminant: There are two real solutions if $b^2 - 4ac > 0$, one real solution if $b^2 - 4ac = 0$, and no real solutions if $b^2 - 4ac < 0$.

Chapter 1 Review Exercises

Name the numbers from the list -12, -6, $-9/10$, $-\sqrt{7}$, $-\sqrt{4}$, 0, $1/8$, $\pi/4$, 6, and $\sqrt{11}$ that are

1. whole numbers

2. integers

3. rational numbers

4. irrational numbers

Identify the properties of real numbers that are illustrated in each of the following.

5. $9[(-3)4] = 9[4(-3)]$

6. $7(4 + 5) = (4 + 5)7$

7. $6(x + y - 3) = 6x + 6y + 6(-3)$

8. $11 + (5 + 3) = (11 + 5) + 3$

Express each statement in symbols.

9. x is at least 9.

10. x is negative.

Write the following numbers in numerical order from smallest to largest.

11. -7, -3, 8, π, -2, 0

12. $\dfrac{5}{6}$, $\dfrac{1}{2}$, $-\dfrac{2}{3}$, $-\dfrac{5}{4}$, $-\dfrac{3}{8}$

13. $|6 - 4|, -|-2|, |8 + 1|, -|3 - (-2)|$

14. $\sqrt{7}, -\sqrt{8}, -|\sqrt{16}|, |-\sqrt{12}|$

Write without absolute value bars.

15. $-|-5| + |3|$

16. $|-6| + |-9|$

17. $7 - |-8|$

18. $|-2| - |-7 + 3|$

Graph each of the following on a number line.

19. $x \geq -3$

20. $-4 < x \leq 6$

21. $x < -2$

22. $x \leq 1$

Use the order of operations to simplify.

23. $(-6 + 3 \cdot 5)(-2)$

24. $-4(-8 - 9 \div 3)$

25. $\dfrac{-8 + (-6)(-3) \div 9}{6 - (-2)}$

26. $\dfrac{20 \div 4 \cdot 2 \div 5 - 1}{-9 - (-3) - 12 \div 3}$

Perform each of the indicated operations.

27. $(3x^4 - x^2 + 5x) - (-x^4 + 3x^2 - 6x)$

28. $(-8y^3 + 8y^2 - 3y) - (2y^3 - 4y^2 - 12)$

29. $-2(q^4 - 3q^3 + 4q^2) + 4(q^4 + 2q^3 + q^2)$

30. $5(3y^4 - 4y^5 + y^6) - 3(2y^4 + y^5 - 3y^6)$

31. $(4z + 2)(3z - 2)$ 32. $(8p - 4)(5p + 2)$

33. $(4k - 3h)(4k + 3h)$ 34. $(2r - 5y)(2r + 5y)$

35. $(6x + 3y)^2$ 36. $(2a - 5b)^2$

Factor as completely as possible.

37. $2kh^2 - 4kh + 5k$ 38. $2m^2n^2 + 6mn^2 + 16n^2$

39. $3a^4 + 13a^3 + 4a^2$ 40. $24x^3 + 4x^2 - 4x$

41. $10y^2 - 11y + 3$ 42. $8q^2 + 3m + 4qm + 6q$

43. $4a^2 - 20a + 25$ 44. $36p^2 + 12p + 1$

45. $144p^2 - 169q^2$ 46. $81z^2 - 25x^2$

47. $8y^3 - 1$ 48. $125a^3 + 216$

Perform each operation.

49. $\dfrac{3x}{5} \cdot \dfrac{35x}{12}$

50. $\dfrac{5k^2}{24} - \dfrac{70k}{36}$

51. $\dfrac{c^2 - 3c + 2}{2c(c - 1)} \div \dfrac{c - 2}{8c}$

52. $\dfrac{p^3 - 2p^2 - 8p}{3p(p^2 - 16)} \div \dfrac{p^2 + 4p + 4}{9p^2}$

53. $\dfrac{2y - 10}{5y} \cdot \dfrac{20y - 25}{14}$

54. $\dfrac{m^2 - 2m}{15m^3} \cdot \dfrac{5}{m^2 - 4}$

55. $\dfrac{2m^2 - 4m + 2}{m^2 - 1} \div \dfrac{6m + 18}{m^2 + 2m - 3}$

56. $\dfrac{x^2 + 6x + 5}{4(x^2 + 1)} \cdot \dfrac{2x(x + 1)}{x^2 - 25}$

57. $\dfrac{6}{15z} + \dfrac{2}{3z} - \dfrac{9}{10z}$

58. $\dfrac{5}{y - 3} - \dfrac{4}{y}$

59. $\dfrac{2}{5q} + \dfrac{10}{7q}$

60. Give two ways to evaluate $125^{2/3}$ and then compare them. Which do you prefer? Why?

Simplify each of the following. Write all answers without negative exponents. Assume that all variables represent positive real numbers.

61. 6^{-3} 62. 10^{-2}

63. -8^0 64. -3^{-1}

65. $\left(-\dfrac{6}{5}\right)^{-2}$ 66. $\left(\dfrac{3}{2}\right)^{-3}$

67. $4^6 \cdot 4^{-3}$ 68. $7^{-5} \cdot 7^{-2}$

69. $\dfrac{9^{-4}}{9^{-3}}$ 70. $\dfrac{5^{-2}}{5^3}$

71. $\dfrac{9^4 \cdot 9^{-5}}{(9^{-2})^2}$ 72. $\dfrac{k^4 \cdot k^{-3}}{(k^{-2})^{-3}}$

73. $4^{-1} + 2^{-1}$ 74. $5^{-2} + 5^{-1}$

75. $125^{2/3}$ 76. $128^{3/7}$

77. $9^{-5/2}$ 78. $\left(\dfrac{144}{49}\right)^{-1/2}$

79. $\dfrac{5^{1/3} \cdot 5^{1/2}}{5^{3/2}}$ 80. $\dfrac{2^{3/4} \cdot 2^{-1/2}}{2^{1/4}}$

81. $(3a^2)^{1/2} \cdot (3^2a)^{3/2}$ 82. $(4p)^{2/3} \cdot (2p^3)^{3/2}$

Simplify each of the following.

83. $\sqrt[3]{27}$ 84. $\sqrt[6]{-64}$

85. $\sqrt{54}$ 86. $\sqrt{63}$

87. $\sqrt[3]{54p^3q^5}$ 88. $\sqrt[4]{64a^5b^3}$

89. $\sqrt{\dfrac{5n^2}{6m}}$ 90. $\sqrt{\dfrac{3x^3}{2z}}$

91. $2\sqrt{3} - 5\sqrt{12}$ 92. $8\sqrt{7} + 2\sqrt{63}$

93. $(\sqrt{6} - 1)(\sqrt{6} + 1)$

94. $(\sqrt{5} - \sqrt{3})(\sqrt{5} + \sqrt{3})$

Rationalize each denominator.

95. $\dfrac{\sqrt{2}}{1 + \sqrt{3}}$

96. $\dfrac{4 + \sqrt{2}}{4 - \sqrt{5}}$

Social Science   *In our system of government, the president is elected by the electoral college and not by individual voters. Because of this, smaller states have a greater voice in the selection of a president than they otherwise would have. Two political scientists have studied the problems of campaigning for president under the current system and have concluded that candidates should allot their money according to the formula*

$$\begin{array}{l} \text{Amount for} \\ \text{large state} \end{array} = \left(\dfrac{E_{large}}{E_{small}}\right)^{3/2} \times \begin{array}{l} \text{amount for} \\ \text{small state.} \end{array}$$

Here E_{large} represents the electoral vote of the large state, and E_{small} represents the electoral vote of the small state. Find the amount that should be spent in each of the following larger states if $1,000,000 is spent in the small state and the following statements are true.

97. Florida has 27 electoral votes, and Vermont has 3.

98. New York has 31 electoral votes, and Kansas has 6.

99. 5 votes in Nevada; 55 in California

100. 10 votes in Wisconsin; 34 in Texas

Solve each equation.

101. $2x - 5(x - 3) = 3x + 9$

102. $4y + 9 = -3(1 - 2y) + 5$

103. $\dfrac{2z}{5} - \dfrac{4z - 3}{10} = \dfrac{-z + 1}{10}$

104. $\dfrac{p}{p + 2} - \dfrac{3}{4} = \dfrac{2}{p + 2}$

105. $\dfrac{2m}{m - 3} = \dfrac{6}{m - 3} + 4$

106. $\dfrac{15}{k + 5} = 4 - \dfrac{3k}{k + 5}$

Solve for x.

107. $6ax - 1 = x$

108. $6x - 5y = 4bx$

109. $\dfrac{2x}{3 - c} = ax + 1$

110. $b^2x - 2x = 4b^2$

Solve each equation.

111. $|m - 5| = 9$

112. $|4 - x| = 12$

113. $\left|\dfrac{2 - y}{5}\right| = 8$

114. $|4k + 1| = |6k - 3|$

115. **Energy**   World energy demand D (in millions of barrels per day equivalent) is approximated by $D = 3.7x + 111$, where $x = 0$ corresponds to 1970.* Find the year in which demand reached or will reach
  (a) 259 millions of barrels per day
  (b) 300 millions of barrels per day

116. **Business**   A laser printer is on sale for 15% off. The sale price is $306. What was the original price?

117. **Health**   The annual per capita medical costs (in thousands of dollars) paid by the Boeing Company in St. Louis are estimated to be given by $C = 1.1953x + 5.343$, where x is the number of years after 2000.† Find the year in which these costs reached
  (a) $10,124
  (b) $13,710.

118. **Finance**   Ellen borrowed $500 from a credit union at 12% annual interest and got $250 in cash with her credit card at 18% annual interest. What single rate of interest on $750 results in the same total amount of interest?

119. **Finance**   A real estate firm invests $100,000 proceeds from a sale in two ways. The first portion is invested in a shopping center that provides an annual return of 8%. The rest is invested in a small apartment building with an annual return of 5%. The firm wants an annual income of $6800 from these investments. How much should be put into each investment?

120. **Business**   To make a special mix for Valentine's Day, the owner of a candy store wants to combine chocolate hearts that sell for $5 per pound with candy kisses that sell for $3.50 per pound. How many pounds of each kind should be used to get 30 pounds of a mix that can be sold for $4.50 per pound?

Determine the number of real solutions of each quadratic equation.

121. $x^2 - 6x = 4$

122. $-3x^2 + 5x + 2 = 0$

123. $4x^2 - 12x + 9 = 0$

124. $5x^2 + 2x + 1 = 0$

125. $x^2 + 3x + 5 = 0$

Find all real solutions of each equation.

126. $(b + 7)^2 = 5$ **127.** $(2p + 1)^2 = 7$

128. $2p^2 + 3p = 2$ **129.** $2y^2 = 15 + y$

130. $x^2 - 2x = 2$ **131.** $r^2 + 4r = 1$

132. $2m^2 - 12m = 11$ **133.** $9k^2 + 6k = 2$

134. $2a^2 + a - 15 = 0$ **135.** $12x^2 = 8x - 1$

---

*Based on data and projections by ExxonMobil.

†Based on a Boeing presentation in 2004.

136. $2q^2 - 11q = 21$

137. $3x^2 + 2x = 16$

138. $6k^4 + k^2 = 1$

139. $21p^4 = 2 + p^2$

140. $2x^4 = 7x^2 + 15$

141. $3m^4 + 20m^2 = 7$

142. $3 = \dfrac{13}{z} + \dfrac{10}{z^2}$

Solve each equation for the specified variable.

143. $p = \dfrac{E^2R}{(r + R)^2}$ for r

144. $p = \dfrac{E^2R}{(r + R)^2}$ for E

145. $K = s(s - a)$ for s

146. $kz^2 - hz - t = 0$ for z

147. **Social Science**   The number N of consumers (in millions) who subscribe to Internet-based phone services is projected to be given by $N = .364x^2 - 2.296x + 3783$, where $x = 0$ corresponds to 2000.*

*Based on data and projections from In-Stat.

(a) How many people subscribed in 2003?

(b) When will the number of subscribers be 9 million?

148. **Business**   A landscaper wants to put a cement walk of uniform width around a rectangular garden that measures 24 by 40 feet. She has enough cement to cover 740 square feet. To the nearest tenth of a foot, how wide should the walk be in order to use up all the cement?

149. **Business**   A recreation director wants to fence off a rectangular playground beside an apartment building. The building forms one boundary, so she needs to fence only the other three sides. The area of the playground is to be 11,250 square meters. She has enough material to build 325 meters of fence. Find the length and width of the playground.

150. **Transportation**   Two cars leave an intersection at the same time. One travels north, and the other heads west traveling 10 mph faster. After 1 hour, they are 50 miles apart. What were their speeds?

C a s e   1

Consumers Often Defy Common Sense*

Imagine two refrigerators in the appliance section of a department store. One sells for $700 and uses $85 worth of electricity a year. The other is $100 more expensive but costs only $25 a year to run. Given that either refrigerator should last at least 10 years without repair, consumers would overwhelmingly buy the second model, right?

Well, not exactly. Many studies by economists have shown that in a wide range of decisions about money—from paying taxes to buying major appliances—consumers consistently make decisions that defy common sense.

*"Consumers Choices About Money Consistently Defy Common Sense." Malcolm Gladwell, *The Washington Post,* February 12, 1990 © 1990, The Washington Post Company. Reprinted with permission.

In some cases—as in the refrigerator example—this means that people are generally unwilling to pay a little more money up front to save a lot of money in the long run. At times, psychological studies have shown, consumers appear to assign entirely whimsical values to money, values that change depending on time and circumstances.

In recent years, these apparently irrational patterns of human behavior have become a subject of intense interest among economists and psychologists, both for what they say about the way the human mind works and because of their implications for public policy.

How, for example, can the United States move toward a more efficient use of electricity if so many consumers refuse to buy energy-efficient appliances even when such a move is in their own best interest?

At the heart of research into the economic behavior of consumers is a concept known as the discount rate. It is a measure of how consumers compare the value of a dollar received today with one received tomorrow.

Consider, for example, if you won $1000 in a lottery. How much more money would officials have to give you before you would agree to postpone cashing the check for a year?

Some people might insist on at least another $100, or 10 percent, since that is roughly how much it would take to make up for the combined effects of a year's worth of inflation and lost interest.

But the studies show that someone who wants immediate gratification might not be willing to postpone receiving the $1000 for 20 percent or 30 percent or even 40 percent more money.

In the language of economists, this type of person has a high discount rate: He or she discounts the value of $1000 so much over a year that it would take hundreds of extra dollars to make waiting as attractive as getting the money immediately.

Of the two alternatives, waiting a year for more money is clearly more rational than taking the check now. Why would people turn down $1400 dollars next year in favor of $1000 today? Even if they needed the $1000 immediately, they would be better off borrowing it from a bank, even at 20 percent or even 30 percent interest. Then, a year later, they could pay off the loan—including the interest—with the $1400 and pocket the difference.

The fact is, however, that economists find numerous examples of such high discount rates implicit in consumer behavior.

While consumers were very much aware of savings to be made at the point of purchase, they so heavily discounted the value of monthly electrical costs that they would pay over the lifetime of their dryer or freezer that they were oblivious to the potential for greater savings.

Gas water heaters, for example, were found to carry an implicit discount rate of 100 percent. This means that in deciding which model was cheapest over the long run, consumers acted as if they valued a $100 gas bill for the first year as if it were really $50. Then, in the second year, they would value the next $100 gas bill as if it were really worth $25, and so on through the life of the appliance.

Few consumers actually make this formal calculation, of course. But there are clearly bizarre behavioral patterns in evidence.

Some experiments, for example, have shown that the way in which consumers make decisions about money depends a great deal on how much is at stake. Few people are willing to give up $10 now for $15 next year. But they are if the choice is between $100 now and $150 next year, a fact that would explain why consumers appear to care less about many small electricity bills—even if they add up to a lot—than one big initial outlay.

◆ Exercises

1. Suppose a refrigerator that sells for $700 costs $85 a year for electricity. Write an expression for the cost to buy and run the refrigerator for x years.

2. Suppose another refrigerator costs $1000 and $25 a year for electricity. Write an expression for the total cost for this refrigerator over x years.

3. Over 10 years, which refrigerator costs the most? By how much?

4. In how many years will the the total costs for the two refrigerators be equal?

2

Graphs, Lines, and Inequalities

Data from current and past events is often a useful tool in business and in the social and health sciences. Gathering data is the first step in developing mathematical models that can be used to analyze a situation and predict future performance. For examples of linear models in business, transportation, and health science, see Exercises 18, 21, and 23–25 on pages 105–107.

Graphical representations of data are commonly used in business and in the health and social sciences. Lines, equations, and inequalities play an important role in developing mathematical models from such data. This chapter presents both algebraic and graphical methods for dealing with these topics.

2.1 Graphs

Just as the number line associates the points on a line with real numbers, a similar construction in two dimensions associates points in the plane with *ordered pairs* of real numbers. A **Cartesian coordinate system,** as shown in Figure 2.1, consists of a horizontal number line (usually called the ***x*-axis**) and a vertical number line (usually called the ***y*-axis**). The point where the number lines meet is called the **origin.** Each point in a Cartesian coordinate system is labeled with an **ordered pair** of real numbers, such as $(-2, 4)$ or $(3, 2)$. Several points and their corresponding ordered pairs are shown in Figure 2.1.

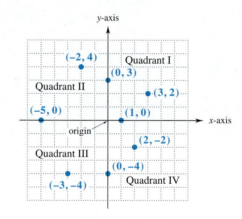

FIGURE 2.1

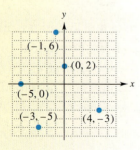
For the point labeled $(-2, 4)$, for example, -2 is the ***x*-coordinate** and 4 is the ***y*-coordinate.** You can think of these coordinates as directions telling you how to move to this point from the origin: you go 2 horizontal units to the left (*x*-coordinate) and 4 vertical units upward (*y*-coordinate). From now on, instead of referring to "the point labeled by the ordered pair $(-2, 4)$," we will say "the point $(-2, 4)$." ①

The *x*-axis and the *y*-axis divide the plane into four parts, or **quadrants,** which are numbered as shown in Figure 2.1. The points on the coordinate axes belong to no quadrant.

Equations and Graphs

A **solution of an equation** in two variables, such as

$$y = -2x + 3$$

or

$$y = x^2 + 7x - 2,$$

is an ordered pair of numbers such that the substitution of the first number for x and the second number for y produces a true statement.

▼ **EXAMPLE 1**   Which of the following are solutions of $y = -2x + 3$?

(a) $(2, -1)$

Solution This is a solution of $y = -2x + 3$ because "$-1 = -2 \cdot 2 + 3$" is a true statement.

(b) $(4, 7)$

Solution Since $-2 \cdot 4 + 3 = -5$ and not 7, the ordered pair $(4, 7)$ is not a solution of $y = -2x + 3$. ▲ ⟨2⟩

Equations in two variables, such as $y = -2x + 3$, typically have an infinite number of solutions. To find one, choose a number for x and then compute the value of y that produces a solution. For instance, if $x = 5$, then $y = -2 \cdot 5 + 3 = -7$, so that the pair $(5, -7)$ is a solution of $y = -2x + 3$. Similarly, if $x = 0$, then $y = -2 \cdot 0 + 3 = 3$, so that $(0, 3)$ is also a solution.

The **graph** of an equation in two variables is the set of points in the plane whose coordinates (ordered pairs) are solutions of the equation. Thus, the graph of an equation is a picture of its solutions. Since a typical equation has infinitely many solutions, its graph has infinitely many points.

▼ **EXAMPLE 2**   Sketch the graph of $y = -2x + 5$.

Solution Since we cannot plot infinitely many points, we construct a table of y-values for a reasonable number of x-values, plot the corresponding points, and make an "educated guess" about the rest. The table of values and points in Figure 2.2 suggests that the graph is a straight line, as shown in Figure 2.3. ▲ ⟨3⟩

x	-1	0	2	4	5
$-2x + 5$	7	5	1	-3	-5

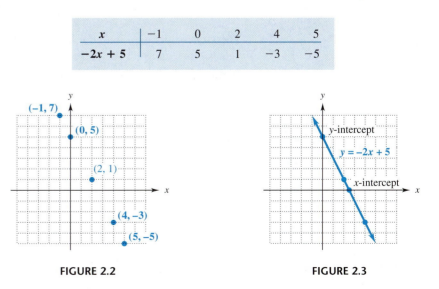

FIGURE 2.2   FIGURE 2.3

An **x-intercept** of a graph is the x-coordinate of a point where the graph intersects the x-axis. The y-coordinate of this point is 0, since it's on the axis.

⟨2⟩

Which of the following are solutions of $y = x^2 + 7x - 2$?

(a) $(1, 6)$

(b) $(-2, -20)$

(c) $(-1, -8)$

Answer:

(a) and (c)

⟨3⟩

Graph $x = 5y$.

Answer:

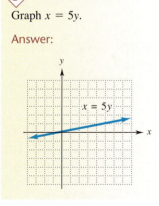

Consequently, to find the *x*-intercepts of the graph of an equation, set $y = 0$ and solve for *x*. For instance, in Example 2, the *x*-intercept of the graph of $y = -2x + 5$ (see Figure 2.3) is found by setting $y = 0$ and solving for *x*.

$$0 = -2x + 5$$

$$2x = 5$$

$$x = \frac{5}{2}$$

Similarly, a **y-intercept** of a graph is the *y*-coordinate of a point where the graph intersects the *y*-axis. The *x*-coordinate of this point is 0. The *y*-intercepts are found by setting $x = 0$ and solving for *y*. For example, the graph of $y = -2x + 5$ in Figure 2.3 has *y*-intercept 5. ◆

⬥ 4

Find the *x*- and *y*-intercepts of the graphs of these equations.

(a) $3x + 4y = 12$

(b) $5x - 2y = 8$

Answers:

(a)   *x*-intercept 4,
      y-intercept 3

(b)   *x*-intercept 8/5,
      y-intercept −4

▼ **EXAMPLE 3**   Find the *x*- and *y*-intercepts of the graph of $y = x^2 - 2x - 8$, and sketch the graph.

Solution   To find the *y*-intercept, set $x = 0$ and solve for *y*.

$$y = x^2 - 2x - 8 = 0^2 - 2 \cdot 0 - 8 = -8.$$

The *y*-intercept is −8. To find the *x*-intercept, set $y = 0$ and solve for *x*.

$$x^2 - 2x - 8 = y$$

$$x^2 - 2x - 8 = 0 \qquad \text{Set } y = 0.$$

$$(x + 2)(x - 4) = 0 \qquad \text{Factor}$$

$$x + 2 = 0 \quad \text{or} \quad x - 4 = 0 \qquad \text{Zero-factor property}$$

$$x = -2 \quad \text{or} \quad x = 4$$

The *x*-intercepts are −2 and 4. Now make a table, using both positive and negative values for *x*, and plot the corresponding points, as in Figure 2.4. These points suggest that the entire graph looks like Figure 2.5. ▲

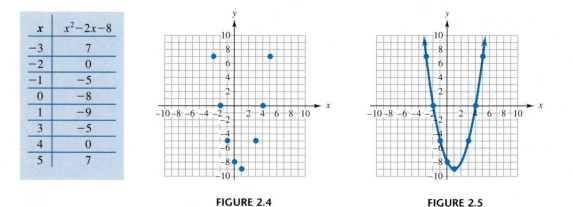

x	x^2-2x-8
−3	7
−2	0
−1	−5
0	−8
1	−9
3	−5
4	0
5	7

FIGURE 2.4   **FIGURE 2.5**

▼ **EXAMPLE 4**   Sketch the graph of $y = \sqrt{x + 2}$.

Solution Notice that $\sqrt{x + 2}$ is a real number only when $x + 2 \geq 0$—that is, when $x \geq -2$. Furthermore, $y = \sqrt{x + 2}$ is always nonnegative. Hence, all points on the graph lie on or above the x-axis and on or to the right of $x = -2$. Computing some typical values, we obtain the graph in Figure 2.6. ▲

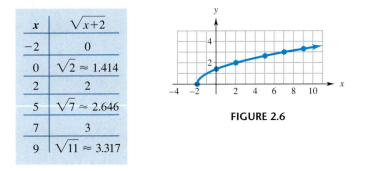

x	$\sqrt{x+2}$
-2	0
0	$\sqrt{2} \approx 1.414$
2	2
5	$\sqrt{7} \approx 2.646$
7	3
9	$\sqrt{11} \approx 3.317$

FIGURE 2.6

Example 2 shows that the solution of the equation $-2x + 5 = 0$ is the x-intercept of the graph of $y = -2x + 5$. Example 3 shows that the solutions of the equation $x^2 - 2x - 8 = 0$ are the x-intercepts of the graph $y = x^2 - 2x - 8$. Similar facts hold in the general case.

Intercepts and Equations

The real solutions of a one-variable equation of the form

expression in x = 0

are the x-intercepts of the graph of

y = same expression in x.

Graph Reading

Information is often given in graphical form, so you must be able to read and interpret graphs—that is, translate graphical information into statements in English.

▼ **EXAMPLE 5**   At various locations, the National Weather Service continuously records the temperature in graphical form. The results for March 20, 2001, in Cleveland, Ohio, are displayed in Figure 2.7 on the next page. The first coordinate of each point on the graph represents the time (measured in hours after midnight) and the second coordinate, the temperature at that time.

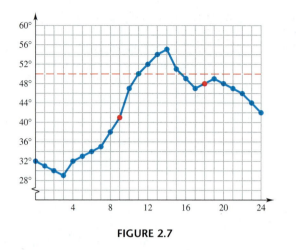

FIGURE 2.7

(a) What was the temperature at 9 A.M. and 6 P.M.?

Solution The point (9, 41) is on the graph, which means that the temperature at 9 A.M. was 41°. Now, 6 P.M. is 18 hours after midnight and the point (18, 48) is on the graph. So the temperature at 6 P.M. was 48°.

(b) At what times during the day was the temperature below 50°?

Solution Look for the points whose second coordinates are less than 50—that is, points which lie below the horizontal line through 50° (shown in red in Figure 2.7). The first coordinates of these points are the times when the temperature was below 50°. The figure shows that these are the points with first coordinates less than 11 or greater than about 15.5. Since 15.5 hours corresponds to 3:30 P.M., the temperature was below 50° from midnight to 11 A.M. and from approximately 3:30 P.M. to midnight. ▲ ⑤

The next example deals with the basic business relationship that was introduced in Section 1.2:

$$\text{Profit} = \text{Revenue} - \text{Cost}.$$

⑤

In Example 5, what were the highest and lowest temperatures during the day? When did they occur?

Answer:

The highest was 55° at 2 P.M. The lowest was 29° at 3 A.M.

▼ **EXAMPLE 6** Monthly revenue and costs for the Fenton Cell Phone Company are determined by the number t of phones produced and sold, as shown in Figure 2.8.

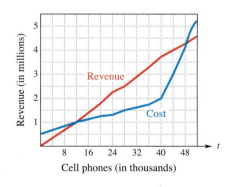

Cell phones (in thousands)

FIGURE 2.8

(a) How many phones should be produced each month if the company is to make a profit (assuming that all phones produced are sold)?

Solution Profit is revenue minus cost, so the company makes a profit whenever revenue is greater than cost—that is, when the revenue graph is above the cost graph. Figure 2.8 shows that this occurs between $t = 12$ and $t = 48$—that is, when 12,000 to 48,000 phones are produced. If the company makes fewer than 12,000 phones, it will lose money (costs will be greater than revenue.) They also lose money by making more than 48,000 phones. (One reason might be that high production levels require large amounts of overtime pay, which drives costs up too much.)

(b) Is it more profitable to make 40,000 or 44,000 phones?

Solution On the revenue graph, the point with first coordinate 40 has second coordinate of approximately 3.7, meaning that the revenue from 40,000 phones is about 3.7 million dollars. The point with first coordinate 40 on the cost graph is (40, 2), meaning that the cost of producing 40,000 phones is 2 million dollars. Therefore, the profit on 40,000 phones is about $3.7 - 2 = 1.7$ million dollars. For 44,000 phones, we have the approximate points (44, 4) on the revenue graph and (44, 3) on the cost graph. So the profit on 44,000 phones is $4 - 3 = 1$ million dollars. Consequently, it is more profitable to make 40,000 phones. ▲ ⟨6⟩

⟨6⟩

In Example 6, find the profit from making

(a) 32,000 phones;

(b) 4000 phones.

Answers:

(a) About $1,200,000 (rounded)

(b) About −$500,000 (that is, a loss of $500,000)

Technology and Graphs

A graphing calculator or computer graphing program follows essentially the same procedure used when graphing by hand: the calculator selects a large number of x-values (95 or more), equally spaced along the x-axis, and plots the corresponding points, simultaneously connecting them with line segments. Calculator-generated graphs are generally quite accurate, although they may not appear as smooth as hand-drawn ones. The next example illustrates the basics of graphing on a graphing calculator. (Computer graphing software operates similarly.)

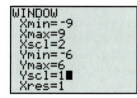

```
WINDOW
Xmin=-9
Xmax=9
Xscl=2
Ymin=-6
Ymax=6
Yscl=1▪
Xres=1
```

FIGURE 2.9

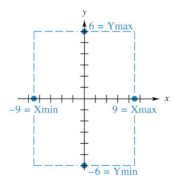

FIGURE 2.10

▼ **EXAMPLE 7** Use a graphing calculator to sketch the graph of the equation $2x^3 - 2y - 10x + 2 = 0$.

Solution *First, set the* **viewing window**—the portion of the coordinate plane that will appear on the screen. Press the WINDOW key (labeled RANGE or PLOT-SETUP on some calculators) and enter the appropriate numbers, as in Figure 2.9 (which shows the screen from a TI-84+; other calculators are similar). Then the calculator will display the portion of the plane inside the dashed lines shown in Figure 2.10—that is, the points (x, y) with $-9 \le x \le 9$ and $-6 \le y \le 6$.

In Figure 2.9 we have set Xscl $= 2$ and Yscl $= 1$, which means the **tick marks** on the x-axis are two units apart and the tick marks on the y-axis are one unit apart (Figure 2.10).

Second, enter the equation to be graphed in the equation memory. To do this, you must first solve the equation for y (because a calculator accepts only equations of the form $y = $ expression in x):

$$2y = 2x^3 - 10x + 2$$
$$y = x^3 - 5x + 1.$$

Now press the Y= key (labeled SYMB on some calculators) and enter the equation, using the "variable key" for x. (This key is labeled X,T,θ,n or X,θ,T or x-VAR, depending on the calculator.) Figure 2.11 shows the equation entered on a TI-84+; other calculators are similar. Now press GRAPH (or PLOT or DRW on some calculators), and obtain Figure 2.12.

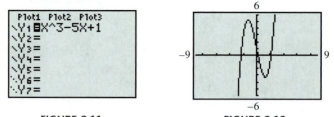

FIGURE 2.11 **FIGURE 2.12**

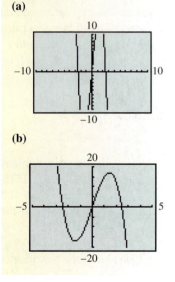

⑦

Use a graphing calculator to graph $y = 18x - 3x^3$ in the following viewing windows.

(a) $-10 \leq x \leq 10$ and $-10 \leq y \leq 10$ with Xscl = 1, Yscl = 1

(b) $-5 \leq x \leq 5$ and $-20 \leq y \leq 20$ with Xscl = 1, Yscl = 5

Answers:

(a)

(b)

Finally, if necessary, change the viewing window to obtain a more readable graph. It is difficult to see the y-intercept in Figure 2.12, so press WINDOW and change the viewing window (Figure 2.13); then press GRAPH to obtain Figure 2.14, in which the y-intercept at $y = 1$ is clearly shown. (It isn't necessary to reenter the equation.) ▲ ⑦

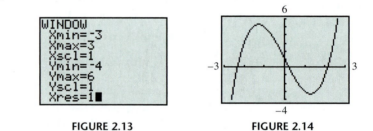

FIGURE 2.13 **FIGURE 2.14**

Technology Tools

In addition to graphing equations, graphing calculators (and graphing software) provide convenient tools for solving equations and reading graphs. For example, when you have graphed an equation, you can readily determine the points the calculator plotted. Press **trace** (a cursor appears on the graph), and use the left and right arrow keys to move the cursor along the graph. The coordinates of the point the cursor is on appear at the bottom of the screen.

Recall that the solutions of an equation, such as $x^3 - 5x + 1 = 0$, are the x-intercepts of the graph of $y = x^3 - 5x + 1$. (See the box on page 75.) A **graphical root finder** enables you to find these x-intercepts and thus to solve the equation.

⑧

Use a graphical root finder to approximate the third solution of the equation in Example 8.

Answer:

$x \approx -2.330059$

▼ **EXAMPLE 8** Use a graphical root finder to solve $x^3 - 5x + 1 = 0$.

Solution First, graph $y = x^3 - 5x + 1$. The x-intercepts of this graph are the solutions of the equation. To find these intercepts, look for "root" or "zero" in the appropriate menu.* Check your instruction manual for the proper syntax. A typical root finder (Figure 2.15) shows that two of the solutions (x-intercepts) are $x \approx .2016$ and $x \approx 2.1284$. For the third solution, see Problem ⑧ at the side. ▲

*CALC on TI-84+, or GRAPH MATH on TI-86, G-SOLV on Casio 9850, or PLOT FNC on HP-39+.

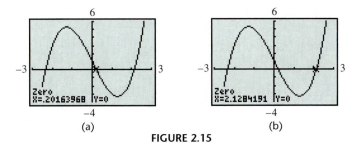

FIGURE 2.15

Many graphs have peaks and valleys (for instance, the graphs in Figures 2.14 and 2.15). A **maximum/minimum finder** provides accurate approximations of the locations of the "tops" of the peaks and the "bottoms" of the valleys.

▼ **EXAMPLE 9**   The percentage y of teens who smoke is approximated by

$$y = -.01285x^3 - .0276x^2 + 2.613x + 23.25,$$

where $x = 1$ corresponds to 1991. In what year was the percentage of teens who smoke the greatest? What was that percentage?

Solution The graph of $y = -.01285x^3 - .0276x^2 + 2.613x + 23.25$ is shown in Figure 2.16. The highest percentage of teen smokers corresponds to the highest point on this graph. To find this point, look for "maximum" or "max" or "extremum" in the same menu as the graphical root finder. Check your instruction manual for the proper syntax. A typical maximum finder (Figure 2.17) shows that the highest point has approximate coordinates (7.55, 35.87). The first coordinate is the year and the second is the percentage. So the largest percentage of teens who smoke (about 35.87%) occured in mid-1997 ($x = 7.55$). ▲

9

Use a minimum finder to locate the approximate coordinates of the lowest point to the left of the y-axis on the graph of $y = 5x - x^3$.

Answer:

$(-1.29, -4.30)$

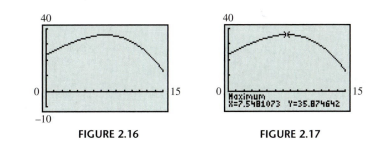

FIGURE 2.16   FIGURE 2.17

◆ **2.1 Exercises**

State the quadrant in which each point lies.

1. $(1, -2), (-2, 1), (3, 4), (-5, -6)$

2. $(\pi, 2), (3, -\sqrt{2}), (4, 0), (-\sqrt{3}, \sqrt{3})$

Determine whether the given ordered pair is a solution of the given equation. (See Example 1.)

3. $(1, -3); 3x - y - 6 = 0$

4. $(2, -1); x^2 + y^2 - 6x + 8y = -15$

5. $(3, 4); (x - 2)^2 + (y + 2)^2 = 6$

6. $(1, -1); \dfrac{x^2}{2} + \dfrac{y^2}{3} = -4$

Sketch the graph of each of these equations. (See Example 2.)

7. $4y + 3x = 12$ **8.** $2x + 7y = 14$

9. $8x + 3y = 12$

10. $9y - 4x = 12$

11. $x = 2y + 3$

12. $x - 3y = 0$

List the x-intercepts and y-intercepts of each graph.

13.

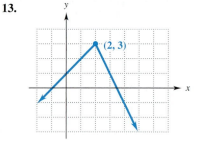

14.

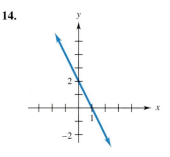

15.

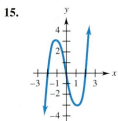

16.

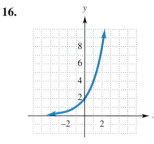

Find the x-intercepts and y-intercepts of the graph of each equation. You need not sketch the graph. (See Example 3.)

17. $3x + 4y = 12$

18. $x - 2y = 5$

19. $2x - 3y = 24$

20. $3x + y = 4$

21. $y = x^2 - 9$

22. $y = x^2 + 4$

23. $y = x^2 + x - 20$

24. $y = 5x^2 + 6x + 1$

25. $y = 2x^2 - 5x + 7$

26. $y = 3x^2 + 4x - 4$

Sketch the graph of the equation. (See Examples 2–4.)

27. $y = x^2$

28. $y = x^2 + 2$

29. $y = x^2 - 3$

30. $y = 2x^2$

31. $y = x^2 - 6x + 5$

32. $y = x^2 + 2x - 3$

33. $y = x^3$

34. $y = x^3 - 3$

35. $y = x^3 + 1$

36. $y = x^3/2$

37. $y = \sqrt{x + 4}$

38. $y = \sqrt{x - 2}$

39. $y = \sqrt{4 - x^2}$

40. $y = \sqrt{9 - x^2}$

Physical Science *The temperature graphs for Fargo and Seattle on the same day are shown in the figure. Use them to do the following exercises. (See Examples 5 and 6.)*

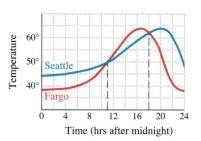

41. Approximately when did the temperature first reach 60° in Fargo? In Seattle?

42. At what times during the day did the two cities have the same temperature?

43. At what times during the day was it warmer in Fargo than in Seattle?

44. Was there any time when it was at least 10° warmer in Seattle than in Fargo?

Business *Use the revenue and cost graphs for the Fenton Cell Phone Company in Example 6 to do Exercises 45–48.*

45. Find the approximate cost of manufacturing the given number of phones.
  (a) 20,000   **(b)** 36,000   **(c)** 48,000

46. Find the approximate revenue from selling the given number of phones.
  (a) 12,000   **(b)** 24,000   **(c)** 36,000

47. Find the approximate profit from manufacturing the given number of phones.
  (a) 20,000   **(b)** 28,000   **(c)** 36,000

48. The company must replace its aging machinery with better, but much more expensive, machines. In addition, raw material prices increase, so that monthly costs go up by $250,000. Owing to competitive pressure, phone prices cannot be increased, so revenue remains the same. Under these new circumstances, find the approximate profit from manufacturing the given number of phones.
  (a) 20,000   **(b)** 36,000   **(c)** 40,000

Physical Science *According to an article in* Scientific American, *the coast-down time for a typical older car as it drops 10 miles per hour from an initial speed depends on variations from the standard condition (automobile in neutral; average drag and tire pressure). The accompanying graph illustrates some of these conditions with coast-down time in seconds and initial speed in miles per hour.*

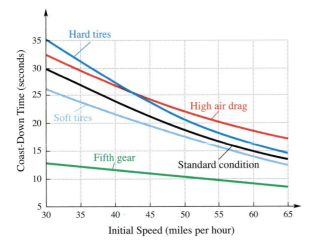

Use the graph to answer the following questions.

49. What is the approximate coast-down time in fifth gear if the initial speed is 40 miles per hour?

50. For what speed is the coast-down time the same for the conditions of high air drag and hard tires?

51. **Finance** Becky Anderson takes out a 30-year mortgage on which her monthly payment is $850. During the early years of the mortgage, most of each payment is for interest and the rather small remainder for principal. As time goes on, the portion of each payment that goes for interest decreases while the portion for principal increases, as shown in the following graph:

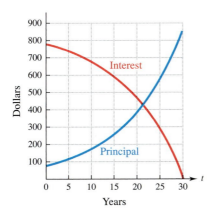

(a) Approximately how much of the $850 monthly payment goes for interest in year 5? In year 15? In year 25?

(b) In what year will the monthly payment be equally divided between interest and principal?

Health *The graph below shows the percentage increase in the cost of prescription drugs over a decade.* Use it to answer the following questions.*

52. What is the approximate percentage increase in 2008?

53. During what periods is the percentage increase below 16%?

54. When is the percentage increase the highest? How high?

55. When is the percentage increase the lowest? How low?

56. During what years between 1998 and 2008 does the graph show the cost of prescription drugs declining?

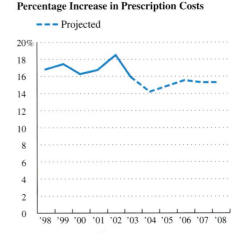

Social Science *The graph on the next page shows the median weekly earnings of men and women. (Half the people earn more than this amount and half less.)† Use this graph to answer the following questions.*

57. What were the approximate median weekly earnings in 2000 for men? For women?

58. What were the highest median earnings for women and when did they occur?

59. What were the lowest median earnings for men and approximately when did they occur?

60. Was the difference between men's median earnings and women's median earnings ever less than $100 per week?

61. By how much did men's and women's median weekly earnings increase from 1979 to 2004?

---

*Data and projections from Express Scripts, Inc., published in the St. Louis *Post-Dispatch,* June 9, 2004.

†Bureau of Labor Statistics, Economy.com, and the Economic Policy Institute; published in the *New York Times,* December 31, 2004.

62. Assuming that the trends suggested in this graph continue, what is a likely scenario for future years?

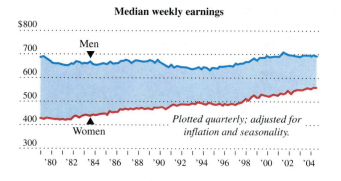

Median weekly earnings

Men

Women

Plotted quarterly; adjusted for inflation and seasonality.

'80 '82 '84 '86 '88 '90 '92 '94 '96 '98 '00 '02 '04

Use a graphing calculator to find the graph of the equation. (See Example 7.)

63. $y = x^2 + x + 1$

64. $y = 2 - x - x^2$

65. $y = (x - 3)^3$

66. $y = x^3 + 2x^2 + 2$

67. $y = x^3 - 3x^2 + x - 1$

68. $y = x^4 - 5x^2 - 2$

Use a graphing calculator for Exercises 69–70.

69. Graph $y = x^4 - 2x^3 + 2x$ in a window with $-3 \le x \le 3$. Is the "flat" part of the graph near $x = 1$ really a horizontal line segment? (*Hint:* Use trace to move along the "flat" part and watch the y-coordinates. Do they remain the same [as they should on a horizontal segment]?)

70. (a) Graph $y = x^4 - 2x^3 + 2x$ in the **standard window** (the one with $-10 \le x \le 10$ and $-10 \le y \le 10$). Use the trace feature to approximate the coordinates of the lowest point on the graph.

   (b) Use a minimum finder to obtain an accurate approximation of the lowest point. How does this compare with your answer in part (a)?

Use a graphing calculator to approximate all real solutions of the equation. (See Example 8.)

71. $x^3 - 3x^2 + 5 = 0$

72. $x^3 + x - 1 = 0$

73. $2x^3 - 4x^2 + x - 3 = 0$

74. $6x^3 - 5x^2 + 3x - 2 = 0$

75. $x^5 - 6x + 6 = 0$

76. $x^3 - 3x^2 + x - 1 = 0$

Use a graphing calculator to work Exercises 77–82. (See Examples 8 and 9.)

77. Physical Science The surface area S of the right circular cone in the figure is given by $S = \pi r \sqrt{r^2 + h^2}$. What

radius should be used to produce a cone of height 5 inches and surface area 100 square inches?

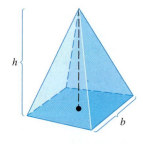

78. Physical Science The surface area of the right square pyramid in the figure is given by $S = b\sqrt{b^2 + 4h^2}$. If the pyramid has height 10 feet and surface area 100 square feet, what is the length of a side b of its base?

79. Health The number of weight-loss surgeries performed in the United States can be approximated by

$$y = 216.9x^4 - 1202.3x^3 + 3223.9x^2 + 2596.8x + 29,087.2,$$

where $x = 0$ corresponds to 1999.* According to this model, in what year did the number of weight-loss surgeries first reach 150,000?

80. Natural Science Carbon monoxide combines with the hemoglobin of the blood to form carboxyhemoglobin (COHb), which reduces the transport of oxygen to tissue. A 4%-to-6% COHb level in the blood (typical for smokers) can cause symptoms such as alterations in blood flow, visual impairment, and poorer vigilance. When $50 \le x \le 100$, the equation $T = .00787x^2 - 1.528x + 75.89$ approximates the exposure time T (in hours) necessary to reach this 4%-to-6% level, where x is the amount of carbon monoxide present in the air in parts per million (ppm).†

   (a) A kerosene heater or a room full of smokers is capable of producing 50 ppm of carbon monoxide. How long would it take for a nonsmoking person to start feeling the aforementioned symptoms?

   (b) What carbon monoxide concentration will cause a person to reach a 4%-to-6% COHb level in three hours?

81. Health The number of doctors per 1000 residents in the United States from 1970 to 2020 can be approximated by

$$y = -.000018x^3 + .000488x^2 + .0366x + 1.6,$$

*Based on data from the American Society for Bariatric Surgery.

†U.S. Department of Energy.

where $x = 0$ corresponds to 1970.* In what year is the number the highest? How many doctors are there per 1000 residents in that year?

82. **Business** During the first part of this decade, the total net assets of U.S. equity mutual funds (in billions of dollars) could be approximated by

$$y = 331.35x^3 - 1099.6x^2 + 224.15x + 3962.3,$$

where $x = 0$ corresponds to 2000.* When were total assets the lowest and when did this occur?

*Based on data from the Investment Company Institute: *Fund Fact Book* (2004).

*Based on data and projections by *Health Affairs* and Dr. Richard Cooper, Institute of Health Policy at the Medical College of Wisconsin.

2.2 Equations of Lines

Straight lines, which are the simplest graphs, play an important role in a wide variety of applications. They are considered here from both a geometric and an algebraic point of view.

The key geometric feature of a nonvertical straight line is how steeply it rises or falls as you move from left to right. The "steepness" of a line can be represented numerically by a number called the *slope* of the line.

To see how the slope is defined, start with Figure 2.18, which shows a line passing through the two different points $(x_1, y_1) = (-3, 5)$ and $(x_2, y_2) = (2, -4)$. The difference in the two x-values,

$$x_2 - x_1 = 2 - (-3) = 5,$$

in this example is called the **change in x.** The Greek letter Δ (delta) is used to denote change. The symbol Δx (read "delta x") represents the change in x. In the same way, Δy represents the **change in y.** In this example,

$$\Delta y = y_2 - y_1 = -4 - 5 = -9.$$

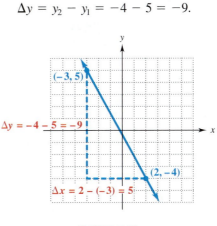

FIGURE 2.18

The **slope** of the line through the two points (x_1, y_1) and (x_2, y_2), where $x_1 \neq x_2$, is defined as the quotient of the change in y and the change in x:

$$\text{slope} = \frac{\text{change in } y}{\text{change in } x} = \frac{\Delta y}{\Delta x} = \frac{y_2 - y_1}{x_2 - x_1}.$$

The slope of the line in Figure 2.18 is

$$\text{slope} = \frac{\Delta y}{\Delta x} = \frac{-4 - 5}{2 - (-3)} = -\frac{9}{5}.$$

Using similar triangles from geometry, we can show that the slope is independent of the choice of points on the line. That is, the same value of the slope will be obtained for *any* choice of two different points on the line.

▼ **EXAMPLE 1**   Find the slope of the line through the points $(-7, 6)$ and $(4, 5)$.

Solution Let $(x_1, y_1) = (-7, 6)$ and $(x_2, y_2) = (4, 5)$. Use the definition of slope.

$$\text{slope} = \frac{\Delta y}{\Delta x} = \frac{y_2 - y_1}{x_2 - x_1} = \frac{5 - 6}{4 - (-7)} = \frac{-1}{11} = -\frac{1}{11}$$

The slope can also be found by letting $(x_1, y_1) = (4, 5)$ and $(x_2, y_2) = (-7, 6)$. In that case,

$$\text{slope} = \frac{6 - 5}{-7 - 4} = \frac{1}{-11} = -\frac{1}{11},$$

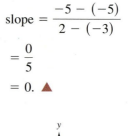

1

Find the slope of the line through

(a) $(6, 11), (-4, -3)$;

(b) $(-3, 5), (-2, 8)$.

Answers:

(a)   7/5

(b)   3

the same answer. ▲ ①

CAUTION When finding the slope of a line, be careful to subtract the *x*-values and the *y*-values in the same order. For example, with the points $(4, 3)$ and $(2, 9)$, if you use $9 - 3$ for the numerator, you must use $2 - 4$ (*not* $4 - 2$) for the denominator.

▼ **EXAMPLE 2**   Find the slope of the horizontal line in Figure 2.19.

Solution Every point on the line has the same *y*-coordinate, -5. Choose any two of them to compute the slope, say, $(x_1, y_1) = (-3, -5)$ and $(x_2, y_2) = (2, -5)$:

$$\text{slope} = \frac{-5 - (-5)}{2 - (-3)}$$

$$= \frac{0}{5}$$

$$= 0. ▲$$

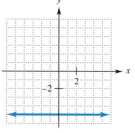

FIGURE 2.19

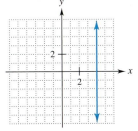

FIGURE 2.20

▼ **EXAMPLE 3**   What is the slope of the vertical line in Figure 2.20?

Solution   Every point on the line has the same x-coordinate, 4. If we attempt to compute the slope with two of these points, say $(x_1, y_1) = (4, -2)$ and $(x_2, y_2) = (4, 1)$, we obtain

$$\text{slope} = \frac{1 - (-2)}{4 - 4}$$

$$= \frac{3}{0}.$$

Division by 0 is not defined, so the slope of this line is undefined. ▲

The arguments used in Examples 2 and 3 work in the general case and lead to the following conclusion.

> The slope of every horizontal line is 0.
>
> The slope of every vertical line is undefined.

Slope-Intercept Form

The slope can be used to develop an algebraic description of nonvertical straight lines. Assume that a line with slope m has y-intercept b, so that it goes through the point $(0, b)$. (See Figure 2.21.) Let (x, y) be any point on the line other than $(0, b)$. Using the definition of slope with the points $(0, b)$ and (x, y) gives

$$m = \frac{y - b}{x - 0}$$

$$m = \frac{y - b}{x}$$

$$mx = y - b \qquad \text{Multiply both sides by } x.$$

$$y = mx + b. \qquad \text{Add } b \text{ to both sides. Reverse the equation.}$$

In other words, the coordinates of any point on the line satisfy the equation $y = mx + b$.

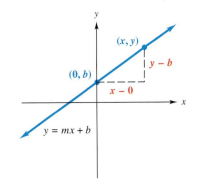

FIGURE 2.21

Slope-Intercept Form

If a line has slope m and y-intercept b, then it is the graph of the equation

$$y = mx + b.$$

This equation is called the **slope-intercept form** of the equation of the line.

▼ **EXAMPLE 4**   Find an equation for the line with y-intercept $7/2$ and slope $-5/2$.

Solution Use the slope-intercept form with $b = 7/2$ and $m = -5/2$.

$$y = mx + b$$
$$y = -\frac{5}{2}x + \frac{7}{2} \quad ▲ \quad \langle 2 \rangle$$

⟨2⟩

Find an equation for the line with

(a) y-intercept -3 and slope $2/3$;

(b) y-intercept $1/4$ and slope $-3/2$.

Answers:

(a) $y = \dfrac{2}{3}x - 3$

(b) $y = -\dfrac{3}{2}x + \dfrac{1}{4}$

▼ **EXAMPLE 5**   Find the equation of the horizontal line with y-intercept 3.

Solution The slope of the line is 0 (why?) and its y-intercept is 3, so its equation is

$$y = mx + b$$
$$y = 0x + 3$$
$$y = 3. \quad ▲$$

The argument in Example 5 also works in the general case.

If k is a constant, then the graph of the equation $y = k$ is the horizontal line with y-intercept k.

▼ **EXAMPLE 6**   Find the slope and y-intercept for each of the following lines.

(a) $5x - 3y = 1$

Solution Solve for y.

$$5x - 3y = 1$$
$$-3y = -5x + 1 \qquad \text{Subtract } 5x \text{ from both sides.}$$
$$y = \frac{5}{3}x - \frac{1}{3} \qquad \text{Divide both sides by } -3.$$

This equation is in the form $y = mx + b$, with $m = 5/3$ and $b = -1/3$. So the slope is $5/3$ and the y-intercept is $-1/3$.

(b) $-9x + 6y = 2$

Solution Solve for y.

3

Find the slope and y-intercept of

(a) $x + 4y = 6$;

(b) $3x - 2y = 1$.

Answers:

(a) Slope $-1/4$; y-intercept $3/2$

(b) Slope $3/2$; y-intercept $-1/2$

$$-9x + 6y = 2$$
$$6y = 9x + 2 \qquad \text{Add } 9x \text{ to both sides.}$$
$$y = \frac{3}{2}x + \frac{1}{3} \qquad \text{Divide both sides by 6.}$$

The slope is $3/2$ (the coefficient of x) and the y-intercept is $1/3$. ▲ **3**

The slope-intercept form can be used to show how the slope measures the steepness of a line. Consider the straight lines A, B, C, and D given by the following equations; each has y-intercept 0 and slope as indicated.

A: $y = .5x$,	B: $y = x$,	C: $y = 3x$,	D: $y = 7x$
Slope .5	Slope 1	Slope 3	Slope 7

For these lines, Figure 2.22 shows that the bigger the slope, the more steeply the line rises from left to right. **4**

4

(a) List the slopes of the following lines:

E: $y = -.3x$, F: $y = -x$,

G: $y = -2x$, H: $y = -5x$.

(b) Graph all four lines on the same set of axes.

(c) How are the slopes of the lines related to their steepness?

Answers:

(a) Slope E $= -.3$; slope F $= -1$; slope G $= -2$; slope H $= -5$.

(b)

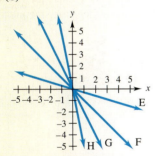

(c) The larger the slope in absolute value, the more steeply the line falls from left to right.

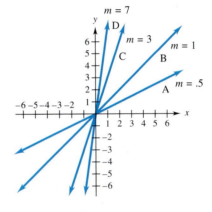

FIGURE 2.22

The preceding discussion and Problem 4 in the margin may be summarized as follows.

Direction of Line (moving from left to right)	Slope
Upward	**Positive** (larger for steeper lines)
Horizontal	**0**
Downward	**Negative** (larger in absolute value for steeper lines)
Vertical	**Undefined**

▼ **EXAMPLE 7**   Sketch the graph of $x + 2y = 5$ and label the intercepts.

Solution Find the x-intercept by setting $y = 0$ and solving for x.

$$x + 2 \cdot 0 = 5$$
$$x = 5$$

The x-intercept is 5, and $(5, 0)$ is on the graph. The y-intercept is found similarly, by setting $x = 0$ and solving for y.

$$0 + 2y = 5$$
$$y = 5/2$$

The y-intercept is $5/2$, and $(0, 5/2)$ is on the graph. The points $(5, 0)$ and $(0, 5/2)$ can be used to sketch the graph (Figure 2.23). ▲ ⑤

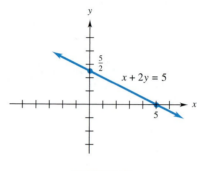

FIGURE 2.23

TECHNOLOGY TIP   To graph a linear equation on a graphing calculator, you must first put the equation in slope-intercept form $y = mx + b$ so that it can be entered in the equation memory (called the Y=list on some calculators). Vertical lines cannot be graphed on most calculators.

Slopes of Parallel and Perpendicular Lines

We shall assume the following facts without proof. The first one is a consequence of the fact that the slope measures steepness and parallel lines have the same steepness.

Two nonvertical lines are parallel whenever they have the same slope.

Two nonvertical lines are perpendicular whenever the product of their slopes is -1.

▼ **EXAMPLE 8**   Determine whether each of the following pairs of lines are *parallel*, *perpendicular*, or *neither*.

(a) $2x + 3y = 5$ and $4x + 5 = -6y$

Solution Put each equation in slope-intercept form by solving for y.

$$3y = -2x + 5 \qquad -6y = 4x + 5$$
$$y = -\frac{2}{3}x + \frac{5}{3} \qquad y = -\frac{2}{3}x - \frac{5}{6}$$

In each case the slope (the coefficient of x) is $-2/3$, so the lines are parallel.

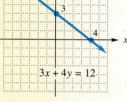
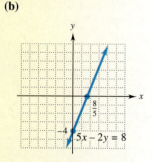

(b) $3x = y + 7$ and $x + 3y = 4$

Solution Put each equation in slope-intercept form to determine the slope of the associated line.

$$3x = y + 7 \qquad\qquad 3y = -x + 4$$

$$y = 3x - 7 \qquad\qquad y = -\frac{1}{3}x + \frac{4}{3}$$

$$\text{slope } 3 \qquad\qquad\qquad \text{slope } -1/3$$

Since $3(-1/3) = -1$, these lines are perpendicular.

(c) $x + y = 4$ and $x - 2y = 3$

Solution Verify that the slope of the first line is -1 and the slope of the second is $1/2$. The slopes are not equal and their product is not -1, so the lines are neither parallel nor perpendicular. ▲ ⑥

⑥

Tell whether the lines in each of the following pairs are *parallel, perpendicular,* or *neither.*

(a) $x - 2y = 6$ and $2x + y = 5$

(b) $3x + 4y = 8$ and $x + 3y = 2$

(c) $2x - y = 7$ and $2y = 4x - 5$

Answers:

(a) Perpendicular

(b) Neither

(c) Parallel

TECHNOLOGY TIP Perpendicular lines may not appear perpendicular on a graphing calculator, unless you use a *square window*—a window in which a one-unit segment on the *y*-axis is the same length as a one-unit segment on the *x*-axis. To obtain such a window on most calculators, use a viewing window in which the *y*-axis is about two-thirds as long as the *x*-axis. The SQUARE (or ZSQUARE) key in the ZOOM menu will change the current window to a square window by automatically adjusting the length of one of the axes.

Point-Slope Form

The slope-intercept form of the equation of a line is usually the most convenient for graphing and for understanding how slopes and lines are related. However, it is not always the best way to *find* the equation of a line. In many situations (particularly in calculus), the slope and a point on the line are known and you must find the equation of the line. In such cases, the best method is to use the *point-slope form,* which we now explain.

Suppose that a line has slope m and that (x_1, y_1) is a point on the line. Let (x, y) represent any other point on the line. Since m is the slope, then, by the definition of slope,

$$\frac{y - y_1}{x - x_1} = m.$$

Multiplying both sides by $x - x_1$ yields

$$y - y_1 = m(x - x_1).$$

Point-Slope Form

If a line has slope m and passes through the point (x_1, y_1), then

$$y - y_1 = m(x - x_1)$$

is the **point-slope form** of the equation of the line.

▼ **EXAMPLE 9** Find the equation of the line satisfying the given conditions.

(a) Slope 2; the point $(5, 3)$ is on the line.

Solution Use the point-slope form with $m = 2$ and $(x_1, y_1) = (5, 3)$. Substitute $x_1 = 5$, $y_1 = 3$, and $m = 2$ into the point-slope form of the equation.

$$y - y_1 = m(x - x_1)$$
$$y - 3 = 2(x - 5)$$

For some purposes, this form of the equation is fine; in other cases, you may want to rewrite it in the slope-intercept form.

(b) Slope -3; the point $(-4, 1)$ is on the line.

Solution Use the point-slope form with $m = -3$ and $(x_1, y_1) = (-4, 1)$.

$$y - y_1 = m(x - x_1)$$
$$y - 1 = -3[x - (-4)] \qquad \text{Point-slope form}$$

Using algebra, we obtain the slope-intercept form of this equation.

$$y - 1 = -3(x + 4)$$
$$y - 1 = -3x - 12 \qquad \text{Distributive property}$$
$$y = -3x - 11 \qquad \text{Slope-intercept form} \; \blacktriangle \; \langle 7 \rangle$$

⟨7⟩

Find both the point-slope and the slope-intercept form of the equation of the line having the given slope and passing through the given point.

(a) $m = -3/5$, $(5, -2)$

(b) $m = 1/3$, $(6, 8)$

Answers:

(a) $y + 2 = -\dfrac{3}{5}(x - 5)$;

$y = -\dfrac{3}{5}x + 1$.

(b) $y - 8 = \dfrac{1}{3}(x - 6)$;

$y = \dfrac{1}{3}x + 6$.

The point-slope form can also be used to find an equation of a line, given two different points on the line. The procedure is shown in the next example.

▼ **EXAMPLE 10** Find an equation of the line through $(5, 4)$ and $(-10, -2)$.

Solution Begin by using the definition of the slope to find the slope of the line that passes through the two points.

$$\text{slope} = m = \frac{-2 - 4}{-10 - 5} = \frac{-6}{-15} = \frac{2}{5}$$

Use $m = 2/5$ and either of the given points in the point-slope form. If $(x_1, y_1) = (5, 4)$, then

$$y - y_1 = m(x - x_1)$$
$$y - 4 = \frac{2}{5}(x - 5) \qquad \text{Let } y_1 = 4, \; m = \frac{2}{5}, \; x_1 = 5.$$
$$5(y - 4) = 2(x - 5) \qquad \text{Multiply both sides by 5.}$$
$$5y - 20 = 2x - 10 \qquad \text{Distributive property}$$
$$5y = 2x + 10.$$

Check that the result is the same when $(x_1, y_1) = (-10, -2)$. $\blacktriangle$ ⟨8⟩

⟨8⟩

Find an equation of the line through

(a) $(2, 3)$ and $(-4, 6)$;

(b) $(-8, 2)$ and $(3, -6)$.

Answers:

(a) $2y = -x + 8$

(b) $11y = -8x - 42$

Vertical Lines

The equation forms developed above do not apply to vertical lines, because slope is not defined for such lines. However, vertical lines can easily be described as graphs of equations.

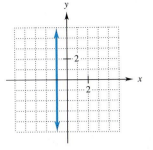

FIGURE 2.24

▼ **EXAMPLE 11**   Find the equation whose graph is the vertical line in Figure 2.24.

Every point on the line has x-coordinate -1 and hence has the form $(-1, y)$. Thus, every point is a solution of the equation $x + 0y = -1$, which is usually written simply as $x = -1$. Note that -1 is the x-intercept of the line.   ▲
The argument in Example 11 also works in the general case.   ▲

If k is a constant, then the graph of the equation $x = k$ is the vertical line with x-intercept k.

Linear Equations

An equation in two variables whose graph is a straight line is called a **linear equation.** Linear equations have a variety of forms, as summarized in the following table. The last form listed is called the **general form** of a linear equation, and every linear equation can be written in this form. For example, $y = 4x - 5$ can be written in general form as $4x - y = 5$, and $x = 6$ can be written in general form as $x + 0y = 6$.

Equation	Description
$x = k$	**Vertical line,** x-intercept k, no y-intercept, undefined slope
$y = k$	**Horizontal line,** y-intercept k, no x-intercept, slope 0
$y = mx + b$	**Slope-intercept form,** slope m, y-intercept b
$y - y_1 = m(x - x_1)$	**Point-slope form,** slope m, the line passes through (x_1, y_1).
$ax + by = c$	**General form.** If $a \neq 0$ and $b \neq 0$, the line has x-intercept c/a and y-intercept c/b, and slope $-a/b$.

Applications

Many relationships are linear or almost linear, so that they can be approximated by linear equations.

▼ **EXAMPLE 12**   In an experiment testing a person's reaction time y (in seconds) after undergoing x hours of stressful activity, the linear equation $y = .1957x + .1243$ was found to be a good approximation of the relationship between stress and reaction time during the first five hours.

(a) Assuming that reaction times continue to follow this pattern, what would be the approximate reaction time after $8\frac{1}{2}$ hours?

Solution Substitute $x = 8.5$ in the equation and use a calculator to compute y.

$$y = .1957x + .1243$$
$$y = .1957(\mathbf{8.5}) + .1243 = 1.78775$$

The reaction time is approximately 1.8 seconds.

(b) When the reaction time is 1.5 seconds, how long has the subject been undergoing stressful activity?

Solution Substitute $y = 1.5$ in the equation and solve for x.

$$1.5 = .1957x + .1243$$
$$1.5 - .1243 = .1957x$$
$$x = \frac{1.5 - .1243}{.1957} \approx 7.0296$$

Hence, the stressful activity has been going on for slightly more than 7 hours. ▲

▼ **EXAMPLE 13** According to data from the National Center for Education Statistics, the average cost of tuition and fees in public four-year colleges was $2987 in the fall of 1996 and grew in an approximately linear fashion to $4061 in the fall of 2002.

(a) Find a linear equation for this data.

Solution Measure time along the x-axis and cost along the y-axis. Then the x-coordinate of each point is a year and the y-coordinate is the average cost of tuition and fees in that year. For convenience, let $x = 0$ correspond to 1990, so that $x = 6$ is 1996 and $x = 12$ is 2002. Then the given data points are $(6, 2987)$ and $(12, 4061)$. The slope of the line joining these points is

$$\frac{4061 - 2987}{12 - 6} = \frac{1074}{6} = 179.$$

Using the point $(6, 2987)$, we obtain the equation of this line.

$$y - 2987 = 179(x - 6) \qquad \text{Point-slope form}$$
$$y - 2987 = 179x - 1074 \qquad \text{Distributive property}$$
$$y = 179x + 1913 \qquad \text{Add 2987 to both sides.}$$

(b) Use this equation to estimate the average cost of tuition and fees in the fall of 2000.

Solution Since 2000 corresponds to $x = 10$, let $x = 10$ in the equation of part (a). Then

$$y = 179(10) + 1913 = \$3703.$$

(c) Assuming the equation remains valid for the next decade, estimate the average cost of tuition and fees in the fall of 2009.

Solution 2009 corresponds to $x = 19$, so the average cost is

$$y = 179(19) + 1913 = \$5314. \; ▲$$

⑨

The average cost of tuition and fees in private four-year colleges was $12,881 in 1996 and $16,949 in 2002.

(a) Let $x = 0$ correspond to 1990, and find a linear equation for the given data.

(b) Assuming that your equation remains accurate, estimate the average cost in 2010.

Answers:

(a) $y = 678x + 8813$

(b) $22,373

◆ ## 2.2 Exercises

Find the slope of the line if it is defined. (See Examples 1–3.)

1. Through $(2, 5)$ and $(0, 8)$

2. Through $(9, 0)$ and $(12, 12)$

3. Through $(-4, 14)$ and $(3, 0)$

4. Through $(-5, -2)$ and $(-4, 11)$

5. Through the origin and $(-4, 10)$

6. Through the origin and $(8, -2)$

7. Through $(-1, 4)$ and $(-1, 6)$

8. Through $(-3, 5)$ and $(2, 5)$

Find an equation of the line with the given y-intercept and slope m. (See Examples 4 and 5.)

9. $5, m = 4$

10. $-3, m = -7$

11. $1.5, m = -2.3$

12. $-4.5, m = 2.5$

13. $4, m = -3/4$

14. $-3, m = 4/3$

Find the slope m and the y-intercept b of the line whose equation is given. (See Example 6.)

15. $2x - y = 9$

16. $x + 2y = 7$

17. $6x = 2y + 4$

18. $4x + 3y = 24$

19. $6x - 9y = 16$

20. $4x + 2y = 0$

21. $2x - 3y = 0$

22. $y = 7$

23. $x = y - 5$

24. On one graph, sketch six straight lines that meet at a single point and satisfy this condition: one line has slope 0, two lines have positive slope, two lines have negative slope, and one line has undefined slope.

25. For which of the line segments in the figure is the slope
   (a) largest?   **(b)** smallest?
   (c) largest in absolute value?   **(d)** closest to 0?

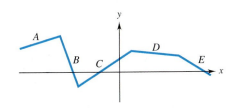

26. Match each equation with the line that most closely resembles its graph. (*Hint:* Consider the signs of m and b in the slope-intercept form.)
   (a) $y = 3x + 2$ **(b)** $y = -3x + 2$
   (c) $y = 3x - 2$ **(d)** $y = -3x - 2$

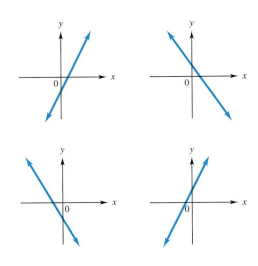

Sketch the graph of the equation and label its intercepts. (See Example 7.)

27. $2x - y = -2$

28. $2y + x = 4$

29. $2x + 3y = 4$

30. $-5x + 4y = 3$

31. $4x - 5y = 2$

32. $3x + 2y = 8$

Determine whether each pair of lines is parallel, perpendicular, or neither. (See Example 8.)

33. $4x - 3y = 6$ and $3x + 4y = 8$

34. $2x - 5y = 7$ and $15y - 5 = 6x$

35. $3x + 2y = 8$ and $6y = 5 - 9x$

36. $x - 3y = 4$ and $y = 1 - 3x$

37. $4x = 2y + 3$ and $2y = 2x + 3$

38. $2x - y = 6$ and $x - 2y = 4$

39. **(a)** Find the slope of each side of the triangle with vertices $(9, 6)$, $(-1, 2)$, and $(1, -3)$.
   (b) Is this triangle a right triangle? (*Hint:* Are two sides perpendicular?)

40. **(a)** Find the slope of each side of the quadrilateral with vertices $(-5, -2)$, $(-3, 1)$, $(3, 0)$, and $(1, -3)$.
   (b) Is this quadrilateral a parallelogram? (*Hint:* Are opposite sides parallel?)

Find an equation of the line that passes through the given point and has the given slope. (See Examples 9 and 11.)

41. $(-1, 2), m = -2/3$

42. $(-4, -3), m = 5/4$

43. $(-2, -2), m = 4$

44. $(-2, 5), m = -1/2$

45. $(8, 2), m = 0$

46. $(2, -4), m = 0$

47. $(6, -5)$, undefined slope

48. $(-8, 9)$, undefined slope

Find an equation of the line that passes through the given points. (See Example 10.)

49. $(-1, 1)$ and $(2, 7)$

50. $(2, 5)$ and $(0, 6)$

51. $(1, 2)$ and $(3, 9)$

52. $(-1, -2)$ and $(2, -1)$

Find an equation of the line satisfying the given conditions.

53. Through the origin with slope 5

54. Through the origin and horizontal

55. Through $(6, 8)$ and vertical

56. Through $(7, 9)$ and parallel to $y = 6$

57. Through $(3, 4)$ and parallel to $4x - 2y = 5$

58. Through $(6, 8)$ and perpendicular to $y = 2x - 3$

59. x-intercept 6; y-intercept -6

60. Through $(-5, 2)$ and parallel to the line through $(1, 2)$ and $(4, 3)$

61. Through $(-1, 3)$ and perpendicular to the line through $(0, 1)$ and $(2, 3)$

62. y-intercept 3 and perpendicular to $2x - y + 6 = 0$

Business *The lost value of equipment over a period of time is called depreciation. The simplest method for calculating depreciation is straight-line depreciation. The annual straight-line depreciation of an item that cost x dollars with a useful life of n years is $D = (1/n)x$. Find the depreciation for items with the following characteristics.*

63. Cost: $12,482; life 10 yr

64. Cost: $39,700; life 12 yr

65. Cost: $145,000; life 28 yr

66. **Business** Ral Corp. has an incentive compensation plan under which a branch manager receives 10% of the branch's income after deduction of the bonus, but before deduction of income tax.* Branch income before the bonus and income tax was $165,000. The tax rate was 30%. The bonus amounted to
   (a) $12,600
   (b) $15,000
   (c) $16,500
   (d) $18,000.

67. **Health** According to data from the U.S. Census Bureau, the number of retail drug prescriptions (in millions) can be approximated by

$$y = 136.25x + 1443.75,$$

where $x = 5$ corresponds to 1995. Find the approximate number of prescriptions in the following years.
   (a) 2003
   (b) 2005
   (c) Assuming that this model remains accurate, in what year will the number of prescriptions be 4,850,000,000?

68. **Natural Sciences** Deer ticks cause concern because they can carry Lyme disease. One study found a relationship between the density of acorns produced in the fall and the density of deer tick larvae the following spring.† The relationship can be approximated by the linear equation

$$y = 34x + 230,$$

where x is the number of acorns per square meter in the fall and y is the number of deer tick larvae per 400 square meters the following spring. According to this formula, approximately how many acorns per square meter would result in 1000 deer tick larvae per 400 square meters?

69. **Business** According to data from the U.S. Department of Commerce, movie-theater box-office receipts (in millions of dollars) in year x can be approximated by

$$y = 535.43x + 5362.21,$$

where $x = 0$ corresponds to 1995. Assume that this equation remains valid until 2012.
   (a) Estimate the box-office receipts in 2004.
   (b) In what year will the box-office receipts be $14 billion?

70. **Business** Over the next 25 years, the number of jobs in the San Diego metropolitan area (in thousands) is projected to be given by

$$y = 55.357x + 1654.29,$$

where $x = 2$ corresponds to 2002.*
   (a) How many jobs are there in 2008?
   (b) In what year will 2,541,000 jobs be available?

71. **Business** Worldwide motor vehicle production was about 60 million in 2000 and is expected to be about 65 million in 2007.†
   (a) Let the x-axis denote time and the y-axis the number of vehicles (in millions). Let $x = 0$ correspond to 2000. Fill in the blanks: the given data is represented by the points (_____ , 60) and (7, _____).
   (b) Find the linear equation determined by the two points in part (a).
   (c) Use the equation in part (b) to estimate the number of vehicles produced in 2003.
   (d) If this model remains accurate, when will vehicle production reach 68 million?

72. **Social Sciences** In 1974, a total of 86,821 people from other countries immigrated to the state of California. In 2002, the number of immigrants was 291,216.‡
   (a) If the change in foreign immigration to California is considered to be linear, write an equation expressing the number y of immigrants in terms of the number x of years after 1974.
   (b) Assuming that your equation in part (a) remains accurate, use it to predict the foreign immigration to California in 2010.
   (c) When will the number of immigrants be 400,800?

73. **Social Sciences** The number of unmarried couples in the United States who are living together has been rising at a roughly linear rate in recent years. The number of cohabitating adults was 1.1 million in 1977 and 5.5 million in 2000.§
   (a) Write an equation expressing the number y of cohabitating adults (in millions) in terms of the number x of years after 1977.

*Based on data in the *New York Times Almanac 2005*.

†*World Almanac and Book of Facts 2005*.

‡*Legal Immigration to California in Federal Fiscal Year 1996*, State of California Demographic Research Unit, June 1999; and U.S. Dept. of Homeland Security.

§*New York Times*, February 15, 2000, p. F8; and U.S. Census Bureau.

*Uniform CPA Examination question

†*Science*, vol. 281, no. 5375, July 17, 1998, pp. 350–351.

(b) Assuming the future accuracy of the equation you found in part (a), predict the number of cohabiting adults in 2010.

(c) When will the number of cohabiting adults reach 6,007,000?

74. Social Science The percentage of people 25 years old and older who have completed four or more years of college was about 7.8 in 1960 and 27.2 in 2002.*

(a) Find a linear equation that gives the percentage of people 25 and over who have completed four or more years of college in terms of time t, where t is the number of years since 1960. Assume that this equations remains valid in the future.

(b) What will the percentage be in 2010?

(c) When will 35% of those 25 and over have completed four or more years of college?

75. Physical Sciences In 2001, the Intergovernmental Panel on Climate Change predicted that the average temperature on the earth would rise about .3°C per decade in the coming century.† Let t measure the time in years since 1970, when the average global temperature was 15°C.

(a) Find a linear equation giving the average global temperature in degrees Celsius in terms of t, the number of years since 1970.

(b) Scientists have estimated that the sea level will rise by 65 cm if the average global temperature rises to 19°C. According to your answer to part (a), when would this occur?

76. Health For you to achieve the maximum benefit for the heart when exercising, your heart rate (in pulse beats per minute) should be in the target heart rate zone. The lower limit of this zone is found by taking 65% of the difference between 220 and your age. The upper limit is found by using 85%.‡

(a) Find linear equations for the upper and lower limits (U and L) in terms of age x.

(b) What is the target heart rate zone for a 20-year-old?

(c) What is the target heart rate zone for a 40-year-old?

(d) Two women in an aerobics class stop to take their pulse rates and are surprised to find that they have the same rate. One woman is 48 years older than the other and is working at the upper limit of her target heart rate zone. The younger woman is working at the lower limit of her target heart rate zone. What are the ages of the two women, and what is their pulse?

(e) Run for 10 minutes, take your pulse, and see if it is in your target heart rate zone. (After all, this is listed as an exercise!)

77. Physical Science The graph shows the winning time (in minutes) at the Olympic Games for the men's 5000-meter run, together with a linear approximation of the data:*

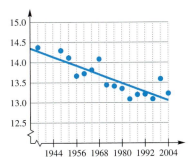

(a) The equation for the linear approximation is

$$y = -.01786x + 48.85.$$

What does the slope of this line represent? Why is the slope negative?

(b) Use the approximation to estimate the winning time in the 2008 Olympics. If possible, check this estimate against the actual time.

(c) Why are there no data points for the years 1940 and 1944?

78. Business The graph shows the number of commercial U.S. radio stations in selected years, along with a linear approximation of the data.†

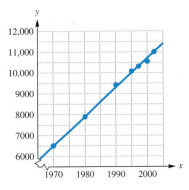

(a) Use the data points (1970, 6519) and (2002, 10965) to estimate the slope of the line shown. Interpret this number.

(b) Use part (a) to find the approximate year in which there will be 12,000 radio stations.

(c) Discuss the accuracy of the linear approximation.

---

*U.S. Census Bureau.

†IPCC, *Climatic Change 2001: the Scientific Basis.*

‡www.drkoop.com.

*United States Olympic Committee.

†Data from the Federal Communications Commission.

2.3 Linear Models

In business and science, it is often necessary to make judgments on the basis of data from the past. For instance, a stock analyst might use a company's profits in previous years to estimate next year's profits. Or a life insurance company might look at life expectancies of people born in various years to predict how much money it should expect to pay out next year.

In such situations, the available data is used to construct a mathematical model, such as an equation or a graph, which is used to approximate the likely outcome in cases where complete data is not available. In this section, we consider applications in which the data can be modeled by a linear equation.

The simplest way to construct a linear model is to use the line determined by two of the data points, as illustrated in the following example.

▼ **EXAMPLE 1**   The profits of the General Electric Company (GE) in billions of dollars, over a five-year period are shown in the following table:*

Year	2000	2001	2002	2003	2004
Profit	13	14	14	15	17

(a) Let $x = 0$ correspond to 2000, and plot the points (x, y), where x is the year and y the profit.

Solution The data points are $(0, 13)$, $(1, 14)$, $(2, 14)$, $(3, 15)$, and $(4, 17)$, as shown in Figure 2.25.

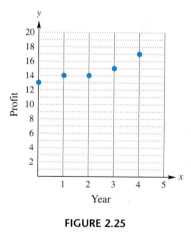

FIGURE 2.25

(b) Use the data points $(0, 13)$ and $(4, 17)$ to find a line that models the data.

---

*GE Annual Report 2004. The profit figures (net earnings) are rounded to the nearest billion.

Solution The slope of the line through $(0, 13)$ and $(4, 17)$ is $\dfrac{17 - 13}{4 - 0} = \dfrac{4}{4} = 1$. Using the point $(13, 0)$ and the slope 1, we find that the equation of this line is

$$y - 13 = 1(x - 0) \qquad \text{Point-slope form for equation of line}$$
$$y - 13 = x \qquad \text{Simplify}$$
$$y = x + 13. \qquad \text{Slope-intercept form}$$

The line and the data points are shown in Figure 2.26. Although the line fits three data points perfectly, it overestimates profits in the middle years.

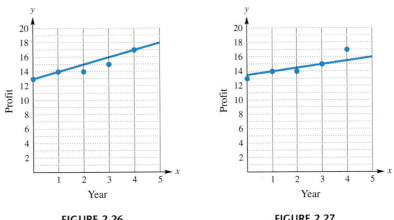

FIGURE 2.26   FIGURE 2.27

(c) Use the points $(1, 14)$ and $(3, 15)$ to find another line that models the data.

Solution The slope is $\dfrac{15 - 14}{3 - 1} = \dfrac{1}{2} = .5$, and the equation is

$$y - 14 = .5(x - 1) \qquad \text{Point-slope form, using (1, 14) and slope .5}$$
$$y - 14 = .5x - .5) \qquad \text{Distributive property}$$
$$y = .5x + 13.5. \qquad \text{Slope-intercept form}$$

The line and the data points are shown in Figure 2.27. This line passes through two data points and is quite close to two others, but significantly underestimates the profit in the last year.

(d) Use the model in part (b) to estimate the profits in 2005.

Solution Substitute $x = 5$ (corresponding to 2005) into the equation.

$$y = x + 13$$
$$y = (5) + 13 = 18 \text{ billion.} \quad \blacktriangle \; \langle 1 \rangle$$

Opinions may vary as to which of the lines in Example 1 best fits the data. To make a decision, we might try to measure the amount of error in each model. One way to do this is to compute the difference between the actual profit p and the amount y given by the model. If the data point is (x, p) and the corresponding point on the line is (x, y), then the difference $p - y$ measure the error in the model for that

⟨1⟩

Use the points $(0, 13)$ and $(3, 15)$ to find another linear model for the data in Example 1.

Answer:

$$y = \dfrac{2}{3}x + 13$$

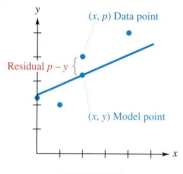

FIGURE 2.28

particular value of x. The number $p - y$ is called a **residual.** As shown in Figure 2.28, the residual $p - y$ is the vertical distance from the data point to the line (positive when the data point is above the line, negative when it is below the line, and 0 when it is on the line).

One way to determine how well a line fits the data points might be to compute the sum of its residuals—that is, the sum of the individual errors. Unfortunately, however, the sum of the residuals of two different line might be equal, thwarting our effort to decide which is the better fit. Furthermore, the residuals may sum to 0, which doesn't mean that there is no error, but only that the positive and negative errors (which might be quite large) cancel each other out. (See Exercise 11 at the end of this section for an example.)

To avoid this difficulty, mathematicians use the sum of the *squares* of the residuals to measure how well a line fits the data points. When the sum of the squares is used, a smaller sum means a smaller overall error and hence a better fit. The error is 0 only when all the data points lie on the line (a perfect fit).

▼ **EXAMPLE 2**   Two linear models for GE's profits were constructed in Example 1:

$$y = x + 13 \quad \text{and} \quad y = .5 + 13.5.$$

For each model, determine the five residuals, the square of each residual, and the sum of the squares of the residuals.

Solution The information for each model is summarized in the following tables.

$$y = x + 13$$

Data Point (x, p)	Model Point (x, y)	Residual $p - y$	Squared Residual $(p - y)^2$
$(0, 13)$	$(0, 13)$	0	0
$(1, 14)$	$(1, 14)$	0	0
$(2, 14)$	$(2, 15)$	-1	1
$(3, 15)$	$(3, 16)$	-1	1
$(4, 17)$	$(4, 17)$	0	0
			Sum: 2

$$y = .5x + 13.5$$

Data Point (x, p)	Model Point (x, y)	Residual $p - y$	Squared Residual $(p - y)^2$
$(0, 13)$	$(0, 13.5)$	$-.5$	$.25$
$(1, 14)$	$(1, 14)$	0	0
$(2, 14)$	$(2, 14.5)$	$-.5$	$.25$
$(3, 15)$	$(3, 15)$	0	0
$(4, 17)$	$(4, 15.5)$	1.5	2.25
			Sum: 2.75

According to this measure of error, the line $y = x + 13$ is a better fit for the data because the sum of the squares of its residuals is smaller than the sum for $y = .5x + 13.5$. ▲ ②

Linear Regression (Optional)*

Mathematical techniques that are introduced in Chapter 14 can be used to prove the following result.

> For any set of data points, there is one and only one line for which the sum of the squares of the residuals is as small as possible.

This *line of best fit* is called the **least squares regression line,** and the computational process for finding its equation is called **linear regression.** Linear regression formulas are quite complicated and require a large amount of computation. Fortunately, most graphing calculators and spreadsheet programs can do linear regression quickly and easily.

▼ **EXAMPLE 3**   Recall that the profits of GE (in billions of dollars) were as follows.

Year	2000	2001	2002	2003	2004
Profit	13	14	14	15	17

Use a graphing calculator to do the following.

(a) Plot the data points, with $x = 0$ corresponding to 2000.†

Solution   The data points are $(0, 13)$, $(1, 14)$, $(2, 14)$, $(3, 15)$, and $(4, 17)$. Press STAT EDIT to bring up the statistics editor. Enter the x-coordinates as a list L_1 and the corresponding y-coordinates as a list L_2, as shown in Figure 2.29. To plot the data points, go to the STAT PLOT menu, choose a plot (here it's Plot 1), choose ON, and enter the lists L_1 and L_2, as shown in Figure 2.30. Then set the viewing window as usual and press GRAPH to produce Figure 2.31.

FIGURE 2.29   FIGURE 2.30   FIGURE 2.31

---

*Examples 3–6 require either a graphing calculator or a spreadsheet program.

†The process outlined here works for most TI graphing calculators. Other graphing calculators and spreadsheet programs operate similarly, but check your instruction manual or see the Graphing Calculator Appendix, which is available at www.aw-bc.com/MWA9.

---

② Another model for the data in Example 1 is $y = .7x + 13$. Use this line to find

(a)   the residuals and

(b)   the sum of the squares of the residuals.

(c)   Does this line fit the data better than the two lines in Examples 1 and 2?

Answers:

(a)   0, .3, −.4, −.1, 1.2

(b)   1.7

(c)   Yes, because the sum of the squares of the residuals is smaller.

(b) Find the least squares regression line for this data.

Solution Go to the STAT CALC menu and choose LIN REG, which returns you to the home screen. As shown in Figure 2.32, enter the list names and the place where the equation of the regression line should be stored (here Y_1 is chosen; it's on the VARS Y-VARS FUNCTION menu); then press ENTER. Figure 2.33 shows that the equation of the regression line is

$$y = .9x + 12.8.$$

(The number r in Figure 2.33 will be discussed below.)

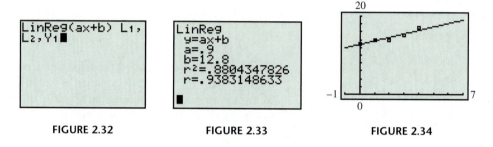

FIGURE 2.32   FIGURE 2.33   FIGURE 2.34

(c) Graph the data points and the regression line on the same screen.

Solution Press GRAPH to see the line plotted with the data points (Figure 2.34). ▲ ③

NOTE The mathematical techniques used to find the least squares regression line are discussed and illustrated with a simple example in the case at the end of Chapter 14. These techniques require multivariable calculus.

③

Use the least squares regression line $y = .9x + 12.8$ and the data points in Example 3 to find

(a) the residuals and

(b) the sum of the squares of the residuals.

(c) How does this line compare with those in Example 1 and side problem 2?

Answers:

(a) .2, .3, −.6, −.5, −.6

(b) 1.1

(c) It fits the data best because the sum of the squares of its residuals is smallest.

▼ **EXAMPLE 4** The following table shows the number of accidental deaths per 100,000 population in the United States over the past half century:*

Year	Death Rate	Year	Death Rate
1960	63.1	1985	38.5
1965	65.8	1990	36.3
1970	62.2	1995	34.9
1975	50.8	2000	34.9
1980	46.4	2003	36.1

(a) Let $x = 0$ correspond to 1960. Use a graphing calculator or spreadsheet program to find the least squares regression line that models the data in the table.

Solution The data points are $(0, 63.1)$, $(5, 65.8)$, ..., $(43, 36.1)$. Enter the x-coordinates as list L_1 and the corresponding y-coordinates as list L_2 in a graphing

*U.S. National Center for Health Statistics; age-adjusted data.

calculator. Then find the regression line, as in Figure 2.35. Rounding the coefficients, the equation of the line is $y = -.8079x + 64.9151$.

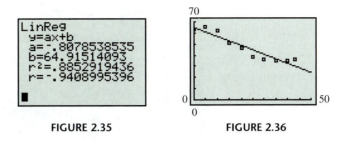

FIGURE 2.35   FIGURE 2.36

(b) Plot the data points and the regression line on the same screen.

Solution See Figure 2.36, which shows that the line is a reasonable model for the data.

(c) Use the regression line to estimate the number of accidental deaths in 2006.

Solution 2006 corresponds to $x = 46$. Substitute $x = 46$ into the regression line equation.

$$y = -.8079(46) + 64.9151 \approx 27.75.$$

So this model estimates that there will be approximately 27.75 accidental deaths per 100,000 population in 2006. ▲ ⬦4

⬦4

Using only the data from 1985 and later in Example 4, find the equation of the least squares regression line. Round the coefficients to three decimal places.

Answers:

$y = -.145x + 41.158$

⬚ Correlation

Although the "best fit" line can always be found by linear regression, it may not be a good model. For instance, if the data points are widely scattered, no straight line will model the data accurately. When the linear regression line was computed in Examples 3 and 4, the calculator also displayed a number r and its square (Figures 2.33 and 2.35). This number is called the **correlation coefficient.** It measures how closely the data points fit the regression line and thus indicates how good the regression line is for predictive purposes.

The correlation coefficient r is always between -1 and 1. When $r = \pm 1$, the data points all lie on the regression line (a perfect fit). When the absolute value of r is close to 1, the line fits the data quite well, and when r is close to 0, the line is a poor fit for the data (but some other curve might be a good fit). Figure 2.37 shows how the value of r varies, depending on the pattern of the data points.

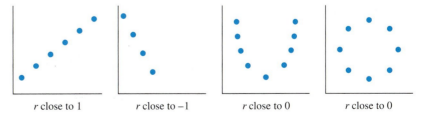

r close to 1 r close to -1 r close to 0 r close to 0

FIGURE 2.37

▼ **EXAMPLE 5**   The number of unemployed people in the U.S. labor force (in millions) in recent years is shown in the table.*

Year	Unemployed	Year	Unemployed	Year	Unemployed
1988	6.701	1994	7.996	2000	5.692
1989	6.528	1995	7.404	2001	6.801
1990	7.047	1996	7.236	2002	8.378
1991	8.628	1997	6.739	2003	8.774
1992	9.613	1998	6.210	2004	8.149
1993	8.940	1999	5.880		

Determine whether or not a linear equation is a good model for this data.

Solution Let $x = 0$ correspond to 1980 and plot the data points $(8, 6.701)$, etc., either by hand or with a graphing calculator, as in Figure 2.38. They do not form a linear pattern (unemployment tends to rise and fall). Alternatively, you could compute the regression equation for the data, as in Figure 2.39. The correlation coefficient is

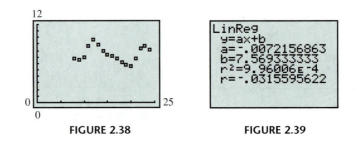

FIGURE 2.38   FIGURE 2.39

*U.S. Department of Labor, Bureau of Labor Statistics.

$r \approx -.03$, a number quite close to 0, which indicates that the regression line is a poor fit for the data. Therefore, a linear equation is not a good model for this data. ▲ ⑤

▼ **EXAMPLE 6**   Enrollment projections (in millions) for all U.S. colleges and universities in selected years are shown in the following table.*

Year	2000	2001	2002	2003	2004	2005	2006
Enrollment	15.313	15.928	16.103	16.360	16.468	16.679	16.887

(a) Let $x = 0$ correspond to 2000. Use a graphing calculator or spreadsheet program to find a linear model for the data and determine how well it fits the data points.

Solution The least squares regression line (with coefficients rounded) is

$$y = .2353x + 15.5423,$$

as shown in Figure 2.40. The correlation coefficient is $r \approx .97$, which is very close to 1, so this line fits the data well.

(b) Predict the enrollment in 2010.

Solution Let $x = 10$ (corresponding to 2010) in the regression equation:

$$y = .2353(10) + 15.5423 = 17.8953.$$

Therefore, the enrollment in 2010 will be approximately 17,895,300 students.

(c) According to this model, in what year will enrollment reach 20 million?

Solution Let $y = 20$ and solve the regression equation for x.

$.2353x + 15.5423 = y$	
$.2353x + 15.5423 = 20$	Let $y = 20$.
$.2353x = 4.4577$	Subtract 15.5423 from both sides.
$x \approx 18.94$	Divide both sides by .2353.

Since these enrollment figures change once a year, use the nearest integer value for x, namely, 19. So enrollment will reach 20 million in 2019. ▲

——————
*As of the fall of each year; *Statistical Abstracts of the United States, 2004–2005.*

⑤

Using only the data from 2000 and later in Example 5, find

(a) the equation of the least squares regression line and

(b) the correlation coefficient.

(c) How well does this line fit the data?

Answers:

(a) $y \approx .69x - 7.59$

(b) $r \approx .85$

(c) It fits reasonably well because $|r|$ is close to 1.

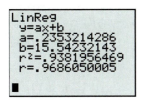

```
LinReg
y=ax+b
a=.2353214286
b=15.54232143
r²=.9381956469
r=.9686050005
■
```

FIGURE 2.40

2.3 Exercises

1. **Physical Science**   The table shows equivalent Fahrenheit and Celsius temperatures.

Degrees Fahrenheit	32	68	104	140	176	212
Degrees Celsius	0	20	40	60	80	100

(a) Choose any two data points and use them to construct a linear equation that models the data, with x being Fahrenheit and y Celsius.

(b) Use the model in part (a) to find the Celsius temperature corresponding to

50° Fahrenheit and 75° Fahrenheit.

Physical Science   *Use the linear equation derived in Exercise 1 to work the following problems.*

2. Convert each temperature.
    (a) 58°F to Celsius
    (b) 50°C to Fahrenheit
    (c) −10°C to Fahrenheit
    (d) −20°F to Celsius

3. According to the *World Almanac and Book of Facts, 2005*, Venus is the hottest planet, with a surface temperature of 867° Fahrenheit. What is this temperature in Celsius degrees?

4. Find the temperature at which Celsius and Fahrenheit temperatures are numerically equal.

In each of the following problems, assume that the data can be modeled by a straight line. Use two data points to find such a line and then answer the question. (See Example 1.)

5. **Business** The Consumer Price Index (CPI), which measures the cost of a typical package of consumer goods, stood at 130.7 in 1990 and 190.7 in early 2005. Let $x = 0$ correspond to 1990, and estimate the CPI in 2004 and 2008.

6. **Finance** The approximate number of amended federal income tax returns was 3.2 million in 2000 and 4 million in 2005.* Let $x = 0$ correspond to 2000, and estimate the number of amended returns in 2008.

7. **Business** The United States is China's largest export market. Imports from China grew from about 82 billion dollars in 1999 to 152 billion dollars in 2003. Let $x = 0$ correspond to 1990, and estimate the amount of imports in 2006.

8. **Physical Science** Suppose a baseball is thrown at 85 miles per hour. The ball will travel 320 feet when hit by a bat swung at 50 miles per hour and will travel 440 feet when hit by a bat swung at 80 miles per hour. Let y be the number of feet traveled by the ball when hit by a bat swung at x miles per hour. (*Note:* The preceding data is valid for $50 \leq x \leq 90$, where the bat is 35 inches long, weighs 32 ounces, and strikes a waist-high pitch so that the place of the swing lies at 10° from the diagonal.)† How much farther will a ball travel for each mile-per-hour increase in the speed of the bat?

9. **Business** Advertising expenditures in the United Sates were $244 billion in 2000 and $245 billion in 2003.‡ Let x be the number of years after 2000 and estimate advertising expenditures in 2007.

10. **Physical Science** Ski resorts require large amounts of water in order to make snow. Snowmass Ski Area in Colorado plans to pump at least 1120 gallons of water per minute for at least 12 hours a day from Snowmass Creek between mid-October and late December.§ Environmentalists are concerned about the effects on the ecosystem. Find the minimum amount of water pumped in 30 days. (*Hint:* Let y be the total number of gallons pumped x days after pumping begins. Note that $(0, 0)$ is on the graph of the equation.)

---

*Internal Revenue Service projections.

†Robert K. Adair, *The Physics of Baseball* (HarperCollins, 1990).

‡*Advertising Age.*

§York Snow, Inc.

In each of the following problems, two linear models are given for the data. For each model,

(a) *Find the residuals and their sum.*
(b) *Find the sum of the squares of the residuals.*
(c) *Decide which model is the better fit. (See Example 2.)*

11. **Business** The weekly amount spent on advertising and the weekly sales revenue of a small store over a five-week period are shown in the following table.

Advertising Expenditure x (in hundreds of dollars)	1	2	3	4	5
Sales Revenue y (in thousands of dollars)	2	2	3	3	5

Two equations that model the data are $y = .5x + 1.5$ and $y = x$.

12. **Business** The table shows the total outstanding consumer debt (excluding home mortgages) in billions of dollars in selected years:*

Year	1985	1990	1995	2000	2003
Consumer debt	585	789	1096	1693	1987

Let $x = 0$ correspond to 1985. Two equations that model the data are $y = 78x + 585$ and $y = 80x + 463$.

In each of the following problems, determine whether a straight line is a good model for the data. You may do this either visually, by plotting the data points, or analytically, by finding the correlation coefficient for the least squares regression line. (See Examples 5 and 6.)

13. **Natural Science** The table shows the average monthly temperature (in degrees Fahrenheit) in Cleveland, Ohio, based on data from 1961 to 1990.† Let $x = 2$ correspond to February, $x = 4$ to April, etc.

Month	Feb	April	June	Aug	Oct	Dec
Temperature	27.3	47.5	67.5	70.3	52.7	30.9

14. **Health** The table that follows shows the number of deaths per 100,000 people from heart disease in selected years.‡ Let $x = 0$ correspond to 1960.

---

*Federal Reserve Bulletin.

†National Climatic Data Center.

‡U.S. National Center for Health Statistics.

Year	1960	1970	1980	1990	2000	2002
Deaths	559	483	412	322	258	240

In the following exercises, find the required linear model as follows: If you do not have a graphing calculator or spreadsheet program, use the first and last data points to determine a line. If you do have a graphing calculator or spreadsheet program, find the least squares regression line. (See Examples 1, 3, 4, and 6.)

15. Health Use the data on death from heart disease in Exercise 14.
   (a) Find a linear model for the data, with $x = 0$ corresponding to 1960.
   (b) Estimate the number of deaths from heart disease in 2008.

16. Social Science The table shows how poverty-level income cutoffs (in dollars) for a family of four have changed over time (in large part because of inflation).*

Year	1980	1985	1990	1995	2000	2003
Income	8414	10,989	13,359	15,569	17,604	18,811

   (a) Find a linear model for the data, with $x = 0$ corresponding to 1980 and income measured in thousands of dollars (so that $8414 is 8.414, etc.).
   (b) What is the approximate poverty level in 2006? In 2008?

17. Physical Science While shopping for an air conditioner, Adam Bryer consulted the following table giving a machine's BTUs and the square footage (ft^2) that it would cool:†

ft^2 (x)	BTUs (y)
150	5000
175	5500
215	6000
250	6500
280	7000
310	7500
350	8000
370	8500
420	9000
450	9500

   (a) Find a linear model for the data.
   (b) To check the fit of the data to the line, use the results from part (a) to find the number of BTUs required to cool a room of 150 ft^2, 280 ft^2, and 420 ft^2. How well do the actual data agree with the predicted values?
   (c) Suppose Adam's room measures 235 ft^2. Use the results from part (a) to decide how many BTUs it requires. If air conditioners are available only with the BTU choices in the table, which should Adam choose?

18. Business The table shows the estimated percentage of all business with Internet phone service:*

Year	2003	2004	2005	2006	2007	2008
%	3.4	11.8	19.2	26.5	32.6	40.5

   (a) Find a linear model for the data, with $x = 3$ corresponding to 2003.
   (b) Estimate the percentage of business with Internet phone service in 2009 and 2012.
   (c) Will this model be accurate until 2020? Explain.

Use technology to do the following regression problems. (See Examples 3, 4 and 6.)

19. Business The graph shows Intel's expenditures on research and development (in billions of dollars) over a ten-year period.†

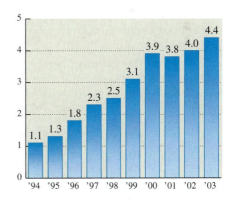

   (a) List the data points, with $x = 4$ corresponding to 1994.
   (b) Find the least squares regression line that models the data.
   (c) If the trend shown continues, what will Intel spend on research and development in 2008?

*U.S. Census Bureau, *Current Population Reports*.

†Morris Carey and James Carey, "On the House," *Sacramento Bee*, July 29, 2000.

*In-Stat.

†Intel Corporation Annual Report, 2003.

20. Natural Science   Biologists have observed a linear relationship between the temperature and the frequency with which a cricket chirps. The following data were measured for the striped ground cricket.*

Temperature °F (x)	Chirps per second (y)
88.6	20.0
71.6	16.0
93.3	19.8
84.3	18.4
80.6	17.1
75.2	15.5
69.7	14.7
82.0	17.1
69.4	15.4
83.3	16.2
79.6	15.0
82.6	17.2
80.6	16.0
83.5	17.0
76.3	14.4

(a) Find the least squares regression line that models the data.

(b) Use the results of part (a) to determine how many chirps per second you would expect to hear from the striped ground cricket if the temperature were 73°F.

(c) Use the results of part (a) to determine what the temperature is when the striped ground crickets are chirping at a rate of 18 times per second.

(d) Find the correlation coefficient.

21. Transportation   The estimated number of scheduled passengers on U.S. commercial airlines (in billions) in selected years is shown in the following table:†

Year	Passengers
2002	.63
2003	.64
2004	.69
2005	.72
2006	.77
2007	.79
2008	.8
2009	.84

(a) Find the least squares regression line that models these data, with $x = 2$ corresponding to 2002.

(b) Estimate the number of passengers in 2012 and in 2016.

22. Health   The following table shows men's and women's life expectancy at birth (in years), selected birth years in the United States:*

Birth Year	Life Expectancy	
	Men	Women
1970	67.1	74.7
1975	68.8	76.6
1980	70	77.4
1985	71.1	78.2
1990	71.8	78.8
1995	72.5	78.9
1998	73.8	79.5
2000	74.3	79.7
2001	74.4	79.8
2005	74.9	80.7

(a) Find the least squares regression line for the men's data, with $x = 70$ corresponding to 1970.

(b) Find the least squares regression line for the women's data, with $x = 70$ corresponding to 1970.

(c) Suppose life expectancy continues to increase as predicted by the equations in parts (a) and (b). Will men's life expectancy ever be the same as women's? If so, in what birth year will this occur?

23. Health   The projected number of new cases of Alzheimer's disease (in thousands) in the United States in selected years is shown in the table.*

Year	2000	2010	2020	2030	2040	2050
New Cases	400	467	489	600	800	956

(a) Find a linear model for this data, with $x = 0$ corresponding to 2000, and use it to estimate the number of new cases in the years 2005, 2015, and 2025.

(b) Use the model to predict the years in which the number of new cases will be 750,000 and 1,000,000.

*Reprinted by permission of the publishers from *The Songs of Insects* by George W. Pierce. Cambridge, MA: Harvard University Press, copyright © 1948 by the President and Fellows of Harvard College.

†Data and projections are from the Federal Aviation Administration.

*U.S. Center for National Health Statistics.

†L. Hebert, L. Beckett, P. Scherr, D. Evans, "Annual Incidence of Alzheimer Disease in the United States Projected to the Years 2000 Through 2050," in *Alzheimer Disease and Associated Disorders,* Vol. 15, No. 4, 2001.

24. Health National expenditures on health care (in billions of dollars) in various years are shown in the table.*

Year	1990	1995	1997	1999
Amount	699.4	993.7	1092.4	1228.5

Year	2000	2001	2002	2003
Amount	1316.2	1403.6	1495.5	1560.4

(a) Find a linear model for these data, with $x = 0$ corresponding to 1990.

(b) Interpret the meaning of the slope and the y-intercept.

(c) If the model remains accurate in the future, what will health care expenditure be in 2007?

---
*Data and projections are from the U.S. Health Care Financing Administration.

25. Business Total corporate profits (after taxes) of U.S. companies (in billions of dollars) are shown in the table.*

Year	1990	1995	1999	2000	2001	2002	2003
Profits	292	487	593	553	569	709	845

(a) Find the least squares regression line that models this data, with $x = 0$ corresponding to 1990.

(b) Does this model fit the data reasonably well? Why?

(c) If this trend continues, what will corporate profits be in 2009?

(d) If this trend continues, when will corporate profits be 1.2 trillion dollars?

---
*U.S. Bureau of Economic Analysis.

2.4 Linear Inequalities

An **inequality** is a statement that one mathematical expression is greater than (or less than) another. Inequalities are very important in applications. For example, a company wants revenue to be *greater than* costs and must use *no more than* the total amount of capital or labor available.

Inequalities may be solved by algebraic or geometric methods. In this section, we shall concentrate on algebraic methods for solving **linear inequalities,** such as

$$4 - 3x \le 7 + 2x \quad \text{and} \quad -2 < 5 + 3m < 20,$$

and absolute value inequalities, such as $|x - 2| < 5$. The following properties are the basic algebraic tools for working with inequalities.

Properties of Inequality

For real numbers a, b, and c,

(a) if $a < b$, then $a + c < b + c$

(b) if $a < b$, and if $c > 0$, then $ac < bc$

(c) if $a < b$, and if $c < 0$, then $ac > bc$.

Throughout this section, definitions are given only for $<$, but they are equally valid for $>$, $\le$, or $\ge$.

CAUTION Pay careful attention to part (c): if both sides of an inequality are multiplied by a negative number, the direction of the inequality symbol must be reversed. For example, starting with the true statement $-3 < 5$ and multiplying both sides by the positive number 2 gives

$$-3 \cdot 2 < 5 \cdot 2,$$

or

$$-6 < 10,$$

still a true statement. However, starting with $-3 < 5$ and multiplying both sides by the negative number -2 gives a true result only if the direction of the inequality symbol is reversed:

$$-3(-2) > 5(-2)$$
$$6 > -10. \quad$$

(a) First multiply both sides of $-6 < -1$ by 4, and then multiply both sides of $-6 < -1$ by -7.

(b) First multiply both sides of $9 \geq -4$ by 2 and then multiply both sides of $9 \geq -4$ by -5.

(c) First add 4 to both sides of $-3 < -1$, and then add -6 to both sides of $-3 < -1$.

Answers:

(a) $-24 < -4; 42 > 7$

(b) $18 \geq -8; -45 \leq 20$

(c) $1 < 3; -9 < -7$

▼ **EXAMPLE 1**   Solve $3x + 5 > 11$. Graph the solution.

Solution   First add -5 to both sides.

$$3x + 5 + (-5) > 11 + (-5)$$
$$3x > 6$$

Now multiply both sides by $1/3$.

$$\frac{1}{3}(3x) > \frac{1}{3}(6)$$
$$x > 2$$

(Why was the direction of the inequality symbol not changed?) In interval notation (introduced in Section 1.1), the solution is the interval $(2, \infty)$, which is graphed on the number line in Figure 2.41. The parenthesis at 2 shows that 2 is not included in the solution.

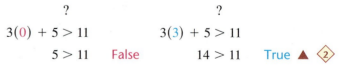

FIGURE 2.41

As a partial check, note that 0, which is not part of the solution, makes the inequality false, while 3, which is part of the solution, makes it true.

$$3(0) + 5 > 11 \qquad\qquad 3(3) + 5 > 11$$
$$5 > 11 \quad \text{False} \qquad\qquad 14 > 11 \quad \text{True} \; \blacktriangle$$

Solve these inequalities. Graph each solution.

(a) $5z - 11 < 14$

(b) $-3k \leq -12$

(c) $-8y \geq 32$

Answers:

(a) $z < 5$

(b) $k \geq 4$

(c) $y \leq -4$

▼ **EXAMPLE 2**   Solve $4 - 3x \leq 7 + 2x$.

Solution   Add -4 to both sides.

$$4 - 3x + (-4) \leq 7 + 2x + (-4)$$
$$-3x \leq 3 + 2x$$

Add $-2x$ to both sides. (Remember that *adding* to both sides never changes the direction of the inequality symbol.)

$$-3x + (-2x) \leq 3 + 2x + (-2x)$$
$$-5x \leq 3$$

Multiply both sides by $-1/5$. Since $-1/5$ is negative, change the direction of the inequality symbol.

$$-\frac{1}{5}(-5x) \geq -\frac{1}{5}(3)$$

$$x \geq -\frac{3}{5}$$

Figure 2.42 shows a graph of the solution, $[-3/5, \infty)$. The bracket in Figure 2.42 shows that $-3/5$ is included in the solution. ▲ ③

FIGURE 2.42

▼ **EXAMPLE 3**   Solve $-2 < 5 + 3m < 20$. Graph the solution.

Solution   The inequality $-2 < 5 + 3m < 20$ says that $5 + 3m$ is *between* -2 and 20. We can solve this inequality with an extension of the properties given at the beginning of this section. Work as follows, first adding -5 to each part.

$$-2 + (-5) < 5 + 3m + (-5) < 20 + (-5)$$
$$-7 < 3m < 15$$

Now multiply each part by $1/3$.

$$-\frac{7}{3} < m < 5$$

A graph of the solution, $(-7/3, 5)$, is given in Figure 2.43. ▲ ④

FIGURE 2.43

▼ **EXAMPLE 4**   The formula for converting from Celsius to Fahrenheit temperature is

$$F = \frac{9}{5}C + 32.$$

What Celsius temperature range corresponds to the range from 32°F to 77°F?

Solution   The Fahrenheit temperature range is $32 < F < 77$. Since $F = (9/5)C + 32$, we have

$$32 < \frac{9}{5}C + 32 < 77.$$

Solve the inequality for C.

$$32 < \frac{9}{5}C + 32 < 77$$

$$0 < \frac{9}{5}C < 45$$

$$\frac{5}{9} \cdot 0 < \frac{5}{9} \cdot \frac{9}{5}C < \frac{5}{9} \cdot 45$$

$$0 < C < 25$$

The corresponding Celsius temperature range is 0°C to 25°C. ▲

In Example 4, what Celsius temperatures correspond to the range from 5°F to 95°F?

Answer:

−15°C to 35°C

A product will break even or produce a profit only if the revenue R from selling the product at least equals the cost C of producing it—that is, if $R \geq C$.

▼ **EXAMPLE 5**   A company analyst has determined that the cost to produce and sell x units of a certain product is $C = 20x + 1000$. The revenue for that product is $R = 70x$. Find the values of x for which the company will break even or make a profit on the product.

Solution Solve the inequality $R \geq C$.

$$R \geq C$$
$$70x \geq 20x + 1000 \qquad \text{Let } R = 70x, \, C = 20x + 1000.$$
$$50x \geq 1000$$
$$x \geq 20$$

The company must produce and sell 20 items to break even and more than 20 to make a profit. ▲

Absolute Value Inequalities

You may wish to review the definition of absolute value in Section 1.1 before reading the following examples, which show how to solve inequalities involving absolute values.

▼ **EXAMPLE 6**   Solve each inequality.

(a) $|x| < 5$

Solution Because absolute value gives the distance from a number to 0, the inequality $|x| < 5$ is true for all real numbers whose distance from 0 is less than 5. This includes all numbers between −5 and 5, or numbers in the interval $(-5, 5)$. A graph of the solution is shown in Figure 2.44.

FIGURE 2.44

(b) $|x| > 5$

Solution In a similar way, the solution of $|x| > 5$ is given by all those numbers whose distance from 0 is *greater* than 5. This includes the numbers satisfying $x < -5$ or $x > 5$. A graph of the solution, all numbers in

$$(-\infty, -5) \qquad \text{or} \qquad (5, \infty),$$

is shown in Figure 2.45. ▲

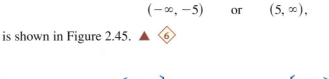

FIGURE 2.45

The preceding examples suggest the following generalizations.

Assume that a and b are real numbers with b positive.

1. Solve $|a| < b$ by solving $-b < a < b$.

2. Solve $|a| > b$ by solving $a < -b$ or $a > b$.

▼ **EXAMPLE 7** Solve $|x - 2| < 5$.

Solution Replace a with $x - 2$ and b with 5 in property (1) in the box above. Now solve $|x - 2| < 5$ by solving the inequality

$$-5 < x - 2 < 5.$$

Add 2 to each part, getting the solution

$$-3 < x < 7,$$

which is graphed in Figure 2.46. ▲

FIGURE 2.46

▼ **EXAMPLE 8** Solve $|2 - 7m| - 1 > 4$.

Solution First add 1 to both sides.

$$|2 - 7m| > 5$$

Now use property (2) from the preceding box to solve $|2 - 7m| > 5$ by solving the inequality

$$2 - 7m < -5 \qquad \text{or} \qquad 2 - 7m > 5.$$

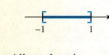

Solve each inequality. Graph each solution.

(a) $|x| \leq 1$

(b) $|y| \geq 3$

Answers:

(a) $[-1, 1]$

(b) All numbers in $(-\infty, -3]$ or $[3, \infty)$

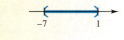

Solve each inequality. Graph each solution.

(a) $|p + 3| < 4$

(b) $|2k - 1| \leq 7$

Answers:

(a) $(-7, 1)$

(b) $[-3, 4]$

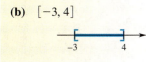

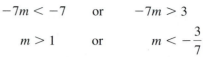

Solve each inequality. Graph
each solution.

(a) $|y - 2| > 5$

(b) $|3k - 1| \geq 2$

(c) $|2 + 5r| - 4 \geq 1$

Answers:

(a) All numbers in
$(-\infty, -3)$ or $(7, \infty)$

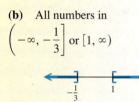

(b) All numbers in
$\left(-\infty, -\dfrac{1}{3}\right]$ or $[1, \infty)$

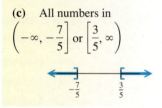

(c) All numbers in
$\left(-\infty, -\dfrac{7}{5}\right]$ or $\left[\dfrac{3}{5}, \infty\right)$

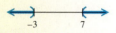

Solve each part separately.

$$-7m < -7 \qquad \text{or} \qquad -7m > 3$$

$$m > 1 \qquad \text{or} \qquad m < -\frac{3}{7}$$

The solution, all numbers in $\left(-\infty, -\dfrac{3}{7}\right)$ or $(1, \infty)$, is graphed in Figure 2.47. ▲ ⑧

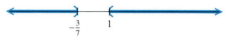

FIGURE 2.47

▼ **EXAMPLE 9** Solve $|2 - 5x| \geq -4$.

Solution The absolute value of a number is always nonnegative. Therefore, $|2 - 5x| \geq -4$ is always true; so the solution is the set of all real numbers. Note that the inequality $|2 - 5x| \leq -4$ has no solution, because the absolute value of a quantity can never be less than a negative number. ▲ ⑨

Absolute value inequalities can be used to indicate how far a number may be from a given number. The next example illustrates this use of absolute value.

▼ **EXAMPLE 10** Write each statement using absolute value notation.

(a) k is at least 4 units from 1.

Solution If k is at least 4 units from 1, then the distance from k to 1 is greater than or equal to 4. (See Figure 2.48(a).) Write this statement in terms of absolute values as follows:

$$|k - 1| \geq 4.$$

⑨

Solve each inequality.

(a) $|5m - 2| > -1$

(b) $|2 + 3a| < -3$

(c) $|6 + r| > 0$

Answers:

(a) All real numbers

(b) No solution

(c) All real numbers except -6

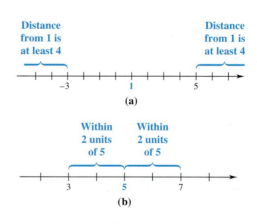

FIGURE 2.48

(b) p is within 2 units of 5.

Solution This statement means that the distance between p and 5 must be less than or equal to 2. (See Figure 2.48(b).) Using absolute value notation, the statement is written as

$$|p - 5| \le 2. \quad \blacktriangle \quad \langle 10 \rangle$$

⟨10⟩

Write each statement using absolute value notation.

(a) m is at least 3 units from 5.

(b) t is within .01 of 4.

Answers:

(a) $|m - 5| \ge 3$

(b) $|t - 4| \le .01$

2.4 Exercises

1. Explain how to determine whether a parenthesis or a bracket is used when graphing the solution of a linear inequality.

2. The three-part inequality $p < x < q$ means "p is less than x and x is less than q." Which one of the following inequalities is not satisfied by any real number x? Explain why.
   (a) $-3 < x < 5$
   (b) $0 < x < 4$
   (c) $-7 < x < -10$
   (d) $-3 < x < -2$

Solve each inequality and graph each solution. (See Examples 1–3.)

3. $-8k \le 32$

4. $-4a \le 36$

5. $-2b > 0$

6. $6 - 6z < 0$

7. $3x + 4 \le 14$

8. $2y - 7 < 9$

9. $-5 - p \ge 3$

10. $5 - 3r \le -4$

11. $7m - 5 < 2m + 10$

12. $6x - 2 > 4x - 10$

13. $m - (4 + 2m) + 3 < 2m + 2$

14. $2p - (3 - p) \le -7p - 2$

15. $-2(3y - 8) \ge 5(4y - 2)$

16. $5r - (r + 2) \ge 3(r - 1) + 6$

17. $3p - 1 < 6p + 2(p - 1)$

18. $x + 5(x + 1) > 4(2 - x) + x$

19. $-7 < y - 2 < 5$

20. $-3 < m + 6 < 2$

21. $8 \le 3r + 1 \le 16$

22. $-6 < 2p - 3 \le 5$

23. $-4 \le \dfrac{2k - 1}{3} \le 2$

24. $-1 \le \dfrac{5y + 2}{3} \le 4$

25. $\dfrac{3}{5}(2p + 3) \ge \dfrac{1}{10}(5p + 1)$

26. $\dfrac{8}{3}(z - 4) \le \dfrac{2}{9}(3z + 2)$

In the following exercises, write a linear inequality that describes the given graph.

27.
$-6 \ -4 \ -2 \quad 0 \quad 2 \quad 4 \quad 6$

28.
$-6 \ -4 \ -2 \quad 0 \quad 2 \quad 4 \quad 6$

29.
$-6 \ -4 \ -2 \quad 0 \quad 2 \quad 4 \quad 6$

30.
$-6 \ -4 \ -2 \quad 0 \quad 2 \quad 4 \quad 6$

31. **Natural Science** Federal guidelines require drinking water to have less than .050 milligram per liter of lead. A test using 21 samples of water in a midwestern city found that the average amount of lead in the samples was .040 milligram per liter. All samples had lead content within 5% of the average.
   (a) Select a variable and write down what it represents.
   (b) Write a three-part inequality to express the results obtained from the sample.
   (c) Did all the samples meet the federal requirement?

32. Natural Science Exposure to radon gas is a known lung cancer risk. According to the Environmental Protection Agency (EPA), the individual lifetime excess cancer risk R for radon exposure is between .0015 and .006, where $R = .01$ represents a 1% increase in the risk of developing cancer.*
  (a) Write the preceding information as an inequality.
  (b) Determine the range of individual annual risk by dividing R by an average life expectancy of 75 years.

33. Finance The following table shows the federal income tax for a single person in 2005:

If Taxable Income Is Over	But Not Over	The Tax Is	Of Amount Over
$0	$7,300	— + 10%	$0
$7,300	$29,700	$730 + 15%	$7,300
$29,700	$71,950	$4,090 + 25%	$29,700
$71,950	$150,150	$14,652.50 + 28%	$71,950
$150,150	$326,450	$36,548.50 + 33%	$150,150
$326,450	no limit	$94,727.50 + 35%	$326,450

  (a) Let x denote the taxable income. Write each of the six income ranges in the table as an inequality.
  (b) Let T denote the income tax. Write an inequality that gives the tax range in dollars for each of the six income ranges in the table.

Solve each inequality. Graph each solution. (See Examples 6–9.)

34. $|p| > 7$ **35.** $|m| < 2$ **36.** $|r| \leq 5$

37. $|a| < -2$ **38.** $|b| > -5$

39. $|2x + 5| < 1$ **40.** $\left|x - \dfrac{1}{2}\right| < 2$

41. $|3z + 1| \geq 4$ **42.** $|8b + 5| \geq 7$

43. $\left|5x + \dfrac{1}{2}\right| - 2 < 5$ **44.** $\left|x + \dfrac{2}{3}\right| + 1 < 4$

Physical Science *The given inequality describes the monthly average temperature T in degrees Fahrenheit in the given location. What range of temperatures corresponds to the inequality?*

45. $|T - 49| \leq 20$; Santa Fe, New Mexico

46. $|T - 43| \leq 24$; Marquette, Michigan

47. $|T - 62| \leq 19$; Memphis, Tennessee

48. $|T - 10| \leq 36$; Chesterfield Inlet, Canada

49. Natural Science Human beings emit carbon dioxide when they breathe. In one study, the emission rates of carbon dioxide by college students were measured both during lectures and during exams. The average individual rate R_L (in grams per hour) during a lecture class satisfied the inequality $|R_L - 26.75| \leq 1.42$, whereas during an exam the rate R_E satisfied the inequality $|R_E - 38.75| \leq 2.17$.*
  (a) Find the range of values for R_L and R_E.
  (b) The class had 225 students. If T_L and T_E represent the total amounts of carbon dioxide (in grams) emitted during a one-hour lecture and one-hour exam, respectively, write inequalities that describe the ranges for T_L and T_E.

50. Social Science When administering a standard intelligence test, we expect about one-third of the scores to be more than 12 units above 100 or more than 12 units below 100. Describe this situation by writing an absolute value inequality.

Work the following problems. (See Example 4.)

51. Business Projections from the U.S. Department of Commerce show that the number of Internet users (in millions) is approximately $16.4x + 121$, where $x = 0$ corresponds to 2000. During what years was the number of Internet users between 203 and 285 million?

52. Business According to data from *Nielson Media Research,* the audience for the ABC nightly news (in millions) can be approximated by $-.458x + 15.375$, where $x = 3$ corresponds to 1993. If this model remains accurate, in what years is the audience between 6,673,000 and 8,505,000?

53. Health At age x, with $0 \leq x \leq 80$, the average remaining life expectancy of a male is approximately $-.8x + 70.5$. For what age range is life expectancy between 54.5 and 42.5 years?

54. Health The blood alcohol level of a 180-pound man after he has imbibed x drinks in one hour can be approximated by $.022x - .0043$. In most states, a person with a blood alcohol level of .08 or more is considered drunk. Coma (and a high probability of death) usually occur at a blood alcohol level of .40 or more. How many drinks does it take for a 180-pound man to be
  (a) legally drunk?
  (b) in a coma?

55. Transportation The cost of a taxi in New York City is $2, plus 30 cents for each $1/5$ mile. How far could you travel for at least $5, but no more than $11?

56. Transportation Anne Kelly went to a conference in Montreal for a week. She decided to rent a car and

---

Indoor-Air-Assessment: A Review of Indoor Air Quality Risk Characterization Studies, Report No. EPA/600/8-90/044, Environmental Protection Agency, 1991.

*T.C. Wang, *ASHRAE Transactions* 81 (Part 1), 32 (1975).

checked with two rental firms. Avery wanted $56 per day, with no mileage fee. Hart wanted $216 per week and $.28 per mile (or part of a mile). How many miles must she drive before the Avery car becomes the better deal?

Business *In Exercises 57–62, find all values of x for which the following products will at least break even. (See Example 5.)*

57. The cost to produce x units of wire is $C = 50x + 6000$, while the revenue is $R = 65x$.

58. The cost to produce x units of squash is $C = 100x + 6000$, while the revenue is $R = 500x$.

59. $C = 85x + 1000$; $R = 105x$

60. $C = 70x + 500$; $R = 60x$

61. $C = 1000x + 5000$; $R = 900x$

62. $C = 25,000x + 21,700,000$; $R = 102,500x$

Write each of the following statements using absolute value notation. (See Example 10.)

63. x is within 3 units of 2.

64. m is no more than 6 units from 9.

65. z is no less than 2 units from 12.

66. p is at least 4 units from 9.

2.5 Polynomial and Rational Inequalities

This section deals with the solution of polynomial and rational inequalities, such as

$$r^2 + 3r - 4 \geq 0, \qquad x^3 - x \leq 0, \qquad \text{and} \qquad \frac{2x - 1}{3x + 4} < 5.$$

We shall concentrate on algebraic solution methods, but to understand why these methods work, we must first look at such inequalities from a graphical point of view.

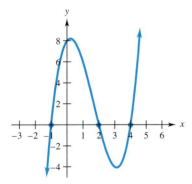

FIGURE 2.49

▼ **EXAMPLE 1**   Use the graph of $y = x^3 - 5x^2 + 2x + 8$ in Figure 2.49 to solve each of the following inequalities.

(a) $x^3 - 5x^2 + 2x + 8 > 0$

Solution Each point on the graph has coordinates of the form $(x, x^3 - 5x^2 + 2x + 8)$. The number x is a solution of the inequality exactly when the second coordinate of this point is positive—that is, when the point lies *above* the x-axis. So to solve the inequality, we need only find the first coordinates of points on the graph that are above the x-axis. This information can be read from Figure 2.49. The graph is above the x-axis when $-1 < x < 2$ and also when $x > 4$. Therefore, the solutions of the inequality are all numbers x in the interval $(-1, 2)$ or the interval $(4, \infty)$.

(b) $x^3 - 5x^2 + 2x + 8 < 0$

The number x is a solution of the inequality exactly when the second coordinate of the point $(x, x^3 - 5x^2 + 2x + 8)$ on the graph is negative—that is, when the point lies *below* the x-axis. Figure 2.49 shows that the graph is below the x-axis when $x < -1$ and also when $2 < x < 4$. Hence, the solutions are all numbers x in the interval $(-\infty, -1)$ or the interval $(2, 4)$. ▲

The solution process in Example 1 depends only on knowing the graph and its x-intercepts (that is, the points where the graph intersects the x-axis). This information can often be obtained algebraically, without doing any graphing, as illustrated in the next example.

▼ **EXAMPLE 2**   Solve each of the following quadratic inequalities.

(a) $x^2 - x < 12$

Solution   First rewrite the inequality as $x^2 - x - 12 < 0$. Now, we don't know what the graph of $y = x^2 - x - 12$ looks like, but we can still find its x-intercepts by setting $y = 0$ and solving for x.

$$x^2 - x - 12 = 0$$
$$(x + 3)(x - 4) = 0.$$
$$x + 3 = 0 \quad \text{or} \quad x - 4 = 0$$
$$x = -3 \qquad\qquad x = 4$$

These numbers divide the x-axis (number line) into three regions, as indicated in Figure 2.50.

Region A	Region B	Region C
$x < -3$	$-3 < x < 4$	$x > 4$

FIGURE 2.50

In each region, the graph of $y = x^2 - x - 12$ is an unbroken curve, so it will be entirely above or entirely below the axis. It can pass from above to below the x-axis only at the x-intercepts. To see whether the graph is above or below the x-axis when x is in region A, choose a value of x in region A, say, $x = -5$, and substitute it into the equation.

$$y = x^2 - x - 12 = (-5)^2 - (-5) - 12 = 18$$

Therefore, the point $(-5, 18)$ is on the graph. Since its y-coordinate 18 is positive, this point lies above the x-axis; hence, the entire graph lies above the x-axis in region A.

Similarly, we can choose a value of x in region B, say, $x = 0$. Then

$$y = x^2 - x - 12 = 0^2 - 0 - 12 = -12,$$

so that $(0, -12)$ is on the graph. Since this point lies below the x-axis (why?), the entire graph in region B must be below the x-axis. Finally, in region C, let $x = 5$. Then $y = 5^2 - 5 - 12 = 8$, so that $(5, 8)$ is on the graph and the entire graph in region C lies above the x-axis. We can summarize the results as follows:

Interval	$x < -3$	$-3 < x < 4$	$x > 4$
Test value in interval	-5	0	5
Value of $x^2 - x - 12$	18	-12	8
Graph	above x-axis	below x-axis	above x-axis
Conclusion	$x^2 - x - 12 > 0$	$x^2 - x - 12 < 0$	$x^2 - x - 12 > 0$

The last row shows that the only region where $x^2 - x - 12 < 0$ is region B, so the solutions of the inequality are all numbers x with $-3 < x < 4$—that is, the interval $(-3, 4)$, as shown in the number line graph in Figure 2.51.

FIGURE 2.51

(b) $x^2 - x - 12 > 0$

Solution Use the chart in part (a). The last row shows that $x^2 - x - 12 > 0$ only when x is in region A or region C. Hence, the solutions of the inequality are all numbers x with $x < -3$ or $x > 4$—that is, all numbers in the interval $(-\infty, -3)$ or the interval $(4, \infty)$. ▲ ①

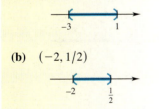

Solve each inequality. Graph the solution on the number line.

(a) $x^2 + 2x - 3 < 0$

(b) $2p^2 + 3p - 2 < 0$

Answers:

(a) $(-3, 1)$

(b) $(-2, 1/2)$

▼ **EXAMPLE 3**   Solve the quadratic inequality $r^2 + 3r \geq 4$.

Solution First rewrite the inequality so that one side is 0.

$$r^2 + 3r \geq 4$$
$$r^2 + 3r - 4 \geq 0 \qquad \text{Add } -4 \text{ to both sides.}$$

Now solve the corresponding equation (which amounts to finding the x-intercepts of $y = r^3 + 3r - 4$).

$$r^2 + 3r - 4 = 0$$
$$(r - 1)(r + 4) = 0$$
$$r = 1 \qquad \text{or} \qquad r = -4$$

These numbers separate the number line into three regions, as shown in Figure 2.52 on the next page. Test a number from each region.

Let $x = -5$ from region **A:** $(-5)^2 + 3(-5) - 4 = 6 > 0$.

Let $x = 0$ from region **B:** $(0)^2 + 3(0) - 4 = -4 < 0$.

Let $x = 2$ from region **C:** $(2)^2 + 3(2) - 4 = 6 > 0$.

We want the inequality to be positive or 0. The solution includes numbers in region A and in region C, as well as -4 and 1, the endpoints. The solution, which

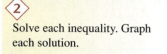

② Solve each inequality. Graph each solution.

(a) $k^2 + 2k - 15 \geq 0$

(b) $3m^2 + 7m \geq 6$

Answers:

(a) All numbers in $(-\infty, -5]$ or $[3, \infty)$

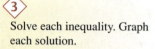

(b) All numbers in $(-\infty, -3]$ or $[2/3, \infty)$

③ Solve each inequality. Graph each solution.

(a) $m^3 - 9m > 0$

(b) $2k^3 - 50k \leq 0$

Answers:

(a) All numbers in $(-3, 0)$ or $(3, \infty)$

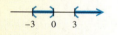

(b) All numbers in $(-\infty, -5]$ or $[0, 5]$

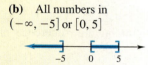

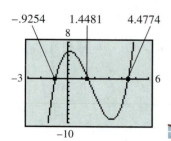

FIGURE 2.54

includes all numbers in the interval $(-\infty, -4]$ or the interval $[1, \infty)$, is graphed in Figure 2.52. ▲ ②

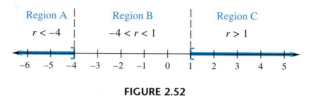

FIGURE 2.52

▼ **EXAMPLE 4**   Solve $q^3 - 4q > 0$.

Solution Solve the corresponding equation by factoring.

$$q^3 - 4q = 0$$
$$q(q^2 - 4) = 0$$
$$q(q + 2)(q - 2) = 0$$

$q = 0$ or $q + 2 = 0$ or $q - 2 = 0$
$q = 0$ $q = -2$ $q = 2$

These three numbers separate the number line into the four regions shown in Figure 2.53. Test a number from each region.

A: If $q = -3$, $(-3)^3 - 4(-3) = -15 < 0$.
B: If $q = -1$, $(-1)^3 - 4(-1) = 3 > 0$.
C: If $q = 1$, $(1)^3 - 4(1) = -3 < 0$.
D: If $q = 3$, $(3)^3 - 4(3) = 15 > 0$.

The numbers that make the polynomial positive are in the interval $(-2, 0)$ or the interval $(2, \infty)$, as graphed in Figure 2.53. ▲ ③

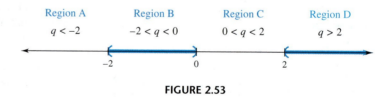

FIGURE 2.53

A graphing calculator can be used to solve inequalities without the need to evaluate at a test number in each interval. A calculator is also useful for finding approximate solutions when the x-intercepts of the graph cannot be found algebraically.

EXAMPLE 5   Use a graphing calculator to solve $x^3 - 5x^2 + x + 6 > 0$.

Solution Begin by graphing $y = x^3 - 5x^2 + x + 6$ (Figure 2.54). Find the x-intercepts by solving $x^3 - 5x^2 + x + 6 = 0$. Since this cannot readily be done algebraically, use the graphical root finder to determine that the solutions (x-intercepts) are approximately $-.9254$, 1.4481, and 4.4774.

The graph is above the x-axis when $-.9254 < x < 1.4481$ and also when $x > 4.4774$. Therefore, the approximate solutions of the inequality are all numbers in the interval $(-.9254, 1.4481)$ or the interval $(4.4774, \infty)$. ▲ ◆4

Use graphical methods to find approximate solutions of these inequalities.

(a) $x^2 - 6x + 2 > 0$

(b) $x^2 - 6x + 2 < 0$

Answers:

(a) All numbers in $(-\infty, .3542)$ or $(5.6458, \infty)$

(b) All numbers in $(.3542, 5.6458)$

Rational Inequalities

Inequalities with quotients of algebraic expressions are called **rational inequalities.** These inequalities can be solved in much the same way as polynomial inequalities can.

▼ **EXAMPLE 6**   Solve the rational inequality $\dfrac{5}{x + 4} \geq 1$.

Solution Write an equivalent inequality with one side equal to 0.

$$\frac{5}{x + 4} \geq 1$$

$$\frac{5}{x + 4} - 1 \geq 0$$

Write the left side as a single fraction.

$$\frac{5}{x + 4} - \frac{x + 4}{x + 4} \geq 0 \qquad \text{Get a common denominator.}$$

$$\frac{5 - (x + 4)}{x + 4} \geq 0 \qquad \text{Subtract fractions.}$$

$$\frac{5 - x - 4}{x + 4} \geq 0 \qquad \text{Distributive property}$$

$$\frac{1 - x}{x + 4} \geq 0$$

The quotient can change sign only at places where the denominator is 0 or the numerator is 0. (In graphical terms, these are the only places where the graph of $y = \dfrac{1 - x}{x + 4}$ can change from above the x-axis to below.) This happens when

$$1 - x = 0 \qquad \text{or} \qquad x + 4 = 0$$
$$x = 1 \qquad \text{or} \qquad x = -4.$$

As in the earlier examples, the numbers -4 and 1 divide the x-axis into three regions. Test a number from each of these regions:

$$x < -4 \qquad \text{Let } x = -5: \quad \frac{1 - (-5)}{-5 + 4} = -6 < 0.$$

$$-4 < x < 1 \qquad \text{Let } x = 0: \quad \frac{1 - 0}{0 + 4} = \frac{1}{4} > 0.$$

$$x > 1 \qquad \text{Let } x = 2: \quad \frac{1 - 2}{2 + 4} = -\frac{1}{6} < 0.$$

Solve each inequality.

(a) $\dfrac{3}{x-2} \geq 4$

(b) $\dfrac{p}{1-p} < 3$

(c) Why is 2 excluded from the solution in part (a)?

Answers:

(a) $(2, 11/4]$

(b) All numbers in $(-\infty, 3/4)$ or $(1, \infty)$

(c) When $x = 2$, the fraction is undefined.

6

Solve each rational inequality.

(a) $\dfrac{3y-2}{2y+5} < 1$

(b) $\dfrac{3c-4}{2-c} \geq -5$

Answers:

(a) $(-5/2, 7)$

(b) All numbers in $(-\infty, 2)$ or $[3, \infty)$

7

(a) Solve the inequality

$$\frac{5}{x+4} \geq 1$$

by first multiplying both sides by $x + 4$.

(b) Show that this method produces a wrong answer by testing $x = -5$.

Answer:

(a) $x \leq 1$

(b) $x = -5$ is a solution of $x \leq 1$, but not of the original inequality $\dfrac{5}{x+4} \geq 1$. For the correct solution see Example 6.

The test shows that numbers in $(-4, 1)$ satisfy the inequality. With a quotient, the endpoints must be considered individually to make sure that no denominator is 0. In this inequality, -4 makes the denominator 0, while 1 satisfies the given inequality. Write the solution as $(-4, 1]$. ▲

CAUTION As suggested by Example 6, be very careful with the endpoints of the intervals in the solution of rational inequalities. 5

▼ **EXAMPLE 7**   Solve $\dfrac{2x-1}{3x+4} < 5$.

Solution Write an equivalent inequality with 0 on one side. Begin by subtracting 5 on both sides and combining the terms on the left into a single fraction.

$$\frac{2x-1}{3x+4} < 5$$

$$\frac{2x-1}{3x+4} - 5 < 0 \qquad \text{Get 0 on one side.}$$

$$\frac{2x-1-5(3x+4)}{3x+4} < 0 \qquad \text{Subtract.}$$

$$\frac{-13x-21}{3x+4} < 0 \qquad \text{Combine terms.}$$

Set the numerator and denominator each equal to 0 and solve the two equations.

$$-13x - 21 = 0 \qquad \text{or} \qquad 3x + 4 = 0$$

$$x = -\frac{21}{13} \quad \text{or} \qquad\qquad x = -\frac{4}{3}$$

Use the values $-21/13$ and $-4/3$ to divide the number line into three intervals. Test a number from each interval in the inequality. The quotient is negative for numbers in $(-\infty, -21/13)$ or $(-4/3, \infty)$. Neither endpoint satisfies the given inequality. ▲ 6

CAUTION In problems like those in Example 6 and 7, you should *not* begin by multiplying both sides by the denominator to simplify the inequality. Doing so will usually produce a wrong answer. For the reason, see Exercise 38. For an example, see side problem 7. 7

TECHNOLOGY TIP   Rational inequalities can also be solved graphically. In Example 6, for instance, after rewriting the original inequality in the form $\dfrac{1-x}{x+4} \geq 0$, determine the values of x that make the numerator and denominator 0 (namely, $x = 1$ and $x = -4$). Then graph $\dfrac{1-x}{x+4}$. Figure 2.55 shows that the graph is above the x-axis when it is between the

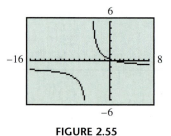

FIGURE 2.55

vertical asymptote at $x = -4$ and the x-intercept at $x = 1$. So the solution of the inequality is the interval $(-4, 1]$. [When the values that make the numerator and denominator 0 cannot be found algebraically, as they were here, you can use the root finder to approximate them.]

◆ 2.5 Exercises

Solve each of these quadratic inequalities. Graph the solutions on the number line. (See Examples 2 and 3.)

1. $(x + 4)(2x - 3) \leq 0$

2. $(5y - 1)(y + 3) > 0$

3. $r^2 + 4r > -3$

4. $z^2 + 6z > -8$

5. $4m^2 + 7m - 2 \leq 0$

6. $6p^2 - 11p + 3 \leq 0$

7. $4x^2 + 3x - 1 > 0$

8. $3x^2 - 5x > 2$

9. $x^2 \leq 36$

10. $y^2 \geq 9$

11. $p^2 - 16p > 0$

12. $r^2 - 9r < 0$

Solve these inequalities. (See Example 4.)

13. $x^3 - 9x \geq 0$

14. $p^3 - 25p \leq 0$

15. $(x + 6)(x + 1)(x - 4) \geq 0$

16. $(2x + 5)(x^2 - 1) \leq 0$

17. $(x + 3)(x^2 - 2x - 3) < 0$

18. $x^3 - 2x^2 - 3x \leq 0$

19. $6k^3 - 5k^2 < 4k$

20. $2m^3 + 7m^2 > 4m$

21. A student solved the inequality $p^2 < 16$ by taking the square root of both sides to get $p < 4$. She wrote the solution as $(-\infty, 4)$. Is her solution correct?

Use a graphing calculator to solve these inequalities. (See Example 5.)

22. $6x + 7 < 2x^2$

23. $.5x^2 - 1.2x < .2$

24. $3.1x^2 - 7.4x + 3.2 > 0$

25. $x^3 - 2x^2 - 5x + 7 \geq 2x + 1$

26. $x^4 - 6x^3 + 2x^2 < 5x - 2$

27. $2x^4 + 3x^3 < 2x^2 + 4x - 2$

28. $x^5 + 5x^4 > 4x^3 - 3x^2 - 2$

Solve the following rational inequalities. (See Examples 6 and 7.)

29. $\dfrac{r - 4}{r - 1} \geq 0$

30. $\dfrac{z + 6}{z + 4} > 1$

31. $\dfrac{a - 2}{a - 5} < -1$

32. $\dfrac{1}{3k - 5} < \dfrac{1}{3}$

33. $\dfrac{1}{p - 2} < \dfrac{1}{3}$

34. $\dfrac{7}{k + 2} \geq \dfrac{1}{k + 2}$

35. $\dfrac{5}{p + 1} > \dfrac{12}{p + 1}$

36. $\dfrac{x^2 - 4}{x} > 0$

37. $\dfrac{x^2 - x - 6}{x} < 0$

38. Determine whether $x + 4$ is positive or negative when
 (a) $x > -4$
 (b) $x < -4$
 (c) If you multiply both sides of the inequality
 $\dfrac{1 - x}{x + 4} \geq 0$ by $x + 4$, should you change the direction of the inequality sign. If so, when?
 (d) Explain how you can use parts (a)–(c) to solve
 $\dfrac{1 - x}{x + 4} \geq 0$ correctly.

Use a graphing calculator to solve these inequalities. You may have to approximate the roots of the numerators or denominators.

39. $\dfrac{2x^2 + x - 1}{x^2 - 4x + 4} \leq 0$

40. $\dfrac{x^3 - 3x^2 + 5x - 29}{x^2 - 7} > 3$

Work these problems.

41. **Business** An analyst has found that his company's profits, in hundreds of thousands of dollars, are given by $P = 3x^2 - 35x + 50$, where x is the amount, in hundreds of dollars, spent on advertising. For what values of x does the company make a profit?

42. **Business** The commodities market is highly unstable; money can be made or lost quickly on investments in soybeans, wheat, and so on. Suppose that an investor kept track of her total profit P at time t, in months, after she began investing, and she found that $P = 4t^2 - 29t + 30$. Find the time intervals during which she has been ahead.

43. **Business** The manager of a large apartment complex has found that the profit is given by $P = -x^2 + 250x - 15,000$, where x is the number of apartments rented. For what values of x does the complex produce a profit?

44. **Health** Annual health care expenditures (in billions of dollars) in the United States can be approximated by $3x^2 + 39x - 900$, where $x = 0$ corresponds to 1990.* In what years since 1990 were health care expenditures $1.5 trillion or less?

45. **Business** The revenue (in millions of dollars) from downloading custom ring tones for cell phones is approximated by $26x^2 - 62x + 60$ $(x \geq 1)$, where x is the number of years since 2000.† If this model remains accurate, in what years will revenue exceed $400,000,000? (*Hint:* $340 = 5 \cdot 68$, which may be helpful in factoring.)

Use a graphing calculator or other technology to do these exercises.

46. **Social Science** According to data from the FBI Uniform Crime Report, the number of violent crimes each year from 1990 to 2005 (in millions) can be approximated by $.0015x^3 - .0287x^2 + .098x + 1.826$, where $x = 0$ corresponds to 1990. During what years were there at least 1.5 million violent crimes?

47. **Health** Projections through 2050 from the Alzheimer's Association suggestion that the number of people with Alzheimer's disease (in millions) can be approximated by $.005x^2 - .052x + 4.7$, where x is the number of years since 2000. In what years will there be more than 5 million Alzheimer's patients?

48. If all planned projects go into operation, China's steel production (in millions of metric tons) from 2000 through 2010 can be approximated by $-1.47x^2 + 48.66x + 113.26$, where x is the number of years since 2000. During what years will production exceed 350 million metric tons? (For comparison purposes, U.S. steel production during this period is expected to remain steady at about 100 million metric tons per year.)

---

*Based on data from the U.S. Centers for Medicare and Medicaid Services.

†Based on data from Jupiter Research.

Chapter 2 Summary

Key Terms and Symbols*

2.1 Cartesian coordinate system
x-axis
y-axis
origin
ordered pair
x-coordinate
y-coordinate
quadrant
solution of an equation
graph
x-intercept
y-intercept
[viewing window]

[trace]
[graphical root finder]
[maximum/minimum finder]
graph reading

2.2 Δx change in x
Δy change in y
slope
slope-intercept form
parallel and perpendicular lines
point-slope form
linear equations
general form

2.3 linear models
residual
[least squares regression line]
[linear regression]
[correlation coefficient]

2.4 linear inequality
properties of inequality
absolute value inequality

2.5 polynomial inequality
algebraic solution methods
[graphical solution methods]
rational inequality

---

*Term in brackets deal with material in which a graphing calculator or other technology is used.

Key Concepts

The **slope** of the line through the points (x_1, y_1) and (x_2, y_2), where $x_1 \neq x_2$, is
$$m = \frac{y_2 - y_1}{x_2 - x_1}.$$

The line with equation $y = mx + b$ has slope m and y-intercept b.

The line with equation $y - y_1 = m(x - x_1)$ has slope m and goes through (x_1, y_1).

The line with equation $ax + by = c$ (with $a \neq 0, b \neq 0$) has x-intercept c/a and y-intercept c/b.

The line with equation $x = k$ is vertical, with x-intercept k, no y-intercept, and undefined slope.

The line with equation $y = k$ is horizontal, with y-intercept k, no x-intercept, and slope 0.

Nonvertical **parallel lines** have the same slope, and **perpendicular lines,** if neither is vertical, have slopes with a product of -1.

Chapter 2 Review Exercises

Which of the ordered pairs $(-2, 3)$, $(0, -5)$, $(2, -3)$, $(3, -2)$, $(4, 3)$, $(7, 2)$ *are solutions of the given equation?*

1. $y = x^2 - 2x - 5$ **2.** $x - y = 5$

Sketch the graph of each equation.

3. $5x - 3y = 15$ **4.** $2x + 7y - 21 = 0$

5. $y + 3 = 0$ **6.** $y - 2x = 0$

7. $y = .25x^2 + 1$ **8.** $y = \sqrt{x + 4}$

9. The following temperature graph was recorded in Bratenahl, Ohio.

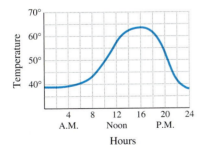

(a) At what times during the day was the temperature over 55°?

(b) When was the temperature below 40°?

10. Greenville, South Carolina, is 500 miles south of Bratenahl, Ohio, and its temperature is 7° higher all day long (see the graph in Exercise 9.) At what time was the temperature in Greenville the same as the temperature at noon in Bratenahl?

11. In your own words, define the slope of a line.

In Exercises 12–21, find the slope of the line

12. Through $(-1, 4)$ and $(2, 3)$

13. Through $(5, -3)$ and $(-1, 2)$

14. Through $(7, -2)$ and the origin

15. Through $(8, 5)$ and $(0, 3)$

16. $2x + 3y = 30$

17. $4x - y = 7$

18. $x + 5 = 0$

19. $y = 3$

20. Parallel to $3x + 8y = 0$

21. Perpendicular to $x = 3y$

22. Graph the line through $(0, 5)$ with slope $m = -2/3$.

23. Graph the line through $(-4, 1)$ with $m = 3$.

24. What information is needed to determine the equation of a line?

Find an equation for each of the following lines.

25. Through $(5, -1)$, slope $2/3$

26. Through $(8, 0)$, slope $-1/4$

27. Through $(5, -2)$ and $(1, 3)$

28. Through $(2, -3)$ and $(-3, 4)$

29. Undefined slope, through $(-1, 4)$

30. Slope 0, through $(-2, 5)$

31. x-intercept -3, y-intercept 5

32. Here is a sample SAT question: which of the following is an equation of the line that has a y-intercept of 2 and an x-intercept of 3?

(a) $-2x + 3y = 4$
(b) $-2x + 3y = 6$
(c) $2x + 3y = 4$
(d) $2x + 3y = 6$
(e) $3x + 2y = 6$

33. **Health** According to the National Center for Health Statistics, 25.3% of adults were smokers in 1990. By 2002, only 22.5% smoked.

(a) Assuming the decline in smoking is linear, write an equation that gives the percentage of smokers in year x, with $x = 0$ corresponding to 1990.
(b) Is the slope of the line positive or negative? Why?
(c) Estimate the percentage of smokers in 2008.

34. **Social Science** In 1990 the percentage of children living with two parents was 72.5, and in 2002 it was 68.7.*

(a) Assuming that the decrease was linear, write an equation relating the percentage y and year x, with 1980 corresponding to $x = 0$.
(b) Graph the equation for the years 1990 to 2010.
(c) Estimate the percentage of children living in two-parent families in 2010.

35. **Business** The following table shows the average hourly earnings of U.S. production workers:†

Year	1998	1999	2001	2002	2003
Earnings	13.00	13.47	14.53	14.95	15.35

(a) Use the first and last data points to find a linear model for the data, with $x = 0$ corresponding to 1990.
(b) If you have access to appropriate technology, find the least squares regression line for the data.
(c) Use the models from part (a) and/or (b) to estimate the hourly earnings in 2000. The actual average in 2000 was $14.00. How far off is the model?
(d) Use the model from part (a) and/or (b) to estimate the hourly earnings in 2008.

36. **Health** Over the next eight decades, Medicare costs as a percentage of Gross Domestic Product (GDP) are pro-

jected to be given by $.14125x + 2.3$, where $x = 0$ corresponds to 2000. In what years will Medicare costs be between 6.82% and 10.21% of GDP?

37. **Transportation** The table, based on data from the American Automobile Manufacturers Association, shows the average expenditure for a new car in various years.

Year	1990	1995	1997	1998
Expenditure	14,371	17,959	19,531	20,370

Year	1999	2000	2001	2002
Expenditure	20,673	20,909	21,258	21,440

(a) Let $x = 0$ correspond to 1990, and find the least squares regression line for the data.
(b) What is the correlation coefficient? Does it indicate that the line is a good fit for the data?
(c) Estimate the expenditure on a new car in 2008.

38. **Business** The following table, from the U.S. Federal Deposit Insurance Corporation, shows the net after-tax income (in billions of dollars) of insured commercial banks:

Year	1990	1995	1997	1998
Income	11.3	56.4	67.9	72.0

Year	1999	2000	2002	2003
Income	82.4	81.7	105.1	120.6

(a) Let $x = 0$ correspond to 1990, and find the least squares regression line for the data.
(b) What is the correlation coefficient? Does it indicate that the line is a good fit for the data?
(c) Estimate the income for the year 2001.

Solve each inequality.

39. $-6x + 3 < 2x$

40. $12z \geq 5z - 7$

41. $2(3 - 2m) \geq 8m + 3$

42. $6p - 5 > -(2p + 3)$

43. $-3 \leq 4x - 1 \leq 7$

44. $0 \leq 3 - 2a \leq 15$

*U.S. Census Bureau.

†U.S. Bureau of Labor Statistics.

45. $|b| \leq 8$

46. $|a| > 7$

47. $|2x - 7| \geq 3$

48. $|4m + 9| \leq 16$

49. $|5k + 2| - 3 \leq 4$

50. $|3z - 5| + 2 \geq 10$

51. Health Here is a sample SAT question: for pumpkin carving, Mr. Sephera will not use pumpkins that weigh less than 2 pounds or more than 10 pounds. If x represents the weight of a pumpkin (in pounds) that he will *not* use, which of the following inequalities represents all possible values of x?

(a) $|x - 2| > 10$

(b) $|x - 4| > 6$

(c) $|x - 5| > 5$

(d) $|x - 6| > 4$

(e) $|x - 10| > 4$

52. Natural Science The number of milligrams of a certain substance per liter in samples of drinking water all tested within .05 of 40 milligrams per liter. Write this information as an inequality, using absolute value.

53. Business Total energy production in the Asia-Pacific region was 415 million metric tons of oil equivalent (Mtoe) in 1990 and 645 Mtoe in 2000 and is increasing linearly.*

(a) Find the linear equation that gives the energy production y in year x, with $x = 0$ corresponding to 1990.

(b) When is energy production less than 599 Mtoe?

(c) When is energy production more than 714 Mtoe?

54. Business One car rental firm charges $75 for a weekend rental (Friday afternoon through Monday morning) and gives unlimited mileage. A second firm charges $50, plus

5 cents per mile. For what range of miles driven is the second firm cheaper?

Solve each inequality.

55. $r^2 + r - 6 < 0$

56. $y^2 + 4y - 5 \geq 0$

57. $2z^2 + 7z \geq 15$

58. $3k^2 \leq k + 14$

59. $(x - 3)(x^2 + 7x + 10) \leq 0$

60. $(x + 4)(x^2 - 1) \geq 0$

61. $\dfrac{m + 2}{m} \leq 0$

62. $\dfrac{q - 4}{q + 3} > 0$

63. $\dfrac{5}{p + 1} > 2$

64. $\dfrac{6}{a - 2} \leq -3$

65. $\dfrac{2}{r + 5} \leq \dfrac{3}{r - 2}$

66. $\dfrac{1}{z - 1} > \dfrac{2}{z + 1}$

67. Social Science During the last century, the population of Cleveland, Ohio (in thousands), in year x was approximately $.06x^2 - 12.5x + 926$, where $x = 0$ corresponds to 1950. During what years since 1950 was the population at least 500,000?

*International Energy Agency, *Global Energy: The Changing Outlook*, 1992.

Case 2

Using Extrapolation to Predict Life Expectancy

One reason for developing a mathematical model is to make predictions. If your model is a least squares line, you can predict the y value corresponding to some new x value by substituting that x into an equation of the form $Y = mx + b$. (We use a capital Y to remind us that we're getting a predicted value rather than an actual data value.) Data analysts distinguish two very different kinds of prediction: *interpolation* and *extrapolation*. An interpolation uses a new x inside the x range of your original data. For example, if you have inflation data at five-year intervals from 1950 to 2000, estimating the rate of inflation in 1957 is an interpolation problem. But if you use the same data to estimate what the inflation rate was in 1920, or what it will be in 2020, you are extrapolating.

In general, interpolation is much safer than extrapolation, because data that are approximately linear over a short interval may be nonlinear over a larger interval. One way to detect nonlinearity is to look at *residuals,* which are the differences between the actual data values and the values predicted by the line of best fit. A simple example is shown in Figure 1.

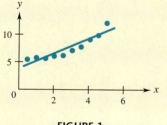

FIGURE 1

The regression equation for the best-fit line in Figure 1 is $Y = 3.431 + 1.334x$. Since the r value for this regression line is 0.93, our linear model fits the data very well. But we might notice that the predictions are a bit low at the ends and high in the middle. We can get a better look at this pattern by plotting the residuals. To find them, we put each value of the independent variable into the regression equation, calculate the predicted value Y, and subtract it from the actual y value. The residual

plot in Figure 2 has the vertical axis rescaled to exaggerate the pattern. The residuals indicate that our data have a nonlinear U-shaped component that isn't captured by the linear fit. Extrapolating from this data set is probably not a good idea; our linear prediction for the value of y when x is 10 may be much too low.

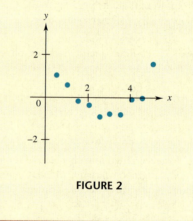

FIGURE 2

◆ Exercises

*The following table gives the life expectancy at birth of females born in the United States in various years from 1950 to 1995:**

Year of Birth	Life Expectancy in Years
1950	71.3
1960	73.1
1970	74.7
1980	77.4
1985	78.2
1990	78.8
1995	78.9

*Health, United States, 1998, Centers for Disease Control.

1. If you have appropriate technology, verify that the least squares regression line that models these data (with coefficients rounded) is $Y = .1827x + 62.2251$, where Y is the life expectancy of a woman born in year x and $x = 0$ corresponds to 1900.

2. Use the regression equation in Exercise 1 to guess a value for the life expectancy of females born in 1900.

3. Compare your answer with the actual life expectancy for females born in 1900, which was 48.3 years. Are you surprised?

4. Find the life expectancy (rounded to one decimal place) predicted by the regression equation for each year in the table, and subtract it from the actual value in the second column. This gives you a table of residuals. Plot your residuals as points on a graph.

5. Now look at the residuals as a fresh data set, and see if you can sketch the graph of a smooth function that fits the residuals well. How easy do you think it will be to predict the life expectancy at birth of females born in 2010?

6. What will happen if you try linear regression on the *residuals?* If you're not sure, use technology to find the regression equation for the residuals. Why does this result make sense?

7. Since most of the females born in 1985 are still alive, how did the Public Health Service come up with a life expectancy of 78.2 years for these women?

Chapter 3   Functions and Graphs

The modern world is overwhelmed with data—from the cost of college, to mortgage rates, to health care expenditures and hundreds of other pieces of information. Functions enable us to construct mathematical models that can sometimes be used to estimate outcomes. Graphs of functions allow us to visualize a situation and to detect trends more easily. See Example 11 on page 145 and Exercises 57 and 58 on page 201.

Functions are an extremely useful way of describing many real-world situations in which the value of one quantity varies with, depends on, or determines the value of another. In this chapter you will be introduced to functions, learn how to use functional notation, develop skill in constructing and interpreting the graphs of functions, and, finally, learn to apply this knowledge in a variety of situations.

3.1 Functions

To understand the origin of the concept of function, we consider some "real-life" situations in which one numerical quantity depends on, corresponds to, or determines another.

▼ **EXAMPLE 1** The amount of income tax you pay depends on the amount of your income. The way in which the income determines the tax is given by the tax law. ▲

▼ **EXAMPLE 2** The graph in Figure 3.1 shows the temperatures in Cleveland, Ohio, on April 11, 2001, as recorded by the U.S. Weather Bureau at Hopkins Airport. The graph shows the temperature that corresponds to each time during the day.

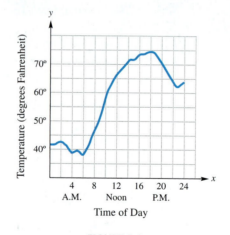

FIGURE 3.1

▼ **EXAMPLE 3** Suppose a rock is dropped straight down from a high point. From physics, we know that the distance traveled by the rock in t seconds is $16t^2$ feet. So the distance depends on the time. ▲

These examples share a couple of features. First, each involves two sets of numbers, which we can think of as inputs and outputs. Second, in each case, there is a rule by which each input determines an output, as summarized here:

	Set of Inputs	Set of Outputs	Rule
Example 1	All incomes	All tax amounts	The tax law
Example 2	Hours since midnight	Temperatures during the day	Time-temperature graph
Example 3	Seconds elapsed after dropping the rock	Distances the rock travels	Distance $= 16t^2$

Each of these examples may be mentally represented by an idealized calculator that has a single operation key: A number is entered [*input*], the rule key is pushed [*rule*], and an answer is displayed [*output*]. The formal definition of a function incorporates these same common features (input-rule-output), with a slight change in terminology.

> A **function** consists of a set of inputs called the **domain,** a set of outputs called the **range,** and a rule by which each input determines *exactly one* output.

▼ **EXAMPLE 4** Find the domains and ranges for the functions in Examples 1 and 2.

Solution In Example 1, the domain consists of all possible income amounts and the range consists of all possible tax amounts.

In Example 2, the domain is the set of hours in the day—that is, all real numbers in the interval [0, 24]. The range consists of the temperatures that actually occur during that day. Figure 3.1 suggests that these are all numbers in the interval [38, 74]. ▲ ①

Be sure that you understand the phrase "exactly one output" in the definition of the rule of a function. In Example 2, for instance, each time of day (input) determines exactly one temperature (output)—you can't have two different temperatures at the same time. However, it is quite possible to have the same temperature (output) at two different times (inputs). In other words,

**In a function, each input produces a single output,
but different inputs may produce the same output.**

▼ **EXAMPLE 5** Which of the following rules describe functions?

(a) Use the optical reader at the checkout counter of the supermarket to convert codes to prices.

Solution For each code, the reader produces exactly one price, so this is a function.

(b) Enter a number in a calculator and press the x^2 key.

Solution This is a function, because the calculator produces just one number x^2 for each number x that is entered.

(c) Assign to each number x the number y given by this table:

x	1	1	2	2	3	3
y	3	−3	−5	−5	8	−8

Solution Since $x = 1$ corresponds to more than one y-value (as does $x = 3$), this table does not define a function.

① Find the domain and range of the function in Example 3, assuming (unrealistically) that the rock can fall forever.

Answer:

The domain consists of all possible times—that is, all nonnegative real numbers. The range consists of all possible distances; thus, the range is also the set of all nonnegative real numbers.

2

Do the following define
functions?

(a) The correspondence defined by the rule $y = x^2 + 5$,
where x is the input and y is the
output.

(b) Enter a nonzero number
in a calculator and press the
$1/x$ key.

(c) The correspondence
between a computer, x, and
several users of the computer, y

Answers:

(a) Yes

(b) Yes

(c) No

3

Do the following define y as a
function of x?

(a) $y = -6x + 1$

(b) $y = x^2$

(c) $x = y^2 - 1$

(d) $y < x + 2$

Answers:

(a) Yes

(b) Yes

(c) No

(d) No

(d) Assign to each number x, the number y given by the equation $y = 3x - 5$.

Solution Because the equation determines a unique value of y for each value
of x, it defines a function. ▲ ②

The equation $y = 3x - 5$ in part (d) of Example 4 defines a function, with x as
input and y as output, because each value of x determines a *unique* value of y. In
such a case, the equation is said to **define y as a function of x.**

▼ **EXAMPLE 6**   Decide whether each of the following equations defines y as a
function of x.

(a) $y = -4x + 11$

Solution For a given value of x, calculating $-4x + 11$ produces exactly one
value of y. (For example, if $x = -7$, then $y = -4(-7) + 11 = 39$.) Because one
value of the input variable leads to exactly one value of the output variable,
$y = -4x + 11$ defines y as a function of x.

(b) $y^2 = x$

Solution Suppose $x = 36$. Then $y^2 = x$ becomes $y^2 = 36$, from which $y = 6$
or $y = -6$. Since one value of x can lead to two values of y, $y^2 = x$ does *not* define
y as a function of x. ▲ ③

Almost all the functions in this book are defined by formulas or equations, as in
part (a) of Example 6. The domain of such a function is determined by the following
agreement on domains.

Unless otherwise stated, assume that the domain of any function defined by a
formula or an equation is the largest set of real numbers (inputs) that each produce a real number as output.

▼ **EXAMPLE 7**   Each of the following equations defines y as a function of x. Find
the domain of each function.

(a) $y = x^4$

Solution Any number can be raised to the fourth power, so the domain is the
set of all real numbers, which is sometimes written as $(-\infty, \infty)$.

(b) $y = \sqrt{x - 6}$

Solution For y to be a real number, $x - 6$ must be nonnegative. This happens
only when $x - 6 \geq 0$—that is, when $x \geq 6$. So the domain is the interval $[6, \infty)$.

(c) $y = \sqrt{4 - x}$

Solution For y to be a real number here, we must have $4 - x \geq 0$, which is
equivalent to $x \leq 4$. So the domain is the interval $(-\infty, 4]$.

(d) $y = \dfrac{1}{x + 3}$

Solution Because the denominator cannot be 0, $x \neq -3$ and the domain consists of all numbers in the interval

$$(-\infty, -3) \quad \text{or} \quad (-3, \infty).$$

(e) $y = \dfrac{\sqrt{x}}{x^2 - 3x + 2}$

Solution The numerator is defined only when $x \geq 0$. The domain cannot contain any numbers that make the denominator 0—that is, the numbers with

$$x^2 - 3x + 2 = 0$$

$$(x - 1)(x - 2) = 0 \qquad \text{Factor}$$

$$x - 1 = 0 \quad \text{or} \quad x - 2 = 0 \qquad \text{Zero-factor property}$$

$$x = 1 \quad \text{or} \qquad x = 2$$

Therefore, the domain consists of all nonnegative real numbers except 1 and 2. ▲

⟨4⟩

Give the domain.

(a) $y = 3x + 1$

(b) $y = x^2$

(c) $y = \sqrt{-x}$

(d) $y = \dfrac{3}{x^2 - 1}$

Answers:

(a) $(-\infty, \infty)$

(b) $(-\infty, \infty)$

(c) $(-\infty, 0]$

(d) All real numbers except 1 and -1

Functional Notation

In actual practice, functions are seldom presented in the style of domain-rule-range, as they have been here. Functions are usually denoted by a letter such as f. If x is an input, then $f(x)$ denotes the output number that the function f produces from the input x. The symbol $f(x)$ is read "f of x." The rule is usually given by a formula, such as $f(x) = \sqrt{x^2 + 1}$. This formula can be thought of as a set of directions.

Name of function   Input number

$$f(x) = \sqrt{x^2 + 1}$$

Output number   Directions that tell you what to do with input x in order to produce the corresponding output f(x); namely, "square it, add 1, and take the square root of the result."

For example, to find $f(3)$ (the output number produced by the input 3), simply replace x by 3 in the formula:

$$f(3) = \sqrt{3^2 + 1}$$

$$= \sqrt{10}.$$

Similarly, replacing x by -5 and 0 shows that

$$f(-5) = \sqrt{(-5)^2 + 1} \quad \text{and} \quad f(0) = \sqrt{0^2 + 1}$$

$$= \sqrt{26} \qquad\qquad\qquad = 1.$$

These directions can be applied to any quantities, such as $a + b$ or c^4 (where a, b, c are real numbers). Thus, to compute $f(a + b)$, the output corresponding to input $a + b$, we square the input [obtaining $(a + b)^2$], add 1 [obtaining $(a + b)^2 + 1$], and take the square root of the result:

$$f(a + b) = \sqrt{(a + b)^2 + 1}$$

$$= \sqrt{a^2 + 2ab + b^2 + 1}.$$

Similarly, the output $f(c^4)$ corresponding to the input c^4 is computed by squaring the input $[(c^4)^2]$, adding 1 $[(c^4)^2 + 1]$, and taking the square root of the result:

$$f(c^4) = \sqrt{(c^4)^2 + 1}$$
$$= \sqrt{c^8 + 1}.$$

▼ **EXAMPLE 8** Let $g(x) = -x^2 + 4x - 5$. Find each of the following.

(a) $g(-2)$

Solution Replace x with -2.

$$g(-2) = -(-2)^2 + 4(-2) - 5$$
$$= -4 - 8 - 5$$
$$= -17$$

(b) $g(x + h)$

Solution Replace x by the quantity $x + h$ in the rule of g.

$$g(x + h) = -(x + h)^2 + 4(x + h) - 5$$
$$= -(x^2 + 2xh + h^2) + (4x + 4h) - 5$$
$$= -x^2 - 2xh - h^2 + 4x + 4h - 5$$

(c) $g(x + h) - g(x)$

Solution Use the result from part (b) and the rule for $g(x)$.

$$g(x + h) - g(x) = \overbrace{(-x^2 - 2xh - h^2 + 4x + 4h - 5)}^{g(x+h)} - \overbrace{(-x^2 + 4x - 5)}^{g(x)}$$
$$= -2xh - h^2 + 4h$$

(d) $\dfrac{g(x + h) - g(x)}{h}$ (assuming that $h \neq 0$)

Solution The numerator was found in part (c). Divide it by h as follows:

$$\frac{g(x + h) - g(x)}{h} = \frac{-2xh - h^2 + 4h}{h}$$
$$= \frac{h(-2x - h + 4)}{h}$$
$$= -2x - h + 4. \; \blacktriangle$$

The quotient found in Example 8(d),

$$\frac{g(x + h) - g(x)}{h},$$

is called the **difference quotient** of the function g. Difference quotients are important in calculus and will appear again in Chapter 11. ⑤

⟨5⟩

Let $f(x) = 5x^2 - 2x + 1$.
Find the following.

(a) $f(1)$

(b) $f(3)$

(c) $f(1 + 3)$

(d) $f(1) + f(3)$

(e) $f(m)$

(f) $f(x + h) - f(x)$

(g)

$$\frac{f(x + h) - f(x)}{h} \quad (h \neq 0)$$

Answers:

(a) 4

(b) 40

(c) 73

(d) 44

(e) $5m^2 - 2m + 1$

(f) $10xh + 5h^2 - 2h$

(g) $10x + 5h - 2$

CAUTION Functional notation is *not* the same as ordinary algebraic notation. You cannot simplify an expression such as $f(x + h)$ by writing $f(x) + f(h)$. To see why, consider the answers to side problems 5(c) and (d) on the opposite page, which show that

$$f(1 + 3) \neq f(1) + f(3).$$

Applications

▼ **EXAMPLE 9**   If you were a single person in Connecticut in 2004 with a taxable income of x dollars, then your state income tax T was determined by the rule

$$T(x) = \begin{cases} .03x & \text{if } 0 \leq x \leq 10{,}000 \\ 300 + .05(x - 10{,}000) & \text{if } x > 10{,}000 \end{cases}$$

Find the income tax paid by a single person with the given taxable income.

(a) $8500

Solution We must find $T(8500)$. Since 8500 is less than 10,000, the first part of the rule applies.

$$T(x) = .03x$$
$$T(8500) = .03(8500) = \$255 \qquad \text{Let } x = 8500.$$

(b) $25,000

Solution Now we must find $T(25{,}000)$. Since 25,000 is greater than 10,000, the second part of the rule applies.

$$T(x) = 300 + .05(x - 10{,}000)$$
$$T(25{,}000) = 300 + .05(25{,}000 - 10{,}000) \qquad \text{Let } x = 25{,}000$$
$$= 300 + .05(15{,}000) \qquad \text{Simplify}$$
$$= 300 + 750 = \$1050 \; \blacktriangle \; \langle 6 \rangle$$

⟨6⟩
In Example 9, find the tax on each of these incomes.

(a) $46,785

(b) $6746

Answers:

(a) $2139.25

(b) $202.38

A function with a multipart rule, as in Example 9, is called a **piecewise-defined function.**

▼ **EXAMPLE 10**   Suppose the projected sales (in thousands of dollars) of a small company over the next ten years are approximated by the function

$$S(x) = .08x^4 - .04x^3 + x^2 + 9x + 54.$$

(a) What are the projected sales this year?

Solution The current year corresponds to $x = 0$ and the sales this year are given by $S(0)$. Substituting 0 for x in the rule of S, we see that $S(0) = 54$. So the current projected sales are $54,000.

(b) What will sales be in three years?

Solution The sales three years from now are given by $S(3)$, which can be computed by hand or with a calculator:

$$S(x) = .08x^4 - .04x^3 + x^2 + 9x + 54$$
$$S(3) = .08(3)^4 - .04(3)^3 + (3)^2 + 9(3) + 54 \qquad \text{Let } x = 3.$$
$$= 95.4.$$

Thus, sales are projected to be $95,400. ▲ ⑦

If you have a graphing calculator, you should learn to use its table feature to evaluate functions.

▼ **EXAMPLE 11** Use the table feature of a graphing calculator to find the projected sales of the company in Example 10 for years 4 through 10.

Enter the sales equation $y = .08x^4 - .04x^3 + x^2 + 9x + 54$ into the equation memory of the calculator (often called the Y= list). Check your instruction manual for how to set the table to start at $x = 4$ and go through $x = 10$. Then display the table, as in Figure 3.2. The figure shows that sales are projected to rise from $123,920 in year 4 to $1,004,000 in year 10. ▲

⑦

A developer estimates that the total cost of building x large apartment complexes in a year is approximated by

$$A(x) = x^2 + 80x + 60,$$

where $A(x)$ represents the cost in hundred thousands of dollars. Find the cost of building

(a) 4 complexes;

(b) 10 complexes.

Answers:

(a) $39,600,000

(b) $96,000,000

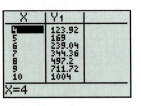

FIGURE 3.2

◆ **3.1 Exercises**

Which of the following rules define y as a function of x? (See Examples 2–6.)

1.

x	3	2	1	0	−1	−2	−3
y	9	4	1	0	1	4	9

2.

x	9	4	1	0	1	4	9
y	3	2	1	0	−1	−2	−3

3. $y = x^3$

4. $y = \sqrt{x - 1}$

5. $x = |y + 2|$

6. $x = y^2 + 3$

7. $y = \dfrac{-1}{x - 1}$

8. $y = \dfrac{4}{2x + 3}$

State the domain of each function. (See Example 7.)

9. $f(x) = 4x - 1$

10. $f(x) = 2x + 7$

11. $f(x) = x^4 - 1$

12. $f(x) = (2x + 5)^2$

13. $f(x) = \sqrt{-x} + 3$

14. $f(x) = \sqrt{5 - x}$

15. $g(x) = \dfrac{1}{x - 2}$

16. $g(x) = \dfrac{x}{x^2 + x - 2}$

17. $g(x) = \dfrac{x^2 + 4}{x^2 - 4}$

18. $g(x) = \dfrac{x^2 - 1}{x^2 + 1}$

19. $h(x) = \dfrac{\sqrt{x + 4}}{x^2 + x - 12}$

20. $h(x) = |5 - 4x|$

21. $g(x) = \begin{cases} 1/x & \text{if } x < 0 \\ \sqrt{x^2 + 1} & \text{if } x \geq 0 \end{cases}$

22. $f(x) = \begin{cases} 2x + 3 & \text{if } x < 4 \\ x^2 - 1 & \text{if } 4 \leq x \leq 10 \end{cases}$

For each of the following functions, find

(a) $f(4)$ **(b)** $f(-3)$ **(c)** $f(2.7)$ **(d)** $f(-4.9)$.

(See Examples 8 and 9.)

23. $f(x) = 8$

24. $f(x) = 0$

25. $f(x) = 2x^2 + 4x$ **26.** $f(x) = x^2 - 2x$
27. $f(x) = \sqrt{x + 3}$ **28.** $f(x) = \sqrt{5 - x}$
29. $f(x) = |x^2 - 6x - 4|$
30. $f(x) = |x^3 - x^2 + x - 1|$
31. $f(x) = \dfrac{\sqrt{x - 1}}{x^2 - 1}$ **32.** $f(x) = \sqrt{-x} + \dfrac{2}{x + 1}$

33. $f(x) = \begin{cases} x^2 & \text{if } x < 2 \\ 5x - 7 & \text{if } x \geq 2 \end{cases}$

34. $f(x) = \begin{cases} -2x + 4 & \text{if } x \leq 1 \\ 3 & \text{if } 1 < x < 4 \\ x + 1 & \text{if } x \geq 4 \end{cases}$

For each of the following functions, find
(a) $f(p)$ **(b)** $f(-r)$ **(c)** $f(m + 3)$.

(See Example 8.)

35. $f(x) = 6 - x$ **36.** $f(x) = 3x + 5$
37. $f(x) = \sqrt{4 - x}$ **38.** $f(x) = \sqrt{-2x}$
39. $f(x) = x^3 + 1$ **40.** $f(x) = 3 - x^3$
41. $f(x) = \dfrac{3}{x - 1}$ **42.** $f(x) = \dfrac{-1}{5 + x}$

For each of the following functions, find the difference quotient
$$\frac{f(x + h) - f(x)}{h} \quad (h \neq 0).$$

(See Example 8.)

43. $f(x) = 2x - 4$ **44.** $f(x) = 2 + 4x$
45. $f(x) = x^2 + 1$ **46.** $f(x) = x^2 - x$

If you have a graphing calculator with table-making ability, display a table showing the (approximate) values of the given function at x = 3.5, 3.9, 4.3, 4.7, 5.1, and 5.5. (See Example 11.)

47. $g(x) = 3x^4 - x^3 + 2x$
48. $f(x) = \sqrt{x^2 - 2.4x + 8}$

Use a calculator to work these exercises. (See Examples 9 and 10.)

49. Finance   The Mississippi state income tax for a single person in 2004 was determined by the rule
$$T(x) = \begin{cases} .03x & \text{if } 0 \leq x \leq 5000 \\ 150 + .04(x - 5000) & \text{if } 5000 < x \leq 10{,}000, \\ 350 + .05(x - 10{,}000) & \text{if } x > 10{,}000 \end{cases}$$
where x is the person's taxable income. Find the tax due on each of these incomes.
(a) \$4750   **(b)** \$27,950   **(c)** \$9320

50. Business   During the first half of this century, the gross domestic product (GDP) of the United States, which measures the overall size of our economy in trillions of dollars, is projected to be given by the function
$$f(x) = .006x^2 + .207x + 10.1,$$

where $x = 0$ corresponds to 2000.* Estimate the GDP in the years
(a) 2005   **(b)** 2010   **(c)** 2025.

51. Health   The table contains incidence ratios by age for death from coronary heart disease (CHD) and lung cancer (LC) when comparing smokers (21–39 cigarettes per day) to nonsmokers.[†]

Age	CHD	LC
55–64	1.9	10
65–74	1.7	9

The incidence ratio of 10 means that smokers are 10 times more likely than nonsmokers to die of lung cancer between the ages of 55 and 64. If the incidence ratio is x, then the percent P (expressed as a decimal) of deaths caused by smoking is given by the function $P(x) = \dfrac{x - 1}{x}$.
(a) What is the percent of lung cancer deaths that can be attributed to smoking between the ages of 65 and 74?
(b) What is the percent of coronary heart disease deaths that can be attributed to smoking between the ages of 55 and 64?

52. Transportation   The number of fliers on commuter airlines (10- to 30-seat planes) between 1975 and 2010 is approximated by $g(x) = .0138x^2 - .172x + 1.4$ (where $x = 0$ corresponds to 1975 and $g(x)$ is in millions).[‡]
(a) How many fliers were there in 1975 and 2000?
(b) How many fliers are projected for 2008?

53. Social Science   The number of Americans (in thousands) that are or are expected to be over 100 years old in year x can be approximated by the function $h(x) = .4018x^2 + 2.039x + 50$ (where $x = 0$ corresponds to 1994).[§]
(a) How many Americans were over 100 in 2000? In 2004?
(b) Predict the number of Americans that will be over 100 years old in the year 2008.

54. Natural Science   High concentrations of zinc ions in water are lethal to rainbow trout. The function
$$f(x) = \left(\frac{x}{1960}\right)^{-.833}$$
gives the approximate average survival time (in minutes) for trout exposed to x milligrams per liter (mg/L) of zinc

---

*Based on data from Goldman Sachs.

[†]Walker, A. *Observations and Inference: An Introduction to the Methods of Epidemiology,* (Newton Lower Falls, MA: Epidemiology Resources, Inc., 1991).

[‡]Based on data from the Federal Aviation Administration (in *USA Today,* March 27, 1995).

[§]Based on data from the U.S. Census Bureau.

ions.[*] Find the survival time (to the nearest minute) for the following concentrations of zinc ions.

(a) 110   (b) 525   (c) 1960   (d) 4500

Use the table feature of a graphing calculator to do these exercises. (See Example 11.)

55. **Health**   The number of surgeons specializing in weight-loss surgery in the United States is approximated by

$$g(x) = -.583x^3 + 38.786x^2 + 12.44x + 363.471,$$

where x is the number of years since 2000.[†]

(a) Create a table that gives the number of these surgeons at the beginning of each year from 2000 through 2005.

(b) Create a table that gives the number of surgeons at the middle of each year from 2000 to 2005. (*Hint:* mid-2000 corresponds to $x = .5$).

56. **Natural Science**   The average monthly temperature in Austin, Texas, can be approximated by the function

$$f(x) = .03x^4 - .716x^3 + 4.38x^2 - 1.07x + 49$$
$$(0 \le x < 12),$$

where $x = 0$ corresponds to January 1, $x = 2$ to February 1, etc. List the average temperature on the first day of each month from April through October.

Work these problems.

57. **Transportation**   The distance from Chicago to Seattle is approximately 2000 miles. A plane flying directly to Seattle passes over Chicago at noon. If the plane travels at 475 mph, find the rule of the function $f(t)$ that gives the distance of the plane from Seattle at time t hours (with $t = 0$ corresponding to noon).

_____

*C. Mason, *Biology of Freshwater Pollution.*

†Based on data from the American Society of Bariatric Surgery.

58. **Business**   A pretzel factory has daily fixed costs of $1800. In addition, it costs 50 cents to produce each bag of pretzels. A bag of pretzels sells for $1.20.

(a) Find the rule of the cost function $c(x)$ that gives the total daily cost of producing x bags of pretzels.

(b) Find the rule of the revenue function $r(x)$ that gives the daily revenue from selling x bags of pretzels.

(c) Find the rule of the profit function $p(x)$ that gives the daily profit from x bags of pretzels.

59. **Business**   A bicycle factory has weekly fixed costs of $36,000. The material and labor costs for each bike are $125.

(a) Express the total weekly cost $C(x)$ as a function of the number x of bicycles that are made.

(b) What is the cost *per bicycle* of producing 600 bikes in a week?

60. **Geometry**   Find an equation that expresses the area y of a square as a function of its

(a) side x

(b) diagonal d

3.2 Graphs of Functions

The **graph** of a function $f(x)$ is defined to be the graph of the *equation* $y = f(x)$. It consists of all points, $(x, f(x))$—that is, every point whose first coordinate is an input number from the domain of f and whose second coordinate is the corresponding output number.

▼ **EXAMPLE 1**   The graph of the function $g(x) = .5x - 3$ is the graph of the equation $y = .5x - 3$. So the graph is a straight line with slope .5 and y-intercept -3, as shown in Figure 3.3. ▲

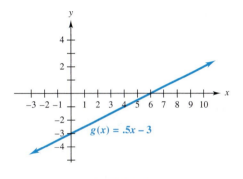

$$g(x) = .5x - 3$$

FIGURE 3.3

A function whose graph is a straight line, as in Example 1, is called a **linear function.** The rule of a linear function can always be put into the form

$$f(x) = ax + b$$

for some constants a and b.

Piecewise Linear Functions

We now consider functions whose graphs consist of straight-line segments. Such functions are called **piecewise linear functions** and are typically defined with different equations for different parts of the domain.

▼ **EXAMPLE 2**   Graph the following function:

$$f(x) = \begin{cases} x + 1 & \text{if } x \le 2 \\ -2x + 7 & \text{if } x > 2 \end{cases}.$$

Solution Consider the two parts of the rule of f. The graphs of $y = x + 1$ and $y = -2x + 7$ are straight lines. The graph of f consists of

the part of the line $y = x + 1$ with $x \le 2$ and

the part of the line $y = -2x + 7$ with $x > 2$.

Each of these line segments can be graphed by plotting two points in the appropriate interval, as shown in Figure 3.4.

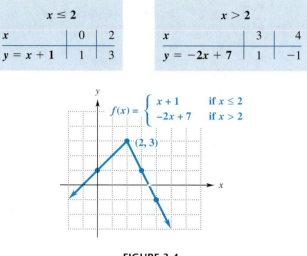

$x \le 2$		
x	0	2
$y = x + 1$	1	3

$x > 2$		
x	3	4
$y = -2x + 7$	1	-1

$$f(x) = \begin{cases} x + 1 & \text{if } x \le 2 \\ -2x + 7 & \text{if } x > 2 \end{cases}$$

FIGURE 3.4

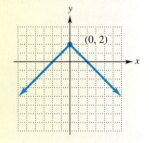

Note that the left and right parts of the graph each extend to the vertical line through $x = 2$, where the two halves of the graph meet at the point $(2, 3)$. ▲ ①

① Graph $f(x) =$
$$\begin{cases} x + 2 & \text{if } x < 0 \\ 2 - x & \text{if } x \ge 0 \end{cases}$$

Answer:

▼ **EXAMPLE 3** Graph the function

$$f(x) = \begin{cases} x - 2 & \text{if } x \le 3 \\ -x + 8 & \text{if } x > 3 \end{cases}.$$

Solution The graph consists of parts of two lines. To find the left side of the graph, choose two values of x with $x \le 3$, say, $x = 0$ and $x = 3$. Then find the corresponding points on $y = x - 2$, namely, $(0, -2)$ and $(3, 1)$. Use these points to draw the line segment to the left of $x = 3$, as in Figure 3.5. Next, choose two values of x with $x > 3$, say, $x = 4$ and $x = 6$, and find the corresponding points on $y = -x + 8$, namely, $(4, 4)$ and $(6, 2)$. Use these points to draw the line segment to the right of $x = 3$, as in Figure 3.5.

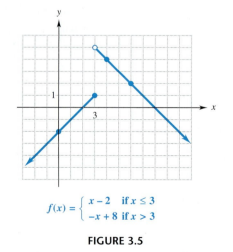

$$f(x) = \begin{cases} x - 2 & \text{if } x \le 3 \\ -x + 8 & \text{if } x > 3 \end{cases}$$

FIGURE 3.5

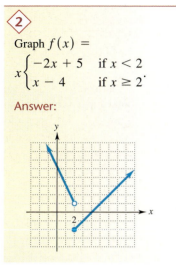

② Graph $f(x) =$
$$x \begin{cases} -2x + 5 & \text{if } x < 2 \\ x - 4 & \text{if } x \geq 2 \end{cases}.$$

Answer:

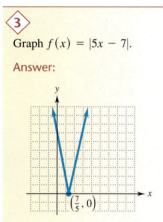

③ Graph $f(x) = |5x - 7|$.

Answer:

Note that both line segments of the graph of f extend to the verticle line through $x = 3$. The closed circle at $(3, 1)$ indicates that this point on the graph of f, whereas the open circle at $(3, 5)$ indicates that this point is *not* on the graph of f (although it is on the graph of the line $y = -x + 8$). ▲ ②

▼ **EXAMPLE 4**   Graph the **absolute value function,** whose rule is $f(x) = |x|$.

Solution The definition of absolute value on page 9 shows that the rule of f can be written

$$f(x) = \begin{cases} x & \text{if } x \geq 0 \\ -x & \text{if } x < 0 \end{cases}.$$

So the right half of the graph (that is, where $x \geq 0$) will consist of a portion of the line $y = x$. It can be graphed by plotting two points, say, $(0, 0)$ and $(1, 1)$. The left half of the graph (where $x < 0$) will consist of a portion of the line $y = -x$, which can be graphed by plotting $(-2, 2)$ and $(-1, 1)$, as shown in Figure 3.6. ▲ ③

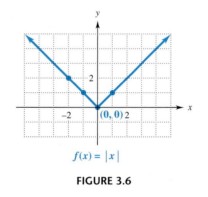

$$f(x) = |x|$$

FIGURE 3.6

TECHNOLOGY TIP   To graph most piecewise linear functions on a graphing calculator, you must use a special syntax. For example, on TI and HP-39+ calculators, the best way to obtain the graph in Example 3 is to graph two separate equations on the same screen:

$$y_1 = (x - 2)/(x \leq 3) \quad \text{and} \quad y_2 = (-x + 8)/(x > 3);$$

the inequality symbols are in the TEST (or CHAR) menu. However, most calculators will graph absolute value functions directly. To graph $f(x) = |x + 2|$, for instance, graph the equation $y = abs(x + 2)$. "Abs" (for absolute value) is on the keyboard or in the MATH menu.

Step Functions

The **greatest integer function,** usually written $f(x) = [x]$, is defined by saying that $[x]$ denotes the largest integer that is less than or equal to x. For example, $[8] = 8, [7.45] = 7, [\pi] = 3, [-1] = -1, [-2.6] = -3$, and so on.

▼ **EXAMPLE 5**   Graph the greatest integer function $f(x) = [x]$.

Solution Consider the values of the function between each two consecutive integers—for instance,

x	$-2 \leq x < -1$	$-1 \leq x < 0$	$0 \leq x < 1$	$1 \leq x < 2$	$2 \leq x < 3$
$[x]$	-2	-1	0	1	2

Thus, between $x = -2$ and $x = -1$, the value of $f(x) = [x]$ is always -2, so the graph there is a horizontal line segment, all of whose points have second coordinate -2. The rest of the graph is obtained similarly (Figure 3.7). An open circle in that figure indicates that the endpoint of the segment is *not* on the graph, whereas a closed circle indicates that the endpoint is on the graph. ▲ ◇④

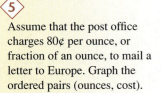

Graph $y = \left[\frac{1}{2}x + 1\right]$.

Answer:

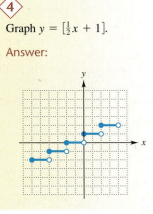

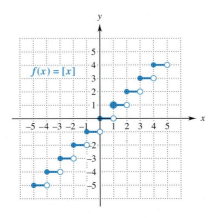

FIGURE 3.7

Functions whose graphs resemble the graph of the greatest integer function are sometimes called **step functions.**

▼ **EXAMPLE 6**   An overnight delivery service charges $25 for a package weighing up to 2 pounds. For each additional pound or fraction of a pound, there is an additional charge of $3. Let $D(x)$ represent the cost to send a package weighing x pounds. Graph $D(x)$ for x in the interval $(0, 6]$.

For x in the interval $(0, 2]$, $y = 25$. For x in $(2, 3]$, $y = 25 + 3 = 28$. For x in $(3, 4]$, $y = 28 + 3 = 31$, and so on. The graph, which is that of a step function, is shown in Figure 3.8. ▲ ◇⑤

⑤

Assume that the post office charges 80¢ per ounce, or fraction of an ounce, to mail a letter to Europe. Graph the ordered pairs (ounces, cost).

Answer:

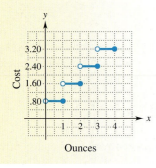

Ounces

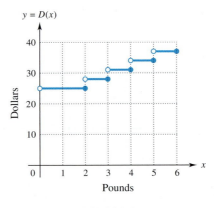

FIGURE 3.8

> **TECHNOLOGY TIP**   On most graphing calculators, the greatest integer function is denoted INT or FLOOR. (Look on the MATH menu or its NUM submenu.) Casio calculators use INTG for the greatest integer function and INT for a different function. When graphing these functions, put your calculator in "dot" graphing mode rather than the usual "connected" mode to avoid erroneous vertical line segments in the graph.

Other Functions

The graphs of many functions do not consist only of straight-line segments. As a general rule when graphing functions by hand, you should follow the procedure introduced in Section 2.1 and summarized here.

This method was used to find the graphs of the functions $f(x) = x^2 - 2x - 8$ and $g(x) = \sqrt{x} + 2$ in Examples 3 and 4 of Section 2.1. Here are some more examples.

Graphing a Function by Plotting Points

1. Determine the domain of the function.

2. Select a few numbers in the domain of f (include both negative and positive ones when possible), and compute the corresponding values of $f(x)$.

3. Plot the points $(x, f(x))$ computed in step 2. Use these points and any other information you may have about the function to make an "educated guess" about the shape of the entire graph.

4. Unless you have information to the contrary, assume that the graph is continuous (unbroken) wherever it is defined.

▼ **EXAMPLE 7**   Graph $g(x) = \sqrt{x - 1}$.

Solution Because the rule of the function is defined only when $x - 1 \geq 0$ (that is, when $x \geq 1$), the domain of g is the interval $[1, \infty)$. Use a calculator to make a table of values, such as the one in Figure 3.9. Plot the corresponding points and connect them to get the graph in Figure 3.9. ▲ ⑥

⑥

Graph $f(x) = \sqrt{4 - x}$.

Answer:

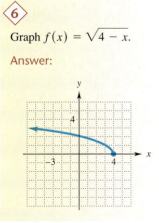

x	$(x) = \sqrt{x - 1}$
1	0
2	1
3	$\sqrt{2} \approx 1.44$
5	2
7	$\sqrt{6} \approx 2.449$
10	3

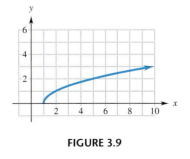

FIGURE 3.9

▼ **EXAMPLE 8**   Graph the function whose rule is $f(x) = 2 - x^3/5$.

Make a table of values and plot the corresponding points. They suggest the graph in Figure 3.10. ▲

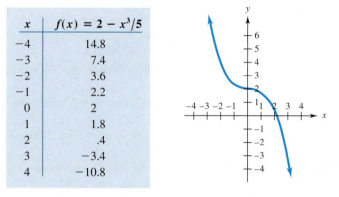

x	$f(x) = 2 - x^3/5$
-4	14.8
-3	7.4
-2	3.6
-1	2.2
0	2
1	1.8
2	.4
3	-3.4
4	-10.8

FIGURE 3.10

▼ **EXAMPLE 9**   Graph the piecewise defined function

$$f(x) = \begin{cases} x^2 & \text{if } x \le 2 \\ \sqrt{x - 1} & \text{if } x > 2 \end{cases}.$$

Solution   When $x \le 2$, the rule of the function is $f(x) = x^2$. Make a table of values, such as the one in Figure 3.11. Plot the corresponding points and connect them to get the left half of the graph in Figure 3.9. When $x > 2$, the rule of the function is $f(x) = \sqrt{x - 1}$, whose graph is shown in Figure 3.9. In Example 7, the entire graph was given, beginning at $x = 1$. Here we use only the part of the graph to the right of $x = 2$, as shown in Figure 3.11. The open circle at $(2, 1)$ indicates that this point is not part of the graph of f (why?). ▲

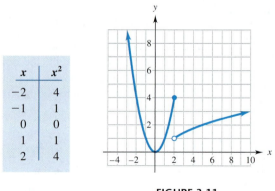

x	x^2
-2	4
-1	1
0	0
1	1
2	4

FIGURE 3.11

Graph Reading

Graphs are often used in business and the social sciences to present data. It is just as important to know how to *read* such graphs as it is to construct them.

▼ **EXAMPLE 10**   The luxury car market in the United States increased significantly after 2000, as shown in Figure 3.12.*

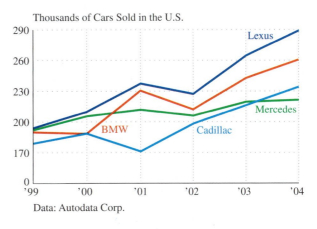

FIGURE 3.12

(a) How do the sales of Cadillac and Mercedes compare over the period shown in the graph?

Solution Sales of the two brands were about the same at the beginning of 1999, but Cadillac sales dropped noticeably until 2001, when they turned around and began a steady increase. Although Mercedes sales increased in all years but 2001, they increased at a slower rate than Cadillac's in 2002 and 2003. Cadillac sales surpassed Mercedes sales in early 2003, and Cadillac's lead appeared to be increasing at the beginning of 2004.

(b) Which brands did best and which did worst during the period shown?

Solution Lexus had the best sales throughout the period. BMW was the worst in 1999 and Cadillac was worst from 2000 to early 2003 when Mercedes took last place. ▲

▼ **EXAMPLE 11**   Figure 3.13 is the graph of the function f whose rule is $f(x) =$ average interest rate on a 30-year fixed-rate mortgage in year x.

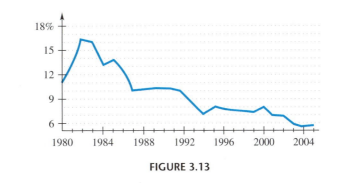

FIGURE 3.13

Business Week, February 28, 2005.

(a) Find the function values of $f(1984)$ and $f(2001)$.

Solution The point $(1984, 13)$ is on the graph, which means that $f(1984) = 13$. Similarly, $f(2001) = 7$ because the point $(2001, 7)$ is on the graph. These values tell us that mortgage rates were 13% in 1984 and 7% in 2001.

(b) During what period were mortgage rates at or above 10%?

Solution Look for points on the graph whose second coordinates are 10 or more—that is, points on or above the horizontal line through 10. These points represent the period from 1980 to 1992.

(c) During what period were mortgage rates below 6%?

Solution Look for points that are below the horizontal line through 6—that is, points with second coordinates less than 6. They occur from 2003 to 2005. ▲

The Vertical-Line Test

The following fact distinguishes function graphs from other graphs.

Vertical-Line Test

No vertical line intersects the graph of a function more than once.

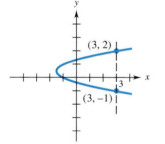

FIGURE 3.14

In other words, if a vertical line intersects a graph at more than one point, the graph is not the graph of a function. To see why this is true, consider the graph in Figure 3.14. The vertical line $x = 3$ intersects the graph at two points. If this were the graph of a function f, it would mean that $f(3) = 2$ (because $(3, 2)$ is on the graph) and that $f(3) = -1$ (because $(3, -1)$ is on the graph). This is impossible, because a *function* can have only one value when $x = 3$ (each input determines exactly one output). Therefore, the graph in Figure 3.14 cannot be the graph of a function. A similar argument works in the general case.

▼ **EXAMPLE 12**   Which of the graphs in Figure 3.15 are the graphs of functions?

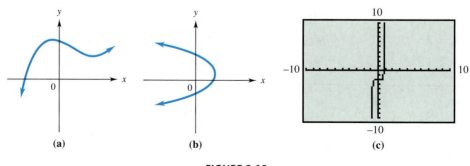

FIGURE 3.15

Solution Every vertical line intersects graph (a) in at most one point, so this graph is the graph of a function. Many vertical lines (including the y-axis) intersect graph (b) twice, so it is not the graph of a function.

Graph (c) appears to fail the Vertical-Line Test near $x = 1$ and $x = -1$, indicating that it is not the graph of a function. But this is misleading because of the low resolution of a calculator screen. The table in Figure 3.16 and the very narrow segment of the graph in Figure 3.17 show that the graph actually rises as it moves to the right. The same happens near $x = -1$. So this graph *does* pass the Vertical-Line Test and *is* the graph of a function. (Its rule is $f(x) = 15x^{11} - 2$.) The moral of this story is that you can't always trust a graphing calculator image. When in doubt, try other viewing windows or a table to see what's really going on. ▲ ⑦

⑦

Find a viewing window that indicates the actual shape of the graph of the function $f(x) = 15x^{11} - 2$ of Example 12 near the point $(-1, -17)$.

Answer:

There are many correct answers, including, $-1.4 \le x \le -.6$ and $-30 \le y \le 0$.

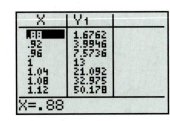

FIGURE 3.16

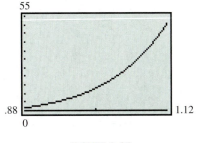

FIGURE 3.17

◆ 3.2 Exercises

Graph each function. (See Examples 1–4.)

1. $f(x) = -.5x + 2$ **2.** $g(x) = 3 - x$

3. $f(x) = \begin{cases} x + 3 & \text{if } x \le 1 \\ 4 & \text{if } x > 1 \end{cases}$

4. $g(x) = \begin{cases} 2x - 1 & \text{if } x < 0 \\ -1 & \text{if } x \ge 0 \end{cases}$

5. $y = \begin{cases} 4 - x & \text{if } x \le 0 \\ 3x + 4 & \text{if } x > 0 \end{cases}$

6. $y = \begin{cases} x + 5 & \text{if } x \le 1 \\ 2 - 3x & \text{if } x > 1 \end{cases}$

7. $f(x) = \begin{cases} |x| & \text{if } x < 2 \\ -2x & \text{if } x \ge 2 \end{cases}$

8. $g(x) = \begin{cases} -|x| & \text{if } x \le 1 \\ 2x & \text{if } x > 1 \end{cases}$

9. $f(x) = |x - 4|$ **10.** $g(x) = |4 - x|$

11. $f(x) = |3 - 3x|$ **12.** $g(x) = -|x|$

13. $y = -|x - 1|$ **14.** $f(x) = |x| - 2$

15. $y = |x - 2| + 3$

16. $|x| + |y| = 1$ (*Hint:* This is not the graph of a function, but is made up of four straight-line segments. Find them by using the definition of absolute value in these four

cases: $x \ge 0$ and $y \ge 0$; $x \ge 0$; and $y < 0$; $x < 0$ and $y \ge 0$; $x < 0$ and $y < 0$.)

Graph each of the following functions. (See Examples 5 and 6.)

17. $f(x) = [x - 3]$ **18.** $g(x) = [x + 3]$

19. $g(x) = [-x]$ **20.** $f(x) = -[x]$

21. $f(x) = [x] + [-x]$ (The graph contains horizontal segments, but is *not* a horizontal line.)

22. Assume that postage rates are 37¢ for the first ounce, plus 23¢ for each additional ounce, and that each letter carries one 37¢ stamp and as many 23¢ stamps as necessary. Graph the *postage stamp function*, whose rule is $p(x) = $ the number of stamps on a letter weighing x ounces.

Graph each function. (See Examples 7–9.)

23. $f(x) = 3 - 2x^2$ **24.** $g(x) = 2 - x^2$

25. $h(x) = x^3/10 + 2$ **26.** $f(x) = x^3/20 - 3$

27. $g(x) = \sqrt{-x}$ **28.** $h(x) = \sqrt{x} - 1$

29. $f(x) = \sqrt[3]{x}$ **30.** $g(x) = \sqrt[3]{x} - 4$

31. $f(x) = \begin{cases} x^2 & \text{if } x < 2 \\ -2x + 2 & \text{if } x \ge 2 \end{cases}$

32. $g(x) = \begin{cases} \sqrt{-x} & \text{if } x \le -4 \\ \dfrac{x^2}{4} & \text{if } x > -4 \end{cases}$

Which of these are graphs of functions? (See Example 12.)

33.

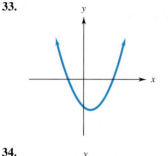

34.

35.

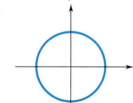

36.

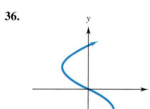

37.

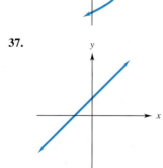

38.

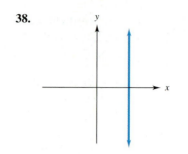

 Use a graphing calculator or other technology to graph each of the following functions. If the graph has any endpoints, indicate whether or not they are part of the graph.

39. $f(x) = .2x^3 - .8x^2 - 4x + 9.6$

40. $g(x) = .1x^4 - .3x^3 - 1.3x^2 + 1.5x$

41. $g(x) = \begin{cases} 2x^2 + x & \text{if } x < 1 \\ x^3 - x - 1 & \text{if } x \ge 1 \end{cases}$ *(Hint:* See the Technology Tip on page 141.)

42. $f(x) = \begin{cases} x|x| & \text{if } x \le 0 \\ -x^2|x| + 2 & \text{if } x > 0 \end{cases}$

 Use a graphical root finder to determine the x-intercepts of the graph of

43. f in Exercise 39   **44.** g in Exercise 40.

 Use a maximum-minimum finder to determine the location of the peaks and valleys in the graph of

45. g in Exercise 40   **46.** f in Exercise 39

Work the following problems. (See Examples 2, 3, 10, and 11.)

47. **Finance** The maximum allowable yearly contribution to an individual retirement account (IRA) was $2000 from 1981 to 2001 and has increased steadily since then. When the maximum is adjusted for inflation, however, the picture is a little different. The approximate maximum IRA contribution in 1981 dollars in year x is given by the function

$$f(x) = \begin{cases} -50x + 2050 & \text{if } 1 \le x < 22 \\ 84x - 348 & \text{if } 22 \le x < 29, \end{cases}$$

where $x = 0$ corresponds to 1980.
(a) Graph this function.
(b) What does the graph say about inflation from 1981 to 2001?

48. **Finance** Example 9 in Section 3.1 gave the rule for the function T that yields the 2004 Connecticut state income tax (for a single person) on a taxable income of x dollars. Graph the function T for taxable incomes between 0 and $50,000.

49. **Natural Science** The depth of snow in Michigan's Isle Royale National Park varies throughout the winter. In a typical winter, the depth in inches is approximated by the function

$$f(x) = \begin{cases} 6.5x & \text{if } 0 \le x \le 4 \\ -5.5x + 48 & \text{if } 4 < x \le 6 \\ -30x + 195 & \text{if } 6 < x \le 6.5 \end{cases},$$

where x represents the time in months, with $x = 0$ representing the beginning of October, $x = 1$ representing the beginning of November, and so on.

(a) Graph $f(x)$.

(b) In what month is the snow deepest? What is the deepest snow depth?

(c) In what months does the snow begin and end?

50. **Natural Science**   A factory begins emitting particulate matter into the atmosphere at 8 A.M. each workday, with the emissions continuing until 4 P.M. The level of pollutants, $P(t)$, measured by a monitoring station 1/2 mile away is approximated as follows, where t represents the number of hours since 8 A.M:

$$P(t) = \begin{cases} 75t + 100 & \text{if } 0 \le t \le 4 \\ 400 & \text{if } 4 < t < 8 \\ -100t + 1200 & \text{if } 8 \le t \le 10 \\ -\dfrac{50}{7}t + \dfrac{1900}{7} & \text{if } 10 < t < 24 \end{cases}.$$

Find the level of pollution at

(a) 9 A.M   (b) 11 A.M.

(c) 5 P.M.   (d) 7 P.M.

(e) Midnight.   (f) Graph $y = P(t)$.

(g) From the graph in part (f), at what time(s) is the pollution level highest? lowest?

51. **Health**   The table shows the consumer price index (CPI) for medical care in selected years.*

Year	Medical CPI
1950	15.1
1975	47.5
2003	297.1

(a) Let $x = 0$ correspond to 1950. Find the rule of a piecewise linear function that models these data— that is, a piecewise linear function f with $f(0) = 15.1$, $f(25) = 47.5$, and $f(53) = 297.1$. Round all coefficients in your final answer to one decimal place.

(b) Graph the function f for $0 \le x \le 60$.

(c) Use the function to estimate the medical CPI in 2000.

(d) Assume that this model remains accurate after 2003, and estimate the medical CPI in 2009.

*U.S. Bureau of Labor Statistics.

52. **Business**   The graph (from the Bureau of Public Debt) shows the federal debt from 1990 to 2004 (in billions of dollars).

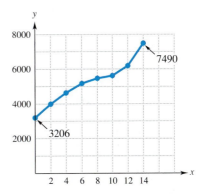

(a) Let $x = 0$ correspond to 1990. Find the rule of a linear function g that passes through the two points corresponding to 1990 and 2004. Draw the graph of g in the preceding figure.

(b) Is the function g a good model for the federal debt? Why?

Finance   *Answer each of the questions in Exercises 53 and 54, and explain how you obtained your answer from the graph, which shows the annual percent change in various consumer price indexes (CPIs). (See Examples 10 and 11.)*

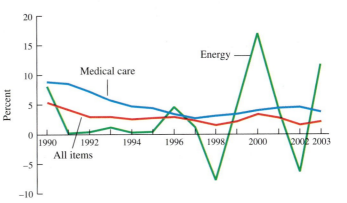

Source: The Conference Board

53. (a) Was there any period between 1990 and 2003 when the CPI for all items was decreasing?

(b) During what years was the CPI for energy decreasing?

(c) During what years was the CPI for energy increasing? How can you tell from the graph?

54. (a) When was the energy CPI decreasing at the fastest rate?

(b) When was the energy CPI increasing at the fastest rate?

(c) Was there ever a time after 2000 when the medical care CPI and the energy CPI were increasing at the same rate? How can you tell from the graph?

55. Natural Science The table gives estimates of the percent of ozone change from 1985 for several years over a hundred-year span if chlorofluorocarbon production is reduced 80% globally.

Year	Percent
1985	0
2005	1.5
2025	3
2065	4
2085	5

(a) Plot these ordered pairs on a grid.
(b) The points from part (a) should lie approximately on a straight line. Use the pairs (2005, 1.5) and (2085, 5) to write an equation of the line.
(c) Letting $f(x)$ represent the percent of ozone change and x represent the year, write your equation from part (b) as a rule that defines a function.
(d) Find $f(2065)$. Does it agree fairly closely with the number in the table that corresponds to 2065? Do you think the expression from part (c) describes this function adequately?

56. Business The graph below shows the number of cellular telephone accounts (in millions) since 1989.
(a) Is this the graph of a function?
(b) What does the domain represent?
(c) Estimate the range.

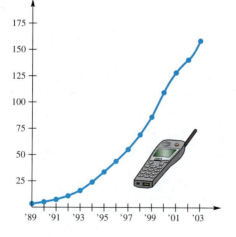

Business *Work these problems. (See Examples 5 and 6.)*

57. Whenever postage rates change, some newspaper publishes a graph like this one, which shows the price of a first-class stamp from 1982 to 2005.

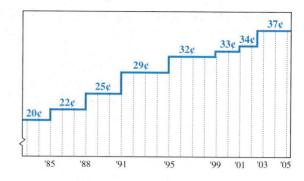

(a) Let f be the function whose rule is

$$f(x) = \text{cost of a first-class stamp in year } x.$$

Find $f(2000)$ and $f(2004)$.
(b) Explain why the graph in the figure is not the graph of the *function f*. What must be done to the figure to make it an accurate graph of the function f?

58. A chain-saw rental firm charges $7 per day or fraction of a day to rent a saw, plus a fixed fee of $4 for resharpening the blade. Let $S(x)$ represent the cost of renting a saw for x days. Find each of the following.

(a) $S\left(\dfrac{1}{2}\right)$ **(b)** $S(1)$ **(c)** $S\left(1\dfrac{1}{4}\right)$ **(d)** $S\left(3\dfrac{1}{2}\right)$

(e) What does it cost to rent a saw for $4\frac{9}{10}$ days?
(f) A portion of the graph of $y = S(x)$ is shown here. Explain how the graph could be continued.

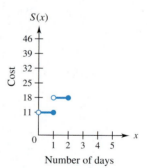

(g) What is the domain variable?
(h) What is the range variable?
(i) Write a sentence or two explaining what part (c) and its answer represent.
(j) We have left $x = 0$ out of the graph. Discuss why it should or shouldn't be included. If it were included, how would you define $S(0)$?

59. The charge to rent a Haul-It-Yourself Trailer is $25, plus $2 per hour or portion of an hour. Find the cost to rent a trailer for
(a) 2 hours;   **(b)** 1.5 hours;
(c) 4 hours;   **(d)** 3.7 hours.
(e) Graph the ordered pairs (hours, cost).

60. A delivery company charges $3, plus 50¢ per mile or part of a mile. Find the cost for a trip of
  (a) 3 miles;   (b) 3.4 miles;
  (c) 3.9 miles;   (d) 5 miles.
  (e) Graph the ordered pairs (miles, cost).
  (f) Is this a function?

Work these problems.

61. Natural Science A laboratory culture contains about 1 million bacteria at midnight. The culture grows very rapidly until noon, when a bactericide is introduced and the bacteria population plunges. By 4 P.M., the bacteria have adapted to the bactericide and the culture slowly increases in population until 9 P.M., when the culture is accidentally destroyed by the cleanup crew. Let $g(t)$ denote the bacteria population at time t (with $t = 0$ corresponding to midnight), and draw a plausible graph of the function g. (Many correct answers are possible.)

62. Transportation A plane flies from Austin, Texas, to Cleveland, Ohio, a distance of 1200 miles. Let f be the function whose rule is

$f(t) = $ distance (in miles) from Austin at time t hours,

with $t = 0$ corresponding to the 4 P.M. takeoff. In each of the following, draw a plausible graph of f under the given circumstances. (There are many correct answers for each part.)
  (a) The flight is nonstop and takes between 3.5 and 4 hours.
  (b) Bad weather forces the plane to land in Dallas (about 200 miles from Austin) at 5 P.M., remain overnight, and leave at 8 A.M. the next morning, flying nonstop to Cleveland.
  (c) The plane flies nonstop, but due to heavy traffic it must fly in a holding pattern for an hour over Cincinnati (about 200 miles from Cleveland) and then go on to Cleveland.

3.3 Applications of Linear Functions

Most of this section deals with the basic business relationships that were introduced in Section 1.2:

$$\textbf{Revenue} = (\textbf{Price per item}) \times (\textbf{Number of items})$$
$$\textbf{Cost} = \textbf{Fixed Costs} + \textbf{Variable Costs}$$
$$\textbf{Profit} = \textbf{Revenue} - \textbf{Cost.}$$

The examples will use only linear functions, but the methods presented here also apply to more complicated functions.

Cost Analysis

Recall that fixed costs are for such things as buildings, machinery, real-estate taxes, and product design. Within broad limits, the fixed cost is constant for a particular product and does not change as more items are made. Variable costs are for labor, materials, shipping, and so on and depend on the number of items made.

If $C(x)$ is the cost of making x items, then the fixed cost (the cost that occurs even when no items are produced) can be found by letting $x = 0$. For example, for the cost function $C(x) = 45x + 250,000$, the fixed cost is

$$C(0) = 45(0) + 250,000 = \$250,000.$$

In this case, the variable cost of making x items is $45x$—that is, $45 per item manufactured.

▼ **EXAMPLE 1** An anticlot drug can be made for $10 per unit. The total cost to produce 100 units is $1500.

(a) Assume that the cost function is linear and find its rule.

Solution Since the cost function $C(x)$ is linear, its rule is of the form $C(x) = mx + b$. We are given that m (the cost per item) is 10, so the rule is $C(x) = 10x + b$. To find b, use the fact that it costs \$1500 to produce 100 units, which means that

$$C(100) = 1500$$
$$10(100) + b = 1500 \qquad C(x) = 10x + b$$
$$1000 + b = 1500$$
$$b = 500.$$

So the rule of the cost function is $C(x) = 10x + 500$.

(b) What are the fixed costs?

Solution The fixed costs are $C(0) = 10(0) + 500 = \$500$. ▲ ◇1

If $C(x)$ is the total cost to produce x items, then the **average cost** per item is given by

$$\overline{C}(x) = \frac{C(x)}{x}.$$

As more and more items are produced, the average cost per item typically decreases.

▼ **EXAMPLE 2** Find the average cost of producing 100 and 1000 units of the anticlot drug in Example 1.

Solution The cost function is $C(x) = 10x + 500$, so the average cost of producing 100 units is

$$\overline{C}(100) = \frac{C(100)}{100} = \frac{10(100) + 500}{100} = \frac{1500}{100} = \$15.00 \text{ per unit.}$$

The average cost of producing 1000 units is

$$\overline{C}(1000) = \frac{C(1000)}{1000} = \frac{10(1000) + 500}{1000} = \frac{10500}{1000} = \$10.50 \text{ per unit.} ▲ ◇2$$

Rates of Change

The rate at which a quantity (such as revenue or profit) is changing can be quite important. For instance, if a company determines that the rate of change of its revenue is decreasing, then sales growth is slowing down, a trend that may require a response.

The rate of change of a linear function is easily determined. For example, suppose $f(x) = 3x + 5$ and consider the table of values in the margin. The table shows that each time x changes by 1, the corresponding value of $f(x)$ changes by 3. Thus, the rate of change of $f(x) = 3x + 5$ with respect to x is 3, which is the slope of the line $y = 3x + 5$. The same thing happens for any linear function:

The rate of change of a linear function $f(x) = mx + b$ is the slope m.

In particular, the rate of change of a linear function is constant. Rates of change of nonlinear functions, which are more complicated (and not constant), will be studied in Section 11.2.

◇1 The total cost of producing 10 calculators is \$100. The variable costs per calculator are \$4. Find the rule of the linear cost function.

Answer:

$C(x) = 4x + 60.$

◇2 In side problem 1, find the average cost per calculator when 100 are produced.

Answer:

\$4.60

x	$f(x) = 3x + 5$
1	8
2	11
3	14
4	17
5	20

The value of a computer, or an automobile, or a machine *depreciates* (decreases) over time. **Linear depreciation** means that the value of the item at time x is given by a linear function $f(x) = mx + b$. The slope m of this line gives the rate of depreciation.

▼ **EXAMPLE 3**   According to the *Kelley Blue Book,* a Ford Focus ZX5 Hatchback that is worth $14,632 today will be worth $10,120 in three years (if it is in good condition with average mileage).

(a) Assuming linear depreciation, find the depreciation function for this car.

Solution We know that the car is worth $14,632 now $(x = 0)$ and will be worth $10,120 in three years $(x = 3)$. So the points $(0, 14632)$ and $(3, 10120)$ are on the graph of the linear depreciation function and can be used to determine its slope:

$$m = \frac{10{,}120 - 14{,}632}{3 - 0} = \frac{-4512}{3} = -1504.$$

Using the point $(0, 14632)$, we find that the equation of the line is

$$y - 14{,}632 = -1504(x - 0) \qquad \text{Point-slope form}$$
$$y = -1504x + 14{,}632$$

Therefore, the rule of the depreciation function is $f(x) = -1504x + 14{,}632$.

(b) What will the car be worth in five years?

Solution Evaluate f when $x = 5$:

$$f(x) = -1504x + 14{,}632$$
$$f(5) = -1504(5) + 14{,}632 = \$7112.$$

(c) At what rate is the car depreciating?

Solution The depreciation rate is given by the slope of $f(x) = -1504x + 14{,}632$, namely, -1504. This negative slope means that the car is decreasing in value an average of $1504 per year. ▲

In economics, the rate of change of the cost function is called the **marginal cost.** Marginal cost is important to management in making decisions in such areas as cost control, pricing, and production planning. When the cost function is linear, say, $C(x) = mx + b$, the marginal cost is the number m (the slope of the graph of C). Marginal cost can also be thought of as the cost of producing one more item, as the next example demonstrates.

▼ **EXAMPLE 4**   An electronics company manufactures handheld PCs. The cost function for one of its models is $C(x) = 160x + 750{,}000$.

(a) What are the fixed costs for this product?

Solution The fixed costs are $C(0) = 160(0) + 750{,}000 = \$750{,}000$.

(b) What is the marginal cost?

Solution The slope of $C(x) = 160x + 750{,}000$ is 160, so the marginal cost is $160.

(c) After 50,000 units have been produced, what is the cost of producing one more?

Solution The cost of producing 50,000 is

$$C(50,000) = 160(50,000) + 750,000 = \$8,750,000.$$

The cost of 50,001 units is

$$C(50,001) = 160(50,001) + 750,000 = \$8,750,160.$$

The cost of the additional unit is the difference

$$C(50,001) - C(50,000) = 8,750,160 - 8,750,000 = \$160.$$

Thus, the cost of one more item is the marginal cost. ▲ ⟨3⟩

Similarly, the rate of change of a revenue function is called the **marginal revenue.** When the revenue function is linear, the marginal cost is the slope of the line, as well as the revenue from producing one more item.

▼ **EXAMPLE 5** The energy company AmerenUE charges each residential customer a basic fee of \$7.25 per month, plus \$0.0764 for every kilowatt hour (kWh) of electricity used.*

(a) Assume that there are 500,000 residential customers, and find the company's revenue function

Solution The monthly revenue from the basic fee is

$$7.25(500,000) = \$3,625,000.$$

If x is the total number of kilowatt hours used by all customers, then the revenue from electricity use is .0764x. So the monthly revenue function is given by

$$R(x) = .0764x + 3,625,000.$$

(b) What is the marginal revenue?

Solution The marginal revenue (the rate at which revenue is changing) is given by the slope of the rate function: \$0.0764 per kWh. ▲ ⟨4⟩

Examples 4 and 5 are typical of the general case, as summarized here.

In a **linear cost function** $C(x) = mx + b$, the marginal cost is m (the slope of the cost line) and the fixed cost is b (the y-intercept of the cost line). The marginal cost is the cost of producing one more item.

Similarly, in a **linear revenue function** $R(x) = kx + d$, the marginal revenue is k (the slope of the revenue line), which is the revenue from producing one more item.

Break-Even Analysis

A typical company must analyze its cost and the potential market for its product to determine when (or even if) it will make a profit.

*Rates in Missouri in summer 2004.

⟨3⟩ The cost in dollars to produce x kilograms of chocolate candy is given by
$C(x) = 3.5x + 800.$
Find each of the following.

(a) The fixed cost

(b) The total cost for 12 kilograms

(c) The marginal cost of the 40th kilogram

(d) The marginal cost per kilogram

Answers:

(a) \$800

(b) \$842

(c) \$3.50

(d) \$3.50

⟨4⟩ If the average customer in Example 5 uses 1500 kWh in a month,

(a) What is the total number of kWh used by all customers?

(b) What is the company's monthly revenue?

Answers:

(a) 750,000,000

(b) \$60,925,000

▼ **EXAMPLE 6**   A company manufactures a particular model of DVD player that sells to retailers for $168. The cost of making x of these DVD players is given by the function $C(x) = 118x + 800,000$.

(a) Find the function R that gives the revenue from selling x players.

Solution Since revenue is the product of the price per item and the number of items, $R(x) = 168x$.

(b) What is the revenue from selling 40,000 players?

Solution Evaluate the revenue function R at 40,000:

$$R(40,000) = 168(40,000) = \$6,720,000.$$

(c) Find the profit function P.

Solution Since profit = revenue − cost,

$$P(x) = R(x) - C(x) = 168x - (118x + 800,000) = 50x - 800,000.$$

(d) What is the profit from selling 10,000 players?

Solution Evaluate the profit function at 10,000:

$$P(10,000) = 50(10,000) - 800,000 = -300,000,$$

that is, a loss of $300,000. ▲

A company can make a profit only if the revenue on a product exceeds the cost of manufacturing it. The number of units at which revenue equals cost (that is, profit is 0) is the **break-even point.**

▼ **EXAMPLE 7**   Find the break-even point for the company in Example 6.

The company will break even when revenue equals cost—that is, when

$$R(x) = C(x)$$
$$168x = 118x + 800,000$$
$$50x = 800,000$$
$$x = 16,000.$$

The company breaks even by selling 16,000 DVD players. The graphs of the revenue and cost functions and the break-even point (where $x = 16,000$) are shown in Figure 3.18. The company must sell more than 16,000 players $(x > 16,000)$ in order to make a profit. ▲ ⑤

⑤

For a certain magazine, the cost equation is

$$c = .70x + 1200,$$

where x is the number of magazines sold. The magazine sells for $1 per copy. Find the break-even point.

Answer:

4000 magazines

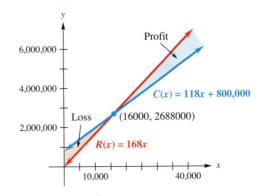

FIGURE 3.18

TECHNOLOGY TIP   The break-even point in Example 7 can be found on a graphing calculator by graphing the cost and revenue functions on the same screen and using the calculator's intersection finder, as shown in Figure 3.19. Depending on the calculator, the intersection finder is in the CALC or G-SOLVE menu or in the MATH or FCN submenu of the GRAPH menu.

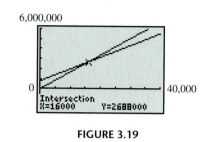

FIGURE 3.19

Supply and Demand

The supply of and demand for an item are usually related to its price. Producers will supply large numbers of the item at a high price, but consumer demand will be low. As the price of the item decreases, consumer demand increases, but producers are less willing to supply large numbers of the item. The curves showing the quantity that will be supplied at a given price and the quantity that will be demanded at a given price are called **supply and demand curves.** In supply and demand problems, we use p for price and q for quantity. We will discuss the economic concepts of supply and demand in more detail in later chapters.

▼ **EXAMPLE 8**   Bill Cornett, an economist, has studied the supply and demand for aluminum siding and has determined that the price per unit,* p, and the quantity demanded, q, are related by the linear equation

$$p = 60 - \frac{3}{4}q.$$

(a) Find the demand at a price of $40 per unit.

Solution Let $p = 40$.

$$p = 60 - \frac{3}{4}q$$

$$40 = 60 - \frac{3}{4}q \qquad \text{Let } p = 40.$$

$$-20 = -\frac{3}{4}q \qquad \text{Add } -60 \text{ on both sides.}$$

$$\frac{80}{3} = q \qquad \text{Multiply both sides by } -\frac{4}{3}.$$

At a price of $40 per unit, $80/3$ (or $26\frac{2}{3}$) units will be demanded.

(b) Find the price if the demand is 32 units.

Solution Let $q = 32$.

$$p = 60 - \frac{3}{4}q$$

$$p = 60 - \frac{3}{4}(32) \qquad \text{Let } q = 32.$$

$$p = 60 - 24$$

$$p = 36$$

---

*An appropriate unit here might be, for example, one thousand square feet of siding.

With a demand of 32 units, the price is $36.

(c) Graph $p = 60 - \dfrac{3}{4}q$.

Solution It is customary to use the horizontal axis for the quantity q and the vertical axis for the price p. In part (a), we saw that 80/3 units would be demanded at a price of $40 per unit; this gives the ordered pair $(80/3, 40)$. Part (b) shows that with a demand of 32 units, the price is $36, which gives the ordered pair $(32, 36)$. Using the points $(80/3, 40)$ and $(32, 36)$ yields the demand graph depicted in Figure 3.20. Only the portion of the graph in Quadrant I is shown, because supply and demand are meaningful only for positive values of p and q. ⬧⁶

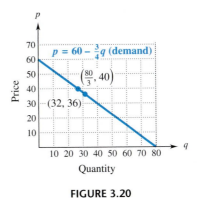

FIGURE 3.20

(d) From Figure 3.20, at a price of $30, what quantity is demanded?

Solution Price is located on the vertical axis. Look for 30 on the p-axis, and read across to where the line $p = 30$ crosses the demand graph. As the graph shows, this occurs where the quantity demanded is 40.

(e) At what price will 60 units be demanded?

Solution Quantity is located on the horizontal axis. Find 60 on the q-axis, and read up to where the vertical line $q = 60$ crosses the demand graph. This occurs where the price is about $15 per unit.

(f) What quantity is demanded at a price of $60 per unit?

Solution The point $(0, 60)$ on the demand graph shows that the demand is 0 at a price of $60 (that is, there is no demand at such a high price). ▲

▼ **EXAMPLE 9** Suppose the economist of Example 8 concludes that the supply q of siding is related to its price p by the equation

$$p = .85q.$$

(a) Find the supply if the price is $51 per unit.

Solution $51 = .85q$ Let $p = 51$.

                          $60 = q$

If the price is $51 per unit, then 60 units will be supplied to the marketplace.

(b) Find the price per unit if the supply is 20 units.

Solution $\qquad p = .85(20) = 17 \qquad$ Let $q = 20$.

If the supply is 20 units, then the price is $17 per unit.

(c) Graph the supply equation $p = .85q$.

Solution As with demand, each point on the graph has quantity q as its first coordinate and the corresponding price p as its second coordinate. Part (a) shows that the ordered pair $(60, 51)$ is on the graph of the supply equation, and part (b) shows that $(20, 17)$ is on the graph. Using these points, we obtain the supply graph in Figure 3.21.

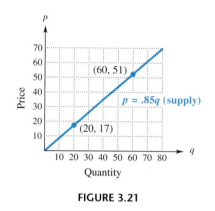

FIGURE 3.21

(d) Use the graph in Figure 3.21 to find the approximate price at which 35 units will be supplied. Then use algebra to find the exact price.

Solution The point on the graph with first coordinate $q = 35$ is approximately $(35, 30)$. Therefore, 35 units will be supplied when the price is approximately $30. To determine the exact price algebraically, substitute $q = 35$ into the supply equation:

$$p = .85q = .85(35) = \$29.75. \; \blacktriangle$$

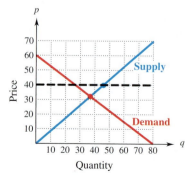

FIGURE 3.22

▼ **EXAMPLE 10** The supply and demand curves of Examples 8 and 9 are shown in Figure 3.22. Determine graphically whether there is a surplus or a shortage of supply at a price of $40 per unit.

Find 40 on the vertical axis in Figure 3.22 and read across to the point where the horizontal line $p = 40$ crosses the supply graph (that is, the point corresponding to a price of $40). This point lies above the demand graph, so supply is greater than demand at a price of $40, and there is a surplus of supply. ▲

Supply and demand are equal at the point where the supply curve intersects the demand curve. This is the **equilibrium point.** Its second coordinate is the **equilibrium price,** the price at which the same quantity will be supplied as is demanded. Its first coordinate is the quantity that will be demanded and supplied at the equilibrium price; this number is called the **equilibrium quantity.**

▼ **EXAMPLE 11**   In the situation described in Examples 8–10, what is the equilibrium quantity? What is the equilibrium price?

The equilibrium point is where the supply and demand curves in Figure 3.22 intersect. To find the quantity q at which the price given by the demand equation $p = 60 - .75q$ (Example 8) is the same as that given by the supply equation $p = .85q$ (Example 9), set these two expressions for p equal and solve the resulting equation.

$$60 - .75q = .85q$$
$$60 = 1.6q$$
$$37.5 = q$$

Therefore, the equilibrium quantity is 37.5 units, the number of units for which supply will equal demand. Substituting $q = 37.5$ into either the demand or supply equation shows that

$$p = 60 - .75(37.5) = 31.875 \quad \text{or} \quad p = .85(37.5) = 31.875.$$

So the equilibrium price is \$31.875 (or \$31.88, rounded). (To avoid error, it's a good idea to substitute into both equations, as we did here, to be sure that the same value of p results; if it doesn't, a mistake has been made.) In this case, the equilibrium point—the point whose coordinates are the equilibrium quantity and price—is $(37.5, 31.875)$. ▲ ⑦

⑦

The demand for a certain commodity is related to the price by $p = 80 - (2/3)q$. The supply is related to the price by $p = (4/3)q$. Find

(a) the equilibrium quantity;

(b) the equilibrium price.

Answers:

(a) 40

(b) $160/3 \approx \$53.33$

TECHNOLOGY TIP   The equilibrium point $(37.5, 31.875)$ can be found on a graphing calculator by graphing the supply and demand curves on the same screen and using the calculator's intersection finder to locate their point of intersection.

◆ **3.3 Exercises**

Business   *Write a cost function for each of the following. Identify all variables used. (See Example 1.)*

1. A chain-saw rental firm charges \$12, plus \$1 per hour.

2. A trailer-hauling service charges \$45, plus \$3 per mile.

3. A parking garage charges \$2.00, plus 50¢ per half hour.

4. For a 1-day rental, a car rental firm charges \$24, plus 16¢ per mile.

Business   *Assume that each of the following can be expressed as a linear cost function. Find the appropriate cost function in each case. (See Examples 1 and 4.)*

5. Fixed cost, \$200; 50 items cost \$2000 to produce.

6. Fixed cost, \$2000; 40 items cost \$5000 to produce.

7. Marginal cost, \$120; 100 items cost \$15,800 to produce.

8. Marginal cost, \$90; 150 items cost \$16,000 to produce.

Business   *In each of the following, a cost function is given. Find the average cost per item when the required number of items is produced. (See Example 2.)*

9. $C(x) = 12x + 1800$; 50 items; 500 items; 1000 items

10. $C(x) = 80x + 12,000$; 100 items; 1000 items; 10,000 items

11. $C(x) = 6.5x + 9800$; 200 items; 2000 items; 5000 items

12. $C(x) = 8.75x + 16,500$; 1000 items; 10,000 items; 75,000 items

Business   *Work these exercises (See Example 3.)*

13. A Honda Civic LX sedan is worth \$15,350 now and is expected to be worth \$9910 in 4 years.
    (a) Find a linear depreciation function for this car.
    (b) Estimate the value of the car 6 years from now.
    (c) At what rate is the car depreciating?

14. A computer that cost $1200 new is expected to depreciate linearly at a rate of $300 per year.
(a) Find the depreciation function f.
(b) Explain why the domain of f is $[0, 4]$.

15. A machine is now worth $120,000 and will be depreciated linearly over an 8-year period, at which time it will be worth $25,000 as scrap.
(a) Find the rule of the depreciation function f.
(b) What is the domain of f?
(c) What will the machine be worth in 6 years?

16. A house increases in value in an approximately linear fashion from $222,000 to $300,000 in 6 years.
(a) Find the *appreciation function* that gives the value of the house in year x.
(b) If the house continues to appreciate at this rate, what will it be worth 12 years from now?

Business *Work these problems. (See Example 4.)*

17. The total cost (in dollars) of producing x college algebra books is $C(x) = 42.5x + 80,000$.
(a) What are the fixed costs?
(b) What is the marginal cost per book?
(c) What is the total cost of producing 1000 books? 32,000 books?
(d) What is the average cost per book when 1000 are produced? When 32,000 are produced?

18. The total cost (in dollars) of producing x compact disks is $C(x) = 6.80x + 450,000$.
(a) What are the fixed costs?
(b) What is the marginal cost per disk?
(c) What is the total cost of producing 50,000 disks? 600,000 disks?
(d) What is the average cost per disk when 50,000 are produced? When 500,000 are produced?

19. The manager of a restaurant found that the cost of producing 100 cups of coffee is $11.02, while the cost of producing 400 cups is $40.12. Assume that the cost $C(x)$ is a linear function of x, the number of cups produced.
(a) Find a formula for $C(x)$.
(b) Find the total cost of producing 1000 cups.
(c) Find the total cost of producing 1001 cups.
(d) Find the marginal cost of producing the 1001st cup.
(e) What is the marginal cost of producing *any* cup?

20. In deciding whether to set up a new manufacturing plant, company analysts have determined that a linear function is a reasonable estimation for the total cost $C(x)$ in dollars of producing x items. They estimate the cost of producing 10,000 items as $547,500 and the cost of producing 50,000 items as $737,500.
(a) Find a formula for $C(x)$.
(b) Find the total cost of producing 100,000 items.
(c) Find the marginal cost of the items to be produced in this plant.

Business *Work these problems. (See Example 5.)*

21. In St. Louis County, the Missouri American Water Company charges a customer with a 5/8" water meter $6.15 per month, plus $2.0337 per 1000 gallons of water used.* If the company has 300,000 such customers, find its monthly revenue function $R(x)$, where the total number of gallons x is measured in thousands.

22. The Laclede Gas Company in St. Louis charges a residential customer $12.62 per month, plus $0.93787 per therm for the first 65 therms of gas used and $0.89722 for each therm above 65.†
(a) How much revenue does the company get from a customer who uses exactly 65 therms of gas in a month.
(b) Find the rule of the function $R(x)$ that gives the company's monthly revenue from one customer, where x is the number of therms of gas used. (*Hint:* $R(x)$ is a piecewise-defined function that has a two-part rule, one part for $x \leq 65$ and the other for $x > 65$.)

Business *Assume that each of the following has a linear cost function. Find (a) the cost function; (b) the revenue function; (c) the profit function; (d) the profit on 100 items. (See Example 6.)*

	Fixed Cost	Marginal Cost per Item	Item Sells For
23.	$500	$10	$35
24.	$180	$11	$20
25.	$250	$18	$28
26.	$15,000	$30	$80
27.	$18,000	$12.50	$25

28. **Business** In the following profit-volume chart, EF and GH represent the profit-volume graphs of a single-product company for 2004 and 2004, respectively.‡

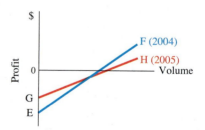

If the 2004 and 2005 unit sales prices are identical, how did the total fixed costs and unit variable costs of 2005 change compared with their values in 2004? Choose one:

---

*Rates in April 2005.

†Rates in March 2005.

‡Adapted from Uniform CPA Examination, American Institute of Certified Public Accountants.

	2004 Total Fixed Costs	2005 Unit Variable Costs
(a)	Decreased	Increased
(b)	Decreased	Decreased
(c)	Increased	Increased
(d)	Increased	Decreased

Use algebra to find the intersection points of the graphs of the given equations. (See Examples 7 and 11.)

29. $2x - y = 7$ and $y = 8 - 3x$

30. $6x - y = 2$ and $y = 4x + 7$

31. $y = 3x - 7$ and $y = 7x + 4$

32. $y = 3x + 5$ and $y = 12 - 2x$

Business *Work the following problems. (See Example 7.)*

33. An insurance company claims that for x thousand policies, its monthly revenue in dollars is given by $R = 125x$ and its monthly cost in dollars is given by $C = 100x + 5000$.
(a) Find the break-even point.
(b) Graph the revenue and cost equations on the same axes.
(c) From the graph, estimate the revenue and cost when $x = 100$ (100,000 policies).

34. The owners of a parking lot have determined that their weekly revenue and cost in dollars are given by $R = 80x$ and $C = 50x + 2400$, where x is the number of long-term parkers.
(a) Find the break-even point.
(b) Graph R and C on the same axes.
(c) From the graph, estimate the revenue and cost when there are 60 long-term parkers.

35. The revenue (in millions of dollars) from the sale of x units at a home supply outlet is given by $r = .21x$. The profit (in millions of dollars) from the sale of x units is given by $p = .084x - 1.5$.
(a) Find the cost equation.
(b) What is the cost of producing 7 units?
(c) What is the break-even point?

36. The profit (in millions of dollars) from the sale of x million units of Blue Glue is given by $p = .7x - 25.5$. The cost is given by $c = .9x + 25.5$.

(a) Find the revenue equation.
(b) What is the revenue from selling 10 million units?
(c) What is the break-even point?

Business *Suppose you are the manager of a firm. The accounting department has provided cost estimates, and the sales department sales estimates, on a new product. You must analyze the data they give you, determine what it will take to break even, and decide whether to go ahead with production of the new product. (See Example 7.)*

37. Cost is estimated by $c = 80x + 7000$ and revenue is estimated by $r = 95x$; no more than 400 units can be sold.

38. Cost is $c = 140x + 3000$ and revenue is $r = 125x$.

39. Cost is $c = 125x + 42,000$ and revenue is $r = 165.5x$; no more than 2000 units can be sold.

40. Cost is $c = 1750x + 95,000$ and revenue is $r = 1975x$; no more than 600 units can be sold.

41. Business The graphs show the hourly compensation (wages and benefits), in U.S. dollars, of production workers in Japan and the United States since 1999.* Estimate the break-even point (the point at which workers in both countries had the same compensation).

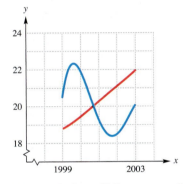

42. Business The graph shows U.S. exports to Ireland and U.S. imports from Ireland (in millions of dollars[†]). Estimate the break-even point (the point at which the values of exports and imports were the same).

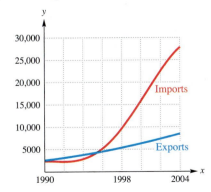

*U.S. Bureau of Labor Statistics.

U.S. Department of Commerce.

43. **Business** The amount of oil produced in the United States (domestic supply) and the (net) amount of imported oil, in millions of barrels per day, in 1990 and 2003 are shown in the table.[‡]

Year	1990	2003
Net Imports	7.2	11.2
Domestic Supply	9.7	8.4

(a) Over the period from 1990 to 2003, net imports and domestic supply each changed in an approximately linear fashion. Let $x = 0$ correspond to 1990, and find two linear functions (one for imports, one for domestic supply) that give the amount of oil for each year x.

(b) Graph the functions in part (a) on the same coordinate axes.

(c) Find the intersection point on the graphs and interpret your answer.

44. **Social Science** The approximate median household income (in thousands of 2002 dollars) for white families in the United States is given by $f(x) = .45x + 40.3$, where x is the number of years since 1990. The approximate median family income for black families is given by $g(x) = .53x + 23.7$.[§]

(a) Graph both functions on the same coordinate axes for $0 \leq x \leq 40$.

(b) Do the graphs intersect in this window? What conclusion do you draw?

(c) Without graphing or doing any algebra, explain why the graphs will eventually intersect.

(d) Use algebra to find the intersection point.

Business *Use the following supply and demand curves to answer Exercises 45–48. (See Examples 8–11.)*

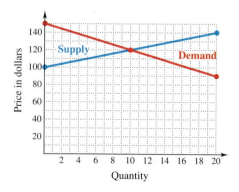

45. At what price are 20 items supplied?

46. At what price are 20 items demanded?

47. Find the equilibrium quantity.

48. Find the equilibrium price.

Business *Work the following exercises. (See Examples 8–11.)*

49. Suppose that the demand and price for a certain brand of shampoo are related by

$$p = 16 - \frac{5}{4}q,$$

where p is price in dollars and q is demand. Find the price for a demand of

(a) 0 units; (b) 4 units; (c) 8 units.

Find the demand for the shampoo at a price of

(d) $6; (e) $11; (f) $16.

(g) Graph $p = 16 - (5/4)q$.

Suppose the price and supply of the shampoo are related by

$$p = \frac{3}{4}q,$$

where q represents the supply and p the price. Find the supply when the price is

(h) $0; (i) $10; (j) $20.

(k) Graph $p = (3/4)q$ on the same axes used for part (g).

(l) Find the equilibrium quantity.

(m) Find the equilibrium price.

50. Let the supply and demand for radial tires in dollars be given by

$$\text{supply: } p = \frac{3}{2}q; \quad \text{demand: } p = 81 - \frac{3}{4}q.$$

(a) Graph these equations on the same axes.

(b) Find the equilibrium quantity.

(c) Find the equilibrium price.

51. Let the supply and demand for bananas in cents per pound be given by

$$\text{supply: } p = \frac{2}{5}q; \quad \text{demand: } p = 100 - \frac{2}{5}q.$$

(a) Graph these equations on the same axes.

(b) Find the equilibrium quantity.

(c) Find the equilibrium price.

(d) On what interval does demand exceed supply?

52. Let the supply and demand for sugar be given by

$$\text{supply: } p = 1.4q - .6$$

and

$$\text{demand: } p = -2q + 3.2,$$

where p is in dollars.

(a) Graph these equations on the same axes.

*U.S. Department of Energy.

†Based on data from the U.S. Census Bureau.

(b) Find the equilibrium quantity.

(c) Find the equilibrium price.

(d) On what interval does supply exceed demand?

53. Explain why the graph of the (total) cost function is always above the x-axis and can never move downward as you go from left to right. Is the same thing true of the graph of the average cost function?

54. Explain why the graph of the profit function can rise or fall (as you go from left to right) and can be below the x-axis.

3.4 Quadratic Functions

A **quadratic function** is a function whose rule is given by a quadratic polynomial, such as

$$f(x) = x^2, \qquad g(x) = 3x^2 + 30x + 67, \qquad \text{and} \quad h(x) = -x^2 + 4x.$$

Thus, a quadratic function is a function whose rule can be written in the form

$$f(x) = ax^2 + bx + c$$

for some constants a, b, c, with $a \neq 0$.

▼ **EXAMPLE 1**   Graph each of these quadratic functions:

$$f(x) = x^2, \qquad g(x) = 4x^2, \qquad h(x) = -.2x^2.$$

Solution In each case, choose several numbers (negative, positive, 0) for x, find the values of the function at these numbers, and plot the corresponding points. Then connect the points with a smooth curve to obtain Figure 3.23. ▲

$f(x) = x^2$					
x	-2	-1	0	1	2
x^2	4	1	0	1	4

$g(x) = 4x^2$					
x	-2	-1	0	1	2
$4x^2$	16	4	0	4	16

$h(x) = -.2x^2$						
x	-5	-3	-1	0	2	4
$-.2x^2$	-5	-1.8	$-.2$	0	$-.8$	-3.2

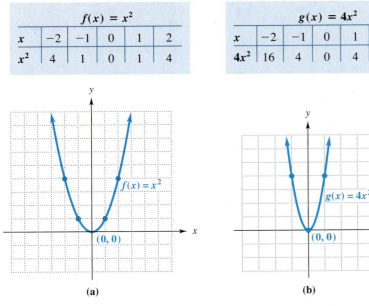

(a)   (b)   (c)

FIGURE 3.23

Each of the curves in Figure 3.23 is a **parabola.** It can be shown that the graph of every quadratic function is a parabola. Parabolas have many useful properties. Cross sections of radar dishes and spotlights form parabolas. Disks often visible on the sidelines of televised football games are microphones having reflectors with parabolic cross sections. These microphones are used by the television networks to pick up the signals shouted by the quarterbacks.

All parabolas have the same basic "cup" shape, although the cup may be broad or narrow and open upward or downward. The general shape of a parabola is determined by the coefficient of x^2 in its rule, as summarized here and illustrated in Example 1. ◇①

①

Graph each quadratic function.

(a) $f(x) = x^2 - 4$

(b) $g(x) = -x^2 + 4$

Answers:

(a)

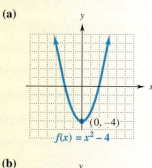

$(0, -4)$
$f(x) = x^2 - 4$

(b)

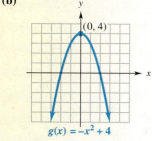

$(0, 4)$
$g(x) = -x^2 + 4$

The graph of a quadratic function $f(x) = ax^2 + bx + c$ is a parabola.

If $a > 0$, the parabola opens upward.   *[Figure 3.23(a) and 3.23(b)]*

If $a < 0$, the parabola opens downward.   *[Figure 3.23(c)]*

If $|a| < 1$, the parabola appears wider than the graph of $y = x^2$. *[Figure 3.23(c)]*

If $|a| > 1$, the parabola appears narrower than the graph of $y = x^2$. *[Figure 3.23(b)]*

When a parabola opens upward (as in Figure 3.23(a), (b)), its lowest point is called the **vertex.** When a parabola opens downward (as in Figure 3.23(c)), its highest point is called the **vertex.** The vertical line through the vertex of a parabola is called the **axis of the parabola.** For example, $(0, 0)$ is the vertex of each of the parabolas in Figure 3.23 and the axis of each parabola is the y-axis. If you were to fold the graph of a parabola along its axis, the two halves of the parabola would match exactly. This means that a parabola is *symmetric* about its axis.

Although the vertex of a parabola can be approximated by a graphing calculator's maximum or minimum finder, its exact coordinates can be found algebraically, as in the following examples.

▼ **EXAMPLE 2**   Consider the function $g(x) = 2(x - 3)^2 + 1$.

(a) Show that g is a quadratic function.

    Solution Multiply out the rule of g to show that it has the required form.

$$g(x) = 2(x - 3)^2 + 1$$
$$= 2(x^2 - 6x + 9) + 1$$
$$= 2x^2 - 12x + 18 + 1$$
$$g(x) = 2x^2 - 12x + 19$$

According to the preceding box, the graph of g is somewhat narrow, upward-opening parabola.

(b) Show that the vertex of the graph of $g(x) = 2(x - 3)^2 + 1$ is $(3, 1)$.

    Solution Since $g(3) = 2(3 - 3)^2 + 1 = 0 + 1 = 1$, the point $(3, 1)$ is on the graph. The vertex of an upward-opening parabola is the lowest point on the graph, so we must show that $(3, 1)$ is the lowest point. Let x be any number except 3 (so that $x - 3 \neq 0$). Then the quantity $2(x - 3)^2$ is positive, and hence

$$g(x) = 2(x - 3)^2 + 1 = \text{(a positive number)} + 1,$$

which means that $g(x) > 1$. Therefore, every point $(x, g(x))$ on the graph with $x \neq 3$ has second coordinate $g(x)$ greater than 1. Hence $(x, g(x))$ lies *above* $(3, 1)$. In other words, $(3, 1)$ is the lowest point on the graph—the vertex of the parabola.

(c) Graph $g(x) = 2(x - 3)^2 + 1$.

Solution Plot some points on both sides of the vertex $(3, 1)$ and obtain the graph in Figure 3.24. The vertical line $x = 3$ through the vertex is the axis of the parabola. ▲

x	y
1	9
2	3
3	1
4	3
5	9

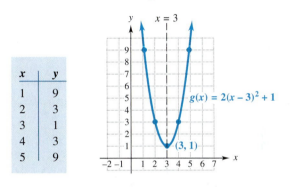

FIGURE 3.24

In Example 2, notice how the rule of the function g is related to the coordinates of the vertex.

$$g(x) = 2(x - 3)^2 + 1 \qquad\qquad (3, 1)$$

Arguments similar to those in Example 2 lead to the following fact.

> The graph of the quadratic function $f(x) = a(x - h)^2 + k$ is a parabola with vertex (h, k). It opens upward when $a > 0$ and downward when $a < 0$.

▼ **EXAMPLE 3**   Determine algebraically whether the parabola opens upward or downward, and find its vertex.

(a) $f(x) = -3(x - 4)^2 - 7$

Solution The rule of the function is in the form $f(x) = a(x - h)^2 + k$ (with $a = -3$, $h = 4$, and $k = -7$). The parabola opens downward $(a < 0)$ and its vertex is $(h, k) = (4, -7)$.

(b) $g(x) = 2(x + 3)^2 + 5$

Solution Be careful here: the vertex is *not* $(3, 5)$. To put the rule of $g(x)$ in the form $a(x - h)^2 + k$, we must rewrite it so that there is a minus sign inside the parentheses:

$$g(x) = 2(x + 3)^2 + 5$$
$$= 2(x - (-3))^2 + 5.$$

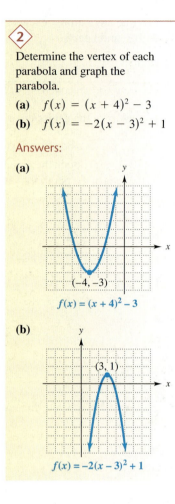

②

Determine the vertex of each parabola and graph the parabola.

(a) $f(x) = (x + 4)^2 - 3$

(b) $f(x) = -2(x - 3)^2 + 1$

Answers:

(a)

$f(x) = (x + 4)^2 - 3$

(b)

$f(x) = -2(x - 3)^2 + 1$

This is the required form, with $a = 2$, $h = -3$, and $k = 5$. The parabola opens upward and its vertex is $(-3, 5)$. ▲ **②**

▼ **EXAMPLE 4** Find the rule of a quadratic function whose graph has vertex $(3, 4)$ and passes through the point $(6, 22)$.

Solution The graph of $f(x) = a(x - h)^2 + k$ has vertex (h, k). We want $h = 3$ and $k = 4$, so that $f(x) = a(x - 3)^2 + 4$. Since $(6, 22)$ is on the graph, we must have $f(6) = 22$. Therefore,

$$f(x) = a(x - 3)^2 + 4$$
$$f(6) = a(6 - 3)^2 + 4$$
$$22 = a(3)^2 + 4$$
$$9a = 18$$
$$a = 2.$$

Thus, the graph of $f(x) = 2(x - 3)^2 + 4$ is a parabola with vertex $(3, 4)$ that passes through $(6, 22)$. ▲

The vertex of each parabola in Examples 2 and 3 was easily determined because the rule of the function had the form

$$f(x) = a(x - h)^2 + k.$$

The rule of *any* quadratic function can be put in this form by using the technique of completing the square, which was discussed in Section 1.7.

▼ **EXAMPLE 5** By completing the square, determine the vertex of the graph of $f(x) = x^2 - 2x + 3$. Then graph the parabola.

Solution In order to get the equation in the form $f(x) = a(x - h)^2 + k$, we first write $f(x) = x^2 - 2x + 3$ as

$$f(x) = (x^2 - 2x \quad) + 3.$$

Now complete the square for the expression in parentheses. Take half the coefficient of x, namely, $(\frac{1}{2})(-2) = -1$, and square the result: $(-1)^2 = 1$. In order to complete the square, we must add 1 inside the parentheses, but in order not to change the rule of the function, we must also subtract 1:

$$f(x) = (x^2 - 2x + 1 - 1) + 3.$$

By the associative property,

$$f(x) = (x^2 - 2x + 1) + (-1 + 3).$$

Factor $x^2 - 2x + 1$ as $(x - 1)^2$, to get

$$f(x) = (x - 1)^2 + 2.$$

With the rule in this form, we can see that the graph is an upward-opening parabola with vertex $(1, 2)$, as shown in Figure 3.25. ▲ **③**

③

Rewrite the rule of each function by completing the square, and use this form to find the vertex of the graph.

(a) $f(x) = x^2 - 6x + 11$

(b) $g(x) = x^2 + 8x + 18$

Answers:

(a) $f(x) = (x - 3)^2 + 2$; $(3, 2)$

(b) $g(x) = (x + 4)^2 + 2$; $(-4, 2)$

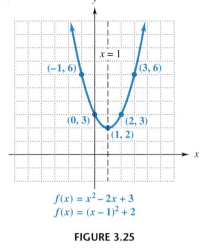

$$f(x) = x^2 - 2x + 3$$
$$f(x) = (x - 1)^2 + 2$$

FIGURE 3.25

CAUTION The technique of completing the square works only when the coefficient of x^2 is 1. To find the vertex of a quadratic function such as

$$f(x) = 2x^2 + 12x - 19,$$

you must first factor the coefficient of x^2 out of the first two terms and write the rule as

$$f(x) = 2(x^2 + 6x) - 19.$$

Now complete the square on the expression in parentheses, and proceed as in Example 5.

The technique of completing the square can be used to rewrite the general equation $f(x) = ax^2 + bx + c$ in the form $f(x) = a(x - h)^2 + k$, as is shown in Exercise 54. When this is done, we obtain a formula for the coordinates of the vertex.

The graph of $f(x) = ax^2 + bx + c$ is a parabola with vertex (h, k), where

$$h = \frac{-b}{2a} \quad \text{and} \quad k = f(h).$$

▼ **EXAMPLE 6** Find the vertex, the axis, and the x- and y-intercepts of the graph of $f(x) = x^2 - x - 6$.

Solution Since $a = 1$ and $b = -1$, the x-value of the vertex is

$$\frac{-b}{2a} = \frac{-(-1)}{2 \cdot 1} = \frac{1}{2}.$$

The y-value of the vertex is

$$f\left(\frac{1}{2}\right) = \left(\frac{1}{2}\right)^2 - \frac{1}{2} - 6 = -\frac{25}{4}.$$

The vertex is $(1/2, -25/4)$ and the axis of the parabola is $x = 1/2$, as shown in Figure 3.26. The intercepts are found by setting each variable equal to 0.

x-intercepts	**y-intercept**
Set $f(x) = y = 0$:	Set $x = 0$:
$0 = x^2 - x - 6$	$f(x) = y = 0^2 - 0 - 6 = -6$
$0 = (x + 2)(x - 3)$	The y-intercept is -6.
$x + 2 = 0$ or $x - 3 = 0$	
$x = -2$ $x = 3$	

The x-intercepts are -2 and 3. ▲ ◇4

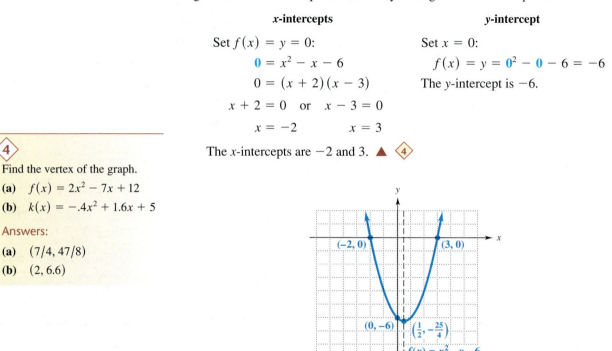

FIGURE 3.26

---

4 ◇

Find the vertex of the graph.

(a) $f(x) = 2x^2 - 7x + 12$

(b) $k(x) = -.4x^2 + 1.6x + 5$

Answers:

(a) $(7/4, 47/8)$

(b) $(2, 6.6)$

TECHNOLOGY TIP   The maximum or minimum finder on a graphing calculator can approximate the vertex of a parabola with a high degree of accuracy. The max-min finder is in the CALC menu or in the MATH or FCN submenu of the GRAPH menu. Similarly, the calculator's graphical root finder can approximate the x-intercepts of a parabola.

◆   **3.4 Exercises**

The graph of each of the following functions is a parabola. Without graphing, determine whether the parabola opens upward or downward. (See Example 1.)

1. $f(x) = x^2 - 3x - 12$

2. $g(x) = -x^2 + 5x + 15$

3. $h(x) = -3x^2 + 14x + 1$

4. $f(x) = 6.5x^2 - 7.2x + 4$

5. $g(x) = 2.9x^2 - 12x - 5$

6. $h(x) = -4x^2 + 4.7x - 6$

Without graphing, determine the vertex of the parabola that is the graph of the given function. State whether the parabola opens upward or downward. (See Examples 2 and 3.)

7. $f(x) = -3(x - 4)^2 + 5$

8. $g(x) = -4(x - 7)^2 - 2$

9. $h(x) = 5(x + 6)^2 - 7$

10. $f(x) = -6(x + 8)^2$

11. $g(x) = 2.1(x - 3.5)^2 - 9$

12. $h(x) = 5.67(x + 4.3)^2 + 11$

Match each function with its graph, which is one of those shown. (See Examples 1–3.)

13. $f(x) = x^2 + 2$

14. $g(x) = x^2 - 2$

15. $g(x) = (x - 2)^2$

16. $f(x) = -(x + 2)^2$

17. $f(x) = 2(x - 2)^2 + 2$

18. $g(x) = -2(x - 2)^2 - 2$

19. $g(x) = -2(x + 2)^2 + 2$

20. $f(x) = 2(x + 2)^2 - 2$

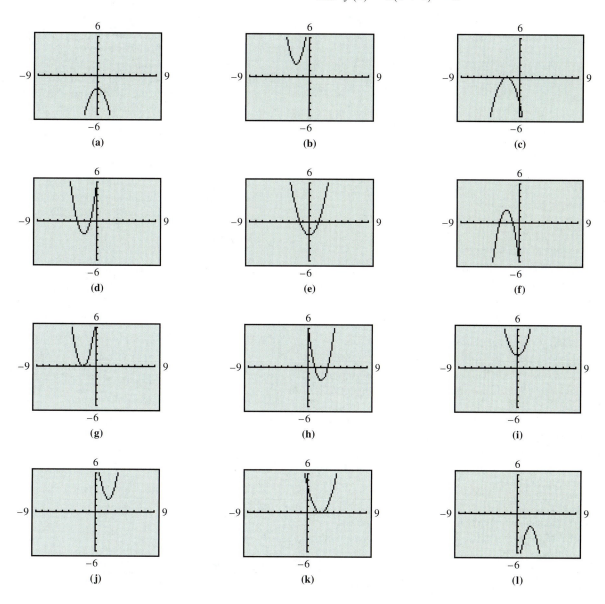

(a) (b) (c)

(d) (e) (f)

(g) (h) (i)

(j) (k) (l)

Find the rule of a quadratic function whose graph has the given vertex and passes through the given point. (See Example 4.)

21. Vertex $(1, 2)$; point $(5, 6)$

22. Vertex $(-3, 2)$; point $(2, 1)$

23. Vertex $(-1, -2)$; point $(1, 2)$

24. Vertex $(2, -4)$; point $(5, 2)$

25. Vertex $(0, 0)$; point $(2, 12)$

26. Vertex $(0, 0)$; point $(-2, -5)$

Without graphing, find the vertex of the parabola that is the graph of the given function. (See Example 5 and 6.)

27. $f(x) = x^2 - 8x + 4$

28. $g(x) = x^2 - 4x + 2$

29. $h(x) = 2x^2 + 12x - 3$

30. $f(x) = 3x^2 + 6x + 1$

31. $g(x) = -x^2 + x$ **32.** $h(x) = -2x^2 - 2x + 1$

33. $f(x) = -3x^2 + 4x + 5$ **34.** $g(x) = 2x^2 - x - 1$

Without graphing, determine the x- and y-intercepts of each of the following parabolas. (See Example 6.)

35. $f(x) = 3(x - 2)^2 - 3$ **36.** $f(x) = x^2 - 4x - 1$

37. $g(x) = 2x^2 + 8x + 6$ **38.** $g(x) = x^2 - 10x + 20$

Graph each parabola. Find the vertex and axis of symmetry of each. (See Examples 1–6.)

39. $f(x) = (x + 2)^2$

40. $f(x) = -(x + 5)^2$

41. $f(x) = (x - 1)^2 - 3$

42. $f(x) = (x - 2)^2 + 1$

43. $f(x) = x^2 - 4x + 6$

44. $f(x) = x^2 + 6x + 3$

45. $f(x) = 2x^2 - 4x + 5$

46. $f(x) = -3x^2 + 24x - 46$

Use a calculator to work these problems.

47. **Transportation** According to data from the National Highway Traffic Safety Administration, the fatal accident rate per 1000 licensed drivers every 100 million miles can be approximated by the function $f(x) = .0031x^2 - .291x + 7.1$, where x is the age of the driver ($16 \leq x \leq 88$). At what age is the rate the lowest?

48. **Social Science** According to data from the U.S. Census Bureau, the population of Cleveland, Ohio (in thousands), in year x can be approximated by $g(x) = -.172x^2 + 16.82x + 487.62$, where $x = 0$ corresponds to 1900. In what year did Cleveland have its largest population?

49. **Physical Science** If an object is thrown upward with an initial velocity of 80 ft/second, then its height after t seconds is given by

$$h = 80t - 16t^2.$$

(a) Find the maximum height attained by the object.

(b) Find the number of seconds it takes the object to hit the ground.

50. **Health** Blood flow to the fetal spleen is of research interest because several diseases are associated with

increased resistance in the splenic artery (the artery that goes to the spleen). Researchers have found that the index of splenic artery resistance in the fetus can be described by the function

$$y = .057x - .001x^2,$$

where x is the number of weeks of gestation.*

(a) At how many weeks is splenic artery resistance a maximum?

(b) What is the maximum splenic artery resistance?

(c) At how many weeks is splenic artery resistance equal to 0, according to the given formula? Is your answer reasonable for this function? Explain.

In Exercises 51–53, graph the functions in parts (a)–(d) on the same set of axes; then answer part (e).

51. **(a)** $k(x) = x^2$ **(b)** $f(x) = x^2 + 2$

    (c) $g(x) = x^2 + 3$ **(d)** $h(x) = x^2 + 5$

    (e) Explain how the graph of $f(x) = x^2 + c$ (where c is a positive constant) can be obtained from the graph of $k(x) = x^2$.

52. **(a)** $k(x) = x^2$ **(b)** $f(x) = x^2 - 1$

    (c) $g(x) = x^2 - 2$ **(d)** $h(x) = x^2 - 4$

    (e) Explain how the graph of $f(x) = x^2 - c$ (where c is a positive constant) can be obtained from the graph of $k(x) = x^2$.

53. **(a)** $k(x) = x^2$ **(b)** $f(x) = (x + 2)^2$

    (c) $g(x) = (x - 2)^2$ **(d)** $h(x) = (x - 4)^2$

    (e) Explain how the graphs of $f(x) = (x + c)^2$ and $f(x) = (x - c)^2$ (where c is a positive constant) can be obtained from the graph of $k(x) = x^2$.

54. Verify that the right side of the equation

$$ax^2 + bx + c = a\left(x - \left(\frac{-b}{2a}\right)\right)^2 + \left(c - \frac{b^2}{4a}\right)$$

equals the left side. Since the right side of the equation is in the form $a(x - h)^2 + k$, we conclude that the vertex of the parabola $f(x) = ax^2 + bx + c$ has x-coordinate $h = -b/2a$.

*Abuhamad, A. Z. et. al., "Doppler Flow Velocimetry of the Splenic Artery in the Human Fetus: Is It a Marker of Chronic Hypoxia?" *American Journal of Obstetrics and Gynecology,* Vol. 172, No. 3, March 1995, pp. 820–825.

3.5 Applications of Quadratic Functions

The fact that the vertex of a parabola $y = ax^2 + bx + c$ is the highest or lowest point on the graph can be used in applications to find a maximum or a minimum value.

▼ **EXAMPLE 1**   Anne Kelly owns and operates Aunt Emma's Blueberry Pies. She has hired a consultant to analyze her business operations. The consultant tells her that her profits, $P(x)$, from the sale of x cases of pies, are given by

$$P(x) = 120x - x^2.$$

How many cases of pies should she sell in order to maximize profit? What is the maximum profit?

Solution The profit function can be rewritten as $P(x) = -x^2 + 120x$. Its graph is a downward-opening parabola. According to the box on page 167, the x-coordinate of the vertex is

$$x = \frac{-b}{2a} = \frac{-120}{2(-1)} = 60$$

and the y-coordinate is $P(60) = -60^2 + 120(60) = 3600$. So the vertex is $(60, 3600)$. Using this fact and plotting some points, we obtain the graph in Figure 3.27. For each point on the graph,

the x-coordinate is the number of cases;

the y-coordinate is the profit on that number of cases.

Maximum profit occurs at the point with the largest y-coordinate, namely, the vertex $(60, 3600)$. So the maximum profit of $3600 is obtained when 60 cases of pies are sold. ▲ ◇①

①

When a company sells x units of a product, its profit is $P(x) = -2x^2 + 40x + 280$. Find

(a) the number of units which should be sold so that maximum profit is received;

(b) the maximum profit.

Answers:

(a) 10 units

(b) $480

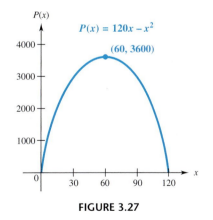

FIGURE 3.27

Supply and demand curves were introduced in Section 3.3. Here is a quadratic example.

▼**EXAMPLE 2**   Suppose that the price of and demand for an item are related by

$$p = 150 - 6q^2, \qquad \text{Demand function}$$

where p is the price (in dollars) and q is the number of items demanded (in hundreds). Suppose also that the price and supply are related by

$$p = 10q^2 + 2q, \qquad \text{Supply function}$$

where q is the number of items supplied (in hundreds). Find the equilibrium quantity and the equilibrium price.

Solution   The graphs of both of these equations are parabolas (Figure 3.28). Only those portions of the graphs that lie in the first quadrant are included, because none of supply, demand, or price can be negative.

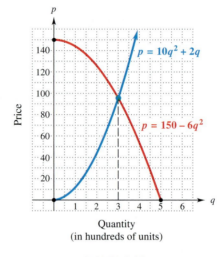

FIGURE 3.28

The point where the demand and supply curves intersect is the equilibrium point. Its first coordinate is the equilibrium quantity, and its second coordinate is the equilibrium price. These coordinates may be found in two ways.

Algebraic Method   At the equilibrium point, the second coordinate of the demand curve must be the same as the second coordinate of the supply curve, so that

$$150 - 6q^2 = 10q^2 + 2q.$$

Write this quadratic equation in standard form as follows:

$$0 = 16q^2 + 2q - 150 \qquad \text{Add } -150 \text{ and } 6q^2 \text{ to both sides.}$$

$$0 = 8q^2 + q - 75. \qquad \text{Multiply both sides by } \frac{1}{2}.$$

This equation can be solved by the quadratic formula given in Section 1.7. Here, $a = 8, b = 1,$ and $c = -75$.

$$q = \frac{-1 \pm \sqrt{1 - 4(8)(-75)}}{2(8)}$$

$$= \frac{-1 \pm \sqrt{1 + 2400}}{16} \qquad -4(8)(-75) = 2400$$

$$= \frac{-1 \pm 49}{16}$$

$\sqrt{1 + 2400} = \sqrt{2401} = 49$

$$q = \frac{-1 + 49}{16} = \frac{48}{16} = 3 \quad \text{or} \quad q = \frac{-1 - 49}{16} = -\frac{50}{16} = -\frac{25}{8}$$

It is not possible to make $-25/8$ units, so discard that answer and use only $q = 3$. Hence, the equilibrium quantity is 300. Find the equilibrium price by substituting 3 for q in either the supply or the demand function (and check your answer by using the other one). Using the supply function gives

$$p = 10q^2 + 2q$$
$$p = 10 \cdot 3^2 + 2 \cdot 3 \quad \text{Let } q = 3.$$
$$= 10 \cdot 9 + 6$$
$$p = \$96.$$

Graphical Method Graph the two functions on a graphing calculator, and use the intersection finder to determine that the equilibrium point is $(3, 96)$, as in Figure 3.29. ▲

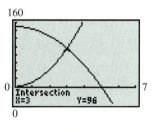

FIGURE 3.29

2

The price and demand for an item are related by
$p = 32 - x^2$, while price and supply are related by $p = x^2$.
Find

(a) the equilibrium quantity;

(b) the equilibrium price.

Answers:

(a) 4

(b) 16

▼ **EXAMPLE 3**   The rental manager of a small apartment complex with 16 units has found from experience that each \$40 increase in the monthly rent results in an empty apartment. All 16 apartments will be rented at a monthly rent of \$500. How many \$40 increases will produce maximum monthly income for the complex?

Solution Let x represent the number of \$40 increases. Then the number of apartments rented will be $16 - x$. Also, the monthly rent per apartment will be $500 + 40x$. (There are x increases of \$40, for a total increase of $40x$.) The monthly income, $I(x)$, is given by the number of apartments rented times the rent per apartment, so

$$I(x) = (16 - x)(500 + 40x)$$
$$= 8000 + 640x - 500x - 40x^2$$
$$= 8000 + 140x - 40x^2.$$

Since x represents the number of \$40 increases and each \$40 increase causes one empty apartment, x must be a whole number. Because there are only 16 apartments, $0 \le x \le 16$. Since there is a small number of possibilities, the value of x that produces maximum income may be found in several ways.

Brute Force Method Use a scientific calculator or the table feature of a graphing calculator (as in Figure 3.30) to evaluate $I(x)$ when $x = 1, 2, \ldots, 16$ and find the largest value.

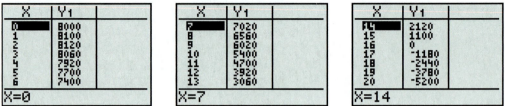

FIGURE 3.30

The tables show that a maximum income of $8120 occurs when $x = 2$. So the manager should charge rent of $500 + 2(40) = \$580$, leaving two apartments vacant.

Algebraic Method The graph of $I(x) = 8000 + 140x - 40x^2$ is a downward-opening parabola (why?), and the value of x that produces maximum income occurs at the vertex. The methods of Section 3.4 show that the vertex is $(1.75, 8122.50)$. Since x must be a whole number, evaluate $I(x)$ at $x = 1$ and $x = 2$ to see which one gives the best result.

$$\text{If } x = 1, \text{ then } I(1) = -40(1)^2 + 140(1) + 8000 = 8100.$$
$$\text{If } x = 2, \text{ then } I(2) = -40(2)^2 + 140(2) + 8000 = 8120.$$

So maximum income occurs when $x = 2$. The manager should charge a rent of $500 + 2(40) = \$580$ and leave two apartments vacant. ▲

Quadratic Models

Real-world data can sometimes be used to construct a quadratic function that approximates the data. Such **quadratic models** can then be used (subject to limitations) to predict future behavior.

▼ **EXAMPLE 4** The number of women employed full-time in civilian jobs has increased dramatically in the past century. The following table shows the number of employed women (in millions) in selected years.*

Year	1910	1930	1940	1950	1960	1970	1980	1990	2000	2003
Working Women	7.4	10.8	12.8	18.6	23.6	31.5	45.5	56.8	65.6	68.3

(a) Display this information graphically.

Solution Let $x = 0$ correspond to 1900. Plot the points given by the table: $(10, 7.4)$, $(30, 10.8)$, and so on, as in Figure 3.31.

*U.S. Bureau of Labor Statistics.

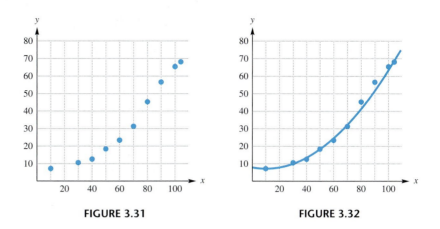

FIGURE 3.31   FIGURE 3.32

(b) The shape of the data points in Figure 3.31 resembles the right half of a parabola. Find a quadratic model $f(x) = a(x - h)^2 + k$ for this data.

Solution Recall that when a quadratic function is written in this form, the vertex of its graph is (h, k). On the basis of Figure 3.31, let $(10, 7.4)$ be the vertex, so that

$$f(x) = a(x - 10)^2 + 7.4.$$

To find a, choose another data point, say, $(103, 68.3)$. If this point is on the parabola, then $f(103) = 68.3$, so that

$$f(x) = a(x - 10)^2 + 7.4$$
$$68.3 = a(103 - 10)^2 + 7.4 \quad \text{Substitute 103 for } x \text{ and 68.3 for } f(x).$$
$$68.3 = 93^2 a + 7.4 \quad \text{Subtract inside parentheses.}$$
$$68.3 = 8649a + 7.4 \quad \text{Square 93.}$$
$$60.9 = 8649a \quad \text{Subtract 7.4 from both sides.}$$
$$a = \frac{60.9}{8649} \approx .00704. \quad \text{Divide both sides by 8649.}$$

Therefore, $f(x) = .00704(x - 10)^2 + 7.4$ is a quadratic model for this data. The graph of $f(x)$ in Figure 3.32 appears to fit the data reasonably well.

(c) Use the quadratic model in part (b) to estimate the number of women in the workforce in 2009.

Solution The year 2009 corresponds to $x = 109$, so the number is approximately

$$f(109) = .00704(109 - 10)^2 + 7.4 \approx 76.4 \text{ million.} \; \blacktriangle \; \text{③}$$

③

Find another quadratic model in Example 4(b) by using $(10, 7.4)$ as the vertex and $(70, 31.4)$ as the other point.

Answer:

$f(x) = .006694(x - 10)^2 + 7.4$

Quadratic Regression

Linear regression was used in Section 2.3 to construct a linear function that modeled a set of data points. When the data points appear to lie on a parabola rather than on a straight line (as in Example 4), a similar least squares regression procedure is available on most graphing calculators and spreadsheet programs to construct a quadratic model for the data. Simply follow the same steps as in linear regression, with one exception: choose quadratic rather than linear regression (they're on the same menu).

▼ **EXAMPLE 5**   Use a graphing calculator to do the following.

(a) Find a quadratic regression model for the data in Example 4.

Solution   Enter the first coordinates of the data points as list L_1 and the second coordinates as list L_2. Performing the quadratic regression, as in Figure 3.33, leads to the model

$$g(x) = .007463x^2 - .16272x + 8.10797.$$

The number R^2 in Figure 3.33 is very close to 1, which indicates that this model fits the data well. The graph of $g(x)$ and the data points in Figure 3.34 visually confirm this agreement.

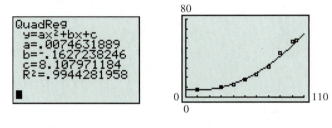

FIGURE 3.33   **FIGURE 3.34**

(b) Use the regression model to estimate the number of women in the workforce in 2009.

Solution   Evaluate $g(x)$ when $x = 109$:

$$g(109) = .007463(109^2) - .16272(109) + 8.10797 \approx 79.0 \text{ million.}$$

This estimate is about 2.6 million higher than the one in Example 4. ▲

◆   ## 3.5 Exercises

Work these problems. (See Example 1.)

1. **Business**   Shannise Cole makes and sells candy. She has found that the cost per box for making x boxes of candy is given by

$$C(x) = x^2 - 40x + 405.$$

(a) How much does it cost per box to make 15 boxes? 18 boxes? 30 boxes?

(b) Graph the cost function $C(x)$, and mark the points corresponding to 15, 18, and 30 boxes.

(c) What point on the graph corresponds to the number of boxes that will make the cost per box as small as possible?

(d) How many boxes should she make in order to keep the cost per box at a minimum? What is the minimum cost per box?

2. **Business**   Greg Tobin sells bottled water. He has found that the average amount of time he spends with each customer is related to his weekly sales volume by the function

$$f(x) = x(60 - x),$$

where x is the number of minutes per customer and $f(x)$ is the number of cases sold per week.

(a) How many cases does he sell if he spends 10 minutes with each customer? 20 minutes? 45 minutes?

(b) Choose an appropriate scale for the axes and sketch the graph of $f(x)$. Mark the points on the graph corresponding to 10, 20, and 45 minutes.

(c) Explain what the vertex of the graph represents.

(d) How long should Greg spend with each customer in order to sell as many cases per week as possible? If he does, how many cases will he sell?

3. Natural Science A researcher in physiology has decided that a good mathematical model for the number of impulses fired after a nerve has been stimulated is given by $y = -x^2 + 20x - 60$, where y is the number of responses per millisecond and x is the number of milliseconds since the nerve was stimulated.
(a) When will the maximum firing rate be reached?
(b) What is the maximum firing rate?

4. Physical Science A bullet is fired upward from ground level. Its height above the ground (in feet) at time t seconds is given by

$$H = -16t^2 + 1000t.$$

Find the maximum height of the bullet and the time at which it hits the ground.

5. Business Colleen Davis owns a factory that manufactures souvenir key chains. Her weekly profit (in hundreds of dollars) is given by $P(x) = -2x^2 + 60x - 120$, where x is the number of cases of key chains sold.
(a) What is the largest number of cases she can sell and still make a profit?
(b) Explain how it is possible for her to lose money if she sells more cases than your answer in part (a).
(c) How many cases should she make and sell in order to maximize her profits?

6. Business The manager of a bicycle shop has found that, at a price (in dollars) of $p(x) = 150 - \dfrac{x}{4}$ per bicycle, x bicycles will be sold.
(a) Find an expression for the total revenue from the sale of x bicycles. (*Hint:* revenue = demand $\times$ price.)
(b) Find the number of bicycle sales that leads to maximum revenue.
(c) Find the maximum revenue.

Work the following problems. (See Example 2.)

7. Business Suppose the supply of and demand for a certain textbook are given by

$$\text{supply: } p = \frac{1}{5}q^2; \qquad \text{demand: } p = -\frac{1}{5}q^2 + 40,$$

where p is price and q is quantity. How many books are demanded at a price of
(a) 10?   (b) 20?   (c) 30?   (d) 40?
How many books are supplied at a price of
(e) 5?   (f) 10?   (g) 20?   (h) 30?
(i) Graph the supply and demand functions on the same axes.

8. Business Find the equilibrium quantity and the equilibrium price in Exercise 7.

9. Business Suppose the price p of widgets is related to the quantity q that is demanded by

$$p = 640 - 5q^2,$$

where q is measured in hundreds of widgets. Find the price when the number of widgets demanded is
(a) 0;   (b) 5;   (c) 10.
Suppose the supply function for widgets is given by $p = 5q^2$, where q is the number of widgets (in hundreds) that are supplied at price p.
(d) Graph the demand function $p = 640 - 5q^2$ and the supply function $p = 5q^2$ on the same axes.
(e) Find the equilibrium quantity.
(f) Find the equilibrium price.

10. Business The supply function for a commodity is given by $p = q^2 + 200$, and the demand function is given by $p = -10q + 3200$.
(a) Graph the supply and demand functions on the same axes.
(b) Find the equilibrium point.
(c) What is the equilibrium quantity? The equilibrium price?

Business *Find the equilibrium quantity and equilibrium price for the commodity whose supply and demand functions are given.*

11. Supply: $p = 45q$; demand: $p = -q^2 + 10,000$

12. Supply: $p = q^2 + q + 10$; demand: $p = -10q + 3060$

13. Supply: $p = q^2 + 20q$; demand: $p = -2q^2 + 10q + 3000$

14. Supply: $p = .2q + 51$; demand: $p = \dfrac{3000}{q + 5}$

Business *The revenue function R and the cost function C for a particular product are given. These functions are valid only for the specified range of values. Find the number of units that must be produced to break even.*

15. $R(x) = 200x - x^2$; $C(x) = 70x + 2200$; $0 \le x \le 100$

16. $R(x) = 300x - x^2$; $C(x) = 65x + 7000$; $0 \le x \le 150$

17. $R(x) = 400x - 2x^2$; $C(x) = -x^2 + 200x + 1900$; $0 \le x \le 100$

18. $R(x) = 500x - 2x^2$; $C(x) = -x^2 + 270x + 5125$; $0 \le x \le 125$

Business *Work each problem. (See Example 3.)*

19. A charter flight charges a fare of \$200 per person, plus \$4 per person for each unsold seat on the plane. If the plane holds 100 passengers and if x represents the number of unsold seats, find the following.
(a) An expression for the total revenue received for the flight. (*Hint:* multiply the number of people flying, $100 - x$, by the price per ticket.)
(b) The graph for the expression of part (a)
(c) The number of unsold seats that will produce the maximum revenue
(d) The maximum revenue

20. The revenue of a charter bus company depends on the number of unsold seats. If 100 seats are sold, the price is $50. Each unsold seat increases the price per seat by $1. Let x represent the number of unsold seats.
  (a) Write an expression for the number of seats that are sold.
  (b) Write an expression for the price per seat.
  (c) Write an expression for the revenue.
  (d) Find the number of unsold seats that will produce the maximum revenue.
  (e) Find the maximum revenue.

21. Farmer Linton wants to find the best time to take her hogs to market. The current price is 88 cents per pound, and her hogs weigh an average of 90 pounds. The hogs gain 5 pounds per week, and the market price for hogs is falling each week, by 2 cents per pound. How many weeks should Ms. Linton wait before taking her hogs to market in order to receive as much money as possible? At that time, how much money (per hog) will she get?

22. The manager of a peach orchard is trying to decide when to arrange for picking the peaches. If they are picked now, the average yield per tree will be 100 pounds, which can be sold for 40¢ per pound. Past experience shows that the yield per tree will increase about 5 pounds per week, while the price will decrease about 2¢ per pound per week.
  (a) Let x represent the number of weeks that the manager should wait. Find the income per pound.
  (b) Find the number of pounds per tree.
  (c) Find the total revenue from a tree.
  (d) When should the peaches be picked in order to produce the maximum revenue?
  (e) What is the maximum revenue?

Work these exercises (See Example 4.)

23. **Business** The yearly spending (in billions of dollars) by U.S. advertisers to promote products to Latinos is shown in the table.*

Year	Spending
2000	2.1
2001	2.2
2002	2.5
2003	2.8
2004	3.1

  (a) Let $x = 0$ correspond to 2000. Use $(0, 2.1)$ as the vertex and the data from 2004 to find a quadratic function $g(x) = a(x - h)^2 + k$ that models this data.
  (b) Use the model to estimate advertisers' spending in 2006 and 2008.

24. **Health** The following table gives the average spending by Medicare for each person enrolled (in 2004 dollars):*

Year	Spending
2000	5000
2001	5900
2002	6200
2003	6600
2004	7400
2005	8000

  (a) Let $x = 0$ correspond to 2000. Use $(0, 5000)$ as the vertex and the data for 2005 to find a quadratic function $f(x) = a(x - h)^2 + k$ that models this data.
  (b) Use the model to estimate the costs per person in 2006 and 2008.

25. **Health** The average health care cost per employee paid by U.S. employers is shown in the table.[†]

Year	Cost
1997	$3820
1999	$4320
2000	$4604
2001	$5162
2002	$5758
2003	$6538

  (a) Let $x = 7$ correspond to 1997. Use $(7, 3820)$ as the vertex and the data from 2003 to find a quadratic function $f(x) = a(x - h)^2 + k$ that models this data.
  (b) Use the model to estimate the costs per employee in 2005 and 2009.

26. **Business** The table shows the China's gross domestic expenditures on applied research (in billions of U.S. dollars).[‡]

Year	Expenditures
1992	2.5
1998	6.5
1999	8.25
2000	10.75
2001	12.5
2002	15.5
2003	18.5

Centers for Medicare and Medicaid Services.

Mercer National Survey of Employer-Sponsored Health Plans.

‡Chinese Ministry of Science and Technology.

Hispanic Business Magazine; St. Louis Post-Dispatch.

(a) Let $x = 2$ correspond to 1992. Use $(2, 2.5)$ and the data from 2003 to find a quadratic function $f(x) = a(x - h)^2 + k$ that models this data.

(b) Use the model to estimate the expenditures on research in 2007 and 2009.

In Exercises 27–30, plot the data points and use quadratic regression to find a function that models the data. (See Example 5.)

27. Business Use the data in Exercise 23, with $x = 0$ corresponding to 2000. What level of spending does this model estimate for 2006 and 2008? Compare your answers with those in Exercise 23.

28. Health Use the data in Exercise 24, with $x = 0$ corresponding to 2000. What per-person costs does this model estimate in 2006 and 2008? Compare your answers with those in Exercise 24.

29. Health Use the data in Exercise 25, with $x = 7$ corresponding to 1997.

(a) On the basis of the value of r^2 in the quadratic regression, how good a model is this?

(b) Use the model to estimate the costs in 2005 and 2009.

30. Business Use the data from Exercise 26, with $x = 2$ corresponding to 1992. Estimate the applied research expenditures in 2007 and 2009.

Work these problems.

31. Geometry A field bounded on one side by a river is to be fenced on three sides to form a rectangular enclosure. There are 320 ft of fencing available. What should the dimensions be to have an enclosure with the maximum possible area?

32. Geometry A rectangular garden bounded on one side by a river is to be fenced on the other three sides. Fencing material for the side parallel to the river costs $30 per foot, and material for the other two sides costs $10 per foot. What are the dimensions of the garden of largest possible area if $1200 is to be spent for fencing material?

Business *Recall that profit equals revenue minus cost. In Exercises 33 and 34, find the following.*

(a) *The break-even point (to the nearest tenth)*
(b) *The x-value that makes profit a maximum*
(c) *The maximum profit*
(d) *For what x-values will a loss occur?*
(e) *For what x-values will a profit occur?*

33. $R(x) = 400x - 2x^2$ and $C(x) = 200x + 2000$, with $0 \le x \le 100$

34. $R(x) = 900x - 3x^2$ and $C(x) = 450x + 5000$, with $20 \le x \le 150$

3.6 Polynomial Functions

A **polynomial function of degree n** is a function whose rule is given by a polynomial of degree n.* For example,

$$f(x) = 3x - 2 \qquad \text{polynomial function of degree 1;}$$
$$g(x) = 3x^2 + 4x - 6 \qquad \text{polynomial function of degree 2;}$$
$$h(x) = 2^4 + 5x^3 - 6x^2 + x - 3 \qquad \text{polynomial function of degree 4.}$$

Basic Graphs

The simplest polynomial functions are those whose rules are of the form $f(x) = ax^n$ (where a is a constant).

▼ **EXAMPLE 1** Graph $f(x) = x^3$.

First, find several ordered pairs belonging to the graph. Be sure to choose some negative x-values, $x = 0$, and some positive x-values to get representative ordered pairs. Find as many ordered pairs as you need in order to see the shape of the graph. Then plot the ordered pairs and draw a smooth curve through them, getting the graph in Figure 3.35 on the next page. ▲ ①

① Graph $f(x) = -\dfrac{1}{2}x^3$

Answer:

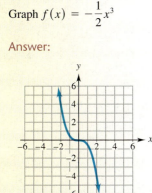

---
*The degree of a polynomial was defined on page 15.

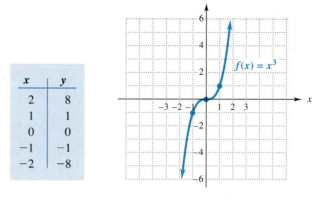

x	y
2	8
1	1
0	0
−1	−1
−2	−8

FIGURE 3.35

▼ **EXAMPLE 2** Graph $f(x) = \dfrac{3}{2}x^4$.

Solution The table below gives some typical ordered pairs and leads to the graph in Figure 3.36. ▲

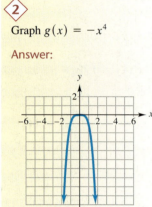

x	f(x)
−2	24
−1	3/2
0	0
1	3/2
2	24

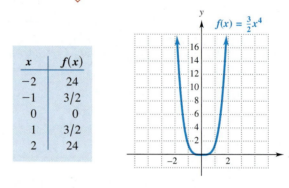

FIGURE 3.36

The graph of $f(x) = ax^n$ has one of the four basic shapes illustrated in Examples 1 and 2 and in side problems 1 and 2, and summarized here.

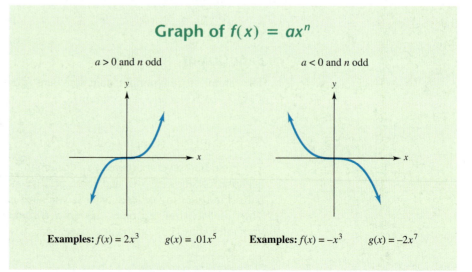

Graph of $f(x) = ax^n$

$a > 0$ and n odd

$a < 0$ and n odd

Examples: $f(x) = 2x^3$ $g(x) = .01x^5$ **Examples:** $f(x) = -x^3$ $g(x) = -2x^7$

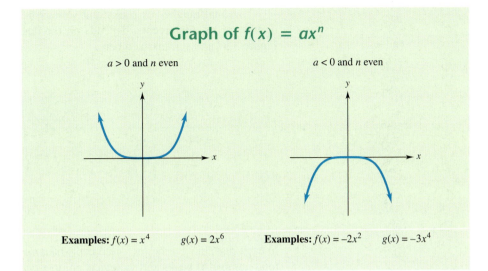

Properties of Polynomial Graphs

Unlike the graphs in the preceding figures, the graphs of more complicated polynomial functions may have several "peaks" and "valleys," as illustrated in Figure 3.37. The locations of the peaks and valleys can be accurately approximated by a maximum or minimum finder on a graphing calculator. Calculus is needed to determine their exact location.

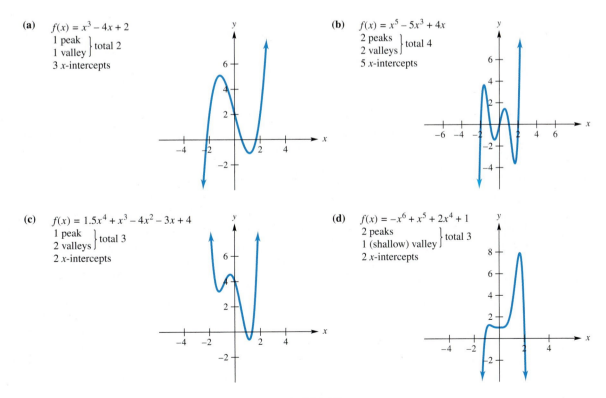

(a) $f(x) = x^3 - 4x + 2$
1 peak
1 valley } total 2
3 x-intercepts

(b) $f(x) = x^5 - 5x^3 + 4x$
2 peaks
2 valleys } total 4
5 x-intercepts

(c) $f(x) = 1.5x^4 + x^3 - 4x^2 - 3x + 4$
1 peak
2 valleys } total 3
2 x-intercepts

(d) $f(x) = -x^6 + x^5 + 2x^4 + 1$
2 peaks
1 (shallow) valley } total 3
2 x-intercepts

FIGURE 3.37

The total number of peaks and valleys in a polynomial graph, as well as the number of its x-intercepts, depends on the degree of the polynomial, as shown in Figure 3.37 and summarized here.

Polynomial	Degree	Number of peaks & valleys	Number of x-intercepts
$f(x) = x^3 - 4x + 2$	3	2	3
$f(x) = x^5 - 5x^3 + 4x$	5	4	5
$f(x) = 1.5x^4 + x^3 - 4x^2 - 3x + 4$	4	3	2
$f(x) = -x^6 + x^5 + 2x^4 + 1$	6	3	2

In each case, the number of x-intercepts is *at most* the degree of the polynomial. The total number of peaks and valleys is at most *one less than* the degree of the polynomial. The same thing is true in every case.

> 1. The total number of peaks and valleys on the graph of a polynomial function of degree n is at most $n - 1$.
>
> 2. The number of x-intercepts on the graph of a polynomial function of degree n is at most n.

The domain of every polynomial function is the set of all real numbers, which means that its graph extends forever to the left and right. We indicate this by the arrows on the ends of polynomial graphs.

Although there may be peaks, valleys, and bends in a polynomial graph, the far ends of the graph are easy to describe: *they look like the graph of the highest-degree term of the polynomial*. Consider, for example, $f(x) = 1.5x^4 + x^3 - 4x^2 - 3x + 4$, whose highest-degree term is $1.5x^4$ and whose graph is shown in Figure 3.37(c). The ends of the graph shoot upward, just as the graph of $y = 1.5x^4$ does in Figure 3.36 on page 180. When $f(x)$ and $y = 1.5x^4$ are graphed in the same large viewing window of a graphing calculator (Figure 3.38), the graphs look almost identical, except near the origin. This is an illustration of the following facts.

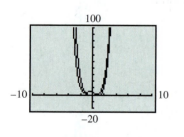

FIGURE 3.38

> The graph of a polynomial function is a smooth, unbroken curve that extends forever to the left and right. When $|x|$ is large, the graph resembles the graph of its highest-degree term and moves sharply away from the x-axis.

▼ **EXAMPLE 3** Let $g(x) = x^3 - 11x^2 - 32x + 24$ and consider the graph in Figure 3.39.

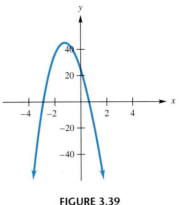

FIGURE 3.39

(a) Is Figure 3.39 a complete graph of $g(x)$; that is, does it show all the important features of the graph?

Solution The far ends of the graph of $g(x)$ should resemble the graph of its highest-degree term x^3. The graph of $f(x) = x^3$ in Figure 3.35 on page 180 moves upward at the far right, but Figure 3.39 does not. So Figure 3.39 is *not* a complete graph.

(b) Use a graphing calculator to find a complete graph of $g(x)$.

Solution Since the graph of $g(x)$ must eventually start rising on the right side (as does the graph of x^3), a viewing window that shows a complete graph must extend beyond $x = 4$. By experimenting with various windows, we obtain Figure 3.40. This graph shows a total of 2 peaks and valleys and 3 x-intercepts (the maximum possible for a polynomial of degree 3). At the far ends, the graph of $g(x)$ resembles the graph of $f(x) = x^3$. Therefore, Figure 3.40 is a complete graph of $g(x)$. ▲ ⟨3⟩

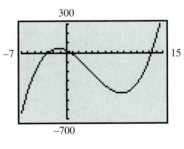

FIGURE 3.40

③ ◤

Find a viewing window on a graphing calculator that shows a complete graph of $f(x) = -.7x^4 + 119x^2 + 400.$ (*Hint:* The graph has two x-intercepts and the maximum possible number of peaks and valleys.)

Answer:

Many correct answers, including $-15 \leq x \leq 15$ and $-2000 \leq y \leq 6000.$

Graphing Techniques

Accurate graphs of first- and second-degree polynomial functions (lines and parabolas) are easily found algebraically, as we saw in Sections 2.2 and 3.4. All polynomial functions of degree 3, and some of higher degree, can be accurately graphed by hand by using calculus and algebra to locate the peaks and valleys. When a polynomial can be completely factored, the general shape of its graph can be determined algebraically by using the basic properties of polynomial graphs, as illustrated in Example 4 on the next page. Accurate graphs of other polynomial functions generally require technology.

▼ **EXAMPLE 4**   Graph $f(x) = (2x + 3)(x - 1)(x + 2)$.

Solution Note that $f(x)$ is a polynomial of degree 3 (if you don't see why, do problem ◄4► at the side). Begin by finding any x-intercepts. Set $f(x) = 0$ and solve for x:

$$f(x) = 0$$
$$(2x + 3)(x - 1)(x + 2) = 0.$$

Solve this equation by placing each of the three factors equal to 0.

$$2x + 3 = 0 \quad \text{or} \quad x - 1 = 0 \quad \text{or} \quad x + 2 = 0$$
$$x = -\frac{3}{2} \qquad\qquad x = 1 \qquad\qquad x = -2$$

The three numbers $-3/2$, 1, and -2 divide the x-axis into four intervals:

$$x < -2, \qquad -2 < x < -\frac{3}{2}, \qquad -\frac{3}{2} < x < 1, \qquad \text{and} \qquad 1 < x.$$

These intervals are shown in Figure 3.41.

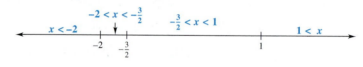

FIGURE 3.41

Since the graph is an unbroken curve, it can change from above the x-axis to below it only by passing through the x-axis. As we have seen, this occurs only at the x-intercepts: $x = -2, -3/2$, and 1. Consequently, in the interval between two intercepts (or to the left of $x = -2$ or to the right of $x = 1$), the graph of $f(x)$ must lie entirely above or entirely below the x-axis.

We can determine where the graph lies over an interval by evaluating $f(x) = (2x + 3)(x - 1)(x + 2)$ at a number in that interval. For example, $x = -3$ is in the interval where $x < -2$, and

$$f(-3) = (2(-3) + 3)(-3 - 1)(-3 + 2)$$
$$= -12.$$

Therefore, $(-3, -12)$ is on the graph. Since this point lies below the x-axis, all points in this interval (that is, all points with $x < -2$) must lie below the x-axis. By testing numbers in the other intervals, we obtain the following table.

Interval	$x < -2$	$-2 < x < -3/2$	$-3/2 < x < 1$	$x > 1$
Test Number	-3	$-7/4$	0	2
Value of f(x)	-12	$11/32$	-6	28
Sign of f(x)	Negative	Positive	Negative	Positive
Graph	Below x-axis	Above x-axis	Below x-axis	Above x-axis

◄4►

Multiply out the expression for $f(x)$ in Example 4 and determine its degree.

Answer:
$f(x) = 2x^3 + 5x^2 - x - 6$;
degree 3.

Since the graph intersects the x-axis at the intercepts   and   and is above the x-axis between these intercepts, there must be at least one peak there. Similarly, there must be at least one valley between $x = -3/2$ and $x = 1$, because the graph is below the x-axis there. However, a polynomial function of degree 3 can have a total of at most $3 - 1 = 2$ peaks and valleys. So there must be exactly one peak and exactly one valley on this graph.

Furthermore, when $|x|$ is large, the graph must resemble the graph of $y = 2x^3$ (the highest-degree term). The graph of $y = 2x^3$, like the graph of $y = x^3$ in Figure 3.35, moves upward to the right and downward to the left. Using these facts and plotting the x-intercepts shows that the graph must have the general shape shown in Figure 3.42. Plotting additional points leads to the reasonably accurate graph in Figure 3.43. We say "reasonably accurate" because we cannot be sure of the exact locations of the peaks and valleys on the graph without using calculus. ▲ ⟨5⟩

⟨5⟩

Graph
$f(x) = .5(x - 1)(x + 2)(x - 3)$.

Answer:

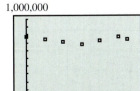

$f(x) = .5(x - 1)(x + 2)(x - 3)$

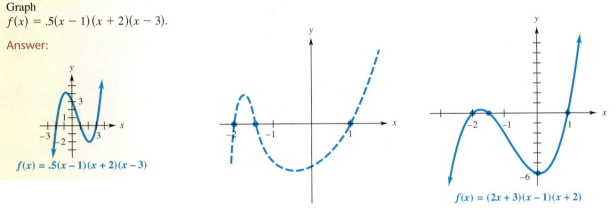

FIGURE 3.42

$f(x) = (2x + 3)(x - 1)(x + 2)$

FIGURE 3.43

Polynomial Models

Regression procedures (similar to linear regression in Section 2.3 and quadratic regression in Section 3.5) can be used to find cubic and quartic (degree 4) polynomial models for appropriate data.

▼ **EXAMPLE 5**   The table shows the population of San Francisco in selected years.

Year	1950	1960	1970	1980	1990	2000	2004
Population	775,357	740,316	715,674	678,974	723,959	776,733	744,230

(a) Plot the data on graphing calculator, with $x = 0$ corresponding to 1950.

Solution The points in Figure 3.44 suggest the general shape of a fourth-degree (quartic) polynomial.

(b) Use quartic regression to obtain a model for this data.

Solution The procedure is the same as for linear regression (just choose "quartic" in place of "linear"). It produces the function

$$f(x) = -.271x^4 + 28.505x^3 - 819.984x^2 + 3440.619x + 772,921.151.$$

FIGURE 3.44

Its graph in Figure 3.45 appears to fit the data well.

(c) Use the model to estimate the population of San Francisco in 1985 and 2005.

Solution 1985 and 2000 correspond to $x = 35$ and $x = 55$, respectively. Verify that

$$f(85) \approx 704{,}345 \quad \text{and} \quad f(55) \approx 744{,}404. \ \blacktriangle$$

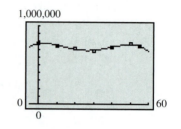

FIGURE 3.45

3.6 Exercises

Graph each of the following polynomial functions. (See Examples 1 and 2.)

1. $f(x) = x^4$

2. $g(x) = -.5x^6$

3. $h(x) = -.2x^5$

4. $f(x) = x^7$

In Exercises 5–8, state whether the graph could possibly be the graph of (a) some polynomial function; (b) a polynomial function of degree 3; (c) a polynomial function of degree 4; (d) a polynomial function of degree 5. (See Example 3.)

5.

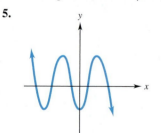

6.

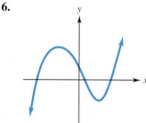

7.

8.

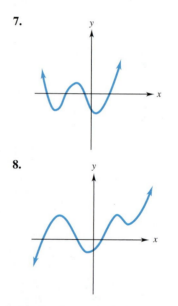

In Exercises 9–14, match the given polynomial function to its graph ((a)–(f)), without using a graphing calculator. (See Example 3 and the two boxes preceding it.)

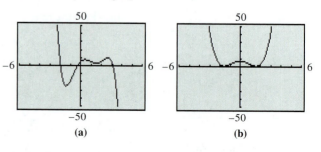

(a)   **(b)**

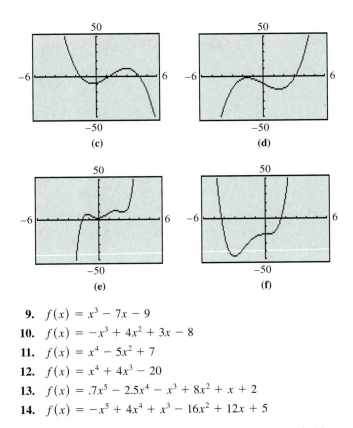

(c)

(d)

(e)

(f)

9. $f(x) = x^3 - 7x - 9$

10. $f(x) = -x^3 + 4x^2 + 3x - 8$

11. $f(x) = x^4 - 5x^2 + 7$

12. $f(x) = x^4 + 4x^3 - 20$

13. $f(x) = .7x^5 - 2.5x^4 - x^3 + 8x^2 + x + 2$

14. $f(x) = -x^5 + 4x^4 + x^3 - 16x^2 + 12x + 5$

Graph each of the following polynomial functions. (See Example 4.)

15. $f(x) = (x + 2)(x - 3)(x + 4)$

16. $f(x) = (x - 3)(x - 1)(x + 1)$

17. $f(x) = x^2(x - 2)(x + 3)$

18. $f(x) = x^2(x + 1)(x - 1)$

19. $f(x) = x^3 + x^2 - 6x$ **20.** $f(x) = x^3 - 2x^2 - 8x$

21. $f(x) = x^3 + 3x^2 - 4x$ **22.** $f(x) = x^4 - 5x^2$

Exercises 23–26 require a graphing calculator. Find a viewing window that shows a complete graph of the polynomial function (that is, a graph which includes all the peaks and valleys and which indicates how the curve moves away from the x-axis at the far left and far right). There are many correct answers. Consider your answer correct if it shows all the features that appear in the window given in the answers. (See Example 3.)

23. $g(x) = x^3 - 3x^2 - 4x - 5$

24. $f(x) = x^4 - 10x^3 + 35x^2 - 50x + 24$

25. $f(x) = 2x^5 - 3.5x^4 - 10x^3 + 5x^2 + 12x + 6$

26. $g(x) = x^5 + 8x^4 + 20x^3 + 9x^2 - 27x - 7$

In Exercises 27–31, use a calculator to evaluate the functions. Generate the graph by plotting points or by using a graphing calculator.

27. **Finance**   An idealized version of the Laffer curve (originated by economist Arthur Laffer) is shown here. According to this theory, decreasing the tax rate, say,

from x_2 to x_1, may actually increase the total revenue to the government. The theory is that people will work harder and earn more if they are taxed at a lower rate, which means higher total tax revenues than would be the case at a higher rate. Suppose that the Laffer curve is given by the function

$$f(x) = \frac{x(x - 100)(x - 160)}{240} \quad (0 \le x \le 100),$$

where $f(x)$ is government revenue (in billions of dollars) from a tax rate of x percent. Find the revenue from the following tax rates.

(a) 20%   **(b)** 40%   **(c)** 50%   **(d)** 70%

(e) Graph $f(x)$

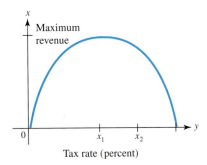

Tax rate (percent)

28. **Health**   A technique for measuring cardiac output depends on the concentration of a dye after a known amount is injected into a vein near the heart. In a normal heart, the concentration of the dye at time x (in seconds) is given by the function defined by

$$g(x) = -.006x^4 + .140x^3 - .053x^2 + 1.79x.$$

(a) Find the following: $g(0)$; $g(1)$; $g(2)$; $g(3)$.

(b) Graph $g(x)$ for $x \ge 0$.

29. **Physical Science**   The pressure of the oil in a reservoir tends to drop with time. By taking sample pressure readings for a particular oil reservoir, petroleum engineers have found that the change in pressure is given by

$$P(t) = t^3 - 18t^2 + 81t,$$

where t is time in years from the date of the first reading.

(a) Find the following: $P(0)$; $P(3)$; $P(7)$; $P(10)$.

(b) Graph $P(t)$.

(c) Over what period is the change (drop) in pressure increasing? decreasing?

30. **Natural Science**   During the early part of the 20th century, the deer population of the Kaibab Plateau in Arizona increased rapidly because hunters had reduced the number of natural predators. The increase in population depleted the food resources and eventually caused the population to decline. For the period from 1905 to 1930, the deer population was approximated by

$$D(x) = -.125x^5 + 3.125x^4 + 4000,$$

where x is time in years from 1905.
(a) Find the following: $D(0)$; $D(5)$; $D(10)$; $D(15)$; $D(20)$; $D(25)$.
(b) Graph $D(x)$.
(c) From the graph, over what period (from 1905 to 1930) was the population increasing? relatively stable? decreasing?

31. Social Science The United Nations projects that the population of China (in millions) will be approximated by

$$g(x) = -.00096x^3 - .1x^2 + 11.3x + 1274 \quad (0 \le x \le 50),$$

where $x = 0$ corresponds to 2000.
(a) What populations does the model predict for 2008 and 2020?
(b) Use your knowledge of polynomial graphs to explain, without graphing, why this model is extremely unlikely to be accurate well beyond 2050.

32. A partial graph of a cubic polynomial function $f(x)$ is shown on the calculator screen. The coefficient of x^3 in the rule of $f(x)$ is negative. Sketch a graph that has the same shape as the entire graph of $f(x)$.

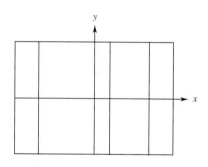

Use a graphing calculator to do the following problems. (See Example 5.)

33. Social Science The table shows the actual and projected enrollment (in millions) in public high schools in selected years.*

Year	Enrollment	Year	Enrollment
1975	14.3	1995	12.5
1980	13.2	2000	13.5
1985	12.4	2005	14.4
1990	11.3	2010	14.1

(a) Plot the data points, with $x = 0$ corresponding to 1975.
(b) Use quartic regression to find a fourth-degree polynomial function $f(x)$ that models this data.
(c) Graph $f(x)$ on the same screen as the data points. Does the graph appear to fit the data well?
(d) According to this model, what is enrollment in 2002 and 2006?
(e) Use your minimum finder to estimate the year between 1975 and 2005 in which enrollment was the lowest.

34. Business The table shows sales in the New York Lottery's numbers game (in million of 2004 dollars).*

Year	Sales	Year	Sales
1981	150	1990	870
1983	400	1994	840
1986	600	1997	780
1988	800	2001	760
1989	880	2004	770

(a) Plot the data points, with $x = 1$ corresponding to 1981, in the viewing window with $0 \le x \le 25$ and $0 \le x \le 1000$.
(b) Use quartic regression to find a fourth-degree polynomial function $g(x)$ that models this data.
(c) Graph $g(x)$ on the same screen as the data points. Does the graph appear to fit the data well?
(d) Estimate lottery sales in 2000 and 2003.
(e) Explain why this model is unlikely to be accurate beyond 2004.

---

*U.S. National Center for Educational Statistics.

*New York Lottery.

3.7 Rational Functions

A **rational function** is a function whose rule is the quotient of two polynomials, such as

$$f(x) = \frac{2}{1 + x}, \qquad g(x) = \frac{3x + 2}{2x + 4}, \qquad h(x) = \frac{x^2 - 2x - 4}{x^3 - 2x^2 + x}.$$

Thus, a rational function is a function whose rule can be written in the form

$$f(x) = \frac{P(x)}{Q(x)},$$

where $P(x)$ and $Q(x)$ are polynomials, with $Q(x) \neq 0$. The function is undefined for any values of x that make $Q(x) = 0$, so there are breaks in the graph at these numbers.

Linear Rational Functions

We begin with rational functions in which both numerator and denominator are first-degree or constant polynomials. Such functions are sometimes called **linear rational functions.**

▼ **EXAMPLE 1** Graph the rational function defined by $y = \dfrac{2}{1 + x}$.

Solution This function is undefined for $x = -1$, since -1 leads to a 0 denominator. For that reason, the graph of this function will not intersect the vertical line $x = -1$. Since x can take on any value except -1, the values of x can approach -1 as closely as desired from either side of -1, as shown in the following table of values.

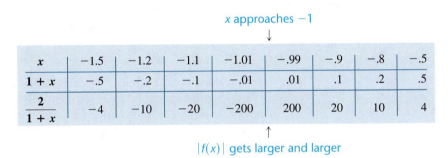

x approaches −1
↓

x	−1.5	−1.2	−1.1	−1.01	−.99	−.9	−.8	−.5
$1 + x$	−.5	−.2	−.1	−.01	.01	.1	.2	.5
$\dfrac{2}{1 + x}$	−4	−10	−20	−200	200	20	10	4

↑
$|f(x)|$ gets larger and larger

The proceeding table suggests that as x gets closer and closer to -1 from either side, $|f(x)|$ gets larger and larger. The part of the graph near $x = -1$ in Figure 3.46 on the next page shows this behavior. The vertical line $x = -1$ that is approached by the curve is called a *vertical asymptote*. For convenience, the vertical asymptote is indicated by a dashed line in Figure 3.46, but this line is *not* part of the graph of the function.

As $|x|$ gets larger and larger, so does the absolute value of the denominator $1 + x$. Hence, $y = 2/(1 + x)$ gets closer and closer to 0, as shown in the following table.

x	−101	−11	−2	0	9	99
$1 + x$	−100	−10	−1	1	10	100
$\dfrac{2}{1 + x}$	−.02	−.2	−2	2	.2	.02

The horizontal line $y = 0$ is called a *horizontal asymptote* for this graph. Using the asymptotes and plotting the intercept and other points gives the graph of Figure 3.46. ▲ ①

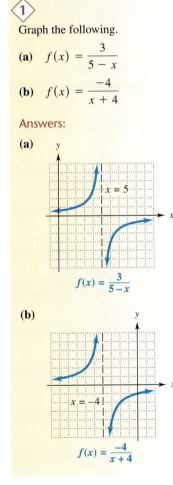

① Graph the following.

(a) $f(x) = \dfrac{3}{5 - x}$

(b) $f(x) = \dfrac{-4}{x + 4}$

Answers:

(a)

$x = 5$

$f(x) = \dfrac{3}{5 - x}$

(b)

$x = -4$

$f(x) = \dfrac{-4}{x + 4}$

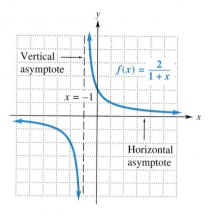

FIGURE 3.46

Example 1 suggests the following conclusion, which applies to all rational functions:

If a number c makes the denominator 0, but the numerator nonzero, in the expression defining a rational function, then the line $x = c$ is a **vertical asymptote** for the graph of the function.

If the graph of a function approaches a horizontal line very closely when x is very large or very small, we say that this line is a **horizontal asymptote** of the graph. In Example 1, the horizontal asymptote was the x-axis. This is not always the case, however, as the next example illustrates.

▼ **EXAMPLE 2**   Graph $f(x) = \dfrac{3x + 2}{2x + 4}$.

Solution Find the vertical asymptote by setting the denominator equal to 0 and then solving for x.

$$2x + 4 = 0$$
$$x = -2$$

In order to see what the graph looks like when $|x|$ is very large, we rewrite the rule of the function. When $x \neq 0$, dividing both numerator and denominator by x does not change the value of the function:

$$f(x) = \frac{3x + 2}{2x + 4} = \frac{\dfrac{3x + 2}{x}}{\dfrac{2x + 4}{x}}$$

$$= \frac{\dfrac{3x}{x} + \dfrac{2}{x}}{\dfrac{2x}{x} + \dfrac{4}{x}} = \frac{3 + \dfrac{2}{x}}{2 + \dfrac{4}{x}}.$$

Now, when $|x|$ is very large, the fractions $2/x$ and $4/x$ are very close to 0 (for instance, when $x = 200$, $4/x = 4/200 = .02$). Therefore, the numerator of $f(x)$ is very close to $3 + 0 = 3$ and the denominator is very close to $2 + 0 = 2$. Hence, $f(x)$ is very close to $3/2$ when $|x|$ is large, so the line $y = 3/2$ is the horizontal asymptote of the graph, as shown in Figure 3.47. ▲ ②

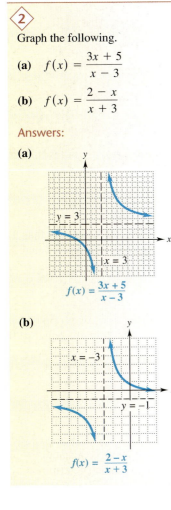

② Graph the following.

(a) $f(x) = \dfrac{3x + 5}{x - 3}$

(b) $f(x) = \dfrac{2 - x}{x + 3}$

Answers:

(a)

$$f(x) = \frac{3x + 5}{x - 3}$$

$y = 3$

$x = 3$

(b)

$x = -3$

$y = -1$

$$f(x) = \frac{2 - x}{x + 3}$$

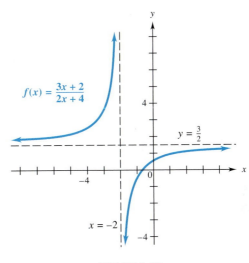

$$f(x) = \frac{3x + 2}{2x + 4}$$

$y = \dfrac{3}{2}$

$x = -2$

FIGURE 3.47

TECHNOLOGY TIP   Depending on the viewing window, a graphing calculator may not accurately represent the graph of a rational function. For example, the graph of $f(x) = \dfrac{3x + 2}{2x + 4}$ in Figure 3.48, which should look like Figure 3.47, has an erroneous vertical line at the place where the graph has a vertical asymptote. This problem can usually be avoided by using a window that has the vertical asymptote at the center of the x-axis, as in Figure 3.49.

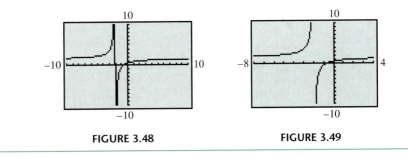

FIGURE 3.48   **FIGURE 3.49**

The horizontal asymptotes of a linear rational function are closely related to the coefficients of the x-terms of the numerator and denominator, as illustrated in Examples 1 and 2:

	Function	Horizontal Asymptote
Example 2:	$f(x) = \dfrac{3x + 2}{2x + 4}$	$y = \dfrac{3}{2}$
Example 1:	$f(x) = \dfrac{2}{1 + x} = \dfrac{0x + 2}{1x + 1}$	$y = \dfrac{0}{1} = 0$ (the x-axis)

The same pattern holds in the general case.

The graph of $f(x) = \dfrac{ax + b}{cx + d}$ (where $c \neq 0$ and $ad \neq bc$) has a vertical asymptote at the root of the denominator and has horizontal asymptote $y = \dfrac{a}{c}$.

Other Rational Functions

When the numerator or denominator of a rational function has degree greater than 1, the graph of the function can be more complicated than those in Examples 1 and 2. The graph may have several vertical asymptotes, as well as peaks and valleys.

▼ **EXAMPLE 3**   Graph $f(x) = \dfrac{2x^2}{x^2 - 4}$.

Solution Find the vertical asymptotes by setting the denominator equal to 0 and solving for x.

$$x^2 - 4 = 0$$
$$(x + 2)(x - 2) = 0$$
$$x + 2 = 0 \quad \text{or} \quad x - 2 = 0$$
$$x = -2 \qquad x = 2$$

Since neither of these numbers makes the numerator 0, the lines $x = -2$ and $x = 2$ are vertical asymptotes of the graph. The horizontal asymptote can be determined by dividing both the numerator and denominator of $f(x)$ by x^2 (the highest power of x that appears in either one).

$$f(x) = \frac{2x^2}{x^2 - 4}$$

$$= \frac{\dfrac{2x^2}{x^2}}{\dfrac{x^2 - 4}{x^2}}$$

$$= \frac{\dfrac{2x^2}{x^2}}{\dfrac{x^2}{x^2} - \dfrac{4}{x^2}}$$

$$= \frac{2}{1 - \dfrac{4}{x^2}}$$

When $|x|$ is very large, the fraction $4/x^2$ is very close to 0, so that the denominator is very close to 1 and $f(x)$ is very close to 2. Hence, the line $y = 2$ is the horizontal asymptote of the graph. Using this information and plotting several points in each of the three regions determined by the vertical asymptotes, we obtain Figure 3.50. ▲

③

List the vertical and horizontal asymptotes of the function.

(a) $f(x) = \dfrac{3x + 5}{x + 5}$

(b) $g(x) = \dfrac{2 - x^2}{x^2 - 4}$

Answers:

(a) Vertical, $x = -5$; horizontal, $y = 3$.

(b) Vertical, $x = -2$ and $x = 2$; horizontal, $y = -1$.

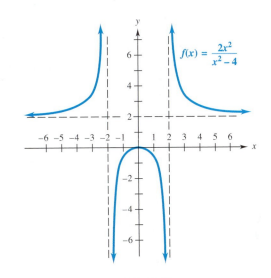

$$f(x) = \frac{2x^2}{x^2 - 4}$$

FIGURE 3.50

TECHNOLOGY TIP   When a function whose graph has more than one vertical asymptote (as in Example 3) is graphed on a graphing calculator, erroneous vertical lines can sometimes be avoided by using a *decimal window* (with the *y*-range adjusted to show the graph). On TI, Sharp, and HP-39+, use (Z)DECIMAL in the ZOOM or VIEWS menu. On Casio, use INIT in the V-WINDOW menu. Figure 3.51 shows the function of Example 3 graphed in a decimal window on a TI-84+. (The *x*-range may be different on other calculators.)

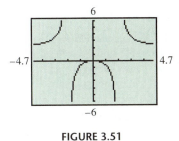

FIGURE 3.51

The arguments used to find the horizontal asymptotes in Examples 1–3 work in the general case and lead to this conclusion.

If the numerator of the rational function $f(x)$ is of *smaller* degree than the denominator, then the *x*-axis (the line $y = 0$) is the horizontal asymptote of the graph. If the numerator and denominator are of the *same* degree, say,

$$f(x) = \frac{ax^n + \cdots}{cx^n + \cdots}, \text{ then the line } y = \frac{a}{c} \text{ is the horizontal asymptote.*}$$

---

*When the numerator is of larger degree than the denominator, the graph has no horizontal asymptote, but may have nonhorizontal lines or other curves as asymptotes; see Exercises 32 and 33 at the end of this section for examples.

Applications

Rational functions have a variety of applications, some of which are explored here.

▼ **EXAMPLE 4** In many situations involving environmental pollution, much of the pollutant can be removed from the air or water at a fairly reasonable cost, but the last, small part of the pollutant can be very expensive to remove.

Cost as a function of the percentage of pollutant removed from the environment can be calculated for various percentages of removal, with a curve fitted through the resulting data points. This curve then leads to a function that approximates the situation. Rational functions often are a good choice for these **cost-benefit functions.**

For example, suppose a cost-benefit function is given by

$$f(x) = \frac{18x}{106 - x},$$

where $f(x)$, or y, is the cost (in thousands of dollars) of removing x percent of a certain pollutant. The domain of x is the set of all numbers from 0 to 100, inclusive; any amount of pollutant from 0% to 100% can be removed. To remove 100% of the pollutant here would cost

$$y = \frac{18(100)}{106 - 100} = 300,$$

or $300,000. Check that 95% of the pollutant can be removed for about $155,000, 90% for about $101,000, and 80% for about $55,000, as shown in Figure 3.52 (in which the displayed y-coordinates are rounded to the nearest integer). ▲ ◈ 4

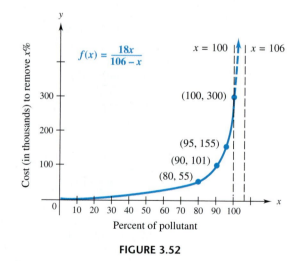

FIGURE 3.52

4

Using the function of Example 4, find the cost to remove the following percentages of pollutants.

(a) 70%

(b) 85%

(c) 98%

Answers:

(a) $35,000

(b) About $73,000

(c) About $221,000

In management, **product-exchange functions** give the relationship between quantities of two items that can be produced by the same machine or factory. For example, an oil refinery can produce gasoline, heating oil, or a combination of the two; a winery can produce red wine, white wine, or a combination of the two. The next example discusses a product-exchange function.

▼ **EXAMPLE 5** The product-exchange function for the Golden Grape Winery for red wine x and white wine y, in tons, is

$$y = \frac{100{,}000 - 50x}{1000 + x}.$$

Graph the function and find the maximum quantity of each kind of wine that can be produced.

Solution Only nonnegative values of x and y make sense in this situation, so we graph the function in the first quadrant (Figure 3.53). Note that the y-intercept of the graph (found by setting $x = 0$) is 100 and the x-intercept (found by setting $y = 0$ and solving for x) is 2000. Since we are interested only in the portion of the graph in Quadrant I, we can find a few more points in that quadrant and complete the graph as shown in Figure 3.53.

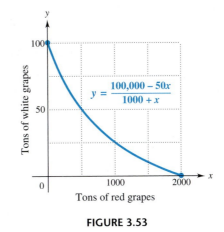

FIGURE 3.53

The maximum value of y occurs when $x = 0$, so the maximum amount of white wine that can be produced is 100 tons, given by the y-intercept. The x-intercept gives the maximum amount of red wine that can be produced, 2000 tons. ▲ ⟨5⟩

⟨5⟩

Rework Example 5 with the product-exchange function

$$y = \frac{60{,}000 - 10x}{60 + x}$$

to find the maximum amount of each wine that can be produced.

Answer:

6000 tons of red, 1000 tons of white

▼ **EXAMPLE 6** A retailer buys 2500 specialty lightbulbs from a distributor each year. In addition to the cost of each bulb, there is an order fee for each order, so she wants to order as few times as possible. However, storage costs are higher when there are fewer orders (and hence more bulbs per order to store). Past experience shows that the total annual cost (for the bulbs, ordering fees, and storage costs) is given by the rational function.

$$C(x) = \frac{.98x^2 + 1200x + 22{,}000}{x},$$

where x is the number of bulbs ordered each time. How many bulbs should be ordered each time in order to have the smallest possible cost?

Graph the cost function $C(x)$ in a window with $0 \le x \le 2500$ (because she can't order a negative number and needs only 2500 for the year).

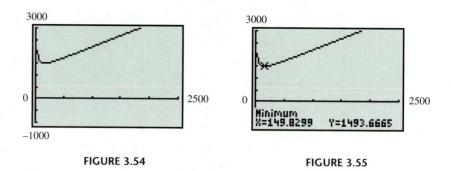

FIGURE 3.54   FIGURE 3.55

Solution For each point on the graph in Figure 3.54,

the x-coordinate is the number of bulbs ordered each time;
the y-coordinate is the annual cost when x bulbs are ordered each time.

Use the minimum finder on a graphing calculator to find the point with the smallest y-coordinate, which is approximately (149.83, 1493.67), as shown in Figure 3.55. Since the retailer can't order part of a lightbulb, she should order 150 bulbs each time, for an approximate annual cost of $1494. ▲

3.7 Exercises

Graph each function. Give the equations of the vertical and horizontal asymptotes. (See Examples 1–3.)

1. $f(x) = \dfrac{1}{x + 5}$

2. $g(x) = \dfrac{-7}{x - 6}$

3. $f(x) = \dfrac{-3}{2x + 5}$

4. $h(x) = \dfrac{-4}{2 - x}$

5. $f(x) = \dfrac{3x}{x - 1}$

6. $g(x) = \dfrac{x - 2}{x}$

7. $f(x) = \dfrac{x + 1}{x - 4}$

8. $f(x) = \dfrac{x - 3}{x + 5}$

9. $f(x) = \dfrac{2 - x}{x - 3}$

10. $g(x) = \dfrac{3x - 2}{x + 3}$

11. $f(x) = \dfrac{2x - 1}{4x + 2}$

12. $f(x) = \dfrac{3x - 6}{6x - 1}$

13. $h(x) = \dfrac{x + 1}{x^2 + 3x - 4}$

14. $g(x) = \dfrac{1}{x(x + 1)^2}$

15. $f(x) = \dfrac{x^2 + 1}{x^2 - 1}$

16. $f(x) = \dfrac{x - 1}{x^2 - x - 6}$

Find the equations of the vertical asymptotes of each of the following rational functions.

17. $f(x) = \dfrac{x - 3}{x^2 + x - 2}$

18. $g(x) = \dfrac{x + 2}{x^2 - 1}$

19. $g(x) = \dfrac{x^2 + 2x}{x^2 - 4x - 5}$

20. $f(x) = \dfrac{x^2 - 2x - 4}{x^3 - 2x^2 + x}$

Work these problems. (See Example 4.)

21. Natural Science Suppose a cost-benefit model is given by

$$f(x) = \frac{4.3x}{100 - x},$$

where $f(x)$ is the cost, in thousands of dollars, of removing x percent of a given pollutant. Find the cost of removing each of the following percentages of pollutants.
(a) 50%
(b) 70%
(c) 80%
(d) 90%
(e) 95%
(f) 98%
(g) 99%
(h) Is it possible, according to this model, to remove *all* the pollutant?
(i) Graph the function.

22. Natural Science Suppose a cost-benefit model is given by

$$f(x) = \frac{4.5x}{101 - x},$$

where $f(x)$ is the cost, in thousands of dollars, of removing x percent of a certain pollutant. Find the cost of removing the following percentages of pollutants.
(a) 0%
(b) 50%

(c) 80%
(d) 90%
(e) 95%
(f) 99%
(g) 100%
(h) Graph the function.

23. **Natural Science** The function

$$f(x) = \frac{\lambda x}{1 + (ax)^b}$$

is used in population models to give the size of the next generation $f(x)$ in terms of the current generation x.*
(a) What is a reasonable domain for this function, considering what x represents?
(b) Graph the function for $\lambda = a = b = 1$ and $x \geq 0$.
(c) Graph the function for $\lambda = a = 1$ and $b = 2$ and $x \geq 0$.
(d) What is the effect of making b larger?

24. **Natural Science** The function

$$f(x) = \frac{Kx}{A + x}$$

is used in biology to give the growth rate of a population in the presence of a quantity x of food. This is called Michaelis-Menten kinetics.[†]
(a) What is a reasonable domain for this function, considering what x represents?
(b) Graph the function for $K = 5$, $A = 2$, and $x \geq 0$.
(c) Show that $y = K$ is a horizontal asymptote.
(d) What do you think K represents?
(e) Show that A represents the quantity of food for which the growth rate is half of its maximum.

25. **Social Science** The average waiting time in a line (or queue) before getting served is given by

$$W = \frac{S(S - A)}{A},$$

where A is the average rate at which people arrive at the line and S is the average service time. At a certain fast-food restaurant, the average service time is 3 minutes. Find W for each of the following average arrival times.
(a) 1 minute   (b) 2 minutes
(c) 2.5 minutes
(d) What is the vertical asymptote?
(e) Graph the equation on the interval (0, 3).
(f) What happens to W when $A > 3$? What does this mean?

Business *Sketch the portion of the graph in Quadrant I of each of the functions defined below, and then estimate the maximum quantities of each product that can be produced. (See Example 5.)*

26. The product-exchange function for gasoline x and heating oil y, in hundreds of gallons per day, is

$$y = \frac{125,000 - 25x}{125 + 2x}.$$

27. A drug factory found that the product-exchange function for a red tranquilizer x and a blue tranquilizer y is

$$y = \frac{900,000,000 - 30,000x}{x + 90,000}.$$

28. **Physical Science** The failure of several O-rings in field joints was the cause of the fatal crash of the *Challenger* space shuttle in 1986. NASA data from 24 successful launches prior to *Challenger* suggested that O-ring failure was related to launch temperature by a function similar to

$$N(t) = \frac{600 - 7t}{4t - 100} \qquad (50 \leq t \leq 85),$$

where t is the temperature (in °F) at launch and N is the approximate number of O-rings that fail. Assume that this function accurately models the number of O-ring failures that would occur at lower launch temperatures (an assumption NASA did not make).
(a) Does $N(t)$ have a vertical asymptote? At what value of t does it occur?
(b) Without graphing, what would you conjecture that the graph would look like just to the right of the vertical asymptote? What does this suggest about the number of O-ring failures that might be expected near that temperature? (The temperature at the *Challenger* launching was 31°.)
(c) Confirm your conjecture by graphing $N(t)$ between the vertical asymptote and $t = 85$.

29. **Business**   A company has fixed costs of $40,000 and a marginal cost of $2.60 per unit.
(a) Find the linear cost function.
(b) Find the average cost function. (Average cost was defined in Section 3.3.)

---

*See J. Maynard Smith, *Models in Ecology* (Cambridge University Press, 1974).

[†]See Leah Edelstein-Keshet, *Mathematical Models in Biology* (Random House, 1988).

(c) Find the horizontal asymptote of the graph of the average cost function. Explain what the asymptote means in this situation. (How low can the average cost be?)

Use a graphing calculator to do Exercises 30–33. (See Example 6.)

30. **Finance**   Another model of a Laffer curve (see Exercise 27 of Section 3.6) is given by

$$f(x) = \frac{300x - 3x^2}{10x + 200},$$

where $f(x)$ is government revenue (in billions of dollars) from a tax rate of x percent. Find the revenue from the following tax rates.
(a) 16%   (b) 25%
(c) 40%   (d) 55%
(e) Graph $f(x)$.
(f) What tax rate produces maximum revenue? What is the maximum revenue?

31. **Business**   When no more than 110 units are produced, the cost of producing x units is given by

$$C(x) = .2x^3 - 25x^2 + 1531x + 25,000.$$

How many units should be produced in order to have the lowest possible average cost?

32. (a) Graph $f(x) = \dfrac{x^3 + 3x^2 + x + 1}{x^2 + 2x + 1}$.

(b) Does the graph appear to have a horizontal asymptote? Does the graph appear to have some nonhorizontal straight line as an asymptote?

(c) Graph $f(x)$ and the line $y = x + 1$ on the same screen. Does this line appear to be an asymptote of the graph of $f(x)$?

33. (a) Graph $g(x) = \dfrac{x^3 - 2}{x - 1}$ in the window with $-4 \le x \le 6$ and $-6 \le y \le 12$.

(b) Graph $g(x)$ and the parabola $y = x^2 + x + 1$ on the same screen. How do the two graphs compare when $|x| \ge 2$?

Chapter 3 Summary

Key Terms and Symbols

3.1   function
domain
range
functional notation
piecewise-defined function

3.2   graph
linear function
piecewise linear function
absolute value function
greatest integer function
step function
graph reading
vertical-line test

3.3   fixed costs
variable cost
average cost
linear depreciation
average rate of change
marginal cost
linear cost function
linear revenue function
break-even point
supply and demand curves
equilibrium point
equilibrium price
equilibrium quantity

3.4   quadratic function
parabola
vertex
axis

3.5   quadratic model

3.6   polynomial function
graph of $f(x) = ax^n$
properties of polynomial graphs
polynomial models

3.7   rational function
linear rational function
vertical asymptote
horizontal asymptote

Key Concepts

A **function** consists of a set of inputs called the **domain,** a set of outputs called the **range,** and a rule by which each input determines exactly one output.

If a vertical line intersects a graph in more than one point, the graph is not that of a function.

A **linear cost function** has equation $C(x) = mx + b$, where m is the **marginal cost** (the cost of producing one more item) and b is the **fixed cost.**

If $p = f(q)$ gives the price per unit when q units can be supplied and $p = g(q)$ gives the price per unit when q units are demanded, then the **equilibrium price** and **equilibrium quantity** occur at the q-value such that $f(q) = g(q)$.

The **quadratic function** defined by $f(x) = a(x - h)^2 + k$ has a graph that is a **parabola** with vertex (h, k) and axis of symmetry $x = h$. The parabola opens upward if $a > 0$, downward if $a < 0$.

If the equation is in the form $f(x) = ax^2 + bx + c$, the vertex is $\left(-\dfrac{b}{2a}, f\left(-\dfrac{b}{2a}\right)\right)$.

When $|x|$ is large, the graph of a **polynomial function** resembles the graph of its highest-degree term ax^n. The graph of $f(x) = ax^n$ is described on pages 180–181.

On the graph of a polynomial function of degree n:

The total number of peaks and valleys is at most $n - 1$.

The total number of x-intercepts is at most n.

If a number c makes the denominator of a **rational function** 0, but the numerator nonzero, then the line $x = c$ is a **vertical asymptote** of the graph.

Whenever the values of y approach, but do not equal, some number k as $|x|$ gets larger and larger, the line $y = k$ is a **horizontal asymptote** of the graph.

If the numerator of a rational function is of *smaller* degree than the denominator, then the x-axis is the horizontal asymptote of the graph.

If the numerator and denominator of a rational function are of the *same* degree, say,
$$f(x) = \frac{ax^n + \cdots}{cx^n + \cdots},$$ then the line $y = \dfrac{a}{c}$ is the horizontal asymptote of the graph.

Chapter 3 Review Exercises

Which of the following rules defines a function?

1.

x	3	2	1	0	1	2
y	8	5	2	0	-2	-5

2.

x	2	1	0	-1	-2
y	5	3	1	-1	-3

3. $y = \sqrt{x}$

4. $x = |y|$

5. $x = y^2 + 1$

6. $y = 5x - 2$

For each function, find

    (a) $f(6)$ **(b)** $f(-2)$ **(c)** $f(p)$ **(d)** $f(r + 1)$.

7. $f(x) = 4x - 1$

8. $f(x) = 3 - 4x$

9. $f(x) = -x^2 + 2x - 4$

10. $f(x) = 8 - x - x^2$

11. Let $f(x) = 5x - 3$ and $g(x) = -x^2 + 4x$. Find each of the following.

    (a) $f(-2)$ **(b)** $g(3)$ **(c)** $g(-k)$

    (d) $g(3m)$ **(e)** $g(k - 5)$ **(f)** $f(3 - p)$

12. Let $f(x) = x^2 + x + 1$. Find each of the following.

    (a) $f(3)$ **(b)** $f(1)$ **(c)** $f(4)$

    (d) Based on your answers in parts (a)–(c), is it true that $f(a + b) = f(a) + f(b)$ for all real numbers a, b?

Graph each function.

13. $f(x) = |x| - 3$

14. $f(x) = -|x| - 2$

15. $f(x) = -|x + 1| + 3$

16. $f(x) = 2|x - 3| - 4$

17. $f(x) = [x - 3]$

18. $f(x) = \left[\dfrac{1}{2}x - 2\right]$

19. $f(x) = \begin{cases} -4x + 2 & \text{if } x \le 1 \\ 3x - 5 & \text{if } x > 1 \end{cases}$

20. $f(x) = \begin{cases} 3x + 1 & \text{if } x < 2 \\ -x + 4 & \text{if } x \ge 2 \end{cases}$

21. $f(x) = \begin{cases} |x| & \text{if } x < 3 \\ 6 - x & \text{if } x \ge 3 \end{cases}$

22. $f(x) = \sqrt{x^2}$

23. $g(x) = x^2/8 - 3$

24. $h(x) = \sqrt{x} + 2$

25. Business Let f be a function that gives the cost to rent a floor polisher for x hours. The cost is a flat $3 for renting the polisher, plus $4 per day or fraction of a day for using the polisher.
(a) Graph f.
(b) Give the domain and range of f.
(c) David Fleming wants to rent a polisher, but he can spend no more than $15. At most how many days can he use it?

26. Business A trailer hauling service charges $45, plus $2 per mile or part of a mile.
(a) Is $90 enough for a 20-mile haul?
(b) Graph the ordered pairs (miles, cost).
(c) Give the domain and range.

27. Social Science The percentage of children born to unmarried mothers since 1970 can be approximated by

$$f(x) = \begin{cases} .9x + 11 & \text{from 1970 to 1994} \\ 32.6 & \text{from 1994 to 2010} \end{cases}.$$

Let $x = 0$ correspond to 1970, and graph the function. What does the graph suggest about births to unmarried mothers?

28. Social Science According to data from the U.S. Department of Agriculture, the annual per capita consumption of red meat in the United States has declined in a linear fashion from 131.7 pounds in 1970 to 107.7 in 2000.
(a) Let $x = 0$ correspond to 1900, and find the rule of a linear function that gives the approximate annual per capita consumption of red meat in year x.
(b) If this trend continues, in what year will per capita consumption be 100 pounds?

Business In Exercises 29–32, find the following:

(a) *the linear cost function*
(b) *the marginal cost*
(c) *the average cost per unit to produce 100 units*

29. Eight units cost $300; fixed cost is $60.

30. Fixed cost is $2000; 36 units cost $8480.

31. Twelve units cost $445; 50 units cost $1585.

32. Thirty units cost $1500; 120 units cost $5640.

33. Business The cost of producing x ink cartridges for a printer is given by $C(x) = 24x + 18{,}000$. Each cartridge can be sold for $28.
(a) What are the fixed costs?
(b) Find the revenue function.
(c) Find the break-even point.
(d) If the company sells exactly the number of cartridges needed to break even, what is its revenue?

34. Business The prime-time viewing share of network TV stations in year x is approximated by

$$f(x) = -1.2x + 68 \qquad (9 \le x \le 22),$$

where $x = 0$ corresponds to 1980. The share of basic cable stations is approximated by $g(x) = 2.3x - 3.6.$* In what year did network and cable stations have the same viewing share?

35. Business Suppose the demand and price for the HBO cable channel are related by $p = -.5q + 30.95$, where p is the monthly price in dollars and q is measured in millions of subscribers. If the price and supply are related by $p = .3q + 2.15$, what are the equilibrium quantity and price?

36. Business Suppose the supply and price for prescription-strength Tylenol are related by $p = .0015q + 1$, where p is the price (in dollars) of a 30-day prescription. If the demand is related to price by $p = -.0025q + 64.36$, what are the equilibrium quantity and price?

Without graphing, determine whether each of the following parabolas opens upward or downward and find its vertex.

37. $f(x) = 3(x - 2)^2 + 6$ **38.** $f(x) = 2(x + 3)^2 - 5$
39. $g(x) = -4(x + 1)^2 + 8$ **40.** $g(x) = -5(x - 4)^2 - 6$

Graph each of the following and label its vertex.

41. $f(x) = x^2 - 4$ **42.** $f(x) = 6 - x^2$
43. $f(x) = x^2 + 2x - 3$ **44.** $f(x) = -x^2 + 6x - 3$
45. $f(x) = -x^2 - 4x + 1$ **46.** $f(x) = 4x^2 - 8x + 3$
47. $f(x) = 2x^2 + 4x - 3$ **48.** $f(x) = -3x^2 - 12x - 8$

Determine whether each of the following functions has a minimum or a maximum value, and find that value.

49. $f(x) = x^2 + 6x - 2$ **50.** $f(x) = x^2 + 4x + 5$
51. $g(x) = -4x^2 + 8x + 3$
52. $g(x) = -3x^2 - 6x + 3$

Solve each problem.

53. Business The commodity market is very unstable; money can be made or lost quickly when investing in soybeans, wheat, pork bellies, and the like. Suppose that an investor kept track of her total profit P at time t, measured in months, after she began investing and found that $P = -4t^2 + 32t - 20$. At what time is her profit largest? (*Hint:* $t > 0$ in this case.)

54. Physical Science The height h (in feet) of a rocket at time t seconds after liftoff is given by $h = -16t^2 + 800t$.
(a) How long does it take the rocket to reach 3200 feet?
(b) What is the maximum height of the rocket?

55. Business The manager of a large apartment complex has found that the profit is given by $P = -x^2 + 250x - 15{,}000$, where x is the number of units rented. For what value of x does the complex produce the largest profit?

---

*Based on data from the National Cable Television Association.

56. Business A rectangular enclosure is to be built with three sides made out of redwood fencing at a cost of $15 per running foot and the fourth side made out of cement blocks at a cost of $30 per running foot. $900 is available for the project. What are the dimensions of the enclosure with maximum possible area, and what is this area?

57. Health The table shows national health care expenditures (in trillions of dollars).*

Year	Expenditures
1985	.38
1990	.7
1995	.94
2000	1.3
2005	2

(a) Let $x = 5$ correspond to 1985. Find a quadratic function $f(x) = a(x - h)^2 + k$ that models this data, using $(5, .38)$ as vertex and the data for 2005.

(b) Estimate health care expenditures in 2008 and 2010.

58. Business The estimated cost of a four-year college education (tuition, fees, books, and room and board) at a public institution is shown in the table.[†]

Enrollment Year	Cost
2002	$58,946
2004	$66,232
2006	$74,418
2008	$83,616
2010	$93,951
2012	$105,564
2014	$118,611

(a) Sketch a scatter plot of the data, with $x = 2$ corresponding to 2002.

(b) Use quadratic regression to find a function g that models this data.

(c) Estimate the costs for a student enrolling in 2009 and for one in 2015.

Graph each of the following polynomial functions.

59. $f(x) = x^4 - 2$ **60.** $g(x) = x^3 - x$

61. $f(x) = x(x - 2)(x + 3)$

62. $f(x) = (x - 1)(x + 2)(x - 3)$

63. $f(x) = 3x(3x + 2)(x - 1)$

64. $f(x) = x^3 - 3x^2 - 4x$

65. $f(x) = x^4 - 5x^2 - 6$ **66.** $f(x) = x^4 - 7x^2 - 8$

Use a graphing calculator to do Problems 67–70.

67. Business The demand equation for automobile oil filters is

$$p = -.000012q^3 - .00498q^2 + .1264q + 1508,$$

where p is in dollars and q is in thousands of items. The supply equation is

$$p = -.000001q^3 + .00097q^2 + 2q.$$

Find the equilibrium quantity and the equilibrium price.

68. Business The average cost (in dollars) per item of manufacturing x thousand cans of spray paint is given by

$$A(x) = -.000006x^4 + .0017x^3 + .03x^2 - 24x + 1110.$$

How many cans should be manufactured if the average cost is to be as low as possible? What is the average cost in that case?

69. Business Plastic racks for holding compact discs sell for $23 each. The cost of manufacturing x racks is given by $C(x) = -.000006x^3 + .07x^2 + 2x + 1200$. At most 600 racks can be produced.

(a) Find the revenue and profit functions.

(b) What is the break-even point?

(c) What is the maximum number of racks that can be made without losing money?

(d) How many racks should be made in order to have as large a profit as possible? What is that profit?

70. Health The table shows the average remaining life expectancy (in years) of a male at selected ages.*

Current Age	Birth	20	40	60	80	90	100
Life Expectancy	73.6	54.7	36.2	19.4	7.5	4	2.4

(a) Let birth $= 0$, and plot the data points.

(b) Use quartic regression to find a fourth-degree polynomial $f(x)$ that models the data.

(c) What is the life expectancy of a man of age 25? Age 35? Age 50?

(d) What is the life expectancy of a man who is exactly your age?

---

*Centers for Medicare and Medicaid Services.

[†]Teachers Insurance and Annuity Association College Retirement Equities Fund.

*National Center for Health Statistics.

List the vertical and horizontal asymptotes of each function, and sketch its graph.

71. $f(x) = \dfrac{1}{x-3}$

72. $f(x) = \dfrac{-2}{x+4}$

73. $f(x) = \dfrac{-3}{2x-4}$

74. $f(x) = \dfrac{5}{3x+7}$

75. $g(x) = \dfrac{5x-2}{4x^2-4x-3}$

76. $g(x) = \dfrac{x^2}{x^2-1}$

77. Business The average cost per carton of producing x cartons of cocoa is given by

$$C(x) = \dfrac{650}{2x+40}.$$

Find the average cost per carton to make the given number of cartons.

(a) 10 cartons   **(b)** 50 cartons

(c) 70 cartons   **(d)** 100 cartons

(e) Graph $C(x)$.

78. Business The cost and revenue functions (in dollars) for a frozen-yogurt shop are given by

$$C(x) = \dfrac{400x+400}{x+4} \quad \text{and} \quad R(x) = 100x,$$

where x is measured in hundreds of units.

(a) Graph $C(x)$ and $R(x)$ on the same axes.

(b) What is the break-even point for this shop?

(c) If the profit function is given by $P(x)$, does $P(1)$ represent a profit or a loss?

(d) Does $P(4)$ represent a profit or a loss?

79. Business The supply and demand functions for the yogurt shop in Exercise 78 are

$$\text{supply: } p = \dfrac{q^2}{4} + 25; \qquad \text{demand: } p = \dfrac{500}{q},$$

where p is the price in dollars for q hundred units of yogurt.

(a) Graph both functions on the same axes, and from the graph, estimate the equilibrium point.

(b) Give the q-intervals where supply exceeds demand.

(c) Give the q-intervals where demand exceeds supply.

80. Business A cost-benefit curve for pollution control is given by

$$y = \dfrac{9.2x}{106-x},$$

where y is the cost in thousands of dollars of removing x percent of a specific industrial pollutant. Find y for each of the following values of x.

(a) $x = 50$ **(b)** $x = 98$

(c) What percent of the pollutant can be removed for $22,000?

Case 3

Architectural Arches

From ancient Roman bridges, to medieval cathedrals, to modern highway tunnels and fast-food restaurants, arches are everywhere. For centuries, builders and architects have used them for both structural and esthetic reasons. There are arches of almost every material, from stone to steel. Some of the most common arch shapes are the parabolic arch, the semicircular arch, and the Norman arch (a semicircular arch set atop a rectangle, as shown here).

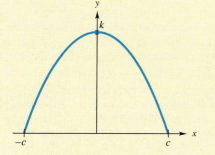

| Parabolic | Semicircular | Norman |

Note that every arch is symmetric around a vertical line through its center. The part of the arch on the left side of the line is mirror image of the part on the right side, with the line being the mirror. To describe these arches mathematically, suppose that each one is located on a coordinate plane with the origin at the intersection of the verticle symmetry line and a horizontal line at the base of the arch. For a parabolic arch, the situation looks like this, for some numbers k and c:

Since $(0, k)$ is the vertex, the arch is the graph of a function of the form

$$f(x) = a(x - 0)^2 + k$$
$$f(x) = ax^2 + k.$$

Note that the point $(c, 0)$ is on the arch, which means that $f(c) = 0$. Therefore,

$$f(x) = ax^2 + k.$$
$$f(c) = ac^2 + k \qquad \text{Let } x = c.$$
$$0 = ac^2 + k \qquad f(c) = 0.$$
$$-k = ac^2 \qquad \text{Subtract } k \text{ from both sides.}$$
$$a = \frac{-k}{c^2} \qquad \text{Divide both sides by } c^2.$$

So the function whose graph is the shape of the parabolic arch is

$$f(x) = \frac{-k}{c^2}x^2 + k \qquad (-c \le x \le c),$$

where k is the height of the arch at its highest point and $2c$ (the distance from $-c$ to c) is the width of the arch at its base. For example, a 12-ft-high arch that is 14 ft wide at its base has $k = 12$ and $c = 7$, so it is the graph of

$$f(x) = \frac{-12}{49}x^2 + 12.$$

In order to describe semicircular and Norman arches, we must first find the equation of a circle or radius r with center at the origin. Consider a point (x, y) on the graph and the right triangle it determines:*

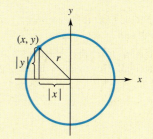

---

*The figure shows a point in the second quadrant, where x is negative and y is positive, but the same argument will work for points in other quadrants.

The horizontal side of the triangle has length $|x|$ (the distance from x to 0 on the x-axis), and the vertical side has length $|y|$ (the distance from y to 0 on the y-axis). The hypotenuse has length r (the radius of the circle). By the Pythagorean theorem,

$$|x|^2 + |y|^2 = r^2,$$

which is equivalent to

$$x^2 + y^2 = r^2,$$

because the absolute value of x is either x or $-x$ (see the definition on page 9), so $|x|^2 = (\pm x)^2 = x^2$, and similarly for y. Solving this equation for y shows that

$$y^2 = r^2 - x^2$$
$$y = \sqrt{r^2 - x^2} \quad \text{or} \quad y = -\sqrt{r^2 - x^2}.$$

In the first equation, y is always positive or 0 (because square roots are nonnegative), so its graph is the top half of the circle. Similarly, the second equation gives the bottom half of the circle.

Now consider a semicircle arch of radius r:

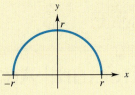

The arch is the top half of a circle with center at the origin. By the preceding paragraph, it is the graph of the function

$$g(x) = \sqrt{r^2 - x^2} \quad (-r \le x \le r).$$

For instance, a semicircular arch that is 8 feet high has $r = 8$ and is the graph of

$$g(x) = \sqrt{8^2 - x^2} = \sqrt{64 - x^2} \quad (-8 \le x \le 8).$$

A Norman arch is not the graph of a function (since its sides are vertical lines), but we can describe the semicircular top of the arch. For example, consider a Norman arch whose top has radius 8 and whose sides are 10 feet high. If the top of the arch were at ground level, then its equation would be $g(x) = \sqrt{64 - x^2}$, as we just saw. But in the actual arch, this semicircular part is raised 10 feet, so it is the graph of

$$h(x) = g(x) + 10 = \sqrt{64 - x^2} + 10 \quad (-8 \le x \le 8),$$

as shown below.

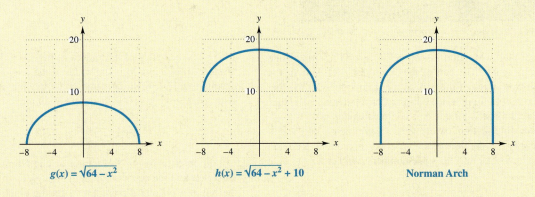

$g(x) = \sqrt{64 - x^2}$ $h(x) = \sqrt{64 - x^2} + 10$ **Norman Arch**

◆ Exercises

1. Write a function $f(x)$ that describes a parabolic arch that is 15 feet tall and 16 feet wide at the base.

2. Write a function $g(x)$ that describes a semicircular arch that is 15 feet tall. How wide is the arch at its base?

3. Write a function $h(x)$ that describes the top part of a Norman arch that is 15 feet tall and 16 feet wide at the base. How high are the vertical sides of this arch?

4. Would a truck that is 12 feet tall and 9 feet wide fit through all of the arches in Exercises 1–3? How could you fix any of the arches that are too small so that the truck would fit through?

Chapter 4

Exponential and Logarithmic Functions

Population growth (of humans, fish, bacteria, etc.), compound interest, radioactive decay, and a host of other phenomena can be described by exponential functions. See Exercises 40–44 on page 214. Archeologists sometimes use carbon-14 dating to determine the approximate age of an artifact (such as a dinosaur skeleton or a mummy). This procedure involves using logarithms to solve an exponential equation. See Exercises 68–70 on page 241. The Richter scale for measuring the magnitude of an earthquake is a logarithmic function. See Exercise 71 on page 241.

Exponential and logarithmic functions play a key role in management, economics, the social and physical sciences, and engineering. We begin with exponential growth and exponential decay functions.

4.1 Exponential Functions

In polynomial functions, the variable is raised to various constant exponents, as in, for instance, $f(x) = x^2 + 3x - 5$. In **exponential functions,** such as

$$f(x) = 10^x \quad \text{or} \quad g(x) = 750(1.05^x) \quad \text{or} \quad h(x) = 3^{.6x} \quad \text{or} \quad k(x) = 2^{-x^2},$$

the variable is in the exponent and the **base** is a positive constant. We begin with the simplest type of exponential function, whose rule is of the form $f(x) = a^x$ with $a > 0$. Exponential functions with the base $a < 0$ are not of interest because, when a is negative, a^x may not be defined; for example, $(-4)^{1/2} = \sqrt{-4}$, which is not a real number.

▼ **EXAMPLE 1**   Graph $f(x) = 2^x$, and estimate the height of the graph when $x = 50$.

Solution   Either use a graphing calculator or graph by hand: make a table of values, plot the corresponding points, and join them by a smooth curve, as in Figure 4.1. The graph has y-intercept 1 and rises steeply to the right. Note that the graph gets very close to the x-axis on the left, but always lies *above* the axis (because *every* power of 2 is positive).

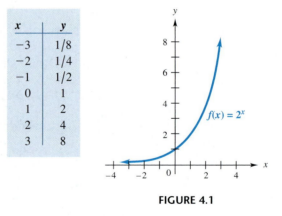

x	y
-3	$1/8$
-2	$1/4$
-1	$1/2$
0	1
1	2
2	4
3	8

FIGURE 4.1

The graph illustrates **exponential growth,** which is far more explosive than polynomial growth. At $x = 50$, the graph would be 2^{50} units high. Since there are approximately 6 units to the inch in Figure 4.1, and since there are 12 inches to the foot and 5280 feet to the mile, the height of the graph at $x = 50$ would be approximately

$$\frac{2^{50}}{6 \times 12 \times 5280} \approx 2{,}961{,}647{,}482 \ \textit{miles!} \quad ▲ \quad \langle 1 \rangle$$

(a)   Fill in this table:

x	$g(x) = 3^x$
-3	
-2	
-1	
0	
1	
2	
3	

(b)   Sketch the graph of $g(x) = 3^x$.

Answers:

(a)   The entries in the second column are $1/27, 1/9, 1/3, 1,$ $3, 9, 27$.

(b)

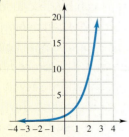

When $a > 1$, the graph of the exponential function $h(x) = a^x$ has the same basic shape as the graph of $f(x) = 2^x$, as illustrated in Figure 4.2 and summarized in the box below.

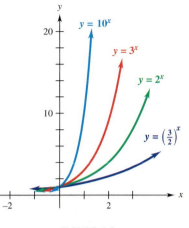

FIGURE 4.2

When $a > 1$, the function $f(x) = a^x$ has the set of all real numbers as its domain. Its graph has the shape shown here and all five of the properties listed below.

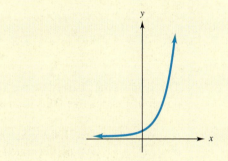

1. The graph is above the x-axis.
2. The y-intercept is 1.
3. The graph climbs steeply to the right.
4. The negative x-axis is a horizontal asymptote.
5. The larger the base a, the more steeply the graph rises to the right.

▼ **EXAMPLE 2**   Consider the function $g(x) = 2^{-x}$.

(a) Rewrite the rule of g so that no minus signs appear in it.

Solution   By the definition of negative exponents,

$$g(x) = 2^{-x} = \frac{1}{2^x} = \left(\frac{1}{2}\right)^x.$$

(b) Graph $g(x)$.

Solution Either use a graphing calculator or graph by hand in the usual way, as shown in Figure 4.3.

x	$y = 2^{-x}$
-3	8
-2	4
-1	2
0	1
1	$1/2$
2	$1/4$
3	$1/8$

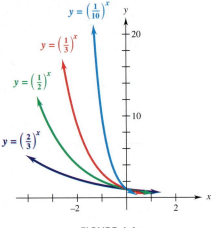

$$g(x) = 2^{-x} = \left(\frac{1}{2}\right)^x$$

FIGURE 4.3

The graph falls sharply to the right, but never touches the x-axis because every power of $\frac{1}{2}$ is positive. This is an example of **exponential decay.** ▲ ⬦2

When $0 < a < 1$, the graph of $k(x) = a^x$ has the same basic shape as the graph of $g(x) = (1/2)^x$, as illustrated in Figure 4.4 and summarized in the box on the next page.

② Graph $h(x) = (1/3)^x$.

Answer:

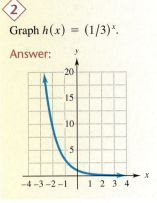

$$y = \left(\frac{1}{10}\right)^x$$
$$y = \left(\frac{1}{3}\right)^x$$
$$y = \left(\frac{1}{2}\right)^x$$
$$y = \left(\frac{2}{3}\right)^x$$

FIGURE 4.4

When $0 < a < 1$, the function $f(x) = a^x$ has the set of all real numbers as its domain. Its graph has the shape shown here and all five of the properties listed below.

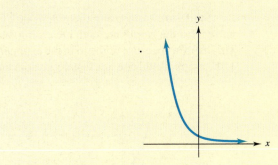

1. The graph is above the x-axis.
2. The y-intercept is 1.
3. The graph falls sharply to the right.
4. The positive x-axis is a horizontal asymptote.
5. The smaller the base a, the more steeply the graph falls to the right.

▼ **EXAMPLE 3**   In each case, graph $f(x)$ and $g(x)$ on the same set of axes and explain how the graphs are related.

(a) $f(x) = 2^x$ and $g(x) = (1/2)^x$.

Solution The graphs of f and g are shown in Figures 4.1 and 4.3, above. Placing them on the same set of axes, we obtain Figure 4.5. It shows that the graph of $g(x) = (1/2)^x$ is the mirror image of the graph of $f(x) = 2^x$, with the y-axis as the mirror. ◇ ③

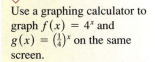

Use a graphing calculator to graph $f(x) = 4^x$ and $g(x) = \left(\frac{1}{4}\right)^x$ on the same screen.

Answer:

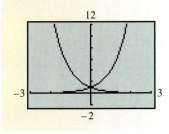

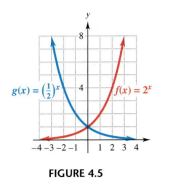

FIGURE 4.5

(b) $f(x) = 3^{1-x}$ and $g(x) = 3^{-x}$

Solution Choose values of x that make the exponent positive, zero, and negative,

and plot the corresponding points. The graphs are shown in Figure 4.6. The graph of $f(x) = 3^{1-x}$ has the same shape as the graph of $g(x) = 3^{-x}$, but is shifted 1 unit to the right, making the y-intercept $(0, 3)$ rather than $(0, 1)$.

(c) $f(x) = 2^{.6x}$ and $g(x) = 2^x$

Solution Comparing the graphs of $f(x) = 2^{.6x}$ and $g(x) = 2^x$ in Figure 4.7, we see that the graphs are both increasing, but the graph of $f(x)$ rises at a slower rate. This happens because of the .6 in the exponent. If the coefficient of x were greater than 1, the graph would rise at a faster rate than the graph of $g(x) = 2^x$. ▲ ⟨4⟩

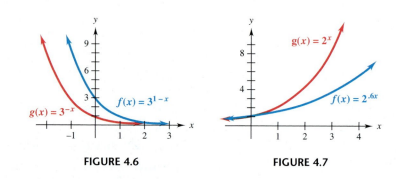

FIGURE 4.6   FIGURE 4.7

When the exponent involves a nonlinear expression in x, an exponential function graph may have a much different shape than the preceding ones have.

▼ **EXAMPLE 4** Graph $f(x) = 2^{-x^2}$.

Solution Either use a graphing calculator or plot points and connect them with a smooth curve, as in Figure 4.8. The graph is symmetric about the y-axis; that is, if the figure were folded on the y-axis, the two halves would match. Like all of the preceding graphs, this graph has the x-axis as a horizontal asymptote. The domain is still all real numbers, but here the range is $0 < y \le 1$. Graphs such as this are important in probability, where the normal curve has an equation similar to $f(x)$ in this example. ▲ ⟨5⟩

x	y
-2	$1/16$
-1.5	$.21$
-1	$1/2$
$-.5$	$.84$
0	1
$.5$	$.84$
1	$1/2$
1.5	$.21$
2	$1/16$

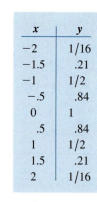

FIGURE 4.8

⟨4⟩

Graph $f(x) = 2^{x+1}$.

Answer:

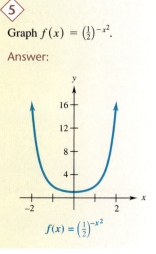

⟨5⟩

Graph $f(x) = \left(\frac{1}{2}\right)^{-x^2}$.

Answer:

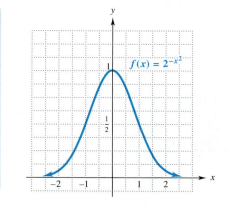

The Number *e*

In Section 5.2, we shall see that a certain irrational number, denoted *e*, plays a central role in the continuous compounding of interest. This number *e* also arises naturally in a variety of other mathematical and scientific contexts. To 12 decimal places,

$$e \approx 2.718281828459.$$

Perhaps the single most useful exponential function is the function defined by $f(x) = e^x$.

> **TECHNOLOGY TIP**   To evaluate powers of *e* with a calculator, use the $\boxed{e^x}$ key. On some calculators, you will need to use the two-keys $\boxed{INV}$ $\boxed{LN}$ or $\boxed{2nd}$ $\boxed{LN}$. For example, your calculator should show that $e^{.14} = 1.150273799$ to 8 decimal places. If these keys do not work, consult the instruction manual for your calculator. ⬧6

⬦6

Evaluate the following powers of *e*.

(a) $e^{.06}$

(b) $e^{-.06}$

(c) $e^{2.30}$

(d) $e^{-2.30}$

Answers:

(a) 1.06184

(b) .94176

(c) 9.97418

(d) .10026

NOTE To display the decimal expansion of *e* on your calculator screen, calculate e^1.

In Figure 4.9, the functions defined by

$$g(x) = 2^x, \qquad f(x) = e^x, \qquad \text{and} \qquad h(x) = 3^x$$

are graphed for comparison.

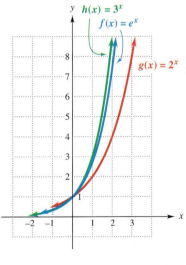

FIGURE 4.9

▼ **EXAMPLE 5**   If the population of the United States continues to grow as it has since 1980, then the approximate population (in millions) in year *t* is given by

$$f(t) = 225e^{.010824t},$$

where $t = 0$ corresponds to 1980.

(a) What was the U.S. population in 2005?

Solution Since 2005 corresponds to $t = 25$, evaluate the function at $t = 25$:

$$f(25) = 225e^{.010824(25)} \approx 294.92.$$

So the 2005 population was about 294,920,000.

(b) What will the U.S. population be in 2010?

Solution Now, 2010 corresponds to $t = 30$ and

$$f(30) = 225e^{.010824(30)} \approx 311.32,$$

which says that the 2010 population will be approximately 311,320,000.

(c) Use a graphing calculator to find the year in which the U.S. population will reach 352,000,000.

Solution Since f measures the population in millions, we must solve the equation $f(t) = 352$—that is, $225e^{.010824t} = 352$. One way to do this is to find the intersection point of the graphs of $y = 225e^{.010824t}$ and $y = 352$. Figure 4.10 shows that the population will reach 352 million when $t \approx 41.35$, in other words, during 2021. ▲ ⑦

---

⑦

The number of organisms present at time t is given by $f(t) = 75e^{.458t}$.

(a) Is this a growth function or a decay function? Find the number of organisms present at

(b) $t = 0$;

(c) $t = 2$;

(d) $t = 4$.

Answers:

(a) A growth function

(b) 75

(c) About 188

(d) About 469

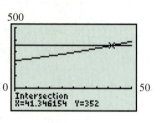

500

0   50.

Intersection
X=41.346154 Y=352

FIGURE 4.10

◆4.1 Exercises

Classify each function as linear, quadratic, or exponential.

1. $f(x) = 6^x$

2. $g(x) = -5x$

3. $h(x) = 4x^2 - x + 5$

4. $k(x) = 4^{x+3}$

5. $f(x) = 675(1.055^x)$

6. $g(x) = 12e^{x^2+1}$

Without graphing,

(a) *describe the shape of the graph of each function;*

(b) *find the second coordinates of the points with first coordinates 0 and 1. (See Examples 1–3.)*

7. $f(x) = .6^x$

8. $g(x) = 4^{-x}$

9. $h(x) = 2^{.5x}$

10. $k(x) = 5(3^x)$

11. $f(x) = e^{-x}$

12. $g(x) = 3(16^{x/4})$

Graph each function. (See Examples 1–3.)

13. $f(x) = 3^x$

14. $g(x) = 3^{.5x}$

15. $f(x) = 2^{x/2}$

16. $g(x) = e^{x/4}$

17. $f(x) = (1/5)^x$

18. $g(x) = 2^{3x}$

19. Graph these functions on the same axes.

(a) $f(x) = 2^x$

(b) $g(x) = 2^{x+3}$

(c) $h(x) = 2^{x-4}$

(d) If c is a positive constant, explain how the graphs of $y = 2^{x+c}$ and $y = 2^{x-c}$ are related to the graph of $f(x) = 2^x$.

20. Graph these functions on the same axes.

(a) $f(x) = 3^x$

(b) $g(x) = 3^x + 2$

(c) $h(x) = 3^x - 4$

(d) If c is a positive constant, explain how the graphs of $y = 3^x + c$ and $y = 3^x - c$ are related to the graph of $f(x) = 3^x$.

The figure shows the graphs of $y = a^x$ for $a = 1.8, 2.3, 3.2, .4, .75,$ and .31. They are identified by letter, but not necessarily in the same order as the values of a just given. Use your knowledge of how the exponential function behaves for various powers of a to match each lettered graph with the correct value of a.

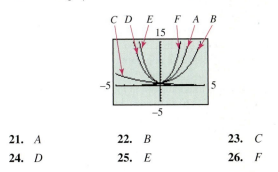

21. A   **22.** B   **23.** C

24. D   **25.** E   **26.** F

In Exercises 27 and 28, the graph of an exponential function with base a is given. Follow the directions in parts (a)–(f) in each exercise.

27.

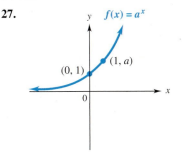

(a) Is $a > 1$ or is $0 < a < 1$?
(b) Give the domain and range of f.
(c) Sketch the graph of $g(x) = -a^x$.
(d) Give the domain and range of g.
(e) Sketch the graph of $h(x) = a^{-x}$.
(f) Give the domain and range of h.

28.

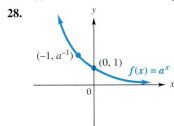

(a) Is $a > 1$ or is $0 < a < 1$?
(b) Give the domain and range of f.
(c) Sketch the graph of $g(x) = a^x + 2$.
(d) Give the domain and range of g.
(e) Sketch the graph of $h(x) = a^{x+2}$.
(f) Give the domain and range of h.

29. If $f(x) = a^x$ and $f(3) = 27$, find the following values of $f(x)$.
(a) $f(1)$ (b) $f(-1)$ (c) $f(2)$ (d) $f(0)$

30. Give a rule of the form $f(x) = a^x$ to define the exponential function whose graph contains the given point.
(a) $(3, 8)$ (b) $(-3, 64)$

Graph each function. (See Example 4.)

31. $f(x) = 2^{-x^2+2}$ **32.** $g(x) = 2^{x^2-2}$

33. $f(x) = x \cdot 2^x$ **34.** $f(x) = x^2 \cdot 2^x$

Work the following exercises.

35. **Finance**   If \$1 is deposited into an account paying 6% per year compounded annually, then after t years the account will contain

$$y = (1 + .06)^t = (1.06)^t$$

dollars.
(a) Use a calculator to complete the following table:

t	0	1	2	3	4	5	6	7	8	9	10
y	1					1.34					1.79

(b) Graph $y = (1.06)^t$.

36. **Finance**   If money loses value at the rate of 3% per year, the value of \$1 in t years is given by

$$y = (1 - .03)^t = (.97)^t.$$

(a) Use a calculator to complete the following table:

t	0	1	2	3	4	5	6	7	8	9	10
y	1					.86					.74

(b) Graph $y = (.97)^t$.

Work these problems. See Example 5.

37. **Finance**   If money loses value, then it takes more dollars to buy the same item. Use the results of Exercise 36(a) to answer the following questions.
(a) Suppose a house costs \$105,000 today. Estimate the cost of the same house in 10 years. (*Hint:* Solve the equation $.74t = \$105,000$.)
(b) Estimate the cost of a \$50 textbook in 8 years.

38. **Natural Science**   Biologists studying salmon have found that the oxygen consumption of yearling salmon is given by $g(x) = 100e^{.7x}$, where x is the speed in feet per second. Find each of the following.
(a) The oxygen consumption when the fish are still
(b) The oxygen consumption when the fish are swimming at a speed of 2 feet per second

39. **Social Science** Cellular phone usage has increased exponentially for the last decade. The number of U.S. cellular phone accounts (in millions) can be approximated by

$$f(x) = 15.76(1.1976^x) \qquad (5 \le x \le 13),$$

where $x = 5$ corresponds to 1995.* Estimate the number of cellular phone accounts in
(a) 2000
(b) 2003.
(c) Will this model continue to be accurate over the long run?

40. **Business** The monthly payment on a car loan at 12% interest per year on the unpaid balance is given by

$$f(n) = \dfrac{P}{\dfrac{1 - 1.01^{-n}}{.01}},$$

where P is the amount borrowed and n is the number of months over which the loan is paid back. Find the monthly payment for each of the following loans.
(a) $8000 for 48 months
(b) $8000 for 24 months
(c) $6500 for 36 months
(d) $6500 for 60 months

41. **Natural Science** The amount of plutonium remaining from 1 kilogram after x years is given by the function $W(x) = 2^{-x/24360}$. How much will be left after
(a) 1000 years?
(b) 10,000 years?
(c) 15,000 years?
(d) Estimate how long it will take for the 1 kilogram to decay to half its original weight. Your answer may help to explain why nuclear waste disposal is a serious problem.

42. **Health** Per capita spending on health care in the United States is approximated by the function

$$g(x) = 1219e^{.0678x} \qquad (0 \le x \le 30),$$

where $x = 0$ corresponds to 1980.* Estimate per capita health care spending in
(a) 2000
(b) 2005
(c) 2009

43. **Social Science** There were fewer than a billion people on earth when Thomas Jefferson died in 1826, and there are now more than 6 billion. If world population continues to grow as expected, the population (in billions) in year t will be given by the function $P(t) = 4.834(1.01^{(t-1980)}).†$ Estimate the world population in the following years.
(a) 2005
(b) 2010
(c) 2030
(d) What will the world population be when you reach 65 years old?

44. **Social Science** Using the function in Exercise 43 and a calculator, estimate the year in which the population will be double what it was in 2005. (*Hint:* If you have a graphing calculator, solve the appropriate equation, as in Example 5(c). If you have a scientific calculator, experiment with various values of t.)

Business *The scrap value of a machine is the value of the machine at the end of its useful life. By one method of calculating scrap value, where it is assumed that a constant percentage of value is lost annually, the scrap value is given by*

$$S = C(1 - r)^n,$$

where C is the original cost, n is the useful life of the machine in years, and r is the constant annual percentage of value lost. Find the scrap value for each of the following machines.

45. Original cost, $54,000; life, 8 years; annual rate of value loss, 12%

46. Original cost, $178,000; life, 11 years; annual rate of value loss, 14%

47. Use the graphs of $f(x) = 2^x$ and $g(x) = 2^{-x}$ (not a calculator) to explain why $2^x + 2^{-x}$ is approximately equal to 2^x when x is very large.

48. **Business** U.S. ethanol fuel production (in billions of gallons) is approximated by

$$f(x) = 1.47(1.225^x),$$

where $x = 0$ corresponds to 2000.‡
(a) What was ethanol production in 2000, 2003, and 2005?
(b) Suppose this trend continues. Use a graphing calculator to estimate the year in which ethanol production reaches 12 billion gallons.

49. **Health** The number of hospitals in the United States is approximated by

$$g(x) = 7311e^{-.00823x},$$

---

*Based on data from the Centers for Medicare and Medicaid Services.

†Based on current data and projections by the U.S. Census Bureau.

‡Based on data from the U.S. Department of Agriculture.

where $x = 0$ corresponds to 1975.

(a) Find the number of hospitals in 2000, 2005, and 2007.

(b) Use a graphing calculator to estimate the year in which the number of hospitals will be less than 5300.

50. The figure at the right is the graph of an exponential growth function $f(x) = Pa^x$.

(a) Find P. (*Hint:* What is $f(0)$?)

(b) Find a. (*Hint:* What is $f(2)$?)

(c) Find $f(5)$.

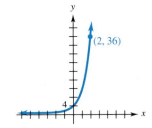

4.2 Applications of Exponential Functions

In many situations in biology, economics, and the social sciences, a quantity changes at a rate proportional to the quantity present. For example, a country's population might be increasing at a rate of 1.3% a year. In such cases, the amount present at time t is given by an **exponential growth function.**

Exponential Growth Function

Under normal conditions, growth can be described by a function of the form

$$f(t) = y_0 e^{kt} \qquad \text{or} \qquad f(t) = y_0 b^t,$$

where $f(t)$ is the amount present at time t, y_0 is the amount present at time $t = 0$, and k and b are constants that depend on the rate of growth.

It is understood that growth can involve either growing larger or growing smaller. Here is an example of growing smaller.

▼ **EXAMPLE 1**   For health reasons, cigarette consumption in the United States has been decreasing for several decades. The number of cigarettes consumed (in millions) in year t is approximated by the exponential function

$$C(t) = y_0 e^{-.018654t},$$

where t is time in years and $t = 0$ corresponds to 1980.

(a) If 635.5 million cigarettes were consumed in 1980, what was cigarette consumption in 2005?

Solution Since $t = 0$ corresponds to 1980, the initial amount y_0 (the amount when $t = 0$) is 635.5. Substitute 635.5 for y_0 and $t = 25$ (corresponding to 2005) in the rule of the function:

$$C(t) = y_0 e^{-.018654t},$$
$$C(25) = 635.5 e^{-.018654(25)} \approx 398.64.$$

Consumption in 2005 will be about 398,640,000 cigarettes.

(b) If this trend continues, what will cigarette consumption be in 2012?

Solution Since 2012 corresponds to $t = 32$, evaluate $C(t)$ at $t = 32$:

$$C(32) = 635.5e^{-.018654(32)} \approx 349.84 \text{ million cigarettes.} \quad \blacktriangle \quad ①$$

When a quantity is known to grow exponentially, it is sometimes possible to find a function that models its growth from a small amount of data.

▼ **EXAMPLE 2**   When money is placed in a bank account that pays compound interest, the amount in the account grows exponentially, as we shall see in Chapter 5. Suppose such an account grows from $1000 to $1316 in seven years.

(a) Find a growth function of the form $f(t) = y_0 b^t$ that gives the amount in the account at time t years.

Solution The value of the account at time $t = 0$ and $t = 7$ are given; that is, $f(0) = 1000$ and $f(7) = 1316$. Solve the first of these equations for y_0:

$$f(0) = 1000$$
$$y_0 b^0 = 1000 \qquad \text{Rule of } f$$
$$y_0 = 1000. \qquad b^0 = 1$$

So the rule of f has the form $f(t) = 1000b^t$. Now solve the equation $f(7) = 1316$ for b:

$$f(7) = 1316$$
$$1000b^7 = 1316 \qquad\qquad \text{Rule of } f$$
$$b^7 = 1.316 \qquad\qquad \text{Divide both sides by 1000.}$$
$$b = (1.316)^{1/7} \approx 1.04. \qquad \text{Take the seventh root of each side.}$$

So the rule of the function is $f(t) = 1000(1.04)^t$.

(b) How much is in the account after 12 years?

Solution $f(12) = 1000(1.04)^{12} = \$1601.03. \quad \blacktriangle \quad ②$

▼ **EXAMPLE 3**   Infant mortality rates in the United States are shown in the following table.*

Year	Rate	Year	Rate
1920	76.7	1980	12.6
1930	60.4	1985	10.6
1940	47.0	1990	9.2
1950	29.2	1995	7.6
1960	26.0	2000	6.9
1970	20.0	2002	7.0

---

*Infants less than 1 year old, deaths per 1000 live births. U.S. National Center for Health Statistics.

①

Suppose the number of bacteria in a culture at time t is

$$y = 500e^{.4t},$$

where t is measured in hours.

(a) How many bacteria are present initially?

(b) How many bacteria are present after 10 hours?

Answers:

(a) 500

(b) About 27,300

②

Suppose an investment grows exponentially from $500 to $587.12 in three years.

(a) Find a function of the form $f(t) = y_0 b^t$ that gives the value of the investment after t years.

(b) How much was the investment worth after 10 years?

Answers:

(a) $f(t) = 500(1.055)^t$

(b) $854.07

(a) Let $t = 0$ correspond to 1920. Use the data for 1920 and 2002 to find a function of the form $f(t) = y_0 b^t$ that models this data.

Solution Since the rate is 76.7 when $t = 0$, we have $y_0 = 76.7$. Hence, $f(t) = 76.7b^t$. Because 2002 corresponds to $t = 82$, we have $f(82) = 7$; that is,

$$76.7b^{82} = 7 \qquad \text{Rule of } f$$

$$b^{82} = \frac{7}{76.7} \qquad \text{Divide both sides by 76.7.}$$

$$b = \left(\frac{7}{76.7}\right)^{\frac{1}{82}} \approx .97123 \qquad \text{Take 82}^{\text{nd}} \text{ roots on both sides.}$$

Therefore, the function is $f(t) = 76.7(.97123^t)$.

 (b) Use exponential regression on a graphing calculator to find another model of the data.

Solution The procedure for entering the data and finding the function is the same as that for linear regression (just choose "exponential" instead of "linear"), as explained in Section 2.3. Depending on the calculator, one of these functions will be produced:

$$g(t) = 81.3143(.96981^t) \qquad \text{or} \qquad h(t) = 81.3143e^{-.03066t}.$$

Both functions give the same values (except for slight differences due to rounding the coefficients displayed above). They fit the data reasonably well, as shown in Figure 4.11.

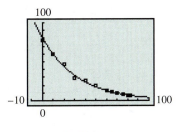

FIGURE 4.11

(c) Use the preceding models to estimate the mortality rate in 2001 and 2008.

Solution Evaluating each of the models at $t = 81$ and $t = 88$ shows that the models in parts (a) and (b) give similar, but not identical, results. ▲

t	$f(t)$	$g(t)$	$h(t)$
81	7.2	6.8	6.8
88	5.9	5.5	5.5

Other Exponential Models

When a quantity changes exponentially, but does not either grow very large or decrease practically to 0, as in Examples 1–3, different functions are needed.

▼ **EXAMPLE 4**   Sales of a new product often grow rapidly at first and then begin to level off with time. Suppose the annual sales of an inexpensive can opener are given by

$$S(x) = 10{,}000(1 - e^{-.5x}),$$

where $x = 0$ corresponds to the time the can opener went on the market.

(a) What were the sales in each of the first three years?

Solution At the end of one year $(x = 1)$, sales were

$$S(1) = 10{,}000(1 - e^{-.5(1)}) \approx 3935.$$

Sales in the next two years were

$$S(2) = 10{,}000(1 - e^{-.5(2)}) \approx 6321 \quad \text{and} \quad S(3) = 10{,}000(1 - e^{-.5(3)}) \approx 7769.$$

(b) What were the sales at the end of the tenth year?

Solution $S(10) = 10{,}000(1 - e^{-.5(10)}) \approx 9933.$

(c) Graph the function S. What does it suggest?

Solution The graph can be obtained by plotting points and connecting them with a smooth curve or by using a graphing calculator, as in Figure 4.12. The graph indicates that sales will level off after the twelfth year, to around 10,000 can openers per year. ▲ ③

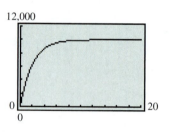

FIGURE 4.12

A variety of activities can be modeled by **logistic functions,** whose rules are of the form

$$f(x) = \frac{c}{1 + ae^{kx}}.$$

The logistic function in the next example is sometimes called a **forgetting curve.**

▼ **EXAMPLE 5** Psychologists have measured people's ability to remember facts that they have memorized. In one such experiment, it was found that the number of facts $N(t)$ remembered after t days was given by

$$N(t) = \frac{10.003}{1 + .0003e^{.8t}}.$$

(a) How many facts were remembered at the beginning of the experiment?

Solution When $t = 0$,

$$N(0) = \frac{10.003}{1 + .0003e^{.8(0)}} = \frac{10.003}{1.0003} = 10.$$

So 10 facts were remembered at the beginning.

③ Suppose the value of the assets (in thousands of dollars) of a certain company after t years is given by

$$V(t) = 100 - 75e^{-.2t}.$$

(a) What is the initial value of the assets?

(b) What is the limiting value of the assets?

(c) Find the value after 10 years.

(d) Graph $V(t)$.

Answers:

(a) $25,000

(b) $100,000

(c) $89,850

(d)

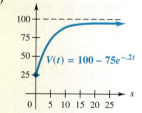

(b) How many facts were remembered after one week? After two weeks?

Solution One and two weeks correspond to $t = 7$ and $t = 14$.

$$N(7) = \frac{10.003}{1 + .003e^{.8(7)}} = \frac{10.003}{1.0811} \approx 9.25$$

$$N(14) = \frac{10.003}{1 + .0003e^{.8(14)}} = \frac{10.003}{22.9391} \approx .44$$

So 9 facts were remembered after one week, but effectively none were remembered after two weeks. (.44 is less than "half a fact"). The graph of the function in Figure 4.13 gives a picture of this forgetting process. ▲ ◆

In Example 5,

(a) find the number of facts remembered after 10 days;

(b) use the graph to estimate when just 1 fact will be remembered.

Answers:

(a) 5

(b) After about 12 days

TECHNOLOGY TIP   Many graphing calculators can find a logistic model for appropriate data.

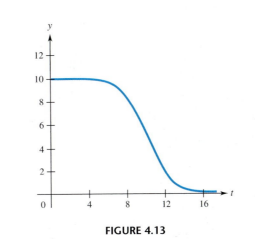

FIGURE 4.13

◆4.2 Exercises

1. **Finance**   Suppose you owe $800 on your credit card and you decide to make no new purchases and to make the minimum monthly payment on the account. Assuming that the interest rate on your card is 1% per month on the unpaid balance and that the minimum payment is 2% of the total (balance plus interest), your balance after t months is given by

$$B(t) = 800(.9898^t).$$

Find your balance at each of these times.
   (a) six months
   (b) one year (remember, t is in months)
   (c) five years
   (d) eight years
   (e) On the basis of your answers to parts (a)–(e), what advice would you give to your friends about minimum payments?

2. **Health**   National health care expenditures (in billions of dollars) are approximated by

$$H(t) = h_0(1.1013^t),$$

Where $t = 0$ corresponds to 1960.*
   (a) Health care expenditures were $29.52 billion in 1960 (that is, when $t = 0$). Find h_0.
   (b) Estimate health care expenditures in 1998 and 2001.

3. **Business**   The average price of an LCD television set is approximated by

$$f(t) = 4295.5e^{-.2294t} \qquad (t \geq 4),$$

where $t = 4$ corresponds to 2004.[†] Find the average cost in the following years.
   (a) 2006   **(b)** 2009

4. **Health**   The graph on the next page shows how the risk of chromosomal abnormality in a child rises with the age of the mother.[‡]

*Based on data from the National Center for Health Statistics.

†Based on data from iSuppli Corp. in *Business Week*, April 4, 2005.

‡*New York Times*, February 5, 1994, p. 24 © 1994, The New York Times Company.

(a) From the graph, read the risk of chromosomal abnormality (per 1000) at ages 20, 35, 42, and 49.

(b) Verify by substitution that the exponential equation $y = .590e^{.061t}$ "fits" the graph for ages 20 and 35.

(c) Does the equation in part (b) also fit the graph for ages 42 and 49? What does this mean?

Maternity

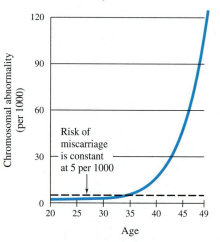

Source: American College of Obstetricians and Gynecologists.

In each of the following problems, find an exponential function of the form $f(t) = y_0 b^t$ to model the data. (See Examples 2 and 3.)

5. Social Sciences The U.S. Census Bureau predicts that the Hispanic population in the United States will increase from 32.5 million in 2000 to 98.2 million in 2050.*

(a) Find a model for this data, in which $t = 0$ corresponds to 2000.

(b) What is the projected Hispanic population in 2010? In 2035?

(c) By experimenting with different values of t (or by using a graphing calculator to solve an appropriate equation), estimate the year in which the Hispanic population will reach 54 million.

6. Social Sciences The U.S. Census Bureau predicts that the African-American population will increase from 35.3 million in 2000 to 59.2 million in 2050.[†]

(a) Find a model for this data, in which $t = 0$ corresponds to 2000.

(b) What is the projected African-American population in 2004? In 2030?

*Statistical Abstract of the United States, 2004–2005.

†Statistical Abstract of the United States, 2004–2005.*

(c) By experimenting with different values of t (or by using a graphing calculator to solve an appropriate equation), estimate the year in which the African-American population will reach 55 million.

7. Business In 1997, China exported $2 billion worth of vehicles and vehicle parts. By 2004, these exports had reached $12 billion.

(a) Let $t = 0$ correspond to 1997 and find a model for this data.

(b) Find the value of Chinese exports in 2003 and 2005?

8. Social Sciences The U.S. Department of Commerce estimates that the number of Internet users in the United States was 84.6 million in 1998 and 188.8 million in 2004.

(a) Find a model f for this data, in which t is measured in years, with $t = 0$ corresponding to 1998, and $f(t)$ is measured in millions.

(b) Use the results of part (a) to determine $f(9)$. Interpret your answer.

In the following exercises, find the exponential model as follows. If you do not have suitable technology, use the first and last data points to find a function. (See Examples 2 and 3.) If you have a graphing calculator or other suitable technology, use exponential regression to find a function. (See Example 3.)

9. Business The table shows the purchasing power of a dollar in various years, with the 1982–1984 average being the base.* For example, the entry .600 for 1999 means that a dollar in 1999 bought what 60 cents did in 1982–1984.

Year	1985	1987	1989	1991
Purchasing Power of $1	.928	.880	.807	.734

Year	1993	1995	1997	1999	2001	2003	2005
Purchasing Power of $1	.692	.656	.623	.600	.562	.541	.515

(a) Find an exponential model for this data, with $t = 0$ corresponding to 1985.

(b) Find the purchasing power of a dollar in 2000, 2006, and 2008.

(c) Use a graphing calculator (or trial and error) to determine the year in which the purchasing power of the 1982–84 dollar will drop to 40 cents.

*U.S. Bureau of Labor Statistics.

10. **Physical Science**   The table shows the atmospheric pressure (in millibars) at various altitudes (in meters).

Altitude	Pressure
0	1,013
1000	899
2000	795
3000	701
4000	617
5000	541
6000	472
7000	411
8000	357
9000	308
10,000	265

(a) Find an exponential model for this data, in which t is measured in thousands (for instance, $t = 2$ is 2000 meters).

(b) Use the function in part (a) to estimate the atmospheric pressure at 1500 meters and 11,000 meters. Compare your results with the actual values of 846 millibars and 227 millibars, respectively.

11. **Health**   The table shows the age-adjusted death rates (per 100,000 population) for heart disease:*

Year	1970	1980	1990	2000	2002
Death Rate	492.7	412.1	321.8	257.9	240.4

(a) Find a model for this data, with $t = 0$ corresponding to 1970.

(b) Estimate the death rate in 1995 and 2005.

(c) Assuming that the model remains accurate, use a graphing calculator (or trial and error) to determine the year in which the death rate will fall to 100.

12. **Business**   The table shows outstanding consumer credit (in billions of dollars) in various years.[†]

Year	1980	1985	1990	1995
Credit	349.4	593.2	789.1	1095.8

Year	2000	2001	2002	2003
Credit	1692.6	1828.8	1905.0	1986.7

---

*U.S. Department of Health and Human Services.

†*Federal Reserve Bulletin.*

(a) Find an exponential model for this data, with $t = 0$ corresponding to 1980.

(b) If this model remains accurate, what will the outstanding consumer credit be in 2007 and 2010?

(c) In what year will consumer credit reach $4500 billion?

Work the following problems. (See Example 4.)

13. **Business**   Assembly-line operations tend to have a high turnover of employees, forcing the companies involved to spend much time and effort in training new workers. It has been found that a worker who is new to the operation of a certain task on the assembly line will produce $P(t)$ items on day t, where

$$P(t) = 25 - 25e^{-.3t}.$$

(a) How many items will be produced on the first day?

(b) How many items will be produced on the eighth day?

(c) According to the function, what is the maximum number of items the worker can produce?

14. **Social Science**   The number of words per minute that an average typist can type is given by

$$W(t) = 60 - 30e^{-.5t},$$

where t is time in months after the beginning of a typing class. Find each of the following.
(a) $W(0)$ **(b)** $W(1)$ **(c)** $W(4)$ **(d)** $W(6)$

Natural Science   *Newton's Law of Cooling says that the rate at which a body cools is proportional to the difference in temperature between the body and an environment into which it is introduced. Using calculus, the temperature $F(t)$ of the body at time t after being introduced into an environment having constant temperature T_0 is*

$$F(t) = T_0 + Cb^t,$$

where C and b are constants. Use this result in Exercises 15 and 16.

15. Boiling water at $100°$ Celsius is placed in a freezer at $-18°$ Celsius. The temperature of the water is $50°$ Celsius after 24 minutes. Find the temperature of the water after 96 minutes.

16. Paisley refuses to drink coffee cooler than $95°$F. She makes coffee with a temperature of $170°$F in a room with a temperature of $70°$F. The coffee cools to $120°$F in 10 minutes. What is the longest time she can let the coffee sit before she drinks it?

17. **Social Science**   A sociologist has shown that the fraction $y(t)$ of people in a group who have heard a rumor after t days is approximated by

$$y(t) = \frac{y_0 e^{kt}}{1 - y_0(1 - e^{kt})},$$

where y_0 is the fraction of people who heard the rumor at time $t = 0$ and k is a constant. A graph of $y(t)$ for a particular value of k is shown in the figure on the next page.

(a) If $k = .1$ and $y_0 = .05$, find $y(10)$.

(b) If $k = .2$ and $y_0 = .10$, find $y(5)$.

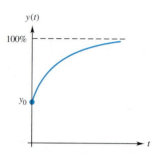 (c) Assume the situation in part (b). How many *weeks* will it take for 65% of the people to have heard the rumor?

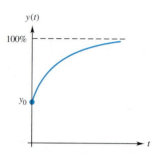

18. **Social Science** Data from the National Highway Traffic Safety Administration indicate that, in year t, the approximate percentage of people in the United States who wear seat belts when driving is given by

$$g(t) = \frac{97}{1 + .823e^{-.1401t}},$$

where $t = 0$ corresponds to 1994. What percentage used seat belts in

(a) 2000   (b) 2003   (c) 2004

Assuming that this function is accurate after 2004, what percentage of people will use seat belts in

(d) 2007;   (e) 2009;   (f) 2011?

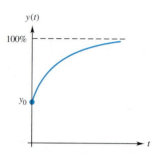 (g) Graph the function. Does the graph suggest that seat belt usage will ever reach 100%?

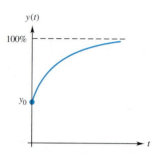 *Use a graphing calculator or other technology to do the following problems. See Example 5.*

19. **Business** The number of people who subscribe to basic cable TV can be approximated by

$$f(x) = \frac{74.22}{1 + 22.34e^{-.21x}},$$

where $x = 0$ corresponds to 1970 and $f(x)$ is in millions.*

*Based on data from *National Cable and Telecommunications Association.*

(a) Estimate the number of subscribers in 2000 and 2005.

(b) Graph $f(x)$ for the period 1970–2020.

(c) Use the graph to determine the year in which the number of subscribers reached 74 million.

(d) Based on the graph, will the number of subscribers reach 90 million in the foreseeable future?

20. **Natural Science** The population of fish in a certain lake at time t months is given by the function

$$p(t) = \frac{20,000}{1 + 24(2^{-.36t})}.$$

(a) Graph the population function from $t = 0$ to $t = 48$ (a four-year period).

(b) What was the population at the beginning of the period?

(c) Use the graph to estimate the one-year period in which the population grew most rapidly.

(d) When do you think the population will reach 25,000? What factors in nature might explain your answer?

21. **Finance** Since 1990, the national debt can be approximated by the logistic model

$$f(x) = \frac{95.8}{1 + 26.1e^{-.0512x}}$$

where $x = 0$ corresponds to 1990 and $f(x)$ is measured in trillions of dollars.

(a) Estimate the national debt in 2000, 2005, and 2008.

(b) Graph the function for the period 1990–2040.

(c) Determine when the debt will reach 10 trillion dollars.

(d) According to this model, will the national debt level off at any point in the future?

22. **Social Science** The probability P percent of having an automobile accident is related to the alcohol level t of the driver's blood by the function $P(t) = e^{21.459t}$.

(a) Graph $P(t)$ in a viewing window with $0 \le t \le .2$ and $0 \le P(t) \le 100$.

(b) At what blood alcohol level is the probability of an accident at least 50%? What is the legal blood alcohol level in your state?

4.3 Logarithmic Functions

Until the development of computers and calculators, logarithms were the only effective tool for large-scale numerical computations. They are no longer needed for this purpose, but logarithmic functions still play a crucial role in calculus and in many applications.

Common Logarithms

Logarithms are simply *a new language for old ideas*—essentially a special case of exponents.

Definition of Common (Base 10) Logarithms

$$y = \log x \qquad \text{means} \qquad 10^y = x.$$

"**Log x,**" which is read "the logarithm of *x*," is the answer to the question

To what exponent must 10 be raised to produce *x*?

▼ **EXAMPLE 1**   To find log 10,000, ask yourself, "To what exponent must 10 be raised to produce 10,000?" Since $10^4 = 10,000$, we see that log 10,000 = 4. Similarly,

$$\log 1 = 0 \qquad \text{because} \qquad 10^0 = 1;$$

$$\log .01 = -2 \qquad \text{because} \qquad 10^{-2} = \frac{1}{10^2} = \frac{1}{100} = .01;$$

$$\log \sqrt{10} = 1/2 \qquad \text{because} \qquad 10^{1/2} = \sqrt{10}. \qquad \blacktriangle \; \langle 1 \rangle$$

▼ **EXAMPLE 2**   Log(-25) is the exponent to which 10 must be raised to produce -25. But every power of 10 is positive! So there is no exponent that will produce -25. *Logarithms of negative numbers and 0 are not defined.* ▲

▼ **EXAMPLE 3**   (a) We know that log 359 must be a number between 2 and 3 because $10^2 = 100$ and $10^3 = 1000$. By using the log key on a calculator, we find that log 359 (to four decimal places) is 2.5551. You can verify this statement by computing $10^{2.5551}$; the result (rounded) is 359. See Figure 4.12.

(b) When 10 is raised to a negative exponent, the result is a number less than 1. Consequently, the logarithms of numbers between 0 and 1 are negative. For instance, log .026 = -1.5850, as shown in Figure 4.14. ▲ ⟨2⟩

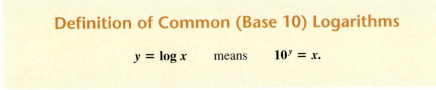

```
log(359)
          2.555094449
10^(2.5551)
          359.004589
log(.026)
         -1.585026652
```

FIGURE 4.14

NOTE On most scientific calculators, enter the number followed by the log key. On graphing calculators, press the log key followed by the number, as in Figure 4.14.

⟨1⟩

Find each common logarithm.

(a)   log 100

(b)   log 1000

(c)   log .1

Answers:

(a)   2

(b)   3

(c)   −1

⟨2⟩

Find each common logarithm.

(a)   log 27

(b)   log 1089

(c)   log .00426

Answers:

(a)   1.4314

(b)   3.0370

(c)   −2.3706

Natural Logarithms

Although common logarithms still have some uses (one of which is discussed in Section 4.4), the most widely used logarithms today are defined in terms of the number e (whose decimal expansion begins $2.71828\dots$) rather than 10. They have a special name and notation.

Definition of Natural (Base e) Logarithms

$$y = \ln x \qquad \text{means} \qquad e^y = x.$$

Thus, the number **ln x** (which is sometimes read "el-en x") is the exponent to which e must be raised to produce the number x. For instance, $\ln 1 = 0$ because $e^0 = 1$. Although logarithms to the base e may not seem as "natural" as common logarithms, there are several reasons for using them, some of which are discussed in Section 4.4.

▼ **EXAMPLE 4** **(a)** To find ln 85, use the $\boxed{\text{LN}}$ key of your calculator. The result is 4.4427. Thus, 4.4427 is the exponent (to four decimal places) to which e must be raised to produce 85. You can verify this result by computing $e^{4.4427}$; the answer (rounded) is 85.

(b) A calculator shows that $\ln .38 = -.9676$ (rounded), which means that $e^{-.9676} \approx .38$.

These results are shown in Figure 4.15. ▲ ⟨3⟩

FIGURE 4.15

▼ **EXAMPLE 5** You don't need a calculator to find $\ln e^8$. Just ask yourself, "To what exponent must e be raised to produce e^8?" The answer, obviously, is 8. So $\ln e^8 = 8$. ▲

Other Logarithms

The procedure used to define common and natural logarithms can be carried out with any positive number $a \neq 1$ as the base in place of 10 or e.

Definition of Logarithms to the Base a

$$y = \log_a x \qquad \text{means} \qquad a^y = x.$$

⟨3⟩

Find the following.

(a) ln 6.1

(b) ln 20

(c) ln .8

(d) ln .1

Answers:

(a) 1.8083

(b) 2.9957

(c) −.2231

(d) −2.3026

Read $y = \log_a x$ as "y is the logarithm of x to the base a." As was the case with common and natural logarithms, $\log_a x$ is an *exponent;* it is the answer to the question

To what power must a be raised to produce x?

For example, the exponential statement $2^4 = 16$ can be translated into the equivalent logarithmic statement $4 = \log_2 16$.

This key definition should be memorized. It is important to remember the location of the base and exponent in each part of the definition.

$$\text{Logarithmic form:} \quad y = \log_a x$$

with Exponent pointing to y and Base pointing to a.

$$\text{Exponential form:} \quad a^y = x$$

with Exponent pointing to y and Base pointing to a.

Common and natural logarithms are the special cases when $a = 10$ and when $a = e$, respectively. Both $\log u$ and $\log_{10} u$ mean the same thing. Similarly, $\ln u$ and $\log_e u$ mean the same thing.

▼ **EXAMPLE 6** This example shows several statements written in both exponential and logarithmic form.

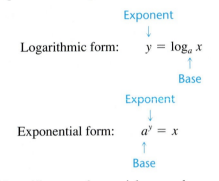

Exponential Form	Logarithmic Form
(a) $3^2 = 9$	$\log_3 9 = 2$
(b) $(1/5)^{-2} = 25$	$\log_{1/5} 25 = -2$
(c) $10^5 = 100{,}000$	$\log_{10} 100{,}000 \ (\text{or } \log 100{,}000) = 5$
(d) $4^{-3} = 1/64$	$\log_4(1/64) = -3$
(e) $2^{-4} = 1/16$	$\log_2(1/16) = -4$
(f) $e^0 = 1$	$\log_e 1 \ (\text{or } \ln 1) = 0$

▲ ⟨4⟩ ⟨5⟩

Properties of Logarithms

Some of the important properties of logarithms arise directly from their definition.

Let x and a be any positive real numbers, with $a \neq 1$, and r be any real number. Then

(a) $\log_a 1 = 0$ (b) $\log_a a = 1$

(c) $\log_a a^r = r$ (d) $a^{\log_a x} = x$

Property (a) was discussed in Example 1 (with $a = 10$). Property (c) was illustrated in Example 5 (with $a = e$ and $r = 8$). Property (b) is property (c) with $r = 1$. To

⟨4⟩

Write the logarithmic form of

(a) $5^3 = 125$;

(b) $3^{-4} = 1/81$;

(c) $8^{2/3} = 4$.

Answers:

(a) $\log_5 125 = 3$

(b) $\log_3(1/81) = -4$

(c) $\log_8 4 = 2/3$

⟨5⟩

Write the exponential form of

(a) $\log_{16} 4 = 1/2$;

(b) $\log_3(1/9) = -2$;

(c) $\log_{16} 8 = 3/4$.

Answers:

(a) $16^{1/2} = 4$

(b) $3^{-2} = 1/9$

(c) $16^{3/4} = 8$

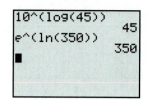

FIGURE 4.16

understand property (d), recall that $\log_a x$ is the exponent to which a must be raised to produce x. Consequently, when you raise a to this exponent, the result is x, as illustrated for common and natural logarithms in Figure 4.16.

The following properties are part of the reason that logarithms are so useful. They will be used in the next section to solve exponential and logarithmic equations.

The Product, Quotient, and Power Properties

Let x, y, and a be any positive real numbers, with $a \neq 1$. Let r be any real number. Then

$$\log_a xy = \log_a x + \log_a y \qquad \text{Product property}$$

$$\log_a \frac{x}{y} = \log_a x - \log_a y \qquad \text{Quotient property}$$

$$\log_a x^r = r \log_a x \qquad \text{Power property}$$

To prove the product property, let

$$m = \log_a x \qquad \text{and} \qquad n = \log_a y.$$

Then, by the definition of logarithm,

$$a^m = x \qquad \text{and} \qquad a^n = y.$$

Multiply to get

$$a^m \cdot a^n = x \cdot y,$$

or, by a property of exponents,

$$a^{m+n} = xy.$$

Use the definition of logarithm to rewrite this last statement as

$$\log_a xy = m + n.$$

Replace m with $\log_a x$ and n with $\log_a y$ to get

$$\log_a xy = \log_a x + \log_a y.$$

The quotient and power properties can be proved similarly. You can also illustrate these properties on a calculator, as in Figure 4.17.

FIGURE 4.17

▼ **EXAMPLE 7** Using the properties of logarithms, we can write each of the following as a single logarithm.*

(a) $\log_a x + \log_a(x-1) = \log_a x(x-1)$;   Product property

(b) $\log_a(x^2 + 4) - \log_a(x+6) = \log_a \dfrac{x^2+4}{x+6}$;   Quotient property

(c) $\log_a 9 + 5\log_a x = \log_a 9 + \log_a x^5 = \log_a 9x^5$.   Power and product properties ▲

CAUTION There is no logarithm property that allows you to simplify the logarithm of a sum, such as $\log_a(x^2+4)$. In particular, $\log_a(x^2+4)$ is *not* equal to $\log_a x^2 + \log_a 4$. The power property of logarithms shows that $\log_a x^2 + \log_a 4 = \log_a 4x^2$.

▼ **EXAMPLE 8** Assume that $\log_6 7 \approx 1.09$ and $\log_6 5 \approx .90$. Use the properties of logarithms to find each of the following.

(a) $\log_6 35 = \log_6(7 \cdot 5) = \log_6 7 + \log_6 5 \approx 1.09 + .90 = 1.99$;

(b) $\log_6 5/7 = \log_6 5 - \log_6 7 \approx .90 - 1.09 = -.19$;

(c) $\log_6 5^3 = 3\log_6 5 \approx 3(.90) = 2.70$;

(d) $\log_6 6 = 1$;

(e) $\log_6 1 = 0$. ▲

In Example 8, several logarithms to the base 6 were given. However, they could have been found by using a calculator and the following formula.

Change of Base Theorem

For any positive numbers a and x (with $a \neq 1$),

$$\log_a x = \frac{\ln x}{\ln a}.$$

▼ **EXAMPLE 9** To find $\log_7 3$, use the theorem with $a = 7$ and $x = 3$:

$$\log_7 3 = \frac{\ln 3}{\ln 7} \approx \frac{1.0986}{1.9459} \approx .5646.$$

You can check this result on your calculator by verifying that $7^{.5646} \approx 3$. ▲

▼ **EXAMPLE 10** Environmental scientists who study the diversity of species in an ecological community use the *Shannon index* to measure diversity. If there are k different species, with n_1 individuals of species 1, n_2 individuals of species 2, and so on, then the Shannon index H is defined as

$$H = \frac{N\log_2 N - [n_1\log_2 n_1 + n_2\log_2 n_2 + \cdots + n_k\log_2 n_k]}{N},$$

6 Write as a single logarithm, using the properties of logarithms.

(a) $\log_a 5x + \log_a 3x^4$

(b) $\log_a 3p - \log_a 5q$

(c) $4\log_a k - 3\log_a m$

Answers:

(a) $\log_a 15x^5$

(b) $\log_a(3p/5q)$

(c) $\log_a(k^4/m^3)$

7 Use the properties of logarithms to rewrite and evaluate each of the following, given that $\log_3 7 \approx 1.77$ and $\log_3 5 \approx 1.46$.

(a) $\log_3 35$

(b) $\log_3 7/5$

(c) $\log_3 25$

(d) $\log_3 3$

(e) $\log_3 1$

Answers:

(a) 3.23

(b) .31

(c) 2.92

(d) 1

(e) 0

*Here and elsewhere, we assume that variable expressions represent positive numbers and that the base a is positive, with $a \neq 1$.

where $N = n_1 + n_2 + n_3 + \cdots + n_k$. A study of the species that barn owls in a particular region typically eat yielded the following data:

Species	Number
Rats	143
Mice	1405
Birds	452

Find the index of diversity of this community.

Solution In this case, $n_1 = 143$, $n_2 = 1405$, $n_3 = 452$, and

$$N = n_1 + n_2 + n_3 = 143 + 1405 + 452 = 2000.$$

So the index of diversity is

$$H = \frac{N \log_2 N - [n_1 \log_2 n_1 + n_2 \log_2 n_2 + \cdots + n_k \log_2 n_k]}{N}$$

$$= \frac{2000 \log_2 2000 - [143 \log_2 143 + 1405 \log_2 1405 + 452 \log_2 452]}{2000}.$$

To compute H, we use the Change of Base Theorem:

$$H = \frac{2000 \dfrac{\ln 2000}{\ln 2} - \left[143 \dfrac{\ln 143}{\ln 2} + 1405 \dfrac{\ln 1405}{\ln 2} + 452 \dfrac{\ln 452}{\ln 2} \right]}{2000}$$

$$\approx 1.1149. \ \blacktriangle$$

Logarithmic Functions

For a given *positive* value of x, the definition of logarithm leads to exactly one value of y, so that $y = \log_a x$ defines a function.

If $a > 0$ and $a \neq 1$, the **logarithmic function** with base a is defined as

$$f(x) = \log_a x.$$

The most important logarithmic function is the natural logarithmic function.

▼ **EXAMPLE 11**   Graph $f(x) = \ln x$ and $g(x) = e^x$ on the same axes.

Solution For each function, use a calculator to compute some ordered pairs. Then plot the corresponding points and connect them with a curve to obtain the graphs in Figure 4.18.

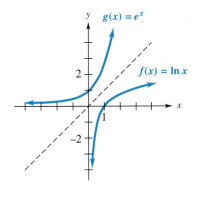

FIGURE 4.18

In Figure 4.18, the dashed straight line is the graph of $y = x$. Observe that the graph of $f(x) = \ln x$ is the mirror image of the graph of $g(x) = e^x$, with the line $y = x$ being the mirror. ▲ ⑧

When the base $a > 1$, the graph of $f(x) = \log_a x$ has the same basic shape as the graph of the natural logarithmic function in Figure 4.18, as summarized below.

⑧

Graph $f(x) = \log x$ and $g(x) = 10^x$ on the same axes.

Answer:

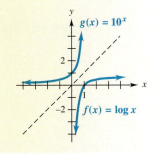

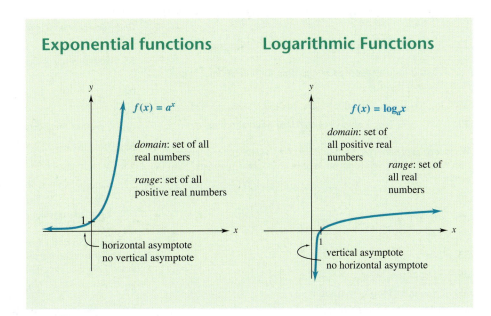

As the information in the box suggests, the graph of $f(x) = \log_a x$ is the mirror image of the graph of $g(x) = a^x$, with the line $y = x$ being the mirror. Functions whose graphs are related in this way are said to be **inverses** of each other. A more complete discussion of inverse functions is given in most standard college algebra books.

◆4.3 Exercises

Complete each statement in Exercises 1–4.

1. $y = \log_a x$ means $x =$ _____.

2. The statement $\log_5 125 = 3$ tells us that _____ is the power of _____ that equals _____.

3. What is wrong with the expression $y = \log_b$?

4. Logarithms of negative numbers are not defined because _____.

Translate each logarithmic statement into an equivalent exponential statement. (See Examples 1, 5, and 6.)

5. $\log 100{,}000 = 5$ **6.** $\log .001 = -3$

7. $\log_9 81 = 2$ **8.** $\log_2 (1/8) = -3$

Translate each exponential statement into an equivalent logarithmic statement. (See Examples 5 and 6.)

9. $10^{1.8751} = 75$ **10.** $e^{3.2189} = 25$

11. $3^{-2} = 1/9$ **12.** $16^{1/2} = 4$

Without using a calculator, evaluate each of the following. (See Examples 1, 5, and 6.)

13. $\log 1000$ **14.** $\log .0001$ **15.** $\log_5 25$

16. $\log_3 81$ **17.** $\log_4 64$ **18.** $\log_5 125$

19. $\log_2 \dfrac{1}{4}$ **20.** $\log_3 \dfrac{1}{27}$ **21.** $\ln \sqrt{e}$

22. $\ln(1/e)$ **23.** $\ln e^{8.77}$ **24.** $\log 10^{74.3}$

Use a calculator to evaluate each logarithm to three decimal places. (See Examples 3 and 4.)

25. $\log 53$ **26.** $\log .005$ **27.** $\ln .452$ **28.** $\ln 423$

29. Why does $\log_a 1$ always equal 0 for any valid base a?

Write each expression as the logarithm of a single number or expression. Assume that all variables represent positive numbers. (See Example 7.)

30. $\log 15 - \log 3$

31. $\log 4 + \log 8 - \log 2$

32. $3 \ln 2 + 2 \ln 3$

33. $2 \ln 5 - \frac{1}{2} \ln 25$

34. $3 \log x - 2 \log y$

35. $2 \log u + 3 \log w - 6 \log v$

36. $\ln(3x + 2) + \ln(x + 4)$

37. $2 \ln(x + 1) - \ln(x + 2)$

Write each expression as a sum and/or a difference of logarithms with all variables to the first degree.

38. $\log 5x^2 y^3$ **39.** $\ln \sqrt{6m^4 n^2}$

40. $\ln \dfrac{3x}{5y}$ **41.** $\log \dfrac{\sqrt{xz}}{z^3}$

42. The calculator-generated table in the figure is for $y_1 = \log(4 - x)$. Why do the values in the y_1 column show ERROR for $x \geq 4$?

X	Y₁
0	.60206
1	.47712
2	.30103
3	0
4	ERROR
5	ERROR
6	ERROR

X=4

Express each of the following in terms of u and v, where $u = \ln x$ and $v = \ln y$. For example, $\ln x^3 = 3(\ln x) = 3u$.

43. $\ln(x^2 y^5)$ **44.** $\ln(\sqrt{x} \cdot y^2)$

45. $\ln(x^3/y^2)$ **46.** $\ln(\sqrt{x}/y)$

Evaluate each expression. (See Example 9.)

47. $\log_6 346$ **48.** $\log_{20} 78$

49. $\log_{35} 7646$ **50.** $\log_6 60 - \log_{60} 6$

Find numerical values for b and c for which the given statement is FALSE.

51. $\log(b + c) = \log b + \log c$

52. $\dfrac{\ln b}{\ln c} = \ln\left(\dfrac{b}{c}\right)$

Graph each of the following. (See Example 11.)

53. $y = \ln(x + 2)$ **54.** $y = \ln x + 2$

55. $y = \log(x - 3)$ **56.** $y = \log x - 3$

57. Graph $f(x) = \log x$ and $g(x) = \log(x/4)$ for $-2 \leq x \leq 8$. How are these graphs related? How does the quotient rule support your answer?

In Exercises 58 and 59, the coordinates of a point on the graph of the indicated function are displayed at the bottom of the screen. Write the logarithmic and exponential equations associated with the display.

58.

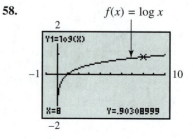

$f(x) = \log x$

Y1=log(X)

X=8   Y=.90308999

59.

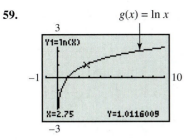

$g(x) = \ln x$

60. Match each equation with its graph. Each tick mark represents one unit.

(a) $y = \log x$ (b) $y = 10^x$
(c) $y = \ln x$ (d) $y = e^x$

(A)

(B)

(C)

(D)

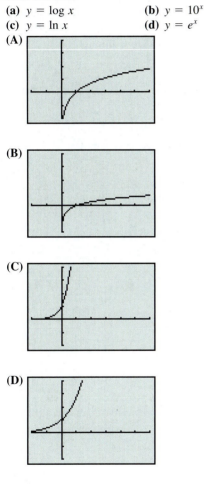

61. Finance The doubling function

$$D(r) = \frac{\ln 2}{\ln(1 + r)}$$

gives the number of years required to double your money when it is invested at interest rate r (expressed as a decimal), compounded annually. How long does it take to double your money at each of the following rates?

(a) 4%   (b) 8%
(c) 18%   (d) 36%

(e) Round each of your answers in parts (a)–(d) to the nearest year, and compare them with these numbers: 72/4, 72/8, 72/18, 72/36. Use this evidence to state a "rule of thumb" for determining approximate doubling time without employing the function D. This rule, which has long been used by bankers, is called the *rule of 72*.

62. Health Two people with the flu visited a college campus. The number of days T that it took for the flu virus to infect n people is given by

$$T = -1.43 \ln\left(\frac{10,000 - n}{4998n}\right).$$

How many days will it take for the virus to infect
(a) 500 people?   (b) 5000 people?

63. Business The number of McDonald's restaurants worldwide (in thousands) is approximated by

$$f(x) = 5.03 + 10.24 \ln x \qquad (x \geq 2),$$

where $x = 2$ corresponds to 1992.*
(a) Estimate the number of restaurants in 1995, 2000, and 2005.
(b) Graph the function f for the period 1992–2020.
(c) Assuming that this model remains accurate, what does the shape of the graph suggest about the number of restaurants.

64. Physical Science The barometric pressure p (in inches of mercury) is related to the height h above sea level (in miles) by the equation

$$h = -5 \ln\left(\frac{p}{29.92}\right).$$

The following pressure readings were made by a weather balloon. At what heights were they made?
(a) 29.92 in.   (b) 20.05 in.   (c) 11.92 in.
(d) Use a graphing calculator to determine the pressure at a height of 3 miles.

*Based on data from McDonald's.

65. Social Science   The percentage of the U.S. population age 60 and older is expected to be given by

$$g(x) = 15.93 + 2.174 \ln x \qquad (x \geq 1),$$

where $x = 1$ corresponds to 2001.*
(a) What is the expected percentage of people who are 60 or older in 2007, 2015, 2030, and 2050?
(b) Graph the function g.
(c) What does the shape of the graph suggest about the percentage of older people as time goes on?

Natural Science   *Do these exercises, which deal with the Shannon index of diversity. (See Example 10.) Note that in two communities with the same number of species, a larger index indicates greater diversity.*

66. A study of barn owl prey in another area produced the following data:

Species	Number
Rats	662
Mice	907
Birds	531

Find the index of diversity of this community. Is this community more or less diverse than the one in Example 10?

67. An eastern forest is composed of the following trees:

Species	Number
Beech	2754
Birch	689
Hemlock	4428
Maple	629

What is the index of diversity of this community?

68. A community has high diversity when all of its species have approximately the same number of individuals. It has low diversity when a few of its species account for most of the total population. Illustrate this fact for the following two communities:

Community 1	Number
Species A	1000
Species B	1000
Species C	1000

Community 2	Number
Species A	2500
Species B	200
Species C	300

*Based on projections by the U.S. Census Bureau.

4.4 Logarithmic and Exponential Equations

Many applications involve solving logarithmic and exponential equations, so we begin with solution methods for such equations.

Logarithmic Equations

When an equation involves only logarithmic terms, use the logarithm properties to write each side as a single logarithm. Then use this fact.

Let u, v, and a be positive real numbers, with $a \neq 1$.

If $\log_a u = \log_a v$, then $u = v$.

▼ **EXAMPLE 1**   Solve $\log x = \log(x + 3) - \log(x - 1)$

Solution First, use the quotient property of logarithms to write the right side as a single logarithm.

$$\log x = \log(x + 3) - \log(x - 1)$$

$$\log x = \log\left(\frac{x + 3}{x - 1}\right)$$

The fact in the preceding box now shows that

$$x = \frac{x + 3}{x - 1}$$

$$x(x - 1) = x + 3$$

$$x^2 - x = x + 3$$

$$x^2 - 2x - 3 = 0$$

$$(x - 3)(x + 1) = 0$$

$$x = 3 \quad \text{or} \quad x = -1.$$

Since $\log x$ is not defined when $x = -1$, the only possible solution is $x = 3$. Use a calculator to verify that 3 actually is a solution. ▲ ⟨1⟩

When an equation involves constants and logarithmic terms, use algebra and the logarithm properties to write one side as a single logarithm and the other as a constant. Then use the following property of logarithms, which was discussed on pages 225–226.

⟨1⟩

Solve each equation.

(a) $\log_2(p + 9) - \log_2 p = \log_2(p + 1)$

(b) $\log_3(m + 1) - \log_3(m - 1) = \log_3 m$

Answers:

(a) 3

(b) $1 + \sqrt{2} \approx 2.414$

If a and u are positive real numbers, with $a \neq 1$, then

$$a^{\log_a u} = u.$$

▼ **EXAMPLE 2** Solve each equation.

(a) $\log_5(2x - 3) = 2$

Solution Since the base of the logarithm is 5, raise 5 to the exponents given by the equation.

$$5^{\log_5(2x-3)} = 5^2$$

On the left side, use the fact in the preceding box (with $a = 5$ and $u = 2x - 3$) to conclude that

$$2x - 3 = 25$$

$$2x = 28$$

$$x = 14.$$

Verify that 14 is a solution of the original equation.

(b) $\log(x - 16) = 2 - \log(x - 1)$

Solution First rearrange the terms to obtain a single logarithm on the left side.

$$\log(x - 16) + \log(x - 1) = 2$$

$$\log[(x - 16)(x - 1)] = 2 \qquad \text{Product property of logarithms}$$

$$\log(x^2 - 17x + 16) = 2$$

Since the base of the logarithm is 10, raise 10 to the given powers.

$$10^{\log(x^2-17x+16)} = 10^2$$

On the left side, apply the logarithm property in the preceding box (with $a = 10$ and $u = x^2 - 17x + 16$).

$$x^2 - 17x + 16 = 100$$
$$x^2 - 17x - 84 = 0$$
$$(x + 4)(x - 21) = 0$$
$$x = -4 \quad \text{or} \quad x = 21$$

In the original equation, when $x = -4$, $\log(x - 16) = \log(-20)$, which is not defined. So -4 is not a solution. You should verify that 21 is a solution of the original equation.

(c) $\log_2 x - \log_2(x - 1) = 1$

Solution Proceed as before, with 2 as the base of the logarithm.

$$\log_2 \frac{x}{x-1} = 1 \qquad \text{Quotient property of logarithms}$$

$$2^{\log_2 x/(x-1)} = 2^1 \qquad \text{Exponentiate to the base 2.}$$

$$\frac{x}{x-1} = 2 \qquad \text{Use the fact in the preceding box.}$$

$$x = 2(x - 1) \qquad \text{Multiply both sides by } x - 1.$$
$$x = 2x - 2$$
$$-x = -2$$
$$x = 2$$

Verify that 2 is a solution of the original equation. ▲ ⟨2⟩

⟨2⟩

Solve each equation.

(a) $\log_5 x + 2 \log_5 x = 3$

(b) $\log_6(a + 2) - \log_6 \dfrac{a - 7}{5} = 1$

Answers:

(a) 5

(b) 52

Exponential Equations

An equation in which all the variables are exponents is called an *exponential equation*. When such an equation can be written as two powers of the same base, it can be solved by using the following fact.

Let a be a positive real number, with $a \neq 1$.

$$\text{If } a^u = a^v, \qquad \text{then} \qquad u = v.$$

▼ **EXAMPLE 3**   Solve $9^x = 27$.

Solution First, write both sides as powers of the same base. Since $9 = 3^2$ and $27 = 3^3$, we have

$$9^x = 27$$
$$(3^2)^x = 3^3$$
$$3^{2x} = 3^3.$$

Apply the fact in the preceding box (with $a = 3$, $u = 2x$, and $v = 3$).

$$2x = 3$$

$$x = \frac{3}{2}$$

Verify that $x = 3/2$ is a solution of the original equation. ▲ ③

Exponential equations involving different bases can often be solved by using the power property of logarithms, which is repeated here for natural logarithms.

If u and r are real numbers, with u positive, then

$$\ln u^r = r \ln u.$$

Although natural logarithms are used in the following examples, logarithms to any base will produce the same solutions.

▼ **EXAMPLE 4** Solve $3^x = 5$.

Solution Take natural logarithms on both sides.

$$\ln 3^x = \ln 5$$

Apply the power property of logarithms in the preceding box (with $r = x$) to the left side.

$$x \ln 3 = \ln 5$$

$$x = \frac{\ln 5}{\ln 3} \approx 1.465 \qquad \text{Divide both sides by the constant ln 3.}$$

To check, evaluate $3^{1.465}$; the answer should be approximately 5, which verifies that the solution of the given equation is 1.465 (to the nearest thousandth). ▲ ④

CAUTION Be careful: $\dfrac{\ln 5}{\ln 3}$ is *not* equal to $\ln\left(\dfrac{5}{3}\right)$ or $\ln 5 - \ln 3$.

▼ **EXAMPLE 5** Solve $3^{2x-1} = 4^{x+2}$.

Solution Taking natural logarithms on both sides gives

$$\ln 3^{2x-1} = \ln 4^{x+2}.$$

Now use the power property of logarithms and the fact that $\ln 3$ and $\ln 4$ are constants to rewrite the equation.

$$(2x - 1)(\ln 3) = (x + 2)(\ln 4) \qquad \text{Power property}$$
$$2x(\ln 3) - 1(\ln 3) = x(\ln 4) + 2(\ln 4) \qquad \text{Distributive property}$$
$$2x(\ln 3) - x(\ln 4) = 2(\ln 4) + 1(\ln 3) \qquad \text{Collect terms with } x \text{ on one side.}$$

③ Solve each equation.
(a) $8^{2x} = 4$
(b) $5^{3x} = 25^4$
(c) $36^{-2x} = 6$

Answers:
(a) $1/3$
(b) $8/3$
(c) $-1/4$

④ Solve each equation. Round solutions to the nearest thousandth.
(a) $2^x = 7$
(b) $5^m = 50$
(c) $3^y = 17$

Answers:
(a) 2.807
(b) 2.431
(c) 2.579

Factor out x on the left side to get

$$[2(\ln 3) - \ln 4]x = 2(\ln 4) + \ln 3.$$

Divide both sides by the coefficient of x:

$$x = \frac{2(\ln 4) + \ln 3}{2(\ln 3) - \ln 4}.$$

Using a calculator to evaluate this last expression, we find that

$$x = \frac{2\ln 4 + \ln 3}{2\ln 3 - \ln 4} \approx 4.774. \; \blacktriangle \; ⑤$$

Recall that $\ln e = 1$ (because 1 is the exponent to which e must be raised to produce e). This fact simplifies the solution of equations involving powers of e.

⑤

Solve each equation. Round solutions to the nearest thousandth.

(a) $6^m = 3^{2m-1}$

(b) $5^{6a-3} = 2^{4a+1}$

Answers:

(a) 2.710

(b) .802

▼ **EXAMPLE 6**   Solve $3e^{x^2} = 600$.

Solution First divide each side by 3 to get

$$e^{x^2} = 200.$$

Now take natural logarithms on both sides; then use the power property of logarithms.

$$e^{x^2} = 200$$
$$\ln e^{x^2} = \ln 200$$
$$x^2 \ln e = \ln 200 \qquad \text{Power property}$$
$$x^2 = \ln 200 \qquad \ln e = 1$$
$$x = \pm\sqrt{\ln 200}$$
$$x \approx \pm 2.302$$

Verify that the solutions are ± 2.302, rounded to the nearest thousandth. (The symbol $\pm$ is used as a shortcut for writing the two solutions 2.302 and -2.302.) ▲ ⑥

⑥

Solve each equation. Round solutions to the nearest thousandth.

(a) $e^{.1x} = 11$

(b) $e^{3+x} = .893$

(c) $e^{2x^2-3} = 9$

Answers:

(a) 23.979

(b) -3.113

(c) ± 1.612

TECHNOLOGY TIP   Logarithmic and exponential equations can be solved on a graphing calculator. To solve $3^x = 5^{2x-1}$, for example, graph $y = 3^x$ and $y = 5^{2x-1}$ on the same screen. Then use the intersection finder to determine the x-coordinates of their intersection points. Alternatively, graph $y = 3^x - 5^{2x-1}$ and use the root finder to determine the x-intercepts of the graph.

Applications

Some of the most important applications of exponential and logarithmic functions arise in banking and finance. They will be thoroughly discussed in Chapter 5. The applications here are from other fields.

▼ **EXAMPLE 7**   According to projections by the U.S. Census Bureau, the world population (in billions) in year x is approximated by the function $P(x) = 4.834(1.01^{x-1980})$. When will the population reach 7 billion?

Solution   You are asked to find the value of x for which $P(x) = 7$—that is, to solve the following equation.

$$4.834(1.01^{x-1980}) = 7$$

$$1.01^{x-1980} = \frac{7}{4.834} \qquad \text{Divide both sides by 4.834.}$$

$$\ln 1.01^{x-1980} = \ln\left(\frac{7}{4.834}\right) \qquad \text{Take logarithms on both sides.}$$

$$(x - 1980)\ln 1.01 = \ln\left(\frac{7}{4.834}\right) \qquad \text{Power property of logarithms}$$

$$x - 1980 = \frac{\ln(7/4.834)}{\ln 1.01} \qquad \text{Divide both sides by ln 1.01.}$$

$$x = \frac{\ln(7/4.834)}{\ln 1.01} + 1980 \approx 2017.2$$

Hence, the world's population will reach 7 billion around 2017. ▲ ⟨7⟩

⟨7⟩

Use the function in Example 7 to determine when the earth's population will be 10 billion.

Answer:

In 2053

The **half-life** of a radioactive substance is the time it takes for a given quantity of the substance to decay to one-half its original mass. The half-life depends only on the substance, not on the size of the sample. It can be shown that the amount of a radioactive substance at time t is given by the function

$$f(t) = y_0\left(\frac{1}{2}\right)^{t/h} = y_0(.5^{t/h}),$$

where y_0 is the initial amount (at time $t = 0$) and h is the half-life of the substance.

Radioactive carbon-14 is found in every living plant and animal. After the plant or animal dies, its carbon-14 decays exponentially, with a half-life of 5730 years. This fact is the basis for a technique called *carbon dating* for determining the age of fossils.

▼ **EXAMPLE 8**   A round wooden table hanging in Winchester Castle (England) was alleged to have belonged to King Arthur, who lived in the fifth century. A recent chemical analysis showed that the wood had lost 9% of its carbon-14.* How old is the table?

Solution   The decay function for carbon-14 is

$$f(t) = y_0(.5^{t/5730}),$$

where $t = 0$ corresponds to the time the wood was cut to make the table. (That's when the tree died.) Since the wood has lost 9% of its carbon-14, the amount now in

_____

*This is done by measuring the ratio of carbon-14 to nonradioactive carbon-12 in the table (a ratio that is approximately constant over long periods) and comparing it with the ratio in living wood.

the table is 91% of the initial amount y_0 (that is, $.91y_0$). We must find the value of t for which $f(t) = .91y_0$. So we must solve the equation

$$y_0(.5^{t/5730}) = .91y_0 \qquad \text{Definition of } f(t)$$

$$.5^{t/5730} = .91 \qquad \text{Divide both sides by } y_0.$$

$$\ln .5^{t/5730} = \ln .91 \qquad \text{Take logarithms on both sides.}$$

$$\left(\frac{t}{5730}\right)\ln .5 = \ln .91 \qquad \text{Power property of logarithms}$$

$$t \ln .5 = 5730 \ln .91 \qquad \text{Multiply both sides by 5730.}$$

$$t = \frac{5730 \ln .91}{\ln .5} \approx 779.63 \qquad \text{Divide both sides by } \ln 5.$$

The table is about 780 years old and therefore could not have belonged to King Arthur. ▲ ⟨8⟩

⟨8⟩
How old is a skeleton that has lost 65% of its carbon-14?

Answer:

About 8679 years

Earthquakes are often in the news. What you may not have known is that the standard method of measuring their size, the **Richter scale,** is a logarithmic function (base 10).

▼ **EXAMPLE 9**   The intensity $R(i)$ of an earthquake, measured on the Richter scale, is given by

$$R(i) = \log\left(\frac{i}{i_0}\right),$$

where i is the intensity of the ground motion of the earthquake and i_0 is the intensity of the ground motion of the so-called *zero earthquake* (the smallest detectable earthquake, against which others are measured). The underwater earthquake that caused the disastrous 2004 tsunami measured 9.1 on the Richter scale.

(a) How did the ground motion of this tsunami compare with that of the zero earthquake?

Solution   In this case,

$$R(i) = 9.1$$

$$\log\left(\frac{i}{i_0}\right) = 9.1.$$

By the definition of logarithms, 9.1 is the exponent to which 10 must be raised to produce i/i_0, which means that

$$10^{9.1} = \frac{i}{i_0}, \qquad \text{or equivalently,} \qquad i = 10^{9.1} i_0.$$

So the earthquake that produced the tsunami had $10^{9.1}$ (about 1.26 *billion*) times more ground motion than the zero earthquake.

(b) What is the Richter-scale intensity of an earthquake with 10 times as much ground motion as the 2004 tsunami earthquake?

Solution From (a), the ground motion of the tsunami quake was $10^{9.1} i_0$. So a quake with 10 times that motion would satisfy

$$i = 10(10^{9.1} i_0) = 10^1 \cdot 10^{9.1} i_0 = 10^{10.1} i_0.$$

Therefore, its Richter scale measure would be

$$R(i) = \log\left(\frac{i}{i_0}\right) = \log\left(\frac{10^{10.1} i_0}{i_0}\right) = \log 10^{10.1} = 10.1.$$

Thus, a tenfold increase in ground motion increases the Richter scale measure by only 1. ▲

⟨9⟩ Find the Richter-scale intensity of an earthquake whose ground motion is 100 times greater than the ground motion of the 2004 tsunami earthquake discussed in Example 9.

Answer:

11.1

▼ **EXAMPLE 10** One action that government could take to reduce carbon emissions into the atmosphere is to place a tax on fossil fuels. This tax would be based on the amount of carbon dioxide that is emitted into the air when such a fuel is burned. The *cost–benefit* equation $\ln(1 - P) = -.0034 - .0053T$ describes the approximate relationship between a tax of T dollars per ton of carbon dioxide and the corresponding percent reduction P (in decimals) in emissions of carbon dioxide.*

(a) Write P as a function of T.

Solution We begin by writing the cost–benefit equation in exponential form.

$$\ln(1 - P) = -.0034 - .0053T$$
$$1 - P = e^{-.0034 - .0053T}$$
$$P = P(T) = 1 - e^{-.0034 - .0053T}$$

A calculator-generated graph of $P(T)$ is shown in Figure 4.19.

(b) Discuss the benefit of continuing to raise taxes on carbon dioxide emissions.

Solution From the graph, we see that initially there is a rapid reduction in carbon dioxide emissions. However, after a while, there is little benefit in raising taxes further. ▲

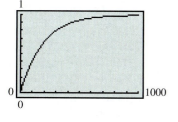

FIGURE 4.19

*W. Nordhause, "To Slow or Not to Slow: The Economics of the Greenhouse Effect." Yale University, New Haven, Connecticut.

◆4.4 Exercises

Solve each logarithmic equation. (See Example 1.)

1. $\ln(3x + 1) - \ln(5 + x) = \ln 2$

2. $\ln(8k - 7) - \ln(3 + 4k) = \ln(9/11)$

3. $\ln(x + 1) = \ln(x - 4)$

4. $\ln(5x + 2) = \ln 4 + \ln(x + 3)$

5. $2\ln(x - 3) = \ln(x + 5) + \ln 4$

6. $\ln(k + 5) + \ln(k + 2) = \ln 14k$

Solve each logarithmic equation. (See Example 2.)

7. $\log_3(6x - 2) = 2$

8. $\log_5(3x - 4) = 1$

9. $\log x - \log(x + 5) = -1$

10. $\log m - \log(m + 4) = -2$

11. $\log_3(y + 2) = \log_3(y - 7) + \log_3 4$

12. $\log_8(z - 6) = 2 - \log_8(z + 15)$

13. $\ln(x + 9) - \ln x = 1$

14. $\ln(2x + 1) - 1 = \ln(x - 2)$

15. $\log x + \log(x - 3) = 1$

16. $\log(x - 1) + \log(x + 2) = 1$

Solve each equation for c.

17. $\log(3 + b) = \log(4c - 1)$

18. $\ln(b + 7) = \ln(6c + 8)$

19. $2 - b = \log(6c + 5)$

20. $8b + 6 = \ln(2c) + \ln c$

21. Suppose you overhear the following statement: "I must reject any negative answer when I solve an equation involving logarithms." Is this correct? Write an explanation of why it is or is not correct.

22. What values of x cannot be solutions of the following equation?

$$\log_a(4x - 7) + \log_a(x^2 + 4) = 0$$

Solve these exponential equations without using logarithms. (See Example 3.)

23. $3^{x-1} = 9$

24. $16^{-x+2} = 8$

25. $25^{-3x} = 3125$

26. $81^{-2x} = 3^{x-1}$

27. $7^{-x} = 49^{x+3}$

28. $16^x = 64$

29. $\left(\dfrac{3}{4}\right)^x = \dfrac{16}{9}$

30. $2^{x^2-4x} = \dfrac{1}{16}$

Use logarithms to solve these exponential equations. (See Examples 4–6.)

31. $2^x = 5$

32. $5^x = 8$

33. $2^x = 3^{x-1}$

34. $4^{x+2} = 2^{x-1}$

35. $3^{1-2x} = 5^{x+5}$

36. $4^{3x-1} = 3^{x-2}$

37. $e^{2x} = 7$

38. $e^{-3x} = 5$

39. $2e^{5a+2} = 8$

40. $10e^{3z-7} = 5$

Solve each equation for c.

41. $10^{4c-3} = d$

42. $3 \cdot 10^{2c+1} = 4d$

43. $e^{2c-1} = b$

44. $3e^{5c-7} = b$

Solve these equations. (See Examples 1–6.)

45. $\log_5(r + 2) + \log_5(r - 2) = 1$

46. $\log_4(z + 3) + \log_4(z - 3) = 2$

47. $\log_3(a - 3) = 1 + \log_3(a + 1)$

48. $\log w + \log(3w - 13) = 1$

49. $\log_2 \sqrt{2y^2} - 1 = 3/2$

50. $\log_2(\log_2 x) = 1$

51. $\log z = \sqrt{\log z}$

52. $\log x^2 = (\log x)^2$

53. $5^{-2x} = \dfrac{1}{25}$

54. $5^{x^2+x} = 1$

55. $2^{|x|} = 16$

56. $5^{-|x|} = \dfrac{1}{25}$

57. $2^{x^2-1} = 10$

58. $3^{2-x^2} = 8$

59. $2(e^x + 1) = 10$

60. $5(e^{2x} - 2) = 15$

61. Explain why the equation $4^{x^2+1} = 2$ has no solutions.

62. Explain why the equation $\log(-x) = -4$ does have a solution, and find that solution.

Work these problems. (See Example 7.)

63. **Business** Worldwide production of automobiles has been steadily increasing. The number of people per automobile is approximated by

$$f(x) = 24.2(.9792^x),$$

where $x = 0$ corresponds to 1960.*
   (a) How many people per automobile were there in 1997 and 2005?
   (b) If this model remains accurate, when will the number of people per car reach 7?

64. **Business** The value of China's exports of automobiles and parts (in billions of dollars) is approximated by

$$g(x) = .1212e^{.3387x} \qquad (x \geq 8),$$

where $x = 8$ corresponds to 1998.†
   (a) What was the value of Chinese exports in 1998?
   (b) When did the level of exports reach $5.1 billion? $12 billion?

65. **Health** A drug's effectiveness decreases over time. If, each hour, a drug is only 90% as effective as the previous hour, at some point the patient will not be receiving enough medication and must receive another dose. This situation can be modeled by an exponential function with $y = y_0(.90)^{t-1}$. In this equation, y_0 is the amount of the initial dose and y is the amount of medication still available t hours after the drug was administered. Suppose 200 mg of the drug is administered. How long will it take for this initial dose to reach the dangerously low level of 50 mg?

66. **Business** The U.S. gross domestic product (GDP), in trillions of dollars, is approximated by

$$f(x) = -1.2155 + 3.7314 \ln x \qquad (x \geq 5),$$

where $x = 5$ corresponds to 1995.‡ According to this model, when will the GDP reach
   (a) $12 trillion
   (b) $13 trillion?

Work these exercises. (See Example 8.)

67. **Natural Science** The amount of cobalt-60 (in grams) in a storage facility at time t is given by

$$C(t) = 25e^{-.14t},$$

where time is measured in years.
   (a) How much cobalt-60 was present initially?
   (b) What is the half-life of cobalt-60? (*Hint:* For what value of t is $C(t) = 12.5$?)

---

*Based on data from Ward's Communications.

†Based on data from Daimler Chrysler and Global Trade Information Services.

‡Based on data from the U.S. Bureau of Economic Analysis.

68. Natural Science A Native American mummy was found recently. It had 73.6% of the amount of radiocarbon present in living beings. Approximately how long ago did this person die?

69. Natural Science How old is a piece of ivory that has lost 36% of its radiocarbon?

70. Natural Science A sample from a refuse deposit near the Strait of Magellan had 60% of the carbon 14 of a contemporary living sample. How old was the sample?

Natural Science *Do the following problems. (See Example 9.)*

71. On July 14, 1991, Peshawar, Pakistan, was shaken by an earthquake that measured 6.6 on the Richter scale.
   (a) Express this reading in terms of i_0.
   (b) In February of the same year, a quake measuring 6.5 on the Richter scale killed about 900 people in the mountains of Pakistan and Afghanistan. Express the intensity of a 6.5 reading in terms of i_0.
   (c) How many times more intense was the July earthquake than the February one?

72. Find the Richter-scale intensity of earthquakes whose ground motion is
   (a) $1000i_0$
   (b) $100,000i_0$
   (c) $10,000,000i_0$.
   (d) Fill in the blank in this statement: increasing the ground motion by a factor of 10^k increases the Richter intensity by _____ units.

73. The loudness of sound is measured in units called decibels. The decibel rating of a sound is given by

$$D(i) = 10 \cdot \log\left(\frac{i}{i_0}\right),$$

where i is the intensity of the sound and i_0 is the minimum intensity detectable by the human ear (the so-called *threshold sound*). Find the decibel rating of each of the following sounds whose intensities are given. Round answers to the nearest whole number.
   (a) Whisper, $115i_0$
   (b) Average sound level in the movie *Godzilla*, $10^{10}i_0$
   (c) Jackhammer, $31,600,000,000i_0$
   (d) Rock music, $895,000,000,000i_0$
   (e) Jetliner at takeoff, $109,000,000,000,000i_0$

74. (a) How much more intense is a sound that measures 100 decibels than the threshold sound?
   (b) How much more intense is a sound that measures 50 decibels than the threshold sound?
   (c) How much more intense is a sound measuring 100 decibels than one measuring 50 decibels?

75. Natural Science Refer to Example 10.
   (a) Determine the percent reduction in carbon dioxide when the tax is $60.

(b) What tax will cause a 50% reduction in carbon dioxide emissions?

76. Social Science The number of years, $N(r)$, since two independently evolving languages split off from a common ancestral language is approximated by

$$N(r) = -5000 \ln r,$$

where r is the proportion of the words from the ancestral language that is common to both languages now. Find each of the following.
   (a) $N(.9)$
   (b) $N(.5)$
   (c) $N(.3)$
   (d) How many years have elapsed since the split if 70% of the words of the ancestral language are common to both languages today?
   (e) If two languages split off from a common ancestral language about 1000 years ago, find r.

77. Natural Science In the central Sierra Nevada of California, the percent of moisture that falls as snow rather than rain is approximated reasonably well by

$$p = 86.3 \ln h - 680,$$

where p is the percent of moisture as snow at an altitude of h feet (with $3000 \leq h < 8500$).
   (a) Graph p.
   (b) At what altitude is 50 percent of the moisture snow?

78. Physical Science The table gives some of the planets' average distances D from the sun and their period P of revolution around the sun in years. The distances have been normalized so that Earth is one unit from the sun. Thus, Jupiter's distance of 5.2 means that Jupiter's distance from the sun is 5.2 times farther than Earth's.*

Planet	D	P
Earth	1	1
Jupiter	5.2	11.9
Saturn	9.54	29.5
Uranus	19.2	84.0

   (a) Plot the points (D, P) for these planets. Would a straight line or an exponential curve fit these points best?
   (b) Plot the points $(\ln D, \ln P)$ for these planets. Do these points appear to lie on a line?
   (c) Determine a linear equation that approximates the data points with $x = \ln D$ and $y = \ln P$. Use the first and last data points (rounded to 2 decimal places). Graph your line and the data on the same coordinate axes.

*C. Ronan, *The Natural History of the Universe* (New York: Macmillan Publishing Co., 1991).

(d) Use the linear equation to predict the period of the planet Pluto if its distance is 39.5. Compare your answer with the true value of 248.5 years.

Use logarithmic regression for these exercises.

79. **Health** The table shows the number of kidney transplants in selected years.*

_____
*U.S. Department of Health and Human Services and the United Network for Organ Sharing.

Year	1990	1995	1996	1997
Transplants	9880	11,901	12,152	12,307

Year	1998	2000	2003
Transplants	13,139	14,149	15,129

(a) Find a logarithmic function that models this data with $x = 10$ corresponding to 1990.

(b) Estimate the number of transplants in 2002.

(c) If this model remains accurate, in what year will there be 16,500 transplants?

80. **Social Science** The table shows the number of visitors to U.S. national parks (in millions) in selected years.*

Year	1960	1980	1994	1998	2000	2002	2003
Visitors	28	47	63	64.5	66.1	66.5	67.5

(a) Find a logarithmic function that models this data with $x = 60$ corresponding to 1960.

(b) Assuming that the model remains accurate, estimate the number of visitors in 2005, 2008, and 2010.

(c) In what year will the number of visitors reach 75 million?

_____
*National Park Service.

Chapter 4 Summary

Key Terms and Symbols

| 4.1 | exponential function
exponential growth and decay
the number $e \approx 2.71828\ldots$ |

| 4.2 | exponential growth function
logistic function |

| 4.3 | $\log x$ common (base 10 logarithm) of x |

$\ln x$ natural (base e logarithm) of x
$\log_a x$ base a logarithm of x
product, quotient, and power properties of logarithms
change of base theorem
logarithmic function
inverses

| 4.4 | logarithmic equations
exponential equations
half-life
Richter scale |

Key Concepts

An important application of exponents is the **exponential growth function,** defined as $f(t) = y_0 e^{kt}$ or $f(t) = y_0 b^t$, where y_0 is the amount of a quantity present at time $t = 0$, $e \approx 2.71828$, and k and b are constants.

The **logarithm** of x to the base a is defined as follows: for $a > 0$ and $a \neq 1$, $y = \log_a x$ means $a^y = x$. Thus, $\log_a x$ is an *exponent,* the power to which a must be raised to produce x.

Properties of Logarithms

Let x, y, and a be positive real numbers, $a \neq 1$, and let r be any real number. Then

$$\log_a 1 = 0 \qquad \log_a a = 1$$

$$\log_a a^r = r \qquad a^{\log_a x} = x$$

Product property: $\log_a xy = \log_a x + \log_a y$

Quotient property: $\log_a \dfrac{x}{y} = \log_a x - \log_a y$

Power property: $\log_a x^r = r \log_a x$

Solving Exponential and Logarithmic Equations

Let $a > 0$, $a \neq 1$.

If $\log_a u = \log_a v$, then $u = v$.

If $a^u = a^v$, then $u = v$.

Chapter 4 Review Exercises

Match each equation with the letter of the graph that most closely resembles the graph of the equation. Assume that $a > 1$.

1. $y = a^{x+2}$

2. $y = a^x + 2$

3. $y = -a^x + 2$

4. $y = a^{-x} + 2$

(a)

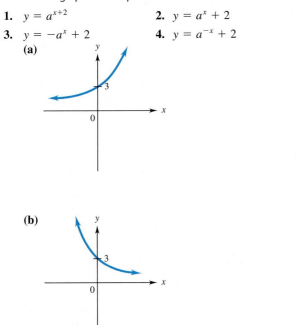

(b)

(c)

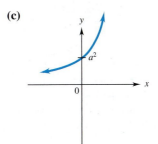

(d)

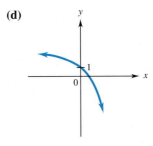

Consider the exponential function $y = f(x) = a^x$ graphed here. Answer each question on the basis of the graph.

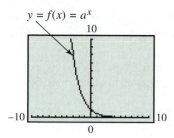

$$y = f(x) = a^x$$

5. What is true about the value of a in comparison to 1?

6. What is the domain of f?

7. What is the range of f?

8. What is the value of $f(0)$?

Graph each function.

9. $f(x) = 4^x$

10. $g(x) = 4^{-x}$

11. $f(x) = \ln x + 5$

12. $g(x) = \log x - 3$

Work these problems.

13. **Social Science**   More people now have digital video recorders (which allow them, among other things, to skip commercials) than ever before. The percentage of U.S. households that have a digital video recorder is projected to be given by

$$f(x) = .789(1.638^x),$$

where $x = 0$ corresponds to 2000.*
   (a) Estimate the percentage of households with digital video recorders in 2006 and 2008.
   (b) According to this model, when will every U.S. household have a digital video recorder? Does this seem likely?

14. **Business**   A person learning certain skills involving repetition tends to learn quickly at first. Then learning tapers off and approaches some upper limit. Suppose the number of symbols per minute a textbook typesetter can produce is given by $p(t) = 250 - 120(2.8)^{-.5t}$, where t is the number of months the typesetter has been in training. Find each of the following.
   (a) $p(2)$ **(b)** $p(4)$
   (c) $p(10)$.   **(d)** Graph $y = p(t)$.

Translate each exponential statement into an equivalent logarithmic one.

15. $10^{1.7404} = 55$

16. $4^5 = 1024$

Based on data in Time, *May 30, 2005.*

17. $e^{3.8067} = 45$

18. $7^{1/2} = \sqrt{7}$

Translate each logarithmic statement into an equivalent exponential one.

19. $\log 10{,}000 = 4$

20. $\log 26.3 = 1.4200$

21. $\ln 81.1 = 4.3957$

22. $\log_2 4096 = 12$

Evaluate these expressions without using a calculator.

23. $\ln e^5$

24. $\log \sqrt[3]{10}$

25. $10^{\log 8.9}$

26. $\ln e^{5t}$

27. $\log_{36} 6$

28. $\log_8 32$

Write these expressions as a single logarithm. Assume all variables represent positive quantities.

29. $\log 3x + \log 4x^4$

30. $5 \log u - 3 \log u^4$

31. $3 \log b - 2 \log c$

32. $7 \ln x - 3(\ln x^3 + 5 \ln x)$

Solve each equation. Round to the nearest thousandth.

33. $\ln(m + 3) - \ln m = \ln 2$

34. $2 \ln(y + 1) = \ln(y^2 - 1) + \ln 5$

35. $\log(m + 2) = 1$

36. $\log x^2 = 2$

37. $\log_2(3k - 2) = 4$

38. $\log_5\left(\dfrac{5z}{z - 2}\right) = 2$

39. $\log x + \log(x + 3) = 1$

40. $\log_2 r + \log_2(r - 2) = 3$

41. $2^{3x} = \dfrac{1}{8}$

42. $\left(\dfrac{9}{16}\right)^x = \dfrac{3}{4}$

43. $9^{2y-1} = 27^y$

44. $\dfrac{1}{2} = \left(\dfrac{b}{4}\right)^{1/4}$

45. $8^p = 19$

46. $3^z = 11$

47. $5 \cdot 2^{-m} = 35$

48. $2 \cdot 15^{-k} = 18$

49. $e^{-5-2x} = 5$

50. $e^{3x-1} = 12$

51. $6^{2-m} = 2^{3m+1}$

52. $5^{3r-1} = 6^{2r+5}$

53. $(1 + .003)^k = 1.089$

54. $(1 + .094)^z = 2.387$

Work these problems.

55. **Business**   Suppose the gross national product (GNP) of a small country (in millions of dollars) is approximated by the function $G(t) = 15 + 2 \log t$, where t is time in years, for $1 \le t \le 6$. Find the GNP at the following times.
   (a) 1 year   **(b)** 2 years   **(c)** 5 years

56. Natural Science A population is increasing according to the growth law $y = 2e^{.02t}$, where y is in millions and t is in years. Match each of the questions (a), (b), (c), and (d) with one of the solutions (A), (B), (C), or (D).

(a) How long will it take for the population to triple?

(A) Evaluate $2e^{.02(1/3)}$.

(b) When will the population reach 3 million?

(B) Solve $2e^{.02t} = 3 \cdot 2$ for t.

(c) How large will the population be in 3 years?

(C) Evaluate $2e^{.02(3)}$.

(d) How large will the population be in 4 months?

(D) Solve $2e^{.02t} = 3$ for t.

57. Natural Science The amount of polonium (in grams) present after t days is given by

$$A(t) = 10e^{-.00495t}.$$

(a) How much polonium was present initially?

(b) What is the half-life of polonium?

(c) How long will it take for the polonium to decay to 3 grams?

58. Natural Science One earthquake measures 4.6 on the Richter scale. A second earthquake has ground motion 1000 times greater than the first. What does the second one measure on the Richter scale?

Natural Science *Here is another form of Newton's Law of Cooling (see Section 4.2, Exercises 15 and 16): $F(t) = T_0 + Ce^{-kt}$, where C and k are constants.*

59. A piece of metal is heated to 300° Celsius and then placed in a cooling liquid at 50° Celsius. After 4 minutes the metal has cooled to 175° Celsius. Find its temperature after 12 minutes.

60. A frozen pizza has a temperature of 3.4° Celsius when it is taken from the freezer and left out in a room at 18° Celsius. After half an hour, its temperature is 7.2° Celsius. How long will it take for the pizza to thaw to 10° Celsius?

In Exercises 61–63, do part (a) and skip part (b) if you do not have a graphing calculator. If you have a graphing calculator, then do part (b) and skip part (a).

61. Business The number of music cassette tapes sold each year (in millions) has been steadily dropping, as shown in the table.

Year	1995	1996	1997	1998	1999
Cassettes Sold	272.6	225.3	172.6	158.5	123.6

Year	2000	2001	2002	2003
Cassettes Sold	76.0	45.0	31.1	17.2

(a) Let $x = 0$ correspond to 1995. Use the data points from 1995 and 2003 to find a function of the form $f(x) = a(b^x)$ that models this data.

(b) Use exponential regression to find a function $g(x)$ that models the data, with $x = 0$ corresponding to 1995.

(c) Estimate the number of cassettes sold in 2004.

(d) According to this model, when will cassette sales drop below 1.5 million?

62. Physical Science The atmospheric pressure (in millibars) at a given altitude (in thousands of meters) is listed in the table.

Altitude	0	2000	4000	6000	8000	10,000
Pressure	1013	795	617	472	357	265

(a) Use the data points for altitudes 0 and 10,000 to find a function of the form $f(x) = a(b^x)$ that models this data, where altitude x is measured in thousands of meters. (For example, $x = 4$ means 4000 meters.)

(b) Use exponential regression to find a function $g(x)$ that models the data, with x measured in thousands of meters.

(c) Estimate the pressure at 1500 m and 11,000 m, and compare the results with the actual values of 846 and 227 millibars.

(d) At what height is the pressure 500 millibars?

63. Business The table shows the number of Southwest Airlines passengers (in millions) in selected years.*

Year	1992	1994	1996	1998	2000	2002	2004
Passengers	27.8	42.7	49.6	52.6	63.7	63.0	70.9

(a) Let $x = 1$ correspond to 1992. Use the data points for 1992 and 2004 to find a function of the form $f(x) = a + b \ln x$ that models this data. [*Hint:* Use $(1, 27.8)$ to find a; then use $(13, 70.9)$ to find b.]

(b) Use logarithmic regression to find a function $g(x)$ that models the data, with $x = 1$ corresponding to 1992.

(c) Estimate the number of passengers in 2008.

(d) According to this model, when will the number of passengers reach 80 million?

64. Physical Science The power of personal computers has grown dramatically as a result of the ability to place an increasing number of transistors on a single processor

---

*Southwest Airlines annual reports, 1992–2004.

chip. The table lists the number of transistors on some popular computer chips by Intel.*

Year	Chip	Transistors
1971	4004	2,300
1986	386DX	275,000
1989	486DX	1,200,000
1993	Pentium	3,100,000
1995	P6	5,500,000
2000	Pentium 4	42,000,000
2003	Itanium 2	220,000,000
2004	Itanium 2 (9-MB cache)	592,000,000

(a) Let x be the year, where $x = 1$ corresponds to 1971, and y be the number of transistors. Plot the data.
(b) Would a linear, exponential, or logarithmic function fit these data best?
(c) Use regression to find a model for the data.
(d) Graph the function from part (c) with the data points. Does the function fit the data?
(e) If this model remains accurate after 2004, when is the number of transistors 1 billion? If possible, check the accuracy of your answer at www.intel.com. (Search for "Moor's Law Transistor Count Chart.")

*www.intel.com

Case 4

Characteristics of the Monkeyface Prickleback*

The monkeyface prickleback (*Cebidichthys violaceus*), known to anglers as the monkeyface "eel," is found in rocky intertidal and subtidal habitats ranging from San Quintin Bay, Baja California, to Brookings, Oregon. Pricklebacks are prime targets of the few sports anglers who "poke pole" in the rocky intertidal zone at low tide. Little is known about the life history of this species. The results of a study of the length, weight, and age of this species are discussed in this case.

Data on standard length (*SL*) and total length (*TL*) were collected. Early in the study only *TL* was measured, so a conversion to *SL* was necessary. The equation relating the two lengths, calculated from 177 observations for which both lengths had been measured, is

$$SL = TL(.931) + 1.416.$$

Ages (determined by standard aging techniques) were used to estimate parameters of the von Bertanfany growth model

$$L_t = L_x(1 - e^{-kt}) \tag{1}$$

where

L_t = length at age *t*,

L_x = asymptotic age of the species,

k = growth completion rate, and

t_0 = theoretical age at zero length.

The constants *a* and *b* in the model

$$W = aL^b \tag{2}$$

where

W = weight in g,

L = standard length in cm,

were determined using 139 fish ranging from 27 cm and 145 g to 60 cm and 195 g.

Growth curves giving length as a function of age are shown in the accompanying figure. For the data marked opercle, the lengths were computed from the ages using equation (1).

*"Characteristics of the Monkeyface Prickleback," by William H. Marshall and Tina Wyllie Echeverria. *California Fish & Game,* Vol 78, No.2, Spring 1992. © 2005, California Department of Fish and Game.

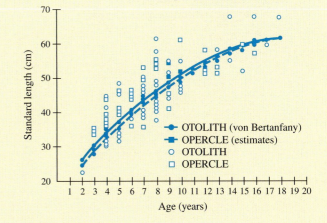

Estimated length from equation (1) at a given age was larger for males than females after age eight. See the table below.

Structure/ Sex	Age (yr)	Length (cm)	L_x	k	t_0	n
Otolith						
Est.	2–18	23–67	72	.10	−1.89	91
S.D.			8	.03	1.08	
Opercle						
Est.	2–18	23–67	71	.10	−2.63	91
S.D.			8	.04	1.31	
Opercle– Females						
Est.	0–18	15–62	62	.14	−1.95	115
S.D.			2	.02	.28	
Opercle– Males						
Est.	0–18	13–67	70	.12	−1.91	74
S.D.			5	.02	.29	

Weight/length relationships found with equation (2) are shown in the accompanying figure, along with data from other studies.

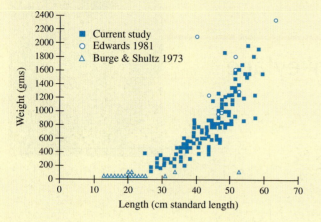

◆ **Exercises**

1. Use equation (1) to estimate the lengths at ages 4, 11, and 17. Let $L_x = 71.5$ and $k = .1$. Compare your answers with the results in the first figure. What do you find?

2. Use equation (2) with $a = .01289$ and $b = 2.9$ to estimate the weights for lengths of 25 cm, 40 cm, and 60 cm. Compare with the results in the second figure. Are your answers reasonable compared with the curve?

Chapter 5 Mathematics of Finance

Most people must take out a loan for a big purchase, such as a car, a major appliance, or a house. People who carry a balance on their credit cards are, in effect, also borrowing money. Loan payments must be accurately determined, and it may take some work to find the "best deal." See Exercise 46 on page 286 and Exercise 47 on page 266. We must all plan for eventual retirement, which usually involves savings accounts and investments in stocks, bonds, and annuities to fund 401K accounts or individual retirement accounts (IRAs). See Exercises 54 and 55 on page 277.

It is important for both businesspersons and consumers to understand the mathematics of finance in order to make sound financial decisions. Interest formulas for borrowing or investing money are introduced in this chapter.

5.1 Simple Interest and Discount

Interest is the fee paid to use someone else's money. Interest on loans of a year or less is frequently calculated as **simple interest,** which is paid only on the amount borrowed or invested and not on past interest. The amount borrowed or deposited is called the **principal.** The **rate** of interest is given as a percent per year, expressed as a decimal. For example, $6\% = .06$ and $11\frac{1}{2}\% = .115$. The time the money is earning interest is calculated in years. Simple interest is the product of the principal, rate, and time.

Simple Interest

The simple interest I on P dollars at a rate of interest r per year for t years is

$$I = Prt.$$

It is customary in financial problems to round interest to the nearest cent.

▼ **EXAMPLE 1**   To furnish her new apartment, Eleanor Chin borrowed $6000 at 9.2% interest for 11 months. How much interest will she pay?

Solution Use the formula $I = Prt$, with $P = 6000$, $r = .092$, and $t = 11/12$ years.

$$I = 6000(.092)(11/12) = 506,$$

or $506.00 ▲ ①

A deposit of P dollars today at a rate of interest r for t years produces interest of $I = Prt$. The interest, added to the original principal P, gives

$$P + Prt = P(1 + rt).$$

This amount is called the **future value** of P dollars at an interest rate r for time t in years. When loans are involved, the future value is often called the **maturity value** of the loan. This idea is summarized as follows.

Future or Maturity Value for Simple Interest

The **future value** or **maturity value,** A, of P dollars for t years at a rate of interest r per year is

$$A = P(1 + rt).$$

① Find the simple interest for each loan.

(a) $2000 at 8.5% for 10 months

(b) $3500 at 10.5% for $1\frac{1}{2}$ years

Answers:

(a) $141.67

(b) $551.25

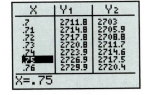

FIGURE 5.1

⟨2⟩

Find each future value.

(a) $1000 at 4.6% for 6 months

(b) $8970 at 11% for 9 months

(c) $95,106 at 9.8% for 76 days

Answers:

(a) $1023

(b) $9710.03

(c) $97,046.68

▼ **EXAMPLE 2** Find each maturity value and the amount of interest paid.

(a) Rick borrows $20,000 from his parents at 5.25% to add a room on his house. He plans to repay the loan in 9 months with a bonus he expects to receive at that time.

Solution The loan is for 9 months, or 9/12 of a year, so $t = .75$, $P = 20{,}000$, and $r = .0525$. Use the formula.

$$A = P(1 + rt)$$
$$= 20{,}000[1 + .0525(.75)]$$
$$\approx 20{,}787.5, \qquad \text{Use a calculator.}$$

or $20,787.50. The maturity value is the sum of the principal and interest, so the interest on the loan is given by

$$\text{Maturity Value} - \text{Principal} = \text{Interest}$$
$$\$20{,}787.50 - \$20{,}000 = \$787.50.$$

(b) A loan of $11,280 for 85 days at 9% interest.

Solution Unless stated otherwise, we assume a 365-day year, so the period in years is 85/365. The maturity value is

$$A = P[1 + rt]$$
$$A = 11{,}280\left[1 + .09\left(\frac{85}{365}\right)\right] \approx 11{,}280[1.020958904] \approx \$11{,}516.42.$$

So the interest is

$$\text{Maturity Value} - \text{Principal} = \$11{,}516.42 - \$11{,}280 = \$236.42. \quad ▲ \ \langle 2 \rangle$$

Present Value

A sum of money that can be deposited today to yield some larger amount in the future is called the **present value** of that future amount. Present value refers to the principal to be invested or loaned, so we use the same variable P as we did for principal. In interest problems, P always represents the amount at the beginning of the period, and A always represents the amount at the end of the period. To find a formula for P, we begin with the future value formula

$$A = P(1 + rt).$$

Dividing each side by $1 + rt$ gives the following formula for the present value.

Present Value for Simple Interest

The **present value** P of a future amount of A dollars at a simple interest rate r for t years is

$$P = \frac{A}{1 + rt}.$$

▼ **EXAMPLE 3**   Find the present value of $32,000 in 4 months at 9% interest.

$$P = \frac{A}{1 + rt} = \frac{32{,}000}{1 + (.09)\left(\dfrac{4}{12}\right)} = \frac{32{,}000}{1.03} = 31{,}067.96$$

Solution A deposit of $31,067.96 today at 9% interest would produce $32,000 in 4 months. These two sums, $31,067.96 today and $32,000.00 in 4 months, are equivalent (at 9%) because the first amount becomes the second amount in 4 months. ▲ ③

<③>
Find the present value of the following future amounts. Assume 6% interest.

(a) $7500 in 1 year
(b) $89,000 in 5 months
(c) $164,200 in 125 days

Answers:

(a) $7075.47
(b) $86,829.27
(c) $160,893.96

▼ **EXAMPLE 4**   Because of a court settlement, Carter Fenton owes $5000 to Jay Beckenstein. The money must be paid in 10 months, with no interest. Suppose Fenton wishes to pay the money today and Beckenstein can invest it at an annual rate of 5%. What amount should he be willing to accept?

Solution The amount that Beckenstein should be willing to accept is given by the present value:

$$P = \frac{5000}{1 + (.05)\left(\dfrac{10}{12}\right)} = 4800.00.$$

Beckenstein should be willing to accept $4800 today in settlement of the obligation. ▲ ④

<④>
Jerrell Davis is owed $19,500 by Christine O'Brien. The money will be paid in 11 months, with no interest. If the current interest rate is 10%, how much should Davis be willing to accept today in settlement of the debt?

Answer:

$17,862.60

▼ **EXAMPLE 5**   Suppose you borrow $40,000 today and are required to pay $41,400 in 4 months to pay off the loan and interest. What is the simple interest rate?

Solution We can use the future value formula with $P = 40{,}000$, $A = 41{,}400$, and $t = 4/12 = 1/3$ and solve for r.

$$A = P(1 + rt)$$

$$41{,}400 = 40{,}000\left(1 + r \cdot \frac{1}{3}\right)$$

$$41{,}400 = 40{,}000 + \frac{40{,}000r}{3}$$

$$1400 = \frac{40{,}000r}{3}$$

$$40{,}000r = 3 \cdot 1400 = 4200$$

$$r = \frac{4200}{40{,}000} = .105$$

<⑤>
You lend a friend $500. She agrees to pay you $520 in 6 months. What is the interest rate?

Answer:

8%

Therefore, the interest rate is 10.5%. ▲ ⑤

Simple Discount Notes

The loans discussed up to this point are called **simple interest notes,** where interest on the face value of the loan is added to the loan and paid at maturity. Another common type of note, called a **simple discount note,** has the interest deducted in advance from the amount of a loan before giving the *balance* to the borrower. The *full* value of the note must be paid at maturity. The money that is deducted is called the **bank discount** or just the **discount,** and the money actually received by the borrower is called the **proceeds.**

For example, consider a loan of $3000 at 6% interest for 9 months. We can compare the two types of loan arrangements as follows.

	Simple Interest Note	Bank Discount Note
Interest on the note	$3000(.06)(9/12) = \$135$	$3000(.06)(9/12) = \$135$
Borrower receives	**$3000**	$2865
Borrower pays back	$3135	**$3000**

▼ **EXAMPLE 6**   Theresa Vasquez needs a loan from her bank and agrees to pay $8500 to her banker in 9 months. The banker subtracts a discount of 12% and gives the balance to Vasquez. Find the amount of the discount and the proceeds.

Solution The discount is found in the same way that simple interest is found, except that it is based on the amount to be repaid.

$$\text{Discount} = 8500(.12)\left(\frac{9}{12}\right) = 765.00$$

The proceeds are found by subtracting the discount from the original amount.

$$\text{Proceeds} = \$8500 - \$765.00 = \$7735.00 \quad ▲ \quad ⟨6⟩$$

In Example 6, the borrower was charged a discount of 12%. However, 12% is *not* the interest rate paid, since 12% applies to the $8500, while the borrower actually received only $7735. In the next example, we find the rate of interest actually paid by the borrower.

▼ **EXAMPLE 7**   Find the actual rate of interest paid by Vasquez in Example 6.

Solution Use the formula for simple interest, $I = Prt$, with r the unknown. Since the borrower received only $7735 and must repay $8500, $I = 8500 - 7735 = 765$. Here, $P = 7735$ and $t = 9/12 = .75$. Substitute these values into $I = Prt$.

$$I = Prt$$
$$765 = 7735(r)(.75)$$
$$\frac{765}{7735(.75)} = r$$
$$.132 \approx r$$

The actual interest rate paid by the borrower is about 13.2%.   ▲   ⟨7⟩

⟨6⟩

Kelly Bell signs an agreement at her bank to pay the bank $25,000 in 5 months. The bank charges a 13% discount rate. Find the amount of the discount and the amount Bell actually receives.

Answer:

$1354.17; $23,645.83

⟨7⟩

Refer to problem 6 and find the actual rate of interest paid by Bell.

Answer:

13.7% (to the nearest tenth)

Let D represent the amount of discount on a loan. Then $D = Art$, where A is the maturity value of the loan (the amount borrowed plus interest) and r is the stated rate of interest. The amount actually received—the proceeds—can be written as $P = A - D$, or $P = A - Art = A(1 - rt)$.

The formulas for discount are summarized below.

Discount

If D is the discount on a loan having a maturity value A at a rate of interest r for t years, and if P represents the proceeds, then

$$P = A - D \qquad \text{or} \qquad P = A(1 - rt).$$

▼ **EXAMPLE 8**   John Cross owes $4250 to Jane Fleming. The loan is payable in 1 year at 10% interest. Fleming needs cash to buy a new car, so 3 months before the loan is payable she goes to the bank to have the loan discounted. That is, she sells the loan (note) to the bank. The bank charges an 11% discount fee. Find the amount of cash she will receive from the bank.

Solution   First find the maturity value of the loan—the amount (with interest) Cross must pay Fleming. By the formula for maturity value,

$$
\begin{aligned}
A &= P(1 + rt) \\
&= 4250[1 + (.10)(1)] \\
&= 4250(1.10) = 4675,
\end{aligned}
$$

or $4675.00.

The bank applies its discount rate to this total:

Amount of discount $= 4675(.11)(3/12) \approx \mathbf{128.56}$.

(Remember that the loan was discounted 3 months before it was due.) Fleming actually receives

$$\$4675 - \$128.56 = \mathbf{\$4546.44}$$

in cash from the bank. Three months later, the bank will get $4675 from Cross. ▲ ⟨8⟩

⟨8⟩

A firm accepts a $21,000 note due in 7 months with interest of 10.5%. Suppose the firm discounts the note at a bank 75 days before it is due. Find the amount the firm would receive if the bank charges a 12.4% discount rate.

Answer:

$21,718.41

◆5.1 Exercises

Unless stated otherwise, assume 365 days in a year and 28 days in February.

1. What factors determine the amount of interest earned on a fixed principal?

Find the simple interest. (See Examples 1 and 2.)

2. $35,000 at 6% for 9 months

3. $2850 at 7% for 8 months

4. $1875 at 5.3% for 7 months

5. $3650 at 6.5% for 11 months

6. $5160 at 7.1% for 58 days

7. $2830 at 8.9% for 125 days

8. $8940 at 9%; loan made on May 7 and due September 19

9. $5328 at 8%; loan made on August 16 and due December 30

10. $7900 at 7%; loan made on July 7 and due October 25

11. $12,000 at 9.5%; loan made on February 19 and due May 31

12. $3468 at 8.7%; loan made on October 4 and due March 15

13. $39,086 at 9.4%; loan made on September 12 and due July 30

14. In your own words, describe the *maturity value* of a loan.

15. What is meant by the *present value* of money?

Find the present value of each future amount. (See Examples 3 and 4.)

16. $15,000 for 9 months; money earns 6%

17. $48,000 for 8 months; money earns 5%

18. $15,402 for 120 days; money earns 6.3%

19. $29,764 for 310 days; money earns 7.2%

Find the proceeds for each loan. (See Example 6.)

20. $7150; discount rate 9%; length of loan 11 months

21. $9450; discount rate 10%; length of loan 7 months

22. $35,800; discount rate 9.1%; length of loan 183 days

23. $50,900; discount rate 8.2%; length of loan 238 days

24. Why is the discount rate charged on a simple discount note different from the actual interest rate paid on the proceeds?

Find the interest rate to the nearest tenth on the proceeds for the following simple discount notes. (See Example 7.)

25. $6200; discount rate 7%; length of loan 8 months

26. $5000; discount rate 8.1%; length of loan 6 months

27. $58,000; discount rate 10.8%; length of loan 9 months

28. $43,000; discount rate 9%; length of loan 4 months

Finance *Work the following applied problems.*

29. Anne Kelly borrowed $25,900 from her father to start a flower shop. She repaid him after 11 months, with interest of 8.4%. Find the total amount she repaid.

30. An accountant for a corporation forgot to pay the firm's income tax of $725,896.15 on time. The government charged a penalty of 9.8% interest for the 34 days the money was late. Find the total amount (tax and penalty) that was paid.

31. Mike Branson invested his summer earnings of $3000 in a savings account for college. The account pays 2.5% interest. How much will this amount to in 9 months?

32. To pay for textbooks, a student borrows $450 dollars from a credit union at 6.5% simple interest. He will repay the loan in 38 days, when he expects to be paid for tutoring. How much interest will he pay?

33. An account invested in a money market fund grew from $67,081.20 to $67,359.39 in a month. What was the interest rate to the nearest tenth?

34. A $100,000 certificate of deposit held for 60 days is worth $101,133.33. To the nearest tenth of a percent, what interest rate was earned?

35. A stock that sold for $22 at the beginning of the year was selling for $24 at the end of the year. If the stock paid a dividend of $.50 per share, what is the simple interest rate on an investment in this stock? (*Hint:* Consider the interest to be the increase in value plus the dividend.)

36. A bond with a face value of $10,000 in 10 years can be purchased now for $5988.02. What is the simple interest rate?

37. Tuition of $1769 will be due when the spring term begins in 4 months. What amount should a student deposit today, at 3.25%, to have enough to pay the tuition?

38. A firm of accountants has ordered 7 new computers at a cost of $5104 each. The machines will not be delivered for 7 months. What amount could the firm deposit in an account paying 6.42% to have enough to pay for the machines?

39. John Sun Yee needs $6000 to pay for remodeling work on his house. A contractor agrees to do the work in 10 months. How much should Yee deposit at 3.6% to accumulate the $6000 at that time?

40. Lorie Reilly decides to go back to college. For transportation, she borrows money from her parents to buy a small car for $7200. She plans to repay the loan in 7 months. What amount can she deposit today at 5.25% to have enough to pay off the loan?

41. John Matthews signs a $4200 note at the bank, which charges a 12.2% discount rate. Find the net proceeds if the note is for 10 months. Find the actual interest rate (to the nearest hundredth) charged by the bank.

42. A building contractor gives a $13,500 note to a plumber. (The plumber loans $13,500 to the contractor.) The note is due in 9 months, with interest of 9%. Three months after the note is signed, the plumber discounts it at the bank. The bank charges a 10.1% discount rate. How much will the plumber receive? Will it be enough to pay a bill for $13,582?

43. Shalia Johnson owes $7000 to the Eastside Music Shop. She has agreed to pay the amount in 7 months at an interest rate of 10%. Two months before the loan is due, the store needs $7350 to pay a wholesaler's bill. The bank will discount the note at a rate of 10.5%. How much will the store receive? Is it enough to pay the bill?

44. Fay, Inc., received a $30,000, six-month, 12% interest-bearing note from a customer.* The note was discounted the same day by Carr National Bank at 15%. The amount of cash received by Fay from the bank was
   (a) $30,000
   (b) $29,550
   (c) $29,415
   (d) $27,750.

45. Many credit card companies target entering college freshmen with attractive interest rates. What they do not emphasize, however, is that the low rate will be in effect for only a short time, changing automatically to a much higher rate, often in just a few months. Freshman Michael Branson accepts a credit card offer at 8% annual interest

on any unpaid charges. The rate will increase to 15% annually after 6 months. To furnish his dorm room, he charges a small refrigerator, some bedding, and personal supplies amounting to $457.80. When he receives his first month's bill, he finds that he can pay just the minimum amount of $87.50.
   (a) If Michael pays the minimum amount and makes no additional charges, how much will he owe the next month?
   (b) Suppose Michael continues to pay the same minimum amount for 4 more months. What will he owe at the end of the next month?
   (c) The minimum amount normally is reduced as the total charge is reduced. Suppose the minimum is reduced by $15 each month and Michael continues to pay only the minimum. Can be pay off the charges in 6 months? If not, what will he owe at the end of 6 months?

*Uniform CPA Examination, May 1989, American Institute of Certified Public Accountants.

5.2 Compound Interest

Simple interest is normally used for loans or investments of a year or less. For longer periods, *compound interest* is used. With **compound interest,** interest is charged (or paid) on interest as well as on principal. For example, if $1000 is deposited at 5% compounded annually, then the interest for the first year is $1000(.05) = $50, just as with simple interest, so that the account balance is $1050 at the end of the year. During the second year, interest is paid on the entire $1050 (not just on the original $1000 as with simple interest), so the amount in the account at the end of the second year is $1050 + $1050(.05) = $1102.50. This is more than simple interest would produce.

Use the formula

$$A = P(1 + rt)$$

to find the amount in the account after 2 years at 5% simple interest.

Answer:

$1100

▼ **EXAMPLE 1** If $7000 is deposited in an account that pays 4% interest compounded annually, how much money is in the account after nine years?

Solution After one year, the account balance is

$$7000 + 4\% \text{ of } 7000 = 7000 + (.04)7000$$
$$= 7000(1 + .04) \qquad \text{Distributive property}$$
$$= 7000(1.04) = \$7280.$$

The initial balance has grown by a factor of 1.04. At the end of the second year, the balance is

$$7280 + 4\% \text{ of } 7280 = 7280 + (.04)7280$$
$$= 7280(1 + .04) \qquad \text{Distributive property}$$
$$= 7280(1.04) = 7571.20.$$

Once again, the amount at the beginning of the year has grown by a factor of 1.04. This is true in general: a balance of P dollars at 4% interest grows to $P(1.04)$ in one year. So the balance grows like this:

Year 1   **Year 2**   **Year 3**

$$7000 \rightarrow 7000(1.04) \rightarrow \underbrace{[7000(1.04)](1.04)}_{7000(1.04)^2} \rightarrow \underbrace{[7000(1.04)(1.040](1.04)}_{7000(1.04)^3} \rightarrow \cdots.$$

At the end of nine years, the balance is

$$7000(1.04)^9 = \$9963.18 \qquad \text{(rounded to the nearest penny).} \ \blacktriangle$$

The argument used in Example 1 applies in the general case and leads to this conclusion.

Compound Interest

If P dollars are invested at interest rate i per period, then the **compound amount** (future value) A after n compounding periods is

$$A = P(1 + i)^n.$$

In Example 1, for instance, we had $P = 7000$, $n = 9$, and $i = .04$ (so that $1 + i = 1 + .04 = 1.04$).

NOTE Compare this formula for compound interest with the formula for simple interest from the previous section.

Compound interest $A = P(1 + r)^t$

Simple interest $A = P(1 + rt)$

The important distinction between the two formulas is that, in the compound interest formula, the number of years t is an *exponent*, so that money grows much more rapidly when interest is compounded.

▼ **EXAMPLE 2**   Suppose $1000 is deposited for 6 years in an account paying 8.31% per year compounded annually.

(a) Find the compound amount.

Solution In the formula above, $P = 1000$, $i = .0831$, and $n = 6$. The compound amount is

$$A = P(1 + i)^n$$
$$A = 1000(1.0831)^6$$
$$A = \$1614.40.$$

(b) Find the amount of interest earned.

2

Suppose $17,000 is deposited at 4% compounded annually for 11 years.

(a) Find the compound amount.

(b) Find the amount of interest earned.

Answers:

(a) $26,170.72

(b) $9170.72

Solution Subtract the initial deposit from the compound amount.

Amount of interest = $1614.40 − $1000 = $614.40 ▲ 2

TECHNOLOGY TIP Spreadsheets are ideal for performing financial calculations. Figure 5.2 shows a Microsoft Excel spreadsheet with the formulas for compound and simple interest used to create columns B and C, respectively, when $1000 is invested at an annual rate of 10%. Notice how rapidly the compound amount increases compared with the maturity value with simple interest. For more details on the use of spreadsheets in the mathematics of finance, see the *Spreadsheet Manual* that is available with this book.

	A	B	C
1	period	compound	simple
2	1	1100	1100
3	2	1210	1200
4	3	1331	1300
5	4	1464.1	1400
6	5	1610.51	1500
7	6	1771.561	1600
8	7	1948.7171	1700
9	8	2143.58881	1800
10	9	2357.947691	1900
11	10	2593.74246	2000
12	11	2853.116706	2100
13	12	3138.428377	2200
14	13	3452.271214	2300
15	14	3797.498336	2400
16	15	4177.248169	2500
17	16	4594.972986	2600
18	17	5054.470285	2700
19	18	5559.917313	2800
20	19	6115.909045	2900
21	20	6727.499949	3000

FIGURE 5.2

▼ **EXAMPLE 3** If a $16,000 investment grows to $50,000 in 18 years, what is the interest rate (assuming annual compounding)?

Solution Use the compound interest formula, with $P = 16{,}000$, $A = 50{,}000$, and $n = 18$, and solve for i.

$$P(1+i)^n = A$$

$$16{,}000(1+i)^{18} = 50{,}000$$

$$(1+i)^{18} = \frac{50{,}000}{16{,}000} = 3.125 \qquad \text{Divide both sides by 16,000.}$$

$$\sqrt[18]{(1+i)^{18}} = \sqrt[18]{3.125} \qquad \text{Take 18th roots on both sides.}$$

$$1 + i = \sqrt[18]{3.125}$$

$$i = \sqrt[18]{3.125} - 1 = .06535 \qquad \text{Subtract 1 from both sides}$$

So the interest rate is about 6.535%. ▲

Interest can be compounded more than once a year. Common **compounding periods** include

semiannually (2 periods per year),

quarterly (4 periods per year),

monthly (12 periods per year), and

daily (usually 365 periods per year).

When the annual interest i is compounded m times per year, the interest rate per period is understood to be i/m.

▼ **EXAMPLE 4**   Determine the amount a $3500 investment is worth after three-and-a-half years at the given interest rate.

(a) 6.4% compounded annually

Solution Use the compound interest formula with $P = 3500$, $i = .064$, and $n = 3.5$

$$A = P(1 + i)^n = 3500(1 + .064)^{3.5} = \$4348.74.$$

(b) 6.4% compounded quarterly

Solution Interest is compounded 4 times a year, so the interest rate per period is $.064/4$. Since there are 4 periods per year, the number of periods in 3.5 years is $n = 4(3.5) = 14$. Hence,

$$A = P(1 + i)^n = 3500\left(1 + \frac{.064}{4}\right)^{14} = \$4370.99.$$

(c) 6.4% compounded monthly

Solution Now interest is compounded 12 times a year, so the interest rate per period is $.064/12$ and the number of periods in 3.5 years is $12(3.5) = 42$. Therefore,

$$A = P(1 + i)^n = 3500\left(1 + \frac{.064}{12}\right)^{42} = \$4376.14.$$

Note that the more often interest is compounded, the larger is the final amount. ▲ ⬦3⬦

⟨3⟩

Find the compound amount.

(a) $10,000 at 8% compounded quarterly for 7 years

(b) $36,000 at 6% compounded monthly for 2 years

Answers:

(a) $17,410.24

(b) $40,577.75

Continuous Compounding

As we saw in Example 4, the more often interest is compounded within a given period, the more interest will be earned. Surprisingly, however, there is a limit on

TECHNOLOGY TIP TI-84+ and Casio 9850GC+ each have a "TVM Solver" for financial computations (in the TI APPS-financial menu or the Casio main menu); a similar one can be downloaded for the TI-86/89. Figure 5.3 shows the solution of Example 4(c) on such a solver. Its use is explained in the next section. Most of the problems in this section can be solved just as quickly with an ordinary calculator.

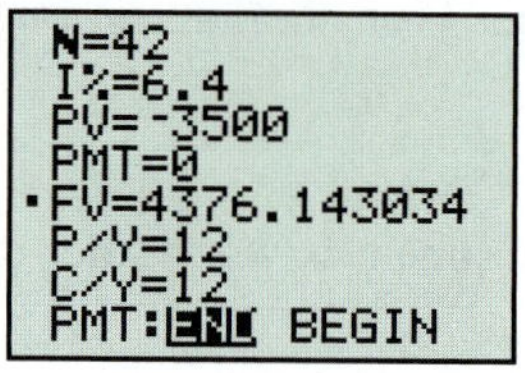

```
N=42
I%=6.4
PV=-3500
PMT=0
•FV=4376.143034
P/Y=12
C/Y=12
PMT:END BEGIN
```

FIGURE 5.3

the amount of interest, no matter how often it is compounded. To see this, suppose that $1 is invested at 100% interest per year, compounded n times per year. Then the interest rate (in decimal form) is 1.00, and the interest rate per period is $1/n$. According to the formula (with $P = 1$), the compound amount at the end of 1 year will be

$$A = \left(1 + \frac{1}{n}\right)^n.$$

A computer gives the following results for various values of n.

Interest Is Compounded	n	$\left(1 + \frac{1}{n}\right)^n$
Annually	1	$\left(1 + \frac{1}{1}\right)^1 = 2$
Semiannually	2	$\left(1 + \frac{1}{2}\right)^2 = 2.25$
Quarterly	4	$\left(1 + \frac{1}{4}\right)^4 \approx 2.4414$
Monthly	12	$\left(1 + \frac{1}{12}\right)^{12} \approx 2.6130$
Daily	365	$\left(1 + \frac{1}{365}\right)^{365} \approx 2.71457$
Hourly	8760	$\left(1 + \frac{1}{8760}\right)^{8760} \approx 2.718127$
Every minute	525,600	$\left(1 + \frac{1}{525,600}\right)^{525,600} \approx 2.7182792$
Every second	31,536,000	$\left(1 + \frac{1}{31,53,600}\right)^{31,536,000} \approx 2.7182818$

Because interest is rounded to the nearest penny, the compound amount never exceeds $2.72, no matter how big n is. (If you try computing the values in the table with your calculator, you may notice that your answers do not agree exactly. This is because of round-off error.)

The preceding table suggests that as n takes larger and larger values, the corresponding values of $\left(1 + \frac{1}{n}\right)^n$ get closer and closer to a specific real number whose decimal expansion begins 2.71828 . . . . This is indeed the case, as is shown in calculus, and the number 2.71828 . . . is denoted e.

The preceding example is typical of what happens when interest is compounded n times per year, with larger and larger values of n. It can be shown that no matter what interest rate or principal is used, there is always an upper limit (involving the number e) on the compound amount, which is called the compound amount from **continuous compounding.**

Continuous Compounding

The compound amount A for a deposit of P dollars at an interest rate r per year compounded continuously for t years is given by

$$A = Pe^{rt}.^*$$

Most calculators have an e^x key for computing powers of e. See page 211 for details on using a calculator to evaluate e^x.

▼ **EXAMPLE 5** Suppose $5000 is invested at an annual rate of 4% compounded continuously for 5 years. Find the compound amount.

   Solution In the formula for continuous compounding, let $P = 5000$, $r = .04$, and $t = 5$. Then a calculator with an e^x key shows that

$$A = 5000e^{(.04)5} = 5000e^{.2} \approx \$6107.01.$$

You can readily verify that daily compounding would have produced a compound amount about 6¢ less. ▲ ④

④

Find the compound amount, assuming continuous compounding.

(a) $12,000 at 10% for 5 years

(b) $22,867 at 7.2% for 9 years

Answers:

(a) $19,784.66

(b) $43,715.15

Effective Rate

If $1 is deposited at 4% compounded quarterly, a calculator can be used to find that at the end of one year the compound amount is $1.0406, an increase of 4.06% over the original $1. The actual increase of 4.06% in the money is somewhat higher than the stated increase of 4%. To differentiate between these two numbers, 4% is called the **nominal** or **stated rate** of interest, while 4.06% is called the **effective rate.** To avoid confusion between stated rates and effective rates, we shall continue to use r for the stated rate, and we will use r_E for the effective rate.[†]

▼ **EXAMPLE 6** Find the effective rate corresponding to a stated rate of 6% compounded semiannually.

   Solution A calculator shows that $100 at 6% compounded semiannually will grow to

$$A = 100\left(1 + \frac{.06}{2}\right)^2 = 100(1.03)^2 = \$106.09.$$

Thus, the actual amount of compound interest is $106.09 - \$100 = \$6.09$. Now, if you earn $6.09 interest on $100 in 1 year with annual compounding, your rate is $6.09/100 = .0609 = 6.09\%$. Thus, the effective rate is $r_E = 6.09\%$. ▲ ⑤

⑤

Find the effective rate corresponding to a nominal rate of

(a) 12% compounded monthly;

(b) 8% compounded quarterly.

Answers:

(a) 12.68%

(b) 8.24%

---

*Other applications of the exponential function $f(x) = e^x$ are discussed in Chapter 4.

[†]When applied to consumer finance, the effective rate is called the annual percentage rate, APR, or annual percentage yield, APY.

If you examine Example 6, you see that the effective rate is $\dfrac{106.09 - 100}{100}$; that is,

$$\frac{100\left(1 + \dfrac{.06}{2}\right)^2 - 100}{100} = \frac{100\left[\left(1 + \dfrac{.06}{2}\right)^2 - 1\right]}{100} = \left(1 + \frac{.06}{2}\right)^2 - 1.$$

The same argument works with any values of P, r, and m in place of 100, .06, and 2, respectively, and leads to this formula.

Effective Rate

The effective rate corresponding to a stated rate of interest r compounded m times per year is

$$r_E = \left(1 + \frac{r}{m}\right)^m - 1.$$

▼ **EXAMPLE 7**   Find the effective rate.

(a) 4.9% compounded quarterly

Solution Use the formula in the box, with $r = .049$ and $m = 4$. The effective rate is

$$r_E = \left(1 + \frac{.049}{4}\right)^4 - 1$$
$$= 1.04990775 - 1 \approx .0499,$$

or about 4.99%

(b) 4.9% compounded monthly

Solution Now $r = .049$ and $m = 12$, so that

$$r_E = \left(1 + \frac{.049}{12}\right)^{12} - 1 = 1.050115575 - 1 \approx .0501,$$

or about 5.01%. Note that the effective rate increases with more frequent compounding. ▲ ⟨6⟩

⟨6⟩

Find the effective rate corresponding to a nominal rate of

(a) 8% compounded monthly;

(b) 10% compounded quarterly.

Answers:

(a) 8.30%

(b) 10.38%

▼ **EXAMPLE 8**   Bank A is now lending money at 10% interest compounded annually. The rate at Bank B is 9.6% compounded monthly, and the rate at Bank C is 9.7% compounded quarterly. If you need to borrow money, at which bank will you pay the least interest?

TECHNOLOGY TIP   Effective rates can be computed on TI-84+ by using "Eff" in the APPS financial menu, as shown in Figure 5.4 for Example 8.

▶Eff(10,1)
                    10
▶Eff(9.6,12)
        10.03386937
▶Eff(9.7,4)
        10.05857629
■

FIGURE 5.4

Solution Compare the effective rates.

Bank A: $\left(1 + \dfrac{.10}{1}\right)^1 - 1 = .10 = 10\%$

Bank B: $\left(1 + \dfrac{.096}{12}\right)^{12} \approx .10034 = 10.034\%$

Bank C: $\left(1 + \dfrac{.097}{4}\right)^4 - 1 \approx .10057 = 10.057\%$

The lowest effective interest rate is at Bank A, which has the highest nominal rate. ▲

Present Value for Compound Interest

The formula for compound interest, $A = P(1 + i)^n$, has four variables: A, P, i, and n. Given the values of any three of these variables, the value of the fourth can be found. In particular, if A (the future amount), i, and n are known, then P can be found. Here, P is the amount that should be deposited today to produce A dollars in n periods.

▼ **EXAMPLE 9**   Keisha Jones must pay a lump sum of $6000 in 5 years. What amount deposited today at 6.2% compounded annually will amount to $6000 in 5 years?

Solution Here, $A = 6000$, $i = .062$, $n = 5$, and P is unknown. Substituting these values into the formula for the compound amount gives

$$6000 = P(1.062)^5$$

$$P = \frac{6000}{(1.062)^5} = 4441.49,$$

or $4441.49. If Jones leaves $4441.49 for 5 years in an account paying 6.2% compounded annually, she will have $6000 when she needs it. To check your work, use the compound interest formula with $P = \$4441.49$, $i = .062$, and $n = 5$. You should get $A = \$6000.00$. ▲ ⑦

As Example 9 shows, $6000 in 5 years is (approximately) the same as $4441.49 today (if money can be deposited at 6.2% annual interest). An amount that can be deposited today to yield a given amount in the future is called the *present value* of the future amount. By solving $A = P(1 + i)^n$ for P, we get the following general formula for present value.

⑦

Find P in Example 9 if the interest rate is

(a) 6%;

(b) 10%.

Answers:

(a) $4483.55

(b) $3725.53

Present Value for Compound Interest

The **present value** of A dollars compounded at an interest rate i per period for n periods is

$$P = \frac{A}{(1 + i)^n} \qquad \text{or} \qquad P = A(1 + i)^{-n}.$$

NOTE Since this is just the compound interest formula solved for P, it is not necessary to remember a new formula for present value. You can use the compound interest formula if you understand what each of the variables represents.

▼ **EXAMPLE 10** Find the present value of $16,000 in 9 years if money can be deposited at 6% compounded semiannually.

Solution In 9 years, there are $2 \cdot 9 = 18$ semiannual periods. A rate of 6% per year is 3% in each semiannual period. Apply the formula with $A = 16,000$, $i = .03$, and $n = 18$.

$$P = \frac{A}{(1 + i)^n} = \frac{16,000}{(1 + .03)^{18}} \approx \frac{16,000}{1.702433} \approx 9398.31$$

A deposit of $9398.31 today at 6% compounded semiannually will produce a total of $16,000 in 9 years. ▲ ⑧

The formula for the compound amount can also be solved for n.

▼ **EXAMPLE 11** Suppose the general level of inflation in the economy averages 8% per year. Find the number of years it would take for the overall level of prices to double.

Solution To find the number of years it will take for $1 worth of goods or services to cost $2, we must solve for n in the equation

$$2 = 1(1 + .08)^n,$$

where $A = 2$, $P = 1$, and $i = .08$. This equation simplifies to

$$2 = (1.08)^n.$$

To solve for n, we will use base 10 logarithms as in Chapter 4.

$$\log 2 = \log (1.08)^n \qquad \text{Take the logarithm of each side.}$$
$$\log 2 = n \log 1.08 \qquad \text{Power property of logarithms}$$
$$n = \frac{\log 2}{\log 1.08} \qquad \text{Divide both sides by } \log 1.08.$$
$$n \approx 9.01$$

To check, use a calculator to get $1.08^{9.01} = 2.00$ to the nearest hundredth. ▲ ⑨

When interest is compounded continuously, the present value can be found by solving the continuous compounding formula $A = Pe^{rt}$ for P.

▼ **EXAMPLE 12** Assuming a continuous inflation rate of 3.5%, how much did a $100 item today cost 4 years ago?

Solution Here the future value is $A = 100$, $r = .035$, and $t = 4$. We want to find the present value P.

$$A = Pe^{rt}$$
$$100 = Pe^{(.035)4}$$

---

⑧

Find the present value in Example 10 if money is deposited at 10% compounded semiannually.

Answer:

$6648.33

---

⑨

Using a calculator, find the number of years it will take for $500 to increase to $750 in an account paying 6% interest compounded semiannually.

Answer:

About 7 years

$$100 = Pe^{.14}$$

$$P = \frac{100}{e^{.14}} \approx \$86.94$$

10

Find the value of a $100 item in 3 years at a continuous inflation rate of 4%.

Answer:

$112.75

The item cost $86.94 four years ago. ▲ **10**

Summary

At this point, it seems helpful to summarize the notation and the most important formulas for simple and compound interest. We use the following variables:

$P = $ principal or present value
$A = $ future or maturity value
$r = $ annual (stated or nominal) interest rate
$t = $ number of years
$m = $ number of compounding periods per year
$i = $ interest rate per period
$n = $ total number of compounding periods
$r_E = $ effective rate.

Simple Interest	Compound Interest	Continuous Compounding
$A = P(1 + rt)$	$A = P(1 + i)^n$	$A = Pe^{rt}$
$P = \dfrac{A}{1 + rt}$	$P = \dfrac{A}{(1 + i)^n} = A(1 + i)^{-n}$	$P = \dfrac{A}{e^{rt}}$
	$r_E = \left(1 + \dfrac{r}{m}\right)^m - 1$	

◆5.2 Exercises

1. In the summary above, what is the difference between r and i? between t and n?

2. Explain the difference between simple interest and compound interest.

3. What factors determine the amount of interest earned on a fixed principal?

4. In your own words, describe the *maturity value* of a loan.

5. What is meant by the *present value* of money?

6. If interest is compounded more than once per year, which rate is higher, the stated rate or the effective rate?

Find the compound amount for each of the following deposits. (See Examples 1, 2, and 4.)

7. $1000 at 4% compounded annually for 8 years

8. $1000 at 6% compounded annually for 10 years

9. $470 at 8% compounded semiannually for 12 years

10. $15,000 at 4.6% compounded semiannually for 11 years

11. $6500 at 5.5% compounded quarterly for 6 years

12. $9100 at 6.1% compounded quarterly for 4 years

Find the amount of interest earned by each of the following deposits. (See Examples 2 and 4.)

13. $26,000 at 6% compounded annually for 5 years

14. $32,000 at 5% compounded annually for 10 years

15. $8000 at 4% compounded semiannually for 6.4 years

16. $2500 at 4.5% compounded semiannually for 8 years

17. $5124.98 at 6.3% compounded quarterly for 5.2 years

18. $27,630.35 at 4.4% compounded quarterly for 3.7 years

Find the interest rate (with annual compounding) that makes the statement true (rounded to the nearest dollar). (See Example 3.)

19. $3000 grows to $3907 in 6 years

20. $2550 grows to $3905 in 11 years

21. $8500 grows to $12,161 in 7 years

22. $9000 grows to $17,118 in 16 years

Find the compound amount if $25,000 is invested at 6% compounded continuously for the following number of years. (See Example 5.)

23. 2   **24.** 5   **25.** 10

26. 12   **27.** 20

28. How do the nominal, or stated, interest rate and the effective interest rate differ?

Find the effective rate corresponding to the following nominal rates. (See Examples 6–8.)

29. 4% compounded semiannually

30. 6% compounded quarterly

31. 6% compounded semiannually

32. 4.7% compounded semiannually

33. 5.2% compounded semiannually

34. 5.2% compounded quarterly

Find the present value of the following future amounts. (See Examples 9 and 10.)

35. $12,000 at 5% compounded annually for 6 years

36. $8500 at 6% compounded annually for 9 years

37. $4253.91 at 5.8% compounded semiannually for 4 years

38. $27,692.53 at 4.6% compounded semiannually for 5 years

39. $17,230 at 4% compounded quarterly for 10 years

40. $5240 at 6% compounded quarterly for 8 years

41. If money can be invested at 8% compounded quarterly, which is larger, $1000 now or $1210 in 5 years? Use present value to decide.

42. If money can be invested at 6% compounded annually, which is larger, $10,000 now or $15,000 in 6 years? Use present value to decide.

Finance *Work the following applied problems.*

43. A small business borrows $50,000 for expansion at 12% compounded monthly. The loan is due in 4 years. How much interest will the business pay?

44. A developer needs $80,000 to buy land. He is able to borrow the money at 10% per year compounded quarterly. How much will the interest amount to if he pays off the loan in 5 years?

45. A company has agreed to pay $2.9 million in 5 years to settle a lawsuit. How much must it invest now in an account paying 8% compounded monthly to have that amount when it is due?

46. Bill Poole wants to have $20,000 available in 5 years for a down payment on a house. He has inherited $15,000. How much of the inheritance should he invest now to accumulate the $20,000 if he can get an interest rate of 8% compounded quarterly?

47. The Flagstar Bank in Michigan offered a 5-year certificate of deposit (CD) at 4.38% interest compounded quarterly.* On the same day on the Internet, Principal Bank offered a 5-year CD at 4.37% interest compounded monthly. Find the APY (effective rate) for each CD. Which bank pays a higher APY?

48. On the Internet, World Savings offered a 13-month CD with an APY (effective rate) of 4.01%.† Given that interest was compounded daily, what was the actual rate? (*Hint:* Solve the effective rate equation for *r*.)

49. The Westfield Bank in Ohio offered the CD rates shown in the table. The APY rates (effective rates) shown assume monthly compounding. Find the corresponding nominal rates to the nearest hundredth. (See the hint for Exercise 48.)

Term	6 mo	1 yr	2 yr	3 yr	5 yr
APY (%)	3.25	3.75	4.00	4.25	4.75

50. The pie graph shows the percent of baby boomers ages 46 to 49 who said they had investments with a total value as shown in each category.‡ Note that 30% say they have saved less than $10,000 and 28% don't know or gave no answer. Assume that the money is invested at an average rate of 8% compounded quarterly for 20 years, when this age group will be ready for retirement. Find the range of amounts each group (except the "don't know or no answer" group) in the graph will have saved for retirement if no more is added.

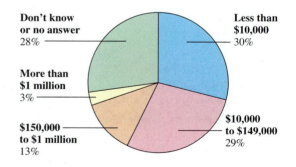

Note: Figures add to more than 100% because of rounding.

Sources: Census Bureau (age distribution); Merrill Lynch Baby Boom Retirement Index (investments); William M. Mercer, Inc. (life expectancy).

---

*June 1, 2005.

†June 3, 2005.

‡*New York Times,* December 31, 1995, Sec. 3, p. 5.

51. Scott Silva borrowed $5200 from his friend Joe Vetere to buy computer equipment. He repaid the loan 10 months later with simple interest at 7%. Vetere then invested the proceeds in a 5-year certificate of deposit paying 6.3% compounded quarterly. How much will he have at the end of 5 years? (*Hint:* You need to use both simple and compound interest.)

52. Lora Reilly has inherited $10,000 from her uncle's estate. She will invest the money for 2 years. She is considering two investments: a money market fund that pays a guaranteed 5.25% interest compounded monthly and a 2-year Treasury note at 6% annual interest. Which investment pays the most interest over the 2-year period?

53. You decide to invest a $16,000 bonus in a money market fund that guarantees a 5.5% annual interest rate compounded monthly for 7 years. A one-time fee of $30 is charged to set up the account. In addition, there is an annual administrative charge of 1.25% of the balance in the account at the end of each year.
(a) How much is in the account at the end of the first year?
(b) How much is in the account at the end of the seventh year?

54. Two partners agree to invest equal amounts in their business. One will contribute $10,000 immediately. The other plans to contribute an equivalent amount in 3 years, when she expects to acquire a large sum of money. How much should she contribute at that time to match her partner's investment now, assuming an interest rate of 6% compounded semiannually?

55. As the prize in a contest, you are offered $1000 now or $1210 in 5 years. If money can be invested at 6% compounded annually, which is larger?

Use the approach in Example 11 to find the time it would take for the general level of prices in the economy to double at the average annual inflation rates in Exercises 56–58.

56. 3%   **57.** 4%   **58.** 5%

59. The consumption of electricity has increased historically at 6% per year. If it continues to increase at this rate indefinitely, find the number of years before the electric utilities will need to double their generating capacity.

60. Suppose a conservation campaign coupled with higher rates causes the demand for electricity to increase at only 2% per year, as it has recently. Find the number of years before the utilities will need to double their generating capacity.

61. In 1995, O. G. McClain of Houston, Texas, mailed a $100 check to a descendant of Texas independence hero Sam Houston to repay a $100 debt of McClain's great-great-grandfather, who died in 1835, to Sam Houston.* A bank

estimated the interest on the loan to be $420 million for the 160 years it was due. Find the interest rate the bank was using, assuming that interest is compounded annually.

62. In the New Testament, Jesus commends a widow who contributed 2 mites to the temple treasury (Mark 12:42–44). A mite was worth roughly 1/8 of a cent. Suppose the temple had invested those 2 mites at 4% interest compounded quarterly. How much would the money be worth 2000 years later?

63. Suppose $10,000 is invested at an annual rate of 5% for 10 years. Find the future value if interest is compounded as follows.
(a) annually
(b) quarterly
(c) monthly
(d) daily

64. In Exercise 63, notice that as the money is compounded more often, the compound amount becomes larger and larger. Is it possible to compound often enough so that the compound amount is $17,000 after 10 years? Explain.

The following exercises are from professional examinations.

65. On January 1, 1998, Jack deposited $1000 into Bank X to earn interest at the rate of j per annum compounded semiannually. On January 1, 2003, he transferred his account to Bank Y to earn interest at the rate of k per annum compounded quarterly. On January 1, 2006, the balance at Bank Y was $1990.76. If Jack could have earned interest at the rate of k per annum compounded quarterly from January 1, 1998, through January 1, 2006, his balance would have been $2203.76. Which of the following represents the ratio k/j?*
(a) 1.25
(b) 1.30
(c) 1.35
(d) 1.40
(e) 1.45

66. On January 1, 2003, Tone Company exchanged equipment for a $200,000 non-interest-bearing note due on January 1, 2006. The prevailing rate of interest for a note of this type at January 1, 2003, was 10%. The present value of $1 at 10% for three periods is 0.75. What amount of interest revenue should be included in Tone's 2004 income statement?[†]
(a) $7500
(b) $15,000
(c) $16,500
(d) $20,000

*Problem adapted from "Course 140 Examination, Mathematics of Compound Interest" of the Education and Examination Committee of the Society of Actuaries. Reprinted by permission of the Society of Actuaries.

†Adapted from the Uniform CPA Examination, American Institute of Certified Public Accountants.

*New York Times, March 30, 1995.

5.3 Future Value of an Annuity and Sinking Funds

So far in this chapter, only lump-sum deposits and payments have been discussed. Many financial situations, however, involve a sequence of equal payments at regular intervals, such as weekly deposits in a savings account or monthly payments on a mortgage or car loan. To develop formulas to deal with periodic payments like these, we first discuss *sequences*.

Geometric Sequences

If a and r are fixed nonzero numbers, then the infinite list of numbers $a, ar, ar^2, ar^3, ar^4, \ldots$ is called a **geometric sequence.** For instance, if $a = 5$ and $r = 2$, we have the sequence

$$5, 5 \cdot 2, 5 \cdot 2^2, 5 \cdot 2^3, 5 \cdot 2^4, \ldots,$$

or

$$5, 10, 20, 40, 80, \ldots.$$

In the sequence $a, ar, ar^2, ar^3, ar^4, \ldots$, the number a is the *first term* of the sequence, ar the *second term,* ar^2 the *third term,* ar^3 the *fourth term,* and so on. Thus, for any $n \geq 1$,

$$ar^{n-1} \text{ is the } n \text{th term of the sequence.}$$

Each term in the sequence is r times the preceding term. The number r is called the **common ratio** of the sequence.

TECHNOLOGY TIP   Some graphing calculators have the ability to produce a list of the terms of a sequence, given the expression for the *n*th term. With other calculators, you can use the TABLE feature by entering the expression for the *n*th term as a function. The first four terms of the sequence discussed in Example 1 are shown in Figure 5.5.

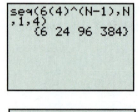

FIGURE 5.5

▼ **EXAMPLE 1**   Consider the geometric sequence with $a = 6$ and $r = 4$.

(a) List the first five terms of the sequence.

Solution The first five terms are

$$6, \quad 6(4), \quad 6(4^2), \quad 6(4^3), \quad 6(4^4)$$

or

$$6, \quad 24, \quad 96, \quad 384, \quad 156.$$

(b) Find the ninth term of the sequence.

Solution Apply the formula for the *n*th term with $n = 9$, $a = 6$, and $r = 4$.

$$ar^{n-1} = 6(4^{9-1}) = 6(4^8) = 393{,}216. \; ▲ \; \text{①}$$

①

Write the first four terms of the geometric sequence with $a = 5$ and $r = -2$. Then find the seventh term.

Answer:

$5, -10, 20, -40; 320$

▼ **EXAMPLE 2** $100 is deposited in an account that pays interest of 5% compounded annually. Show that the account balances at the beginning of each year form a geometric series.

Solution With $P = 100$ and $i = .05$, the compound interest formula $A = P(1 + i)^n$, which gives the account balance at the *end* of year n, becomes

$$A = 100(1 + .05)^n = 100(1.05^n).$$

Rounding to the nearest penny at each step, we have the following:

Time	Account balance
Beginning of year 1	$100
Beginning of year 2 = end of year 1	$100(1.05^1) = \$105$
Beginning of year 3 = end of year 2	$100(1.05^2) = \$110.25$
Beginning of year 4 = end of year 3	$100(1.05^3) = \$115.76$

and so on. Thus, the account balances at the beginning of each year form a geometric sequence with $a = 100$ and $r = 1.05$:

$$100, \qquad 100(1.05), \qquad 100(1.05^2), \qquad 100(1.05^3), \qquad 100(1.05^4), \dots,$$

or

$$\$100, \qquad \$105, \qquad \$110.25, \qquad \$115.76, \qquad \$121.55, \dots \ \blacktriangle$$

The sum S_n of the first n terms of a geometric sequence can be written as

$$S_n = a + ar + ar^2 + ar^3 + \cdots + ar^{n-1}. \tag{1}$$

To derive a formula for S_n when $r \neq 1$, multiply both sides of equation (1) by r, obtaining

$$rS_n = ar + ar^2 + ar^3 + ar^4 + \cdots + ar^n. \tag{2}$$

Now subtract corresponding sides of equation (1) from equation (2), and solve the equation for S_n.

$$
\begin{aligned}
rS_n &= ar + ar^2 + ar^3 + \cdots + ar^{n-1} + ar^n \\
-S_n &= -(a + ar + ar^2 + ar^3 + \cdots + ar^{n-1}) \\
\hline
rS_n - S_n &= ar^n - a \\
S_n(r - 1) &= a(r^n - 1) \\
S_n &= \frac{a(r^n - 1)}{r - 1}
\end{aligned}
$$

Sum of Terms

If a geometric sequence has first term a and common ratio r, then the **sum of the first n terms** is given by

$$S_n = \frac{a(r^n - 1)}{r - 1}, \qquad r \neq 1.$$

2

(a)   Find S_4 and S_7 for the geometric sequence 5, 15, 45, 135, ...

(b)   Find S_5 for the geometric sequence having $a = -3$ and $r = -5$.

Answers:

(a) $S_4 = 200$, $S_7 = 5465$

(b) $S_5 = -1563$

▼ **EXAMPLE 3**   Find the sum of the first six terms of the geometric sequence 3, 12, 48, ...

Solution   Here, $a = 3$ and $r = 4$. Find S_6 by the preceding result.

$$
\begin{aligned}
S_6 &= \frac{3(4^6 - 1)}{4 - 1} \qquad \text{Let } n = 6,\ a = 3,\ r = 4. \\
&= \frac{3(4096 - 1)}{3} \\
&= 4095 \ \blacktriangle \ \langle 2 \rangle
\end{aligned}
$$

FIGURE 5.6

TECHNOLOGY TIP   Graphing calculators with sequence capability can also find the sum of the first *n* terms of a sequence, given the expression for the *n*th term. In Example 3, for instance, the sequence is given by $ar^{n-1} = 3(4^{n-1})$. The sum of the first six terms is shown in Figure 5.6.

Ordinary Annuities

A sequence of equal payments made at equal periods of time is called an **annuity.** If the payments are made at the end of the period, and if the frequency of payments is the same as the frequency of compounding, the annuity is called an **ordinary annuity.** The time between payments is the **payment period,** and the time from the beginning of the first payment period to the end of the last period is called the **term of the annuity.** The **future value of the annuity**—the final sum on deposit—is defined as the sum of the compound amounts of all the payments, compounded to the end of the term.

Two common uses of annuities are to accumulate funds for some goal or to withdraw funds from an account. For example, an annuity may be used to save money for a large purchase, such as an automobile, an expensive trip, or a down payment on a home. An annuity also may be used to provide monthly payments for retirement. We explore these options in this and the next section.

▼ **EXAMPLE 4** $1500 is deposited at the end of each year for the next 6 years in an account paying 8% interest compounded annually. Find the future value of this annuity.

Solution   Figure 5.7 shows the situation schematically.

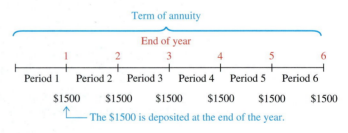

FIGURE 5.7

To find the future value of this annuity, look separately at each of the $1500 payments. The first $1500 is deposited at the end of period 1 and earns interest for the remaining 5 periods. From the formula in the box on page 257, the compound amount produced by this payment is

$$1500(1 + .08)^5 = 1500(1.08)^5.$$

The second $1500 payment is deposited at the end of period 2 and earns interest for the remaining 4 periods. So the compound amount produced by the second payment is

$$1500(1 + .08)^4 = 1500(1.08^4).$$

Continue to compute the compound amount for each subsequent payment, as shown in Figure 5.8. Note that the last payment earns no interest.

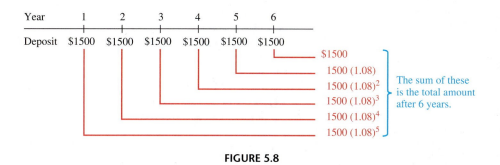

FIGURE 5.8

The last column of Figure 5.8 shows that the total amount after 6 years is the sum

$$1500 + 1500 \cdot 1.08^1 + 1500 \cdot 1.08^2 + 1500 \cdot 1.08^3$$
$$+ 1500 \cdot 1.08^4 + 1500 \cdot 1.08^5,$$

which is the sum of the first six terms of a geometric sequence with $a = 1500$ and $r = 1.08$. Applying the sum formula with $n = 6$, we see that the sum is

$$\frac{a(r^n - 1)}{r - 1} = \frac{1500[(1.08)^6 - 1]}{1.08 - 1} = \$11,003.089. \ \blacktriangle \ \boxed{3}$$

We can generalize the preceding example to an annuity, as follows. Suppose that a payment of R dollars is deposited into an account at the end of each period for n periods at an interest rate of i per period. [In Example 4, $R = \$1500$, $n = 6$, and $i = .08$.] The first payment is made at the end of period 1 and earns interest for the $n - 1$ remaining periods. By the formula in the box on page 257, the compound amount produced by this payment is $R(1 + i)^{n-1}$ dollars. The second payment is made at the end of period 2 and earns interest for the remaining $n - 2$ periods. It produces a compound amount of $R(1 + i)^{n-2}$ dollars. The compound amounts for each subsequent payment are computed similarly, as indicated in Figure 5.9. Note that the last payment earns no interest and contributes R dollars to the total.

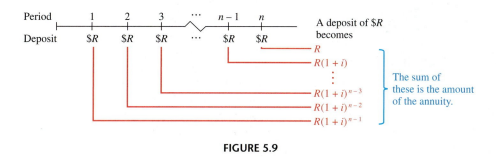

FIGURE 5.9

If S is the future value of the annuity, then the last column of Figure 5.9 shows that

$$S = R + R(1 + i) + R(1 + i)^2 + \cdots + R(1 + i)^{n-2} + R(1 + i)^{n-1}.$$

Hence, S is the sum of the first n terms of a geometric sequence having first term R and common ratio $1 + i$. Using the formula for the sum of the first n terms of a geometric sequence with $a = R$ and $r = i + 1$, we have

$$S = \frac{R[(1 + i)^n - 1]}{(1 + i) - 1} = \frac{R[(1 + i)^n - 1]}{i}$$

$$= R\left[\frac{(1 + i)^n - 1}{i}\right].$$

The quantity in brackets in the preceding line is sometimes written $s_{\overline{n}|i}$ (read "s-angle-n at i"), so that

$$S = R \cdot s_{\overline{n}|i}.$$

We summarize the preceding work as follows.*

Future Value of an Ordinary Annuity

The future value of an ordinary annuity is given by

$$S = R\left[\frac{(1 + i)^n - 1}{i}\right] \quad \text{or} \quad S = R \cdot s_{\overline{n}|i},$$

where

  S is the future value,

  R is the payment at the end of each period,

  i is the interest rate per period, and

  n is the number of periods.

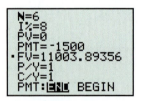

FIGURE 5.10

TECHNOLOGY TIP   Most computations with annuities can be done quickly with a spreadsheet program or a graphing calculator. On a calculator, use the TVM solver if there is one (see the Tip on page 260); otherwise, use the programs in the Program Appendix at www.aw.com/MWA9. (TI programs may be downloaded; others must be entered by hand).

Figure 5.10 shows how to do Example 4 on a TI-84+ TVM solver. First, enter the known quantities: N = number of payments, I% = annual interest rate, PV = present value, PMT = payment per period (entered as a negative amount), P/Y = number of payments per year, and C/Y = number of compoundings per year. Then put the cursor next to the unknown amount FV (future value), and press SOLVE.

Note: P/Y and C/Y should always be the same for problems in this book. If you use the solver for ordinary compound interest problems, set PMT = 0 and enter either PV or FV (whichever is known) as a negative amount.

▼ **EXAMPLE 5**   Chris Webber is an athlete who feels that his playing career will last 7 years. To prepare for his future, he deposits $122,000 at the end of each year for 7 years in an account paying 6% compounded annually. How much will he have on deposit after 7 years?

---

*We use S for the future value here, instead of A as in the compound interest formula, to help avoid confusing the two formulas.

Solution His payments form an ordinary annuity with $R = 122,000$, $n = 7$, and $i = .06$. The future value of this annuity (by the formula above) is

$$S = 122,000\left[\frac{(1.06)^7 - 1}{.06}\right] = \$1,024,048.19. \; \blacktriangle$$

Sinking Funds

A fund set up to receive periodic payments is called a **sinking fund.** The periodic payments, together with the interest earned by them, are designed to produce a given sum at some time in the future. For example, a sinking fund might be set up to receive money that will be needed to pay off the principal on a loan at some future time. If the payments are all the same amount and are made at the end of a regular period, they form an ordinary annuity.

▼ **EXAMPLE 6**   Experts say that the baby-boom generation (born between 1946 and 1960) cannot count on a company pension or Social Security to provide a comfortable retirement, as their parents did. It is recommended that they start to save early and regularly. Michael Boezi, a baby boomer, has decided to deposit $200 at the end of each month in a sinking fund that pays interest of 7.2% compounded monthly for retirement in 20 years.

(a)   How much will be in the account at that time?

Solution This savings plan is an annuity with $R = 200$, $i = .072/12$, and $n = 12(20)$. The future value is

$$S = 200\left[\frac{(1 + .072/12)^{12(20)} - 1}{.072/12}\right] = 106,752.47,$$

or $106,752.47.

(b)   Michael believes he needs to accumulate $130,000 in the 20-year period to have enough for retirement. What interest rate would provide that amount?

Solution Once again, we have an annuity with $R = 200$ and $n = 12(20)$. Now the future value $S = 130,000$ is given, and we must find the interest rate. If x is the annual interest rate in decimal form, then the interest rate per month is $i = x/12$ and we have

$$R\left[\frac{(1 + i)^n - 1}{i}\right] = S$$

$$200\left[\frac{(1 + x/12)^{12(20)} - 1}{x/12}\right] = 130,000. \tag{1}$$

This last equation is difficult to solve algebraically. We can get a rough approximation by using a calculator to compute the left side of the equation for various values of x, to see which one comes closest to producing $130,000. For example, $x = .08$ produces 117,804 and $x = .09$ produces 133,577, so the solution is between .08 and .09. Further experimentation of this type leads to the approximate solution $x = .0879$, that is, 8.79%. A more exact approximation can be obtained with a graphing calculator (see the following Technology Tip). ▲

⟨4⟩

Johnson Building Materials deposits $2500 at the end of each year into an account paying 8% per year compounded annually. Find the total amount on deposit after

(a)   6 years;

(b)   10 years.

Answers:

(a) $18,339.82

(b) $36,216.41

⟨5⟩

Pete's Pizza deposits $5800 at the end of each quarter for 4 years.

(a)   Find the final amount on deposit if the money earns 6.4% compounded quarterly.

(b)   Pete wants to accumulate $110,000 in the 4-year period. What interest rate (to the nearest tenth) will be required?

Answers:

(a) $104,812.44

(b)   8.9%

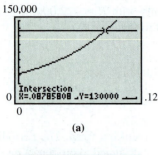

(a)

(b)

FIGURE 5.11

TECHNOLOGY TIP To solve equation (1) in Example 6 with a graphing calculator, graph the equations

$$y = 200\left[\frac{[(1 + x/12)^{12(20)} - 1]}{x/12}\right] \quad \text{and} \quad y = 130,000$$

on the same screen and find the intersection of their graphs as in Figure 5.11(a).

This result can also be obtained on a TVM solver, as shown in Figure 5.11(b): Enter all the information except the interest rate (with the monthly payment PMT as a negative amount); then put the cursor next to I% and press SOLVE.

▼ **EXAMPLE 7** The Chinns are close to retirement. They agree to sell an antique urn to the local museum for $17,000. Their tax adviser suggests that they defer receipt of this money until they retire, 5 years in the future. (At that time, they might well be in a lower tax bracket.) Find the amount of each payment the museum must make into a sinking fund so that it will have the necessary $17,000 in 5 years. Assume that the museum can earn 6% compounded annually on its money. Also, assume that the payments are made annually.

Solution These payments are the periodic payments into an ordinary annuity that will amount to $17,000 in 5 years at 6% compounded annually. Using the formula and a calculator yields

$$17,000 = R\left[\frac{(1.06)^5 - 1}{.06}\right]$$

$$17,000 \approx R(5.637093)$$

$$R \approx \frac{17,000}{5.637093} = 3015.74,$$

or $3015.74. If the museum management deposits $3015.74 at the end of each year for 5 years in an account paying 6% compounded annually, it will have the total amount needed. This result is shown in the following sinking-fund table.

Payment Number	Amount of Deposit	Interest Earned	Total in Account
1	$3015.74	$0	$3,015.74
2	3015.74	180.94	6,212.42
3	3015.74	372.75	9,600.91
4	3015.74	576.05	13,192.70
5	3015.74	791.56	17,000.00

To construct the table, notice that the first payment does not earn interest until the second payment is made. Line 2 of the table shows the second payment, 6% interest of $180.94 on the first payment, and the sum of these amounts added to the total in line 1. Line 3 shows the third payment, 6% interest of $372.75 on the total

6

Francisco Arce needs $8000 in 6 years so that he can go on an archaeological dig. He wants to deposit equal payments at the end of each quarter so that he will have enough to go on the dig. Find the amount of each payment if the bank pays

(a) 12% compounded quarterly;

(b) 8% compounded quarterly.

Answers:

(a) $232.38

(b) $262.97

from line 2, and the new total found by adding these amounts to the total in line 2. This procedure is continued to complete the table.* ▲ ⬦6

Annuities Due

The formula developed above is for *ordinary annuities*—those with payments at the *end* of each period. The results can be modified slightly to apply to **annuities due**— annuities where payments are made at the *beginning* of each period.

An example will illustrate how this is done. Consider an annuity due, in which payments of $100 are made for 3 years, and an ordinary annuity, in which payments of $100 are made for 4 years, both with 5% interest, compounded annually. Figure 5.12 computes the growth of each payment separately (as was done in Example 4).

Annuity Due (payments at beginning of year for 3 years)

Year 1 Year 2 Year 3

$100 $100 $100

$100(1.05^1)$
$100(1.05^2)$
$100(1.05^3)$

Future value of is the sum of this column ⟶

Ordinary Annuity (payments at end of year for 4 years)

Year 1 Year 2 Year 3 Year 4

$100 $100 $100 $100

100
$100(1.05^1)$
$100(1.05^2)$
$100(1.05^3)$

Future value of is the sum of this column ⟶

FIGURE 5.12

Figure 5.12 shows that the future values are the same, *except* for one $100 payment on the ordinary annuity (shown in red). So we can use the formula on page 272 to find the future value of the 4-year ordinary annuity and then subtract one $100 payment to get the future value of the 3-year annuity due:

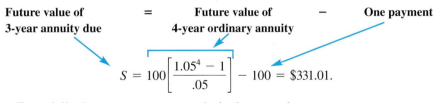

Future value of = **Future value of** − **One payment**
3-year annuity due   **4-year ordinary annuity**

$$S = 100\left[\frac{1.05^4 - 1}{.05}\right] - 100 = \$331.01.$$

Essentially the same argument works in the general case.

_____

*In some sinking-fund tables, the last payment may differ slightly from the others because of rounding in the preceding lines.

The **future value S of an annuity due of n payments** of R dollars each at the beginning of consecutive interest periods with interest compounded at the rate of i per period is

$$S = R\left[\frac{(1 + i)^{n+1} - 1}{i}\right] - R.$$

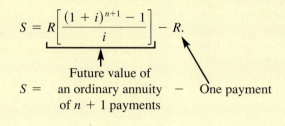

$S =$ Future value of an ordinary annuity $-$ One payment of $n + 1$ payments

7

(a) Ms. Black deposits $800 at the beginning of each 6-month period for 5 years. Find the final amount if the account pays 6% compounded semiannually.

(b) Find the final amount if this account were an ordinary annuity.

Answers:

(a) $9446.24

(b) $9171.10

▼ **EXAMPLE 8** Payments of $500 are made at the beginning of each quarter for 7 years in an account paying 12% compounded quarterly. Find the future value of this annuity due.

Solution In 7 years, there are 28 quarterly periods. Add one period to get $n = 29$, and use the formula with $i = 12\%/4 = 3\%$.

$$S = 500\left[\frac{(1.03)^{29} - 1}{.03}\right] - 500$$

$$S \approx 500(45.21885) - 500 = \$22,109.43$$

The account will contain a total of $22,109.43 after 7 years. ▲ ⑦

◆5.3 Exercises

🖉 **1.** In your own words, describe a geometric sequence.

2. What is the nth term of a geometric sequence with first term a and common ratio r?

3. According to T. Rowe Price Associates, a person with a moderate investment strategy and n years to retirement should have accumulated savings of a_n percent of his or her annual salary. The geometric sequence with

$$a_n = 1276(.916)^n$$

gives the appropriate percent for each year n.
(a) Find a_1 and r.
(b) Find and interpret the terms a_{10} and a_{20}.

4. Refer to Exercise 3. For someone who has a conservative investment strategy with n years to retirement, the geometric sequence is*

$$a_n = 1278(.935)^n.$$

(a) Repeat part (a) of Exercise 3.
(b) Repeat part (b) of Exercise 3.
(c) Why are the answers in parts (a) and (b) larger than in Exercise 3?

Find the fourth term of each of the following geometric sequences. (See Example 1.)

5. $a = 4, r = 3$ **6.** $a = 15, r = 2$

7. $a = 24, r = .5$ **8.** $a = 80, r = .1$

9. $a = 2000, r = 1.05$ **10.** $a = 10,000, r = 1.01$

Find the sum of the first four terms for each of the following geometric sequences. (See Example 3.)

11. $a = 3, r = 2$ **12.** $a = 4, r = 3$

13. $a = 5, r = .2$ **14.** $a = 7, r = .5$

15. $a = 128, r = 1.1$ **16.** $a = 100, r = 1.05$

Find each of the following values.

17. $s_{\overline{12}|.05}$ **18.** $s_{\overline{25}|.06}$

19. $s_{\overline{16}|.04}$ **20.** $s_{\overline{30}|.02}$

21. $s_{\overline{40}|.01}$ **22.** $s_{\overline{18}|.015}$

Find the future value of the following ordinary annuities. Payments are made and interest is compounded as given. (See Examples 4, 5, and 6(a).)

23. $R = \$12,000$, 6.2% interest compounded annually for 8 years

24. $R = \$20{,}000$, 4.5% interest compounded annually for 12 years

25. $R = \$865$, 6% interest compounded semiannually for 10 years

26. $R = \$7300$, 9% interest compounded semiannually for 6 years

27. $R = \$1200$, 8% interest compounded · quarterly for 10 years

28. $R = \$20{,}000$, 6% interest compounded quarterly for 12 years

Find the periodic payment that will amount to the given sums under the given conditions if payments are made at the end of each period. (See Example 7.)

29. $S = \$14{,}500$; interest is 4% compounded semiannually for 8 years.

30. $S = \$43{,}000$; interest is 6% compounded semiannually for 10 years.

31. $S = \$62{,}000$; interest is 6% compounded quarterly for 6 years.

32. $S = \$12{,}800$; interest is 6% compounded monthly for 5 years.

33. What is meant by a sinking fund? List some reasons for establishing a sinking fund.

Find the amount of each payment to be made into a sinking fund to accumulate the following amounts. Payments are made at the end of each period. (See Example 7.)

34. $11,000; money earns 5% compounded semiannually for 6 years

35. $65,000; money earns 6% compounded semiannually for $4\frac{1}{2}$ years

36. $50,000; money earns 8% compounded quarterly for $2\frac{1}{2}$ years

37. $25,000; money earns 9% compounded quarterly for $3\frac{1}{2}$ years

38. $6000; money earns 6% compounded monthly for 3 years

39. $9000; money earns 7% compounded monthly for $2\frac{1}{2}$ years

40. Explain the difference between an ordinary annuity and an annuity due.

Find the future value of each annuity due. (See Example 8.)

41. Payments of $500 for 10 years at 5% compounded annually

42. Payments of $1050 for 8 years at 3.5% compounded annually

43. Payments of $16,000 for 11 years at 4.7% compounded annually

44. Payments of $25,000 for 12 years at 6% compounded annually

45. Payments of $1000 for 9 years at 8% compounded semiannually

46. Payments of $750 for 15 years at 6% compounded semiannually

47. Payments of $100 for 7 years at 9% compounded quarterly

48. Payments of $1500 for 11 years at 7% compounded quarterly

Finance   *Work the following applied problems.*

49. A typical pack-a-day smoker spends about $100 per month on cigarettes. Suppose the smoker invests that amount at the end of each month in a savings account at 4.8% compounded monthly. What would the account be worth after 40 years?

50. Ron Hampton is saving for a computer. At the end of each month, he puts $60 in a savings account that pays 4.5% interest compounded monthly. How much is in the account after 3 years?

51. Hassi is paid on the first day of the month, and $80 is automatically deducted from his pay and deposited in a savings account. If the account pays 7.5% interest compounded monthly, how much will be in the account after 3 years and 9 months?

52. A father opened a savings account for his daughter on the day she was born, depositing $1000. Each year on her birthday, he deposits another $1000, making the last deposit on her 21st birthday. If the account pays 6.5% interest compounded annually, how much is in the account at the end of the day on the daughter's 21st birthday?

53. Becky Anderson deposits $12,000 at the end of each year for 9 years in an account paying 6% interest compounded annually.
  (a) Find the final amount she will have on deposit.
  (b) Becky's brother-in-law works in a bank that pays 5% compounded annually. If she deposits money in this bank instead of the other one, how much will she have in her account?
  (c) How much would Becky lose over 9 years by using her brother-in-law's bank?

54. Raul Vasquez, a 25-year-old professional, puts $750 in a retirement fund at the end of each quarter until he reaches age 60. The account pays 8% interest compounded quarterly.
  (a) How much will be in the account when he is 60?
  (b) If Raul makes no further deposits after age 60, how much will he have for retirement at age 65?

55. At the end of each quarter, a 50-year-old woman puts $1200 in a retirement account that pays 7% interest compounded quarterly. When she reaches age 60, she withdraws the entire amount and places it in a mutual fund that pays 9% interest compounded monthly. From then on, she deposits $300 in the mutual fund at the end of each month. How much is in the account when she reaches age 65?

56. Jasspreet Kaur deposits $2435 at the beginning of each semiannual period for 8 years in an account paying 6% compounded semiannually. She then leaves that money alone, with no further deposits, for an additional 5 years. Find the final amount on deposit after the entire 13-year period.

57. Chuck Hickman deposits $10,000 at the beginning of each year for 12 years in an account paying 5% compounded annually. He then puts the total amount on deposit in another account paying 6% compounded semiannually for another 9 years. Find the final amount on deposit after the entire 21-year period.

58. Suppose Michael, in Example 6, cannot get the higher interest rate to produce $130,000 in 20 years. To meet that goal at an interest rate of 7.2%, he must increase his monthly payment. What payment should he make each month?

59. David Horwitz needs $10,000 in 8 years.
   (a) What amount should he deposit at the end of each quarter at 5% compounded quarterly so that he will have his $10,000?
   (b) Find Horwitz's quarterly deposit if the money is deposited at 5.8% compounded quarterly.

60. Harv's Meats knows that it must buy a new deboner machine in 4 years. The machine costs $12,000. In order to accumulate enough money to pay for the machine, Harv decides to deposit a sum of money at the end of each 6 months in an account paying 6% compounded semiannually. How much should each payment be?

61. Barbara Margolius wants to buy an $18,000 car in 6 years. How much money must she deposit at the end of each quarter in an account paying 6.1% compounded quarterly so that she will have enough to pay for her car?

In Exercises 62 and 63, use a graphing calculator to find the value of i that produces the given value of S. (See the Technology Tip following Example 6.)

62. To save for retirement, Karla Harby put $300 each month into an ordinary annuity for 20 years. Interest was compounded monthly. At the end of the 20 years, the annuity was worth $147,126. What annual interest rate did she receive?

63. Jennifer Wall made payments of $250 per month at the end of each month to purchase a piece of property. After 30 years, she owned the property, which she sold for $330,000. What annual interest rate would she need to earn on an ordinary annuity for a comparable rate of return?

64. In a 1992 Virginia lottery, the jackpot was $27 million. An Australian investment firm tried to buy all possible combinations of numbers, which would have cost $7 million. In fact, the firm ran out of time and was unable to buy all combinations, but ended up with the only winning ticket anyway. The firm received the jackpot in 20 equal annual payments of $1.35 million.* Assume these payments meet the conditions of an ordinary annuity.
   (a) Suppose the firm can invest money at 8% interest compounded annually. How many years would it take until the investors would be further ahead than if they had simply invested the $7 million at the same rate? (*Hint:* Experiment with different values of n, the number of years, or use a graphing calculator to plot the value of both investments as a function of the number of years.)
   (b) How many years would it take in part (a) at an interest rate of 12%?

65. Diane Gray sells some land in Nevada. She will be paid a lump sum of $60,000 in 7 years. Until then, the buyer pays 8% simple interest quarterly.
   (a) Find the amount of each quarterly interest payment.
   (b) The buyer sets up a sinking fund so that enough money will be present to pay off the $60,000. The buyer wants to make semiannual payments into the sinking fund; the account pays 6% compounded semiannually. Find the amount of each payment into the fund.
   (c) Prepare a table showing the amount in the sinking fund after the first 6 deposits.

66. Joe Seniw bought a rare stamp for his collection. He agreed to pay a lump sum of $4000 after 5 years. Until then, he pays 6% simple interest semiannually.
   (a) Find the amount of each semiannual interest payment.
   (b) Seniw sets up a sinking fund so that enough money will be present to pay off the $4000. He wants to make annual payments into the fund. The account pays 8% compounded annually. Find the amount of each payment.
   (c) Prepare a table showing the amount in the sinking fund after each deposit.

---

*Washington Post, March 10, 1992, p. A1.

5.4 Present Value of an Annuity and Amortization

Suppose that at the end of each year, for the next 10 years, $500 is deposited in a savings account paying 7% interest compounded annually. This is an example of an ordinary annuity. The **present value** of this annuity is the amount that would have to be deposited in one lump sum today (at the same compound interest rate) in order to produce exactly the same balance at the end of 10 years. We can find a formula for the present value of an annuity as follows.

Suppose deposits of R dollars are made at the end of each period for n periods at interest rate i per period. Then the amount in the account after n periods is the future value of this annuity:

$$S = R\left[\frac{(1 + i)^n - 1}{i}\right].$$

On the other hand, if P dollars are deposited today at the same compound interest rate i, then, at the end of n periods, the amount in the account is $P(1 + i)^n$. This amount must be the same as the amount S in the preceding formula; that is,

$$P(1 + i)^n = R\left[\frac{(1 + i)^n - 1}{i}\right].$$

To solve this equation for P, multiply both sides by $(1 + i)^{-n}$.

$$P = R(1 + i)^{-n}\left[\frac{(1 + i)^n - 1}{i}\right]$$

Use the distributive property; also, recall that $(1 + i)^{-n}(1 + i)^n = (1 + i)^0 = 1$.

$$P = R\left[\frac{(1 + i)^{-n}(1 + i)^n - (1 + i)^{-n}}{i}\right]$$

$$P = R\left[\frac{1 - (1 + i)^{-n}}{i}\right]$$

The amount P is called the *present value of the annuity*. The quantity in brackets is abbreviated as $a_{\overline{n}|i}$, so

$$a_{\overline{n}|i} = \frac{1 - (1 + i)^{-n}}{i}.$$

(The symbol $a_{\overline{n}|i}$ is read "*a*-angle-*n* at *i*." Compare this quantity with $s_{\overline{n}|i}$ in the previous section.) The formula for the present value of an annuity is summarized next.

Present Value of an Annuity

The present value P of an annuity of n payments of R dollars each at the end of consecutive interest periods with interest compounded at a rate of interest i per period is

$$P = R\left[\frac{1 - (1 + i)^{-n}}{i}\right] \qquad \text{or} \qquad P = R \cdot a_{\overline{n}|i}.$$

CAUTION Don't confuse the formula for the present value of an annuity with the one for the future value of an annuity. Notice the difference: the numerator of the fraction in the present-value formula is $1 - (1 + i)^{-n}$; but in the future-value formula, it is $(1 + i)^n - 1$.

▼ **EXAMPLE 1** Surinder Sinah and Maria Gonzalez are both graduates of the Forestvire Institute of Technology. They both agree to contribute to the endowment fund of FIT. Sinah says that he will give $500 at the end of each year for 9 years. Gonzalez prefers to give a lump sum today. What lump sum can she give that will equal the present value of Sinah's annual gifts if the endowment fund earns 7.5% compounded annually?

Solution Here, $R = 500$, $n = 9$, and $i = .075$, and we have

$$P = 500\left[\frac{1 - (1.075)^{-9}}{.075}\right]$$

$$= 3189.44.$$

Therefore, Gonzalez must donate a lump sum of $3189.44 today. ▲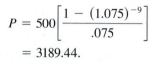

1

What lump sum deposited today would be equivalent to equal payments of

(a) $650 at the end of each year for 9 years at 4% compounded annually?

(b) $1000 at the end of each quarter for 4 years at 4% compounded quarterly?

Answers:

(a) $4832.97

(b) $14,717.87

▼ **EXAMPLE 2** A car costs $22,000. After a down payment of $4000, the balance will be paid off in 48 equal monthly payments with interest of 12% per year on the unpaid balance. Find the amount of each payment.

Solution A single lump-sum payment of $18,000 today would pay off the loan. So $18,000 is the present value of an annuity of 48 monthly payments with interest of $12\%/12 = 1\%$ per month. Thus, $P = 18,000$, $n = 48$, and $i = .01$, and we must find the monthly payment R in the formula

$$P = R\left[\frac{1 - (1 + i)^{-n}}{i}\right].$$

$$18,000 = R\left[\frac{1 - (1.01)^{-48}}{.01}\right]$$

$$R \approx 474.0090378.$$

A monthly payment of $474.01 will be needed. ▲ **2**

Amortization

A loan is **amortized** if both the principal and interest are paid by a sequence of equal periodic payments. In Example 2, a loan of $18,000 at 12% interest compounded monthly could be amortized by paying $474.01 per month for 48 months.

The periodic payment needed to amortize a loan may be found, as in Example 2, by solving the present-value equation for R.

2

Kelly Erin buys a small business for $174,000. She agrees to pay off the cost in payments at the end of each semiannual period for 7 years, with interest of 8% compounded semiannually on the unpaid balance. Find the amount of each payment.

Answer:

$16,472.40

Amortization Payments

A loan of P dollars at interest rate i per period may be amortized in n equal periodic payments of R dollars made at the end of each period, where

$$R = \frac{P}{a_{\overline{n}|i}} = \frac{P}{\left[\dfrac{1 - (1 + i)^{-n}}{i}\right]} = \frac{Pi}{1 - (1 + i)^{-n}}.$$

TECHNOLOGY TIP   A TVM
solver on a graphing calculator
can find the present value of an
annuity or the payment on a
loan: fill in the known information, put the cursor next to
the unknown item (PV or PMT),
and press SOLVE. Figure 5.13
shows the solution to Example 3(a) on a TVM solver. Alternatively, you can use the program
in the Program Appendix at
www.aw.com/MWA9.

FIGURE 5.13

▼ **EXAMPLE 3**   The Rechtien family buys a house for $140,000 with a down payment of $30,000. The family takes out a 30-year, $110,000 mortgage at an annual interest rate of 6.6%.

(a) Find the amount of the monthly payment needed to amortize this loan.

Solution Here, $P = \$110,000$. Since the annual interest rate is .066, the monthly interest rate is $i = .066/12 = .0055.*$ The number of payments is $n = 12 \cdot 30 = 360$. Hence,

$$R = \frac{P}{\left[\dfrac{1 - (1 + i)^{-n}}{i}\right]} = \frac{110,000}{\left[\dfrac{1 - (1 + .0055)^{-360}}{.0055}\right]}$$

$$= \frac{110,000}{\left[\dfrac{1 - (1.0055)^{-360}}{.0055}\right]} \approx \frac{110,000}{156.5781} \approx \$702.52.$$

Monthly payments of $702.52 are required to amortize the loan.

(b) Find the total amount of interest paid when the loan is amortized over 30 years.

Solution The Rechtein family makes 360 payments of $702.52, for a total of

$$360 \cdot 702.52 = \$252,907.20.$$

Since the amount of the loan was $110,000, the total interest paid is

$$\$252,907.20 - \$110,000 = \$142,907.20.$$

The large amount of interest is typical of what happens with a long mortgage. A 15-year mortgage would have somewhat higher payments, but would involve signifi-cantly less interest. ③

(c) Find the part of the first payment that is interest and the part that is applied to reducing the debt.

Solution The monthly interest rate is .0055. During the first month, the entire $110,000 is owed. Interest on this amount for one month is found by the formula for simple interest:

$$I = Prt = \$110,000(.0055) = \$605.$$

At the end of the month, a payment of $702.52 is made. Since $605 of this is inter-est, the amount applied to the reduction of the original debt is

$$\$702.52 - \$605 = \$97.52.$$

So the remaining balance after one payment is $110,000 - \$97.52 = \$109,902.48$. ▲

Amortization Schedules

In Example 3, 360 payments are made to amortize a $110,000 loan balance after the first payment is reduced by only $97.52, which is much less than

③

If the mortgage in Example 3 runs for 15 years, find

(a) the monthly payment;

(b) the total amount of interest paid.

Answers:

(a) $964.28

(b) $63,570.40

---
*Mortgage rates are quoted in terms of annual interest, but it is always understood that the monthly rate is 1/12 of the annual rate and that interest is compounded monthly.

$\left(\dfrac{1}{360}\right)(110{,}000) \approx \$305.56.$ Therefore, even though equal *payments* are made to amortize a loan, the loan *balance* does not decrease in equal steps. This fact is very important if a loan is paid off early.

▼ **EXAMPLE 4** Beth Hill borrows $1000 for one year at 12% annual interest compounded monthly. Her monthly interest rate is $.12/12 = .01$, so her monthly loan payment is

$$R = \dfrac{P}{\left[\dfrac{1 - (1 + i)^{-n}}{i}\right]} = \dfrac{1000}{\left[\dfrac{1 - (1 + .01)^{-12}}{.01}\right]} = \dfrac{1000}{\left[\dfrac{1 - (1.01)^{-12}}{.01}\right]} \approx \$88.85.$$

After making three payments, she decides to pay off the remaining balance all at once. Approximately how much must she pay?

Solution Since nine payments remain to be paid, they can be thought of as an annuity consisting of nine payments of $88.85 at 1% interest per period. The present value of this annuity is

$$P = R\left[\dfrac{1 - (1 + i)^{-n}}{i}\right] = 88.85\left[\dfrac{1 - (1.01)^{-9}}{.01}\right] = \$761.09.$$

So Beth's remaining balance is about $761.09. The actual balance due may differ slightly from this figure because earlier payments and interest amounts are rounded to the nearest penny. ▲ ◇4

The method used to compute the remaining balance in Example 4 applies in the general case. After x payments, there are $n - x$ payments remaining. So the balance is the present value of an annuity of $n - x$ payments of the same amount at the same interest rate. As in Example 4 (where $n = 12$ and $x = 3$), this can be found by using $n - x$ in place of n in the present-value formula.

Remaining Balance

If a loan can be amortized by n payments of R dollars each at an interest rate i per period, then the *approximate* remaining balance after x payments is

$$y = R\left[\dfrac{1 - (1 + i)^{-(n-x)}}{i}\right].$$

▼ **EXAMPLE 5** In Example 3, we saw that the monthly payment on a $110,000 mortgage for 30 years with an annual interest rate of 6.6% was $702.52.

(a) Use the formula in the preceding box to find the approximate remaining balance after one payment

Solution Apply the formula with $R = 702.52$, $n = 360$, $x = 1$, and $i = .066/12 = .0055$ (monthly rate of interest). The approximate remaining balance is

$$y = 702.52\left[\frac{1 - (1 + .0055)^{-(360-1)}}{.0055}\right]$$

$$= 702.52\left[\frac{1 - (1.0055)^{-359}}{.0055}\right] \approx \$109{,}901.74.$$

This differs by 74 cents from the actual remaining balance of $109,902.48 found in Example 3(c).

(b) About how long it will take to repay half the loan?

Solution Half the loan is $55,000. Let $y = 55{,}000$, and solve the remaining balance equation for x.

$$55{,}000 = 702.52\left[\frac{1 - (1 + .0055)^{-(360-x)}}{.0055}\right]$$

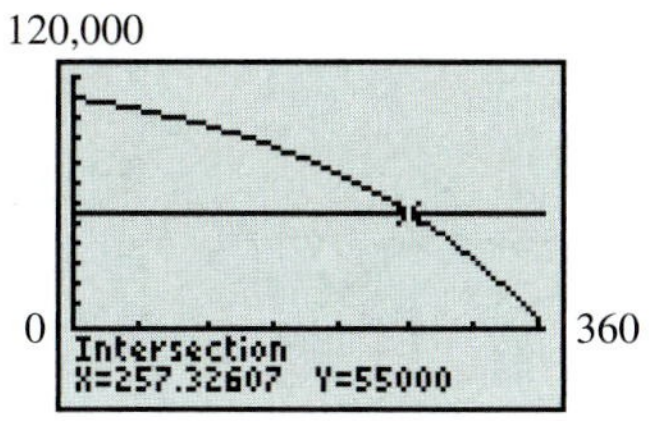

This can be done with a graphing calculator by graphing $y_1 = 55{,}000$ and $y_2 = 702.52\left[\dfrac{1 - (1.0055)^{-(360-x)}}{.0055}\right]$ on the same screen and finding their intersection. Figure 5.14 shows that the intersection occurs when $x \approx 257.33$—that is, after approximately 257 monthly payments. This means that it takes more than 21 years to pay off half of the 30-year mortgage. ▲

The remaining-balance formula is a quick and convenient way to get a reasonable estimate of the remaining balance on a loan, but it is not accurate enough for a bank or business, which must keep its books exactly. To determine the exact remaining balance after each loan payment, financial institutions normally use an **amortization schedule,** which lists how much of each payment is interest, how much goes to reduce the balance, and how much is still owed after each payment.

▼ **EXAMPLE 6** Determine the exact amount Beth Hill in Example 4 owes after three monthly payments.

Solution An amortization table for the loan is shown on the next page; it was obtained as follows. The annual interest rate is 12% compounded monthly, so the interest rate per month is $12\%/12 = 1\% = .01$. When the first payment is made, 1 month's interest, namely, $.01(1000) = \$10$, is owed. Subtracting this from the $88.85 payment leaves $78.85 to be applied to repayment. Hence, the principal at the end of the first payment period is $1000 - 78.85 = \$921.15$, as shown in "payment 1" line of the table.

When payment 2 is made, 1 month's interest on $921.15 is owed, namely, $.01(921.15) = \$9.21$. Subtracting this from the $88.85 payment leaves $79.64 to reduce the principal. So the principal at the end of payment 2 is $921.15 - 79.64 = \$841.51$. The interest portion of payment 3 is based on this amount, and the remaining lines of the table are found in a similar fashion.

120,000

0   360

Intersection
X=257.32607   Y=55000

FIGURE 5.14

Payment Number	Amount of Payment	Interest for Period	Portion to Principal	Principal at End of Period
0	—	—	—	$1000.00
1	$88.85	$10.00	$78.85	921.15
2	88.85	9.21	79.64	841.51
3	88.85	8.42	80.43	761.08
4	88.85	7.61	81.24	679.84
5	88.85	6.80	82.05	597.79
6	88.85	5.98	82.87	514.92
7	88.85	5.15	83.70	431.22
8	88.85	4.31	84.54	346.68
9	88.85	3.47	85.38	261.30
10	88.85	2.61	86.24	175.06
11	88.85	1.75	87.10	87.96
12	88.84	.88	87.96	0

The schedule shows that after three payments, she still owes $761.08, an amount that differs slightly from that obtained in Example 4. ▲

The amortization schedule in Example 6 is typical. In particular, note that all payments are the same except the last one. It is often necessary to adjust the amount of the final payment to account for rounding off earlier and to ensure that the final balance is exactly 0.

An amortization schedule also shows how the periodic payments are applied to interest and principal. The amount going to interest decreases with each payment, while the amount going to reduce the principal owed increases with each payment.

TECHNOLOGY TIP   Graphing calculator programs to produce amortization schedules are in the Program Appendix (at www.aw.com/MWA9). Spreadsheets are another useful tool for creating amortization tables. Microsoft Excel has a built-in feature for calculating monthly payments. Figure 5.15 shows an Excel amortization table for Example 6. For more details, see the *Spreadsheet Manual,* also available with this book.

	A	B	C	D	E	F
1	Prnt#	Payment	Interest	Principal	End Principal	
2	0				1000	
3	1	88.85	10.00	78.85	921.15	
4	2	88.85	9.21	79.64	841.51	
5	3	88.85	8.42	80.43	761.08	
6	4	88.85	7.61	81.24	679.84	
7	5	88.85	6.80	82.05	597.79	
8	6	88.85	5.98	82.87	514.92	
9	7	88.85	5.15	83.70	431.22	
10	8	88.85	4.31	84.54	346.68	
11	9	88.85	3.47	85.38	261.30	
12	10	88.85	2.61	86.24	175.06	
13	11	88.85	1.75	87.10	87.96	
14	12	88.85	0.88	87.97	-0.01	

FIGURE 5.15

5.4 Exercises

1. Which of the following is represented by $a_{\overline{n}|i}$?

   (a) $\dfrac{(1+i)^{-n}-1}{i}$ (b) $\dfrac{(1+i)^{n}-1}{i}$

   (c) $\dfrac{1-(1+i)^{-n}}{i}$ (d) $\dfrac{1-(1+i)^{n}}{i}$

2. Which of the choices in Exercise 1 represents $s_{\overline{n}|i}$?

Find each of the following.

3. $a_{\overline{15}|.05}$ 4. $a_{\overline{10}|.03}$ 5. $a_{\overline{18}|.04}$

6. $a_{\overline{30}|.01}$ 7. $a_{\overline{16}|.015}$ 8. $a_{\overline{32}|.027}$

9. Explain the difference between the present value of an annuity and the future value of an annuity. For a given annuity, which is larger? Why?

Find the present value of each ordinary annuity. (See Example 1.)

10. Payments of $890 each year for 16 years at 6% compounded annually

11. Payments of $1400 each year for 8 years at 6% compounded annually

12. Payments of $10,000 semiannually for 15 years at 7.5% compounded semiannually

13. Payments of $50,000 quarterly for 10 years at 5% compounded quarterly

14. Payments of $15,806 quarterly for 3 years at 6.8% compounded quarterly

15. Payments of $18,579 every 6 months for 8 years at 7.4% compounded semiannually

Find the lump sum deposited today that will yield the same total amount as payments of $10,000 at the end of each year for 15 years at each of the given interest rates. (See Example 1.)

16. 3% compounded annually

17. 4% compounded annually

18. 6% compounded annually

19. What does it mean to amortize a loan?

Find the payment necessary to amortize each of the following loans. (See Example 2.)

20. $2500; 8% compounded quarterly; 6 quarterly payments

21. $41,000; 9% compounded semiannually; 10 semiannual payments

22. $90,000; 7% compounded annually; 12 annual payments

23. $140,000; 12% compounded quarterly; 15 quarterly payments

24. $7400; 8.2% compounded semiannually; 18 semiannual payments

25. $5500; 9.5% compounded monthly; 24 monthly payments

Find the monthly house payment necessary to amortize the following loans. (See Example 3.)

26. $149,560 at 6.75% for 25 years

27. $170,892 at 7.11% for 30 years

28. $153,762 at 5.45% for 30 years

29. $96,511 at 8.57% for 25 years

Use the amortization table in Example 6 to answer the questions in Exercises 30–33.

30. How much of the 5th payment is interest?

31. How much of the 10th payment is used to reduce the debt?

32. How much interest is paid in the first 5 months of the loan?

33. How much interest is paid in the last 5 months of the loan?

34. What sum deposited today at 5% compounded annually for 8 years will provide the same amount as $1000 deposited at the end of each year for 8 years at 6% compounded annually?

35. What lump sum deposited today at 8% compounded quarterly for 10 years will yield the same final amount as deposits of $4000 at the end of each 6-month period for 10 years at 6% compounded semiannually?

Finance *Work the following applied problems.*

36. An auto stereo dealer sells a stereo system for $600 down and monthly payments of $30 for the next 3 years. If the interest rate is 1.25% per month on the unpaid balance, find
   (a) the cost of the stereo system;
   (b) the total amount of interest paid.

37. John Kushida buys a used car costing $6000. He agrees to make payments at the end of each monthly period for 4 years. He pays 12% interest, compounded monthly.
   (a) What is the amount of each payment?
   (b) Find the total amount of interest Kushida will pay.

38. A speculator agrees to pay $15,000 for a parcel of land; this amount, with interest, will be paid over 4 years with semiannual payments at an interest rate of 10% compounded semiannually. Find the amount of each payment.

39. In a recent Powerball lottery, the prize was about $34 million. The winner could choose to receive either 30 payments of $1,133,334 — one payment now and one payment at the end of each year for 29 years — or a single "cash value" payment now. Assuming that money earns 4.215% interest compounded annually, what was the cash value? (*Hint:* The first option is equivalent to $1,133,334 plus an ordinary annuity of 29 payments of $1,133,334 per year. So the cash value should be $1,133,334 plus the present value of the annuity.)

40. If the interest rate in Exercise 39 is 5%, what is the cash value of the $34 million prize?

41. Refer to Exercise 39. Suppose you win a Powerball prize of $54 million.
(a) If you choose to receive 30 equal payments, how much will each payment be?
(b) If money earns interest at 4.8%, what is the cash value of the lottery prize?

42. Do Exercise 41 if the prize is $18 million and money earns interest at 6%.

Finance *Student borrowers now have more options to choose from when selecting repayment plans.* The standard plan repays the loan in 10 years with equal monthly payments. The extended plan allows from 12 to 30 years to repay the loan. A student borrows $35,000 at 7.43% compounded monthly.*

43. Find the monthly payment and total interest paid under the standard plan.

44. Find the monthly payment and total interest paid under the extended plan with 20 years to pay off the loan.

Finance *Use the formula for the approximate remaining balance to work each problem. (See Example 5.)*

45. When Teresa Flores opened her law office, she bought $14,000 worth of law books and $7200 worth of office furniture. She paid $1200 down and agreed to amortize the balance with semiannual payments for 5 years at 12% compounded semiannually.
(a) Find the amount of each payment.
(b) When her loan had been reduced below $5000, Flores received a large tax refund and decided to pay off the loan. How many payments were left at this time?

46. Kareem Adams buys a house for $285,000. He pays $60,000 down and takes out a mortgage at 6.9% on the balance. Find his monthly payment and the total amount of interest he will pay if the length of the mortgage is

(a) 15 years;
(b) 20 years;
(c) 25 years.
(d) When will half the 20-year loan be paid off?

47. The Beyes plan to purchase a home for $212,000. They will pay 20% down and finance the remainder for 30 years at 7.2% interest, compounded monthly.*
(a) How large are their monthly payments?
(b) What will be their approximate loan balance right after they have made their 96th payment?
(c) How much interest will they pay during the 7th year of the loan?
(d) If they were to increase their monthly payments by $150, how long would it take to pay off the loan?

Finance *Work each problem.*

48. Sandi Goldstein has inherited $25,000 from her grandfather's estate. She deposits the money in an account offering 6% interest compounded annually. She wants to make equal annual withdrawals from the account so that the money (principal and interest) lasts exactly 8 years.
(a) Find the amount of each withdrawal.
(b) Find the amount of each withdrawal if the money must last 12 years.

 49. Jeni Ramirez plans to retire in 20 years. She will make 240 equal monthly contributions to her retirement account. One month after her last contribution, she will begin the first of 120 monthly withdrawals from the account. She expects to withdraw $3500 per month. How large must her monthly contributions be in order to accomplish her goal if her account is assumed to earn interest of 10.5% compounded monthly for the duration of her contributions and the 120 months of withdrawals?

50. Ron Okimura also plans to retire in 20 years. Ron will make 120 equal monthly contributions to his account. Ten years after his last contribution (120 months) he will begin the first of 120 monthly withdrawals from the account. Ron also expects to withdraw $3500 per month. His account also earns interest of 10.5% compounded monthly for the duration of his contributions and the 120 months of withdrawals. How large must Ron's monthly contributions be in order to accomplish his goals?

51. Madeline and Nick Swenson took out a 30-year mortgage for $160,000 at 9.8% interest compounded monthly. After they had made 12 years of payments (144 payments), they decided to refinance the remaining loan balance for 25 years at 7.2% interest, compounded monthly. What will be the balance on their loan 5 years after they refinance?

New York Times, April 2, 1995, "Money and College," Saul Hansell, p. 28.

*Exercises 47–51 were supplied by Norman Lindquist at Western Washington University.

Finance   In Exercises 52–55, prepare an amortization schedule showing the first 4 payments for each loan. (See Example 5.)

52. An insurance firm pays $4000 for a new printer for its computer. It amortizes the loan for the printer in 4 annual payments at 8% compounded annually.

53. Large semitrailer trucks cost $72,000 each. Ace Trucking buys such a truck and agrees to pay for it by a loan that will be amortized with 9 semiannual payments at 6% compounded semiannually.

54. One retailer charges $1048 for a certain computer. A firm of tax accountants buys 8 of these computers. They make a down payment of $1200 and agree to amortize the balance with monthly payments at 12% compounded monthly for 4 years.

55. Joan Varozza plans to borrow $20,000 to stock her small boutique. She will repay the loan with semiannual payments for 5 years at 7% compounded semiannually.

Chapter 5 Summary

Key Terms and Symbols

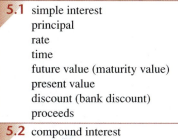

5.1 simple interest
principal
rate
time
future value (maturity value)
present value
discount (bank discount)
proceeds

5.2 compound interest
compound amount

compounding period
continuous compounding
nominal rate (stated rate)
effective rate

5.3 geometric sequence
common ratio
annuity
ordinary annuity
payment period

term of an annuity
future value of an annuity
sinking fund
annuity due

5.4 present value of an annuity
amortize a loan
amortization schedule

Key Concepts

A Strategy for Solving Finance Problems

We have presented a lot of new formulas in this chapter. By answering the following questions, you can decide which formula to use for a particular problem.

1. Is simple or compound interest involved?

   Simple interest is normally used for investments or loans of a year or less; compound interest is normally used in all other cases.

2. If simple interest is being used, what is being sought—interest amount, future value, present value, or discount?

3. If compound interest is being used, does it involve a lump sum (single payment) or an annuity (sequence of payments)?

   (a) For a lump sum,

   (i) Is ordinary compound interest or continuous interest involved?

   (ii) What is being sought—present value, future value, number of periods, or effective rate?

(b) For an annuity,

　(i) Is it an ordinary annuity (payment at the end of each period) or an annuity due (payment at the beginning of each period)?

　(ii) What is being sought—present value, future value, or payment amount?

Once you have answered these questions, choose the appropriate formula and work the problem. As a final step, consider whether the answer you get makes sense. For instance, the amount of interest or the payments in an annuity should be fairly small compared with the total future value.

Key Formulas

List of Variables

r is the annual interest rate.

m is the number of periods per year.

i is the interest rate per period.　　　　$i = \dfrac{r}{m}$

t is the number of years.

n is the number of periods.　　　　$n = tm$

P is the principal or present value.

A is the future value of a lump sum.

S is the future value of an annuity.

R is the periodic payment in an annuity.

Interest

	Simple Interest	Compound Interest
Interest	$I = Prt$	$I = A - P$
Future value	$A = P(1 + rt)$	$A = P(1 + i)^n$
Present value	$P = \dfrac{A}{1 + rt}$	$P = \dfrac{A}{(1 + i)^n} = A(1 + i)^{-n}$

Effective rate　$r_E = \left(1 + \dfrac{r}{m}\right)^m - 1$

Discount

If D is the **discount** on a loan having maturity value A at simple interest rate r for t years, then $D = Art$. If D is the discount and P the **proceeds** of a loan having maturity value A at simple interest rate r for t years, then $P = A - D$　　or　　$P = A(1 - rt)$.

Continuous Interest

If P dollars are deposited for t years at interest rate r per year, compounded continuously, the **compound amount (future value)** is $A = Pe^{rt}$.

The **present value** P of A dollars at interest rate r per year compounded continuously for t years is

$$P = \dfrac{A}{e^{rt}}.$$

Annuities

Ordinary annuity	Future value	$S = R\left[\dfrac{(1 + i)^n - 1}{i}\right] = R \cdot s_{\overline{n}|i}$
	Present value	$P = R\left[\dfrac{1 - (1 + i)^{-n}}{i}\right] = R \cdot a_{\overline{n}|i}$
Annuity due	Future value	$S = R\left[\dfrac{(1 + i)^{n+1} - 1}{i}\right] - R$

Chapter 5 Review Exercises

Find the simple interest for the following loans.

1. $4902 at 6.5% for 11 months
2. $42,368 at 9.22% for 5 months
3. $3478 at 7.4% for 88 days
4. $2390 at 8.7% from May 3 to July 28
5. What is meant by the present value of an amount A?

Find the present value of the following future amounts; use simple interest.

6. $459.57 in 7 months; money earns 5.5%
7. $80,612 in 128 days; money earns 6.77%
8. Explain what happens when a borrower is charged a discount. What are the proceeds?

Find the proceeds in Exercises 9 and 10.

9. $802.34; discount rate 8.2%; length of loan 11 months
10. $12,000; discount rate 7.09%; length of loan 145 days
11. For a given amount of money at a given interest rate for a given period greater than 1 year, does simple interest or compound interest produce more interest? Explain.

Find the compound amount and the amount of interest earned in each of the following.

12. $2800 at 6% compounded annually for 12 years
13. $57,809.34 at 4% compounded quarterly for 6 years
14. $12,903.45 at 6.37% compounded quarterly for 29 quarters
15. $4677.23 at 4.57% compounded monthly for 32 months

Find the present value of the following amounts.

16. $42,000 in 7 years; 12% compounded monthly
17. $17,650 in 4 years; 8% compounded quarterly
18. $1347.89 in 3.5 years; 6.2% compounded semiannually
19. $2388.90 in 44 months; 5.75% compounded monthly
20. Write the first five terms of the geometric sequence with $a = 5$ and $r = 3$.
21. Write the first four terms of the geometric sequence with $a = 16$ and $r = 1/2$.
22. Find the sixth term of the geometric sequence with $a = -3$ and $r = 2$.
23. Find the fifth term of the geometric sequence with $a = -2$ and $r = -2$.
24. Find the sum of the first four terms of the geometric sequence with $a = -3$ and $r = 3$.
25. Find the sum of the first five terms of the geometric sequence with $a = 8000$ and $r = -1/2$.

26. Find $s_{\overline{30}|.02}$.
27. What is meant by the future value of an annuity?

Find the future value of each annuity.

28. $1288 deposited at the end of each year for 14 years; money earns 7% compounded annually
29. $4000 deposited at the end of each quarter for 8 years; money earns 6% compounded quarterly
30. $233 deposited at the end of each month for 4 years; money earns 6% compounded monthly
31. $672 deposited at the beginning of each quarter for 7 years; money earns 5% compounded quarterly
32. $11,900 deposited at the beginning of each month for 13 months; money earns 7% compounded monthly
33. What is the purpose of a sinking fund?

Find the amount of each payment that must be made into a sinking fund to accumulate the following amounts. Assume payments are made at the end of each period.

34. $6500; money earns 5% compounded annually; 6 annual payments
35. $57,000; money earns 6% compounded semiannually for $8\frac{1}{2}$ years
36. $233,188; money earns 5.7% compounded quarterly for $7\frac{3}{4}$ years
37. $56,788; money earns 6.12% compounded monthly for $4\frac{1}{2}$ years

Find the present value of each ordinary annuity.

38. Payments of $850 annually for 4 years at 5% compounded annually
39. Payments of $1500 quarterly for 7 years at 8% compounded quarterly
40. Payments of $4210 semiannually for 8 years at 5.6% compounded semiannually
41. Payments of $877.34 monthly for 17 months at 6.4% compounded monthly
42. Give two examples of the types of loans that are commonly amortized.

Find the amount of the payment necessary to amortize each of the following loans.

43. $32,000 at 8.4% compounded quarterly; 10 quarterly payments
44. $5607 at 7.6% compounded monthly; 32 monthly payments

Find the monthly house payments for the following mortgages.

45. $56,890 at 6.74% for 25 years

46. $77,110 at 8.45% for 30 years

A portion of an amortization table is given below for a $127,000 loan at 8.5% interest compounded monthly for 25 years.

Payment Number	Amount of Payment	Interest for Period	Portion to Principal	Principal at End of Period
0	—	—	—	$127,000.00
1	$1022.64	$899.58	$123.06	126,876.94
2	1022.64	898.71	123.93	126,753.02
3	1022.64	897.83	124.80	126,628.21
4	1022.64	896.95	125.69	126,502.53
5	1022.64	896.06	126.58	126,375.95
6	1022.64	895.16	127.48	126,248.47

Use the table to answer the following questions.

47. How much of the fifth payment is interest?

48. How much of the sixth payment is used to reduce the debt?

49. How much interest is paid in the first 3 months of the loan?

50. How much has the debt been reduced at the end of the first 6 months?

Finance *Work the following applied problems.*

51. Larry DiCenso needs $9812 to buy new equipment for his business. The bank charges a discount rate of 12%. Find the amount of DiCenso's loan if he borrows the money for 7 months.

52. Tom Wilson owes $5800 to his mother. He has agreed to pay back the money in 10 months at an interest rate of 6%. Three months before the loan is due, Tom's mother discounts the loan at the bank to get cash for $5900 worth of new furniture. The bank charges a 10.45% discount rate. How much money does Tom's mother receive? Is it enough to buy the furniture?

53. A firm of attorneys deposits $15,000 of profit-sharing money in an account at 6% compounded semiannually for $7\frac{1}{2}$ years. Find the amount of interest earned.

54. Tom, a graduate student, is considering investing $500 now, when he is 23, or waiting until he is 40 to invest $500. How much more money will he have at age 65 if he invests now, given that he can earn 5% interest compounded quarterly?

55. The graph shows the growth of a $10,000 investment in the Calamos Growth Fund (Class A shares, before sales charges) for a 10-year period from March 31, 1995, to March 31, 2005.*
   (a) What was the rate of return for the entire period?
   (b) If the investment was worth $60,000 on March 31, 2001, what was the rate of return from then until March 31, 2005?

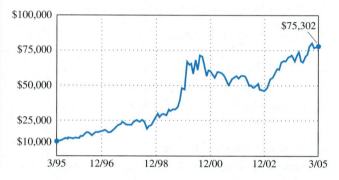

56. According to a financial website, on June 15, 2005, Frontenac Bank of Earth City, Missouri, paid 3.94% interest, compounded quarterly, on a 2-year CD, while E*-TRADE Bank of Arlington, Virginia, paid 3.93% compounded daily.[†] What was the effective rate for the two CDs, and which bank pays a higher effective rate?

57. The Consumer Price Index in May 2005 was approximately double that for May 1982.[‡] What was the inflation rate for that period? (*Hint:* Use continuous compounding.)

58. In 2005, an item that cost $172 in 2000 sold for $194. What inflation rate does this represent? (*Hint:* Use continuous compounding.)

59. Chalon Bridges deposits semiannual payments of $3200, received in payment of a debt, in an ordinary annuity at 6.8% compounded semiannually. Find the final amount in the account and the interest earned at the end of 3.5 years.

60. Each year, a firm must set aside enough funds to provide employee retirement benefits of $52,000 in 20 years. If the firm can invest money at 7.5% compounded monthly, what amount must be invested at the end of each month for this purpose?

61. In 1995, Oseola McCarty donated $150,000 to the University of Southern Mississippi to establish a scholarship fund.[§] What is unusual about her is that the entire amount

*www.calamos.com.

†www.bankrate.com.

‡U.S. Bureau of Labor Statistics.

§New York Times, November 12, 1996, pp. A1, A22.

came from what she was able to save each month from her work as a washerwoman, a job she began in 1916 at the age of 8, when she dropped out of school. How much would Ms. McCarty have to put into her savings account at the end of every 3 months to accumulate $150,000 over 79 years? Assume that she received an interest rate of 5.25% compounded quarterly.

62. Pension experts recommend that you start drawing at least 40% of your full pension as early as possible.* Suppose you have built up a pension with $12,000 annual payments by working 10 years for a company when you leave to accept a better job. The company gives you the option of collecting half the full pension when you reach age 55 or the full pension at age 65. Assume an interest rate of 8% compounded annually. By age 75, how much will each plan produce? Which plan would produce the larger amount?

63. In 3 years, Ms. Nguyen must pay a pledge of $7500 to her favorite charity. What lump sum can she deposit today at 10% compounded semiannually so that she will have enough to pay the pledge?

64. To finance the $15,000 cost of their kitchen remodeling, the Chews will make equal payments at the end of each month for 36 months. They will pay interest at the rate of 7.2% compounded monthly. Find the amount of each payment.

65. To expand her business, the owner of a small restaurant borrows $40,000. She will repay the money in equal payments at the end of each semiannual period for 8 years at 9% interest compounded semiannually. What payments must she make?

66. The Taggart family bought a house for $91,000. They paid $20,000 down and took out a 30-year mortgage for the balance at 9%.

*Smart Money, October 1994, "Pocket That Pension," p. 33.

(a) Find their monthly payment.

(b) How much of the first payment is interest?

After 180 payments, the family sold their house for $136,000. They paid closing costs of $3700 plus 2.5% of the sale price.

(c) Estimate the current mortgage balance at the time of the sale, using one of the methods from Examples 4 and 5 in Section 5.4.

(d) Find the total closing costs.

(e) Find the amount of money they receive from the sale after paying off the mortgage.

67. The proceeds of a $10,000 death benefit are left on deposit with an insurance company for 7 years at an annual effective interest rate of 5%.* The balance at the end of 7 years is paid to the beneficiary in 120 equal monthly payments of X, with the first payment made immediately. During the payout period, interest is credited at an annual effective interest rate of 3%. Calculate X.

(a) 117

(b) 118

(c) 129

(d) 135

(e) 158

68. Gene deposits $500 per quarter to his nest-egg account. The account earns interest of 9%, compounded quarterly. Right after Gene makes his 16th deposit, he loses his job and cannot make any deposits for the next 5 years (20 quarters). Eventually, Gene gets another job and again begins making deposits to his account. Since he missed so many deposits while he was out of work, Gene now deposits $750 per quarter. His first $750 deposit comes

*Problem from "Course 140 Examination, Mathematics of Compound Interest" of the Education and Examination Committee of the Society of Actuaries. Reprinted by permission of the Society of Actuaries.

exactly 20 quarters after his last $500 deposit. What will be Gene's account balance right after he has made his 32nd $750 deposit?*

69. Cathy wants to retire on $55,000 per year for her life expectancy of 20 years. She estimates that she will be able to earn interest of 9%, compounded annually, throughout her lifetime. To reach her retirement goal, Cathy will make annual contributions to her account for the next 25 years. One year after making her last contribution, she will take her first retirement check. How large must her yearly contributions be?

70. The *New York Times* posed a scenario with two individuals, Sue and Joe, who each have $1200 a month to spend on housing and investing. Each takes out a mortgage for $140,000. Sue gets a 30-year mortgage at a rate of 6.625%. Joe gets a 15-year mortgage at a rate of 6.25%. Whatever money is left after the mortgage payment is invested in a mutual fund with a return of 10% annually.[†]

(a) What annual interest rate, compounded monthly, gives an effective annual rate of 10%?

(b) What is Sue's monthly payment?

(c) If, after the payment in part (b), Sue invests the remainder of her $1200 each month in a mutual fund with the interest rate found in part (a), how much money will she have in the fund at the end of 30 years?

(d) What is Joe's monthly payment?

(e) You found in part (d) that Joe has nothing left to invest until his mortgage is paid off. If he then invests the entire $1200 monthly in a mutual fund with the interest rate in part (a), how much money will he have at the end of 30 years (i.e., after 15 years of paying the mortgage and 15 years of investing)?

(f) Who is ahead at the end of the 30 years and by how much?

(g) Discuss to what extent the difference found in part (f) is due to the different interest rates or to the different amounts of time.

*Exercises 68 and 69 supplied by Norman Lindquist, Western Washington University.

[†]*New York Times*, September 27, 1998, p. BU 10.

Case 5

Time, Money, and Polynomials*

A *time line* is often helpful for evaluating complex investments. For example, suppose you buy a $1000 CD at time t_0. After one year, $2500 is added to the CD at t_1. By time t_2, after another year, your money has grown to $3851 with interest. What rate of interest, called *yield to maturity* (*YTM*), did your money earn? A time line for this situation is shown in the next figure.

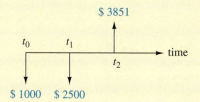

Assuming that interest is compounded annually at a rate i and using the compound interest formula gives the following description of the YTM:

$$1000(1 + i)^2 + 2500(1 + i) = 3851.$$

To determine the yield to maturity, we must solve this equation for i. Since the quantity $1 + i$ is repeated, let $x = 1 + i$ and first solve the second-degree (quadratic) polynomial equation for x.

$$1000x^2 + 2500x - 3851 = 0$$

We can use the quadratic formula with $a = 1000$, $b = 2500$, and $c = -3851$.

$$x = \frac{-2500 \pm \sqrt{2500^2 - 4(1000)(-3851)}}{2(1000)}$$

We get $x = 1.0767$ and $x = -3.5767$. Since $x = 1 + i$, the two values for i are $.0767 = 7.67\%$ and $-4.5767 = -457.67\%$. We reject the negative value because the final accumulation is greater than the sum of the deposits. In some applications, however, negative rates may be meaningful. By checking in the first equation, we see that the yield to maturity for the CD is 7.67%.

---

*From *Time, Money, and Polynomials*, COMAP.

Now let us consider a more complex, but realistic, problem. Suppose Curt Reynolds has contributed for 4 years to a retirement fund. He contributed $6000 at the beginning of the first year. At the beginning of the next 3 years, he contributed $5840, $4000, and $5200, respectively. At the end of the fourth year, he had $29,912.38 in his fund. The interest rate earned by

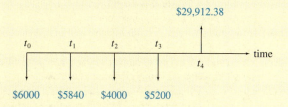

the fund varied between 21% and -3%, so Reynolds would like to know the YTM $= i$ for his hard-earned retirement dollars. Look at the time line in the figure above.

We can set up the following in $1 + i$ for Reynold's savings program:

$$6000(1 + i)^4 + 5840(1 + i)^3 + 4000(1 + i)^2$$
$$+ 5200(1 + i) = 29,912.38.$$

Let $x = 1 + i$. We need to solve the fourth-degree polynomial equation

$$f(x) = 6000x^4 + 5840x^3 + 4000x^2 + 5200x$$
$$- 29,912.38 = 0.$$

If you have a graphing calculator or computer algebra system, you can solve this equation graphically or by using the equation solver. If the only technology you have is an ordinary calculator, you can solve the equation by the guess-and-check method, as follows.

We expect that $0 < i < 1$, so that $1 < x < 2$. Let us calculate $f(1)$ and $f(2)$. If there is a change of sign, we will know that there is a solution to $f(x) = 0$ between 1 and 2. We find that

$$f(1) = -8872.38 \qquad \text{and} \qquad f(2) = 139,207.62.$$

There is a change in sign, as expected. Now we find $f(1.1)$, $f(1.2)$, $f(1.3)$, and so on, and look for another change in sign. Right away we find that

$$f(1.1) = -2794.74 \quad \text{and} \quad f(1.2) = 4620.74.$$

This process can be repeated for values of x between 1.1 and 1.2 to get $f(1.11) = -2116.59$, $f(1.12) = -1424.88$, $f(1.13) = -719.42$, and $f(1.14) = 0$. We were lucky; the solution for $f(x) = 0$ is exactly 1.14, so $i = $ YTM $= .14 = 14\%$.

◆ Exercises

1. Brenda Bravener received $50 on her 16th birthday and $70 on her 17th birthday, both of which she immediately invested in the bank with interest compounded annually. On her 18th birthday, she had $127.40 in her account. Draw a time line, set up a polynomial equation, and calculate the YTM.

2. At the beginning of the year, Jay Beckenstein invested $10,000 at 5% for the first year. At the beginning of the second year, he added $12,000 to the account. The total account earned 4.5% for the second year.
   (a) Draw a time line for this investment.
   (b) How much was in the fund at the end of the second year?
   (c) Set up and solve a polynomial equation, and determine the YTM. What do you notice about the YTM?

3. On January 2 each year for 3 years, Greg Tobin deposited bonuses of $1025, $2200, and $1850, respectively, into an account. He received no bonus the following year, so he made no deposit. At the end of the fourth year, there was $5864.17 in the account.
   (a) Draw a time line for these investments.
   (b) Write a polynomial equation in $x (x = 1 + i)$, and use technology or the guess-and-check method to find the YTM for Greg's investments.

4. Pat McCutcheon invested yearly in a fund for his children's college education. At the beginning of the first year, he invested $1000; at the beginning of the second year, $2000; at the third through the sixth, $2500 each year; and at the beginning of the seventh, $5000. At the beginning of the eighth year, there was $21,259 in the fund.
   (a) Draw a time line for this investment program.
   (b) Write a seventh-degree polynomial equation in $1 + i$ that gives the YTM for this investment program.
   (c) Use a graphing calculator to show that the YTM is less than 5.07% and greater than 5.05%.
   (d) Use a graphing calculator to calculate the solution for $1 + i$ and find the YTM.

5. People often lose money on investments. Christine O'Brien invested $50 at the beginning of each of 2 years in a mutual fund, and at the end of 2 years her investment was worth $90.
   (a) Draw a time line and set up a polynomial equation in $1 + i$. Solve for i.
   (b) Examine each negative solution (rate of return on the investment) to see if it has a reasonable interpretation in the context of the problem. To do this, use the compound interest formula on each value of i to trace each $50 payment to maturity.

Chapter

6

Systems of Linear Equations and Matrices

The structure of some crystals can be described by a large system of linear equations. A variety of resource allocation problems involving many variables can be handled by sovling an appropriate system of linear equations. Technology (such as computers and graphing calculators) is very helpful for handling large systems. Smaller ones can easily be solved by hand. See Exercises 40 and 41 on page 324.

Many applications of mathematics require finding the solution of a *system* of first-degree equations. This chapter presents methods for solving such systems, including matrix methods. Matrix algebra and other applications of matrices are also discussed.

6.1 Systems of Linear Equations

This section deals with **linear** (or **first-degree**) **equations** such as

$$2x + 3y = 14 \qquad \text{linear equation in two variables,}$$
$$4x - 2y + 5z = 8 \qquad \text{linear equation in three variables,}$$

and so on. A **solution** of such an equation is an ordered list of numbers that, when substituted for the variables in the order they appear, produces a true statement. For instance, $(1, 4)$ is a solution of the equation $2x + 3y = 14$ because substituting $x = 1$ and $y = 4$ produces the true statement $2(1) + 3(4) = 14$. Similarly, $(0, -4, 0)$ is a solution of $4x - 2y + 5z = 8$ because $4(0) - 2(-4) + 5(0) = 8$.

Many applications involve **systems of linear equations,** such as these two:

Two equations in two variables

Three equations in four variables

$$
\begin{array}{ll}
5x - 3y = 7 & 2x + y + z = 3 \\
2x + 4y = 8 & x + y + z + w = 5 \\
& -4x + z + w = 0.
\end{array}
$$

A **solution of a system** is a solution that satisfies *all* the equations in the system. For instance, in the right-hand system of equations above, $(1, 0, 1, 3)$ is a solution of all three equations (check it) and hence is a solution of the system. By contrast, $(1, 1, 0, 3)$ is a solution of the first two equations but not the third. Hence, $(1, 1, 0, 3)$ is not a solution of the system. ①

Systems of Two Equations in Two Variables

The graph of a linear equation in two variables is a straight line. Earlier we saw that the coordinates of each point on the graph represent a solution of the equation. Thus, the solution of a system of two such equations is represented by the point or points where the two lines intersect. There are exactly three geometric possibilities for two lines: they intersect at a single point, or they coincide, or they are distinct and parallel. As illustrated in Figure 6.1, each of these geometric possibilities corresponds to an algebraic outcome for which special terminology is used.

In theory, every system of two equations in two variables can be solved by graphing the lines and finding their intersection points (if any). In practice, however, algebraic techniques are easily implemented and usually needed to determine the solutions precisely. One algebraic solution method is **substitution,** which is explained in Example 1.

---

① Which of $(-8, 3, 0)$, $(8, -2, -2)$, $(-6, 5, -2)$ are solutions of the system

$$
\begin{array}{l}
x + 2y + 3z = -2 \\
2x + 6y + z = 2 \\
3x + 3y + 10z = -2?
\end{array}
$$

Answer:

Only $(8, -2, -2)$

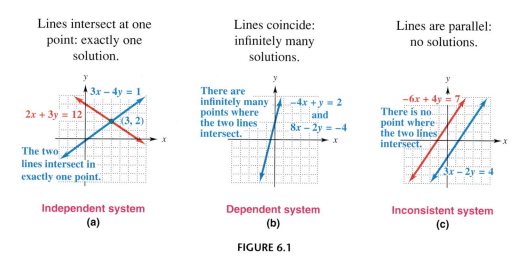

Lines intersect at one point: exactly one solution.

The two lines intersect in exactly one point.

$3x - 4y = 1$

$2x + 3y = 12$

$(3, 2)$

Independent system
(a)

Lines coincide: infinitely many solutions.

There are infinitely many points where the two lines intersect.

$-4x + y = 2$
and
$8x - 2y = -4$

Dependent system
(b)

Lines are parallel: no solutions.

There is no point where the two lines intersect.

$-6x + 4y = 7$

$3x - 2y = 4$

Inconsistent system
(c)

FIGURE 6.1

▼ **EXAMPLE 1**   Solve the system

$$2x - y = 1$$
$$3x + 2y = 4.$$

Solution Begin by solving the first equation for y.

$$2x - y = 1$$
$$-y = -2x + 1 \qquad \text{Subtract } 2x \text{ from both sides.}$$
$$y = 2x - 1 \qquad \text{Multiply both sides by } -1.$$

Substitute this expression for y in the second equation and solve for x.

$$3x + 2y = 4$$
$$3x + 2(2x - 1) = 4 \qquad \text{Substitute } 2x - 1 \text{ for } y.$$
$$3x + 4x - 2 = 4 \qquad \text{Multiply out the left side.}$$
$$7x - 2 = 4 \qquad \text{Combine like terms.}$$
$$7x = 6 \qquad \text{Add 2 to both sides.}$$
$$x = 6/7 \qquad \text{Divide both sides by 7.}$$

Therefore, every solution of the system must have $x = 6/7$. To find the corresponding solution for y, substitute $x = 6/7$ in one of the two original equations and solve for y. We shall use the first equation.

$$2x - y = 1$$
$$2\left(\frac{6}{7}\right) - y = 1 \qquad \text{Substitute 6/7 for } x.$$
$$\frac{12}{7} - y = 1 \qquad \text{Multiply out the left side.}$$
$$-y = -\frac{12}{7} + 1 \qquad \text{Subtract 12/7 from both sides.}$$
$$y = \frac{12}{7} - 1 = \frac{5}{7} \qquad \text{Multiply both sides by } -1.$$

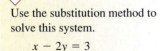

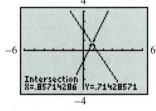

Use the substitution method to
solve this system.

$$x - 2y = 3$$
$$2x + 3y = 13$$

Answer:

$(5, 1)$

Hence, the solution of the original system is $x = 6/7$ and $y = 5/7$. We would have obtained the same solution if we had substituted $x = 6/7$ in the second equation of the original system, as you can easily verify. ▲ ②

The substitution method is useful when at least one of the equations has a variable with coefficient 1. That is why we solved for y in the first equation of Example 1. If we had solved for x in the first equation or for x or y in the second, we would have had a lot more fractions to deal with.

TECHNOLOGY TIP   A graphing calculator can be used to solve systems of two equations in two variables. Solve each of the given equations for y and graph them on the same screen. In Example 1, for instance we would graph

$$y_1 = 2x - 1 \quad \text{and} \quad y_2 = \frac{-3x + 4}{2}.$$

Use the intersection finder to determine the point where the graphs intersect (the solution of the system), as shown in Figure 6.2. Note that this is an *approximate* solution, rather than the exact solution found algebraically in Example 1.

FIGURE 6.2

The **elimination method** of solving systems of linear equations is often more convenient than substitution, as illustrated in the next four examples.

▼ **EXAMPLE 2**   Solve the system

$$5x + y = 4$$
$$3x + 2y = 1.$$

Solution Multiply the first equation by -2, so that the coefficients of y in the two equations are negatives of each other.

$$-10x - 2y = -8 \qquad \text{First equation multiplied by } -2.$$
$$3x + 2y = 1$$

Any solution of this system of equations must also be a solution of the sum of the two equations.

$$
\begin{array}{rl}
-10x - 2y = & -8 \\
\underline{3x + 2y = 1} & \\
-7x = & -7 \qquad \text{Sum; variable } y \text{ is eliminated.} \\
x = & 1 \qquad \text{Divide both sides by } -7.
\end{array}
$$

To find the corresponding value of y, substitute $x = 1$ in one of the original equations, say the first one:

$$5x + y = 4$$
$$5(\mathbf{1}) + y = 4 \qquad \text{Substitute 1 for } x.$$
$$y = -1. \qquad \text{Subtract 5 from both sides.}$$

3

Use the elimination method to solve this system.

$$x + 2y = 4$$
$$3x - 4y = -8$$

Answer:

$(0, 2)$

Therefore, the solution of the original system is $(1, -1)$. ▲ **3**

▼ **EXAMPLE 3** Solve the system

$$3x - 4y = 1$$
$$2x + 3y = 12.$$

Solution Multiply the first equation by 2 and the second equation by -3 to get

$$6x - 8y = 2$$
$$-6x - 9y = -36.$$

The multipliers 2 and -3 were chosen so that the coefficients of x in the two equations would be negatives of each other. Any solution of both these equations must also be a solution of their sum:

$$6x - 8y = 2$$
$$\underline{-6x - 9y = -36}$$
$$-17y = -34 \qquad \text{Sum; variable } x \text{ is eliminated.}$$
$$y = 2$$

To find the corresponding value of x, substitute 2 for y in either of the original equations. We choose the first equation.

$$3x - 4(2) = 1$$
$$3x - 8 = 1$$
$$x = 3$$

Therefore, the solution of the system is $(3, 2)$. The graphs of both equations of the system are shown in Figure 6.1(a). They intersect at the point $(3, 2)$, the solution of the system. ▲ **4**

4

Solve the system of equations

$$3x + 2y = -1$$
$$5x - 3y = 11.$$

Draw the graph of each equation on the same axes.

Answer:

$(1, -2)$

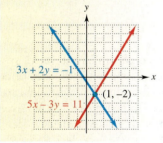

▼ **EXAMPLE 4** Solve the system

$$-4x + y = 2$$
$$8x - 2y = -4.$$

Solution Eliminate x by multiplying both sides of the first equation by 2 and adding the results to the second equation.

$$-8x + 2y = 4$$
$$\underline{8x - 2y = -4}$$
$$0 = 0 \qquad \text{Both variables are eliminated.}$$

Although both variables have been eliminated, the resulting statement "$0 = 0$" is true, which is the algebraic indication that the two equations have the same graph, as shown in Figure 6.1(b). Therefore, the system is dependent and has an infinite number of solutions: every ordered pair that is a solution of the equation $-4x + y = 2$ is a solution of the system. ▲ **5**

5

Solve the following system.

$$3x - 4y = 13$$
$$12x - 16y = 52$$

Answer:

All ordered pairs that satisfy the equation $3x - 4y = 13$ (or $12x - 16y = 52$)

▼ **EXAMPLE 5** Solve the system

$$3x - 2y = 4$$
$$-6x + 4y = 7.$$

Solution The graphs of these equations are parallel lines (each has slope 3/2), as shown in Figure 6.1(c). Therefore, the system has no solution. However, you do not need the graphs to discover this fact. If you try to solve the system algebraically by multiplying both sides of the first equation by 2 and adding the results to the second equation, you obtain

$$
\begin{array}{r}
6x - 4y = 8 \\
-6x + 4y = 7 \\
\hline
0 = 15.
\end{array}
$$

The false statement "$0 = 15$" is the algebraic signal that the system is inconsistent and has no solution. ▲ ⟨6⟩

Larger Systems of Linear Equations

Two systems of equations are said to be **equivalent** if they have the same solutions. The basic procedure for solving a large system of equations is to use properties of algebra to transform the system into a simpler, equivalent system and then solve this simpler system

Performing any one of the following **elementary operations** on a system of linear equations produces an equivalent system:

1. Interchange any two equations.

2. Multiply both sides of an equation by a nonzero constant.

3. Replace an equation by the sum of itself and a constant multiple of another equation in the system.

Performing either of the first two elementary operations produces an equivalent system because rearranging the order of the equations or multiplying both sides of an equation by a constant does not affect the solutions of the individual equations, and hence it does not affect the solutions of the system. No formal proof will be given here that performing the third elementary operation produces an equivalent system, but side problem 7 illustrates this fact. ⟨7⟩

Example 6 shows how to use elementary operations on a system to eliminate certain variables and produce an equivalent system that is easily solved. The italicized statements provide an outline of the procedure. Here and below we use R_1 to denote the first equation in a system, R_2 the second equation, and so on.

▼ **EXAMPLE 6** Use the elimination method to solve the system

$$2x + y - z = 2$$
$$x + 3y + 2z = 1$$
$$x + y + z = 2.$$

⟨6⟩

Solve the system

$$x - y = 4$$
$$2x - 2y = 3.$$

Draw the graph of each equation on the same axes.

Answer:

No solution

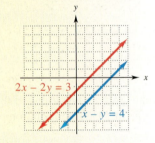

⟨7⟩

Verify that $x = 2$, $y = 1$ is the solution of the system

$$x - 3y = -1$$
$$3x + 2y = 8.$$

(a) Replace the second equation by the sum of itself and -3 times the first equation.

(b) What is the solution of the system in part (a)?

Answers:

(a) The system becomes
$$x - 3y = -1$$
$$11y = 11.$$

(b) $x = 2$, $y = 1$

Solution First, *use elementary operations to produce an equivalent system in which 1 is the coefficient of x in the first equation.* One way to do this is to interchange the first two equations (another would be to multiply both sides of the first equation by $1/2$).

$$x + 3y + 2z = 1 \qquad \text{Interchange } R_1, R_2.$$
$$2x + y - z = 2$$
$$x + y + z = 2$$

Next, *use elementary operations to produce an equivalent system in which the x-term has been eliminated from the second and third equations.* To eliminate the *x*-term from the second equation, replace the second equation by the sum of itself and -2 times the first equation.

$$\begin{array}{r|rcl} -2R_1 & -2x - 6y - 4z & = & -2 \\ R_2 & 2x + y - z & = & 2 \\ \hline -2R_1 + R_2 & -5y - 5z & = & 0 \end{array}$$

$$x + 3y + 2z = 1$$
$$-5y - 5z = 0 \qquad -2R_1 + R_2 \leftarrow$$
$$x + y + z = 2$$

To eliminate the *x*-term from the third equation of this last system, replace the third equation by the sum of itself and -1 times the first equation.

$$\begin{array}{r|rcl} -1R_1 & -x - 3y - 2z & = & -1 \\ R_3 & x + y + z & = & 2 \\ \hline -1R_1 + R_3 & -2y - z & = & 1 \end{array}$$

$$x + 3y + 2z = 1$$
$$-5y - 5z = 0$$
$$-2y - z = 1 \qquad -1R_1 + R_3 \leftarrow$$

Now that *x* has been eliminated from all but the first equation, we ignore the first equation and work on the remaining ones. *Use elementary operations to produce an equivalent system in which 1 is the coefficient of y in the second equation.* This can be done by multiplying the second equation in the system by $-1/5$.

$$x + 3y + 2z = 1$$
$$y + z = 0 \qquad -\tfrac{1}{5}R_2$$
$$-2y - z = 1$$

Then *use elementary operations to obtain an equivalent system in which y has been eliminated from the third equation.* Replace the third equation by the sum of itself and 2 times the second equation:

$$x + 3y + 2z = 1$$
$$y + z = 0$$
$$z = 1. \qquad 2R_2 + R_3$$

The solution of the third equation is obvious: $z = 1$. Now work backward in the system. Substitute 1 for z in the second equation and solve for y, obtaining $y = -1$. Finally, substitute 1 for z and -1 for y in the first equation and solve for x, obtaining $x = 2$. This process is known as **back substitution.** When it is finished, we have the solution of the original system, namely, $(2, -1, 1)$. It is always wise to check the solution by substituting the values for x, y, and z in *all* equations of the original system. ▲

The procedure used in Example 6 to eliminate variables and produce a system in which back substitution works can be carried out with any system, as summarized below. In this summary, the first variable that appears in an equation with nonzero coefficient is called the **leading variable** of that equation, and its nonzero coefficient is called the **leading coefficient.**

The Elimination Method for Solving Large Systems of Linear Equations

Use elementary operations to transform the given system into an equivalent one as follows:

1. Make the leading coefficient of the first equation 1 either by interchanging equations or by multiplying the first equation by a suitable constant.

2. Eliminate the leading variable of the first equation from each later equation by replacing the later equation by the sum of itself and a suitable multiple of the first equation.

3. Repeat Steps 1 and 2 for the second equation: make its leading coefficient 1 and eliminate its leading variable from each later equation by replacing the later equation by the sum of itself and a suitable multiple of the second equation.

4. Repeat Steps 1 and 2 for the third equation, fourth equation, and so on, until it is not possible to go any further.

Then solve the resulting system by back substitution.

At various stages in the elimination process, you may have a choice of elementary operations that can be used. As long as the final result is a system in which back substitution can be used, the choice does not matter. To avoid unnecessary errors, choose elementary operations that minimize the amount of computation and, as far as possible, avoid complicated fractions. ⟨8⟩

⟨8⟩

Use the elimination method to solve each system.

(a) $2x + y = -1$

$x + 3y = 2$

(b) $2x - y + 3z = 2$

$x + 2y - z = 6$

$-x - y + z = -5$

Answers:

(a) $(-1, 1)$

(b) $(3, 1, -1)$

Matrix Methods

You may have noticed that the variables in a system of equations remain unchanged during the solution process. So we need to keep track of only the coefficients and the constants. For instance, consider the system in Example 6:

$$2x + y - z = 2$$
$$x + 3y + 2z = 1$$
$$x + y + z = 2.$$

This system can be written in an abbreviated form as

$$\begin{bmatrix} 2 & 1 & -1 & 2 \\ 1 & 3 & 2 & 1 \\ 1 & 1 & 1 & 2 \end{bmatrix}.$$

Such a rectangular array of numbers, consisting of horizontal **rows** and vertical **columns,** is called a **matrix** (plural, **matrices**). Each number in the array is an **element** or **entry.** To separate the constants in the last column of the matrix from the coefficients of the variables, we sometimes use a vertical line, producing the following **augmented matrix:**

$$\left[\begin{array}{ccc|c} 2 & 1 & -1 & 2 \\ 1 & 3 & 2 & 1 \\ 1 & 1 & 1 & 2 \end{array}\right].$$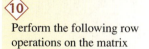

The rows of the augmented matrix can be transformed in the same way as the equations of the system, since the matrix is just a shortened form of the system. The following **row operations** on the augmented matrix correspond to the elementary operations used on systems of equations.

Performing any one of the following **row operations** on the augmented matrix of a system of linear equations produces the augmented matrix of an equivalent system:

1. Interchange any two rows.

2. Multiply each element of a row by a nonzero constant.

3. Replace a row by the sum of itself and a constant multiple of another row of the matrix.

Row operations on a matrix are indicated by the same notation we used for elementary operations on a system of equations. For example, $2R_3 + R_1$ indicates the sum of 2 times row 3 and row 1.

▼ **EXAMPLE 7** Use matrices to solve the system

$$x - 2y = 6 - 4z$$
$$x + 13z = 6 - y$$
$$-2x + 6y - z = -10.$$

Solution First, put the system in the required form, with the constants on the right side of the equals sign and the terms with variables *in the same order* in each equation on the left side of the equals sign. Then write the augmented matrix of the system.

$$\begin{array}{c} x - 2y + 4z = 6 \\ x + y + 13z = 6 \\ -2x + 6y - z = -10 \end{array} \qquad \left[\begin{array}{ccc|c} 1 & -2 & 4 & 6 \\ 1 & 1 & 13 & 6 \\ -2 & 6 & -1 & -10 \end{array}\right]$$

The matrix method is the same as the elimination method, except that row operations are used on the augmented matrix instead of elementary operations on the corresponding system of equations, as shown in this side-by-side comparison.

9

(a) Write the augmented matrix of this system.

$$4x - 2y + 3z = 4$$
$$3x + 5y + z = -7$$
$$5x - y + 4z = 6$$

(b) Write the system of equations associated with this augmented matrix.

$$\left[\begin{array}{cc|c} 2 & -2 & -2 \\ 1 & 1 & 4 \\ 3 & 5 & 8 \end{array}\right]$$

Answers:

(a) $\left[\begin{array}{ccc|c} 4 & -2 & 3 & 4 \\ 3 & 5 & 1 & -7 \\ 5 & -1 & 4 & 6 \end{array}\right]$

(b) $2x - 2y = -2$
$x + y = 4$
$3x + 5y = 8$

10

Perform the following row operations on the matrix

$$\left[\begin{array}{cc} -1 & 5 \\ 3 & -2 \end{array}\right].$$

(a) Interchange R_1 and R_2.

(b) $2R_1$

(c) Replace R_2 by $-3R_1 + R_2$.

(d) Replace R_1 by $2R_2 + R_1$.

Answers:

(a) $\left[\begin{array}{cc} 3 & -2 \\ -1 & 5 \end{array}\right]$

(b) $\left[\begin{array}{cc} -2 & 10 \\ 3 & -2 \end{array}\right]$

(c) $\left[\begin{array}{cc} -1 & 5 \\ 6 & -17 \end{array}\right]$

(d) $\left[\begin{array}{cc} 5 & 1 \\ 3 & -2 \end{array}\right]$

Equation Method	**Matrix Method**

Replace the second equation by the sum of itself and -1 times the first equation.

$$\begin{aligned} x - 2y + 4z &= 6 \\ 3y + 9z &= 0 \\ -2x + 6y - z &= -10 \end{aligned}$$

$\leftarrow -1R_1 + R_2 \rightarrow$

$$\begin{bmatrix} 1 & -2 & 4 & 6 \\ 0 & 3 & 9 & 0 \\ -2 & 6 & -1 & -10 \end{bmatrix}$$

Replace the second row by the sum of itself and -1 times the first row.

Replace the third equation by the sum of itself and 2 times the first equation.

$$\begin{aligned} x - 2y + 4z &= 6 \\ 3y + 9z &= 0 \\ 2y + 7z &= 2 \end{aligned}$$

$\leftarrow 2R_1 + R_3 \rightarrow$

$$\begin{bmatrix} 1 & -2 & 4 & 6 \\ 0 & 3 & 9 & 0 \\ 0 & 2 & 7 & 2 \end{bmatrix}$$

Replace the third row by the sum of itself and 2 times the first row.

Multiply both sides of the second equation by $1/3$.

$$\begin{aligned} x - 2y + 4z &= 6 \\ y + 3z &= 0 \\ 2y + 7z &= 2 \end{aligned}$$

$\leftarrow \frac{1}{3}R_2 \rightarrow$

$$\begin{bmatrix} 1 & -2 & 4 & 6 \\ 0 & 1 & 3 & 0 \\ 0 & 2 & 7 & 2 \end{bmatrix}$$

Multiply each element of row 2 by $1/3$.

Replace the third equation by the sum of itself and -2 times the second equation.

$$\begin{aligned} x - 2y + 4z &= 6 \\ y + 3z &= 0 \\ z &= 2 \end{aligned}$$

$\leftarrow -2R_2 + R_3 \rightarrow$

$$\begin{bmatrix} 1 & -2 & 4 & 6 \\ 0 & 1 & 3 & 0 \\ 0 & 0 & 1 & 2 \end{bmatrix}$$

Replace the third row by the sum of itself and -2 times the second row.

Now use back substitution.

$$z = 2 \qquad y + 3(2) = 0 \qquad x - 2(-6) + 4(2) = 6$$
$$y = -6 \qquad x + 20 = 6$$
$$x = -14$$

The solution of the system is $(-14, -6, 2)$. ▲

11 Complete the matrix solution of the system with this augmented matrix:

$$\begin{bmatrix} 1 & 1 & 1 & 2 \\ 1 & -2 & 1 & -1 \\ 0 & 3 & 1 & 5 \end{bmatrix}.$$

Answer:

$(-1, 1, 2)$

A matrix, such as the last one in Example 7, is said to be in **row echelon form** when

All rows consisting entirely of zeros (if any) are at the bottom.

The first nonzero entry in each row is 1 (called a *leading* 1).

Each leading 1 appears to the right of the leading 1's in any preceding rows.

When a row echelon form matrix is the augmented matrix of a system of equations, as in Example 7, the system can readily be solved by back substitution. So the matrix solution method amounts to transforming the augmented matrix of a system of equations into a row echelon form matrix.

Virtually all graphing calculators have matrix capabilities that make it easy to solve systems of linear equations.

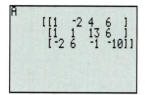

FIGURE 6.3

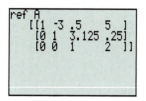

FIGURE 6.4

▼ **EXAMPLE 8**   Use a graphing calculator to solve the system in Example 7.

Solution Enter the augmented matrix of the system into the calculator (Figure 6.3) Then use the REF key (in the MATH or OPS submenu of the MATRIX menu) to put this matrix into row echelon form. Figure 6.4 shows the resulting row echelon form matrix; the corresponding system of equations is:

$$x - 3y + \quad .5z = 5$$
$$y + 3.125z = .25$$
$$z = 2$$

Because the calculator used a different sequence of row operations than was used in Example 7, it produced a different row echelon form matrix (and corresponding system). However, back substitution shows that the solutions of this system are the same ones found in Example 5.

$z = 2$	$y + 3.125(2) = .25$	$x - 3(-6) + .5(2) = 5$
	$y + 6.25 = .25$	$x + 18 + 1 = 5$
	$y = -6$	$x = -14$ ▲

Dependent and Inconsistent Systems

Recall that a system of equations in two variables may have exactly one solution, or an infinite number of solutions, or no solutions at all. This fact was illustrated geometrically in Figure 6.1. The same thing is true for systems with three or more variables (and the same terminology is used): a system has exactly one solution (an **independent system**), or an infinite number of solutions (a **dependent system**), or no solutions at all (an **inconsistent system**). Both the equation method and the matrix method always produce the unique solution of an independent system. The matrix method also provides a useful way of describing the infinitely many solutions of a dependent system, as we now see.

▼ **EXAMPLE 9**   Solve the system

$$2x - 3y + 4z = 6$$
$$x - 2y + \quad z = 9.$$

Solution Use the steps of the matrix method as far as possible, beginning with the augmented matrix of the system.

$$\begin{bmatrix} 2 & -3 & 4 & | & 6 \\ 1 & -2 & 1 & | & 9 \end{bmatrix}$$

$$\begin{bmatrix} 1 & -2 & 1 & | & 9 \\ 2 & -3 & 4 & | & 6 \end{bmatrix} \qquad \text{Interchange } R_1 \text{ and } R_2.$$

$$\begin{bmatrix} 1 & -2 & 1 & | & 9 \\ 0 & 1 & 2 & | & -12 \end{bmatrix} \qquad -2R_1 + R_2. \qquad \begin{aligned} x - 2y + \quad z &= 9 \\ y + 2z &= -12 \end{aligned}$$

The last augmented matrix above represents the system shown to its right. Since there are only two rows in the matrix, it is not possible to continue the process. The fact that the corresponding system has one variable (namely, z) that is not the leading variable of an equation indicates a dependent system. Its solutions can be found as follows. Solve the second equation for y.

$$y = -2z - 12$$

Now substitute the result for y in the first equation and solve for x.

$$x - 2y + z = 9$$
$$x - 2(-2z - 12) + z = 9$$
$$x + 4z + 24 + z = 9$$
$$x + 5z = -15$$
$$x = -5z - 15$$

Each choice of a value for z leads to values for x and y. For example,

$$\text{if } z = 1, \quad \text{then } x = -20 \quad \text{and} \quad y = -14;$$
$$\text{if } z = -6, \quad \text{then } x = -15 \quad \text{and} \quad y = 0;$$
$$\text{if } z = 0, \quad \text{then } x = -15 \quad \text{and} \quad y = -12.$$

There are infinitely many solutions of the original system, since z can take on infinitely many values. The solutions are all ordered triples of the form

$$(-5z - 15, -2z - 12, z),$$

where z is any real number. ▲ ⟨12⟩

⟨12⟩

Use the following values of z to find additional solutions for the system of Example 9.

(a) $z = 7$

(b) $z = -14$

(c) $z = 5$

Answers:

(a) $(-50, -26, 7)$

(b) $(55, 16, -14)$

(c) $(-40, -22, 5)$

Since both x and y in Example 9 were expressed in terms of z, the variable z is called a **parameter.** If we solved the system in a different way, x or y could be the parameter. The system in Example 9 had one more variable than equations. If there are two more variables than equations, there usually will be two parameters, and so on.

▼ EXAMPLE 10   Row operations were used to reduce the augmented matrix of a system of three equations in four variables $(x, y, z, \text{and } w)$ to the row echelon form matrix in Figure 6.5. Solve the system.

Solution   First write out the system represented by the matrix.

$$x + 2y + 4z - w = 0$$
$$y - 2z + 3w = 2$$
$$z - 2w = 1$$

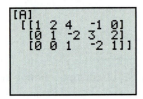

FIGURE 6.5

Let w be the parameter and use back substitution.

$$z = 2w + 1 \quad y - 2(2w + 1) + 3w = 2 \qquad x + 2(w + 4) + 4(2w + 1) - w = 0$$
$$y - 4w - 2 + 3w = 2 \qquad\qquad x + 2w + 8 + 8w + 4 - w = 0$$
$$y = w + 4 \qquad\qquad\qquad x = -9w - 12$$

The solutions are given by $(-9w - 12, w + 4, 2w + 1, w)$, where w is any real number. ▲

When a system is inconsistent, the matrix method will indicate this fact, too, as in the next example.

▼ **EXAMPLE 11** Solve the system

$$4x + 12y + 8z = -4$$
$$2x + 8y + 5z = 0$$
$$3x + 9y + 6z = 2.$$

Solution Write the augmented matrix and go through the steps of the matrix method.

$$\begin{bmatrix} 4 & 12 & 8 & | & -4 \\ 2 & 8 & 5 & | & 0 \\ 3 & 9 & 6 & | & 2 \end{bmatrix}$$

$$\begin{bmatrix} 1 & 3 & 2 & | & -1 \\ 2 & 8 & 5 & | & 0 \\ 3 & 9 & 6 & | & 2 \end{bmatrix} \quad (1/4)R_1$$

$$\begin{bmatrix} 1 & 3 & 2 & | & -1 \\ 0 & 2 & 1 & | & 2 \\ 3 & 9 & 6 & | & 2 \end{bmatrix} \quad -2R_1 + R_2$$

$$\begin{bmatrix} 1 & 3 & 2 & | & -1 \\ 0 & 2 & 1 & | & 2 \\ 0 & 0 & 0 & | & 5 \end{bmatrix} \quad -3R_1 + R_3$$

Stop! The last row has all zeros to the left of the vertical bar, so the corresponding equation is $0x + 0y + 0z = 5$. This equation has no solution, because its left side is 0 and its right side is 5, resulting in the false statement "$0 = 5$." Therefore, the entire system cannot have any solutions and is inconsistent. ▲ ⟨13⟩

⟨13⟩

Complete the matrix solution of the system.

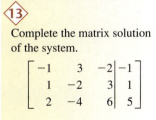

$$\begin{bmatrix} -1 & 3 & -2 & | & -1 \\ 1 & -2 & 3 & | & 1 \\ 2 & -4 & 6 & | & 5 \end{bmatrix}$$

Answer:

No solution

CAUTION It is possible for the elimination method to result in an equation that is 0 on *both* sides, such as $0x + 0y + 0z = 0$. Unlike the situation in Example 11, such an equation has infinitely many solutions, so a system that contains it may also have solutions.

Applications

To apply the mathematical techniques in this text to practical problems, begin by reading the problem carefully. Next, identify what must be found. Let each unknown quantity be represented by a variable. (It is a good idea to *write down* exactly what each variable represents.) Now reread the problem, looking for all necessary data. Write that down, too. Finally, look for one or more sentences that lead to equations or inequalities.

▼ **EXAMPLE 12** Kelly Karpet Kleaners sells rug-cleaning machines. The EZ model weighs 10 pounds and comes in a 10-cubic-foot box. The compact model weighs 20 pounds and comes in an 8-cubic-foot box. The commercial model weighs 60 pounds and comes in a 28-cubic-foot box. Each of Kelly's delivery vans has 248 cubic feet of space and can hold a maximum of 440 pounds. In order for a van to be fully loaded, how many of each model should it carry?

Solution Let x be the number of EZ, y the number of compact, and z the number of commercial models carried by a van. Then we can summarize the given information in this table.

Model	Number	Weight	Volume
EZ	x	10	10
Compact	y	20	8
Commercial	z	60	28
Total for a load		440	248

Since a fully loaded van can carry 440 pounds and 248 cubic feet, we must solve this system of equations:

$$10x + 20y + 60z = 440 \quad \text{Weight equation}$$
$$10x + 8y + 28z = 248. \quad \text{Volume equation}$$

The augmented matrix of the system is

$$\begin{bmatrix} 10 & 20 & 60 & | & 440 \\ 10 & 8 & 28 & | & 248 \end{bmatrix}.$$

Transform it into row echelon form, either by using a graphing calculator or by performing row operations by hand, as in problem 14 at the side. ⬦14⬦

A graphing calculator or problem 14 at the side shows that the original system is equivalent to the following one.

$$x + 2y + 6z = 44$$
$$y + \frac{8}{3}z = 16.$$

Solving this dependent system by back substitution, we get

$$y = 16 - \frac{8}{3}z$$

$$x = 44 - 2y - 6z = 44 - 2\left(16 - \frac{8}{3}z\right) - 6z = 12 - \frac{2}{3}z,$$

so that all solutions of the system are given by $\left(12 - \frac{2}{3}z, 16 - \frac{8}{3}z, z\right)$. The only solutions that apply in this situation, however, are those given by $z = 0, 3,$ or 6, because all other values of z lead to fractions or negative numbers. (You can't deliver part of a box or a negative number of boxes). Hence, there are three ways to have a fully loaded van.

Solution	Van Load
$(12, 16, 0)$	12 EZ, 16 compact, 0 commercial
$(10, 8, 3)$	10 EZ, 8 compact, 3 commercial
$(8, 0, 6)$	8 EZ, 0 compact, 6 commercial ▲

⬦14⬦

List a sequence of row operations that will transform the augmented matrix in Example 10 into row echelon form.

Answer:

Many sequences are possible, including this one:

replace R_1 by $\frac{1}{10}R_1$;

replace R_2 by $\frac{1}{2}R_2$.

replace R_2 by $-5R_1 + R_2$;

replace R_2 by $-\frac{1}{6}R_2$.

$$\begin{bmatrix} 1 & 2 & 6 & | & 44 \\ 0 & 1 & \frac{8}{3} & | & 16 \end{bmatrix}$$

◆6.1 Exercises

Determine whether the given ordered list of numbers is a solution of the system of equations.

1. $(-1, 3)$
$$2x + y = 1$$
$$-3x + 2y = 9$$

2. $(2, 1.5, -.5)$
$$3x + 4y - 2z = -.5$$
$$.5x \qquad + 8z = -3$$
$$x - 3y + 5z = -5$$

Use substitution to solve each system. (See Example 1.)

3. $3x - y = 1$
$\quad x + 2y = -9$

4. $x + y = 7$
$\quad x - 2y = -5$

5. $3x - 2y = 4$
$\quad 2x + y = -1$

6. $5x - 3y = -2$
$\quad -x - 2y = 3$

7. $r + s = 0$
$\quad r - s = 5$

8. $t = 3u + 5$
$\quad t = u + 5$

Use elimination to solve each system. (See Examples 2–5.)

9. $x - 2y = 5$
$\quad 2x + y = 3$

10. $3x - y = 1$
$\quad -x + 2y = 4$

11. $2x - 2y = 12$
$\quad -2x + 3y = 10$

12. $3x + 2y = -4$
$\quad 4x - 2y = -10$

13. $x + 3y = -1$
$\quad 2x - y = 5$

14. $4x - 3y = -1$
$\quad x + 2y = 19$

15. $2x + 3y = 15$
$\quad 8x + 12y = 40$

16. $2x + 5y = 8$
$\quad 6x + 15y = 18$

17. $2x - 8y = 2$
$\quad 3x - 12y = 3$

18. $3x - 2y = 4$
$\quad 6x - 4y = 8$

19. $3x + 2y = 5$
$\quad 6x + 4y = 8$

20. $9x - 5y = 1$
$\quad -18x + 10y = 1$

21. Only one of the three screens below gives the correct graphs for the system in Exercise 14. Which is it? (*Hint:* Solve for y first in each equation and use the slope-intercept form to help you answer the question.)

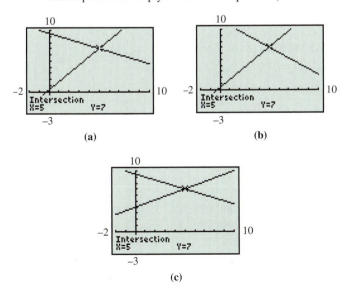

(a)

(b)

(c)

In Exercises 22 and 23, multiply both sides of each equation by a common denominator to eliminate the fractions. Then solve the system.

22. $\dfrac{x}{2} + \dfrac{y}{3} = 8$

$\dfrac{2x}{3} + \dfrac{3y}{2} = 17$

23. $\dfrac{x}{5} + 3y = 31$

$2x - \dfrac{y}{5} = 8$

24. **Business** When Neil Simon opens a new play, he has to decide whether to open the show on Broadway or off Broadway. For example, he decided to open his play *London Suite* off Broadway. From information provided by Emanuel Azenberg, his producer, the following equations were developed:

$$43,500x - y = 1,295,000$$
$$27,000x - y = 440,000,$$

where x represents the number of weeks that the show has run and y represents the profit or loss from the show (first equation is Broadway and second equation is off Broadway).*

(a) Solve this system of equations to determine when the profit/loss from the show will be equal for each venue. What is the profit?

(b) Discuss which venue is favorable for the show.

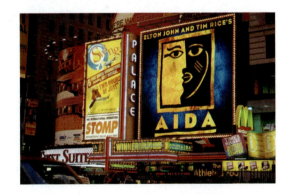

25. **Social Sciences** One of the factors that contribute to the success or failure of a particular army during war is its ability to get new troops ready for service. It is possible to analyze the rate of change in the number of troops of two hypothetical armies with the simplified model

Rate of increase (Red Army) $= 200,000 - .5r - .3b$

Rate of increase (Blue Army) $= 350,000 - .5r - .7b,$

where r is the number of soldiers in the Red Army at a given time and b is the number of soldiers in the Blue

Army at the same time. The factors .5 and .7 represent each army's efficiency at bringing new soldiers into the fight.*

(a) Solve this system of equations to determine the number of soldiers in each army when the rate of increase for each is zero.

(b) Describe what might be going on in a war when the rate of increase is zero.

26. **Social Science**   The population y in year x of Long Beach and New Orleans is approximated by the equations

$$Long\ Beach: \qquad -9.92x + 2y = 722$$
$$New\ Orleans: \qquad 3.87x + \ \ y = 558,$$

where $x = 0$ corresponds to 1980 and y is in thousands.[†] In what year do the two cities have the same population?

27. **Finance**   On the basis of data from 1990 to 2002, the median income y in year x for men and women is approximated by the equations

$$Men: \qquad -1490x + 2y = 40,586$$
$$Women: \qquad -1686x + 3y = 30,210,$$

where $x = 0$ corresponds to 1990 and y is in constant 2001 dollars.[‡] If these equations remain valid in the future, when will the median income of men and women be the same?

28. **Health**   The death rate per 100,000 population y in year x from heart disease and cancer is approximated by the equations

$$Heart\ Disease: \qquad 13.6x + 2y = 644$$
$$Cancer: \qquad 1.8x + \ \ y = 216,$$

where $x = 0$ corresponds to 1990.[§] If these equations remain accurate, when will the death rates for heart disease and cancer be the same?

29. **Finance**   Shirley Cicero has \$16,000 invested in Boeing and GE stock. The Boeing stock currently sells for \$30 a share and the GE stock for \$70 a share. If GE stock triples in value and Boeing stock goes up 50%, her stock will be worth \$34,500. How many shares of each stock does she own?

30. **Business**   An apparel shop sells skirts for \$45 and blouses for \$35. Its entire stock is worth \$51,750, but sales are slow and only half the skirts and two-thirds of the blouses are sold, for a total of \$30,600. How many skirts and blouses are left in the store?

*Ian Bellamy, "Modeling War," *Journal of Peace Research* 36, no. 6 (1999): 729–739. ©1999 Sage Publications, Ltd.

†U.S. Census Bureau.

‡U.S. Census Bureau.

§U.S. Department of Health and Human Services.

31. **Business**   A company produces two models of bicycles: model 201 and model 301. Model 201 requires 2 hours of assembly time, and model 301 requires 3 hours of assembly time. The parts for model 201 cost \$25 per bike, and the parts for model 301 cost \$30 per bike. If the company has a total of 34 hours of assembly time and \$365 available per day for these two models, how many of each can be made in a day?

32. **Social Science**   The relationship between a professional basketball player's height H (in inches) and weight W (in pounds) was modeled using two different samples of players. The resulting equations that modeled each sample were $W = 7.46H - 374$ and $W = 7.93H - 405$.

(a) Use both equations to predict the weight of a 6′11″ professional basketball player.

(b) According to each model, what change in weight is associated with a 1-inch increase in height?

(c) Determine the weight and height at which the two models agree.

Obtain an equivalent system by performing the stated elementary operation on the system.

33. Interchange equations 1 and 2.

$$2x - 4y + 5z = 1$$
$$x \qquad\quad - 3z = 2$$
$$5x - 8y + 7z = 6$$
$$3x - 4y + 2z = 3$$

34. Interchange equations 1 and 3.

$$2x - 2y + \ \ z = -6$$
$$3x + \ \ y + 2z = \ \ 2$$
$$x + \ \ y - 2z = \ \ 0$$

35. Multiply the second equation by -1.

$$3x \qquad + z + 2w + 18v = 0$$
$$4x - y + \qquad w + 24v = 0$$
$$7x - y + z + 3w + 42v = 0$$
$$4x \qquad + z + 2w + 24v = 0$$

36. Multiply the third equation by $1/2$.

$$x + 2y + 4z = 3$$
$$x \qquad + 2z = 0$$
$$2x + 4y + \ \ z = 3$$

37. Replace the second equation by the sum of itself and -2 times the first equation.

$$x + y + 2z + 3w = 1$$
$$2x + y + 3z + 4w = 1$$
$$3x + y + 4z + 5w = 2$$

38. Replace the third equation by the sum of itself and -1 times the first equation.

$$x + 2y + 4z = \ \ 6$$
$$y + \ \ z = \ \ 1$$
$$x + 3y + 5z = 10$$

39. Replace the third equation by the sum of itself and -2 times the second equation.

$$x + 12y - 3z + 4w = 10$$
$$2y + 3z + w = 4$$
$$4y + 5z + 2w = 1$$
$$6y - 2z - 3w = 0$$

40. Replace the third equation by the sum of itself and 3 times the second equation.

$$2x + 2y - 4z + w = -5$$
$$2y + 4z - w = 2$$
$$-6y - 4z + 2w = 6$$
$$2y + 5z - 3w = 7$$

Solve the system by back substitution.

41.
$$x + 3y - 4z + 2w = 1$$
$$y + z - w = 4$$
$$2z + 2w = -6$$
$$3w = 9$$

42.
$$x + 5z + 6w = 10$$
$$y + 3z - 2w = 4$$
$$z - 4w = -6$$
$$2w = 4$$

43.
$$2x + 2y - 4z + w = -5$$
$$3y + 4z - w = 0$$
$$2z - 7w = -6$$
$$5w = 15$$

44.
$$3x - 2y - 4z + 2w = 6$$
$$2y + 5z - 3w = 7$$
$$3z + 4w = 0$$
$$3w = 15$$

Write the augmented matrix of each of the following systems. Do not solve the systems.

45.
$$2x + y + z = 3$$
$$3x - 4y + 2z = -5$$
$$x + y + z = 2$$

46.
$$3x + 4y - 2z - 3w = 0$$
$$x - 3y + 7z + 4w = 9$$
$$2x + 5z - 6w = 0$$

Write the system of equations associated with the following augmented matrices. Do not solve the systems.

47. $\begin{bmatrix} 2 & 3 & 8 & | & 20 \\ 1 & 4 & 6 & | & 12 \\ 0 & 3 & 5 & | & 10 \end{bmatrix}$

48. $\begin{bmatrix} 3 & 2 & 6 & | & 18 \\ 2 & -2 & 5 & | & 7 \\ 1 & 0 & 5 & | & 20 \end{bmatrix}$

Use the indicated row operation to transform each matrix.

49. Interchange R_2 and R_3.

$\begin{bmatrix} 1 & 2 & 3 & | & -1 \\ 6 & 5 & 4 & | & 6 \\ 2 & 0 & 7 & | & -4 \end{bmatrix}$

50. Replace R_3 by $-3R_1 + R_3$.

$\begin{bmatrix} 1 & 5 & 2 & 0 & | & -1 \\ 8 & 5 & 4 & 6 & | & 6 \\ 3 & 0 & 7 & 1 & | & -4 \end{bmatrix}$

51. Replace R_2 by $2R_1 + R_2$.

$\begin{bmatrix} -4 & -3 & 1 & -1 & | & 2 \\ 8 & 2 & 5 & 0 & | & 6 \\ 0 & -2 & 9 & 4 & | & 5 \end{bmatrix}$

52. Replace R_3 by $\dfrac{1}{4}R_3$.

$\begin{bmatrix} 2 & 5 & 1 & | & -1 \\ -4 & 0 & 4 & | & 6 \\ 6 & 0 & 8 & | & -4 \end{bmatrix}$

In Exercises 53–58, perform row operations on the augmented matrix as far as necessary to determine whether the system is independent, dependent, or inconsistent. (See Examples 9–11.)

53.
$$x + 2y = 0$$
$$y - z = 2$$
$$x + y + z = -2$$

54.
$$x + 2y + z = 0$$
$$y + 2z = 0$$
$$x + y - z = 0$$

55.
$$x + 2y + 4z = 6$$
$$y + z = 1$$
$$x + 3y + 5z = 10$$

56.
$$x + y + 2z + 3w = 1$$
$$2x + y + 3z + 4w = 1$$
$$3x + y + 4z + 5w = 2$$

57.
$$a - 3b - 2c = -3$$
$$3a + 2b - c = 12$$
$$-a - b + 4c = 3$$

58.
$$2x + 2y + 2z = 6$$
$$3x - 3y - 4z = -1$$
$$x + y + 3z = 11$$

Write the augmented matrix of the system and use the matrix method to solve the system. If the system has an infinite number of solutions, express them in terms of the parameter z. (See Examples 6–11.)

59.
$$-x + 3y + 2z = 0$$
$$2x - y - z = 3$$
$$x + 2y + 3z = 0$$

60.
$$3x + 7y + 9z = 0$$
$$x + 2y + 3z = 2$$
$$x + 4y + z = 2$$

61.
$$x - 2y + 4z = 6$$
$$x + y + 13z = 6$$
$$-2x + 6y - z = -10$$

62.
$$x - y + 5z = -6$$
$$3x + 3y - z = 10$$
$$x + 3y + 2z = 5$$

63.
$$x + y + z = 200$$
$$x - 2y = 0$$
$$2x + 3y + 5z = 600$$
$$2x - y + z = 200$$

64.
$$2x - y + 2z = 3$$
$$-x + 2y - z = 0$$
$$3y - 2z = 1$$
$$x + y - z = 1$$

65.
$$x + y + z = 2$$
$$2x + y - z = 5$$
$$x - y + z = -2$$

66.
$$2x + y + z = 9$$
$$-x - y + z = 1$$
$$3x - y + z = 9$$

67.
$$x + 3y + 4z = 14$$
$$2x - 3y + 2z = 10$$
$$3x - y + z = 9$$
$$4x + 2y + 5z = 23$$

68.
$$4x - y + 3z = -2$$
$$3x + 5y - z = 15$$
$$-2x + y + 4z = 14$$
$$x + 6y + 3z = 29$$

69.
$$x + 2y + 3z = 8$$
$$3x - y + 2z = 5$$
$$-2x - 4y - 6z = 5$$

70.
$$3x - 2y - 8z = 1$$
$$9x - 6y - 24z = -2$$
$$x - y + z = 1$$

71.
$$2x - 4y + z = -4$$
$$x + 2y - z = 0$$
$$-x + y + z = 6$$
$$2x - y + z = 2$$

72.
$$4x - 3y + z = 9$$
$$3x + 2y - 2z = 4$$
$$x - y + 3z = 5$$
$$2x + 3y - 5z = -1$$

73.
$$5x + 3y + 4z = 19$$
$$3x - y + z = -4$$

74.
$$3x + y - z = 0$$
$$2x - y + 3z = -7$$

75. $11x + 10y + 9z = 5$
$x + 2y + 3z = 1$
$3x + 2y + z = 1$

76. $x + y = 3$
$5x - y = 3$
$9x - 4y = 1$

77. (a) Find the equation of the straight line through $(1, 2)$ and $(3, 4)$.

(b) Find the equation of the line through $(-1, 1)$ with slope 3.

(c) Find a point that lies on both of the lines in (a) and (b).

78. Find constants a, b, and c such that the points $(2, 3)$, $(-1, 0)$, and $(-2, 2)$ lie on the graph of the equation $y = ax^2 + bx + c$. (*Hint:* Since $(2, 3)$ is on the graph, we must have $3 = a(2^2) + b(2) + c$; that is, $4a + 2b + c = 3$. Similarly, the other two points lead to two more equations. Solve the resulting system for a, b, and c.)

79. Graph the equations in the given system. Then explain why the graphs show that the system is inconsistent.

$$2x + 3y = 8$$
$$x - y = 4$$
$$5x + y = 7$$

80. Explain why a system with more variables than equations cannot have a unique solution (that is, be an independent system). (*Hint:* When you apply the elimination method to such a system, what must happen?)

Work the following problems by writing and solving a system of equations. (See Example 12.)

81. **Finance** An investor wants to invest $30,000 in corporate bonds that are rated AAA, A, and B. The lower rated ones pay higher interest but pose a higher risk as well. The average yield is 5% on AAA bonds, 6% on A bonds and 10% on B bonds. Being conservative, the investor wants to have twice as much in AAA bonds as in B bonds. How much should she invest in each type of bond to have an interest income of $2000?

82. **Business** A candy store sells cashews for $4.40 per pound and peanuts for $1.20 per pound. If you want to buy exactly 3 pounds of nuts for $6, how many pounds of each kind should you buy? (*Hint:* If you buy x pounds of cashews and y pounds of peanuts, then $x + y = 3$. Find a second equation by considering cost; then solve the resulting system.)

83. **Business** The Crunchy Company wants to make a 100-pound mixture of corn chips, nuts, and pretzels that will cost $4 per pound. Corn chips cost $2 per pound, nuts cost $6 per pound, and pretzels cost $3 per pound. If the mixture is to have three times as many corn chips as pretzels (by weight), how many pounds of each ingredient should be used?

84. **Business** A winemaker has two large casks of wine. One wine is 8% alcohol, the other 18% alcohol. How many gallons of each wine should be mixed to produce 30 gallons of wine that is 12% alcohol? (*Hint:* If x is the number of gallons of the 8% wine and y the number of gallons of the 18% wine, then $x + y = ?$ Use the fact that the amount of alcohol in the mixture is "8% of $x + 18\%$ of y" to find a second equation.)

85. **Business** A minor league baseball park has 7000 seats. Box seats cost $6, grandstand cost $4, and bleacher seats cost $2. When all seats are sold, the revenue is $26,400. If the number of box seats is one-third the number of bleacher seats, how many seats of each type are there?

86. **Business** Shipping charges at the online bookstore Heracles.com are $4 for one book, $6 for two books, and $7 for three to five books. Last week, there were 6400 orders of five or fewer books, and total shipping charges for these orders were $33,600. The number of shipments with $7 charges was 1000 less than the number with $6 charges. How many shipments were made in each category (one book, two books, three-to-five books)?

87. **Health** Computer-aided tomography (CAT) scanners take X-rays of a part of the body from different directions and put the information together to create a picture of a cross section of the body.* The amount by which the energy of the X-ray decreases, measured in linear-attenuation units, tells whether the X-ray has passed through healthy tissue, tumorous tissue, or bone, on the basis of the following table.

Type of Tissue	Linear-Attenuation Values
Healthy tissue	.1625–.2977
Tumorous tissue	.2679–.3930
Bone	.3857–.5108

The part of the body to be scanned is divided into cells. If an X-ray passes through more than one cell, the total linear-attenuation value is the sum of the values for the cells. For example, in the figure, let a, b, and c be the values for cells A, B, and C, respectively. Then the attenuation value for beam 1 is $a + b$ and for beam 2 is $a + c$.

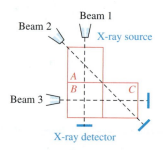

*Exercises 87 and 88 are based on the article "Medical Applications of Linear Equations," by David Jabon, Gail Nord, Bryce W. Wilson, and Penny Coffman, *Mathematics Teacher* 89, no. 5 (May 1996) p. 98: 398.

(a) Find the attenuation value for beam 3.

(b) Suppose that the attenuation values are .8, .55, and .65 for beams 1, 2, and 3, respectively. Set up and solve the system of three equations for a, b, and c. What can you conclude about cells A, B, and C?

88. Health (Refer to Exercise 87.) Four X-ray beams are aimed at four cells, as shown in the following figure.

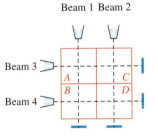

Beam 1 Beam 2

(a) Suppose the attenuation values for beams 1, 2, 3, and 4 are .60, .75, .65, and .70, respectively. Do we have enough information to determine the values of a, b, c, and d? Explain.

(b) Suppose we have the data from part (a), as well as the following values for d. Find the values for a, b, and c, and make conclusions about cells A, B, C, and D in each case.

  (i) .33   (ii) .43

(c) Two X-ray beams are added, as shown in the figure. In addition to the data in part (a), we now have attenuation values of .85 and .50 for beams 5 and 6, respectively. Find the values for a, b, c, and d, and make conclusions about cells A, B, C, and D.

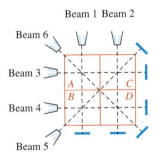

Beam 1 Beam 2

89. Transportation At rush hours, substantial traffic congestion is encountered at the traffic intersections shown in the figure at the top of the next column. (The arrows indicate one-way streets.)

The city wishes to improve the signals at these corners to speed the flow of traffic. The traffic engineers first gather data. As the figure shows, 700 cars per hour come down M Street to intersection A, and 300 cars per hour come down 10th Street to intersection A. x_1 of these cars leave A on M Street, and x_4 cars leave A on 10th Street.

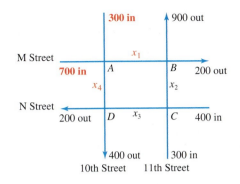

The number of cars entering A must equal the number leaving, so that

$$x_1 + x_4 = 700 + 300,$$

or

$$x_1 + x_4 = 1000.$$

For intersection B, x_1 cars enter on M street and x_2 on 11th Street. The figure shows that 900 cars leave B on 11th and 200 on M. We have

$$x_1 + x_2 = 900 + 200$$
$$x_1 + x_2 = 1100.$$

(a) Write two equations representing the traffic entering and leaving intersections C and D.

(b) Solve the system of four equations, using x_4 as the parameter.

(c) On the basis of your solution to part (b), what are the largest and smallest possible values for the number of cars leaving intersection A on 10th Street?

(d) Answer the question in part (c) for the other three variables.

(e) Verify that you could have discarded any one of the four original equations without changing the solution. What does this tell you about the original problem?

90. Transportation The diagram shows the traffic flow at four intersections during rush hour, as in Exercise 89.

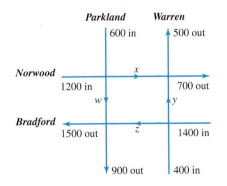

(a) What are the possible values of x, y, z, and w in order to avoid any congestion? (Use w as the parameter.)

(b) What are the possible values of w?

6.2 The Gauss–Jordan Method

In Example 7 of the previous section, we used matrix methods to rewrite the system

$$x - 2y = 6 - 4z$$
$$x + 13z = 6 - y$$
$$-2x + 6y - z = -10$$

as an augmented matrix. We carried out the steps of the matrix method until the final matrix was

$$\begin{bmatrix} 1 & -2 & 4 & | & 6 \\ 0 & 1 & 3 & | & 0 \\ 0 & 0 & 1 & | & 2 \end{bmatrix}.$$

We then used back substitution to solve it. In the **Gauss–Jordan method,** additional elimination of variables replaces back substitution, as follows.

$$\begin{bmatrix} 1 & -2 & 4 & | & 6 \\ 0 & 1 & 0 & | & -6 \\ 0 & 0 & 1 & | & 2 \end{bmatrix} \quad -3R_3 + R_2$$

$$\begin{bmatrix} 1 & -2 & 0 & | & -2 \\ 0 & 1 & 0 & | & -6 \\ 0 & 0 & 1 & | & 2 \end{bmatrix} \quad -4R_3 + R_1$$

$$\begin{bmatrix} 1 & 0 & 0 & | & -14 \\ 0 & 1 & 0 & | & -6 \\ 0 & 0 & 1 & | & 2 \end{bmatrix}. \quad 2R_2 + R_1$$

The solution of the system is now obvious: $(-14, -6, 2)$. Note that this solution is the last column of the final augmented matrix. ◇①

A matrix such as the last one shown is said to be in **reduced row echelon form** if it is in echelon form *and* every column containing a leading 1 has zeros in all its other entries. In the Gauss–Jordan method, row operations may be performed in any order, but it is best to transform the matrix systematically. Either follow the procedure in the preceding example (which first puts the system into a form in which back substitution can be used and then eliminates additional variables), or work column by column from left to right, as in the next example.

▼ **EXAMPLE 1** Use the Gauss–Jordan method to solve the system

$$x \quad + 5z = -6 + y$$
$$3x + 3y \quad = 10 + z$$
$$x + 3y + 2z = 5.$$

Solution The system must first be rewritten in proper form as follows:

$$x - y + 5z = -6$$
$$3x + 3y - z = 10$$
$$x + 3y + 2z = 5.$$

---

① Use the Gauss–Jordan method to solve the system

$$x + 2y = 11$$
$$-4x + y = -8$$

as instructed. Give the shorthand notation and the new matrix in (b)–(d).

(a) Set up the augmented matrix.

(b) Get 0 in row two, column one.

(c) Get 1 in row two, column two.

(d) Finally, get 0 in row one, column two.

(e) The solution of the system is _____.

Answers:

(a) $\begin{bmatrix} 1 & 2 & | & 11 \\ -4 & 1 & | & -8 \end{bmatrix}$

(b) $4R_1 + R_2;$ $\begin{bmatrix} 1 & 2 & | & 11 \\ 0 & 9 & | & 36 \end{bmatrix}$

(c) $\frac{1}{9}R_2;$ $\begin{bmatrix} 1 & 2 & | & 11 \\ 0 & 1 & | & 4 \end{bmatrix}$

(d) $-2R_2 + R_1;$ $\begin{bmatrix} 1 & 0 & | & 3 \\ 0 & 1 & | & 4 \end{bmatrix}$

(e) $(3, 4)$

Begin the solution by writing the augmented matrix of the linear system.

$$\begin{bmatrix} 1 & -1 & 5 & | & -6 \\ 3 & 3 & -1 & | & 10 \\ 1 & 3 & 2 & | & 5 \end{bmatrix}$$

The first element in column one is already 1. Get 0 for the second element in column one by multiplying each element in the first row by -3 and adding the results to the corresponding elements in row two.

$$\begin{bmatrix} 1 & -1 & 5 & | & -6 \\ 0 & 6 & -16 & | & 28 \\ 1 & 3 & 2 & | & 5 \end{bmatrix} \quad -3R_1 + R_2$$

Now change the first element in row three to 0 by multiplying each element of the first row by -1 and adding the results to the corresponding elements of the third row.

$$\begin{bmatrix} 1 & -1 & 5 & | & -6 \\ 0 & 6 & -16 & | & 28 \\ 0 & 4 & -3 & | & 11 \end{bmatrix} \quad -1R_1 + R_3$$

This transforms the first column. Transform the second column in a similar manner, as directed in side problem 2.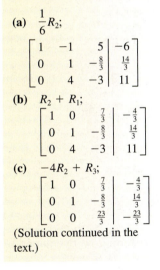

Complete the solution by transforming the third column of the matrix in part (c) of side problem 2.

$$\begin{bmatrix} 1 & 0 & \frac{7}{3} & | & -\frac{4}{3} \\ 0 & 1 & -\frac{8}{3} & | & \frac{14}{3} \\ 0 & 0 & 1 & | & 1 \end{bmatrix} \quad \frac{3}{23}R_3$$

$$\begin{bmatrix} 1 & 0 & 0 & | & 1 \\ 0 & 1 & -\frac{8}{3} & | & \frac{14}{3} \\ 0 & 0 & 1 & | & -1 \end{bmatrix} \quad -\frac{7}{3}R_3 + R_1$$

$$\begin{bmatrix} 1 & 0 & 0 & | & 1 \\ 0 & 1 & 0 & | & 2 \\ 0 & 0 & 1 & | & -1 \end{bmatrix} \quad \frac{8}{3}R_3 + R_2$$

The linear system associated with this last augmented matrix is

$$\begin{aligned} x & & & = 1 \\ & y & & = 2 \\ & & z & = -1, \end{aligned}$$

and the solution is $(1, 2, -1)$. ▲ ③

Most graphing calculators can transform a matrix into reduced row echelon form in a single keystroke, as illustrated in the next example.

2

Continue the solution of the system in Example 1 as instructed. Give the shorthand notation and the matrix for each step.

(a) Get 1 in row two, column two.

(b) Get 0 in row one, column two.

(c) Now get 0 in row three, column two.

Answers:

(a) $\frac{1}{6}R_2$;

$$\begin{bmatrix} 1 & -1 & 5 & | & -6 \\ 0 & 1 & -\frac{8}{3} & | & \frac{14}{3} \\ 0 & 4 & -3 & | & 11 \end{bmatrix}$$

(b) $R_2 + R_1$;

$$\begin{bmatrix} 1 & 0 & \frac{7}{3} & | & -\frac{4}{3} \\ 0 & 1 & -\frac{8}{3} & | & \frac{14}{3} \\ 0 & 4 & -3 & | & 11 \end{bmatrix}$$

(c) $-4R_2 + R_3$;

$$\begin{bmatrix} 1 & 0 & \frac{7}{3} & | & -\frac{4}{3} \\ 0 & 1 & -\frac{8}{3} & | & \frac{14}{3} \\ 0 & 0 & \frac{23}{3} & | & -\frac{23}{3} \end{bmatrix}$$

(Solution continued in the text.)

3

Use the Gauss–Jordan method to solve

$x + y - z = 6$
$2x - y + z = 3$
$-x + y + z = -4.$

Answer:

$(3, 1, -2)$

EXAMPLE 2 Use a graphing calculator to solve the system

$$x + 2y - z = 0$$
$$3x - y + z = 6$$
$$7x + 28y - 8z = -5$$
$$5x + 3y - z = 6.$$

Solution Enter the augmented matrix into the calculator (Figure 6.6). Then use the RREF key (in the MATH or OPS submenu of the MATRIX menu) or a REFF program (in the Graphing Calculator Appendix that is available at www.aw.com/MWA9) to put this matrix in reduced row echelon form (Figure 6.7). In Figure 6.7, you must use the arrow key to scroll over to see the full decimal expansions in the right hand column. This can be avoided on TI calculators by using the FRAC key (in the MATH menu), as in Figure 6.8.

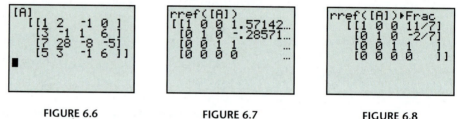

FIGURE 6.6   FIGURE 6.7   FIGURE 6.8

The answers can now be read from the last column of the matrix in Figure 6.8: $x = 11/7, y = -2/7, z = 1$. ▲

4

Use a graphing calculator to solve the system.

$$x + 3y = 4$$
$$4x + 8y = 4$$
$$6x + 12y = 6$$

Answer:

$(-5, 3)$

EXAMPLE 3 Use the Gauss–Jordan method to solve the system

$$2x + 4y = 4$$
$$3x + 6y = 8$$
$$2x + y = 7.$$

Solution There are several possible ways to proceed.

Manual Method Write the augmented matrix and perform row operations to obtain a first column whose entries (from top to bottom) are $1, 0, 0$.

$$\begin{bmatrix} 2 & 4 & | & 4 \\ 3 & 6 & | & 8 \\ 2 & 1 & | & 7 \end{bmatrix}$$

$$\begin{bmatrix} 1 & 2 & | & 2 \\ 3 & 6 & | & 8 \\ 2 & 1 & | & 7 \end{bmatrix} \quad \frac{1}{2}R_1$$

$$\begin{bmatrix} 1 & 2 & | & 2 \\ 0 & 0 & | & 2 \\ 2 & 1 & | & 7 \end{bmatrix} \quad -3R_1 + R_2$$

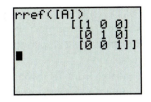

FIGURE 6.9

⟨5⟩

Solve each system.

(a)
$$x - y = 4$$
$$-2x + 2y = 1$$

(b)
$$3x - 4y = 0$$
$$2x + y = 0$$

Answers:

(a) No solution

(b) $(0, 0)$

Stop! The second row of the matrix denotes the equation $0x + 0y = 2$. Since the left side of this equation is always 0 and the right side is 2, it has no solution. Therefore, the original system has no solution.

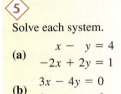**Calculator Method**   Enter the augmented matrix into a graphing calculator and use the RREF key to put it into reduced row echelon form, as in Figure 6.9. The last row of that matrix corresponds to $0x + 0y = 1$, which has no solution. Hence, the original system has no solution. ▲ ⟨5⟩

NOTE Whenever the Gauss–Jordan method produces a row whose elements are all 0 except the last one, as in Example 3, the system is inconsistent and has no solutions. In contrast, if a row with *every* element 0 is produced, the system may have solutions. In that case, continue with the Gauss–Jordan method.

▼ **EXAMPLE 4**   Use the Gauss–Jordan method to solve the system

$$x + 2y - 3z = 0$$
$$3x - 2y + z = 6.$$

Solution   When working manually, start with the augmented matrix and use row operations to obtain a first column whose entries (from top to bottom) are 1, 0.

$$\begin{bmatrix} 1 & 2 & -3 & 0 \\ 3 & -2 & 1 & 6 \end{bmatrix}$$

$$\begin{bmatrix} 1 & 2 & -3 & 0 \\ 0 & -8 & 10 & 6 \end{bmatrix} \quad -3R_1 + R_2$$

Now use row operations to obtain a second column whose entries (from top to bottom) are 0, 1.

$$\begin{bmatrix} 1 & 2 & -3 & 0 \\ 0 & 1 & -\frac{5}{4} & -\frac{3}{4} \end{bmatrix} \quad -\frac{1}{8}R_2$$

$$\begin{bmatrix} 1 & 0 & -\frac{1}{2} & \frac{3}{2} \\ 0 & 1 & -\frac{5}{4} & -\frac{3}{4} \end{bmatrix} \quad -2R_2 + R_1$$

The last matrix (which is the same one produced by the RREF key on a graphing calculator) is the augmented matrix of the system

$$x - \frac{1}{2}z = \frac{3}{2}$$
$$y - \frac{5}{4}z = -\frac{3}{4}.$$

Solving the first equation for x and the second for y gives the solution

$$x = \frac{1}{2}z + \frac{3}{2} = .5z + 1.5$$

$$y = \frac{5}{4}z - \frac{3}{4} = 1.25z - .75$$

z arbitrary,

⟨6⟩

Use the Gauss–Jordan method to solve the following.

(a) $3x + 9y = -6$
       $-x - 3y = 2$

(b) $2x + 9y = 12$
       $4x + 18y = 5$

Answers:

(a) y arbitrary,
      $x = -3y - 2$;
      or $(-3y - 2, y)$

(b) No solution

or $(z/2 + 3/2, 5z/4 - 3/4, z)$. ▲ ⟨6⟩

TECHNOLOGY TIP   Many graphing calculators have solvers for systems of linear equations. TI-86 and Casio solvers work only for systems that have the same number of variables as equations *and* have a unique solution. When an error message is displayed, the system may have no solutions or it may have infinitely many solutions.

The TI-84+ PolySmlt solver (in the APPS menu or downloadable from TI) can handle virtually any system. Figure 6.10 shows its solution for Example 4. When the message "no solution found" is displayed, as in Figure 6.11, press the RREF key at the bottom of the screen to display the reduced row echelon matrix of the system. From that, you can tell whether the system has no solution or has infinitely many solutions.

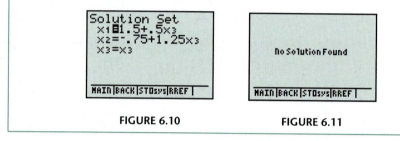

FIGURE 6.10   FIGURE 6.11

The techniques used in Examples 1–4 can be summarized as follows.

The Gauss–Jordan Method for Solving a System of Linear Equations

1. Arrange the equations with the variable terms in the same order on the left of the equals sign and the constants on the right.

2. Write the augmented matrix of the system.

3. Use row operations or a graphing calculator to transform the augmented matrix into reduced row echelon form.

4. Stop the process in Step 3 if you obtain a row whose elements are all zero except the last one. In that case, the system is inconsistent and has no solutions. Otherwise, finish Step 3 and read the solutions of the system from the final matrix.

Applications

The key to solving applied problems is to determine the system of equations that must be solved.

▼ **EXAMPLE 5**   An animal feed is to be made from corn, soybeans, and cottonseed. Determine how many units of each ingredient are needed to make a feed that supplies 1800 units of fiber, 2800 units of fat, and 2200 units of protein, given that 1 unit of each ingredient provides the numbers of units shown in the table at the top of the next page. The table states, for example, that a unit of corn provides 10 units of fiber, 30 units of fat, and 20 units of protein.

	Corn	Soybeans	Cottonseed	Totals
Units of Fiber	10	20	30	1800
Units of Fat	30	20	40	2800
Units of Protein	20	40	25	2200

Solution Let x represent the required number of units of corn, y the number of units of soybeans, and z the number of units of cottonseed. Since the total amount of fiber is to be 1800,

$$10x + 20y + 30z = 1800.$$

The feed must supply 2800 units of fat, so

$$30x + 20y + 40z = 2800.$$

Finally, since 2200 units of protein are required,

$$20x + 40y + 25z = 2200.$$

Thus, we must solve this system of equations:

$$10x + 20y + 30z = 1800$$
$$30x + 20y + 40z = 2800 \qquad (1)$$
$$20x + 40y + 25z = 2200.$$

Now solve the system, either manually or with technology.

Manual Method Write the augmented matrix and use row operations to transform it into row echelon form, as in problem 7 at the side. The resulting matrix represents the following system: ⟨7⟩

$$x + 2y + 3z = 180$$
$$y + \frac{5}{4}z = 65 \qquad (2)$$
$$z = 40.$$

Back substitution now shows that

$$z = 40, \quad y = 65 - \frac{5}{4}(40) = 15, \quad \text{and} \quad x = 180 - 2(15) - 3(40) = 30.$$

Thus, the feed should contain 30 units of corn, 15 of soybeans, and 40 of cottonseed.

Calculator Method Enter the augmented matrix of the system into the calculator (top of Figure 6.12). Use the RREF key or a REFF program to transform it into reduced row echelon form (bottom of Figure 6.12), which shows that $x = 30$, $y = 15$, and $z = 40$. ▲

▼ **EXAMPLE 6** The U-Drive Rent-a-Truck Company plans to spend 3 million dollars on 200 new vehicles. Each van will cost $10,000, each small truck $15,000, and each large truck $25,000. Past experience shows that U-Drive needs twice as many vans as small trucks. How many of each kind of vehicle can the company buy?

⟨7⟩

List a sequence of row operations that will transform the augmented matrix of system (1) in Example 5 into the augmented matrix of system (2).

Answer:

Many sequences are possible, including this one:

replace R_1 by $\frac{1}{10}R_1$;

replace R_2 by $\frac{1}{10}R_2$;

replace R_3 by $\frac{1}{5}R_3$;

replace R_2 by $-3R_1 + R_2$;
replace R_3 by $-4R_1 + R_3$;

replace R_2 by $-\frac{1}{4}R_2$;

replace R_3 by $-\frac{1}{7}R_3$.

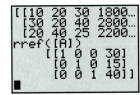

FIGURE 6.12

Solution Let x be the number of vans, y the number of small trucks, and z the number of large trucks. Then

$$\begin{pmatrix} \text{Number of} \\ \text{vans} \end{pmatrix} + \begin{pmatrix} \text{Number of} \\ \text{small trucks} \end{pmatrix} + \begin{pmatrix} \text{Number of} \\ \text{large trucks} \end{pmatrix} = \text{Total number of vehicles}$$

$$x + y + z = 200. \tag{1}$$

Similarly,

$$\begin{pmatrix} \text{Cost of } x \\ \text{vans} \end{pmatrix} + \begin{pmatrix} \text{Cost of } y \\ \text{small trucks} \end{pmatrix} + \begin{pmatrix} \text{Cost of } z \\ \text{large trucks} \end{pmatrix} = \text{Total cost}$$

$$10{,}000x + 15{,}000y + 25{,}000z = 3{,}000{,}000$$

Dividing both sides by 5000 produces the equivalent equation

$$2x + 3y + 5z = 600. \tag{2}$$

Finally, the number of vans is twice the number of small trucks; that is, $x = 2y$, or equivalently,

$$x - 2y = 0. \tag{3}$$

We must solve the system given by equations (1)–(3).

$$\begin{aligned} x + y + z &= 200 \\ 2x + 3y + 5z &= 600 \\ x - 2y \phantom{{}+ 5z} &= 0 \end{aligned}$$

Form the augmented matrix and transform it into reduced row echelon form (either manually, as shown below, or by using technology).

$$\left[\begin{array}{ccc|c} 1 & 1 & 1 & 200 \\ 2 & 3 & 5 & 600 \\ 1 & -2 & 0 & 0 \end{array}\right]$$

$$\left[\begin{array}{ccc|c} 1 & 1 & 1 & 200 \\ 0 & 1 & 3 & 200 \\ 0 & -3 & -1 & -200 \end{array}\right] \quad \begin{array}{l} -2R_1 + R_2 \\ -R_1 + R_3 \end{array}$$

$$\left[\begin{array}{ccc|c} 1 & 0 & -2 & 0 \\ 0 & 1 & 3 & 200 \\ 0 & 0 & 8 & 400 \end{array}\right] \quad \begin{array}{l} -R_2 + R_1 \\ \\ 3R_2 + R_3 \end{array}$$

$$\left[\begin{array}{ccc|c} 1 & 0 & -2 & 0 \\ 0 & 1 & 3 & 200 \\ 0 & 0 & 1 & 50 \end{array}\right] \quad \begin{array}{l} \\ \\ \frac{1}{8}R_3 \end{array}$$

$$\left[\begin{array}{ccc|c} 1 & 0 & 0 & 100 \\ 0 & 1 & 0 & 50 \\ 0 & 0 & 1 & 50 \end{array}\right] \quad \begin{array}{l} 2R_3 + R_1 \\ -3R_3 + R_2 \end{array}$$

The final matrix corresponds to the system

$$x = 100$$
$$y = 50$$
$$z = 50.$$

Therefore, U-Drive should buy 100 vans, 50 small trucks, and 50 large trucks. ▲

8

In Example 6, suppose that U-Drive can spend only 2 million dollars on 150 new vehicles and the company needs three times as many vans as small trucks. Write a system of equations to express these conditions.

Answer:

$$x + y + z = 150$$
$$2x + 3y + 5z = 400$$
$$x - 3y = 0$$

▼ **EXAMPLE 7** The concentrations (in parts per million) of carbon dioxide (a greenhouse gas) have been measured at Mauna Loa, Hawaii, since 1959. The concentrations are known to have increased quadratically. The table lists readings for three years.*

Year	1964	1984	2004
Carbon Dioxide	319	344	377

(a) Use the given data to construct a quadratic function that gives the concentration in year x.

Solution Let $x = 0$ correspond to 1959. Then the table is represented by the data points $(5, 319)$, $(25, 344)$, and $(45, 377)$. We must find a function of the form

$$f(x) = ax^2 + bx + c$$

whose graph contains these three points. If $(5, 319)$ is to be on the graph, we must have $f(5) = 319$; that is,

$$a(5^2) + b(5) + c = 319$$
$$25a + 5b + c = 319.$$

The other two points lead to these equations:

$$f(25) = 344 \qquad\qquad f(45) = 377$$
$$a(25^2) + b(25) + c = 344 \qquad a(45^2) + b(45) + c = 377$$
$$625a + 25b + c = 344 \qquad\qquad 2025a + 45b + c = 377.$$

Now work by hand or use technology to solve the system.

$$25a + 5b + c = 319$$
$$625a + 25b + c = 344$$
$$2025a + 45b + c = 377.$$

The reduced row echelon form of the augmented matrix in Figure 6.13 shows that the solution is $a = .01$, $b = .95$, and $c = 314$. So the function is

$$f(x) = .01x^2 + .95x + 314.$$

```
[[25    5   1  319…
 [625  25   1  344…
 [2025 45   1  377…
rref([A])
    [[1  0  0  .01]
     [0  1  0  .95]
     [0  0  1  314]]
```

FIGURE 6.13

*C. D. Keeling and T. P. Whorf, Scripps Institution of Oceanography.

(b) Use this model to estimate the carbon dioxide concentrations in 2006 and 2010.

Solution The year 2006 corresponds to $x = 47$, so the concentration is

$$f(47) = .01(47^2) + .95(47) + 314 = 380.74.$$

Similarly, the concentration in 2010 $(x = 51)$ is

$$f(51) = .01(51^2) + .95(51) + 314 = 388.46. \; \blacktriangle$$

◆6.2 Exercises

In Exercises 1–4, the reduced row echelon form of the augmented matrix of a system of equations is given. Find the solutions of the system.

1. $\begin{bmatrix} 1 & 0 & 0 & 0 & 3/2 \\ 0 & 1 & 0 & 0 & 7 \\ 0 & 0 & 1 & 0 & -3 \\ 0 & 0 & 0 & 1 & 0 \end{bmatrix}$

2. $\begin{bmatrix} 1 & 0 & 0 & 0 & 0 & 6 \\ 0 & 1 & 0 & 0 & 0 & 4 \\ 0 & 0 & 1 & 0 & 0 & 5 \\ 0 & 0 & 0 & 0 & 1 & 2 \\ 0 & 0 & 0 & 0 & 0 & 1 \end{bmatrix}$

3. $\begin{bmatrix} 1 & 0 & 0 & 1 & 2 \\ 0 & 1 & 0 & 2 & -3 \\ 0 & 0 & 1 & 0 & 5 \\ 0 & 0 & 0 & 0 & 0 \end{bmatrix}$

4. $\begin{bmatrix} 1 & 0 & 0 & 0 & 7 \\ 0 & 1 & 0 & 0 & 2 \\ 0 & 0 & 1 & 0 & -5 \\ 0 & 0 & 0 & 1 & 3 \\ 0 & 0 & 0 & 0 & 0 \\ 0 & 0 & 0 & 0 & 0 \end{bmatrix}$

Use the Gauss–Jordan method to solve each of the following systems of equations. (See Examples 1–4.)

5. $\begin{aligned} x + 2y + z &= 5 \\ 2x + y - 3z &= -2 \\ 3x + y + 4z &= -5 \end{aligned}$

6. $\begin{aligned} 3x - 2y + z &= 6 \\ 3x + y - z &= -4 \\ -x + 2y - 2z &= -8 \end{aligned}$

7. $\begin{aligned} x + 3y - 6z &= 7 \\ 2x - y + 2z &= 0 \\ x + y + 2z &= -1 \end{aligned}$

8. $\begin{aligned} x &= 1 - y \\ 2x &= z \\ 2z &= -2 - y \end{aligned}$

9. $\begin{aligned} x - 2y + 4z &= 6 \\ x + y + 13z &= 6 \\ -2x + 6y - z &= -10 \end{aligned}$

10. $\begin{aligned} x - y + 5z &= -6 \\ 3x + 3y - z &= 10 \\ x + 3y + 2z &= 5 \end{aligned}$

11. $\begin{aligned} 3x + 5y - z &= 0 \\ 4x - y + 2z &= 1 \\ -6x - 10y + 2z &= 0 \end{aligned}$

12. $\begin{aligned} x + y &= -1 \\ y + z &= 4 \\ x + z &= 1 \end{aligned}$

13. $\begin{aligned} x + y - z &= 6 \\ 2x - y + z &= -9 \\ x - 2y + 3z &= 1 \end{aligned}$

14. $\begin{aligned} y &= x - 1 \\ y &= 6 + z \\ z &= -1 - x \end{aligned}$

15. $\begin{aligned} x + y + z &= 1 \\ x - 2y + 2z &= 4 \\ 2x - y + 3z &= 5 \end{aligned}$

16. $\begin{aligned} 2x - y + z &= 1 \\ 3x + y + z &= 0 \\ 7x - y + 3z &= 2 \end{aligned}$

Solve the system by any method.

17. $\begin{aligned} x - 2y + z &= 5 \\ 2x + y - z &= 2 \\ -2x + 4y - 2z &= 2 \end{aligned}$

18. $\begin{aligned} 2x + 3y + z &= 9 \\ 4x + y - 3z &= -7 \\ 6x + 2y - 4z &= -8 \end{aligned}$

19. $\begin{aligned} -8x - 9y &= 11 \\ 24x + 34y &= 2 \\ 16x + 11y &= -57 \end{aligned}$

20. $\begin{aligned} 2x + y &= 7 \\ x - y &= 3 \\ x + 3y &= 4 \end{aligned}$

21. $\begin{aligned} x + 2y &= 3 \\ 2x + 3y &= 4 \\ 3x + 4y &= 5 \\ 4x + 5y &= 6 \end{aligned}$

22. $\begin{aligned} x - y &= 2 \\ x + y &= 4 \\ 2x + 3y &= 9 \\ 3x - 2y &= 6 \end{aligned}$

23. $\begin{aligned} x + y - z &= -20 \\ 2x - y + z &= 11 \end{aligned}$

24. $\begin{aligned} 4x + 3y + z &= 1 \\ -2x - y + 2z &= 0 \end{aligned}$

25. $\begin{aligned} 2x + y + 3z - 2w &= -6 \\ 4x + 3y + z - w &= -2 \\ x + y + z + w &= -5 \\ -2x - 2y + 2z + 2w &= -10 \end{aligned}$

26. $\begin{aligned} x + y + z + w &= -1 \\ -x + 4y + z - w &= 0 \\ x - 2y + z - 2w &= 11 \\ -x - 2y + z + 2w &= -3 \end{aligned}$

27. $\begin{aligned} x + 2y - z &= 3 \\ 3x + y + w &= 4 \\ 2x - y + z + w &= 2 \end{aligned}$

28. $\begin{aligned} x - 2y - z - 3w &= -3 \\ -x + y + z &= 2 \\ 4y + 3z - 6w &= -2 \end{aligned}$

29. $\begin{aligned} \frac{3}{x} - \frac{1}{y} + \frac{4}{z} &= -13 \\ \frac{1}{x} + \frac{2}{y} - \frac{1}{z} &= 12 \\ \frac{4}{x} - \frac{1}{y} + \frac{3}{z} &= -7 \end{aligned}$

[*Hint:* Let $u = 1/x, v = 1/y, w = 1/z$ and solve the resulting system.]

30. $\dfrac{1}{x+1} - \dfrac{2}{y-3} + \dfrac{3}{z-2} = 4$

$\dfrac{5}{y-3} - \dfrac{10}{z-2} = -5$

$\dfrac{-3}{x+1} + \dfrac{4}{y-3} - \dfrac{1}{z-2} = -2$

[*Hint:* Let $u = 1/(x+1)$, $v = 1/(y-3)$, $w = 1/(z-2)$.]

Use systems of equations to answer these questions. (See Examples 5–7.)

31. **Business** McFrugal Snack Shops plan to hire two public-relations firms to survey 500 customers by phone, 750 by mail, and 250 by in-person interviews. The Garcia firm has personnel to do 10 phone surveys, 30 mail surveys, and 5 interviews per hour. The Wong firm can handle 20 phone surveys, 10 mail surveys, and 10 interviews per hour. For how many hours should each firm be hired to produce the exact number of surveys needed?

32. **Finance** Kate borrows $10,000, some from her friend at 8% annual interest, twice as much as that from her bank at 9%, and the remainder from her insurance company at 5%. She pays a total of $830 in interest for the first year. How much did she borrow from each source?

33. **Business** Pretzels cost $3 per pound, dried fruit $4 per pound, and nuts $8 per pound. How many pounds of each should be used to produce 140 pounds of trail mix costing $6 per pound in which there are twice as many pretzels (by weight) as dried fruit?

34. **Business** An auto manufacturer sends cars from two plants, I and II, to dealerships A and B, located in a midwestern city. Plant I has a total of 28 cars to send, and plant II has 8. Dealer A needs 20 cars, and dealer B needs 16. Transportation costs based on the distance of each dealership from each plant are $220 from I to A, $300 from I to B, $400 from II to A, and $180 from II to B. The manufacturer wants to limit transportation costs to $10,640. How many cars should be sent from each plant to each of the two dealerships?

35. **Physical Sciences** The stopping distance for a car traveling 25 mph is 61.7 feet, and for a car traveling 35 mph it is 106 feet.* The stopping distance in feet can be described by the equation $y = ax^2 + bx$, where x is the speed in mph.
   (a) Find the values of a and b.
   (b) Use your answers from part (a) to find the stopping distance for a car traveling 55 mph.

36. **Natural Science** An animal breeder can buy four types of tiger food. Each case of Brand A contains 25 units of fiber, 30 units of protein, and 30 units of fat. Each case of Brand B contains 50 units of fiber, 30 units of protein, and 20 units of fat. Each case of Brand C contains 75 units of fiber, 30 units of protein, and 20 units of fat. Each case of Brand D contains 100 units of fiber, 60 units of protein, and 30 units of fat. How many cases of each brand should the breeder mix together to obtain a food that provides 1200 units of fiber, 600 units of protein, and 400 units of fat?

37. **Finance** An investor plans to invest $70,000 in a mutual fund, corporate bonds, and a fast-food franchise. She plans to put twice as much in bonds as in the mutual fund. On the basis of past performance, she expects the mutual fund to pay a 2% dividend, the bonds 10%, and the franchise 6%. She would like a dividend income of $4800. How much should she put in each of three investments?

38. **Business** According to data from a Texas agricultural report, the amount of nitrogen (lb/acre), phosphate (lb/acre), and labor (hr/acre) needed to grow honeydews, yellow onions, and lettuce is given by the following table.*

	Honeydews	Yellow Onions	Lettuce
Nitrogen	120	150	180
Phosphate	180	80	80
Labor	4.97	4.45	4.65

   (a) If a farmer has 220 acres, 29,100 pounds of nitrogen, 32,600 pounds of phosphate, and 480 hours of labor, can he use all of his resources completely? If so, how many acres should he allot for each crop?
   (b) Suppose everything is the same as in part (a), except that 1061 hours of labor are available. Is it possible to use all of his resources completely? If so, how many acres should he allot for each crop?

A graphing calculator or other technology is recommended for the following exercises.

39. **Health** The table shows the calories, sodium, and protein in one cup of various kinds of soup.

	Progresso™ Roasted Chicken Rotini	Healthy Choice™ Hearty Chicken	Campbell's™ Chunky Chicken Noodle
Calories	100	130	130
Sodium (mg)	970	480	880
Protein (g)	6	8	8

How many cups of each kind of soup should be mixed together to produce ten servings of soup, each of which

---

*Miguel Paredes, Mohammad Fatehi, and Richard Hinthorn, "The Transformation of an Inconsistent Linear System into a Consistent System," *AMATYC Review*, 13, no. 2 (spring 1992).

provides 203 calories, 1190 milligrams of sodium, and 12.4 grams of protein? What is the serving size (in cups)? (*Hint:* In ten servings, there must be 2030 calories, 11,900 milligrams of sodium, and 124 grams of protein.)

40. Health The table shows the calories, sodium, and fat in 1 ounce of various snack foods (all produced by Planters™).

	Cashews	Dry Roasted Honey Peanuts	Cajun Crunch Trail Mix
Calories	170	150	130
Sodium (mg)	120	95	10
Fat (g)	6	7	3

How many ounces of each kind of snack should be combined to produce ten servings, each of which provides 288 calories, 115 milligrams of sodium, and 9.6 grams of fat? What is the serving size?

41. Finance An investment firm recommends that a client invest in bonds rated AAA, A, and B. The average yield on AAA bonds is 6%, on A bonds 7%, and on B bonds 10%. The client wants to invest twice as much in AAA bonds as in B bonds. How much should be invested in each type of bond under the following conditions?
(a) The total investment is $25,000, and the investor wants an annual return of $1810 on the three investments.
(b) The values in part (a) are changed to $30,000 and $2150, respectively.
(c) The values in part (a) are changed to $40,000 and $2900, respectively.

42. Business An electronics company produces transistors, resistors, and computer chips. Each transistor requires 3 units of copper, 1 unit of zinc, and 2 units of glass. Each resistor requires 3, 2, and 1 units of the three materials, and each computer chip requires 2, 1, and 2 units of these materials, respectively. How many of each product can be made with the following amounts of materials?
(a) 810 units of copper, 410 of zinc, and 490 of glass
(b) 765 units of copper, 385 of zinc, and 470 of glass
(c) 1010 units of copper, 500 of zinc, and 610 of glass

43. Business At a pottery factory, fuel consumption for heating the kilns varies with the size of the order being fired. In the past, the company recorded the figures in the table at the top of the next column.
(a) Find an equation of the form $y = ax^2 + bx + c$ whose graph contains the three points corresponding to the data in the table.
(b) How many platters should be fired at one time in order to minimize the fuel cost per platter? What is the minimum fuel cost per platter?

x = Number of Platters	y = Fuel Cost per Platter
6	$2.80
8	2.48
10	2.24

44. Business The business analyst for Melcher Manufacturing wants to find an equation that can be used to project sales of a relatively new product. For the years 2000, 2001, and 2002, sales were $15,000, $32,000, and $123,000, respectively.
(a) Graph the sales for the years 2000, 2001, and 2002, letting the year 2000 equal 0 on the *x*-axis. Let the values on the vertical axis be in thousands. (For example, the point (2001, 32,000) will be graphed as (1, 32).)
(b) Find the equation of the straight line $ax + by = c$ through the points for 2000 and 2002.
(c) Find the equation of the parabola $y = ax^2 + bx + c$ through the three given points.
(d) Find the projected sales for 2005 first by using the equation from part (b) and then by using the equation from part (c). If you were to estimate sales of the product in 2005, which result would you choose? Why?

45. Business The gross domestic product (GDP) of the United States was $11 trillion in 2003 and is projected to be $20 trillion in 2028 and $30 trillion in 2044.*
(a) Let $x = 0$ correspond to 2000. Find a quadratic function $f(x) = ax^2 + bx + c$ that gives the GDP (in trillions of dollars) in year *x*.
(b) Estimate the GDP in 2009 and 2015.
(c) In what year will the GDP reach $25 trillion?

46. Health The number of Alzheimer's cases in people 85 and older was 2 million in 2004 and is projected to be 3 million in 2025 and 8 million in 2050.[†]
(a) Let $x = 0$ correspond to 2000, and find a quadratic function that models the given data.
(b) How many people 85 or older will have Alzheimer's disease in 2020 and in 2034?
(c) In the year you turn 85, how many people your age or older are expected to have Alzheimer's disease?

47. Physical Science For certain aircraft, there exists a quadratic relationship between an airplane's maximum speed *S* (in knots) and its ceiling *C*—its highest altitude possible (in thousands of feet).[‡] The table lists three airplanes that conform to this relationship.

---

*Goldman Sachs.

[†]Alzheimer'Association.

[‡]D. Sanders, *Statistics: A First Course*, Fifth Edition (McGraw Hill, 1995).

Airplane	Maximum Speed	Ceiling
Hawkeye	320	33
Corsair	600	40
Tomcat	1283	50

(a) If the relationship between C and S is written as $C = aS^2 + bS + c$, use a linear system of equations to determine the constants a, b, and c.

(b) A new aircraft of this type has a ceiling of 45,000 feet. Predict its top speed.

6.3 Basic Matrix Operations

Until now, we have used matrices only as a convenient shorthand to solve systems of equations. However, matrices are also important in the fields of management, natural science, engineering, and social science as a way to organize data, as Example 1 demonstrates.

▼ **EXAMPLE 1** The EZ Life Company manufactures sofas and armchairs in three models: A, B, and C. The company has regional warehouses in New York, Chicago, and San Francisco. In its August shipment, the company sends 10 model A sofas, 12 model B sofas, 5 model C sofas, 15 model A chairs, 20 model B chairs, and 8 model C chairs to each warehouse.

This data might be organized by first listing it as follows.

Sofas	10 model A	12 model B	5 model C
Chairs	15 model A	20 model B	8 model C

Alternatively, we might tabulate the data.

			MODEL	
		A	B	C
FURNITURE	Sofa	10	12	5
	Chair	15	20	8

With the understanding that the numbers in each row refer to the type of furniture (sofa, chair) and the numbers in each column refer to the model (A, B, C), the same information can be given by a matrix, as follows.

$$M = \begin{bmatrix} 10 & 12 & 5 \\ 15 & 20 & 8 \end{bmatrix} \quad ▲ \quad ⬦ \; 1$$

A matrix with m horizontal rows and n vertical columns has dimension, or size, $m \times n$. The number of rows is always given first.

⬦ 1

Rewrite this information in a matrix with three rows and two columns.

Answer:

$$\begin{bmatrix} 10 & 15 \\ 12 & 20 \\ 5 & 8 \end{bmatrix}$$

▼ **EXAMPLE 2** (a) The matrix $\begin{bmatrix} 6 & 5 \\ 3 & 4 \\ 5 & -1 \end{bmatrix}$ is a 3×2 matrix.

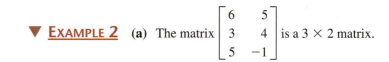

(b) $\begin{bmatrix} 5 & 8 & 9 \\ 0 & 5 & -3 \\ -4 & 0 & 5 \end{bmatrix}$ is a 3 × 3 matrix.

(c) $\begin{bmatrix} 1 & 6 & 5 & -2 & 5 \end{bmatrix}$ is a 1 × 5 matrix.

(d) A graphing calculator displays a 4 × 1 matrix like this.

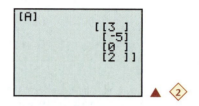

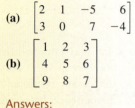

2

Give the size of each of the following matrices.

(a) $\begin{bmatrix} 2 & 1 & -5 & 6 \\ 3 & 0 & 7 & -4 \end{bmatrix}$

(b) $\begin{bmatrix} 1 & 2 & 3 \\ 4 & 5 & 6 \\ 9 & 8 & 7 \end{bmatrix}$

Answers:

(a) 2 × 4

(b) 3 × 3

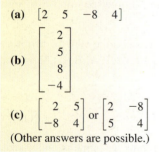

3

Use the numbers 2, 5, −8, 4 to write

(a) a row matrix;

(b) a column matrix;

(c) a square matrix.

Answers:

(a) $\begin{bmatrix} 2 & 5 & -8 & 4 \end{bmatrix}$

(b) $\begin{bmatrix} 2 \\ 5 \\ 8 \\ -4 \end{bmatrix}$

(c) $\begin{bmatrix} 2 & 5 \\ -8 & 4 \end{bmatrix}$ or $\begin{bmatrix} 2 & -8 \\ 5 & 4 \end{bmatrix}$

(Other answers are possible.)

A matrix with only one row, as in Example 2(c), is called a **row matrix** or **row vector**. A matrix with only one column, as in Example 2(d), is called a **column matrix** or **column vector**. A matrix with the same number of rows as columns is called a **square matrix**. The matrix in Example 2(b) is a square matrix, as are

$$A = \begin{bmatrix} -5 & 6 \\ 8 & 3 \end{bmatrix} \quad \text{and} \quad B = \begin{bmatrix} 0 & 0 & 0 & 0 \\ -2 & 4 & 1 & 3 \\ 0 & 0 & 0 & 0 \\ -5 & -4 & 1 & 8 \end{bmatrix}.$$

3

When a matrix is denoted by a single letter, such as the matrix A above, the element in row i and column j is denoted a_{ij}. For example, $a_{21} = 8$ (the element in row 2, column 1). Similarly, in matrix B, $b_{42} = -4$ (the element in row 4, column 2).

Addition

The matrix given in Example 1,

$$M = \begin{bmatrix} 10 & 12 & 5 \\ 15 & 20 & 8 \end{bmatrix},$$

shows the August shipment from the EZ Life plant to each of its warehouses. If matrix N below gives the September shipment to the New York warehouse, what is the total shipment for each item of furniture to the New York warehouse for the two months?

$$N = \begin{bmatrix} 45 & 35 & 20 \\ 65 & 40 & 35 \end{bmatrix}$$

If 10 model A sofas were shipped in August and 45 in September, then altogether 55 model A sofas were shipped in the two months. Adding the other corresponding entries gives a new matrix, Q, that represents the total shipment to the New York warehouse for the two months.

$$Q = \begin{bmatrix} 55 & 47 & 25 \\ 80 & 60 & 43 \end{bmatrix}$$

It is convenient to refer to Q as the *sum* of M and N.

The way these two matrices were added illustrates the following definition of addition of matrices:

> The **sum** of two $m \times n$ matrices X and Y is the $m \times n$ matrix $X + Y$ in which each element is the sum of the corresponding elements of X and Y.

It is important to remember that only matrices that are the same size can be added.

▼ **EXAMPLE 3**   Find each sum if possible.

(a) $\begin{bmatrix} 5 & -6 \\ 8 & 9 \end{bmatrix} + \begin{bmatrix} -4 & 6 \\ 8 & -3 \end{bmatrix} = \begin{bmatrix} 5 + (-4) & -6 + 6 \\ 8 + 8 & 9 + (-3) \end{bmatrix} = \begin{bmatrix} 1 & 0 \\ 16 & 6 \end{bmatrix}$

(b) The matrices

$$A = \begin{bmatrix} 5 & 8 \\ 6 & 2 \end{bmatrix} \quad \text{and} \quad B = \begin{bmatrix} 3 & 9 & 1 \\ 4 & 2 & 5 \end{bmatrix}$$

are different sizes, so it is not possible to find the sum $A + B$. ▲ ④

TECHNOLOGY TIP   Spreadsheets and graphing calculators can find matrix sums, as illustrated in Figure 6.14.

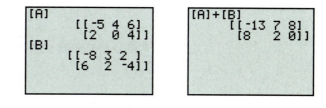

FIGURE 6.14

▼ **EXAMPLE 4**   The September shipments of the three models of sofas and chairs from the EZ Life Company to the New York, San Francisco, and Chicago warehouses are given in matrices N, S, and C as follows:

$$N = \begin{bmatrix} 45 & 35 & 20 \\ 65 & 40 & 35 \end{bmatrix}, \quad S = \begin{bmatrix} 30 & 32 & 28 \\ 43 & 47 & 30 \end{bmatrix}, \quad C = \begin{bmatrix} 22 & 25 & 38 \\ 31 & 34 & 35 \end{bmatrix}.$$

What was the total amount shipped to the three warehouses in September?

Solution   The total of the September shipments is represented by the sum of the three matrices N, S, and C.

$$N + S + C = \begin{bmatrix} 45 & 35 & 20 \\ 65 & 40 & 35 \end{bmatrix} + \begin{bmatrix} 30 & 32 & 28 \\ 43 & 47 & 30 \end{bmatrix} + \begin{bmatrix} 22 & 25 & 38 \\ 31 & 34 & 35 \end{bmatrix}$$

$$= \begin{bmatrix} 97 & 92 & 86 \\ 139 & 121 & 100 \end{bmatrix}$$

For example, from this sum, the total number of model C sofas shipped to the three warehouses in September was 86. ▲ ⑤

④

Find each sum when possible.

(a) $\begin{bmatrix} 2 & 5 & 7 \\ 3 & -1 & 4 \end{bmatrix}$ $+ \begin{bmatrix} -1 & 2 & 0 \\ 10 & -4 & 5 \end{bmatrix}$

(b) $\begin{bmatrix} 1 \\ 2 \\ 3 \end{bmatrix} + \begin{bmatrix} 2 & -1 \\ 4 & 5 \\ 6 & 0 \end{bmatrix}$

(c) $\begin{bmatrix} 5 & 4 & -1 \end{bmatrix} + \begin{bmatrix} -5 & 2 & 3 \end{bmatrix}$

Answers:

(a) $\begin{bmatrix} 1 & 7 & 7 \\ 13 & -5 & 9 \end{bmatrix}$

(b) Not possible

(c) $\begin{bmatrix} 0 & 6 & 2 \end{bmatrix}$

⑤

From the result of Example 4, find the total number of the following shipped to the three warehouses.

(a) Model A chairs

(b) Model B sofas

(c) Model C chairs

Answers:

(a) 139

(b) 92

(c) 100

As mentioned in Section 1.1, the additive inverse of the real number a is $-a$; a similar definition is given for the additive inverse of a matrix.

The **additive inverse** (or *negative*) of a matrix X is the matrix $-X$ in which each element is the additive inverse of the corresponding element of X.

TECHNOLOGY TIP A graphing calculator gives the additive inverse of a matrix by preceding the matrix with a negative sign. See Figure 6.15.

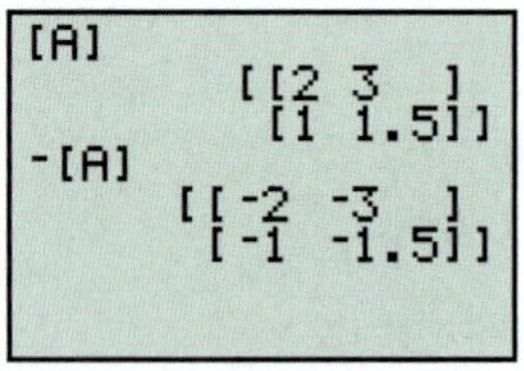

If

$$A = \begin{bmatrix} 1 & 2 & 3 \\ 0 & -1 & 5 \end{bmatrix} \quad \text{and} \quad B = \begin{bmatrix} -2 & 3 & 0 \\ 1 & -7 & 2 \end{bmatrix},$$

then, by the definition of the additive inverse of a matrix,

$$-A = \begin{bmatrix} -1 & -2 & -3 \\ 0 & 1 & -5 \end{bmatrix} \quad \text{and} \quad -B = \begin{bmatrix} 2 & -3 & 0 \\ -1 & 7 & -2 \end{bmatrix}.$$

```
[A]
        [[2 3 ]
         [1 1.5]]
-[A]
        [[-2 -3 ]
         [-1 -1.5]]
```

FIGURE 6.15

By the definition of matrix addition, for each matrix X, the sum $X + (-X)$ is a **zero matrix,** O, whose elements are all zeros. There is an $m \times n$ zero matrix for each pair of values of m and n.

$$\begin{bmatrix} 0 & 0 \\ 0 & 0 \end{bmatrix} \qquad \begin{bmatrix} 0 & 0 & 0 & 0 \\ 0 & 0 & 0 & 0 \end{bmatrix}$$

2×2 zero matrix $\qquad 2 \times 4$ zero matrix

Zero matrices have the following *identity property.*

If O is the $m \times n$ zero matrix, and A is any $m \times n$ matrix, then

$$A + O = O + A = A.$$

Compare this with the identity property for real numbers: for any real number a, $a + 0 = 0 + a = a$.

Subtraction

Subtraction of matrices can be defined in a manner comparable to subtraction for real numbers.

For two $m \times n$ matrices X and Y, the **difference** of X and Y is the $m \times n$ matrix $X - Y$ in which each element is the difference of the corresponding elements of X and Y, or, equivalently,

$$X - Y = X + (-Y).$$

According to this definition, matrix subtraction can be performed by subtracting corresponding elements. For example,

$$\begin{bmatrix} 1 & 2 & 3 \\ 0 & -1 & 5 \end{bmatrix} - \begin{bmatrix} -2 & 3 & 0 \\ 1 & -7 & 2 \end{bmatrix} = \begin{bmatrix} 1 - (-2) & 2 - 3 & 3 - 0 \\ 0 - 1 & -1 - (-7) & 5 - 2 \end{bmatrix}$$

$$= \begin{bmatrix} 3 & -1 & 3 \\ -1 & 6 & 3 \end{bmatrix}.$$

▼ **EXAMPLE 5** (a) $\begin{bmatrix} 8 & 6 & -4 \end{bmatrix} - \begin{bmatrix} 3 & 5 & -8 \end{bmatrix} = \begin{bmatrix} 5 & 1 & 4 \end{bmatrix}$

(b) The matrices

$$\begin{bmatrix} -2 & 5 \\ 0 & 1 \end{bmatrix} \quad \text{and} \quad \begin{bmatrix} 3 \\ 5 \end{bmatrix}$$

are different sizes and cannot be subtracted.

(c) Spreadsheets and graphing calculators can do matrix subtraction, as shown in Figure 6.16. ▲ ⬦6⬦

⬦6⬦

Find each of the following differences when possible.

(a) $\begin{bmatrix} 2 & 5 \\ -1 & 0 \end{bmatrix} - \begin{bmatrix} 6 & 4 \\ 3 & -2 \end{bmatrix}$

(b) $\begin{bmatrix} 1 & 5 & 6 \\ 2 & 4 & 8 \end{bmatrix} - \begin{bmatrix} 2 & 1 \\ 10 & 3 \end{bmatrix}$

(c) $\begin{bmatrix} 5 & -4 & 1 \end{bmatrix} - \begin{bmatrix} 6 & 0 & -3 \end{bmatrix}$

Answers:

(a) $\begin{bmatrix} -4 & 1 \\ -4 & 2 \end{bmatrix}$

(b) Not possible

(c) $\begin{bmatrix} -1 & -4 & 4 \end{bmatrix}$

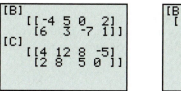

FIGURE 6.16

▼ **EXAMPLE 6** During September, the Chicago warehouse of the EZ Life Company shipped out the following numbers of each model:

$$K = \begin{bmatrix} 5 & 10 & 8 \\ 11 & 14 & 15 \end{bmatrix}.$$

What was the Chicago warehouse inventory on October 1, taking into account only the number of items received and sent out during the month?

Solution The number of each kind of item received during September is given by matrix C from Example 4; the number of each model sent out during September is given by matrix K above. The October 1 inventory will be represented by the matrix $C - K$ as shown below.

$$C - K = \begin{bmatrix} 22 & 25 & 38 \\ 31 & 34 & 35 \end{bmatrix} - \begin{bmatrix} 5 & 10 & 8 \\ 11 & 14 & 15 \end{bmatrix} = \begin{bmatrix} 17 & 15 & 30 \\ 20 & 20 & 20 \end{bmatrix}. \quad ▲$$

▼ **EXAMPLE 7** A drug company is testing 200 patients to see if Painoff (a new headache medicine) is effective. Half the patients receive Painoff and half receive a placebo. The data on the first 50 patients is summarized in this matrix.

Pain Relief Obtained
Yes   No

Patient took Painoff $\begin{bmatrix} 22 & 3 \\ Patient took placebo & 8 & 17 \end{bmatrix}$

For example, row 2 shows that, of the people who took the placebo, 8 got relief but 17 did not. The test was repeated on three more groups of 50 patients each, with the results summarized by these matrices:

$$\begin{bmatrix} 21 & 4 \\ 6 & 19 \end{bmatrix} \begin{bmatrix} 19 & 6 \\ 10 & 15 \end{bmatrix} \begin{bmatrix} 23 & 2 \\ 3 & 22 \end{bmatrix}.$$

The total results of the test can be obtained by adding these four matrices:

$$\begin{bmatrix} 22 & 3 \\ 8 & 17 \end{bmatrix} + \begin{bmatrix} 21 & 4 \\ 6 & 19 \end{bmatrix} + \begin{bmatrix} 19 & 6 \\ 10 & 15 \end{bmatrix} + \begin{bmatrix} 23 & 2 \\ 3 & 22 \end{bmatrix} = \begin{bmatrix} 85 & 15 \\ 27 & 73 \end{bmatrix}.$$

Because 85 of 100 patients got relief with Painoff and only 27 of 100 with the placebo, it appears that Painoff is effective. ▲ ⟨7⟩

Suppose one of the EZ Life Company warehouses receives the following order, written in matrix form, where the entries have the same meaning as given earlier:

$$\begin{bmatrix} 5 & 4 & 1 \\ 3 & 2 & 3 \end{bmatrix}.$$

Later, the store that sent the order asks the warehouse to send six more of the same order. The six new orders can be written as one matrix by multiplying each element in the matrix by 6, giving the product

$$6\begin{bmatrix} 5 & 4 & 1 \\ 3 & 2 & 3 \end{bmatrix} = \begin{bmatrix} 30 & 24 & 6 \\ 18 & 12 & 18 \end{bmatrix}.$$

In work with matrices, a real number, like the 6 in the preceding multiplication, is called a **scalar.**

The **product** of a scalar k and a matrix X is the matrix kX, each of whose elements is k times the corresponding element of X.

Here are two examples, one done by hand and the other on a graphing calculator (Figure 6.17).

$$(-3)\begin{bmatrix} 2 & -5 \\ 1 & 7 \\ 4 & -6 \end{bmatrix} = \begin{bmatrix} -6 & 15 \\ -3 & -21 \\ -12 & 18 \end{bmatrix}$$

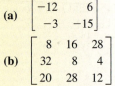

FIGURE 6.17 ⟨8⟩

⟨7⟩

Later, it was discovered that the data in the last group of 50 patients in Example 7 was invalid. Use a matrix to represent the total test results after that data was eliminated.

Answer:

$$\begin{bmatrix} 62 & 13 \\ 24 & 51 \end{bmatrix}$$

⟨8⟩

Find each product.

(a) $-3\begin{bmatrix} 4 & -2 \\ 1 & 5 \end{bmatrix}$

(b) $4\begin{bmatrix} 2 & 4 & 7 \\ 8 & 2 & 1 \\ 5 & 7 & 3 \end{bmatrix}$

Answers:

(a) $\begin{bmatrix} -12 & 6 \\ -3 & -15 \end{bmatrix}$

(b) $\begin{bmatrix} 8 & 16 & 28 \\ 32 & 8 & 4 \\ 20 & 28 & 12 \end{bmatrix}$

◆6.3 Exercises

Find the size of each of the given matrices. Identify any square, column, or row matrices. (See Example 2.) Give the additive inverse of each matrix.

1. $\begin{bmatrix} 7 & -8 & 4 \\ 0 & 13 & 9 \end{bmatrix}$ **2.** $\begin{bmatrix} -7 & 23 \\ 5 & -6 \end{bmatrix}$

3. $\begin{bmatrix} -3 & 0 & 11 \\ 1 & \frac{1}{4} & -7 \\ 5 & -3 & 9 \end{bmatrix}$ **4.** $\begin{bmatrix} 6 & -4 & \frac{2}{3} & 12 & 2 \end{bmatrix}$

5. $\begin{bmatrix} 7 \\ 11 \end{bmatrix}$ **6.** $\begin{bmatrix} -5 \end{bmatrix}$

7. If A is a 5×3 matrix and $A + B = A$, what do you know about B?

8. If C is a 3×3 matrix and D is a 3×4 matrix, then $C + D$ is _____.

Perform the indicated operations where possible. (See Examples 3–6.)

9. $\begin{bmatrix} 1 & 2 & 5 & -1 \\ 3 & 0 & 2 & -4 \end{bmatrix} + \begin{bmatrix} 8 & 12 & -5 & 5 \\ -2 & -1 & 0 & 0 \end{bmatrix}$

10. $\begin{bmatrix} 1 & 5 \\ 2 & -3 \\ 3 & 7 \end{bmatrix} + \begin{bmatrix} 2 & 8 \\ 8 & 6 \\ -1 & 9 \end{bmatrix}$

11. $\begin{bmatrix} -1 & 5 & 9 \\ 2 & 2 & 3 \end{bmatrix} + \begin{bmatrix} 4 & 8 & -7 \\ 1 & -1 & 5 \end{bmatrix}$

12. $\begin{bmatrix} 2 & 4 \\ -8 & 2 \end{bmatrix} + \begin{bmatrix} 9 & -5 \\ 8 & 5 \end{bmatrix}$

13. $\begin{bmatrix} 4 & -2 & 5 \\ 3 & 7 & 0 \end{bmatrix} - \begin{bmatrix} 1 & 5 & -2 \\ -3 & 3 & 8 \end{bmatrix}$

14. $\begin{bmatrix} 9 & 1 \\ 0 & -3 \\ 4 & 10 \end{bmatrix} - \begin{bmatrix} 1 & 9 & -4 \\ -1 & 1 & 0 \end{bmatrix}$

Let $A = \begin{bmatrix} -2 & 0 \\ 5 & 3 \end{bmatrix}$ and $B = \begin{bmatrix} 0 & 2 \\ 4 & -6 \end{bmatrix}$. Find each of the following.

15. $2A$ **16.** $-3B$ **17.** $-4B$

18. $5A$ **19.** $-4A + 5B$ **20.** $3A - 10B$

Let $A = \begin{bmatrix} 1 & -2 \\ 4 & 3 \end{bmatrix}$ and $B = \begin{bmatrix} 2 & -1 \\ 0 & 5 \end{bmatrix}$. Find a matrix X satisfying the given equation.

21. $2X = 2A + 3B$ **22.** $3X = A - 3B$

Using matrices

$$O = \begin{bmatrix} 0 & 0 \\ 0 & 0 \end{bmatrix}, P = \begin{bmatrix} m & n \\ p & q \end{bmatrix}, T = \begin{bmatrix} r & s \\ t & u \end{bmatrix}, \text{ and } X = \begin{bmatrix} x & y \\ z & w \end{bmatrix},$$

verify that the statements in Exercises 23–28 are true.

23. $X + T$ is a 2×2 matrix.

24. $X + T = T + X$ (commutative property of addition of matrices)

25. $X + (T + P) = (X + T) + P$ (associative property of addition of matrices)

26. $X + (-X) = O$ (inverse property of addition of matrices)

27. $P + O = P$ (identity property of addition of matrices)

28. Which of the preceding properties are valid for matrices that are not square?

Work the following exercises. (See Examples 1 and 7.)

29. **Business** When ticket holders fail to attend, major league sports teams lose the money these fans would have spent on refreshments, souvenirs, etc. The percentage of fans who don't show up is 16% in basketball and hockey, 20% in football, and 18% in baseball. The lost revenue per fan is $18.20 in basketball, $18.25 in hockey, $19 in football, and $15.40 in baseball. The total annual lost revenue is $22.7 million in basketball, $35.8 million in hockey, $51.9 million in football, and $96.3 million in baseball.* Express this information in matrix form; specify what the rows and columns represent.

30. **Finance** 78% of undergraduate students had credit cards in 2000 and 76% in 2004. Average credit card debt was $2748 in 2000 and $2169 in 2004. The percentage of students with balances of $7000 or more was 9% in 2000 and 7% in 2004. The average number of credit cards per student was 3 in 2000 and 4.09 in 2004.† Write this information first as a 4×2 matrix and then as a 2×4 matrix.

31. **Health** The shortage of organs for transplants is a continuing problem in the United States. In 1998, there were 4121 people waiting for a heart transplant, 3171 for a lung transplant, 12,070 for a liver transplant, and 38,270 for a kidney transplant. Corresponding figures for 2000 were 4143, 3614, 15,359, and 45,273. In mid-2005, the figures

_____
*American Demographics.

†Nellie Mae.

were 3140, 3601, 17,376, and 62,130.* Express this information as a matrix, labeling rows and columns.

Work these exercises. (See Examples 1, 4, 6, and 7.)

32. Management There are three convenience stores in Gambier. This week, Store I sold 88 loaves of bread, 48 quarts of milk, 16 jars of peanut butter, and 112 pounds of cold cuts. Store II sold 105 loaves of bread, 72 quarts of milk, 21 jars of peanut butter, and 147 pounds of cold cuts. Store III sold 60 loaves of bread, 40 quarts of milk, no peanut butter, and 50 pounds of cold cuts.
(a) Use a 3 × 4 matrix to express the sales information for the three stores.
(b) During the following week, sales on these products at Store I increased by 25%, sales at Store II increased by 1/3, and sales at Store III increased by 10%. Write the sales matrix for that week.
(c) Write a matrix that represents total sales over the two-week period.

33. Health Heart disease death rates (per 100,000 population) vary in different age groups. The rates for people 15–24 years old were 2.9 in 1995, 2.6 in 2000, and 2.5 in 2001 and 2002. In 1995, the death rates were 111 for people 45–54 years old and 799.9 for people 65–74 years old. The death rates for people 45–54 years old were 94.2 in 2000, 92.9 in 2001, and 90.7 in 2002. For people 65–74, the death rates were 666.6 in 2000, 635.1 in 2001, and 615.9 in 2002.[†] Write this information in matrix form, labling rows and columns.

34. Social Sciences The following table gives the educational attainment of the U.S. population 25 years and older in various years.[‡]

	MALE		FEMALE	
	Four Years of High School or More	Four Years of College or More	Four Years of High School or More	Four Years of College or More
1950	32.6	7.3	36.0	5.2
1960	39.5	9.7	42.5	5.8
1970	51.9	13.5	52.8	8.1
1980	67.3	20.1	65.8	12.8
1990	77.7	24.4	77.5	18.4
2000	84.2	27.8	84.0	23.6
2003	84.1	28.9	85.0	25.7

(a) Write a 2 × 7 matrix for the educational attainment of males.
(b) Write a 2 × 7 matrix for the educational attainment of females.
(c) Use the matrices from parts (a) and (b) to write a matrix showing how much more (or less) education males have attained than females.

35. Transportation The tables give the death rates (per million person trips) for male and female drivers for various ages and numbers of passengers.*

MALE DRIVERS				
Age	Number of Passengers			
	0	1	2	≥3
16	2.61	4.39	6.29	9.08
17	1.63	2.77	4.61	6.92
30–59	.92	.75	.62	.54

FEMALE DRIVERS				
Age	Number of Passengers			
	0	1	2	≥3
16	1.38	1.72	1.94	3.31
17	1.26	1.48	2.82	2.28
30–59	.41	.33	.27	.40

(a) Write a matrix A for the death rate of male drivers.
(b) Write a matrix B for the death rate of female drivers.
(c) Use the matrices from parts (a) and (b) to write a matrix showing the difference between the death rate of males and females.

36. Transportation Use matrix operations on the matrices found in Exercise 35(a)(b) to find one matrix that gives the combined death rates for males and females (per million person trips) of drivers of various ages, with varying numbers of passengers. (*Hint:* Consider $\frac{1}{2}(A + B)$.)

---

*United Network of Organ Sharing and National Organ Procurement and Transplantation Network.

†U.S. Department of Halth and Human Services, center for Health Statistics.

‡U.S. Census Bureau.

*Li-Hui Chen, Susan Baker, Elisa Braver, and Guohua Li, "Carrying Passengers as a Risk Factor for Crashes Fatal to 16- and 17-Year-Old Drivers," *JAMA* 283, no. 12 (March 22/29, 2000): 1578–1582.

6.4 Matrix Products and Inverses

In the previous section, we showed how to multiply a matrix by a scalar. Finding the product of two matrices is more involved, but is important in solving practical problems. To understand the reasoning behind the definition of matrix multiplication, look again at the EZ Life Company. Suppose sofas and chairs of the same model are often sold as sets, with matrix W showing the number of each model set in each warehouse.

$$
\begin{array}{c}
\begin{array}{ccc} A & B & C \end{array} \\
\begin{array}{c} \text{New York} \\ \text{Chicago} \\ \text{San Francisco} \end{array}
\begin{bmatrix} 10 & 7 & 3 \\ 5 & 9 & 6 \\ 4 & 8 & 2 \end{bmatrix} = W
\end{array}
$$

If the selling price of a model A set is $800, of a model B set $1000, and of a model C set $1200, find the total value of the sets in the New York warehouse as follows.

Type	Number of Sets		Price of Set		Total
A	10	×	$ 800	=	$ 8,000
B	7	×	1000	=	7,000
C	3	×	1200	=	3,600
			Total for New York		$18,600

The total value of the three kinds of sets in New York is $18,600. ◇①

The work done in the preceding table is summarized as

$$10(\$800) + 7(\$1000) + 3(\$1200) = \$18,600.$$

In the same way, the Chicago sets have a total value of

$$5(\$800) + 9(\$1000) + 6(\$1200) = \$20,200,$$

and in San Francisco, the total value of the sets is

$$4(\$800) + 8(\$1000) + 2(\$1200) = \$13,600.$$

The selling prices can be written as a column matrix P and the total value in each location as a column matrix V.

$$
P = \begin{bmatrix} 800 \\ 1000 \\ 1200 \end{bmatrix} \quad \text{and} \quad V = \begin{bmatrix} 18,600 \\ 20,200 \\ 13,600 \end{bmatrix}
$$

①

In this example of the EZ Life Company, find the total value of the New York sets if model A sets sell for $1200, model B for $1600, and model C for $1300.

Answer:

$27,100

Consider how the first row of the matrix W and the single column P lead to the first entry of V.

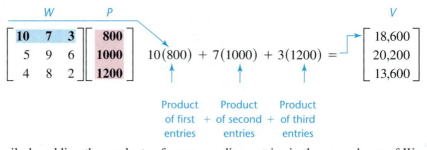

Similarly, adding the products of corresponding entries in the second row of W and the column P produces the second entry in V. The third entry in V is obtained in the same way by using the third row of W and column P. This suggests that it is reasonable to *define* the product WP to be V.

$$WP = \begin{bmatrix} 10 & 7 & 3 \\ 5 & 9 & 6 \\ 4 & 8 & 2 \end{bmatrix} \begin{bmatrix} 800 \\ 1000 \\ 1200 \end{bmatrix} = \begin{bmatrix} 18{,}600 \\ 20{,}200 \\ 13{,}600 \end{bmatrix} = V$$

Note the sizes of the matrices here: the product of a 3×3 matrix and a 3×1 matrix is a 3×1 matrix.

Multiplying Matrices

We first define the **product of a row of a matrix and a column of a matrix** (with the same number of entries in each) to be the *number* obtained by multiplying the corresponding entries (first by first, second by second, etc.) and adding the results. For instance,

$$\begin{bmatrix} 3 & -2 & 1 \end{bmatrix} \cdot \begin{bmatrix} 4 \\ 5 \\ 0 \end{bmatrix} = 3 \cdot 4 + (-2) \cdot 5 + 1 \cdot 0 = 12 - 10 + 0 = 2.$$

Now **matrix multiplication** is defined as follows.

Let A be an $m \times n$ matrix and let B be an $n \times k$ matrix. The **product matrix** AB is the $m \times k$ matrix whose entry in the ith row and jth column is

the product of the ith row of A and the jth column of B.

CAUTION Be careful when multiplying matrices. Remember that the number of *columns* of A must equal the number of *rows* of B in order to get the product matrix AB. The final product will have as many rows as A and as many columns as B.

▼ **EXAMPLE 1**   Suppose matrix A is 2×2 and matrix B is 2×4. Can the product AB be calculated? What is the size of the product?

Solution The following diagram helps decide the answers to these questions.

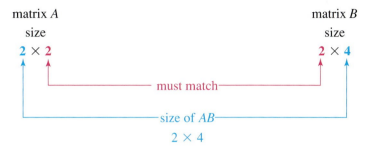

The product AB can be calculated because A has two columns and B has two rows. The product will be a 2×4 matrix. ▲ ②

▼ **EXAMPLE 2** Find the product CD when

$$C = \begin{bmatrix} -3 & 4 & 2 \\ 5 & 0 & 4 \end{bmatrix} \quad \text{and} \quad D = \begin{bmatrix} -6 & 4 \\ 2 & 3 \\ 3 & -2 \end{bmatrix}.$$

Solution Here, matrix C is 2×3 and matrix D is 3×2, so matrix CD can be found and will be 2×2.

Step 1   row 1, column 1

$$\begin{bmatrix} -3 & 4 & 2 \\ 5 & 0 & 4 \end{bmatrix} \begin{bmatrix} -6 & 4 \\ 2 & 3 \\ 3 & -2 \end{bmatrix} \qquad (-3) \cdot (-6) + 4 \cdot 2 + 2 \cdot 3 = 32$$

Hence, 32 is the entry in row 1, column 1, of CD, as shown in Step 5 below.

Step 2   row 1, column 2

$$\begin{bmatrix} -3 & 4 & 2 \\ 5 & 0 & 4 \end{bmatrix} \begin{bmatrix} -6 & 4 \\ 2 & 3 \\ 3 & -2 \end{bmatrix} \qquad (-3) \cdot 4 + 4 \cdot 3 + 2 \cdot (-2) = -4$$

So -4 is the entry in row 1, column 2, of CD, as shown in Step 5. Continue in this manner to find the remaining entries of CD.

Step 3   row 2, column 1

$$\begin{bmatrix} -3 & 4 & 2 \\ 5 & 0 & 4 \end{bmatrix} \begin{bmatrix} -6 & 4 \\ 2 & 3 \\ 3 & -2 \end{bmatrix} \qquad 5 \cdot (-6) + 0 \cdot 2 + 4 \cdot 3 = -18$$

Step 4   row 2, column 2

$$\begin{bmatrix} -3 & 4 & 2 \\ 5 & 0 & 4 \end{bmatrix} \begin{bmatrix} -6 & 4 \\ 2 & 3 \\ 3 & -2 \end{bmatrix} \qquad 5 \cdot 4 + 0 \cdot 3 + 4 \cdot (-2) = 12$$

Step 5   The product is

$$CD = \begin{bmatrix} -3 & 4 & 2 \\ 5 & 0 & 4 \end{bmatrix} \begin{bmatrix} -6 & 4 \\ 2 & 3 \\ 3 & -2 \end{bmatrix} = \begin{bmatrix} 32 & -4 \\ -18 & 12 \end{bmatrix}. \quad \blacktriangle \quad \langle 3 \rangle$$

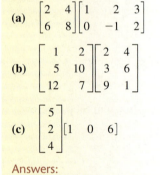

③

Find the product CD, given that

$$C = \begin{bmatrix} 1 & 3 & 5 \\ 2 & -4 & -1 \end{bmatrix}$$

and

$$D = \begin{bmatrix} 2 & -1 \\ 4 & 3 \\ 1 & -2 \end{bmatrix}.$$

Answer:

$$CD = \begin{bmatrix} 19 & -2 \\ -13 & -12 \end{bmatrix}$$

▼ **EXAMPLE 3**   Find BA, given that

$$A = \begin{bmatrix} 1 & 7 \\ -3 & 2 \end{bmatrix} \quad \text{and} \quad B = \begin{bmatrix} 1 & 0 & -1 \\ 3 & 1 & 4 \end{bmatrix}.$$

Since B is a 2×3 matrix and A is a 2×2 matrix, the product BA cannot be found. ▲ ④

> **TECHNOLOGY TIP**   Graphing calculators can find matrix products. However, if you use a graphing calculator to try to find the product in Example 3, the calculator will display an error message.

④

Give the size of each of the following products that can be found.

(a) $\begin{bmatrix} 2 & 4 \\ 6 & 8 \end{bmatrix} \begin{bmatrix} 1 & 2 & 3 \\ 0 & -1 & 2 \end{bmatrix}$

(b) $\begin{bmatrix} 1 & 2 \\ 5 & 10 \\ 12 & 7 \end{bmatrix} \begin{bmatrix} 2 & 4 \\ 3 & 6 \\ 9 & 1 \end{bmatrix}$

(c) $\begin{bmatrix} 5 \\ 2 \\ 4 \end{bmatrix} \begin{bmatrix} 1 & 0 & 6 \end{bmatrix}$

Answers:

(a) 2×3

(b) Not possible

(c) 3×3

Matrix multiplication has some similarities to the multiplication of numbers.

> For any matrices A, B, and C such that all the indicated sums and products exist, matrix multiplication is associative and distributive.
>
> $$A(BC) = (AB)C \qquad A(B + C) = AB + AC \qquad (B + C)A = BA + CA$$

However, there are important differences between matrix multiplication and the multiplication of numbers. (See Exercises 19–22 at the end of this section.) In particular, matrix multiplication is *not* commutative.

> If A and B are matrices such that the products AB and BA exist,
>
> $$AB \text{ may not equal } BA.$$

Figure 6.18 shows an example of this situation.

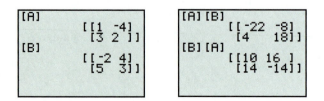

FIGURE 6.18

▼ **EXAMPLE 4**   A contractor builds three kinds of houses, models A, B, and C, with a choice of two styles, Spanish or contemporary. Matrix P shows the number of each kind of house planned for a new 100-home subdivision.

$$
\begin{array}{c}
 \\
\text{Model A} \\
\text{Model B} \\
\text{Model C}
\end{array}
\begin{array}{cc}
\text{Spanish} & \text{Contemporary} \\
\end{array}
\left[
\begin{array}{cc}
0 & 30 \\
10 & 20 \\
20 & 20
\end{array}
\right] = P
$$

The amounts for each of the exterior materials used depend primarily on the style of the house. These amounts are shown in matrix Q. (Concrete is in cubic yards, lumber in units of 1000 board feet, brick in thousands, and shingles in units of 100 square feet.)

$$
\begin{array}{c}
 \\
\text{Spanish} \\
\text{Contemporary}
\end{array}
\begin{array}{cccc}
\text{Concrete} & \text{Lumber} & \text{Brick} & \text{Shingles} \\
\end{array}
\left[
\begin{array}{cccc}
10 & 2 & 0 & 2 \\
50 & 1 & 20 & 2
\end{array}
\right] = Q
$$

Matrix R gives the cost for each kind of material.

$$
\begin{array}{c}
 \\
\text{Concrete} \\
\text{Lumber} \\
\text{Brick} \\
\text{Shingles}
\end{array}
\begin{array}{c}
\text{Cost per Unit} \\
\end{array}
\left[
\begin{array}{c}
20 \\
180 \\
60 \\
25
\end{array}
\right] = R
$$

(a) What is the total cost for each model house?

Solution   First find the product PQ, which shows the amount of each material needed for each model house.

$$
PQ = \begin{bmatrix} 0 & 30 \\ 10 & 20 \\ 20 & 20 \end{bmatrix} \begin{bmatrix} 10 & 2 & 0 & 2 \\ 50 & 1 & 20 & 2 \end{bmatrix}
$$

$$
PQ =
\begin{array}{cccc}
\text{Concrete} & \text{Lumber} & \text{Brick} & \text{Shingles} \\
\end{array}
\left[
\begin{array}{cccc}
1500 & 30 & 600 & 60 \\
1100 & 40 & 400 & 60 \\
1200 & 60 & 400 & 80
\end{array}
\right]
\begin{array}{c}
\text{Model A} \\
\text{Model B} \\
\text{Model C}
\end{array}
$$

Now multiply PQ and R, the cost matrix, to get the total cost for each model house.

$$
\begin{bmatrix} 1500 & 30 & 600 & 60 \\ 1100 & 40 & 400 & 60 \\ 1200 & 60 & 400 & 80 \end{bmatrix} \begin{bmatrix} 20 \\ 180 \\ 60 \\ 25 \end{bmatrix} =
\begin{array}{c}
\text{Cost} \\
\end{array}
\left[
\begin{array}{c}
72{,}900 \\
54{,}700 \\
60{,}800
\end{array}
\right]
\begin{array}{c}
\text{Model A} \\
\text{Model B} \\
\text{Model C}
\end{array}
$$

(b) How much of each of the four kinds of material must be ordered?

Solution   The totals of the columns of matrix PQ will give a matrix whose elements represent the total amounts of each material needed for the subdivision. Call this matrix T and write it as a row matrix.

$$
T = \begin{bmatrix} 3800 & 130 & 1400 & 200 \end{bmatrix}
$$

(c) What is the total cost for material?

Solution Find the total cost of all the materials by taking the product of matrix T, the matrix showing the total amounts of each material, and matrix R, the cost matrix. (To multiply these and get a 1×1 matrix representing total cost, we must multiply a 1×4 matrix by a 4×1 matrix. This is why T was written as a row matrix in (b).

$$TR = \begin{bmatrix} 3800 & 130 & 1400 & 200 \end{bmatrix} \begin{bmatrix} 20 \\ 180 \\ 60 \\ 25 \end{bmatrix} = \begin{bmatrix} 188{,}400 \end{bmatrix}$$

(d) Suppose the contractor builds the same number of homes in five subdivisions.

Solution Calculate the total amount of each material for each model for all five subdivisions. Multiply PQ by the scalar 5 as follows.

$$5 \begin{bmatrix} 1500 & 30 & 600 & 60 \\ 1100 & 40 & 400 & 60 \\ 1200 & 60 & 400 & 80 \end{bmatrix} = \begin{bmatrix} 7500 & 150 & 3000 & 300 \\ 5500 & 200 & 2000 & 300 \\ 6000 & 300 & 2000 & 400 \end{bmatrix} \blacktriangle$$

We can introduce notation to help keep track of the quantities a matrix represents. For example, we can say that matrix P from Example 4 represents models/styles, matrix Q represents styles/materials, and matrix R represents materials/cost. In each case, the meaning of the rows is written first and the columns second. When we found the product PQ in Example 4, the rows of the matrix represented models and the columns represented materials. Therefore, we can say that the matrix product PQ represents models/materials. The common quantity, styles, in both P and Q was eliminated in the product PQ. Do you see that the product $(PQ)R$ represents models/cost?

In practical problems, this notation helps decide in what order to multiply two matrices so that the results are meaningful. In Example 4(c), we could have found either product RT or product TR. However, since T represents subdivisions/materials and R represents materials/cost, the product TR gives subdivisions/cost.

Identity and Inverse Matrices

Recall from Section 1.1 that the real number 1 is the identity element for multiplication of real numbers: for any real number a, $a \cdot 1 = 1 \cdot a = a$. In this section, an **identity matrix** I is defined that has properties similar to those of the number 1.

If I is to be the identity matrix, the products AI and IA must both equal A. The 2×2 identity matrix that satisfies these conditions is

$$I = \begin{bmatrix} 1 & 0 \\ 0 & 1 \end{bmatrix}.$$

To check that I is really the 2×2 identity matrix, let

$$A = \begin{bmatrix} a & b \\ c & d \end{bmatrix}.$$

⬦5

Let matrix A be

$$\begin{array}{c} & \textit{Vitamin} \\ & \begin{array}{ccc} \text{C} & \text{E} & \text{K} \end{array} \\ \textit{Brand} \begin{array}{c} \text{X} \\ \text{Y} \end{array} & \begin{bmatrix} 2 & 7 & 5 \\ 4 & 6 & 9 \end{bmatrix} \end{array}$$

and matrix B be

$$\begin{array}{c} & \textit{Cost} \\ & \begin{array}{cc} \text{X} & \text{Y} \end{array} \\ \textit{Vitamin} \begin{array}{c} \text{C} \\ \text{E} \\ \text{K} \end{array} & \begin{bmatrix} 12 & 14 \\ 18 & 15 \\ 9 & 10 \end{bmatrix}. \end{array}$$

(a) What quantities do matrices A and B represent?

(b) What quantities does the product AB represent?

(c) What quantities does the product BA represent?

Answers:

(a) $A =$ brand/vitamin, $B =$ vitamin/cost

(b) $AB =$ brand/cost

(c) Not meaningful, although the product BA can be found

⬦6

Let $A = \begin{bmatrix} 3 & -2 \\ 4 & -1 \end{bmatrix}$ and

$I = \begin{bmatrix} 1 & 0 \\ 0 & 1 \end{bmatrix}.$

Find IA and AI.

Answer:

$IA = \begin{bmatrix} 3 & -2 \\ 4 & -1 \end{bmatrix} = A$ and

$AI = \begin{bmatrix} 3 & -2 \\ 4 & -1 \end{bmatrix} = A$

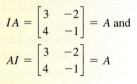

Then AI and IA should both equal A.

$$AI = \begin{bmatrix} a & b \\ c & d \end{bmatrix} \begin{bmatrix} 1 & 0 \\ 0 & 1 \end{bmatrix} = \begin{bmatrix} a(1) + b(0) & a(0) + b(1) \\ c(1) + d(0) & c(0) + d(1) \end{bmatrix} = \begin{bmatrix} a & b \\ c & d \end{bmatrix} = A$$

$$IA = \begin{bmatrix} 1 & 0 \\ 0 & 1 \end{bmatrix} \begin{bmatrix} a & b \\ c & d \end{bmatrix} = \begin{bmatrix} 1(a) + 0(c) & 1(b) + 0(d) \\ 0(a) + 1(c) & 0(b) + 1(d) \end{bmatrix} = \begin{bmatrix} a & b \\ c & d \end{bmatrix} = A$$

This verifies that I has been defined correctly. (It can also be shown that I is the only 2×2 identity matrix.)

The identity matrices for 3×3 matrices and 4×4 matrices are, respectively,

$$I = \begin{bmatrix} 1 & 0 & 0 \\ 0 & 1 & 0 \\ 0 & 0 & 1 \end{bmatrix} \quad \text{and} \quad I = \begin{bmatrix} 1 & 0 & 0 & 0 \\ 0 & 1 & 0 & 0 \\ 0 & 0 & 1 & 0 \\ 0 & 0 & 0 & 1 \end{bmatrix}.$$

TECHNOLOGY TIP An $n \times n$ identity matrix can be displayed on most graphing calculators by using IDENTITY n or IDENT n or IDENMAT(n). Look in the MATH or OPS submenu of the TI MATRIX menu, or the OPTN MAT menu of Casio, or the MATRIX submenu of the HP-39+ MATH menu.

By generalizing, an identity matrix can be found for any n by n matrix: this identity matrix will have 1's on the main diagonal from upper left to lower right, with all other entries equal to 0.

Recall that for every nonzero real number a, the equation $ax = 1$ has a solution, namely, $x = 1/a = a^{-1}$. Similarly, for a square matrix A, we consider the matrix equation $AX = I$. This equation does not always have a solution, but when it does, we use special terminology. If there is a matrix A^{-1} satisfying

$$AA^{-1} = I,$$

(that is, A^{-1} is a solution of $AX = I$), then A^{-1} is called the **inverse matrix** of A. In this case, it can be proved that $A^{-1}A = I$ and that A^{-1} is unique (that is, a square matrix has no more than one inverse). When a matrix has an inverse, it can be found by using the row operations given in Section 6.2, as we shall see below.

CAUTION Only square matrices have inverses, but not every square matrix has one. A matrix that does not have an inverse is called a **singular matrix.** Note that the symbol A^{-1} (read A-inverse) does *not* mean $1/A$ or I/A; the symbol A^{-1} is just the notation for the inverse of matrix A. There is no such thing as matrix division.

▼ **EXAMPLE 5** Given matrices A and B below, determine whether B is the inverse of A.

$$A = \begin{bmatrix} 1 & 2 \\ 4 & 6 \end{bmatrix} \qquad B = \begin{bmatrix} -3 & 1 \\ 2 & -\frac{1}{2} \end{bmatrix}$$

⑦

Given $A = \begin{bmatrix} 2 & 1 \\ 3 & 8 \end{bmatrix}$ and

$B = \begin{bmatrix} -1 & 3 \\ 1 & -2 \end{bmatrix}$, determine if they are inverses.

Answer:

No, because $AB \neq I$.

Solution B is the inverse of A if $AB = I$ and $BA = I$, so we find those products.

$$AB = \begin{bmatrix} 1 & 2 \\ 4 & 6 \end{bmatrix} \begin{bmatrix} -3 & 1 \\ 2 & -\frac{1}{2} \end{bmatrix} = \begin{bmatrix} 1 & 0 \\ 0 & 1 \end{bmatrix} = I$$

$$BA = \begin{bmatrix} -3 & 1 \\ 2 & -\frac{1}{2} \end{bmatrix} \begin{bmatrix} 1 & 2 \\ 4 & 6 \end{bmatrix} = \begin{bmatrix} 1 & 0 \\ 0 & 1 \end{bmatrix} = I$$

Therefore, B is the inverse of A; that is, $A^{-1} = B$. (It is also true that A is the inverse of B, or $B^{-1} = A$.) ▲ ⑦

To see how to find the multiplicative inverse of a matrix, let us look for the inverse of

$$A = \begin{bmatrix} 2 & 4 \\ 1 & -1 \end{bmatrix}.$$

Let the unknown inverse matrix be

$$A^{-1} = \begin{bmatrix} x & y \\ z & w \end{bmatrix}.$$

By the definition of matrix inverse, $AA^{-1} = I$, or

$$AA^{-1} = \begin{bmatrix} 2 & 4 \\ 1 & -1 \end{bmatrix}\begin{bmatrix} x & y \\ z & w \end{bmatrix} = \begin{bmatrix} 1 & 0 \\ 0 & 1 \end{bmatrix}.$$

Use matrix multiplication to get

$$\begin{bmatrix} 2x + 4z & 2y + 4w \\ x - z & y - w \end{bmatrix} = \begin{bmatrix} 1 & 0 \\ 0 & 1 \end{bmatrix}.$$

Setting corresponding elements equal gives the system of equations

$$2x + 4z = 1 \tag{1}$$
$$2y + 4w = 0 \tag{2}$$
$$x - z = 0 \tag{3}$$
$$y - w = 1. \tag{4}$$

Since equations (1) and (3) involve only x and z, while equations (2) and (4) involve only y and w, these four equations lead to two systems of equations:

$$\begin{matrix} 2x + 4z = 1 \\ x - z = 0 \end{matrix} \quad \text{and} \quad \begin{matrix} 2y + 4w = 0 \\ y - w = 1. \end{matrix}$$

Writing the two systems as augmented matrices gives

$$\left[\begin{array}{cc|c} 2 & 4 & 1 \\ 1 & -1 & 0 \end{array}\right] \quad \text{and} \quad \left[\begin{array}{cc|c} 2 & 4 & 0 \\ 1 & -1 & 1 \end{array}\right].$$

Note that the row operations needed to transform both matrices are the same because the first two columns of both matrices are identical. Consequently, we can save time by combining these matrices into the single matrix

$$\left[\begin{array}{cc|cc} 2 & 4 & 1 & 0 \\ 1 & -1 & 0 & 1 \end{array}\right]. \tag{5}$$

Columns 1–3 represent the first system and columns 1, 2, and 4 represent the second system. Now use row operations as follows.

$$\left[\begin{array}{cc|cc} 1 & -1 & 0 & 1 \\ 2 & 4 & 1 & 0 \end{array}\right] \qquad \text{Interchange } R_1, R_2$$

$$\left[\begin{array}{cc|cc} 1 & -1 & 0 & 1 \\ 0 & 6 & 1 & -2 \end{array}\right] \qquad -2R_1 + R_2$$

$$\left[\begin{array}{cc|cc} 1 & -1 & 0 & 1 \\ 0 & 1 & \frac{1}{6} & -\frac{1}{3} \end{array}\right] \qquad \frac{1}{6}R_2$$

$$\left[\begin{array}{cc|cc} 1 & 0 & \frac{1}{6} & \frac{2}{3} \\ 0 & 1 & \frac{1}{6} & -\frac{1}{3} \end{array}\right] \qquad R_2 + R_1 \tag{6}$$

The left half of the augmented matrix (6) is the identity matrix, so the Gauss–Jordan process is finished and the solutions can be read from the right half of the augmented matrix. The numbers in the first column to the right of the vertical bar give the values of x and z. The second column to the right of the bar gives the values of y and w. That is,

$$\begin{bmatrix} 1 & 0 & | & x & y \\ 0 & 1 & | & z & w \end{bmatrix} = \begin{bmatrix} 1 & 0 & | & \frac{1}{6} & \frac{2}{3} \\ 0 & 1 & | & \frac{1}{6} & -\frac{1}{3} \end{bmatrix},$$

so that

$$A^{-1} = \begin{bmatrix} x & y \\ z & w \end{bmatrix} = \begin{bmatrix} \frac{1}{6} & \frac{2}{3} \\ \frac{1}{6} & -\frac{1}{3} \end{bmatrix}.$$

Check by multiplying A and A^{-1}. The result should be I.

$$AA^{-1} = \begin{bmatrix} 2 & 4 \\ 1 & -1 \end{bmatrix}\begin{bmatrix} \frac{1}{6} & \frac{2}{3} \\ \frac{1}{6} & -\frac{1}{3} \end{bmatrix} = \begin{bmatrix} \frac{1}{3}+\frac{2}{3} & \frac{4}{3}-\frac{4}{3} \\ \frac{1}{6}-\frac{1}{6} & \frac{2}{3}+\frac{1}{3} \end{bmatrix} = \begin{bmatrix} 1 & 0 \\ 0 & 1 \end{bmatrix} = I \quad \diamond \boxed{8}$$

Thus, the original augmented matrix (5) has A as its left half and the identity matrix as its right half, while the final augmented matrix (6), at the end of the Gauss–Jordan process, has the identity matrix as its left half and the inverse matrix A^{-1} as its right half.

$$[A \mid I] \rightarrow [I \mid A^{-1}]$$

This procedure for finding the inverse of a matrix can be generalized as follows.

To obtain an **inverse matrix** A^{-1} for any $n \times n$ matrix A for which A^{-1} exists, follow these steps:

1. Form the augmented matrix $[A \mid I]$, where I is the $n \times n$ identity matrix.
2. Perform row operations on $[A \mid I]$ to get a matrix of the form $[I \mid B]$.
3. Matrix B is A^{-1}.

▼ **EXAMPLE 6** Find A^{-1} if $A = \begin{bmatrix} 1 & 0 & 1 \\ 2 & -2 & -1 \\ 3 & 0 & 0 \end{bmatrix}$.

Solution First write the augmented matrix $[A \mid I]$.

$$[A \mid I] = \begin{bmatrix} 1 & 0 & 1 & | & 1 & 0 & 0 \\ 2 & -2 & -1 & | & 0 & 1 & 0 \\ 3 & 0 & 0 & | & 0 & 0 & 1 \end{bmatrix}$$

The augmented matrix already has 1 in the upper left-hand corner as needed, so begin by selecting the row operation that will result in a 0 for the first element in row two. Multiply row one by -2 and add the result to row two, giving

$$\begin{bmatrix} 1 & 0 & 1 & | & 1 & 0 & 0 \\ 0 & -2 & -3 & | & -2 & 1 & 0 \\ 3 & 0 & 0 & | & 0 & 0 & 1 \end{bmatrix}. \quad -2R_1 + R_2$$

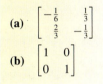

⟨8⟩

(a) Find A^{-1} if

$A = \begin{bmatrix} 2 & 2 \\ 4 & 1 \end{bmatrix}$.

(b) Check your answer by finding AA^{-1}.

Answers:

(a) $\begin{bmatrix} -\frac{1}{6} & \frac{1}{3} \\ \frac{2}{3} & -\frac{1}{3} \end{bmatrix}$

(b) $\begin{bmatrix} 1 & 0 \\ 0 & 1 \end{bmatrix}$

9

(a) Complete this step.

(b) Write this row transformation as _____.

Answers:

(a)

$$\begin{bmatrix} 1 & 0 & 1 & | & 1 & 0 & 0 \\ 0 & -2 & -3 & | & -2 & 1 & 0 \\ 0 & 0 & -3 & | & -3 & 0 & 1 \end{bmatrix}$$

(b) $-3R_1 + R_3$

10

(a) Complete these steps.

(b) Write these row transformations as _____.

Answers:

(a)

$$\begin{bmatrix} 1 & 0 & 0 & | & 0 & 0 & \frac{1}{3} \\ 0 & 1 & 0 & | & -\frac{1}{2} & -\frac{1}{2} & \frac{1}{2} \\ 0 & 0 & 1 & | & 1 & 0 & -\frac{1}{3} \end{bmatrix}$$

(b) $-1R_3 + R_1;\ -\frac{3}{2}R_3 + R_2$

Get 0 for the first element in row three by multiplying row one by -3 and adding to row three as directed in Side Problem 9. **9**

Get 1 for the second element in row two by multiplying row two of the matrix found in Problem 9 at the side by $-1/2$, obtaining the new matrix

$$\begin{bmatrix} 1 & 0 & 1 & | & 1 & 0 & 0 \\ 0 & 1 & \frac{3}{2} & | & 1 & -\frac{1}{2} & 0 \\ 0 & 0 & -3 & | & -3 & 0 & 1 \end{bmatrix}. \quad -\frac{1}{2}R_2$$

Get 1 for the third element in row three by multiplying row three by $-1/3$, with the result

$$\begin{bmatrix} 1 & 0 & 1 & | & 1 & 0 & 0 \\ 0 & 1 & \frac{3}{2} & | & 1 & -\frac{1}{2} & 0 \\ 0 & 0 & 1 & | & 1 & 0 & -\frac{1}{3} \end{bmatrix}. \quad -\frac{1}{3}R_3$$

Now do side problem 10 to get 0's for the third elements in rows one and two. **10**

The answer to Side Problem 10(a) gives the desired inverse:

$$A^{-1} = \begin{bmatrix} 0 & 0 & \frac{1}{3} \\ -\frac{1}{2} & -\frac{1}{2} & \frac{1}{2} \\ 1 & 0 & -\frac{1}{3} \end{bmatrix}.$$

Verify that AA^{-1} is I. ▲

A graphing calculator provides a fast way to find matrix inverses.

▼ EXAMPLE 7 Use a graphing calculator to find the inverse of the following matrices (if they have inverses):

$$A = \begin{bmatrix} 2 & 1 & -1 \\ 1 & 3 & 2 \\ 1 & 1 & 1 \end{bmatrix} \quad \text{and} \quad B = \begin{bmatrix} 2 & 4 \\ 3 & 6 \end{bmatrix}.$$

Solution Enter matrix A into the calculator (Figure 6.19(a)). Then use the x^{-1} key to find the inverse matrix, as in Figure 6.19(b). (Using $\wedge$ and -1 for the inverse results in an error message on most calculators.)

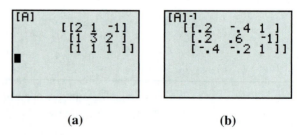

(a)   (b)

FIGURE 6.19

Use a graphing calculator to find the inverses of these matrices (if they exist).

(a) $A = \begin{bmatrix} 1 & 2 & 3 \\ 4 & -1 & 0 \\ 5 & 1 & 3 \end{bmatrix}$

(b) $B = \begin{bmatrix} 2 & 3 \\ -1 & 4 \end{bmatrix}$

Answers:

(a) No inverse

(b) Use the FRAC key (if you have one) to simplify the answer, which is

$$B^{-1} = \begin{bmatrix} 4/11 & -3/11 \\ 1/11 & 2/11 \end{bmatrix}.$$

Now enter matrix B into the calculator and use the x^{-1} key. The result is an error message (Figure 6.20), which indicates that the matrix is singular; it does not have an inverse. (If you were working by hand, you would have found that the appropriate system of equations has no solution.) ▲ ⑪

```
ERR:SINGULAR MAT
1▮Quit
2:Goto
```

FIGURE 6.20

TECHNOLOGY TIP   Because of round-off error, a graphing calculator may sometimes display an "inverse" for a matrix that doesn't actually have one. So always verify your results by multiplying A and A^{-1}. If the product is not the identity matrix, then A does not have an inverse.

◆6.4 Exercises

In Exercises 1–6, the sizes of two matrices A and B are given. Find the sizes of the product AB and the product BA whenever these products exist. (See Example 1.)

1. A is 2×2 and B is 2×2.

2. A is 3×3 and B is 3×2.

3. A is 3×5 and B is 5×3.

4. A is 4×3 and B is 3×6.

5. A is 4×2 and B is 3×4.

6. A is 7×3 and B is 2×7.

7. To find the product matrix AB, the number of _____ of A must be the same as the number of _____ of B.

8. The product matrix AB has the same number of _____ as A and the same number of _____ as B.

Find each of the following matrix products. (See Examples 2–4.)

9. $\begin{bmatrix} 1 & 2 \\ 3 & 4 \end{bmatrix}\begin{bmatrix} -1 \\ 3 \end{bmatrix}$

10. $\begin{bmatrix} -2 & 5 \\ 7 & 0 \end{bmatrix}\begin{bmatrix} 6 \\ 2 \end{bmatrix}$

11. $\begin{bmatrix} 2 & 2 & -1 \\ 3 & 0 & 1 \end{bmatrix}\begin{bmatrix} 0 & 2 \\ -1 & 5 \\ 0 & 2 \end{bmatrix}$

12. $\begin{bmatrix} -9 & 3 & 1 \\ 3 & 0 & 0 \end{bmatrix}\begin{bmatrix} 2 \\ -1 \\ 4 \end{bmatrix}$

13. $\begin{bmatrix} -4 & 1 \\ 2 & -3 \end{bmatrix}\begin{bmatrix} 1 & 0 \\ 0 & 1 \end{bmatrix}$

14. $\begin{bmatrix} 1 & 0 \\ 0 & 1 \end{bmatrix}\begin{bmatrix} 3 & -2 \\ 1 & -5 \end{bmatrix}$

15. $\begin{bmatrix} 1 & 0 & 0 \\ 0 & 1 & 0 \\ 0 & 0 & 1 \end{bmatrix}\begin{bmatrix} 3 & -5 & 7 \\ -2 & 1 & 6 \\ 0 & -3 & 4 \end{bmatrix}$

16. $\begin{bmatrix} -8 & 9 \\ 3 & -4 \\ -1 & 6 \end{bmatrix}\begin{bmatrix} 1 & 0 & 0 \\ 0 & 1 & 0 \end{bmatrix}$

17. $\begin{bmatrix} 1 & 2 & 3 \\ 4 & 5 & 6 \\ 7 & 8 & 9 \end{bmatrix}\begin{bmatrix} -1 & 5 \\ 7 & 0 \\ 1 & 3 \end{bmatrix}$

18. $\begin{bmatrix} -2 & 0 & 3 \\ 5 & -3 & -1 \end{bmatrix}\begin{bmatrix} 2 & 0 & -1 & 3 \\ 0 & 1 & 0 & -1 \\ 4 & 2 & 5 & -4 \end{bmatrix}$

In Exercises 19–21, use the matrices

$$A = \begin{bmatrix} -3 & -9 \\ 2 & 6 \end{bmatrix} \quad \text{and} \quad B = \begin{bmatrix} 4 & 6 \\ 2 & 3 \end{bmatrix}.$$

19. Show that $AB \neq BA$. Hence, matrix multiplication is not commutative.

20. Show that $(A + B)^2 \neq A^2 + 2AB + B^2$.

21. Show that $(A + B)(A - B) \neq A^2 - B^2$.

22. Show that $D^2 = D$, where

$$D = \begin{bmatrix} 1 & 0 & 0 \\ \frac{1}{2} & 0 & \frac{1}{2} \\ 0 & 0 & 1 \end{bmatrix}.$$

Given matrices

$$P = \begin{bmatrix} m & n \\ p & q \end{bmatrix}, \quad X = \begin{bmatrix} x & y \\ z & w \end{bmatrix}, \quad T = \begin{bmatrix} r & s \\ t & u \end{bmatrix},$$

verify that the statements in Exercises 23–26 are true.

23. $(PX)T = P(XT)$ (associative property)

24. $P(X + T) = PX + PT$ (distributive property)

25. $k(X + T) = kX + kT$ for any real number k

26. $(k + h)P = kP + hP$ for any real numbers k and h

Determine whether the given matrices are inverses of each other by computing their product. (See Example 5.)

27. $\begin{bmatrix} 5 & 2 \\ 3 & -1 \end{bmatrix}$ and $\begin{bmatrix} -1 & 2 \\ 3 & -4 \end{bmatrix}$

28. $\begin{bmatrix} 3 & 5 \\ 7 & 9 \end{bmatrix}$ and $\begin{bmatrix} -\frac{9}{8} & \frac{5}{8} \\ \frac{7}{8} & -\frac{3}{8} \end{bmatrix}$

29. $\begin{bmatrix} 3 & -1 \\ -4 & 2 \end{bmatrix}$ and $\begin{bmatrix} 1 & \frac{1}{2} \\ 2 & \frac{3}{2} \end{bmatrix}$

30. $\begin{bmatrix} 1 & 1 \\ .1 & .2 \end{bmatrix}$ and $\begin{bmatrix} 2 & -10 \\ -1 & 10 \end{bmatrix}$

31. $\begin{bmatrix} 1 & 1 & 1 \\ 2 & 3 & 0 \\ 1 & 2 & 1 \end{bmatrix}$ and $\begin{bmatrix} 1.5 & .5 & -1.5 \\ -1 & 0 & 1 \\ .5 & -.5 & .5 \end{bmatrix}$

32. $\begin{bmatrix} 2 & 5 & 4 \\ 1 & 4 & 3 \\ 1 & 3 & 2 \end{bmatrix}$ and $\begin{bmatrix} 1 & 2 & 1 \\ -5 & 8 & 2 \\ 7 & -11 & -3 \end{bmatrix}$

Find the inverse, if it exists, for each of the following matrices. (See Example 6.)

33. $\begin{bmatrix} 2 & -3 \\ -1 & 2 \end{bmatrix}$
34. $\begin{bmatrix} 1 & 2 \\ 1 & 1 \end{bmatrix}$

35. $\begin{bmatrix} 2 & 4 \\ 3 & 6 \end{bmatrix}$
36. $\begin{bmatrix} -3 & -5 \\ 6 & 10 \end{bmatrix}$

37. $\begin{bmatrix} 2 & 6 \\ 1 & 4 \end{bmatrix}$
38. $\begin{bmatrix} 1 & 2 \\ 3 & 4 \end{bmatrix}$

39. $\begin{bmatrix} 1 & -1 & 0 \\ -1 & 2 & 3 \\ 1 & 0 & 2 \end{bmatrix}$
40. $\begin{bmatrix} 0 & 1 & -1 \\ 2 & -2 & -1 \\ -1 & 1 & 1 \end{bmatrix}$

41. $\begin{bmatrix} 1 & 4 & 3 \\ 1 & -3 & -2 \\ 2 & 5 & 4 \end{bmatrix}$
42. $\begin{bmatrix} 1 & 2 & 0 \\ 3 & -1 & 2 \\ -2 & 3 & -2 \end{bmatrix}$

43. $\begin{bmatrix} 1 & -3 & 4 \\ 2 & -5 & 7 \\ 0 & -1 & 1 \end{bmatrix}$
44. $\begin{bmatrix} 5 & 0 & 2 \\ 2 & 2 & 1 \\ -3 & 1 & -1 \end{bmatrix}$

Use a graphing calculator to find the inverse of each matrix.

45. $\begin{bmatrix} 2 & 4 & 6 \\ 1 & 4 & 2 \\ 0 & 1 & -1 \end{bmatrix}$
46. $\begin{bmatrix} 2 & 2 & -4 \\ 2 & 6 & 0 \\ -3 & -3 & 5 \end{bmatrix}$

47. $\begin{bmatrix} 1 & 0 & -2 & 0 \\ -2 & 1 & 2 & 2 \\ 3 & -1 & -2 & -3 \\ 0 & 1 & 4 & 1 \end{bmatrix}$
48. $\begin{bmatrix} 1 & 1 & 0 & 2 \\ 2 & -1 & 1 & -1 \\ 3 & 3 & 2 & -2 \\ 1 & 2 & 1 & 0 \end{bmatrix}$

A graphing calculator or other technology is recommended for part (c) of Exercises 49–51.

49. **Social Sciences** The average birth and death rates per million for several regions of the world and their populations (in millions) are shown.*

Region	Births	Deaths
Asia	.024	.008
Latin America	.025	.007
North America	.015	.009
Europe	.011	.011

Year	Asia	Latin America	North America	Europe
1970	1996	286	226	460
1980	2440	365	252	484
1990	2906	455	277	499
2000	3683	519	310	729
2025 (projected)	4723	697	364	702

(a) Write the information in the first table as a 4×2 matrix R.

(b) Write the information in the second table as a 5×4 matrix P.

(c) Find the product PR.

(d) Explain what PR represents.

(e) From matrix PR, what was the total number of births in 2000? What total number of deaths is projected for 2025?

*U.S. Census Bureau and the United Nations Population Fund.

50. Business The first table shows the number of telecommunications devices sold (in millions) over a four-year period. The second gives the average price of each such device (in dollars) over the same period.*

Year	2001	2002	2003	2004
Cordless Phone	35.0	36.6	39.8	37.9
Answering Machine	20.1	20.7	21.5	20.3
Home Fax Machine	2.8	2.3	2.1	1.6

Year	2001	2002	2003	2004
Cordless Phone	64.51	43.06	36.06	34.04
Answering Machine	60.20	58.89	59.76	59.11
Home Fax Machine	219.64	154.78	138.10	120.00

(a) Write the information in the first table as a 3 × 4 matrix *A*.

(b) Write the information in the second table as a 4 × 3 matrix *B*.

(c) Find the product *AB*.

(d) Explain what the following entries in *AB* represent: row 1, column 1; row 2, column 2; row 3, column 3. What do the other entries represent?

(e) From matrix *AB*, what were the total sales (in dollars) of answering machines over the four-year period?

51. Social Science The following table shows the population (in millions) of four countries in various years.[†]

	2001	2010
United States	278.1	300.1
Canada	31.6	34.3
Argentina	37.4	41.1
Japan	126.8	127.3

The next table gives the average birth and death rates for these countries.[†]

	Birth Rate	Death Rate
United States	.01425	.00865
Canada	.01145	.00775
Argentina	.0175	.00755
Japan	.00945	.00925

(a) Write the information in the first table as a 2 × 4 matrix *A*.

(b) Write the information in the second table as a 4 × 2 matrix *B*.

(c) Find the product *AB*.

(d) Explain what *AB* represents.

(e) According to matrix *AB*, what is the total number of people born in thse four countries combined in 2001?

52. Health The first table shows the number of live births (in thousands). The second shows the infant mortality rates (deaths per 1000 live births) in those years.*

Year	2000	2001	2002	2003
Black	623	606	594	576
White	3194	3178	3175	3233

Year	2000	2001	2002	2003
Black	14.1	14.0	14.4	14.1
White	5.7	5.7	5.8	5.8

(a) Write the information in the first table as a 4 × 2 matrix *C*.

(b) Write the information in the second table as a 2 × 4 matrix *D*.

(c) Find the product *DC*.

(d) Which entry in *DC* gives the total number of black infant deaths from 2000 to 2003?

(e) Which entry in *DC* gives the total number of white infant deaths from 2000 to 2003?

53. Business The first table shows the number of cell phone accounts and the number of basic cable accounts (in millions). The second shows the average monthly cost of each service (in dollars) over the same period.

Year	2001	2002	2003	2004
Cell Phones	128.4	140.8	158.7	182.1
Basic Cable	73.0	73.5	73.4	73.2

---

*Multimedia Telecommunications Association.

†U.S. Census Bureau, U.S. Department of Commerce.

*U.S. National Center for Health Statistics; data for 2003 is preliminary.

Year	2001	2002	2003	2004
Cell Phones	47.37	48.40	49.91	50.64
Basic Cable	35.33	36.47	37.64	38.79

(a) Write the information in the first table as a 4 × 2 matrix *A*.

(b) Write the information in the second table as a 2 × 4 matrix *B*.

(c) Find *AB* and round each entry to two decimal places.

(d) Find *BA* and round each entry to two decimal places.

(e) Explain the meaning (if any) of each entry in *AB* and in *BA*.

54. Business Burger Barn's three locations sell hamburgers, fries, and soft drinks. Barn I sells 900 burgers, 600 orders of fries, and 750 soft drinks each day. Barn II sells 1500 burgers a day and Barn III sells 1150. Soft drink sales number 900 a day at Barn II and 825 a day at Barn III. Barn II sells 950 orders of fries per day and Barn III sells 800.

(a) Write a 3 × 3 matrix *S* that displays daily sales figures for all locations.

(b) Burgers cost $1.50 each, fries $.90 an order, and soft drinks $.60 each. Write a 1 × 3 matrix *P* that displays the prices.

(c) What matrix product displays the daily revenue at each of the three locations?

(d) What is the total daily revenue from all locations?

55. Business The four departments of Stagg Enterprises need to order the following amounts of the same products.

	Paper	Tape	Printer Ribbon	Memo Pads	Pens
Department 1	10	4	3	5	6
Department 2	7	2	2	3	8
Department 3	4	5	1	0	10
Department 4	0	3	4	5	5

The unit price (in dollars) of each product is as follows for two suppliers.

Product	Supplier A	Supplier B
Paper	2	3
Tape	1	1
Printer Ribbon	4	3
Memo Pads	3	3
Pens	1	2

(a) Use matrix multiplication to get a matrix showing the comparative costs for each department for the products from the two suppliers.

(b) Find the total cost to buy products from each supplier. From which supplier should the company make the purchase?

56. Business The Perulli Candy Company makes three types of chocolate candy: Cheery Cherry, Mucho Mocha, and Almond Delight. The company produces its products in San Diego, Mexico City, and Managua, using two main ingredients: chocolate and sugar.

(a) Each kilogram of Cheery Cherry requires .5 kilogram of sugar and .2 kilogram of chocolate; each kilogram of Mucho Mocha requires .4 kilogram of sugar and .3 kilogram of chocolate; and each kilogram of Almond Delight requires .3 kilogram of sugar and .3 kilogram of chocolate. Put this information into a 2 × 3 matrix, labeling the rows and columns.

(b) The cost of 1 kilogram of sugar is $3 in San Diego, $2 in Mexico City, and $1 in Managua. The cost of 1 kilogram of chocolate is $3 in San Diego, $3 in Mexico City, and $4 in Managua. Put this information into a matrix in such a way that when you multiply it with your matrix from part (a), you get a matrix representing the cost by ingredient of producing each type of candy in each city.

(c) Multiply the matrices in parts (a) and (b), labeling the product matrix.

(d) From part (c), what is the combined cost of sugar and chocolate required to produce 1 kilogram of Mucho Mocha in Managua?

(e) Perulli Candy needs to produce quickly a special shipment of 100 kilograms of Cheery Cherry, 200 kilograms of Mucho Mocha, and 500 kilograms of Almond Delight, and it decides to select one factory to fill the entire order. Use matrix multiplication to determine in which city the total cost of sugar and chocolate combined required to produce the order is the smallest.

6.5 Applications of Matrices

This section gives a variety of applications of matrices.

Solving Systems with Matrices

Consider this system of linear equations:

$$2x - 3y = 4$$
$$x + 5y = 2.$$

Let

$$A = \begin{bmatrix} 2 & -3 \\ 1 & 5 \end{bmatrix}, \qquad X = \begin{bmatrix} x \\ y \end{bmatrix}, \qquad B = \begin{bmatrix} 4 \\ 2 \end{bmatrix}.$$

Since

$$AX = \begin{bmatrix} 2 & -3 \\ 1 & 5 \end{bmatrix} \begin{bmatrix} x \\ y \end{bmatrix} = \begin{bmatrix} 2x - 3y \\ x + 5y \end{bmatrix} \qquad \text{and} \qquad B = \begin{bmatrix} 4 \\ 2 \end{bmatrix},$$

the original system is equivalent to the single matrix equation $AX = B$. Similarly, any system of linear equations can be written as a matrix equation $AX = B$. The matrix A is called the **coefficient matrix.** ◁1▷

A matrix equation $AX = B$ can be solved if A^{-1} exists. Assuming that A^{-1} exists and using the facts that $A^{-1}A = I$ and $IX = X$, along with the associative property of multiplication of matrices, gives

$$AX = B$$
$$A^{-1}(AX) = A^{-1}B \qquad \text{Multiply both sides by } A^{-1}.$$
$$(A^{-1}A)X = A^{-1}B \qquad \text{Associative property}$$
$$IX = A^{-1}B \qquad \text{Inverse property}$$
$$X = A^{-1}B. \qquad \text{Identity property}$$

When multiplying by matrices on both sides of a matrix equation, be careful to multiply in the same order on both sides, since multiplication of matrices is not commutative (unlike the multiplication of real numbers). This discussion is summarized below.

Suppose that a system of equations with the same number of equations as variables is written in matrix form as $AX = B$, where A is the square matrix of coefficients, X is the column matrix of variables, and B is the column matrix of constants. If A has an inverse, then the unique solution of the system is $X = A^{-1}B$.*

*If A does not have an inverse, then the system either has no solution or has an infinite number of solutions. Use the methods of Sections 6.1 or 6.2.

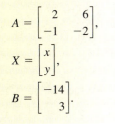

1 Write the matrix of coefficients, the matrix of variables, and the matrix of constants for the system

$$2x + 6y = -14$$
$$-x - 2y = 3.$$

Answers:

$$A = \begin{bmatrix} 2 & 6 \\ -1 & -2 \end{bmatrix},$$

$$X = \begin{bmatrix} x \\ y \end{bmatrix},$$

$$B = \begin{bmatrix} -14 \\ 3 \end{bmatrix}.$$

▼ **EXAMPLE 1** Consider this system of equations:

$$x + y + z = 2$$
$$2x + 3y = 5$$
$$x + 2y + z = -1.$$

(a) Write the system as a matrix equation.

Solution We have these three matrices:

$$\text{Coefficient Matrix} \qquad \text{Matrix of Variables} \qquad \text{Matrix of Constants}$$

$$A = \begin{bmatrix} 1 & 1 & 1 \\ 2 & 3 & 0 \\ 1 & 2 & 1 \end{bmatrix}, \qquad A = \begin{bmatrix} x \\ y \\ z \end{bmatrix}, \qquad B = \begin{bmatrix} 2 \\ 5 \\ -1 \end{bmatrix}.$$

So the matrix equation is

$$AX = B$$

$$\begin{bmatrix} 1 & 1 & 1 \\ 2 & 3 & 0 \\ 1 & 2 & 1 \end{bmatrix} \begin{bmatrix} x \\ y \\ z \end{bmatrix} = \begin{bmatrix} 2 \\ 5 \\ -1 \end{bmatrix}.$$

(b) Find A^{-1} and solve the equation.

Solution Use Exercise 31 of Section 6.4, or technology, or the method of Section 6.4 to find that

$$A^{-1} = \begin{bmatrix} 1.5 & .5 & -1.5 \\ -1 & 0 & 1 \\ .5 & -.5 & .5 \end{bmatrix}.$$

Hence,

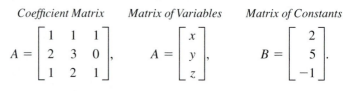

$$X = A^{-1}B = \begin{bmatrix} 1.5 & .5 & -1.5 \\ -1 & 0 & 1 \\ .5 & -.5 & .5 \end{bmatrix} \begin{bmatrix} 2 \\ 5 \\ -1 \end{bmatrix} = \begin{bmatrix} 7 \\ -3 \\ -2 \end{bmatrix}.$$

Thus, the solution of the original system is $(7, -3, -2)$. ▲ ◇2

② Use the inverse matrix to solve the system in Example 1 if the constants for the three equations are 12, 0, and 8, respectively.

Answer:

$(6, -4, 10)$

▼ **EXAMPLE 2** Use the inverse of the coefficient matrix to solve the system

$$x + 1.5y = 8$$
$$2x + 3y = 10.$$

Solution The coefficient matrix is $A = \begin{bmatrix} 1 & 1.5 \\ 2 & 3 \end{bmatrix}$. A graphing calculator will indicate that A^{-1} does not exist. If we try to carry out the row operations, we see why.

$$\left[\begin{array}{cc|cc} 1 & 1.5 & 1 & 0 \\ 2 & 3 & 0 & 1 \end{array} \right]$$

$$\left[\begin{array}{cc|cc} 1 & 1.5 & 1 & 0 \\ 0 & 0 & -2 & 1 \end{array} \right]$$

The next step cannot be performed because of the zero in the second row, second column. Verify that the original system has no solution. ▲ ③

Input–Output Analysis

An interesting application of matrix theory to economics was developed by Nobel Prize winner Wassily Leontief. His application of matrices to the interdependencies in an economy is called **input–output** analysis. In practice, input–output analysis is very complicated with many variables. We shall discuss only simple examples with a few variables.

Input–output models are concerned with the production and flow of goods and services. A typical economy is composed of a number of different sectors (such as manufacturing, energy, transportation, agriculture, etc.). Each sector requires input from other sectors (and possibly from itself) to produce its output. For instance, manufacturing output requires energy, transportation, and manufactured items (such as tools and machinery). If an economy has n sectors, then the inputs required by the various sectors from each other to produce their outputs can be described by an $n \times n$ matrix called the **input–output matrix** (or the **technological matrix**).

▼ **EXAMPLE 3** Suppose a simplified economy involves just three sectors—agriculture, manufacturing, and transportation—all in appropriate units. The production of 1 unit of agriculture requires $1/2$ unit of manufacturing and $1/4$ unit of transportation. The production of 1 unit of manufacturing requires $1/4$ unit of agriculture and $1/4$ unit of transportation. The production of 1 unit of transportation requires $1/3$ unit of agriculture and $1/4$ unit of manufacturing. Write the input–output matrix of this economy.

Solution Since there are three sectors in the economy, the input–output matrix A is 3×3. Each row and each column is labeled by a sector of the economy, as shown below. The first column lists the units from each sector of the economy that are required to produce one unit of agriculture. The second column lists the units required from each sector to produce 1 unit of manufacturing, and the last column lists the units required from each sector to produce 1 unit of transportation.

$$Input \quad \begin{array}{c} \text{Agriculture} \\ \text{Manufacturing} \\ \text{Transportation} \end{array} \begin{bmatrix} 0 & \frac{1}{4} & \frac{1}{3} \\ \frac{1}{2} & 0 & \frac{1}{4} \\ \frac{1}{4} & \frac{1}{4} & 0 \end{bmatrix} = A$$

Output: Agriculture, Manufacturing, Transporting ▲ ④

Example 3 is a bit unrealistic in that no sector of the economy requires any input from itself. In an actual economy, most sectors require input from themselves as well as from other sectors to produce their output (as discussed in the paragraph preceding Example 3). Nevertheless, it is easier to learn the basic concepts from simplified examples, so we shall continue to use them.

The input–output matrix gives only a partial picture of an economy. We also need to know the amount produced by each sector and the amount of the economy's output that is used up by the sectors themselves in the production process. The remainder of the total output is available to satisfy the needs of consumers and others outside the production system.

③ Solve the system in Example 2 if the constants are, respectively, 3 and 6.

Answer:

$(3 - 1.5y, y)$ for all real numbers y

④ Write a 2×2 input–output matrix in which 1 unit of electricity requires $1/2$ unit of water and $1/3$ unit of electricity, while 1 unit of water requires no water but $1/4$ unit of electricity.

Answer:

$$\begin{array}{c} \text{Elec.} \\ \text{Water} \end{array} \begin{bmatrix} \frac{1}{3} & \frac{1}{4} \\ \frac{1}{2} & 0 \end{bmatrix}$$

▼ **EXAMPLE 4** Consider the economy whose input–output matrix A was found in Example 3.

(a) Suppose this economy produces 60 units of agriculture, 52 units of manufacturing, and 48 units of transportation. Write this information as a column matrix.

Solution Listing the sectors in the same order as the rows of input–output matrix, we have

$$X = \begin{bmatrix} 60 \\ 52 \\ 48 \end{bmatrix}.$$

The matrix X is called the **production matrix.**

(b) How much from each sector is used up in the production process?

Solution Since 1/4 unit of agriculture is used to produce each unit of manufacturing and there are 52 units of manufacturing output, the amount of agriculture used up by manufacturing is $1/4 \times 52 = 13$ units. Similarly 1/3 unit of agriculture is used to produce a unit of transportation, so $1/3 \times 48 = 16$ units of agriculture are used up by transportation. Therefore $13 + 16 = 29$ units of agriculture are used up in the economy's production process.

A similar analysis shows that the economy's production process uses up

$$\underset{\substack{\text{agricul-} \\ \text{ture}}}{1/2 \times 60} + \underset{\substack{\text{transpor-} \\ \text{tation}}}{1/4 \times 48} = 30 + 12 = 42 \text{ units of manufacturing}$$

and

$$\underset{\substack{\text{agricul-} \\ \text{ture}}}{1/4 \times 60} + \underset{\substack{\text{transpor-} \\ \text{tation}}}{1/4 \times 52} = 15 + 13 = 28 \text{ units of transportation.}$$

(c) Describe the conclusions of part (b) in terms of the input–output matrix A and the production matrix X.

Solution The matrix product AX gives the amount from each sector that is used up in the production process, as shown here (with selected entries color coded as in part (b)):

$$AX = \begin{bmatrix} 0 & \frac{1}{4} & \frac{1}{3} \\ \frac{1}{2} & 0 & \frac{1}{4} \\ \frac{1}{4} & \frac{1}{4} & 0 \end{bmatrix} \begin{bmatrix} 60 \\ 52 \\ 48 \end{bmatrix} = \begin{bmatrix} 0 \cdot 60 + \frac{1}{4} \cdot 52 + \frac{1}{3} \cdot 48 \\ \frac{1}{2} \cdot 60 + 0 \cdot 52 + \frac{1}{4} \cdot 48 \\ \frac{1}{4} \cdot 60 + \frac{1}{4} \cdot 52 + 0 \cdot 48 \end{bmatrix}$$

$$= \begin{bmatrix} \frac{1}{4} \cdot 52 + \frac{1}{3} \cdot 48 \\ \frac{1}{2} \cdot 60 + \frac{1}{4} \cdot 48 \\ \frac{1}{4} \cdot 60 + \frac{1}{4} \cdot 52 \end{bmatrix} = \begin{bmatrix} 29 \\ 42 \\ 28 \end{bmatrix}. \ \blacktriangle$$

(d) Find the matrix $D = X - AX$ and explain what its entries represent.

Solution From parts (a) and (c), we have

$$D = X - AX = \begin{bmatrix} 60 \\ 52 \\ 48 \end{bmatrix} - \begin{bmatrix} 29 \\ 42 \\ 28 \end{bmatrix} = \begin{bmatrix} 31 \\ 10 \\ 20 \end{bmatrix}.$$

The matrix D lists the amount of each sector that is *not* used up in the production process and hence is available to groups outside the production process (such as consumers). For example, 60 units of agriculture are produced and 29 units are used up in the process, so the difference $60 - 29 = 31$ is the amount of agriculture that is available to groups outside the production process. Similar remarks apply to manufacturing and transportation. The matrix D is called the **demand matrix.** Matrices A and D show that the production of 60 units of agriculture, 52 units of manufacturing, and 48 units of transportation would satisfy an outside demand of 31 units of agriculture, 10 units of manufacturing, and 20 units of transportation. ▲ ⑤

⑤

(a)   Write a 2×1 matrix X to represent the gross production of 9000 units of electricity and 12,000 units of water.

(b)   Find AX, using A from side problem 4.

(c)   Find D, using $D = X - AX$.

Answers:

(a) $\begin{bmatrix} 9000 \\ 12,000 \end{bmatrix}$

(b) $\begin{bmatrix} 6000 \\ 4500 \end{bmatrix}$

(c) $\begin{bmatrix} 3000 \\ 7500 \end{bmatrix}$

Example 4 illustrates the general situation. In an economy with n sectors, the input–output matrix A is $n \times n$. The production matrix X is a column matrix whose n entries are the outputs of each sector of the economy. The demand matrix D is also a column matrix with n entries. This matrix is defined by

$$D = X - AX.$$

D lists the amount from each sector that is available to meet the demands of consumers and other groups outside the production process.

In Example 4, we knew the input–output matrix A and the production matrix X and used them to find the demand matrix D. In practice, however, this process is reversed: the input–output matrix A and the demand matrix D are known, and we must find the production matrix X needed to satisfy the required demands. Matrix algebra can be used to solve the equation $D = X - AX$ for X.

$$D = X - AX$$
$$D = IX - AX \qquad \text{Identity property}$$
$$D = (I - A)X \qquad \text{Distributive property}$$

If the matrix $I - A$ has an inverse, then

$$X = (I - A)^{-1}D.$$

▼ **EXAMPLE 5**   Suppose, in the three-sector economy of Examples 3 and 4, there is a demand for 516 units of agriculture, 258 units of manufacturing, and 129 units of transportation. What should production be for each sector?

Solution   The demand matrix is

$$D = \begin{bmatrix} 516 \\ 258 \\ 129 \end{bmatrix}.$$

Find the production matrix by first calculating $I - A$.

$$I - A = \begin{bmatrix} 1 & 0 & 0 \\ 0 & 1 & 0 \\ 0 & 0 & 1 \end{bmatrix} - \begin{bmatrix} 0 & \frac{1}{4} & \frac{1}{3} \\ \frac{1}{2} & 0 & \frac{1}{4} \\ \frac{1}{4} & \frac{1}{4} & 0 \end{bmatrix} = \begin{bmatrix} 1 & -\frac{1}{4} & -\frac{1}{3} \\ -\frac{1}{2} & 1 & -\frac{1}{4} \\ -\frac{1}{4} & -\frac{1}{4} & 1 \end{bmatrix}$$

Using a calculator with matrix capability or row operations, find the inverse of $I - A$.

$$(I - A)^{-1} = \begin{bmatrix} 1.3953 & .4961 & .5891 \\ .8372 & 1.3643 & .6202 \\ .5581 & .4651 & 1.3023 \end{bmatrix}$$

(The entries are rounded to four decimal places.)* Since $X = (I - A)^{-1}D$,

$$X = \begin{bmatrix} 1.3953 & .4961 & .5891 \\ .8372 & 1.3643 & .6202 \\ .5581 & .4651 & 1.3023 \end{bmatrix} \begin{bmatrix} 516 \\ 258 \\ 129 \end{bmatrix} = \begin{bmatrix} 924 \\ 864 \\ 576 \end{bmatrix}$$

(rounded to the nearest whole numbers).

From the last result, we see that the production of 924 units of agriculture, 864 units of manufacturing, and 576 units of transportation is required to satisfy demands of 516, 258, and 129 units, respectively. ▲

▼ **EXAMPLE 6** An economy depends on two basic products: wheat and oil. To produce 1 metric ton of wheat requires .25 metric ton of wheat and .33 metric ton of oil. The production of 1 metric ton of oil consumes .08 metric ton of wheat and .11 metric ton of oil. Find the production that will satisfy a demand of 500 metric tons of wheat and 1000 metric tons of oil.

Solution The input–output matrix is

$$\begin{array}{cc} & \text{Wheat} \quad \text{Oil} \\ A = & \begin{bmatrix} .25 & .08 \\ .33 & .11 \end{bmatrix} \begin{array}{l} \text{Wheat} \\ \text{Oil} \end{array} \end{array}$$

and we also have

$$I - A = \begin{bmatrix} 1 & 0 \\ 0 & 1 \end{bmatrix} - \begin{bmatrix} .25 & .08 \\ .33 & .11 \end{bmatrix} = \begin{bmatrix} .75 & -.08 \\ -.33 & .89 \end{bmatrix}.$$

Next, use technology or the methods of Section 6.4 to calculate $(I - A)^{-1}$.

$$(I - A)^{-1} = \begin{bmatrix} 1.3882 & .1248 \\ .5147 & 1.1699 \end{bmatrix} \quad \text{(rounded)}$$

The demand matrix is

$$D = \begin{bmatrix} 500 \\ 1000 \end{bmatrix}.$$

Consequently, the production matrix is

$$X = (I - A)^{-1}D = \begin{bmatrix} 1.3882 & .1248 \\ .5147 & 1.1699 \end{bmatrix} \begin{bmatrix} 500 \\ 1000 \end{bmatrix} = \begin{bmatrix} 819 \\ 1427 \end{bmatrix},$$

where the production numbers have been rounded to the nearest whole numbers. The production of 819 metric tons of wheat and 1427 metric tons of oil is required to satisfy the indicated demand. ▲ ⟨6⟩

Code Theory

Governments need sophisticated methods of coding and decoding messages. One example of such an advanced code uses matrix theory. This code takes the letters in the words and divides them into groups. (Each space between words is treated as

---

⟨6⟩
A simple economy depends on just two products: beer and pretzels.

(a) Suppose 1/2 unit of beer and 1/2 unit of pretzels are needed to make 1 unit of beer, and 3/4 unit of beer is needed to make 1 unit of pretzels. Write the technological matrix A for the economy.

(b) Find $I - A$.

(c) Find $(I - A)^{-1}$.

(d) Find the gross production X that will be needed to get a net production of

$$D = \begin{bmatrix} 100 \\ 1000 \end{bmatrix}.$$

Answers:

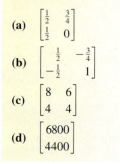

(a) $\begin{bmatrix} \frac{1}{2} & \frac{3}{4} \\ \frac{1}{2} & 0 \end{bmatrix}$

(b) $\begin{bmatrix} \frac{1}{2} & -\frac{3}{4} \\ -\frac{1}{2} & 1 \end{bmatrix}$

(c) $\begin{bmatrix} 8 & 6 \\ 4 & 4 \end{bmatrix}$

(d) $\begin{bmatrix} 6800 \\ 4400 \end{bmatrix}$

TECHNOLOGY TIP   If you are using a graphing calculator to determine X, you can calculate $(I - A)^{-1}D$ in one step without finding the intermediate matrices $I - A$ and $(I - A)^{-1}$.

---

*Although we show the matrix $(I - A)^{-1}$ with entries rounded to four decimal places, we did not round off in calculating $(I - A)^{-1}D$. If the rounded figures are used, the numbers in the product may vary slightly in the last digit.

a letter; punctuation is disregarded.) Then, numbers are assigned to the letters of the alphabet. For our purposes, let the letter *a* correspond to 1, *b* to 2, and so on. Let the number 27 correspond to a space.

For example, the message

mathematics is for the birds

can be divided into groups of three letters each.

mat   hem   ati   cs−   is−   for   −th   e−b   ird   s−−

(We used − to represent a space.) We now write a column matrix for each group of three symbols, using the corresponding numbers, as determined above, instead of letters. For example, the first four groups can be written as

$$
\begin{matrix} mat \\ \begin{bmatrix} 13 \\ 1 \\ 20 \end{bmatrix} \end{matrix},
\begin{matrix} hem \\ \begin{bmatrix} 8 \\ 5 \\ 13 \end{bmatrix} \end{matrix},
\begin{matrix} ati \\ \begin{bmatrix} 1 \\ 20 \\ 9 \end{bmatrix} \end{matrix},
\begin{matrix} cs- \\ \begin{bmatrix} 3 \\ 19 \\ 27 \end{bmatrix} \end{matrix}.
$$

The entire message consists of ten 3×1 column matrices:

$$
\begin{bmatrix} 13 \\ 1 \\ 20 \end{bmatrix},
\begin{bmatrix} 8 \\ 5 \\ 13 \end{bmatrix},
\begin{bmatrix} 1 \\ 20 \\ 9 \end{bmatrix},
\begin{bmatrix} 3 \\ 19 \\ 27 \end{bmatrix},
\begin{bmatrix} 9 \\ 19 \\ 27 \end{bmatrix},
\begin{bmatrix} 6 \\ 15 \\ 18 \end{bmatrix},
\begin{bmatrix} 27 \\ 20 \\ 8 \end{bmatrix},
\begin{bmatrix} 5 \\ 27 \\ 2 \end{bmatrix},
\begin{bmatrix} 9 \\ 18 \\ 4 \end{bmatrix},
\begin{bmatrix} 19 \\ 27 \\ 27 \end{bmatrix}.
$$

⟨7⟩

Although you could transmit these matrices, a simple substitution code such as this is very easy to break.

To get a more reliable code, we choose a 3×3 matrix M that has an inverse. Suppose we choose

$$
M = \begin{bmatrix} 1 & 3 & 3 \\ 1 & 4 & 3 \\ 1 & 3 & 4 \end{bmatrix}.
$$

Then encode each message group by multiplying by M:

$$
\begin{bmatrix} 1 & 3 & 3 \\ 1 & 4 & 3 \\ 1 & 3 & 4 \end{bmatrix}
\begin{bmatrix} 13 \\ 1 \\ 20 \end{bmatrix} =
\begin{bmatrix} 76 \\ 77 \\ 96 \end{bmatrix},
\qquad
\begin{bmatrix} 1 & 3 & 3 \\ 1 & 4 & 3 \\ 1 & 3 & 4 \end{bmatrix}
\begin{bmatrix} 8 \\ 5 \\ 13 \end{bmatrix} =
\begin{bmatrix} 62 \\ 67 \\ 75 \end{bmatrix},
$$

$$
\begin{bmatrix} 1 & 3 & 3 \\ 1 & 4 & 3 \\ 1 & 3 & 4 \end{bmatrix}
\begin{bmatrix} 1 \\ 20 \\ 9 \end{bmatrix} =
\begin{bmatrix} 88 \\ 108 \\ 97 \end{bmatrix},
$$

and so on. The coded message consists of the ten 3×1 column matrices

$$
\begin{bmatrix} 76 \\ 77 \\ 96 \end{bmatrix},
\begin{bmatrix} 62 \\ 67 \\ 75 \end{bmatrix},
\begin{bmatrix} 88 \\ 108 \\ 97 \end{bmatrix},
\ldots,
\begin{bmatrix} 181 \\ 208 \\ 208 \end{bmatrix}.
$$

The message would be sent as a string of numbers

$$76, 77, 96, 62, 67, 75, 88, 108, 97, \ldots, 181, 208, 208.$$

Note that the same letter may be encoded by different numbers. For instance, the first *a* in "mathematics" is 77 and the second *a* is 88. This makes the code harder to break. ⟨8⟩

7

Write the message *"when"* using 2×1 matrices.

Answer:

$$\begin{bmatrix} 23 \\ 8 \end{bmatrix}, \begin{bmatrix} 5 \\ 14 \end{bmatrix}$$

8

Use the following matrix to find the 2×1 matrices to be transmitted for the message in side problem 7:

$$\begin{bmatrix} 2 & 1 \\ 5 & 0 \end{bmatrix}$$

Answer:

$$\begin{bmatrix} 54 \\ 115 \end{bmatrix}, \begin{bmatrix} 24 \\ 25 \end{bmatrix}$$

The receiving agent rewrites the message as the ten 3×1 column matrices shown in color above. The agent then decodes the message by multiplying each of these column matrices by the matrix M^{-1}. Verify that

$$M^{-1} = \begin{bmatrix} 7 & -3 & -3 \\ -1 & 1 & 0 \\ -1 & 0 & 1 \end{bmatrix}.$$

So the first two matrices of the coded message are decoded as

$$\begin{bmatrix} 7 & -3 & -3 \\ -1 & 1 & 0 \\ -1 & 0 & 1 \end{bmatrix} \begin{bmatrix} 76 \\ 77 \\ 96 \end{bmatrix} = \begin{bmatrix} 13 \\ 1 \\ 20 \end{bmatrix} \begin{matrix} m \\ a \\ t \end{matrix}$$

and

$$\begin{bmatrix} 7 & -3 & -3 \\ -1 & 1 & 0 \\ -1 & 0 & 1 \end{bmatrix} \begin{bmatrix} 62 \\ 67 \\ 75 \end{bmatrix} = \begin{bmatrix} 8 \\ 5 \\ 13 \end{bmatrix} \begin{matrix} h \\ e \\ m \end{matrix}.$$

The other blocks are decoded similarly.

Routing

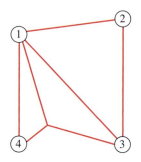

FIGURE 6.21

The diagram in Figure 6.21 shows the roads connecting four cities. Another way of representing this information is via matrix A, where the entries represent the number of roads connecting two cities without passing through another city.* For example, from the diagram, we see that there are two roads connecting city 1 to city 4 without passing through either city 2 or 3. This information is entered in row one, column four, and again in row four, column one, of matrix A.

$$A = \begin{bmatrix} 0 & 1 & 2 & 2 \\ 1 & 0 & 1 & 0 \\ 2 & 1 & 0 & 1 \\ 2 & 0 & 1 & 0 \end{bmatrix}$$

Note that there are no roads connecting each city to itself. Also, there is one road connecting cities 3 and 2.

How many ways are there to go from city 1 to city 2 by going, for example, through exactly one other city? Because we must go through one other city, we must go through either city 3 or city 4. On the diagram in Figure 6.21, we see that we can go from city 1 to city 2 through city 3 in two ways. We can go from city 1 to city 3 in two ways and then from city 3 to city 2 in one way, so there are $2 \cdot 1 = 2$ ways to get from city 1 to city 2 through city 3. It is not possible to go from city 1 to city 2 through city 4, because there is no direct route between cities 4 and 2.

The matrix A^2 gives the number of ways to travel between any two cities by passing through exactly one other city. Multiply matrix A by itself to get A^2. Let the

---

*Campbell, *Matrices with Applications*, 1st ed., section 2.3, example 5, pp. 50–51. © 1968. Electronically reproduced by permission of Pearson Education, Inc., Upper Saddle River, New Jersey.

entry in the first row, second column, of A^2 be b_{12}. (We use a_{ij} to denote the entry in the ith row and jth column of matrix A.) The entry b_{12} is found as follows:

$$b_{12} = a_{11}a_{12} + a_{12}a_{22} + a_{13}a_{32} + a_{14}a_{42}$$
$$= 0 \cdot 1 + 1 \cdot 0 + 2 \cdot 1 + 2 \cdot 0$$
$$= 2.$$

The first product, $0 \cdot 1$, in this calculation represents the number of ways to go from city 1 to city 1 (that is, 0) and then from city 1 to city 2 (that is, 1). The 0 result indicates that such a trip does not involve a third city. The only nonzero product $(2 \cdot 1)$ represents the two routes from city 1 to city 3 and the one route from city 3 to city 2, which result in the $2 \cdot 1$, or 2, routes from city 1 to city 2 by going through city 3.

Similarly, A^3 gives the number of ways to travel between any two cities by passing through exactly two cities. Also, $A + A^2$ represents the total number of ways to travel between two cities with at most one intermediate city. ⑨

The diagram can be given many other interpretations. For example, the lines could represent lines of mutual influence between people or nations, or they could represent communication lines, such as telephone lines.

⑨ 〰

Use a graphing calculator to find the following.

(a) A^3

(b) $A + A^2$

Answers:

(a)

```
[A]^3
 [[12 12 22 21]
  [12  4  9  6]
  [22  9 12 12]
  [21  6 12  8]]
■
```

(b)

```
[A]+[A]²
 [[9  3  5  4]
  [3  4  3  3]
  [5  3  6  5]
  [4  3  5  5]]
```

6.5 Exercises

Solve the matrix equation $AX = B$ for X. (See Example 1.)

1. $A = \begin{bmatrix} 1 & -1 \\ 5 & 6 \end{bmatrix}$, $B = \begin{bmatrix} -4 \\ 2 \end{bmatrix}$

2. $A = \begin{bmatrix} 1 & 2 \\ 1 & 3 \end{bmatrix}$, $B = \begin{bmatrix} -3 \\ 5 \end{bmatrix}$

3. $A = \begin{bmatrix} 3 & 1 \\ 4 & 2 \end{bmatrix}$, $B = \begin{bmatrix} 3 & 5 \\ 4 & 6 \end{bmatrix}$

4. $A = \begin{bmatrix} 1 & 3 \\ 2 & 7 \end{bmatrix}$, $B = \begin{bmatrix} 0 & 8 \\ 4 & 1 \end{bmatrix}$

5. $A = \begin{bmatrix} 2 & 1 & 0 \\ -4 & -1 & 3 \\ 3 & 1 & -2 \end{bmatrix}$, $B = \begin{bmatrix} 2 \\ 7 \\ 4 \end{bmatrix}$

6. $A = \begin{bmatrix} 3 & -1 & 0 \\ 0 & 1 & 2 \\ 6 & 0 & 5 \end{bmatrix}$, $B = \begin{bmatrix} -3 \\ 6 \\ 12 \end{bmatrix}$

Use the inverse of the coefficient matrix to solve each system of equations. (The inverses for Exercises 9–14 were found in Exercises 41 and 44–48 of Section 6.4.) (See Example 1.)

7. $\begin{aligned} x + 2y + 3z &= 10 \\ 2x + 3y + 2z &= 6 \\ -x - 2y - 4z &= -1 \end{aligned}$

8. $\begin{aligned} -x + y - 3z &= 3 \\ 2x + 4y - 4z &= 6 \\ -x + y + 4z &= -1 \end{aligned}$

9. $\begin{aligned} x + 4y + 3z &= -12 \\ x - 3y - 2z &= 0 \\ 2x + 5y + 4z &= 7 \end{aligned}$

10. $\begin{aligned} 5x \phantom{{}+2y} + 2z &= 3 \\ 2x + 2y + z &= 4 \\ -3x + y - z &= 5 \end{aligned}$

11. $\begin{aligned} 2x + 4y + 6z &= 4 \\ x + 4y + 2z &= 8 \\ y - z &= -4 \end{aligned}$

12. $\begin{aligned} 2x + 2y - 4z &= 12 \\ 2x + 6y \phantom{{}- 4z} &= 16 \\ -3x - 3y + 5z &= -20 \end{aligned}$

13. $\begin{aligned} x \phantom{{}+ y} - 2z \phantom{{}+ 2w} &= 4 \\ -2x + y + 2z + 2w &= -8 \\ 3x - y - 2z - 3w &= 12 \\ y + 4z + w &= -4 \end{aligned}$

14. $\begin{aligned} x + y \phantom{{}+ z} + 2w &= 3 \\ 2x - y + z - w &= 3 \\ 3x + 3y + 2z - 2w &= 5 \\ x + 2y + z \phantom{{}- 2w} &= 3 \end{aligned}$

Use matrix algebra to solve the following matrix equations for X.
Then use the given matrices to find X and check your work.

15. $N = X - MX, N = \begin{bmatrix} 8 \\ -12 \end{bmatrix}, M = \begin{bmatrix} 0 & 1 \\ -2 & 1 \end{bmatrix}$

16. $A = BX + X, A = \begin{bmatrix} 4 & 6 \\ -2 & 2 \end{bmatrix}, B = \begin{bmatrix} -2 & -2 \\ 3 & 3 \end{bmatrix}$

Write a system of equations and use the inverse of the coefficient matrix to solve the system.

17. **Business** Felsted Furniture makes dining room furniture. A buffet requires 30 hours for construction and 10 hours for finishing. A chair requires 10 hours for construction and 10 hours for finishing. A table requires 10 hours for construction and 30 hours for finishing. The construction department has 350 hours of labor and the finishing department 150 hours of labor available each week. How many pieces of each type of furniture should be produced each week if the factory is to run at full capacity?

18. **Natural Science** (a) A hospital dietician is planning a special diet for a certain patient. The total amount per meal of food groups A, B, and C must equal 400 grams. The diet should include one-third as much of group A as of group B, and the sum of the amounts of group A and group C should equal twice the amount of group B. How many grams of each food group should be included?

(b) Suppose we drop the requirement that the diet include one-third as much of group A as of group B. Describe the set of all possible solutions.

(c) Suppose that, in addition to the conditions given in part (a), foods A and B cost 2 cents per gram and food C costs 3 cents per gram and that a meal must cost $8. Is a solution possible?

19. **Natural Science** Three species of bacteria are fed three foods: I, II, and III. A bacterium of the first species consumes 1.3 units each of foods I and II and 2.3 units of food III each day. A bacterium of the second species consumes 1.1 units of food I, 2.4 units of food II, and 3.7 units of food III each day. A bacterium of the third species consumes 8.1 units of I, 2.9 units of II, and 5.1 units of III each day. If 16,000 units of I, 28,000 units of II, and 44,000 units of III are supplied each day, how many of each species can be maintained in this environment?

20. **Business** A company produces three combinations of mixed vegetables, which sell in 1-kilogram packages. Italian style combines .3 kilogram of zucchini, .3 of broccoli, and .4 of carrots. French style combines .6 kilogram of broccoli and .4 of carrots. Oriental style combines .2 kilogram of zucchini, .5 of broccoli, and .3 of carrots. The company has a stock of 16,200 kilograms of zucchini, 41,400 kilograms of broccoli, and 29,400 kilograms of carrots. How many packages of each style should the company prepare to use up its supplies?

21. **Business** A national chain of casual clothing stores recently sent shipments of jeans, jackets, sweaters, and shirts to its stores in various cities. The number of items shipped to each city and their total wholesale cost are shown in the table. Find the wholesale price of one pair of jeans, one jacket, one sweater, and one shirt.

City	Jeans	Jackets	Sweaters	Shirts	Total Cost
Cleveland	3000	3000	2200	4200	$507,650
St. Louis	2700	2500	2100	4300	459,075
Seattle	5000	2000	1400	7500	541,225
Phoenix	7000	1800	600	8000	571,500

22. **Health** A 100-bed nursing home provides two levels of long-term care: regular and maximum. Patients at each level have a choice of a private room or a less expensive semiprivate room. The tables show the number of patients in each category at various times last year. The total daily costs for all patients were $18,824 in January, $18,738 in April, $18,606 in July, and $18,824 in October. Find the daily cost of each of the following: a private room (regular care), a private room (maximum care), a semiprivate room (regular care), and a semiprivate room (maximum care).

	REGULAR-CARE PATIENTS	
Month	Semiprivate	Private
January	22	8
April	26	8
July	24	14
October	20	10

	MAXIMUM-CARE PATIENTS	
Month	Semiprivate	Private
January	60	10
April	54	12
July	56	6
October	62	8

Find the production matrix for the given input–output and demand matrices. (See Examples 3–6.).

23. $A = \begin{bmatrix} \frac{1}{2} & \frac{2}{5} \\ \frac{1}{4} & \frac{1}{5} \end{bmatrix}, D = \begin{bmatrix} 2 \\ 4 \end{bmatrix}$

24. $A = \begin{bmatrix} .1 & .03 \\ .07 & .6 \end{bmatrix}$, $D = \begin{bmatrix} 5 \\ 10 \end{bmatrix}$

Exercises 25 and 26 refer to Example 6.

25. Business If the demand is changed to 690 metric tons of wheat and 920 metric tons of oil, how many units of each commodity should be produced?

26. Business Change the technological matrix so that the production of 1 metric ton of wheat requires 1/5 metric ton of oil (and no wheat) and the production of 1 metric ton of oil requires 1/3 metric ton of wheat (and no oil). To satisfy the same demand matrix, how many units of each commodity should be produced?

Work these problems. (See Examples 3–6.)

27. Business A simplified economy has only two industries: the electric company and the gas company. Each dollar's worth of the electric company's output requires $.40 of its own output and $.50 of the gas company's output. Each dollar's worth of the gas company's output requires $.25 of its own output and $.60 of the electric company's output. What should the production of electricity and gas be (in dollars) if there is a $12 million demand for gas and a $15 million demand for electricity?

28. Business A two-segment economy consists of manufacturing and agriculture. To produce 1 unit of manufacturing output requires .40 unit of its own output and .20 unit of agricultural output. To produce 1 unit of agricultural output requires .30 unit of its own output and .40 unit of manufacturing output. If there is a demand of 240 units of manufacturing and 90 units of agriculture, what should be the output of each segment?

29. Business A primitive economy depends on two basic goods: yams and pork. The production of 1 bushel of yams requires 1/4 bushel of yams and 1/2 of a pig. To produce 1 pig requires 1/6 bushel of yams. Find the amount of each commodity that should be produced to get
(a) 1 bushel of yams and 1 pig;
(b) 100 bushels of yams and 70 pigs.

30. Business A simplified economy is based on agriculture, manufacturing, and transportation. Each unit of agricultural output requires .4 unit of its own output, .3 unit of manufacturing, and .2 unit of transportation output. One unit of manufacturing output requires .4 unit of its own output, .2 unit of agricultural, and .3 unit of transportation output. One unit of transportation output requires .4 unit of its own output, .1 unit of agricultural, and .2 unit of manufacturing output. There is demand for 35 units of agricultural, 90 units of manufacturing, and 20 units of transportation output. How many units should each segment of the economy produce?

31. Business In his work *Input–Output Economics*, Leontief provides an example of a simplified economy with just three sectors: agriculture, manufacturing, and households (that is, the sector of the economy that produces labor.)* It has the following input–output matrix.

	Agriculture	Manufacturing	Households
Agriculture	.25	.40	.133
Manufacturing	.14	.12	.100
Households	.80	3.60	.133

(a) How many units from each sector does the manufacturing sector require to produce 1 unit?
(b) What production levels are needed to meet a demand of 35 units of agriculture, 38 units of manufacturing, and 40 unit of households?
(c) How many units of agriculture are used up in the economy's production process?

32. Business A much-simplified version of Leontief's 42-sector analysis of the 1947 American economy has the following input–output matrix.[†]

	Agriculture	Manufacturing	Households
Agriculture	.245	.102	.051
Manufacturing	.099	.291	.279
Households	.433	.372	.011

(a) What information about the needs of agricultural production is given by column 1 of the matrix?
(b) Suppose the demand matrix (in billions of dollars) is
$$D = \begin{bmatrix} 2.88 \\ 31.45 \\ 30.91 \end{bmatrix}$$
Find the amount of each commodity that should be produced.

33. Business An analysis of the 1958 Israeli economy is simplified here by grouping the economy into three sectors, with the following input–output matrix.[‡]

	Agriculture	Manufacturing	Energy
Agriculture	.293	0	0
Manufacturing	.014	.207	.017
Energy	.044	.010	.216

(a) How many units from each sector does the energy sector require to produce one unit?
(b) If the economy's production (in thousands of Israeli pounds) is 175,000 of agriculture, 22,000 of manufacturing, and 12,000 of energy, how much is available from each sector to satisfy the demand from consumers and others outside the production process?

*Wassily Leontief, *Input–Output Economics*, 2d ed. (Oxford University Press, 1986), pp. 19–27.

[†]Ibid, pp. 6–9.

[‡]Ibid, pp. 174–175.

(c) The actual 1958 demand matrix is

$$D = \begin{bmatrix} 138,213 \\ 17,597 \\ 1786 \end{bmatrix}.$$

How much must each sector produce to meet this demand?

34. **Business** The 1981 Chinese economy can be simplified to three sectors: agriculture, industry and construction, and transportation and commerce.* The input–output matrix is

	Agri.	Industry/Constr.	Trans./Commerce
Agri.	.158	.156	.009
Industry/Constr.	.136	.432	.071
Trans./Commerce	.013	.041	.011

The demand (in 100,000 RMB, the unit of money in China) is

$$D = \begin{bmatrix} 106,674 \\ 144,739 \\ 26,725 \end{bmatrix}.$$

(a) Find the amount each sector should produce.

(b) Interpret the economic value of an increase in demand of 1 RMB in agriculture exports.

35. **Business** The 1987 economy of the state of Washington has been simplified to four sectors: natural resource, manufacturing, trade and services, and personal consumption. The input–output matrix is[†]

	natural resources	manufacturing	trade and services	personal consumption
natural resources	.1045	.0428	.0029	.0031
manufacturing	.0826	.1087	.0584	.0321
trade and services	.0867	.1019	.2032	.3555
personal consumption	.6253	.3448	.6106	.0798

Suppose the demand (in millions of dollars) is

$$D = \begin{bmatrix} 450 \\ 300 \\ 125 \\ 100 \end{bmatrix}.$$

Find the amount each sector should produce.

36. **Business** The 1963 economy of the state of Nebraska has been condensed to six sectors; livestock, crops, food

products, mining and manufacturing, households, and other. The input–output matrix is*

$$\begin{bmatrix} .178 & .018 & .411 & 0 & .005 & 0 \\ .143 & .018 & .088 & 0 & .001 & 0 \\ .089 & 0 & .035 & 0 & .060 & .003 \\ .001 & .010 & .012 & .063 & .007 & .014 \\ .141 & .252 & .088 & .089 & .402 & .124 \\ .188 & .156 & .103 & .255 & .008 & .474 \end{bmatrix}.$$

(a) Find the matrix $(I - A)^{-1}$ and interpret the value in row two, column one, of this matrix.

(b) Suppose the demand (in millions of dollars) is

$$D = \begin{bmatrix} 1980 \\ 650 \\ 1750 \\ 1000 \\ 2500 \\ 3750 \end{bmatrix}.$$

Find the dollar amount each sector should produce.

37. **Business** Input–output analysis can also be used to model how changes in one city can affect cities that are connected with it in some way.[†] For example, if a large manufacturing company shuts down in one city, it is very likely that the economic welfare of all of the cities around it will suffer. Consider three Pennsylvania communities: Sharon, Farrell, and Hermitage. Due to their proximity to each other, residents of these three communities regularly spend time and money in the other communities. Suppose that we have gathered information in the form of an input–output matrix

$$\begin{bmatrix} .2 & .1 & .1 \\ .1 & .1 & 0 \\ .5 & .6 & .7 \end{bmatrix}.$$

This matrix can be thought of as the likelihood that a person from a particular community will spend money in each of the communities.

(a) Treat this matrix like an input–output matrix and calculate $(I - A)^{-1}$, where A is the given input–output matrix.

(b) Interpret the entries of this inverse matrix.

38. **Social Science** Use the method discussed in the text to encode the message

Anne is home.

*Input–Output Tables of China, 1981, China Statistical Information and Consultancy Service Centre, 1987, pp. 17–19.

†Robert Chase, Philip Bourque, and Richard Conway, Jr., *The 1987 Washington State Input–Output Study*, Report to the Graduate School of Business Administration, University of Washington, September 1993.

*F. Charles Lamphear and Theodore Roesler, "1970 Nebraska Input–Output Tables," *Nebraska Economic and Business Report No. 10*, Bureau of Business Research, University of Nebraska, Lincoln.

†The idea for this problem came from an example created by Thayer Watkins, Department of Economics, San Jose State University.

Break the message into groups of two letters and use the matrix

$$M = \begin{bmatrix} 1 & 3 \\ 2 & 7 \end{bmatrix}.$$

39. Social Science   Use the matrix of Exercise 38 to encode the message

Head for the hills!

40. Social Science   Decode the following message, which was encoded by using the matrix M of Exercise 38:

$$\begin{bmatrix} 90 \\ 207 \end{bmatrix}, \begin{bmatrix} 39 \\ 87 \end{bmatrix}, \begin{bmatrix} 26 \\ 57 \end{bmatrix}, \begin{bmatrix} 66 \\ 145 \end{bmatrix}, \begin{bmatrix} 61 \\ 142 \end{bmatrix}, \begin{bmatrix} 89 \\ 205 \end{bmatrix}.$$

Work these routing problems.

41. Social Science   Use matrix A in the discussion on routing in the text to find A^2. Then answer the following questions. How many ways are there to travel from
   (a) City 1 to city 3 by passing through exactly one city?
   (b) City 2 to city 4 by passing through exactly one city?
   (c) City 1 to city 3 by passing through at most one city?
   (d) City 2 to city 4 by passing through at most one city?

42. Social Sciences   The matrix A^3 (See Exercise 41) was found in Side Problem 9 of the text. Use it to answer the following questions.
   (a) How many ways are there to travel between cities 1 and 4 by passing through exactly two cities?
   (b) How many ways are there to travel between cities 1 and 4 by passing through at most two cities?

43. Business   A small telephone system connects three cities. There are four lines between cities 3 and 2, three lines connecting city 3 with city 1, and two lines between cities 1 and 2.
   (a) Write a matrix B to represent this information.
   (b) Find B^2.
   (c) How many lines that connect cities 1 and 2 go through exactly one other city (city 3)?
   (d) How many lines that connect cities 1 and 2 go through at most one other city?

44. Transportation   The figure shows four southern cities served by Supersouth Airlines.

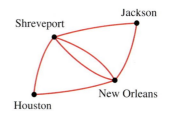

   (a) Write a matrix to represent the number of nonstop routes between cities.
   (b) Find the number of one-stop flights between Houston and Jackson.
   (c) Find the number of flights between Houston and Shreveport that require at most one stop.
   (d) Find the number of one-stop flights between New Orleans and Houston.

45. Natural Science   The figure shows a food web. The arrows indicate the food sources of each population. For example, cats feed on rats and on mice.

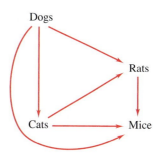

   (a) Write a matrix C in which each row and corresponding column represent a population in the food chain. Enter a 1 when the population in a given row feeds on the population in the given column.
   (b) Calculate and interpret C^2.

Chapter 6 Summary

Key Terms and Symbols

6.1 linear equation
system of linear equations
solution of a system
independent system
dependent system
inconsistent system
substitution method
elimination method
equivalent systems
elementary operations
row
column
matrix (matrices)
element (entry)

augmented matrix
row operations
row echelon form
parameter

6.2 Gauss–Jordan method
reduced row echelon form

6.3 row matrix (row vector)
column matrix (column vector)
square matrix
additive inverse of a matrix
zero matrix
scalar
product of a scalar and a matrix

6.4 product matrix
identity matrix
inverse matrix
singular matrix

6.5 coefficient matrix
input–output model
input–output matrix
production matrix
demand matrix
code theory
routing theory

Key Concepts

Solving Systems of Equations

The following **elementary operations** are used to transform a system of equations into a simpler equivalent system:

1. Interchange any two equations.

2. Multiply both sides of an equation by a nonzero constant.

3. Replace an equation by the sum of itself and a constant multiple of another equation in the system.

The **elimination method** is a systematic way of using elementary operations to transform a system into an equivalent one that can be solved by **back substitution.** See Section 6.1 for details.

The matrix version of the elimination method uses the following **matrix row operations,** which correspond to using elementary row operations with back substitution on a system of equations:

1. Interchange any two rows.

2. Multiply each element of a row by a nonzero constant.

3. Replace a row by the sum of itself and a constant multiple of another row in the matrix.

The **Gauss–Jordan method** is an extension of the elimination method for solving a system of linear equations. It uses row operations on the augmented matrix of the system. See Section 6.2 for details.

Operations on Matrices

The **sum** of two $m \times n$ matrices X and Y is the $m \times n$ matrix $X + Y$ in which each element is the sum of the corresponding elements of X and Y. The **difference** of two $m \times n$ matrices X and Y is the $m \times n$ matrix $X - Y$ in which each element is the difference of the corresponding elements of X and Y.

The **product** of a scalar k and a matrix X is the matrix kX, with each element k times the corresponding element of X.

The **product matrix** AB of an $m \times n$ matrix A and an $n \times k$ matrix B is the $m \times k$ matrix whose entry in the ith row and jth column is the product of the ith row of A and the jth column of B.

The **inverse matrix** A^{-1} for any $n \times n$ matrix A for which A^{-1} exists is found as follows: form the augmented matrix $[A|I]$; perform row operations on $[A|I]$ to get the matrix $[I|A^{-1}]$.

Chapter 6 Review Exercises

Use the substitution, elimination, or matrix method to solve each of the following systems. Identify any dependent or inconsistent systems.

1. $-5x - 3y = -3$
$\quad\ \ 2x + \ y = 4$

2. $3x - \ y = 8$
$\quad\ \ 2x + 3y = 6$

3. $4x - 5y = 6$
$\quad\ \ 3x - 3y = 9$

4. $\dfrac{1}{4}x - \dfrac{1}{3}y = -\dfrac{1}{4}$
$\quad\ \dfrac{1}{10}x + \dfrac{2}{5}y = \dfrac{2}{5}$

5. $\quad\ x - 2y = 1$
$\quad\ 4x + 4y = 2$
$\quad 10x + 8y = 4$

6. $\quad x + y - 4z = 0$
$\quad 2x + y - 3z = 2$

7. $\quad 3x + y - \ z = 3$
$\quad\ x \quad\ + 2z = 6$
$\quad -3x - y + 2z = 9$

8. $4x - y - 2z = 4$
$\quad x - y - \dfrac{1}{2}z = 1$
$\quad 2x - y - \ z = 8$

9. Business An office supply manufacturer makes two kinds of paper clips: standard and extra large. To make 1000 standard paper clips requires 1/4 hour on a cutting machine and 1/2 hour on a machine that shapes the clips. Making 1000 extra-large paper clips requires 1/3 hour on each machine. The manager of paper clip production has 4 hours per day available on the cutting machine and 6 hours per day on the shaping machine. How many of each kind of clip can he make?

10. Business Gretchen Schmidt plans to buy shares of two stocks. One costs $32 per share and pays dividends of $1.20 per share. The other costs $23 per share and pays dividends of $1.40 per share. She has $10,100 to spend and wants to earn dividends of $540. How many shares of each stock should she buy?

11. Business Joyce Pluth has money in two investment funds. Last year the first fund paid a dividend of 8% and the second a dividend of 2%, and Joyce received a total of $780. This year the first fund paid a 10% dividend and the second only 1%, and Joyce received $810. How much does she have invested in each fund?

12. Finance You are given $144 in one-, five-, and ten-dollar bills. There are 35 bills. There are two more ten-dollar bills than five-dollar bills. How many bills of each type are there?

13. Social Science A social service agency provides counseling, meals, and shelter to clients referred by sources I, II, and III. Clients from source I require an average of $100 for food, $250 for shelter, and no counseling. Source II clients require an average of $100 for counseling, $200 for food, and nothing for shelter. Source III clients require an average of $100 for counseling, $150 for food, and $200 for shelter. The agency has funding of $25,000 for counseling, $50,000 for food, and $32,500 for shelter. How many clients from each source can be served?

14. Business The Waputi Indians make woven blankets, rugs, and skirts. Each blanket requires 24 hours for spinning the yarn, 4 hours for dying the yarn, and 15 hours for weaving. Rugs require 30, 5, and 18 hours and skirts 12, 3, and 9 hours, respectively. If there are 306, 59, and 201 hours available for spinning, dying, and weaving, respectively, how many of each item can be made? (*Hint:* Simplify the equations you write, if possible, before solving the system.)

Use the Gauss–Jordan method to solve the following systems.

15. $\quad x + \ z = -3$
$\quad\quad\ y - \ z = \ 6$
$\quad 2x + 3z = \ 5$

16. $\quad 3x + 2y + 4z = -1$
$\quad -2x + \ y - 2z = \ 6$
$\quad 3x + 3y + 6z = \ 3$

17. $\quad 5x - \ 8y + \ z = 1$
$\quad 3x - \ 2y + 4z = 3$
$\quad 10x - 16y + 2z = 3$

18. $\quad x - 2y + 3z = \ 4$
$\quad 2x + \ y - 4z = \ 3$
$\quad -3x + 4y - \ z = -2$

19. $3x + 2y - 6z = 3$
$\quad x + \ y + 2z = 2$
$\quad 2x + 2y + 5z = 0$

20. Business Each week at a furniture factory, 2000 work hours are available in the construction department, 1400 work hours in the painting department, and 1300 work hours in the packing department. Producing a chair requires 2 hours of construction, 1 hour of painting, and 2 hours for packing. Producing a table requires 4 hours of construction, 3 hours of painting, and 3 hours for packing. Producing a chest requires 8 hours of construction, 6 hours of painting, and 4 hours for packing. If all available time is used in every department, how many of each item are produced each week?

For each of the following, find the dimensions of the matrix and identify any square, row, or column matrices.

21. $\begin{bmatrix} 2 & 3 \\ 5 & 9 \end{bmatrix}$ 22. $\begin{bmatrix} 2 & -1 \\ 4 & 6 \\ 5 & 7 \end{bmatrix}$

23. $\begin{bmatrix} 12 & 4 & -8 & -1 \end{bmatrix}$ 24. $\begin{bmatrix} -7 & 5 & 6 & 4 \\ 3 & 2 & -1 & 2 \\ -1 & 12 & 8 & -1 \end{bmatrix}$

25. $\begin{bmatrix} 6 & 8 & 10 \\ 5 & 3 & -2 \end{bmatrix}$ 26. $\begin{bmatrix} -9 \\ 15 \\ 4 \end{bmatrix}$

27. **Natural Science**   The activities of a grazing animal can be classified roughly into three categories: grazing, moving, and resting. Suppose horses spend 8 hours grazing, 8 moving, and 8 resting; cattle spend 10 grazing, 5 moving, and 9 resting; sheep spend 7 grazing, 10 moving, and 7 resting; and goats spend 8 grazing, 9 moving, and 7 resting. Write this information as a 4×3 matrix.

28. **Business**   The New York Stock Exchange reports in the daily newspapers give the dividend, price-to-earnings ratio, sales (in hundreds of shares), last price, and change in price for each company. Write the following stock reports as a 4×5 matrix: American Telephone & Telegraph, 5, 7, 2532, $52\frac{3}{8}$, $-\frac{1}{4}$; General Electric, 3, 9, 1464, 56, $+\frac{1}{8}$; Gulf Oil, 2.50, 5, 4974, 41, $-1\frac{1}{2}$; Sears, 1.36, 10, 1754, 18, $+\frac{1}{2}$.

Given the matrices

$A = \begin{bmatrix} 4 & 10 \\ -2 & -3 \\ 6 & 9 \end{bmatrix}$, $B = \begin{bmatrix} 1 & 3 & -2 \\ 2 & 3 & 0 \\ 0 & 1 & 5 \end{bmatrix}$, $C = \begin{bmatrix} 5 & 0 \\ -1 & 3 \\ 4 & 7 \end{bmatrix}$,

$D = \begin{bmatrix} 6 \\ 1 \\ 0 \end{bmatrix}$, $E = \begin{bmatrix} 1 & 3 & -4 \end{bmatrix}$, $F = \begin{bmatrix} -1 & 2 \\ 8 & 7 \end{bmatrix}$,

$G = \begin{bmatrix} 2 & 5 \\ 1 & 6 \end{bmatrix}$,

Find each of the following (if possible).

29. $-B$ 30. $-D$

31. $3A - 2C$ 32. $F + 3G$

33. $2B - 5C$ 34. $G - 2F$

35. **Business**   Refer to Exercise 28. Write a 4×2 matrix using the sales and price changes for the four companies.

The next day's sales and price changes for the same four companies were 2310, 1258, 5061, 1812 and $-1/4$, $-1/4$, $+1/2$, $+1/2$, respectively. Write a 4×2 matrix using these new sales and price change figures. Use matrix addition to find the total sales and price changes for the two days.

36. **Business**   An oil refinery in Tulsa sent 110,000 gallons of oil to a Chicago distributor, 73,000 to a Dallas distributor, and 95,000 to an Atlanta distributor. Another refinery in New Orleans sent the following amounts to the same three distributors: 85,000, 108,000, 69,000. The next month, the two refineries sent the same distributors new shipments of oil as follows: from Tulsa, 58,000 to Chicago, 33,000 to Dallas, and 80,000 to Atlanta; from New Orleans, 40,000, 52,000, and 30,000, respectively.
   (a) Write the monthly shipments from the two distributors to the three refineries as 3×2 matrices.
   (b) Use matrix addition to find the total amounts sent to the refineries from each distributor.

Use the matrices shown before Exercise 29 to find each of the following (if possible).

37. AG 38. EB 39. GF

40. CA 41. AGF

42. **Health**   In a study, the numbers of head and neck injuries among hockey players wearing full face shields and half face shields were compared. The following table provides the rates per 1000 athlete exposures for specific injuries that caused a player wearing either shield to miss one or more events.*

	Half Shield	Full Shield
Head and Face Injuries, Excluding Concussions	3.54	1.41
Concussions	1.53	1.57
Neck Injuries	.34	.29
Other	7.53	6.21

If an equal number of players in a large league wear each type of shield and the total number of athlete exposures for the league in a season is 8000, use matrix operations to estimate the total number of injuries of each type.

43. **Business**   An office supply manufacturer makes two kinds of paper clips: standard and extra large. To make a unit of standard paper clips requires 1/4 hour on a cutting machine and 1/2 hour on a machine that shapes the clips. A unit of extra-large paper clips requires 1/3 hour on each machine.

---

*Brian Benson, Nicholas Nohtadi, Sarah Rose, and Willem Meeuwisse, "Head and Neck Injuries among Ice Hockey Players Wearing Full Face Shields vs. Half Face Shields," *JAMA*, 282, no. 24, (December 22/29, 1999): 2328–2332.

(a) Write this information as a 2 × 2 matrix (size/machine).

(b) If 48 units of standard and 66 units of extra-large clips are to be produced, use matrix multiplication to find out how many hours each machine will operate. (*Hint:* Write the units as a 1 × 2 matrix.)

44. **Business**   Theresa DePalo buys shares of three stocks. Their cost per share and dividend earnings per share are $32, $23, and $54, and $1.20, $1.49, and $2.10, respectively. She buys 50 shares of the first stock, 20 shares of the second, and 15 shares of the third.

(a) Write the cost per share and earnings per share of the stocks as a 3 × 2 matrix.

(b) Write the number of shares of each stock as a 1 × 3 matrix.

(c) Use matrix multiplication to find the total cost and total dividend earnings of these stocks.

45. If $A = \begin{bmatrix} 3 & 0 \\ 2 & 1 \end{bmatrix}$, find a matrix B such that both AB and BA are defined and $AB \neq BA$.

46. Is it possible to do Exercise 45 if $A = \begin{bmatrix} 4 & 0 \\ 0 & 4 \end{bmatrix}$? Explain.

Find the inverse of each of the following matrices that has an inverse.

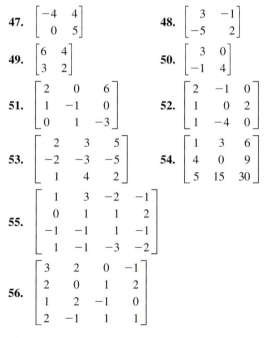

47. $\begin{bmatrix} -4 & 4 \\ 0 & 5 \end{bmatrix}$

48. $\begin{bmatrix} 3 & -1 \\ -5 & 2 \end{bmatrix}$

49. $\begin{bmatrix} 6 & 4 \\ 3 & 2 \end{bmatrix}$

50. $\begin{bmatrix} 3 & 0 \\ -1 & 4 \end{bmatrix}$

51. $\begin{bmatrix} 2 & 0 & 6 \\ 1 & -1 & 0 \\ 0 & 1 & -3 \end{bmatrix}$

52. $\begin{bmatrix} 2 & -1 & 0 \\ 1 & 0 & 2 \\ 1 & -4 & 0 \end{bmatrix}$

53. $\begin{bmatrix} 2 & 3 & 5 \\ -2 & -3 & -5 \\ 1 & 4 & 2 \end{bmatrix}$

54. $\begin{bmatrix} 1 & 3 & 6 \\ 4 & 0 & 9 \\ 5 & 15 & 30 \end{bmatrix}$

55. $\begin{bmatrix} 1 & 3 & -2 & -1 \\ 0 & 1 & 1 & 2 \\ -1 & -1 & 1 & -1 \\ 1 & -1 & -3 & -2 \end{bmatrix}$

56. $\begin{bmatrix} 3 & 2 & 0 & -1 \\ 2 & 0 & 1 & 2 \\ 1 & 2 & -1 & 0 \\ 2 & -1 & 1 & 1 \end{bmatrix}$

Refer again to the matrices shown before Exercise 29 to find each of the following (if possible).

57. F^{-1}

58. G^{-1}

59. $(G - F)^{-1}$

60. $(F + G)^{-1}$

61. B^{-1}

62. Explain why the matrix $\begin{bmatrix} a & 0 \\ c & 0 \end{bmatrix}$, where a and c are nonzero constants, cannot have an inverse.

Solve each of the following matrix equations $AX = B$ for X.

63. $A = \begin{bmatrix} 2 & 4 \\ -1 & -3 \end{bmatrix}$, $B = \begin{bmatrix} 8 \\ 3 \end{bmatrix}$

64. $A = \begin{bmatrix} 1 & 3 \\ -2 & 4 \end{bmatrix}$, $B = \begin{bmatrix} 9 \\ 6 \end{bmatrix}$

65. $A = \begin{bmatrix} 1 & 0 & 2 \\ -1 & 1 & 0 \\ 3 & 0 & 4 \end{bmatrix}$, $B = \begin{bmatrix} 8 \\ 4 \\ -6 \end{bmatrix}$

66. $A = \begin{bmatrix} 2 & 4 & 0 \\ 1 & -2 & 0 \\ 0 & 0 & 3 \end{bmatrix}$, $B = \begin{bmatrix} 72 \\ -24 \\ 48 \end{bmatrix}$

Use the method of matrix inverses to solve each of the following systems.

67. $\begin{aligned} x + y &= 2 \\ 2x + 3y &= 8 \end{aligned}$

68. $\begin{aligned} 5x - 3y &= -2 \\ 2x + 7y &= -9 \end{aligned}$

69. $\begin{aligned} 2x + y &= 10 \\ 3x - 2y &= 8 \end{aligned}$

70. $\begin{aligned} x - 2y &= 7 \\ 3x + y &= 7 \end{aligned}$

71. $\begin{aligned} x + y + z &= 1 \\ 2x - y &= -2 \\ 3y + z &= 2 \end{aligned}$

72. $\begin{aligned} x &= -3 \\ y + z &= 6 \\ 2x - 3z &= -9 \end{aligned}$

73. $\begin{aligned} 3x - 2y + 4z &= 4 \\ 4x + y - 5z &= 2 \\ -6x + 4y - 8z &= -2 \end{aligned}$

74. $\begin{aligned} x + 2y &= -1 \\ 3y - z &= -5 \\ x + 2y - z &= -3 \end{aligned}$

Solve each of the following problems by any method.

75. **Business**   A wine maker has two large casks of wine. One is 9% alcohol and the other is 14% alcohol. How many liters of each wine should be mixed to produce 40 liters of wine that is 12% alcohol?

76. **Business**   A gold merchant has some 12-carat gold (12/24 pure gold), and some 22-carat gold (22/24 pure). How many grams of each could be mixed to get 25 grams of 15-carat gold?

77. **Natural Science**   A chemist has a 40% acid solution and a 60% solution. How many liters of each should be used to get 40 liters of a 45% solution?

78. **Business**   How many pounds of tea worth $4.60 a pound should be mixed with tea worth $6.50 a pound to get 10 pounds of a mixture worth $5.74 a pound?

79. **Business**   A machine in a pottery factory takes 3 minutes to form a bowl and 2 minutes to form a plate. The material for a bowl costs $.25 and the material for a plate costs $.20. If the machine runs for 8 hours and exactly $44 is spent for material, how many bowls and plates can be produced?

80. Transportation   A boat travels at a constant speed a distance of 57 kilometers downstream in 3 hours and then turns around and travels 55 kilometers upstream in 5 hours. What are the speeds of the boat and the current?

81. Business   Ms. Tham invests $50,000 three ways—at 8%, $8\frac{1}{2}\%$, and 11%. In total, she receives $4436.25 per year in interest. The interest on the 11% investment is $80 more than the interest on the 8% investment. Find the amount she has invested at each rate.

82. Business   Tickets to a band concert cost $2 for children, $3 for teenagers, and $5 for adults. 570 people attended the concert and total ticket receipts were $1950. Three-fourths as many teenagers as children attended. How many children, teenagers, and adults were at the concert?

83. Given the input–output matrix $A = \begin{bmatrix} 0 & \frac{1}{4} \\ \frac{1}{2} & 0 \end{bmatrix}$ and the

demand matrix $D = \begin{bmatrix} 2100 \\ 1400 \end{bmatrix}$, find each of the following.

(a) $I - A$

(b) $(I - A)^{-1}$

(c) the production matrix X

84. Business   An economy depends on two commodities: goats and cheese. It takes 2/3 unit of goats to produce 1 unit of cheese and 1/2 unit of cheese to produce 1 unit of goats.

(a) Write the input–output matrix for this economy.

(b) Find the production required to satisfy a demand of 400 units of cheese and 800 units of goats.

85. Business   In a simple economic model, a country has two industries: agriculture and manufacturing. To produce $1 of agricultural output requires $.10 of agricultural output and $.40 of manufacturing output. To produce $1 of manufacturing output requires $.70 of agricultural output and $.20 of manufacturing output. If agricultural demand is $60,000 and manufacturing demand is $20,000, what must each industry produce? (Round answers to the nearest whole number.)

86. Business   Here is the input–output matrix for a small economy.

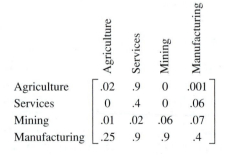

	Agriculture	Services	Mining	Manufacturing
Agriculture	.02	.9	0	.001
Services	0	.4	0	.06
Mining	.01	.02	.06	.07
Manufacturing	.25	.9	.9	.4

(a) How many units from each sector does the service sector require to produce 1 unit?

(b) What production levels are needed to meet a demand for 760 units of agriculture, 1600 units of

services, 1000 units of mining, and 2000 units of manufacturing?

(c) How many units of manufacturing production are used up in the economy's production process?

87. Use this input–output matrix to answer the questions below.

	Agriculture	Construction	Energy	Manufacturing	Transportation
Agriculture	.18	.017	.4	.005	0
Construction	.14	.018	.09	.001	0
Energy	.9	0	.4	.06	.002
Manufacturing	.19	.16	.1	.008	.5
Transportation	.14	.25	.9	.4	.12

(a) How many units from each sector does the energy sector require to produce 1 unit?

(b) If the economy produces 28,067 units of agriculture, 9383 units of construction, 51,372 units of energy, 61,364 units of manufacturing, and 90,403 units of transportation, how much is available from each sector to satisfy the demand from consumers and others outside the production system?

(c) A new demand matrix for the economy is

$$\begin{bmatrix} 2400 \\ 850 \\ 1400 \\ 3200 \\ 1800 \end{bmatrix}.$$

How much must each sector produce to meet this demand?

88. Business   The following matrix represents the number of direct flights between four cities:

$$\begin{array}{c} \\ A \\ B \\ C \\ D \end{array} \begin{array}{c} \begin{array}{cccc} A & B & C & D \end{array} \\ \begin{bmatrix} 0 & 1 & 0 & 1 \\ 1 & 0 & 0 & 1 \\ 0 & 0 & 0 & 1 \\ 1 & 1 & 1 & 0 \end{bmatrix} \end{array}.$$

(a) Find the number of one-stop flights between cities A and C.

(b) Find the total number of flights between cities B and C that are either direct or one stop.

(c) Find the matrix that gives the number of two-stop flights between these cities.

89. Social Science   (a) Use the matrix $M = \begin{bmatrix} 2 & 6 \\ 1 & 4 \end{bmatrix}$ to encode the message "leave now."

(b) What matrix should be used to decode this message?

Case 6

Matrix Operations and Airline Route Maps

Airline route maps are usually published on airline websites, as well as in in-flight magazines. The purpose of these maps is to show what cities are connected to each other by nonstop flights provided by the airline. We can think of these maps as another type of **graph,** and we can use matrix operations to answer questions of interest about the graph. In order to study these graphs, a bit of terminology will be helpful. A **graph** is a set of points called **vertices** or **nodes** and a set of lines called **edges** connecting some pairs of vertices. Two vertices connected by an edge are said to be **adjacent.** Consider, for example, the northern route map for Cape Air from May 2001 (Figure 1).* Here, the vertices are the cities to which Cape Air flies, and two vertices are connected if there is a nonstop flight between them.

FIGURE 1 Cape Air Northern Route Map—May 2001

Some natural questions arise about graphs. It might be important to know if two vertices are connected by a sequence of two edges, even if they are not connected by a single edge. In the route map, Provincetown and Hyannis are connected by a two-edge sequence, meaning that a passenger would have to stop in

Boston while flying between those cities on Cape Air. It might be important to know if it is possible to get from a vertex to another vertex in a given number of flights. In the example, a passenger on Cape Air can get from any city in the company's network to any other city, given enough flights. But how many flights are enough? This is another issue of interest: what is the minimum number of steps required to get from one vertex to another? What is the minimum number of steps required to get from **any** vertex on the graph to any other? While these questions are relatively easy to answer for a small graph, as the number of vertices and edges grows, it becomes harder to keep track of all the different ways the vertices are connected. Matrix notation and computation can help to answer these questions.

The **adjacency matrix** for a graph with n vertices is an $n \times n$ matrix whose (i, j) entry is 1 if the ith and jth vertices are connected and 0 if they are not. If the vertices in the Cape Air graph respectively correspond to Boston (B), Hyannis (H), Martha's Vineyard (M), Nantucket (N), New Bedford (NB), Providence (P), and Provincetown (PT), then the adjacency matrix for Cape Air is as follows.

$$A = \begin{array}{c} \\ \\ \\ \\ \\ \\ \\ \\ \end{array} \begin{array}{ccccccc} B & H & M & N & NB & P & PT \\ \left[\begin{array}{ccccccc} 0 & 1 & 1 & 1 & 0 & 0 & 1 \\ 1 & 0 & 1 & 1 & 0 & 0 & 0 \\ 1 & 1 & 0 & 1 & 1 & 1 & 0 \\ 1 & 1 & 1 & 0 & 1 & 1 & 0 \\ 0 & 0 & 1 & 1 & 0 & 0 & 0 \\ 0 & 0 & 1 & 1 & 0 & 0 & 0 \\ 1 & 0 & 0 & 0 & 0 & 0 & 0 \end{array}\right] & \begin{array}{c} B \\ H \\ M \\ N \\ NB \\ P \\ PT \end{array} \end{array}$$

Adjacency matrices can be used to address the questions about graphs raised earlier. Which vertices are connected by a two-edge sequence? How many different two-edge sequences connect each pair of vertices? Consider the matrix A^2, which is A multiplied by itself. For example, let the $(6, 2)$ entry in the matrix A^2 be named b_{62}. This entry in A^2 is the product of the 6th row of A and the 2nd column of A, or

$$b_{62} = a_{61}a_{12} + a_{62}a_{22} + a_{63}a_{32} + a_{64}a_{42} + a_{65}a_{52} + a_{66}a_{62} + a_{67}a_{72}$$

$$= 0 \cdot 1 + 0 \cdot 0 + 1 \cdot 1 + 1 \cdot 1 + 0 \cdot 0 + 0 \cdot 0 + 0 \cdot 0$$

$$= 2,$$

*Compliments of Cape Air. See www.flycapeair.com/rout_cc_frameset.htm

which happens to be the number of two-flight sequences that connect city 6 (Providence) and city 2 (Hyannis). A careful look at Figure 1 confirms this fact. This calculation works because, in order for a two-flight sequence to occur between Providence and Hyannis, Providence and Hyannis must each connect to an intermediate city. Since Providence connects to Martha's Vineyard (city 3) and Martha's Vineyard connects to Hyannis, $a_{63}a_{32} = 1 \cdot 1 = 1$. Thus, there is 1 two-flight sequence from Providence to Hyannis that passes through Martha's Vineyard. Since Providence does not connect to Boston (city 1), but Boston does connect with Hyannis, $a_{61}a_{12} = 0 \cdot 1 = 0$. Hence, there is no two-flight sequence from Providence to Hyannis that passes through Boston. To find the total number of two-flight sequences between Providence and Hyannis, simply sum over all intermediate points. Notice that this sum, which is

$$a_{61}a_{12} + a_{62}a_{22} + a_{63}a_{32} + a_{64}a_{42}$$
$$+ a_{65}a_{52} + a_{66}a_{62} + a_{67}a_{72},$$

is just b_{62}, the (6, 2) entry in the matrix A^2. So it is seen that the number of two-step sequences between vertex i and vertex j in a graph with adjacency matrix A is the (i, j) entry in A^2. A more general result is the following:

The number of k-step sequences between vertex i and vertex j in a graph with adjacency matrix A is the (i, j) entry in A^k.

If A is the adjacency matrix for Figure 1, then

$$A^2 = \begin{bmatrix} 4 & 2 & 2 & 2 & 2 & 2 & 0 \\ 2 & 3 & 2 & 2 & 2 & 2 & 1 \\ 2 & 2 & 5 & 4 & 1 & 1 & 1 \\ 2 & 2 & 4 & 5 & 1 & 1 & 1 \\ 2 & 2 & 1 & 1 & 2 & 2 & 0 \\ 2 & 2 & 1 & 1 & 2 & 2 & 0 \\ 0 & 1 & 1 & 1 & 0 & 0 & 1 \end{bmatrix} \begin{matrix} B \\ H \\ M \\ N \\ NB \\ P \\ PT \end{matrix}$$

with column headers B H M N NB P PT

and

$$A^3 = \begin{bmatrix} 6 & 8 & 12 & 12 & 4 & 4 & 4 \\ 8 & 6 & 11 & 11 & 4 & 4 & 2 \\ 12 & 11 & 10 & 11 & 9 & 9 & 2 \\ 12 & 11 & 11 & 10 & 9 & 9 & 2 \\ 4 & 4 & 9 & 9 & 2 & 2 & 2 \\ 4 & 4 & 9 & 9 & 2 & 2 & 2 \\ 4 & 2 & 2 & 2 & 2 & 2 & 0 \end{bmatrix} \begin{matrix} B \\ H \\ M \\ N \\ NB \\ P \\ PT \end{matrix}$$

with column headers B H M N NB P PT

Since the (6,3) entry in A^2 is 1, there is 1 two-step sequence from Providence to Martha's Vineyard. Likewise, there are 4 three-step sequences between Hyannis and New Bedford since the (2,5) entry in A^3 is 4.

In observing the figures, note that some two-step or three step sequences may not be meaningful. On the Cape Air route map, Nantucket is reachable in two steps from Boston (via Hyannis or Martha's Vineyard), but in reality this does not mat-

ter, since there is a nonstop flight between the two cities. A better question to ask of a graph might be "What is the least number of edges that must be traversed to go from vertex i to vertex j?"

To answer this question, consider the matrix $S_k = A + A^2 + \cdots + A^k$. The (i, j) entry in this matrix tallies the number of ways to get from vertex i to vertex j in k steps or less. If such a trip is impossible, this entry will be zero. Thus, to find the shortest number of steps between the vertices, continue to compute S_k as k increases; the first k for which the (i, j) entry in S_k is nonzero is the shortest number of steps between i and j. Note that although the shortest number of steps may be computed, the method does not determine what those steps are.

If you are interested in other airlines' route maps, visit the website dir.yahoo.com/Business_and_Economy/Shopping_and_Services/Travel_and_Transportation/Airlines/. This site includes links to many obscure and smaller airlines, as well as to the more well-known carriers.

◆ **Exercises**

1. Which Cape Air cities may be reached by a two-flight sequence from New Bedford? Which may be reached by a three-flight sequence?

2. It was shown above that there are 4 three-step sequences between Hyannis and New Bedford. Describe these three-step sequences.

3. Which trips in the Cape Air network take the greatest number of flights?

4. The route map for the northern routes of Big Sky Airlines for May 2001 is given in Figure 2.* Produce an adjacency matrix for this map. The current route map is available at hosts3.in-tch.com/www.bigskyair.com/body_northern_routes.html.

FIGURE 2 Big Sky Airlines Northern Route Map—May 2001

5. Which Big Sky cities may be reached by a three-flight sequence from Helena? For which cities does it take *at least* three flights to get from them to Helena?

6. Which trips in the Big Sky network take the largest number of flights? How many flights do these trips take?

_____

*Airline route map courtesy of Big Sky Airlines, Billings, MT, www.bigskyair.com.

Chapter 7 Linear Programming

Linear programming is one of the most remarkable (and useful) mathematical techniques developed in the last 60 years. It is used to deal with a variety of issues faced by businesses, financial planners, medical personnel, sports leagues, and others. Typical applications include maximizing company profits by adjusting production schedules, minimizing shipping costs by locating warehouses efficiently, and maximizing pension income by choosing the best mix of financial products. See Exercise 13 on page 410, Exercise 25 on page 434, and Exercise 19 on page 390.

Many realistic problems involve inequalities. For example, a factory may have no more than 200 workers on a shift and must manufacture at least 3000 units at a cost of no more than $35 each. How many workers should it have per shift in order to produce the required units at minimal cost? *Linear programming* is a method for finding the optimal (best possible) solution for such problems—if there is one.

In this chapter, we shall study two methods of solving linear programming problems: the graphical method and the simplex method. The graphical method requires a knowledge of **linear inequalities,** those involving only first-degree polynomials in x and y. So we begin with a study of such inequalities.

7.1   Graphing Linear Inequalities in Two Variables

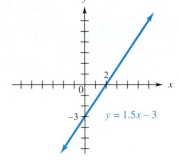

FIGURE 7.1

Examples of linear inequalities in two variables include

$$x + 2y < 4, \qquad 3x + 2y > 6, \qquad \text{and} \qquad 2x - 5y \geq 10.$$

A solution of a linear inequality is an ordered pair that satisfies the inequality. For example $(4, 4)$ is a solution of

$$3x - 2y \leq 6.$$

(Check by substituting 4 for x and 4 for y.) A linear inequality has an infinite number of solutions, one for every choice of a value for x. The best way to show these solutions is to sketch the **graph of the inequality,** which consists of all points in the plane whose coordinates satisfy the inequality.

▼ **EXAMPLE 1**   Graph the inequality $3x - 2y \leq 6$.

    Solution   First, solve the inequality for y.

$$3x - 2y \leq 6$$
$$-2y \leq -3x + 6$$
$$y \geq \frac{3}{2}x - 3 \qquad \text{{\color{blue}Multiply by } } -1/2\text{{\color{blue}; reverse the inequality.}}$$
$$y \geq 1.5x - 3$$

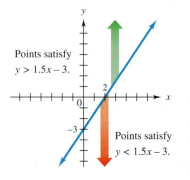

FIGURE 7.2

This inequality has the same solutions as the original one. To solve it, note that the points on the line $y = 1.5x - 3$ certainly satisfy $y \geq 1.5x - 3$. Plot some points and graph this line, as in Figure 7.1.

The points on the line satisfy "y *equals* $1.5x - 3$." The points satisfying "y is *greater than* $1.5x - 3$" are the points *above* the line (because they have larger second coordinates than the points on the line; see Figure 7.2). Similarly, the points satisfying

$$y < 1.5x - 3$$

lie below the line (because they have smaller second coordinates), as shown in Figure 7.2. The line $y = 1.5x - 3$ is the **boundary line.**

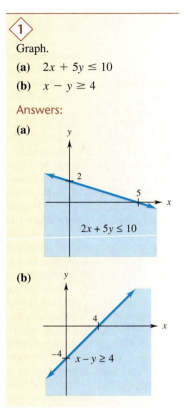

1

Graph.

(a) $2x + 5y \leq 10$

(b) $x - y \geq 4$

Answers:

(a)

(b)

2

Graph.

(a) $2x + 3y > 12$

(b) $3x - 2y < 6$

Answers:

(a)

(b)

In sum, the solutions of $y \geq 1.5x - 3$ are all points *on or above* the line $y = 1.5x - 3$. The line and the shaded region of Figure 7.3 make up the graph of the inequality $y \geq 1.5x - 3$. ▲ **1**

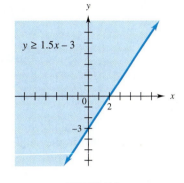

FIGURE 7.3

▼ **EXAMPLE 2**   Graph $x + 4y < 4$.

Solution First obtain an equivalent inequality by solving for y.

$$4y < -x + 4$$

$$y < -\frac{1}{4}x + 1$$

$$y < -.25x + 1$$

The boundary line is $y = -.25x + 1$, but *it is not* part of the solution, since points on the line do not satisfy $y < -.25x + 1$. To indicate this, the line is drawn dashed in Figure 7.4. The points *below* the boundary line are the solutions of $y < -.25x + 1$ because they have smaller second coordinates than the points on the line $y = -.25x + 1$. The shaded region in Figure 7.4 (excluding the dashed line) is the graph of the inequality $y < -.25x + 1$. ▲ **2**

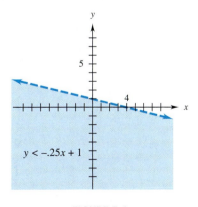

FIGURE 7.4

Examples 1 and 2 show that the solutions of a linear inequality form a **half plane** consisting of all points on one side of the boundary line (and possibly the line itself). When an inequality is solved for y, the inequality symbol immediately tells

TECHNOLOGY TIP   To shade the area above or below the graph of Y_1 on TI-84+, go to the Y= menu and move the cursor to the left of Y_1. Press ENTER until the correct shading pattern appears (◥ for above the line and ◣ for below the line). Then press GRAPH. On TI-86/89, use the STYLE key in the Y= menu instead of the ENTER key. For other calculators, consult your instruction manual.

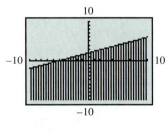

FIGURE 7.5

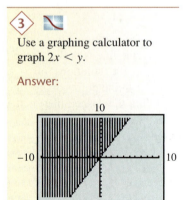

whether the points above ($>$), on ($=$), or below ($<$) the boundary line satisfy the inequality, as summarized here.

Inequality	Solution Consists of All Points
$y \geq mx + b$	*on or above* the line $y = mx + b$
$y > mx + b$	*above* the line $y = mx + b$
$y \leq mx + b$	*on or below* the line $y = mx + b$
$y < mx + b$	*below* the line $y = mx + b$

When graphing by hand, draw the boundary line $y = mx + b$ solid when it is included in the solution ($\geq$ or $\leq$ inequalities) and dashed when it is not part of the solution ($>$ or $<$ inequalities).

▼ **EXAMPLE 3**   Graph $5y - 2x \leq 10$.

Solution Solve the inequality for y.

$$5y \leq 2x + 10$$

$$y \leq \frac{2}{5}x + 2$$

$$y \leq .4x + 2$$

The graph consists of all points on or below the boundary line $y \leq .4x + 2$, as shown in Figure 7.5 (see the Technology Tip). ▲ ③

CAUTION You cannot tell from a calculator-produced graph whether the boundary line is included. It is included in Figure 7.5, but not in the answer to problem 3 at the side.

▼ **EXAMPLE 4**   Graph each of the following.

(a) $y \geq 2$

Solution The boundary line is the horizontal line $y = 2$. The graph consists of all points on or above this line (Figure 7.6).

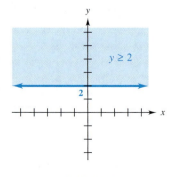

FIGURE 7.6

(b) $x \leq -1$

Solution This inequality does not fit the pattern discussed above, but it can be solved by a similar technique. Here, the boundary line is the vertical line $x = -1$, and it is included in the solution. The points satisfying $x < -1$ are all points to the left of this line (because they have x-coordinates smaller than -1). So the graph consists of the points that are *on or to the left of* the vertical line $x = -1$, as shown in Figure 7.7. ▲ ◇4◇

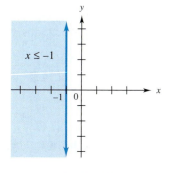

FIGURE 7.7

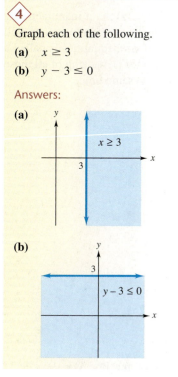

Graph each of the following.

(a) $x \geq 3$

(b) $y - 3 \leq 0$

Answers:

(a)

(b)

An alternative technique for solving inequalities that does not require solving for y is illustrated in the next example. Feel free to use it if you find it easier than the technique presented in Examples 1–4.

▼ **EXAMPLE 5** Graph $4y - 2x > 6$.

Solution The boundary line is $4y - 2x = 6$, which can be graphed by finding its intercepts.

x-intercept: let $y = 0$. y-intercept: let $x = 0$.

$4(0) - 2x = 6$ $4y - 2(0) = 6$

$x = -3$ $y = \dfrac{6}{4} = 1.5$

The graph is the half plane above or below this line. To determine which, choose a test point—any point not on the boundary line, say, $(0, 0)$. Letting $x = 0$ and $y = 0$ in the inequality produces

$$4(0) - 2(0) > 6, \qquad \text{a } \textit{false} \text{ statement.}$$

Therefore, $(0, 0)$ is not in the solution. So the solution is the half-plane that does *not* include $(0, 0)$, as shown in Figure 7.8. If a different test point is used, say, $(3, 5)$, then substituting $x = 3$ and $y = 5$ in the inequality produces

$$4(5) - 2(3) > 6, \qquad \text{a } \textit{true} \text{ statement.}$$

Therefore, the solution of the inequality is the half-plane containing $(3, 5)$, as shown in Figure 7.8. ▲

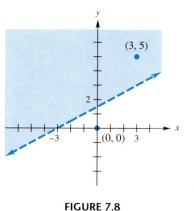

FIGURE 7.8

NOTE When using the method of Example 5, $(0, 0)$ is the best choice for the test point because it makes the calculation very easy. The only time that $(0, 0)$ cannot be used is for inequalities of the form $ax + by > 0$ (or $\geq$ or $<$ or $\leq$); in such cases, $(0, 0)$ is on the line $ax + by = 0$.

Systems of Inequalities

Realistic problems often involve many inequalities. For example, a manufacturing problem might produce inequalities resulting from production requirements, as well as inequalities about cost requirements. A set of at least two inequalities is called a **system of inequalities.** The **graph** of a system of inequalities is made up of all those points which satisfy all the inequalities of the system at the same time.

▼ **EXAMPLE 6**   Graph the system

$$3x + y < 12$$
$$x < 2y.$$

Solution First, solve each inequality for y.

$$3x + y < 12 \qquad\qquad x < 2y$$
$$y < -3x + 12 \qquad\qquad y > \frac{x}{2}$$

Then the original system is equivalent to this one:

$$y < -3x + 12$$
$$y > \frac{x}{2}.$$

The solutions of the first inequality are the points *below* the line $y = -3x + 12$ (Figure 7.9). The solutions of the second inequality are the points *above* the line $y = x/2$ (Figure 7.10). So the solutions of the *system* are the points that satisfy both of these conditions, as shown in Figure 7.11. ▲ ⑤

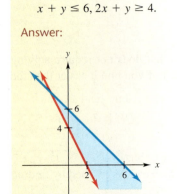

⑤

Graph the system

$$x + y \leq 6, 2x + y \geq 4.$$

Answer:

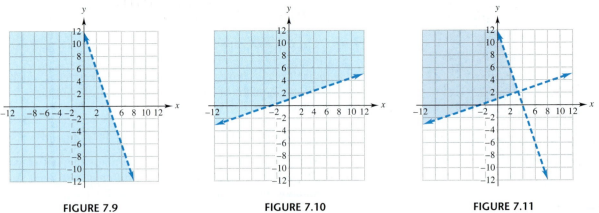

FIGURE 7.9   FIGURE 7.10   FIGURE 7.11

The shaded region in Figure 7.11 is sometimes called the **region of feasible solutions,** or just the **feasible region,** since it consists of all the points that satisfy (are feasible for) every inequality of the system.

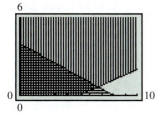

FIGURE 7.12

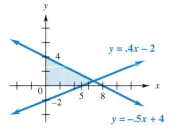

FIGURE 7.13

⟨**6**⟩

Graph the feasible region of the system

$$x + 4y \leq 8$$
$$x - y \geq 3$$
$$x \geq 0, y \geq 0.$$

Answer:

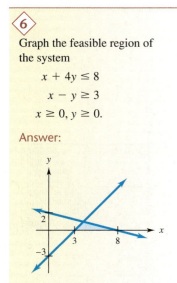

▼ **EXAMPLE 7**   Graph the feasible region for the system

$$2x - 5y \leq 10$$
$$x + 2y \leq 8$$
$$x \geq 0, y \geq 0.$$

Solution   Begin by solving the first two inequalities for y.

$$2x - 5y \leq 10 \qquad\qquad x + 2y \leq 8$$
$$-5y \leq -2x + 10 \qquad\qquad 2y \leq -x + 8$$
$$y \geq .4x - 2 \qquad\qquad y \leq -.5x + 4.$$

Then the original system is equivalent to this one:

$$y \geq .4x - 2$$
$$y \leq .5x + 4$$
$$x \geq 0, y \geq 0.$$

The inequalities $x \geq 0$ and $y \geq 0$ restrict the graph to the first quadrant. So the feasible region consists of all points in the first quadrant that are on or above the line $y = .4x - 2$ *and* on or below the line $y = -.5x + 4$. In the calculator-generated graph of Figure 7.12, the feasible region is the region with both vertical and horizontal shading. This is confirmed by the hand-drawn graph of the feasible region in Figure 7.13. ▲ ⟨6⟩

Applications

As we shall see in the rest of this chapter, many realistic problems lead to systems of linear inequalities. The next example is typical of such problems.

▼ **EXAMPLE 8**   Midtown Manufacturing Company makes plastic plates and cups, both of which require time on two machines. Producing a unit of plates requires 1 hour on machine A and 2 on machine B, while producing a unit of cups requires 3 hours on machine A and 1 on machine B. Each machine is operated for at most 15 hours per day. Write a system of inequalities expressing these conditions, and graph the feasible region.

Solution   Let x represent the number of units of plates to be made and y represent the number of units of cups. Then make a chart that summarizes the given information.

	Number of Units	Time on Machine	
		A	B
Plates	x	1	2
Cups	y	3	1
Maximum Time Available		15	15

We must have $x \geq 0$ and $y \geq 0$ because they can't produce a negative number of cups or plates. On machine A, producing x units of plates requires a total of

$1 \cdot x = x$ hours, while producing y units of cups requires $3 \cdot y = 3y$ hours. Since machine A is available no more than 15 hours a day,

$$x + 3y \le 15$$

$$y \le -\frac{x}{3} + 5.$$

Similarly, the requirement that machine B be used no more than 15 hours a day gives

$$2x + y \le 15.$$

$$y \le -2x + 15.$$

So we must solve the system

$$y \le -\frac{x}{3} + 5$$

$$y \le -2x + 15$$

$$x \ge 0, y \ge 0.$$

The feasible region is shown in Figure 7.14. ▲

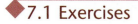

FIGURE 7.14

◆7.1 Exercises

Match the inequality with its graph, which is one of the ones shown.

1. $y \ge -x - 2$ **2.** $y \le 2x - 2$ **3.** $y \le x + 2$

4. $y \ge x + 1$ **5.** $6x + 4y \ge -12$

6. $3x - 2y \ge -4$

A.

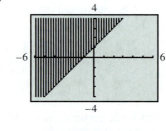

B.

C.

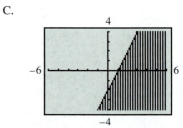

D.

E.

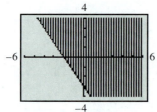

F.

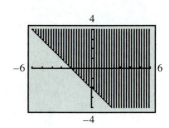

Graph each of the following linear inequalities. (See Examples 1–5.)   **13.** $y \le -4$ **14.** $x \ge -2$

7. $y < 5 - 2x$ **8.** $y < x + 3$ **15.** $3x - 2y \ge 18$ **16.** $3x + 2y \ge -4$

9. $3x - 2y \ge 18$ **10.** $2x + 5y \ge 10$ **17.** $3x + 4y > 12$ **18.** $4x - 3y > 9$

11. $2x - y \le 4$ **12.** $4x - 3y \le 24$ **19.** $2x - 4y < 3$ **20.** $4x - 3y < 12$

21. $x \le 5y$ **22.** $2x \ge y$

23. $-3x < y$ **24.** $-x > 6y$

25. $y < x$ **26.** $y > -2x$

27. In your own words, explain how to determine whether the boundary of an inequality is solid or dashed.

28. When graphing $y \le 3x - 6$, would you shade above or below the line $y = 3x - 6$? Explain your answer.

Graph the feasible region for the following systems of inequalities. (See Examples 6 and 7.)

29. $y \ge 3x - 6$
$y \ge -x + 1$

30. $x + y \le 4$
$x - y \ge 2$

31. $2x + y \le 5$
$x + 2y \le 5$

32. $x - y \ge 1$
$x \le 3$

33. $2x + y > 8$
$4x - y < 3$

34. $4x + y \ge 9$
$2x + 3y \le 7$

35. $2x - y < 1$
$3x + y < 6$

36. $x + 3y \le 6$
$2x + 4y \ge 7$

37. $-x - y < 5$
$2x - y < 4$

38. $6x - 4y > 8$
$3x + 2y > 4$

39. $3x + y \ge 6$
$x + 2y \ge 7$
$x \ge 0$
$y \ge 0$

40. $2x + 3y \ge 12$
$x + y \ge 4$
$x \ge 0$
$y \ge 0$

41. $-2 < x < 3$
$-1 \le y \le 5$
$2x + y < 6$

42. $-2 < x < 2$
$y > 1$
$x - y > 0$

43. $2y - x \ge -5$
$y \le 3 + x$
$x \ge 0$
$y \ge 0$

44. $2x + 3y \le 12$
$2x + 3y > -6$
$3x + y < 4$
$x \ge 0$
$y \ge 0$

45. $3x + 4y > 12$
$2x - 3y < 6$
$0 \le y \le 2$
$x \ge 0$

46. $0 \le x \le 9$
$x - 2y \ge 4$
$3x + 5y \le 30$
$y \ge 0$

Find a system of inequalities that has the given graph.

47.

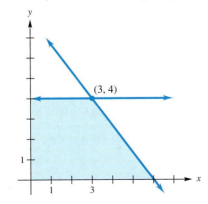

48.

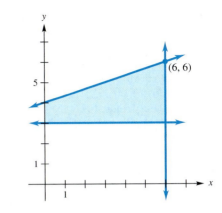

In Exercises 49 and 50, find a system of inequalities whose feasible region is the interior of the given polygon.

49. Rectangle with vertices $(2, 3)$, $(2, -1)$, $(7, 3)$, $(7, -1)$

50. Triangle with vertices $(2, 4)$, $(-4, 0)$, $(2, -1)$

51. Business Cindi Herring and Kent Merrill produce hand-made shawls and afghans. They spin the yarn, dye it, and then weave it. A shawl requires 1 hour of spinning, 1 hour of dyeing, and 1 hour of weaving. An afghan needs 2 hours of spinning, 1 of dyeing, and 4 of weaving. Together, they spend at most 8 hours spinning, 6 hours dyeing, and 14 hours weaving.

(a) Complete the following table.

	Number	Hours Spinning	Hours Dyeing	Hours Weaving
Shawls	x			
Afghans	y			
Maximum Number of Hours Available		8	6	14

(b) Use the table to write a system of inequalities that describes the situation.

(c) Graph the feasible region of this system of inequalities.

52. Business An electric shaver manufacturer makes two models: the regular and the flex. Because of demand, the number of regular shavers made is never more than half the number of flex shavers. The factory's production cannot exceed 1200 shavers per week.

(a) Write a system of inequalities that describes the possibilities for making x regular and y flex shavers per week.

(b) Graph the feasible region of this system of inequalities.

In each of the following, write a system of inequalities that describes all the given conditions, and graph the feasible region of the system. (See Example 8.)

53. Business Southwestern Oil supplies two distributors located in the Northwest. One distributor needs at least 3000 barrels of oil, the other at least 5000 barrels. Southwestern can send out at most 10,000 barrels. Let x = the number of barrels of oil sent to distributor 1 and y = the number sent to distributor 2.

54. Business The California Almond Growers have 2400 boxes of almonds to be shipped from their plant in Sacramento to Des Moines and San Antonio. The Des Moines market needs at least 1000 boxes, while the San Antonio market must have at least 800 boxes. Let x = the number of boxes to be shipped to Des Moines and y = the number of boxes to be shipped to San Antonio.

55. Business A cement manufacturer produces at least 3.2 million barrels of cement annually. He is told by the Environmental Protection Agency that his operation emits 2.5 pounds of dust for each barrel produced. The EPA has ruled that annual emissions must be reduced to 1.8 million pounds. To do this, the manufacturer plans to replace the present dust collectors with two types of electronic precipitators. One type would reduce emissions to .5 pound per barrel and would cost 16¢ per barrel. The other would reduce the dust to .3 pound per barrel and would cost 20¢ per barrel. The manufacturer does not want to spend more than .8 million dollars on the precipitators. He needs to know how many barrels he should produce with each type. Let x = the number of barrels (in millions produced) with the first type and y = the number of barrels (in millions produced) with the second type.

56. Health A dietician is planning a snack package of fruit and nuts. Each ounce of fruit will supply 1 unit of protein, 2 units of carbohydrates, and 1 unit of fat. Each ounce of nuts will supply 1 unit of protein, 1 unit of carbohydrates, and 1 unit of fat. Every package must provide at least 7 units of protein, at least 10 units of carbohydrates, and no more than 9 units of fat. Let x be the number of ounces of fruit and y the number of ounces of nuts to be used in each package.

7.2 Linear Programming: The Graphical Method

Many problems in business, science, and economics involve finding the optimal value of a function (for instance, the maximum value of the profit function or the minimum value of the cost function), subject to various **constraints** (such as transportation costs, environmental protection laws, availability of parts, and interest rates). **Linear programming** deals with such situations. In linear programming, the function to be optimized, called the **objective function,** is linear and the constraints are given by linear inequalities. Linear programming problems that involve only two variables can be solved by the graphical method, explained in Example 1.

▼ **EXAMPLE 1** Find the maximum and minimum values of the objective function $z = 2x + 5y$, subject to the following constraints:

$$3x + 2y \le 6$$
$$-2x + 4y \le 8$$
$$x + y \ge 1$$
$$x \ge 0, y \ge 0.$$

Solution First, graph the feasible region of the system of inequalities (Figure 7.15). The points in this region or on its boundaries are the only ones that satisfy all the constraints. However, each such point may produce a different value of the objective function. For instance, the points $(.5, 1)$ and $(1, 0)$ in the feasible region lead to the values

$$z = 2(.5) + 5(1) = 6 \quad \text{and} \quad z = 2(1) + 5(0) = 2.$$

We must find the points that produce the maximum and minimum values of z.

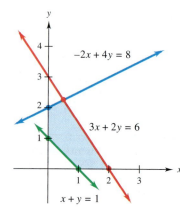

FIGURE 7.15

To find the maximum value, consider various possible values for z. For instance, when $z = 0$, the objective function is $0 = 2x + 5y$, whose graph is a straight line. Similarly, when z is 5, 10, and 15, the objective function becomes (in turn)

$$5 = 2x + 5y, \qquad 10 = 2x + 5y, \qquad 15 = 2x + 5y.$$

These four lines are graphed in Figure 7.16. (All the lines are parallel because they have the same slope.) The figure shows that z cannot take on the value 15, because the graph for $z = 15$ is entirely outside the feasible region. The maximum possible value of z will be obtained from a line parallel to the others and between the lines representing the objective function when $z = 10$ and $z = 15$. The value of z will be as large as possible, and all constraints will be satisfied, if this line just touches the feasible region. This occurs with the green line through point A.

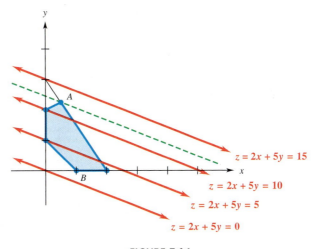

FIGURE 7.16

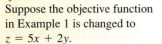

1

Suppose the objective function in Example 1 is changed to $z = 5x + 2y$.

(a) Sketch the graphs of the objective function when $z = 0$, $z = 5$, and $z = 10$ on the region of feasible solutions given in Figure 7.15.

(b) From the graph, decide what values of x and y will maximize the objective function.

Answers:

(a)

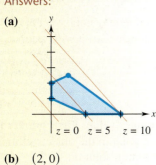

(b) $(2, 0)$

The point A is the intersection of the graphs of $3x + 2y = 6$ and $-2x + 4y = 8$ (see Figure 7.15). Its coordinates can be found either algebraically or graphically (using a graphing calculator).

Algebraic Method	**Graphical Method**
Solve the system	Solve the two equations for y.

$$3x + 2y = 6$$ $$-2x + 4y = 8,$$	$$y = -1.5x + 3$$ $$y = .5x + 2$$
as in Section 6.1, to get $x = 1/2$ and $y = 9/4$. Hence, A has coordinates $(1/2, 9/4) = (.5, 2.25)$.	Graph both equations on the same screen and use the intersection finder to find that the coordinates of the intersection point A are $(.5, 2.25)$.

The value of z at the point A is

$$z = 2x + 5y = 2(.5) + 5(2.25) = 12.25.$$

Thus, the maximum possible value of z is 12.25. Similarly, the minimum value of z occurs at point B, which has coordinates $(1, 0)$. The minimum value of z is $2(1) + 5(0) = 2$. ▲

Points such as A and B in Example 1 are called corner points. A **corner point** is a point in the feasible region where the boundary lines of two constraints cross. The feasible region in Figure 7.15 is **bounded** because the region is enclosed by boundary lines on all sides. Linear programming problems with bounded regions always have solutions. However, if Example 1 did not include the constraint $3x + 2y \le 6$, the feasible region would be **unbounded,** and there would be no way to *maximize* the value of the objective function.

Some general conclusions can be drawn from the method of solution used in Example 1. Figure 7.17 shows various feasible regions and the lines that result from various values of z. (Figure 7.17 shows the situation in which the lines are in order from left to right as z increases.) In part (a) of the figure, the objective function takes on its minimum value at corner point Q and its maximum value at P. The minimum is again at Q in part (b), but the maximum occurs at P_1 or P_2, or any point on the line segment connecting them. Finally, in part (c), the minimum value occurs at Q, but the objective function has no maximum value because the feasible region is unbounded.

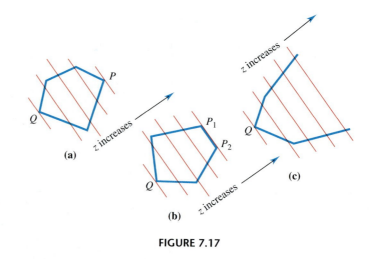

FIGURE 7.17

The preceding discussion suggests the truth of the **corner point theorem.**

Corner Point Theorem

If the feasible region is bounded, then the objective function has both a maximum and a minimum value and each occurs at one or more corner points.

If the feasible region is unbounded, the objective function may not have a maximum or minimum. But if a maximum or minimum value exists, it will occur at one or more corner points.

This theorem simplifies the job of finding an optimum value. First, graph the feasible region and find all corner points. Then test each point in the objective function. Finally, identify the corner point producing the optimum solution.

With the theorem, the problem in Example 1 could have been solved by identifying the five corner points of Figure 7.15: $(0, 1)$, $(0, 2)$, $(.5, 2.25)$, $(2, 0)$, and $(1, 0)$. Then, substituting each of these points into the objective function $z = 2x + 5y$ would identify the corner points that produce the maximum and minimum values of z.

Corner Point	Value of $z = 2x + 5y$	
$(0, 1)$	$2(0) + 5(1) = 5$	
$(0, 2)$	$2(0) + 5(2) = 10$	
$(.5, 2.25)$	$2(.5) + 5(2.25) = 12.25$	(maximum)
$(2, 0)$	$2(2) + 5(0) = 4$	
$(1, 0)$	$2(1) + 5(0) = 2$	(minimum)

From these results, the corner point $(.5, 2.25)$ yields the maximum value of 12.25 and the corner point $(1, 0)$ gives the minimum value of 2. These are the same values found earlier. ◇2

A summary of the steps for solving a linear programming problem by the graphical method is given here.

2

(a) Identify the corner points in the graph.

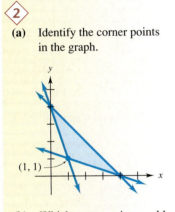

(b) Which corner point would minimize $z = 2x + 3y$?

Answers:

(a) $(0, 4)$, $(1, 1)$, $(4, 0)$

(b) $(1, 1)$

Solving a Linear Programming Problem Graphically

1. Write the objective function and all necessary constraints.
2. Graph the feasible region.
3. Determine the coordinates of each of the corner points.
4. Find the value of the objective function at each corner point.
5. If the feasible region is bounded, the solution is given by the corner point producing the optimum value of the objective function.
6. If the feasible region is an unbounded region in the first quadrant and both coefficients of the objective function are positive,* then the minimum value of the objective function occurs at a corner point and there is no maximum value.

▼ **EXAMPLE 2**   Sketch the feasible region for the following set of constraints:

$$3y - 2x \geq 0$$
$$y + 8x \leq 52$$
$$y - 2x \leq 2$$
$$x \geq 3.$$

Then find the maximum and minimum values of the objective function $z = 5x + 2y$.

*This is the only case of an unbounded region that occurs in the applications considered here.

Solution The graph in Figure 7.18(a) shows that the feasible region is bounded. The corner points are found by solving systems of two equations either algebraically, using the methods of Chapter 6, or graphically, using the intersection finder on a graphing calculator. Figure 7.18(b) shows the calculator graphs of the first three inequalities. With the graphical method, corner points on the line $x = 3$ are found by observation.

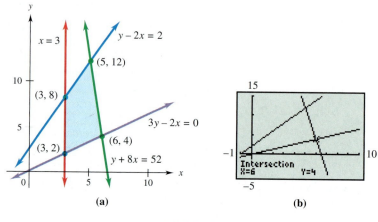

FIGURE 7.18

Use the corner points from the graph to find the maximum and minimum values of the objective function.

Corner Point	Value of $z = 5x + 2y$	
$(3, 2)$	$5(3) + 2(2) = 19$	**(minimum)**
$(6, 4)$	$5(6) + 2(4) = 38$	
$(5, 12)$	$5(5) + 2(12) = 49$	**(maximum)**
$(3, 8)$	$5(3) + 2(8) = 31$	

The minimum value of $z = 5x + 2y$ is 19, at the corner point $(3, 2)$. The maximum value is 49, at $(5, 12)$. ▲ ③

▼ **EXAMPLE 3** Solve the following linear programming problem.

$$\text{Minimize} \quad z = x + 2y$$
$$\text{subject to} \quad x + y \le 10$$
$$3x + 2y \ge 6$$
$$x \ge 0, y \ge 0.$$

Solution The feasible region is shown in Figure 7.19. From the figure, the corner points are $(0, 3)$, $(0, 10)$, $(10, 0)$, and $(2, 0)$. These corner points give the following values of z.

③

Use the region of feasible solutions in the sketch to find the following.

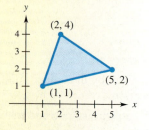

(a) The values of x and y that maximize $z = 2x - y$.

(b) The maximum value of $z = 2x - y$.

(c) The values of x and y that minimize $z = 4x + 3y$.

(d) The minimum value of $z = 4x + 3y$.

Answers:

(a) $(5, 2)$

(b) 8

(c) $(1, 1)$

(d) 7

Corner Point	Value of $z = x + 2y$
$(0, 3)$	$0 + 2(3) = 6$
$(0, 10)$	$0 + 2(10) = 20$
$(10, 0)$	$10 + 2(0) = 10$
$(2, 0)$	$2 + 2(0) = 2$ (minimum)

The minimum value of z is 2; it occurs at $(2, 0)$. ▲

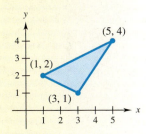

4

The sketch shows a feasible region. Let $z = x + 3y$. Use the sketch to find the values of x and y that

(a) minimize z;

(b) maximize z.

Answers:

(a) $(3, 1)$

(b) $(5, 4)$

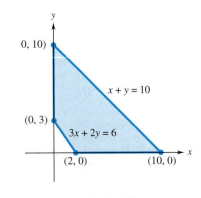

FIGURE 7.19

▼ **EXAMPLE 4**   Solve the following linear programming problem.

$$\text{Minimize} \quad z = 2x + 4y$$
$$\text{subject to} \quad x + 2y \geq 10$$
$$3x + y \geq 10$$
$$x \geq 0, y \geq 0.$$

Solution Figure 7.20 shows the hand-drawn graph with corner points $(0, 10)$, $(2, 4)$, and $(10, 0)$, as well as the calculator graph with the corner point $(2, 4)$. Find the value of z for each point.

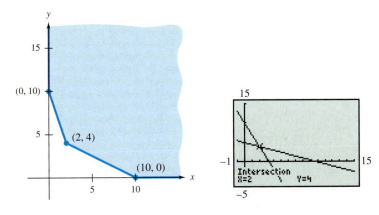

FIGURE 7.20

⟨5⟩

The sketch shows a region of feasible solutions. From the sketch, decide what ordered pair would minimize $z = 2x + 4y$.

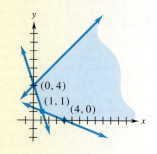

Corner Point	Value of $z = 2x + 4y$	
$(0, 10)$	$2(0) + 4(10) = 40$	
$(2, 4)$	$2(2) + 4(4) = 20$	(minimum)
$(10, 0)$	$2(10) + 4(0) = 20$	(minimum)

In this case, both $(2, 4)$ and $(10, 0)$, as well as all the points on the boundary line between them, give the same optimum value of z. There is an infinite number of equally "good" values of x and y that give the same minimum value of the objective function $z = 2x + 4y$. The minimum value is 20. ▲ ⟨5⟩

Answer:

$(1, 1)$

◆7.2 Exercises

Exercises 1–6 show regions of feasible solutions. Use these regions to find maximum and minimum values of each given objective function. (See Examples 1 and 2.)

1. $z = 3x + 5y$

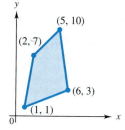

2. $z = x + 6y$

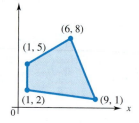

3. $z = .75x + .40y$

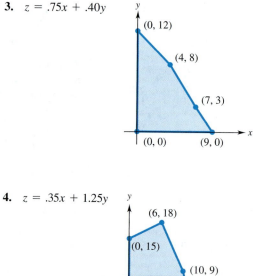

4. $z = .35x + 1.25y$

5. (a) $z = x + 5y$
(b) $z = 2x + 3y$
(c) $z = 2x + 4y$
(d) $z = 4x + y$

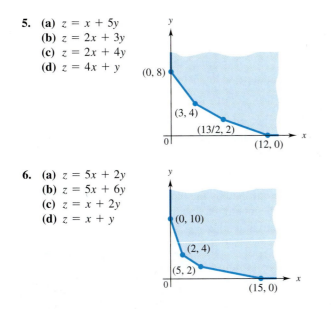

(0, 8)
(3, 4)
(13/2, 2)
(12, 0)

6. (a) $z = 5x + 2y$
(b) $z = 5x + 6y$
(c) $z = x + 2y$
(d) $z = x + y$

(0, 10)
(2, 4)
(5, 2)
(15, 0)

Use graphical methods to solve Exercises 7–12. (See Examples 2–4.)

7. Maximize $z = 2x + 5y$
subject to $2x + 3y \leq 6$
$4x + y \leq 6$
$x \geq 0, y \geq 0.$

8. Minimize $z = x + 3y$
subject to $2x + y \leq 10$
$5x + 2y \geq 20$
$-x + 2y \geq 0$
$x \geq 0, y \geq 0.$

9. Minimize $z = 2x + y$
subject to $3x - y \geq 12$
$x + y \leq 15$
$x \geq 2, y \geq 3.$

10. Maximize $z = x + 3y$
subject to $2x + 3y \leq 100$
$5x + 4y \leq 200$
$x \geq 10, y \geq 20.$

11. Maximize $z = 2x + 4y$
subject to $x - y \leq 10$
$5x + 3y \leq 75$
$x \geq 0, y \geq 0.$

12. Maximize $z = 4x + 5y$
subject to $10x - 5y \leq 100$
$20x + 10y \geq 150$
$x \geq 0, y \geq 0.$

Find the minimum and maximum values of $z = 3x + 4y$ (if possible) for each of the following sets of constraints. (See Examples 2–4.)

13. $3x + 2y \geq 6$
$x + 2y \geq 4$
$x \geq 0, y \geq 0$

14. $2x + y \leq 20$
$10x + y \geq 36$
$2x + 5y \geq 36$

15. $x + y \leq 6$
$-x + y \leq 2$
$2x - y \leq 8$

16. $-x + 2y \leq 6$
$3x + y \geq 3$
$x \geq 0, y \geq 0$

17. Find values of $x \geq 0$ and $y \geq 0$ that maximize $z = 10x + 12y$, subject to each of the following sets of constraints.
(a) $x + y \leq 20$
$x + 3y \leq 24$
(b) $3x + y \leq 15$
$x + 2y \leq 18$
(c) $x + 2y \geq 10$
$2x + y \geq 12$
$x - y \leq 8$

18. Find values of $x \geq 0$ and $y \geq 0$ that minimize $z = 3x + 2y$, subject to each of the following sets of constraints.
(a) $10x + 7y \leq 42$
$4x + 10y \geq 35$
(b) $6x + 5y \geq 25$
$2x + 6y \geq 15$
(c) $2x + 5y \geq 22$
$4x + 3y \leq 28$
$2x + 2y \leq 17$

19. Explain why it is impossible to maximize the function $z = 3x + 4y$ subject to the constraints
$x + y \geq 8$
$2x + y \leq 10$
$x + 2y \leq 8$
$x \geq 0, y \geq 0.$

20. You are given the following linear programming problem:*

Maximize $z = c_1x_1 + c_2x_2$
subject to $2x_1 + x_2 \leq 11$
$-x_1 + 2x_2 \leq 2$
$x_1 \geq 0, x_2 \geq 0.$

If $c_2 > 0$, determine the range of c_1/c_2 for which $(x_1, x_2) = (4, 3)$ is an optimal solution.
(a) $[-2, 1/2]$ (b) $[-1/2, 2]$
(c) $[-11, -1]$ (d) $[1, 11]$
(e) $[-11, 11]$

*Problem from "Course 130 Examination Operations Research" of the *Education and Examination Committee of the Society of Actuaries.* Reprinted by permission of the Society of Actuaries.

7.3 Applications of Linear Programming

In this section, we show several applications of linear programming with two variables.

▼ **EXAMPLE 1**   A 4-H Club member raises geese and pigs. She wants to raise no more than 16 animals, including no more than 10 geese. She spends $15 to raise a goose and $45 to raise a pig, and she has $540 available for the project. Find the maximum profit she can make if each goose produces a profit of $7 and each pig a profit of $20.

Solution   The total profit is determined by the number of geese and pigs. So let x be the number of geese to be produced and let y be the number of pigs. Then summarize the information of the problem in a table.

	Number	Cost to Raise	Profit Each
Geese	x	$15	$7
Pigs	y	$45	$20
Maximum Available	16	$540	

Use this table to write the necessary constraints. Since the total number of animals cannot exceed 16, the first constraint is

$$x + y \leq 16.$$

"No more than 10 geese" leads to

$$x \leq 10.$$

The cost to raise x geese at $15 per goose is $15x$ dollars, while the cost for y pigs at $45 each is $45y$ dollars. Only $540 is available, so

$$15x + 45y \leq 540.$$

Dividing both sides by 15 gives the equivalent inequality

$$x + 3y \leq 36.$$

The number of geese and pigs cannot be negative, so

$$x \geq 0, \qquad y \geq 0.$$

The 4-H Club member wants to know the number of geese and the number of pigs that should be raised for maximum profit. Each goose produces a profit of $7, each pig $20. If z represents total profit, then

$$z = 7x + 20y$$

is the objective function that is to be maximized.

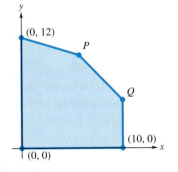

FIGURE 7.21

Find the corner points P and Q in Figure 7.19.

Answer:

$P = (6, 10)$
$Q = (10, 6)$

We must solve the following linear programming problem:

Maximize $z = 7x + 20y$ Objective function

subject to $\begin{aligned} x + y &\le 16 \\ x &\le 10 \\ x + 3y &\le 36 \\ x \ge 0, y &\ge 0. \end{aligned}$ Constraints

Using the methods of the previous section, graph the feasible region for the system of inequalities given by the constraints, as in Figure 7.21.

The corner points $(0, 12)$, $(0, 0)$, and $(10, 0)$ can be read directly from the graph. Find the coordinates of the other corner points by solving a system of equations or with a graphing calculator. ①

Test each corner point in the objective function to find the maximum profit.

Corner Point	$z = 7x + 20y$
$(0, 12)$	$7(0) + 20(12) = 240$
(6, 10)	**$7(6) + 20(10) = 242$** **(maximum)**
$(10, 6)$	$7(10) + 20(6) = 190$
$(10, 0)$	$7(10) + 20(0) = 70$
$(0, 0)$	$7(0) + 20(0) = 0$

The maximum value for z of 242 occurs at $(6, 10)$. Thus, 6 geese and 10 pigs will produce a maximum profit of $242. ▲

▼ **EXAMPLE 2**   An office manager needs to purchase new filing cabinets. He knows that Ace cabinets cost $40 each, require 6 square feet of floor space, and hold 8 cubic feet of files. On the other hand, each Excello cabinet costs $80, requires 8 square feet of floor space, and holds 12 cubic feet. His budget permits him to spend no more than $560 on files, while the office has room for no more than 72 square feet of cabinets. The manager desires the greatest storage capacity within the limitations imposed by funds and space. How many of each type of cabinet should he buy?

Solution Let x represent the number of Ace cabinets to be bought, and let y represent the number of Excello cabinets. The information given in the problem can be summarized as follows.

	Number	Cost of Each	Space Required	Storage Capacity
Ace	x	$ 40	6 sq ft	8 cu ft
Excello	y	$ 80	8 sq ft	12 cu ft
Maximum Available		$560	72 sq ft	

The constraints imposed by cost and space are

$$40x + 80y \leq 560 \qquad \text{Cost}$$
$$6x + 8y \leq 72. \qquad \text{Floor space}$$

The number of cabinets cannot be negative, so $x \geq 0$ and $y \geq 0$. The objective function to be maximized gives the amount of storage capacity provided by some combination of Ace and Excello cabinets. From the information in the chart, the objective function is

$$z = \text{Storage space} = 8x + 12y.$$

In sum, the given problem has produced the following linear programming problem:

$$\text{Maximize} \quad z = 8x + 12y$$
$$\text{subject to} \quad 40x + 80y \leq 560$$
$$6x + 8y \leq 72$$
$$x \geq 0, y \geq 0.$$

A graph of the feasible region is shown in Figure 7.22. Three of the corner points can be identified from the graph as $(0, 0)$, $(0, 7)$, and $(12, 0)$. The fourth corner point, labeled Q in the figure, can be found algebraically or with a graphing calculator to be $(8, 3)$. **2**

2

Find the corner point labeled P on the region of feasible solutions.

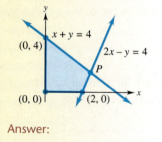

Answer:

$(8/3, 4/3)$

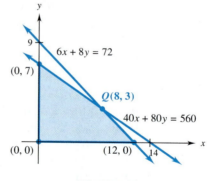

FIGURE 7.22

Use the corner point theorem to find the maximum value of z.

Corner Point	Value of $z = 8x + 12y$	
$(0, 0)$	0	
$(0, 7)$	84	
$(12, 0)$	96	
$(8, 3)$	**100**	**(maximum)**

The objective function, which represents storage space, is maximized when $x = 8$ and $y = 3$. The manager should buy 8 Ace cabinets and 3 Excello cabinets. ▲

▼ **EXAMPLE 3** Certain laboratory animals must have at least 30 grams of protein and at least 20 grams of fat per feeding period. These nutrients come from food A, which costs 18¢ per unit and supplies 2 grams of protein and 4 of fat, and food B, with 6 grams of protein and 2 of fat, costing 12¢ per unit. Food B is bought under a long-term contract requiring that at least 2 units of B be used per serving. How much of each food must be bought to produce the minimum cost per serving?

Solution Let x represent the amount of food A needed and y the amount of food B. Use the given information to produce the following table.

Food	Number of Units	Grams of Protein	Grams of Fat	Cost
A	x	2	4	18¢
B	y	6	2	12¢
Minimum Required		30	20	

The linear programming problem can be stated as follows:

$$\text{Minimize} \quad z = .18x + .12y \quad \text{Cost}$$
$$\text{subject to} \quad 2x + 6y \geq 30 \quad \text{Protein}$$
$$4x + 2y \geq 20 \quad \text{Fat}$$
$$y \geq 2 \quad \text{Contract}$$
$$x \geq 0, y \geq 0.$$

(The constraint $y \geq 0$ is redundant because of the constraint $y \geq 2$.) A graph of the feasible region with the corner points identified is shown in Figure 7.23.

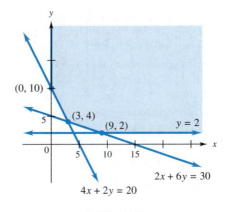

FIGURE 7.23

Use the corner point theorem to find the minimum value of z as shown in the table.

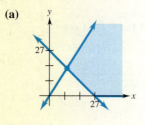

Corner Points	$z = .18x + .12y$
$(0, 10)$	$.18(0) + .12(10) = 1.20$
$(3, 4)$	$.18(3) + .12(4) = 1.02$ **(minimum)**
$(9, 2)$	$.18(9) + .12(2) = 1.86$

4

Use the information in Side Problem 3 to do the following.

(a) Graph the feasible region and find the corner points.

(b) Determine the minimum value of the objective function and the point where it occurs.

(c) Is there a maximum cost?

Answers:

(a)

Corner points: $(27, 0)$, $(9, 18)$

(b) $6300 at $(9, 18)$

(c) No

The minimum value of 1.02 occurs at $(3, 4)$. Thus, 3 units of A and 4 units of B will produce a minimum cost of $1.02 per serving. ▲

The feasible region in Figure 7.23 is an unbounded one: the region extends indefinitely to the upper right. With this region, it would not be possible to *maximize* the objective function, because the total cost of the food could always be increased by encouraging the animals to eat more.

◆7.3 Exercises

Write the constraints in Exercises 1–5 as linear inequalities and identify all variables used. In some instances, not all of the information is needed to write the constraints. (See Examples 1–3.)

1. A canoe requires 8 hours of fabrication and a rowboat 5 hours. The fabrication department has at most 110 hours of labor available each week.

2. Doug Gilbert needs at least 2800 miligrams of vitamin C per day. Each Supervite pill provides 250 milligrams and each Vitahealth pill provides 350 milligrams.

3. A candidate can afford to spend no more than $9500 on radio and TV advertising. Each radio spot costs $250 and each TV ad costs $750.

4. A hospital dietician has two meal choices: one for patients on solid food that costs $2.75 and one for patients on liquids that costs $3.75. There is a maximum of 600 patients in the hospital.

Solve these linear programming problems, which are somewhat simpler than the examples in the text.

5. **Business** A chain saw requires 4 hours of assembly and a wood chipper 6 hours. A maximum of 48 hours of assembly time is available. The profit is $150 on a chain saw and $220 on a chipper. How many of each should be assembled for maximum profit?

6. **Health** Don Garcia needs at least 24 units of vitamin A per day. Each green pill provides 4 units and each red pill provides 1 unit. Green pills cost 3.5¢ each and red pills 1¢ each. How many of each kind of pill should he take to minimize his cost?

7. **Business** Deluxe coffee is to be mixed with regular coffee to make at least 50 pounds of a blended coffee. The mixture must contain at least 10 pounds of deluxe coffee. Deluxe coffee costs $6 per pound and regular coffee $5 per pound. How many pounds of each kind of coffee should be used to minimize costs?

8. **Business** Pauline Wong spends 4 hours selling a used car and 6 hours selling a new car. She works no more than 40 hours per week. In order to receive a bonus, she must sell at least one used car and four new cars each week. In that case, she receives a bonus of $180 for each used car and $290 for each new car. How many used and how many new cars should she try to sell to maximize her bonus?

Solve the following linear programming problems. (See Examples 1–3.)

9. **Business** A company is considering two insurance plans with the types of coverage and premiums shown in the table.

	Policy A	Policy B
Fire/Theft	$10,000	$15,000
Liability	$180,000	$120,000
Premium per unit	$50	$40

(For example, $50 buys one unit of plan A, consisting of $10,000 of fire and theft insurance and $180,000 of liability insurance.)

(a) The company wants at least $300,000 of fire/theft insurance and at least $3,000,000 of liability insurance from these plans. How many units should be purchased from each plan to minimize the cost of the premiums? What is the minimum premium?

(b) Suppose the premium for policy A is reduced to $25. Now how many units should be purchased from each plan to minimize the cost of the premiums? What is the minimum premium?

10. **Business** A manufacturer of refrigerators must ship at least 100 refrigerators to its two West Coast warehouses. Each warehouse holds a maximum of 100 refrigerators. Warehouse A holds 25 refrigerators already, and warehouse B has 20 on hand. It costs $12 to ship a refrigerator to warehouse A and $10 to ship one to warehouse B. Union rules require that at least 300 workers be hired. Shipping a refrigerator to Warehouse A requires 4 workers, while shipping a refrigerator to Warehouse B requires 2 workers. How many refrigerators should be shipped to each warehouse to minimize costs? What is the minimum cost?

11. **Health** Sing, who is dieting, requires two food supplements: I and II. He can get these supplements from two different products, A and B, as shown in the table.

		Grams of Supplement per Serving	
		I	II
Product	A	3	2
	B	2	4

Sing's physician has recommended that he include at least 15 g of each supplement in his daily diet. If product A costs 25¢ per serving and product B costs 40¢ per serving, how can he satisfy his dietetic requirements most economically? Find the minimum cost.

12. **Business** The manufacturing process requires that oil refineries manufacture at least 2 gallons of gasoline for every gallon of fuel oil. To meet the winter demand for fuel oil, at least 3 million gallons a day must be produced. The demand for gasoline is no more than 12 million gallons per day. It takes .25 hour to ship each million gallons of gasoline and 1 hour to ship each million gallons of fuel

oil out of the warehouse. No more than 6.6 hours are available for shipping. If the refinery sells gasoline for $1.25 per gallon and fuel oil for $1 per gallon, how much of each should be produced to maximize revenue? Find the maximum revenue.

13. **Business** A machine shop manufactures two types of bolts. The bolts require time on each of three groups of machines, but the time required on each group differs, as shown in the table.

		Machine Group		
		I	II	III
Bolts	Type 1	.1 min	.1 min	.1 min
	Type 2	.1 min	.4 min	.02 min

Production schedules are made up one day at a time. In a day, there are 240, 720, and 160 minutes available, respectively, on these machines. Type 1 bolts sell for 10¢ and type 2 bolts for 12¢. How many of each type of bolt should be manufactured per day to maximize revenue? What is the maximum revenue?

14. **Health** Kim Walrath has a nutritional deficiency and is told to take at least 2400 mg of iron, 2100 mg of vitamin B-1, and 1500 mg of vitamin B-2. One Maxivite pill contains 40 mg of iron, 10 mg of B-1, and 5 mg of B-2 and costs 6¢. One Healthovite pill provides 10 mg of iron, 15 mg of B-1, and 15 mg of B-2 and costs 8¢. What combination of Maxivite and Healthovite pills will meet the requirement at lowest cost? What is the minimum cost?

15. **Social Sciences** An anthropology article recounts a hypothetical situation that could be described by a linear programming model.* Suppose a population gathers plants and animals for survival. They need at least 360 units of energy, 300 units of protein, and 8 hides during some time period. One unit of plants provides 30 units of energy, 10 units of protein, and no hides. One animal provides 20 units of energy, 25 units of protein, and 1 hide. Only 25 units of plants and 25 animals are available. It costs the population 30 hours of labor to gather one unit of a plant and 15 hours for an animal. Find how many units of plants and how many animals should be gathered to meet the requirements with a minimum number of hours of labor.

16. **Business** The Miers Company produces small engines for several manufacturers. The company receives orders from two assembly plants for their Topflight engine. Plant I needs at least 50 engines, and plant II needs at least 27 engines. The company can send at most 105 engines to these two assembly plants. It costs $20 per engine to ship

---

*Van A. Reidhead, "Linear Programming Models in Archaeology," *Annual Review of Anthropology* 8 (1979): 543–578.

to plant I and $35 per engine to ship to plant II. Plant I gives Miers $15 in rebates toward its products for each engine Miers buys, while plant II gives similar $10 rebates. Miers estimates that it needs at least $1200 in rebates to cover products it plans to buy from the two plants. How many engines should be shipped to each plant to minimize shipping costs? What is the minimum cost?

17. **Business** A greeting card manufacturer has 500 boxes of a particular card in warehouse I and 290 boxes of the same card in warehouse II. A greeting card shop in San Jose orders 350 boxes of the card, and another shop in Memphis orders 250 boxes. The shipping costs per box to these shops from the two warehouses are shown in the following table.

		DESTINATION	
		San Jose	Memphis
Warehouse	I	$.25	$.22
	II	$.23	$.21

How many boxes should be shipped to each city from each warehouse to minimize shipping costs? What is the minimum cost? (*Hint:* Use x, $350 - x$, y, and $250 - y$ as the variables.)

18. **Business** *Hotnews Magazine* publishes a U.S. and a Canadian edition each week. There are 30,000 subscribers in the United States and 20,000 in Canada. Other copies are sold at newsstands. Postage and shipping costs average $80 per thousand copies for the United States and $60 per thousand copies for Canada. Surveys show that no more than 120,000 copies of each issue can be sold (including subscriptions) and that the number of copies of the Canadian edition should not exceed twice the number of copies of the U.S. edition. The publisher can spend at most $8400 a month on postage and shipping. If the profit is $200 for each thousand copies of the U.S. edition and $150 for each thousand copies of the Canadian edition, how many copies of each version should be printed to earn as large a profit as possible? What is that profit?

19. **Finance** A pension fund manager decides to invest at most $50 million in U.S. Treasury Bonds paying 4% annual interest and in mutual funds paying 6% annual interest. He plans to invest at least $20 million in bonds and at least $6 million in mutual funds. Bonds have an initial fee of $300 per million dollars, while the fee for mutual funds is $100 per million. The fund manager is allowed to spend no more than $8400 on fees. How much should be invested in each to maximize annual interest? What is the maximum annual interest?

20. **Natural Science** A certain predator requires at least 10 units of protein and 8 units of fat per day. One prey of Species I provides 5 units of protein and 2 units of fat; one prey of Species II provides 3 units of protein and 4 units of fat. Capturing and digesting each Species II prey requires 3 units of energy, and capturing and digesting each Species I prey requires 2 units of energy. How many of each prey would meet the predator's daily food requirements with the least expenditure of energy? Are the answers reasonable? How could they be interpreted?

21. **Social Science** Students at Upscale U. are required to take at least 4 humanities and 4 science courses. The maximum allowable number of science courses is 12. Each humanities course carries 4 credits and each science course 5 credits. The total number of credits in science and humanities cannot exceed 92. Quality points for each course are assigned in the usual way: the number of credit hours times 4 for an A grade, times 3 for a B grade, and times 2 for a C grade. Susan Katz expects to get B's in all her science courses. She expects to get C's in half her humanities courses, B's in one-fourth of them, and A's in the rest. Under these assumptions, how many courses of each kind should she take in order to earn the maximum possible number of quality points?

22. **Social Science** In Exercise 21, find Susan's grade point average (the total number of quality points divided by the total number of credit hours) at each corner point of the feasible region. Does the distribution of courses that produces the highest number of quality points also yield the highest grade point average? Is this a contradiction?

*The importance of linear programming is shown by the inclusion of linear programming problems on most qualification examinations for Certified Public Accountants. Exercises 23–25 are reprinted from one such examination.**

The Random Company manufactures two products: Zeta and Beta. Each product must pass through two processing operations. All materials are introduced at the start of Process No. 1. There are no work-in-process inventories. Random may produce either one product exclusively or various combinations of both products, subject to the following constraints.

	Process No. 1	Process No. 2	Contribution Margin per Unit
Hours required to produce 1 unit of:			
Zeta	1 hour	1 hour	$4.00
Beta	2 hours	3 hours	$5.25
Total capacity in hours per day	1000 hours	1275 hours	

*Material from *Uniform CPA Examinations and Unofficial Answers,* copyright © 1973, 1974, 1975 by the American Institute of Certified Public Accountants, Inc., is reprinted with permission.

A shortage of technical labor has limited Beta production to 400 units per day. There are no constraints on the production of Zeta other than the hour constraints shown in the schedule. Assume that all the relationships between capacity and production are linear.

23. Given the objective to maximize total contribution margin, what is the production constraint for Process No. 1?
  (a) Zeta + Beta ≤ 1000   **(b)** Zeta + 2 Beta ≤ 1000
  (c) Zeta + Beta ≥ 1000   **(d)** Zeta + 2 Beta ≥ 1000

24. Given the objective to maximize total contribution margin, what is the labor constraint for production of Beta?
  (a) Beta ≤ 400   **(b)** Beta ≥ 400

(c) Beta ≤ 425   **(d)** Beta ≥ 425

25. What is the objective function of the data presented?
  (a) Zeta + 2 Beta = $9.25
  (b) $4.00 Zeta + 3($5.25) Beta = total contribution margin
  (c) $4.00 Zeta + $5.25 Beta = total contribution margin
  (d) 2($4.00) Zeta + 3($5.25) Beta = total contribution margin

7.4 The Simplex Method: Maximization

For linear programming problems with more than two variables or with two variables and many constraints, the graphical method is usually inefficient or impossible, so the **simplex method** is used. The simplex method, which is introduced here, was developed for the U.S. Air Force by George B. Danzig in 1947. It is now used in industrial planning, factory design, product distribution networks, sports scheduling, truck routing, resource allocation, and a variety of other ways.

Because the simplex method is used for problems with many variables, it usually is not convenient to use letters such as x, y, z, or w as variable names. Instead, the symbols x_1 (read "x-sub-one"), x_2, x_3, and so on, are used. In the simplex method, all constraints must be expressed in the linear form

$$a_1x_1 + a_2x_2 + a_3x_3 + \cdots \le b,$$

where $x_1, x_2, x_3, \ldots$ are variables, $a_1, a_2, a_3, \ldots$ are coefficients, and b is a constant.

We first discuss the simplex method for linear programming problems in *standard maximum form*.

Standard Maximum Form

A linear programming problem is in **standard maximum form** if

1. the objective function is to be maximized;

2. all variables are nonnegative $(x_i \ge 0, i = 1, 2, 3, \ldots)$;

3. all constraints involve ≤;

4. the constants on the right side in the constraints are all nonnegative $(b \ge 0)$.

Problems that do not meet all of these conditions are considered in Sections 7.6 and 7.7.

The "mechanics" of the simplex method are demonstrated in Examples 1–5. Although the procedures to be followed will be made clear, as will the fact that they result in an optimal solution, the reasons these procedures are used may not be immediately apparent. Examples 6 and 7 will supply these reasons and explain the connection between the simplex method and the graphical method used in Section 7.3.

Setting Up the Problem

The first step is to convert each constraint, a linear inequality, into a linear equation. This is done by adding a nonnegative variable, called a **slack variable,** to each constraint. For example, convert the inequality $x_1 + x_2 \leq 10$ into an equation by adding the slack variable x_3, to get

$$x_1 + x_2 + x_3 = 10, \qquad \text{where } x_3 \geq 0.$$

The inequality $x_1 + x_2 \leq 10$ says that the sum $x_1 + x_2$ is less than or perhaps equal to 10. The variable x_3 "takes up any slack" and represents the amount by which $x_1 + x_2$ fails to equal 10. For example, if $x_1 + x_2$ equals 8, then x_3 is 2. If $x_1 + x_2 = 10$, the value of x_3 is 0.

CAUTION A different slack variable must be used for each constraint.

▼ **EXAMPLE 1** Restate the following linear programming problem by introducing slack variables:

$$\begin{aligned} \text{Maximize} \quad & z = 2x_1 + 3x_2 + x_3 \\ \text{subject to} \quad & x_1 + x_2 + 4x_3 \leq 100 \\ & x_1 + 2x_2 + x_3 \leq 150 \\ & 3x_1 + 2x_2 + x_3 \leq 320 \\ \text{with } & x_1 \geq 0, x_2 \geq 0, x_3 \geq 0. \end{aligned}$$

Solution Rewrite the three constraints as equations by introducing nonnegative slack variables x_4, x_5, and x_6, one for each constraint. Then the problem can be restated as

$$\begin{aligned} \text{Maximize} \quad & z = 2x_1 + 3x_2 + x_3 & \\ \text{subject to} \quad & x_1 + x_2 + 4x_3 + x_4 & = 100 & \quad \text{Constraint 1} \\ & x_1 + 2x_2 + x_3 + x_5 & = 150 & \quad \text{Constraint 2} \\ & 3x_1 + 2x_2 + x_3 + x_6 & = 320 & \quad \text{Constraint 3} \\ \text{with} \quad & x_1 \geq 0, x_2 \geq 0, x_3 \geq 0, x_4 \geq 0, x_5 \geq 0, x_6 \geq 0. \; \blacktriangle \end{aligned}$$

Adding slack variables to the constraints converts a linear programming problem into a system of linear equations. These equations should have all variables on the left of the equals sign and all constants on the right. All the equations of Example 1 satisfy this condition except for the objective function, $z = 2x_1 + 3x_2 + x_3$, which may be written with all variables on the left as

$$-2x_1 - 3x_2 - x_3 + z = 0. \qquad \text{Objective Function}$$

Now the equations of Example 1 (with the constraints listed first and the objective function last) can be written as the following augmented matrix.

①

Rewrite the following set of constraints as equations by adding nonnegative slack variables.

$$\begin{aligned} x_1 + x_2 + x_3 & \leq 12 \\ 2x_1 + 4x_2 & \leq 15 \\ x_2 + 3x_3 & \leq 10 \end{aligned}$$

Answer:

$$\begin{aligned} x_1 + x_2 + x_3 + x_4 & = 12 \\ 2x_1 + 4x_2 + x_5 & = 15 \\ x_2 + 3x_3 + x_6 & = 10 \end{aligned}$$

$$
\begin{array}{ccccccc}
x_1 & x_2 & x_3 & x_4 & x_5 & x_6 & z \\
\end{array}
$$

$$
\left[\begin{array}{ccccccc|c}
1 & 1 & 4 & 1 & 0 & 0 & 0 & 100 \\
1 & 2 & 1 & 0 & 1 & 0 & 0 & 150 \\
3 & 2 & 1 & 0 & 0 & 1 & 0 & 320 \\
\hline
-2 & -3 & -1 & 0 & 0 & 0 & 1 & 0
\end{array}\right]
\begin{array}{l}
\text{Constraint 1} \\
\text{Constraint 2} \\
\text{Constraint 3} \\
\text{Objective Function}
\end{array}
$$

Indicators

This matrix is the initial **simplex tableau.** Except for the last entries—the 1 and 0 on the right end—the numbers in the bottom row of a simplex tableau are called **indicators.** ②

This simplex tableau represents a system of four linear equations in seven variables. Since there are more variables than equations, the system is dependent and has infinitely many solutions. Our goal is to find a solution in which all the variables are nonnegative and z is as large as possible. This will be done by using row operations to replace the given system by an equivalent one in which certain variables are eliminated from some of the equations. The process will be repeated until the optimum solution can be read from the matrix, as explained below.

Selecting the Pivot

Recall how row operations are used to eliminate variables in the Gauss–Jordan method. A particular nonzero entry in the matrix is chosen and changed to a 1; then all other entries in that column are changed to zeros. A similar process is used in the simplex method. The chosen entry is called the **pivot.** If we were interested only in solving the system, we could choose the various pivots in many different ways, as in Chapter 6. Here, however, it's not enough just to find a solution. We must find one that is nonnegative (so that it satisfies all the constraints) *and* makes z as a large as possible. Consequently, the pivot must be chosen carefully, as explained in the next example. The reasons this procedure is used and why it works are discussed in Example 7.

②

Set up the initial simplex tableau for the following linear programming problem:

Maximize $z = 2x_1 + 3x_2$

subject to $x_1 + 2x_2 \le 85$

$2x_1 + x_2 \le 92$

$x_1 + 4x_2 \le 104$

with $x_1 \ge 0, x_2 \ge 0$.
Locate and label the indicators.

Answer:

$$
\begin{array}{cccccc}
x_1 & x_2 & x_3 & x_4 & x_5 & z \\
\end{array}
$$

$$
\left[\begin{array}{cccccc|c}
1 & 2 & 1 & 0 & 0 & 0 & 85 \\
2 & 1 & 0 & 1 & 0 & 0 & 92 \\
1 & 4 & 0 & 0 & 1 & 0 & 104 \\
\hline
-2 & -3 & 0 & 0 & 0 & 1 & 0
\end{array}\right]
$$

Indicators

▼ **EXAMPLE 2**   Determine the pivot in the simplex tableau for the problem in Example 1.

Solution Look at the indicators (the last row of the tableau) and choose the most negative one.

$$
\begin{array}{ccccccc}
x_1 & x_2 & x_3 & x_4 & x_5 & x_6 & z \\
\end{array}
$$

$$
\left[\begin{array}{ccccccc|c}
1 & 1 & 4 & 1 & 0 & 0 & 0 & 100 \\
1 & 2 & 1 & 0 & 1 & 0 & 0 & 150 \\
3 & 2 & 1 & 0 & 0 & 1 & 0 & 320 \\
\hline
-2 & \boxed{-3} & -1 & 0 & 0 & 0 & 1 & 0
\end{array}\right]
$$

↑ Most negative indicator

The most negative indicator identifies the variable that is to be eliminated from all but one of the equations (rows), in this case x_2. The column containing the most negative indicator is called the **pivot column.** Now, for each *positive* entry in the pivot column, divide the number in the far right column of the same row by the positive number in the pivot column.

x_1	x_2	x_3	x_4	x_5	x_6	z		Quotients
1	**1**	4	1	0	0	0	**100**	$100/1 = 100$
1	**2**	1	0	1	0	0	**150**	$150/2 = 75 \leftarrow$ Smallest
3	**2**	1	0	0	1	0	**320**	$320/2 = 160$
-2	-3	-1	0	0	0	1	0	

The row with the smallest quotient (in this case, the second row) is called the **pivot row.** The entry in the pivot row and pivot column is the pivot.

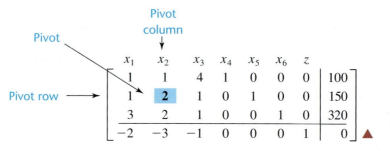

Pivot

Pivot column

Pivot row →

x_1	x_2	x_3	x_4	x_5	x_6	z	
1	1	4	1	0	0	0	100
1	**2**	1	0	1	0	0	150
3	2	1	0	0	1	0	320
-2	-3	-1	0	0	0	1	0

CAUTION In some simplex tableaus, the pivot column may contain zeros or negative entries. Only the positive entries in the pivot column should be used to form the quotients and determine the pivot row. If there are no positive entries in the pivot column (so that a pivot row cannot be chosen), then no maximum solution exists. ◇ ③

⟨3⟩

Find the pivot for the following tableau.

x_1	x_2	x_3	x_4	x_5	z	
0	1	1	0	0	0	50
-2	3	0	1	0	0	78
2	4	0	0	1	0	65
-5	-3	0	0	0	1	0

Answer:

2 (in first column)

Pivoting

Once the pivot has been selected, row operations are used to replace the initial simplex tableau by another simplex tableau in which the pivot column variable is eliminated from all but one of the equations. Since this new tableau is obtained by row operations, it represents an equivalent system of equations (that is, a system with the same solutions as the original system). This process, which is called **pivoting,** is explained in the next example.

▼ **EXAMPLE 3**   Use the indicated pivot, 2, to perform the pivoting on the simplex tableau of Example 2:

x_1	x_2	x_3	x_4	x_5	x_6	z	
1	1	4	1	0	0	0	100
1	**2**	1	0	1	0	0	150
3	2	1	0	0	1	0	320
-2	-3	-1	0	0	0	1	0

Solution   Start by multiplying each entry of row two by $1/2$ in order to change the pivot to 1.

$$\begin{array}{cccccccc} & x_1 & x_2 & x_3 & x_4 & x_5 & x_6 & z \\ \left[\begin{array}{ccccccc|c} 1 & 1 & 4 & 1 & 0 & 0 & 0 & 100 \\ \frac{1}{2} & 1 & \frac{1}{2} & 0 & \frac{1}{2} & 0 & 0 & 75 \\ 3 & 2 & 1 & 0 & 0 & 1 & 0 & 320 \\ -2 & -3 & -1 & 0 & 0 & 0 & 1 & 0 \end{array}\right] & & & & & & & \frac{1}{2}R_2 \end{array}$$

Now use row operations to make the entry in row one, column two, a 0.

$$\begin{array}{cccccccc} & x_1 & x_2 & x_3 & x_4 & x_5 & x_6 & z \\ \left[\begin{array}{ccccccc|c} \frac{1}{2} & 0 & \frac{7}{2} & 1 & -\frac{1}{2} & 0 & 0 & 25 \\ \frac{1}{2} & 1 & \frac{1}{2} & 0 & \frac{1}{2} & 0 & 0 & 75 \\ 3 & 2 & 1 & 0 & 0 & 1 & 0 & 320 \\ -2 & -3 & -1 & 0 & 0 & 0 & 1 & 0 \end{array}\right] & & & & & & & -R_2 + R_1 \end{array}$$

Change the 2 in row three, column two, to a 0 by a similar process.

$$\begin{array}{cccccccc} & x_1 & x_2 & x_3 & x_4 & x_5 & x_6 & z \\ \left[\begin{array}{ccccccc|c} \frac{1}{2} & 0 & \frac{7}{2} & 1 & -\frac{1}{2} & 0 & 0 & 25 \\ \frac{1}{2} & 1 & \frac{1}{2} & 0 & \frac{1}{2} & 0 & 0 & 75 \\ 2 & 0 & 0 & 0 & -1 & 1 & 0 & 170 \\ -2 & -3 & -1 & 0 & 0 & 0 & 1 & 0 \end{array}\right] & & & & & & & -2R_2 + R_3 \end{array}$$

Finally, add 3 times row 2 to the last row in order to change the indicator -3 to 0.

$$\begin{array}{cccccccc} & x_1 & x_2 & x_3 & x_4 & x_5 & x_6 & z \\ \left[\begin{array}{ccccccc|c} \frac{1}{2} & 0 & \frac{7}{2} & 1 & -\frac{1}{2} & 0 & 0 & 25 \\ \frac{1}{2} & 1 & \frac{1}{2} & 0 & \frac{1}{2} & 0 & 0 & 75 \\ 2 & 0 & 0 & 0 & -1 & 1 & 0 & 170 \\ -\frac{1}{2} & 0 & \frac{1}{2} & 0 & \frac{3}{2} & 0 & 1 & 225 \end{array}\right] & & & & & & & -3R_2 + R_4 \end{array}$$

The pivoting is now complete, because the pivot column variable x_2 has been eliminated from all equations except the one represented by the pivot row. The initial simplex tableau has been replaced by a new simplex tableau, which represents an equivalent system of equations. ▲

CAUTION During pivoting, do not interchange rows of the matrix. Make the pivot entry 1 by multiplying the pivot row by an appropriate constant, as in Example 3. ◆ 4

When at least one of the indicators in the last row of a simplex tableau is negative (as is the case with the tableau obtained in Example 3), the simplex method requires that a new pivot be selected and the pivoting be performed again. This procedure is repeated until a simplex tableau with no negative indicators in the last row is obtained or a tableau is reached in which no pivot row can be chosen.

▼ **EXAMPLE 4** In the simplex tableau obtained in Example 3, select a new pivot and perform the pivoting.

Solution First, locate the pivot column by finding the most negative indicator in the last row. Then locate the pivot row by computing the necessary quotients and finding the smallest one, as shown here:

◇ 4

For the following simplex tableau,

(a) find the pivot.

(b) Perform the pivoting and write the new tableau.

$$\begin{array}{cccccc} x_1 & x_2 & x_3 & x_4 & x_5 & z \\ \left[\begin{array}{cccccc|c} 1 & 2 & 6 & 1 & 0 & 0 & 16 \\ 1 & 3 & 0 & 0 & 1 & 0 & 25 \\ -1 & -4 & -3 & 0 & 0 & 1 & 0 \end{array}\right] \end{array}$$

Answers:

(a) 2

(b)

$$\begin{array}{cccccc} x_1 & x_2 & x_3 & x_4 & x_5 & z \\ \left[\begin{array}{cccccc|c} \frac{1}{2} & 1 & 3 & \frac{1}{2} & 0 & 0 & 8 \\ -\frac{1}{2} & 0 & -9 & -\frac{3}{2} & 1 & 0 & 1 \\ 1 & 0 & 9 & 2 & 0 & 1 & 32 \end{array}\right] \end{array}$$

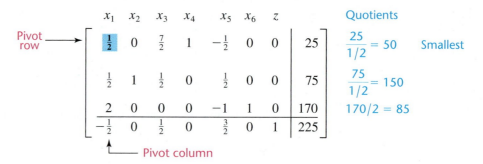

So the pivot is the number $1/2$ in row one, column one. Begin the pivoting by multiplying every entry in row one by 2. Then continue as indicated to obtain the following simplex tableau.

$$
\begin{array}{ccccccc}
x_1 & x_2 & x_3 & x_4 & x_5 & x_6 & z \\
\end{array}
$$

x_1	x_2	x_3	x_4	x_5	x_6	z		
1	0	7	2	−1	0	0	50	$2R_1$
0	1	−3	−1	1	0	0	50	$-\frac{1}{2}R_1 + R_2$
0	0	−14	−4	1	1	0	70	$-2R_1 + R_3$
0	0	4	1	1	0	1	250	$\frac{1}{2}R_1 + R_4$

Since there are no negative indicators in the last row, no further pivoting is necessary and we call this the **final simplex tableau.** ▲

Reading the Solution

The next example shows how to read an optimal solution of the original linear programming problem from the final simplex tableau.

▼ **EXAMPLE 5** Solve the linear programming problem introduced in Example 1.

Solution Look at the final simplex tableau for this problem, which was obtained in Example 4.

x_1	x_2	x_3	x_4	x_5	x_6	z	
1	0	7	2	−1	0	0	50
0	1	−3	−1	1	0	0	50
0	0	−14	−4	1	1	0	70
0	0	4	1	1	0	1	250

The last row of this matrix represents the equation

$$4x_3 + x_4 + x_5 + z = 250, \qquad \text{or equivalently,} \qquad z = 250 - 4x_3 - x_4 - x_5.$$

If x_3, x_4, and x_5 are all 0, then the value of z is 250. If any one of x_3, x_4, or x_5 is positive, then z will have a smaller value than 250 (why?). Consequently, since we want a solution for this system in which all the variables are nonnegative and z is as large as possible, we must have

$$x_3 = 0, \qquad x_4 = 0, \qquad x_5 = 0.$$

When these values are substituted into the first equation (represented by the first row of the final simplex tableau), the result is

$$x_1 + 7 \cdot 0 + 2 \cdot 0 - 1 \cdot 0 = 50; \qquad \text{that is,} \qquad x_1 = 50.$$

Similarly, substituting 0 for x_3, x_4, and x_5 in the last three equations represented by the final simplex tableau shows that

$$x_2 = 50, \qquad x_6 = 70, \qquad z = 250.$$

Therefore, the maximum value of $z = 2x_1 + 3x_2 + x_3$ occurs when

$$x_1 = 50, \qquad x_2 = 50, \qquad x_3 = 0,$$

in which case $z = 2 \cdot 50 + 3 \cdot 50 + 0 = 250$. (The values of the slack variables are irrelevant in stating the solution of the original problem.) ▲

In any simplex tableau, some columns look like columns of an identity matrix (one entry is 1 and the rest are 0). The variables corresponding to these columns are called **basic variables,** the variables corresponding to the other columns **non-basic variables.** In the tableau of Example 5, for instance, the basic variables are x_1, x_2, x_6, and z (shown in blue), and the nonbasic variables are x_3, x_4, and x_5.

$$
\begin{array}{ccccccc}
x_1 & x_2 & x_3 & x_4 & x_5 & x_6 & z \\
\end{array}
$$

$$
\left[
\begin{array}{ccccccc|c}
1 & 0 & 7 & 2 & -1 & 0 & 0 & 50 \\
0 & 1 & -3 & -1 & 1 & 0 & 0 & 50 \\
0 & 0 & -14 & -4 & 1 & 1 & 0 & 70 \\
\hline
0 & 0 & 4 & 1 & 1 & 0 & 1 & 250 \\
\end{array}
\right]
$$

The optimal solution in Example 5 was obtained from the final simplex tableau by setting the nonbasic variables equal to 0 and solving for the basic variables. Furthermore, the values of the basic variables are easy to read from the matrix: find the 1 in the column representing a basic variable; the last entry in that row is the value of that basic variable in the optimal solution. In particular, *the entry in the lower right-hand corner of the final simplex tableau is the maximum value of z.* ⬦5⬦

CAUTION If there are two identical columns in a tableau, each of which is a column in an identity matrix, only one of the variables corresponding to these columns can be a basic variable. The other is treated as a nonbasic variable. You may choose either one to be the basic variable, unless one of them is z, in which case z must be the basic variable.

The steps involved in solving a standard maximum linear programming problem by the simplex method have been illustrated in Examples 1–5 and are summarized here.

Simplex Method

1. Determine the objective function.

2. Write all necessary constraints.

3. Convert each constraint into an equation by adding slack variables.

4. Set up the initial simplex tableau.

5. Locate the most negative indicator. If there are two such indicators, choose one. This indicator determines the pivot column.

continued

⬦5⬦

A linear programming problem with slack variables x_4 and x_5 has the following final simplex tableau:

$$
\begin{array}{cccccc}
x_1 & x_2 & x_3 & x_4 & x_5 & z \\
\end{array}
$$

$$
\left[
\begin{array}{cccccc|c}
0 & 3 & 1 & 5 & 2 & 0 & 9 \\
1 & -2 & 0 & 4 & 1 & 0 & 6 \\
0 & 5 & 0 & 1 & 0 & 1 & 21 \\
\end{array}
\right].
$$

What is the optimal solution?

Answer:

$z = 21$ when $x_1 = 6$, $x_2 = 0$, $x_3 = 9$.

6. Use the positive entries in the pivot column to form the quotients necessary for determining the pivot. If there are no positive entries in the pivot column, no maximum solution exists. If two quotients are equally the smallest, let either determine the pivot.*

7. Multiply every entry in the pivot row by the reciprocal of the pivot to change the pivot to 1. Then use row operations to change all other entries in the pivot column to 0 by adding suitable multiples of the pivot row to the other rows.

8. If the indicators are all positive or 0, you have found the final tableau. If not, go back to Step 5 and repeat the process until a tableau with no negative indicators is obtained.[†]

9. In the final tableau, the *basic* variables correspond to the columns that have one entry 1 and the rest 0. The *nonbasic* variables correspond to the other columns. Set each nonbasic variable equal to 0 and solve the system for the basic variables. The maximum value of the objective function is the number in the lower right-hand corner of the final tableau.

The solution found by the simplex method may not be unique, especially when choices are possible in steps 5, 6, or 9. There may be other solutions that produce the same maximum value of the objective function. (See Exercises 37 and 38 at the end of this section) ⬥6

6

A linear programming problem has the following initial tableau:

$$\begin{array}{ccccc|c} x_1 & x_2 & x_3 & x_4 & z & \\ 1 & 1 & 1 & 0 & 0 & 40 \\ 2 & 1 & 0 & 1 & 0 & 24 \\ \hline -300 & -200 & 0 & 0 & 1 & 0 \end{array}$$

Use the simplex method to solve the problem.

Answer:

$x_1 = 0, x_2 = 24, x_3 = 16,$
$x_4 = 0, z = 4800$

The Simplex Method with Technology

In the preceding discussion and in most of the examples that follow, the simplex method is carried out by hand. This allows the reader to become familiar with the procedure and to understand how it works.

Unfortunately, it's easy to make an error in manual computations, and even a small error can make a big difference in the final result. Furthermore, most real-life applications involve a large number of variables and constraints, which makes the manual approach impractical. Consequently,

we recommend that you use some kind of technology in applying the simplex method.

Unless marked otherwise, the exercises can be done manually, but using technology will make your life much easier and your answers more accurate. Readily available technology includes the following.

GRAPHING CALCULATORS Any graphing calculator that has matrix capabilities can do row operations. Doing the simplex method this way is an improvement on manual computation but still rather cumbersome. It is much more effective to use the simplex method programs in the Graphing Calculator Appendix. You can install these programs manually in your calculator or use a computer to download them from www.aw.com/MWA9 if you have the appropriate hardware for transferring the programs to the calculator.

---

*It may be that the first choice of a pivot does not produce a solution. In that case try the other choice.
†Some special circumstances are noted at the end of Section 7.7.

SPREADSHEETS Most spreadsheets have a built-in simplex method program. Figure 7.24 shows the Solver of Microsoft Excel. Spreadsheets also provide a "sensitivity analysis," which allows you to see how much the constraints can be varied without changing the maximal solution.

FIGURE 7.24

Geometric Interpretation of the Simplex Method

Although it may not be immediately apparent, the simplex method is based on the same geometrical considerations as the graphical method. This can be seen by looking at a problem that can be readily solved by both methods.

▼ **EXAMPLE 6** In Example 2 of Section 7.3, the following problem was solved graphically (using x, y instead of x_1, x_2):

$$\text{Maximize} \quad z = 8x_1 + 12x_2$$
$$\text{subject to} \quad 40x_1 + 80x_2 \leq 560$$
$$6x_1 + 8x_2 \leq 72$$
$$x_1 \geq 0, x_2 \geq 0.$$

Graphing the feasible region (Figure 7.25) and evaluating z at each corner point shows that the maximum value of z occurs at $(8, 3)$.

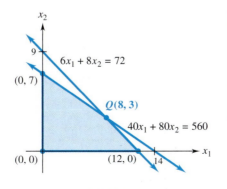

Corner Point	Value of $z = 8x_1 + 12x_2$	
$(0, 0)$	0	
$(0, 7)$	84	
$(12, 0)$	96	
$(8, 3)$	100	(maximum)

FIGURE 7.25

To solve the same problem by the simplex method, add a slack variable to each constraint.

$$40x_1 + 80x_2 + x_3 \qquad = 560$$
$$6x_1 + 8x_2 \qquad + x_4 = 72$$

Then write the initial simplex tableau.

$$
\begin{array}{ccccc}
x_1 & x_2 & x_3 & x_4 & z \\
\left[\begin{array}{ccccc|c}
40 & 80 & 1 & 0 & 0 & 560 \\
6 & 8 & 0 & 1 & 0 & 72 \\
\hline
-8 & -12 & 0 & 0 & 1 & 0
\end{array}\right]
\end{array}
$$

In this tableau, the basic variables are x_3, x_4, and z (why?). By setting the non-basic variables (namely, x_1 and x_2) equal to 0 and solving for the basic variables, we obtain the following solution (which will be called a **basic feasible solution**):

$$x_1 = 0, \qquad x_2 = 0, \qquad x_3 = 560, \qquad x_4 = 72, \qquad z = 0.$$

Since $x_1 = 0$ and $x_2 = 0$, this solution corresponds to the corner point at the origin in the graphical solution (Figure 7.25). The value $z = 0$ at the origin is obviously not maximal, and pivoting in the simplex method is designed to improve it.

The most negative indicator in the initial tableau is -12, and it determines the pivot column. Then we form the necessary quotients and determine the pivot row.

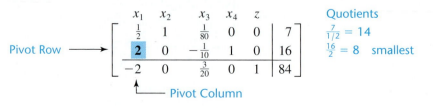

Thus, the pivot is 80 in row 1, column 2. Performing the pivoting leads to this tableau:

$$
\begin{array}{ccccc}
x_1 & x_2 & x_3 & x_4 & z \\
\left[\begin{array}{ccccc|c}
\frac{1}{2} & 1 & \frac{1}{80} & 0 & 0 & 7 \\
2 & 0 & -\frac{1}{10} & 1 & 0 & 16 \\
\hline
-2 & 0 & \frac{3}{20} & 0 & 1 & 84
\end{array}\right]
\end{array}
\qquad
\begin{array}{l}
\frac{1}{80}R_1 \\
-8R_1 + R_2 \\
12R_1 + R_3
\end{array}
$$

The basic variables here are x_2, x_4, and z, and the basic feasible solution (found by setting the nonbasic variables equal to 0 and solving for the basic variables) is

$$x_1 = 0, \qquad x_2 = 7, \qquad x_3 = 0, \qquad x_4 = 16, \qquad z = 84,$$

which corresponds to the corner point $(0, 7)$ in Figure 7.25. Note that the new value of the pivot variable x_2 is precisely the smallest quotient, 7, that was used to select the pivot row. Although this value of z is better, further improvement is possible.

Now the most negative indicator is -2. We form the necessary quotients and determine the pivot as usual.

Pivot Row $\longrightarrow$

$$
\begin{array}{ccccc}
x_1 & x_2 & x_3 & x_4 & z \\
\left[\begin{array}{ccccc|c}
\frac{1}{2} & 1 & \frac{1}{80} & 0 & 0 & 7 \\
2 & 0 & -\frac{1}{10} & 1 & 0 & 16 \\
\hline
-2 & 0 & \frac{3}{20} & 0 & 1 & 84
\end{array}\right]
\end{array}
\qquad
\begin{array}{l}
\text{Quotients} \\
\frac{7}{1/2} = 14 \\
\frac{16}{2} = 8 \quad \text{smallest}
\end{array}
$$

Pivot Column

The pivot is 2 in row 2, column 1. Pivoting now produces the final tableau.

$$
\begin{array}{ccccc|c}
x_1 & x_2 & x_3 & x_4 & z & \\
0 & 1 & \frac{3}{80} & -\frac{1}{4} & 0 & 3 \\
1 & 0 & -\frac{1}{20} & \frac{1}{2} & 0 & 8 \\
\hline
0 & 0 & \frac{1}{20} & 1 & 1 & 100
\end{array}
\qquad
\begin{array}{l}
-\frac{1}{2}R_2 + R_1 \\
\frac{1}{2}R_2 \\
2R_2 + R_3
\end{array}
$$

Here, the basic feasible solution is

$$x_1 = 8, \qquad x_2 = 3, \qquad x_3 = 0, \qquad x_4 = 0, \qquad z = 100,$$

which corresponds to the corner point $(8, 3)$ in Figure 7.25. Once again, the new value of the pivot variable x_1 is the smallest quotient, 8, that was used to select the pivot. From the graphical method, we know that this solution provides the maximum value of the objective function. This fact can also be seen algebraically by an algebraic argument similar to the one used in Example 5. Thus, there is no need to move to another corner point, and the simplex method ends. ▲

As illustrated in Example 6, the basic feasible solution obtained from a simplex tableau corresponds to a corner point of the feasible region. Pivoting, which replaces one tableau with another, is a systematic way of moving from one corner point to another, each time improving the value of the objective function. The simplex method ends when a corner point that produces the maximum value of the objective function is reached (or when it becomes clear that the problem has no maximum solution).

When there are three or more variables in a linear programming problem, it may be difficult or impossible to draw a picture, but it can be proved that the optimal value of the objective function occurs at a basic feasible solution (corresponding to a corner point in the two-variable case). The simplex method provides a means of moving from one basic feasible solution to another until one that produces the optimal value of the objective function is reached.

Explanation of Pivoting

The rules for selecting the pivot in the simplex method can be understood by examining how the first pivot was chosen in Example 6.

▼ **EXAMPLE 7**   The initial simplex tableau of Example 6 provides a basic feasible solution with $x_1 = 0$, $x_2 = 0$.

$$
\begin{array}{ccccc|c}
x_1 & x_2 & x_3 & x_4 & z & \\
40 & 80 & 1 & 0 & 0 & 560 \\
6 & 8 & 0 & 1 & 0 & 72 \\
\hline
-8 & -12 & 0 & 0 & 1 & 0
\end{array}
$$

This solution certainly does not give a maximum value for the objective function $z = 8x_1 + 12x_2$. Since x_2 has the largest coefficient, z will be increased most if x_2 is increased. In other words, the most negative indicator in the tableau (which corresponds to the largest coefficient in the objective function) identifies the variable that will provide the greatest change in the value of z.

To determine how much x_2 can be increased without leaving the feasible region, look at the first two equations,

$$40x_1 + 80x_2 + x_3 \qquad = 560$$
$$6x_1 + 8x_2 \qquad + x_4 = 72,$$

and solve for the basic variables x_3 and x_4.

$$x_3 = 560 - 40x_1 - 80x_2$$
$$x_4 = 72 - 6x_1 - 8x_2$$

Now x_2 is to be increased while x_1 is to keep the value 0. Hence,

$$x_3 = 560 - 80x_2$$
$$x_4 = 72 - 8x_2.$$

Since $x_3 \geq 0$ and $x_4 \geq 0$, we must have

$$0 \leq x_3 \qquad\qquad 0 \leq x_4$$
$$0 \leq 560 - 80x_2 \quad \text{and} \quad 0 \leq 72 - 8x_2$$
$$80x_2 \leq 560 \qquad\qquad 8x_2 \leq 72$$
$$x_2 \leq \frac{560}{80} = 7 \qquad\qquad x_2 \leq \frac{72}{8} = 9.$$

The right sides of these last inequalities are the quotients used to select the pivot row. Since x_2 must satisfy both inequalities, x_2 can be at most 7. In other words, the smallest quotient formed from positive entries in the pivot column identifies the value of x_2 that produces the largest change in z while remaining in the feasible region. By pivoting with the pivot determined in this way, we obtain the second tableau and a basic feasible solution in which $x_2 = 7$, as was shown in Example 6. ▲

An analysis similar to that in Example 7 applies to each occurrence of pivoting in the simplex method. The idea is to improve the value of the objective function by adjusting one variable at a time. The most negative indicator identifies the variable that will account for the largest increase in z. The smallest quotient determines the largest value of that variable which will produce a feasible solution. Pivoting leads to a solution in which the selected variable has this largest value.

The simplex method is easily implemented on a computer and some graphing calculators. Technology is essential for the simplex method in any situation where there are a large number of variables and constraints (and hence, an enormous number of corner points to check).

◆7.4 Exercises

In Exercises 1–4, (a) determine the number of slack variables needed; (b) name them; (c) use the slack variables to convert each constraint into a linear equation. (See Example 1.)

1. Maximize $z = 32x_1 + 9x_2$
   subject to $4x_1 + 2x_2 \leq 20$
   $5x_1 + x_2 \leq 50$
   $2x_1 + 3x_2 \leq 25$
   $x_1 \geq 0, x_2 \geq 0.$

2. Maximize $z = 3.7x_1 + 4.3x_2$
   subject to $2.4x_1 + 1.5x_2 \leq 10$
   $1.7x_1 + 1.9x_2 \leq 15$
   $x_1 \geq 0, x_2 \geq 0.$

3. Maximize $z = 8x_1 + 3x_2 + x_3$
   subject to $3x_1 - x_2 + 4x_3 \leq 95$
   $7x_1 + 6x_2 + 8x_3 \leq 118$
   $4x_1 + 5x_2 + 10x_3 \leq 220$
   $x_1 \geq 0, x_2 \geq 0, x_3 \geq 0.$

4. Maximize $z = 12x_1 + 15x_2 + 10x_3$
    subject to $2x_1 + 2x_2 + x_3 \leq 8$
    $x_1 + 4x_2 + 3x_3 \leq 12$
    $x_1 \geq 0, x_2 \geq 0, x_3 \geq 0.$

Introduce slack variables as necessary and then write the initial simplex tableau for each of these linear programming problems.

5. Maximize $z = 5x_1 + x_2$
    subject to $2x_1 + 5x_2 \leq 6$
    $4x_1 + x_2 \leq 6$
    $5x_1 + 3x_2 \leq 15$
    $x_1 \geq 0, x_2 \geq 0.$

6. Maximize $z = 5x_1 + 3x_2 + 7x_3$
    subject to $4x_1 + 3x_2 + 2x_3 \leq 60$
    $3x_1 + 4x_2 + x_3 \leq 24$
    $x_1 \geq 0, x_2 \geq 0, x_3 \geq 0.$

7. Maximize $z = x_1 + 5x_2 + 10x_3$
    subject to $x_1 + 2x_2 + 3x_3 \leq 10$
    $2x_1 + x_2 + x_3 \leq 8$
    $3x_1 + 4x_3 \leq 6$
    $x_1 \geq 0, x_2 \geq 0, x_3 \geq 0.$

8. Maximize $z = 5x_1 - x_2 + 3x_3$
    subject to $3x_1 + 2x_2 + x_3 \leq 36$
    $x_1 + 6x_2 + x_3 \leq 24$
    $x_1 - x_2 - x_3 \leq 32$
    $x_1 \geq 0, x_2 \geq 0, x_3 \geq 0.$

Find the pivot in each of the following simplex tableaus. (See Example 2.)

9.

x_1	x_2	x_3	x_4	x_5	z	
2	2	0	3	1	0	15
3	4	1	6	0	0	20
−2	−3	0	1	0	1	10

10.

x_1	x_2	x_3	x_4	x_5	z	
0	2	1	1	3	0	5
1	−5	0	1	2	0	8
0	−2	0	−3	1	1	10

11.

x_1	x_2	x_3	x_4	x_5	x_6	z	
6	2	1	3	0	0	0	8
0	2	0	1	0	1	0	7
6	1	0	3	1	0	0	6
−3	−2	0	2	0	0	1	12

12.

x_1	x_2	x_3	x_4	x_5	x_6	z	
0	2	0	1	2	2	0	3
0	3	1	0	1	2	0	4
1	4	0	0	3	5	0	5
0	−4	0	0	4	3	1	20

In Exercises 13–16, use the indicated entry as the pivot and perform the pivoting. (See Examples 3 and 4.)

13.

x_1	x_2	x_3	x_4	x_5	z	
1	2	4	1	0	0	56
2	**2**	1	0	1	0	40
−1	−3	−2	0	0	1	0

14.

x_1	x_2	x_3	x_4	x_5	x_6	z	
2	2	**1**	1	0	0	0	12
1	2	3	0	1	0	0	45
3	1	1	0	0	1	0	20
−2	−1	−3	0	0	0	1	0

15.

x_1	x_2	x_3	x_4	x_5	x_6	z	
1	1	1	1	0	0	0	60
3	1	**2**	0	1	0	0	100
1	2	3	0	0	1	0	200
−1	−1	−2	0	0	0	1	0

16.

x_1	x_2	x_3	x_4	x_5	x_6	z	
4	2	3	1	0	0	0	22
2	2	**5**	0	1	0	0	28
1	3	2	0	0	1	0	45
−3	−2	−4	0	0	0	1	0

For each simplex tableau in Exercises 17–20, (a) list the basic and the nonbasic variables, (b) find the basic feasible solution determined by setting the nonbasic variables equal to 0, and (c) decide whether this is a maximum solution. (See Examples 5 and 6.)

17.

x_1	x_2	x_3	x_4	x_5	z	
3	2	0	−3	1	0	29
4	0	1	−2	0	0	16
−5	0	0	−1	0	1	11

18.

x_1	x_2	x_3	x_4	x_5	x_6	z	
−3	0	$\frac{1}{2}$	1	−2	0	0	22
2	0	−3	0	1	1	0	10
4	1	4	0	$\frac{3}{4}$	0	0	17
−1	0	0	0	1	0	1	120

19.

x_1	x_2	x_3	x_4	x_5	x_6	z	
1	0	2	$\frac{1}{2}$	0	$\frac{1}{3}$	0	6
0	1	−1	5	0	−1	13	13
0	0	1	$\frac{3}{2}$	1	$-\frac{1}{3}$	0	21
0	0	2	$\frac{1}{2}$	0	3	1	18

20.

x_1	x_2	x_3	x_4	x_5	x_6	x_7	z	
−1	0	0	1	0	3	−2	0	47
2	0	1	0	0	2	$-\frac{1}{2}$	0	37
3	0	0	0	1	−1	6	0	43
4	1	0	0	0	6	0	1	86

Use the simplex method to solve Exercises 21–36.

21. Maximize $z = x_1 + 3x_2$
subject to $x_1 + x_2 \le 10$
$5x_1 + 2x_2 \le 20$
$x_1 + 2x_2 \le 36$
$x_1 \ge 0, x_2 \ge 0.$

22. Maximize $z = 5x_1 + x_2$
subject to $2x_1 + 3x_2 \le 8$
$4x_1 + 8x_2 \le 12$
$5x_1 + 2x_2 \le 30$
$x_1 \ge 0, x_2 \ge 0.$

23. Maximize $z = 2x_1 + x_2$
subject to $x_1 + 3x_2 \le 12$
$2x_1 + x_2 \le 10$
$x_1 + x_2 \le 4$
$x_1 \ge 0, x_2 \ge 0.$

24. Maximize $z = 4x_1 + 2x_2$
subject to $-x_1 - x_2 \le 12$
$3x_1 - x_2 \le 15$
$x_1 \ge 0, x_2 \ge 0.$

25. Maximize $z = 5x_1 + 4x_2 + x_3$
subject to $-2x_1 + x_2 + 2x_3 \le 3$
$x_1 - x_2 + x_3 \le 1$
$x_1 \ge 0, x_2 \ge 0, x_3 \ge 0.$

26. Maximize $z = 3x_1 + 2x_2 + x_3$
subject to $2x_1 + 2x_2 + x_3 \le 10$
$x_1 + 2x_2 + 3x_3 \le 15$
$x_1 \ge 0, x_2 \ge 0, x_3 \ge 0.$

27. Maximize $z = 2x_1 + x_2 + x_3$
subject to $x_1 - 3x_2 + x_3 \le 3$
$x_1 - 2x_2 + 2x_3 \le 12$
$x_1 \ge 0, x_2 \ge 0, x_3 \ge 0.$

28. Maximize $z = 4x_1 + 5x_2 + x_3$
subject to $x_1 + 2x_2 + 4x_3 \le 10$
$2x_1 + 2x_2 + x_3 \le 10$
$x_1 \ge 0, x_2 \ge 0, x_3 \ge 0.$

29. Maximize $z = 2x_1 + 2x_2 - 4x_3$
subject to $3x_1 + 3x_2 - 6x_3 \le 51$
$5x_1 + 5x_2 + 10x_3 \le 99$
$x_1 \ge 0, x_2 \ge 0, x_3 \ge 0.$

30. Maximize $z = 4x_1 + x_2 + 3x_3$
subject to $x_1 + 3x_3 \le 6$
$6x_1 + 3x_2 + 12x_3 \le 40$
$x_1 \ge 0, x_2 \ge 0, x_3 \ge 0.$

31. Maximize $z = 300x_1 + 200x_2 + 100x_3$
subject to $x_1 + x_2 + x_3 \le 100$
$2x_1 + 3x_2 + 4x_3 \le 320$
$2x_1 + x_2 + x_3 \le 160$
$x_1 \ge 0, x_2 \ge 0, x_3 \ge 0.$

32. Maximize $z = x_1 + 5x_2 - 10x_3$
subject to $8x_1 + 4x_2 + 12x_3 \le 18$

$x_1 + 6x_2 + 2x_3 \le 45$
$5x_1 + 7x_2 + 3x_3 \le 60$
$x_1 \ge 0, x_2 \ge 0, x_3 \ge 0.$

33. Maximize $z = 4x_1 - 3x_2 + 2x_3$
subject to $2x_1 - x_2 + 8x_3 \le 40$
$4x_1 - 5x_2 + 6x_3 \le 60$
$2x_1 - 2x_2 + 6x_3 \le 24$
$x_1 \ge 0, x_2 \ge 0, x_3 \ge 0.$

34. Maximize $z = 3x_1 + 2x_2 - 4x_3$
subject to $x_1 - x_2 + x_3 \le 10$
$2x_1 - x_2 + 2x_3 \le 30$
$-3x_1 + x_2 + 3x_3 \le 40$
$x_1 \ge 0, x_2 \ge 0, x_3 \ge 0.$

35. Maximize $z = x_1 + 2x_2 + x_3 + 5x_4$
subject to $x_1 + 2x_2 + x_3 + x_4 \le 50$
$3x_1 + x_2 + 2x_3 + x_4 \le 100$
$x_1 \ge 0, x_2 \ge 0, x_3 \ge 0, x_4 \ge 0.$

36. Maximize $z = x_1 + x_2 + 4x_3 + 5x_4$
subject to $x_1 + 2x_2 + 3x_3 + x_4 \le 115$
$2x_1 + x_2 + 8x_3 + 5x_4 \le 200$
$x_1 + x_3 \le 50$
$x_1 \ge 0, x_2 \ge 0, x_3 \ge 0, x_4 \ge 0.$

37. The initial simplex tableau of a linear programming problem is

x_1	x_2	x_3	x_4	x_5	z	
1	1	1	1	0	0	12
2	1	2	0	1	0	30
-2	-2	-1	0	0	1	0

(a) Use the simplex method to solve the problem with column one as the first pivot column.
(b) Now use the simplex method to solve the problem with column two as the first pivot column.
(c) Does this problem have a unique maximum solution? Why?

38. The final simplex tableau of a linear programming problem is

x_1	x_2	x_3	x_4	z	
1	1	2	0	0	24
2	0	2	1	0	8
4	0	0	0	1	40

(a) What is the solution given by this tableau?
(b) Even though all the indicators are nonnegative, perform one more round of pivoting on this tableau, using column three as the pivot column and choosing the pivot row by forming quotients in the usual way.
(c) Show that there is more than one solution to the linear programming problem by comparing your answer in part (a) with the basic feasible solution given by the tableau found in part (b). Does it give the same value of z as the solution in part (a)?

7.5 Maximization Applications

Applications of the simplex method are considered in this section. First, however, we make a slight change in notation. You have noticed that the column representing the variable z in a simplex tableau never changes during pivoting. (Since all the entries except the last one in this column are 0, performing row operations has no effect on these entries—they remain 0.) Consequently, this column is unnecessary and can be omitted without causing any difficulty.

> Hereafter in this text, the column corresponding to the variable z (representing the objective function) will be omitted from all simplex tableaus.

▼ **EXAMPLE 1**   A farmer has 110 acres of available land he wishes to plant with a mixture of potatoes, corn, and cabbage. It costs him $400 to produce an acre of potatoes, $160 to produce an acre of corn, and $280 to produce an acre of cabbage. He has a maximum of $20,000 to spend. He makes a profit of $120 per acre of potatoes, $40 per acre of corn, and $60 per acre of cabbage.

(a) How many acres of each crop should he plant to maximize his profit?

   Solution Let the number of acres alloted to each of potatoes, corn, and cabbage be x_1, x_2, and x_3, respectively. Then summarize the given information as follows.

Crop	Number of Acres	Cost per Acre	Profit per Acre
Potatoes	x_1	$400	$120
Corn	x_2	$160	$ 40
Cabbage	x_3	$280	$ 60
Maximum Available	110	$20,000	

The constraints can be expressed as

$$x_1 + \quad x_2 + \quad x_3 \le 110 \qquad \text{Number of acres}$$
$$400x_1 + 160x_2 + 280x_3 \le 20{,}000, \qquad \text{Production costs}$$

where x_1, x_2, and x_3 are all nonnegative. The first of these constraints says that $x_1 + x_2 + x_3$ is less than or perhaps equal to 110. Use x_4 as the slack variable, giving the equation

$$x_1 + x_2 + x_3 + x_4 = 110.$$

Here, x_4 represents the amount of the farmer's 110 acres that will not be used. (x_4 may be 0 or any value up to 110.)

In the same way, the constraint $400x_1 + 160x_2 + 280x_3 \leq 20,000$ can be converted into an equation by adding a slack variable x_5:

$$400x_1 + 160x_2 + 280x_3 + x_5 = 20,000.$$

The slack variable x_5 represents any unused portion of the farmer's $20,000 capital. (Again, x_5 may have any value from 0 to 20,000.)

The farmer's profit on potatoes is the product of the profit per acre ($120) and the number x_1 of acres, that is, $120x_1$. His profits on corn and cabbage are computed similarly. Hence, his total profit is given by

$$z = \text{profit on potatoes} + \text{profit on corn} + \text{profit on cabbage}$$
$$z = 120x_1 + 40x_2 + 60x_3.$$

The linear programming problem can now be stated as follows:

$$\text{Maximize} \quad z = 120x_1 + 40x_2 + 60x_3$$
$$\text{subject to} \quad x_1 + x_2 + x_3 + x_4 = 110$$
$$400x_1 + 160x_2 + 280x_3 + x_5 = 20,000$$
$$\text{with} \quad x_1 \geq 0, x_2 \geq 0, x_3 \geq 0, x_4 \geq 0, x_5 \geq 0.$$

The initial simplex tableau (without the z column) is

$$\begin{array}{ccccc}
x_1 & x_2 & x_3 & x_4 & x_5 \\
\end{array}$$
$$\left[\begin{array}{ccccc|c}
1 & 1 & 1 & 1 & 0 & 110 \\
400 & 160 & 280 & 0 & 1 & 20,000 \\
\hline
-120 & -40 & -60 & 0 & 0 & 0
\end{array}\right].$$

The most negative indicator is -120; column one is the pivot column. The quotients needed to determine the pivot row are $110/1 = 110$ and $20,000/400 = 50$. So the pivot is 400 in row two, column one. Multiplying row two by $1/400$ and completing the pivoting leads to the final simplex tableau.

$$\begin{array}{ccccc}
x_1 & x_2 & x_3 & x_4 & x_5 \\
\end{array}$$
$$\left[\begin{array}{ccccc|c}
0 & .6 & .3 & 1 & -.0025 & 60 \\
1 & .4 & .7 & 0 & .0025 & 50 \\
\hline
0 & 8 & 24 & 0 & .3 & \boxed{6000}
\end{array}\right] \quad \begin{array}{l} -1R_2 + R_1 \\ \frac{1}{400}R_2 \\ 120R_2 + R_3 \end{array}$$

Setting the nonbasic variables x_2, x_3, and x_5 equal to 0, solving for the basic variables x_1 and x_4, and remembering that the value of z is in the lower right-hand corner leads to this maximum solution:

$$x_1 = 50, \quad x_2 = 0, \quad x_3 = 0, \quad x_4 = 60, \quad x_5 = 0, \quad z = 6000.$$

Therefore, the farmer will make a maximum profit of $6000 by planting 50 acres of potatoes and no corn or cabbage.

(b) If the farmer maximizes his profit, how much land will remain unplanted? What is the explanation for this?

Solution Since 50 of 110 acres are planted, 60 acres will remain unplanted. Alternatively, note that the unplanted acres of land are represented by x_4, the slack variable in the "number of acres" constraint. In the maximal solution found in part (a), $x_4 = 60$, which means that 60 acres are left unplanted.

The amount of unused cash is represented by x_5, the slack variable in the "production costs" constraint. Since $x_5 = 0$, all the available money has been used. By using the maximal solution in part (a), the farmer has used his $20,000 most effectively. If he had more cash, he would plant more crops and make a larger profit. ▲

▼ **EXAMPLE 2**   Ana Pott, who is a candidate for the state legislature, has $96,000 to buy TV advertising time. Ads cost $400 per minute on a local cable channel, $4000 per minute on a regional independent channel, and $12,000 per minute on a national network channel. Because of existing contracts, the TV stations can provide at most 30 minutes of advertising time, with a maximum of 6 minutes on the national network channel. At any given time during the evening, approximately 100,000 people watch the cable channel, 200,000 the independent channel, and 600,000 the network channel. To get maximum exposure, how much time should Ana buy from each station?

(a) Set up the initial simplex tableau for this problem.

Solution   Let x_1 be the number of minutes of ads on the cable channel, x_2 the number of minutes on the independent channel, and x_3 the number of minutes on the network channel. Exposure is measured in viewer-minutes. For instance, 100,000 people watching x_1 minutes of ads on the cable channel produces $100,000x_1$ viewer-minutes. The amount of exposure is given by the total number of viewer-minutes for all three channels, namely,

$$100,000x_1 + 200,000x_2 + 600,000x_3.$$

Since 30 minutes are available,

$$x_1 + x_2 + x_3 \le 30.$$

The fact that only 6 minutes can be used on the network channel means that

$$x_3 \le 6.$$

Expenditures are limited to $96,000, so

$$\text{Cable cost} + \text{independent cost} + \text{network cost} \le 96,000$$
$$400x_1 \quad + \quad 4000x_2 \quad + \quad 12,000x_3 \quad \le 96,000.$$

Therefore, Ana must solve the following linear programming problem:

$$\text{Maximize} \quad z = 100,000x_1 + 200,000x_2 + 600,000x_3$$
$$\text{subject to} \quad x_1 + x_2 + x_3 \le 30$$
$$x_3 \le 6$$
$$400x_1 + 4000x_2 + 12,000x_3 \le 96,000$$

with $x_1 \ge 0$, $x_2 \ge 0$, $x_3 \ge 0$.

Introducing slack variables x_4, x_5, and x_6 (one for each constraint), rewriting the constraints as equations, and expressing the objective function as

$$-100,000x_1 - 200,000x_2 - 600,000x_3 + z = 0$$

leads to the initial simplex tableau:

x_1	x_2	x_3	x_4	x_5	x_6	
1	1	1	1	0	0	30
0	0	1	0	1	0	6
400	4000	12,000	0	0	1	96,000
−100,000	−200,000	−600,000	0	0	0	0

(b) Use the simplex method to find the final simplex tableau.

Solution Work by hand, or use a graphing calculator simplex program or a spreadsheet, to obtain this final tableau:

$$
\begin{array}{cccccc}
x_1 & x_2 & x_3 & x_4 & x_5 & x_6 \\
\end{array}
$$

$$
\left[
\begin{array}{cccccc|c}
1 & 0 & 0 & \frac{10}{9} & \frac{20}{9} & -\frac{25}{90,000} & 20 \\
0 & 0 & 1 & 0 & 1 & 0 & 6 \\
0 & 1 & 0 & -\frac{1}{9} & -\frac{29}{9} & \frac{25}{90,000} & 4 \\
\hline
0 & 0 & 0 & \frac{800,000}{9} & \frac{1,600,000}{9} & \frac{250}{9} & 6,400,000
\end{array}
\right].
$$

Therefore, the optimal solution is

$$x_1 = 20, \quad x_2 = 4, \quad x_3 = 6, \quad x_4 = 0, \quad x_5 = 0, \quad x_6 = 0.$$

Ana should buy 20 minutes of time on the cable channel, 4 minutes on the independent channel, and 6 minutes on the network channel.

(c) What do the values of the slack variables in the optimal solution tell you?

Solution All three slack variables are 0. This means that all the available minutes have been used ($x_4 = 0$ in the first constraint), the maximum possible six minutes on the national network have been used ($x_5 = 0$ in the second constraint), and all of the $96,000 has been spent ($x_6 = 0$ in the third constraint). ▲ ①

In Example 2, what is the number of viewer-minutes in the optimal solution?

Answer:

$z = 6,400,000$

▼ **EXAMPLE 3** A chemical plant makes three products—glaze, solvent, and clay—each of which brings in different revenue per truckload. Production is limited, first by the number of air pollution units the plant is allowed to produce each day and second by the time available in the evaporation tank. The plant manager wants to maximize the daily revenue. Using information not given here, he sets up an initial simplex tableau and uses the simplex method to produce the following final simplex tableau:

$$
\left[
\begin{array}{ccccc|c}
-10 & -25 & 0 & 1 & -1 & 60 \\
3 & 4 & 1 & 0 & .1 & 24 \\
\hline
7 & 13 & 0 & 0 & .4 & 96
\end{array}
\right].
$$

The three variables represent the number of truckloads of glaze, solvent, and clay, respectively. The first slack variable comes from the air pollution constraint and the second slack variable from the time constraint on the evaporation tank. The revenue function is given in hundreds of dollars.

(a) What is the optimal solution?

Solution $x_1 = 0, \quad x_2 = 0, \quad x_3 = 24, \quad x_4 = 60, \quad x_5 = 0, \quad$ and $\quad z = 96.$

(b) Interpret this solution. What do the variables represent and what does the solution mean?

Solution The variable x_1 is the number of truckloads of glaze, x_2 the number of truckloads of solvent, x_3 the number of truckloads of clay to be produced, and z the revenue produced (in hundreds of dollars). The plant should produce 24 truckloads of clay and no glaze or solvent, for a maximum revenue of $9600. The first slack variable, x_4, represents the number of air pollution units below the maximum number allowed. Since $x_4 = 60$, the number of air pollution units will be 60 less than the allowable maximum. The second slack variable, x_5, represents the unused time in the evaporation tank. Since $x_5 = 0$, the evaporation tank is fully used. ▲

7.5 Exercises

Set up the initial simplex tableau for each of the following problems.

1. **Business** A cat breeder has the following amounts of cat food: 90 units of tuna, 80 units of liver, and 50 units of chicken. To raise a Siamese cat, the breeder must use 2 units of tuna, 1 of liver, and 1 of chicken per day, while raising a Persian cat requires 1, 2, and 1 units, respectively, per day. If a Siamese cat sells for $12 while a Persian cat sells for $10, how many of each should be raised in order to obtain maximum gross income? What is the maximum gross income?

2. **Business** Banal, Inc., produces art for motel rooms. Its painters can turn out mountain scenes, seascapes, and pictures of clowns. Each painting is worked on by three different artists: T, D, and H. Artist T works only 25 hours per week, while D and H work 45 and 40 hours per week, respectively. Artist T spends 1 hour on a mountain scene, 2 hours on a seascape, and 1 hour on a clown. Corresponding times for D and H are 3, 2, and 2 hours and 2, 1, and 4 hours, respectively. Banal makes $20 on a mountain scene, $18 on a seascape, and $22 on a clown. The head painting packer can't stand clowns, so that no more than 4 clown paintings may be done in a week. Find the number of each type of painting that should be made weekly in order to maximize profit. Find the maximum possible profit.

3. **Health** A biologist has 500 kilograms of nutrient A, 600 kilograms of nutrient B, and 300 kilograms of nutrient C. These nutrients will be used to make 4 types of food, whose contents (in percent of nutrient per kilogram of food) and whose "growth values" are as shown in the table.

	P	Q	R	S
A	0	0	37.5	62.5
B	0	75	50	37.5
C	100	25	12.5	0
Growth Value	90	70	60	50

How many kilograms of each food should be produced in order to maximize total growth value? Find the maximum growth value.

4. **Natural Science** A lake is stocked each spring with three species of fish: A, B, and C. The average weights of the fish are 1.62, 2.12, and 3.01 kilograms for species A, B, and C, respectively. Three foods—I, II, and III—are available in the lake. Each fish of species A requires 1.32 units of food I, 2.9 units of food II, and 1.75 units of food III, on the average, each day. Species B fish each require 2.1 units of food I, .95 units of food II,

and .6 units of food III daily. Species C fish require .86, 1.52, and 2.01 units of I, II, and III per day, respectively. If 490 units of food I, 897 units of food II, and 653 units of food III are available daily, how should the lake be stocked to maximize the weight of the fish it supports?

In each of the following exercises, (a) use the simplex method to solve the problem and (b) explain what the values of the slack variables in the optimal solution mean in the context of the problem. (See Examples 1–3).

5. **Business** A manufacturer of bicycles builds one-, three-, and ten-speed models. The bicycles are made of both aluminum and steel. The company has available 91,800 units of steel and 42,000 units of aluminum. The one-, three-, and ten-speed models need, respectively, 20, 30, and 40 units of steel and 12, 21, and 16 units of aluminum. How many of each type of bicycle should be made in order to maximize profit if the company makes $8 per one-speed bike, $12 per three-speed, and $24 per ten-speed? What is the maximum possible profit?

6. **Social Science** Jayanta is working to raise money for the homeless by sending information letters and making follow-up calls to local labor organizations and church groups. She discovered that each church group requires 2 hours of letter writing and 1 hour of follow-up, while, for each labor union, she needs 2 hours of letter writing and 3 hours of follow-up. Jayanta can raise $100 from each church group and $200 from each union local, and she has a maximum of 16 hours of letter-writing time and a maximum of 12 hours of follow-up time available per month. Determine the most profitable mixture of groups she should contact and the most money she can raise in a month.

7. **Social Science** A political party is planning a half-hour television show. The show will have 3 minutes of direct requests for money from viewers. Three of the party's politicians will be on the show: a senator, a congresswoman, and a governor. The senator, a party "elder statesman," demands that he be on screen at least twice as long

as the governor. The total time taken by the senator and the governor must be at least twice the time taken by the congresswoman. On the basis of a preshow survey, it is believed that 40, 60, and 50 (in thousands) viewers will watch the program for each minute the senator, congresswoman, and governor, respectively, are on the air. Find the time that should be allotted to each politician in order to get the maximum number of viewers. Find the maximum number of viewers.

8. **Business** The Cut-Right Company sells sets of kitchen knives. The Basic Set consists of 2 utility knives and 1 chef's knife. The Regular Set consists of 2 utility knives, 1 chef's knife, and 1 slicer. The Deluxe Set consists of 3 utility knives, 1 chef's knife, and 1 slicer. The profit is $30 on a Basic Set, $40 on a Regular Set, and $60 on a Deluxe Set. The factory has on hand 800 utility knives, 400 chef's knives, and 200 slicers. Assuming that all sets will be sold, how many of each type should be made up in order to maximize profit? What is the maximum profit?

Use the simplex method to solve the following problems. (See Examples 1–3.)

9. **Business** The Fancy Fashions Store has $8000 available each month for advertising. Newspaper ads cost $400 each, and no more than 20 can be run per month. Radio ads cost $200 each, and no more than 30 can run per month. TV ads cost $1200 each, with a maximum of 6 available each month. Approximately 2000 women will see each newspaper ad, 1200 will hear each radio commercial, and 10,000 will see each TV ad. How much of each type of advertising should be used if the store wants to maximize its ad exposure?

10. **Business** Caroline's Quality Candy Confectionery is famous for fudge, chocolate cremes, and pralines. Its candy-making equipment is set up to make 100-pound batches at a time. Currently there is a chocolate shortage, and the company can get only 120 pounds of chocolate in the next shipment. On a week's run, the confectionery's cooking and processing equipment is available for a total of 42 machine hours. During the same period, the employees have a total of 56 work hours available for packaging. A batch of fudge requires 20 pounds of chocolate, while a batch of cremes uses 25 pounds of chocolate. The cooking and processing take 120 minutes for fudge, 150 minutes for chocolate cremes, and 200 minutes for pralines. The packaging times, measured in minutes per 1-pound box, are 1, 2, and 3, respectively, for fudge, cremes, and pralines. Determine how many batches of each type of candy the confectionery should make, assuming that the profit per 1-pound box is 50¢ on fudge, 40¢ on chocolate cremes, and 45¢ on pralines. Also, find the maximum profit for the week.

11. **Finance** A political party is planning its fund-raising activities for a coming election. It plans to raise money through large fund-raising parties, letters requesting funds, and dinner parties where people can meet the candidate personally. Each large fund-raising party costs $3000, each mailing costs $1000, and each dinner party costs $12,000. The party can spend up to $102,000 for these activities. From experience, it is known that each large party will raise $200,000, each letter campaign will raise $100,000, and each dinner party will raise $600,000. The party is able to carry out as many as 25 of these activities.

   (a) How many of each should the party plan in order to raise the maximum amount of money? What is the maximum amount?

   (b) Dinner parties are more expensive than letter campaigns, yet the optimum solution found in part (a) includes dinner parties but no letter campaigns. Explain how this is possible.

12. **Business** A baker has 60 units of flour, 132 units of sugar, and 102 units of raisins. A loaf of raisin bread requires 1 unit of flour, 1 unit of sugar, and 2 units of raisins, while a raisin cake needs 2, 4, and 1 units, respectively. If raisin bread sells for $3 a loaf and a raisin cake for $4, how many of each should be baked so that the gross income is maximized? What is the maximum gross income?

Business *The next two problems come from past CPA examinations.* *Select the appropriate answer for each question.*

13. The Ball Company manufactures three types of lamps, labeled A, B, and C. Each lamp is processed in two departments: I and II. Total available person-hours per day for departments I and II are 400 and 600, respectively. No additional labor is available. Time requirements and profit per unit for each type of lamp are as follows.

	A	B	C
Person-Hours in I	2	3	1
Person-Hours in II	4	2	3
Profit per Unit	$5	$4	$3

The company has assigned you as the accounting member of its profit-planning committee to determine the numbers of types of A, B, and C lamps that it should produce in order to maximize its total profit from the sale of lamps. The following questions relate to a linear programming model that your group has developed.

(a) The coefficients of the objective function would be
   (1) 4, 2, 3;
   (2) 2, 3, 1;
   (3) 5, 4, 3;
   (4) 400,600.
(b) The constraints in the model would be
   (1) 2, 3, 1;
   (2) 5, 4, 3;
   (3) 4, 2, 3;
   (4) 400,600.
(c) The constraint imposed by the available number of person-hours in department I could be expressed as
   (1) $4X_1 + 2X_2 + 3X_3 \leq 400$;
   (2) $4X_1 + 2X_2 + 3X_3 \geq 400$;
   (3) $2X_1 + 3X_2 + 1X_3 \leq 400$;
   (4) $2X_1 + 3X_2 + 1X_3 \geq 400$.

14. The Golden Hawk Manufacturing Company wants to maximize the profits on products A, B, and C. The contribution margin for each product is as follows.

Product	Contribution Margin
A	$2
B	5
C	4

The production requirements and the departmental capacities are as follows.

	PRODUCTION REQUIREMENTS BY PRODUCT (HOURS)			DEPARTMENTAL CAPACITY
Department	A	B	C	(TOTAL HOURS)
Assembling	2	3	2	30,000
Painting	1	2	2	38,000
Finishing	2	3	1	28,000

(a) What is the profit-maximization formula for the Golden Hawk Company?
   (1) $\$2A + \$5B + \$4C = X$ (where X = profit)
   (2) $5A + 8B + 5C \leq 96,000$
   (3) $\$2A + \$5B + \$4C \leq X$
   (4) $\$2A + \$5B + \$4C = 96,000$
(b) What is the constraint for the Painting Department of the Golden Hawk Company?
   (1) $1A + 2B + 2C \geq 38,000$
   (2) $\$2A + \$5B + \$4C \geq 38,000$
   (3) $1A + 2B + 2C \leq 38,000$
   (4) $2A + 3B + 2C \leq 30,000$

15. Solve the problem in Exercise 1.

Use a graphing calculator or a computer program for the simplex method to solve the following linear programming problems.

16. Exercise 2. Your final answer should consist of whole numbers (Banal can't sell half a painting).

17. Exercise 3

18. Exercise 4

7.6 The Simplex Method: Duality and Minimization

This section and the next are independent of each other, so either one may be read first. Here, we present a method for solving linear programming problems in which the objective function is to be *minimized*. An alternative approach to such problems is given in the next section, which also considers mixed constraints ($\geq$, $=$, and $\leq$).

We begin with a necessary tool from matrix algebra: if A is a matrix, then the **transpose** of A is the matrix obtained by interchanging the rows and columns of A.

▼ EXAMPLE 1   Find the transpose of each matrix.

(a) $A = \begin{bmatrix} 2 & -1 & 5 \\ 6 & 8 & 0 \\ -3 & 7 & -1 \end{bmatrix}$

Solution Write the rows of matrix A as the columns of the transpose.

$$\text{Transpose of } A = \begin{bmatrix} 2 & 6 & -3 \\ -1 & 8 & 7 \\ 5 & 0 & -1 \end{bmatrix}$$

(b) Solution The transpose of $\begin{bmatrix} 1 & 2 & 4 & 0 \\ 2 & 1 & 7 & 6 \end{bmatrix}$ is $\begin{bmatrix} 1 & 2 \\ 2 & 1 \\ 4 & 7 \\ 0 & 6 \end{bmatrix}$. ▲ ◇ 1

---

1 ⟨1⟩

Give the transpose of each matrix.

(a) $\begin{bmatrix} 2 & 4 \\ 6 & 3 \\ 1 & 5 \end{bmatrix}$

(b) $\begin{bmatrix} 4 & 7 & 10 \\ 3 & 2 & 6 \\ 5 & 8 & 12 \end{bmatrix}$

Answers:

(a) $\begin{bmatrix} 2 & 6 & 1 \\ 4 & 3 & 5 \end{bmatrix}$

(b) $\begin{bmatrix} 4 & 3 & 5 \\ 7 & 2 & 8 \\ 10 & 6 & 12 \end{bmatrix}$

TECHNOLOGY TIP   Most graphing calculators can find the transpose of a matrix. Look for this feature in the MATRIX MATH menu (TI), or the OPTN MAT menu (Casio), or the MATH MATRIX menu (HP-39+). The transpose of matrix A from Example 1(a) is shown in Figure 7.26.

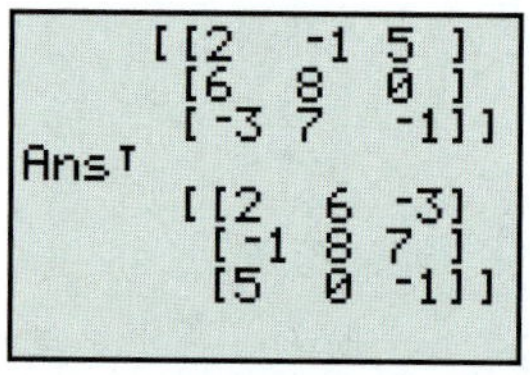

```
[[2   -1   5 ]
 [6    8   0 ]
 [-3   7  -1]]
Ans T
        [[2    6  -3]
         [-1   8   7 ]
         [5    0  -1]]
```

FIGURE 7.26

We now consider linear programming problems satisfying the following conditions:

1. The objective function is to be minimized.
2. All the coefficients of the objective function are nonnegative.
3. All constraints involve $\geq$.
4. All variables are nonnegative.

The method of solving minimization problems presented here is based on an interesting connection between maximization and minimization problems: any solution of a maximizing problem produces the solution of an associated minimizing problem, and vice versa. Each of the associated problems is called the **dual** of the other. Thus, duals enable us to solve minimization problems of the type just described by the simplex method introduced in Section 7.4.

When dealing with minimization problems, we use y_1, y_2, y_3, etc., as variables and denote the objective function by w. The next two examples show how to construct the dual problem. Later examples will show how to solve both the dual problem and the original one.

▼ **EXAMPLE 2**   Construct the dual of this problem:

$$\text{Minimize} \quad w = 8y_1 + 16y_2$$
$$\text{subject to} \quad y_1 + 5y_2 \geq 9$$
$$2y_1 + 2y_2 \geq 10$$
$$y_1 \geq 0, \ y_2 \geq 0.$$

Solution Write the augmented matrix of the system of inequalities *and* include the coefficients of the objective function (not their negatives) as the last row of the matrix.

Constants

$$\begin{bmatrix} 1 & 5 & 9 \\ 2 & 2 & 10 \\ 8 & 16 & 0 \end{bmatrix}$$

Objective function ⟶

Now form the transpose of the preceding matrix.

$$\begin{bmatrix} 1 & 2 & 8 \\ 5 & 2 & 16 \\ 9 & 10 & 0 \end{bmatrix}$$

In this last matrix, think of the first two rows as constraints and the last row as the objective function. Then the dual maximization problem is as follows:

$$\text{Maximize} \quad z = 9x_1 + 10x_2$$
$$\text{subject to} \quad x_1 + 2x_2 \le 8$$
$$5x_1 + 2x_2 \le 16$$
$$x_1 \ge 0, x_2 \ge 0. \quad \blacktriangle$$

▼ **EXAMPLE 3**   Write the duals of the following minimization linear programming problems.

(a) Minimize $w = 10y_1 + 8y_2$
subject to $y_1 + 2y_2 \ge 2$
$y_1 + y_2 \ge 5$
$y_1 \ge 0, y_2 \ge 0.$

Solution Begin by writing the augmented matrix for the given problem.

$$\begin{bmatrix} 1 & 2 & 2 \\ 1 & 1 & 5 \\ 10 & 8 & 0 \end{bmatrix}$$

Form the transpose of this matrix to get

$$\begin{bmatrix} 1 & 1 & 10 \\ 2 & 1 & 8 \\ 2 & 5 & 0 \end{bmatrix}.$$

The dual problem is stated from this second matrix as follows (using x instead of y):

$$\text{Maximize} \quad z = 2x_1 + 5x_2$$
$$\text{subject to} \quad x_1 + x_2 \le 10$$
$$2x_1 + x_2 \le 8$$
$$x_1 \ge 0, x_2 \ge 0.$$

(b) Minimize $\quad w = 7y_1 + 5y_2 + 8y_3$

$\qquad$ subject to $\qquad 3y_1 + 2y_2 + y_3 \geq 10$

$\qquad\qquad\qquad\qquad y_1 + y_2 + y_3 \geq 8$

$\qquad\qquad\qquad\quad 4y_1 + 5y_2 \geq 25$

$\qquad\qquad\qquad y_1 \geq 0, y_2 \geq 0, y_3 \geq 0.$

Solution The dual problem is stated as follows:

Maximize $\quad z = 10x_1 + 8x_2 + 25x_3$

$\qquad$ subject to $\qquad 3x_1 + x_2 + 4x_3 \leq 7$

$\qquad\qquad\qquad\quad 2x_1 + x_2 + 5x_3 \leq 5$

$\qquad\qquad\qquad\quad x_1 + x_2 \leq 8$

$\qquad\qquad\quad x_1 \geq 0, x_2 \geq 0, x_3 \geq 0.$ ▲ ◈2

In Example 3, all the constraints of the minimization problems were $\geq$ inequalities, while all those in the dual maximization problems were $\leq$ inequalities. This is generally the case; inequalities are reversed when the dual problem is stated.

The following table shows the close connection between a problem and its dual.

Given Problem	Dual Problem
m variables	n variables
n constraints	m constraints (m slack variables)
Coefficients from objective function	Constraint constants
Constraint constants	Coefficients from objective function

Now that you know how to construct the dual problem, we examine how it is related to the original problem and how both may solved.

▼ **EXAMPLE 4** Solve this problem and its dual:

Minimize $\quad w = 8y_1 + 16y_2$

$\qquad$ subject to $\qquad y_1 + 5y_2 \geq 9$

$\qquad\qquad\qquad\quad 2y_1 + 2y_2 \geq 10$

$\qquad\qquad\qquad y_1 \geq 0, y_2 \geq 0.$

Solution In Example 2, we saw that the dual problem is

Maximize $\quad z = 9x_1 + 10x_2$

$\qquad$ subject to $\qquad x_1 + 2x_2 \leq 8$

$\qquad\qquad\qquad\quad 5x_1 + 2x_2 \leq 16$

$\qquad\qquad\qquad x_1 \geq 0, x_2 \geq 0.$

In this case, both the original problem and the dual may be solved geometrically, as in Section 7.2. Figure 7.27(a) shows the region of feasible solutions for the original minimization problem, and Problem 3 at the side shows that

the minimum value of w is 48 at the vertex $(4, 1)$. ◈3

◈2

Write the dual of the following linear programming problem:

Minimize

$\qquad w = 2y_1 + 5y_2 + 6y_3$

subject to

$\qquad 2y_1 + 3y_2 + y_3 \geq 15$

$\qquad\quad y_1 + y_2 + 2y_3 \geq 12$

$\qquad 5y_1 + 3y_2 \geq 10$

$\qquad y_1 \geq 0, y_2 \geq 0, y_3 \geq 0.$

Answer:

Maximize

$\qquad z = 15x_1 + 12x_2 + 10x_3$

subject to

$\qquad 2x_1 + x_2 + 5x_3 \leq 2$

$\qquad 3x_1 + x_2 + 3x_3 \leq 5$

$\qquad\quad x_1 + 2x_2 \leq 6$

$\qquad x_1 \geq 0, x_2 \geq 0, x_3 \geq 0.$

◈3

Use the corner points in Figure 7.27(a) to find the minimum value of $w = 8y_1 + 16y_2$ and where it occurs.

Answer:

48 when $y_1 = 4$, $y_2 = 1$

Figure 7.27(b) shows the region of feasible solutions for the dual maximization problem, and Problem 4 at the side shows that

the maximum value of z is 48 at the vertex $(2, 3)$.

Even though the regions and the corner points are different, the minimization problem and its dual have the same solution, 48. ▲

4

Use Figure 7.27(b) to find the maximum value of $z = 9x_1 + 10x_2$ and where it occurs.

Answer:

48 when $x_1 = 2, x_2 = 3$

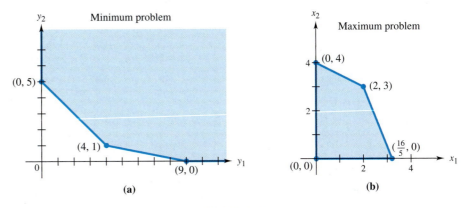

FIGURE 7.27

The next theorem, whose proof requires advanced methods, guarantees that what happened in Example 4 happens in the general case as well.

Theorem of Duality

The objective function w of a minimizing linear programming problem takes on a minimum value if and only if the objective function z of the corresponding dual maximizing problem takes on a maximum value. The maximum value of z equals the minimum value of w.

Geometric solution methods were used in Example 4, but the simplex method can also be used. In fact, the final simplex tableau shows the solutions for both the original minimization problem and the dual maximization problem, as illustrated in the next example.

▼ **EXAMPLE 5**   Use the simplex method to solve the minimization problem in Example 4.

Solution   First, set up the dual problem, as in Example 4:

$$
\begin{aligned}
\text{Maximize} \quad & z = 9x_1 + 10x_2 \\
\text{subject to} \quad & x_1 + 2x_2 \le 8 \\
& 5x_1 + 2x_2 \le 16 \\
& x_1 \ge 0, x_2 \ge 0.
\end{aligned}
$$

This is a maximization problem in standard form, so it can be solved by the simplex method as follows.

Maximization problem

$$
\begin{array}{cccc}
x_1 & x_2 & x_3 & x_4 \\
\end{array}
$$

$$
\left[
\begin{array}{cccc|c}
1 & \boxed{2} & 1 & 0 & 8 \\
5 & 2 & 0 & 1 & 16 \\
\hline
-9 & -10 & 0 & 0 & 0
\end{array}
\right]
$$

$$
\begin{array}{cccc}
x_1 & x_2 & x_3 & x_4 \\
\end{array}
$$

$$
\left[
\begin{array}{cccc|c}
\frac{1}{2} & 1 & \frac{1}{2} & 0 & 4 \\
\boxed{4} & 0 & -1 & 1 & 8 \\
\hline
-4 & 0 & 5 & 0 & 40
\end{array}
\right]
\begin{array}{l}
\frac{1}{2}R_1 \\
-2R_1 + R_2 \\
10R_1 + R_3
\end{array}
$$

$$
\begin{array}{cccc}
x_1 & x_2 & x_3 & x_4 \\
\end{array}
$$

$$
\left[
\begin{array}{cccc|c}
0 & 1 & \frac{5}{8} & -\frac{1}{8} & 3 \\
1 & 0 & -\frac{1}{4} & \frac{1}{4} & 2 \\
\hline
0 & 0 & 4 & 1 & 48
\end{array}
\right]
\begin{array}{l}
-\frac{1}{2}R_2 + R_1 \\
\frac{1}{4}R_2 \\
4R_2 + R_3
\end{array}
$$

The final simplex tableau shows that the maximum value of 48 occurs when $x_1 = 2$ and $x_2 = 3$. In Example 4, we saw that the minimum value of 48 occurs when $y_1 = 4$ and $y_2 = 1$. Note that this information appears in the last row (shown in blue). The minimum 48 is in the lower right-hand corner, and the values where this occurs $(y_1 = 4, y_2 = 1)$ are in the last row at the bottom of the slack variable columns. ▲

A minimization problem that meets the conditions listed after Example 1 can be solved by the method used in Example 5 and summarized here.

Solving Minimization Problems with Duals

1. Find the dual standard maximization problem.*
2. Use the simplex method to solve the dual maximization problem.
3. Read the optimal solution of the original minimization problem from the final simplex tableau:

   y_1 is the last entry in the column corresponding to the first slack variable;

   y_2 is the last entry in the column corresponding to the second slack variable; and so on.

These values of y_1, y_2, y_3, etc., produce the minimum value of w, which is the entry in the lower right-hand corner of the tableau.

---

*The coefficients of the objective function in the minimization problem are the constants on the right side of the constraints in the dual maximization problem. So when all these coefficients are nonnegative (condition 2), the dual problem is in standard maximum form.

▼ **EXAMPLE 6**

$$\text{Minimize} \quad w = 3y_1 + 2y_2$$
$$\text{subject to} \quad y_1 + 3y_2 \geq 6$$
$$2y_1 + y_2 \geq 3$$
$$y_1 \geq 0, y_2 \geq 0.$$

Solution Use the given information to write the matrix.

$$\begin{bmatrix} 1 & 3 & | & 6 \\ 2 & 1 & | & 3 \\ \hline 3 & 2 & | & 0 \end{bmatrix}$$

Transpose to get the following matrix for the dual problem:

$$\begin{bmatrix} 1 & 2 & | & 3 \\ 3 & 1 & | & 2 \\ \hline 6 & 3 & | & 0 \end{bmatrix}.$$

Write the dual problem from this matrix as follows:

$$\text{Maximize} \quad z = 6x_1 + 3x_2$$
$$\text{subject to} \quad x_1 + 2x_2 \leq 3$$
$$3x_1 + x_2 \leq 2$$
$$x_1 \geq 0, x_2 \geq 0.$$

Solve this standard maximization problem by the simplex method. Start by introducing slack variables to give the system

$$x_1 + 2x_2 + x_3 \qquad = 3$$
$$3x_1 + x_2 \qquad + x_4 \qquad = 2$$
$$-6x_1 - 3x_2 - 0x_3 - 0x_4 + z = 0,$$

with $x_1 \geq 0, x_2 \geq 0, x_3 \geq 0, x_4 \geq 0$.

The initial tableau for this system is

$$\begin{array}{cccc} x_1 & x_2 & x_3 & x_4 \\ \end{array}$$
$$\begin{bmatrix} 1 & 2 & 1 & 0 & | & 3 \\ \mathbf{3} & 1 & 0 & 1 & | & 2 \\ \hline -6 & -3 & 0 & 0 & | & 0 \end{bmatrix}$$

Quotients
$3/1 = 3$
$2/3$

with the pivot as indicated. The simplex method gives the following final tableau:

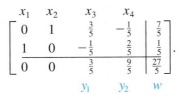

As indicated in blue below the final tableau, the last entries in the columns corresponding to the slack variables (x_3 and x_4) give the values of the original variables y_1 and y_2 that produce the minimal value of w. This minimal value of w appears in the lower right-hand corner (and is the same as the maximal value of z in the dual problem). So the solution of the given minimization problems is as follows:

The minimum value of $w = 3y_1 + 2y_2$, subject to the given constraints, is $27/5$ and occurs when $y_1 = 3/5$ and $y_2 = 9/5$. ▲ ⑤

⑤

Minimize $w = 10y_1 + 8y_2$
subject to

$$y_1 + 2y_2 \geq 2$$
$$y_1 + y_2 \geq 5$$
$$y_1 \geq 0, y_2 \geq 0.$$

Answer:

$y_1 = 0, y_2 = 5$, for a minimum of 40

▼ **EXAMPLE 7**   A minimization problem in three variables was solved by the use of duals. The final simplex tableau for the dual maximization problem is shown here:

$$\begin{bmatrix} x_1 & x_2 & x_3 & x_4 & x_5 & x_6 & \\ 3 & 1 & 1 & 0 & 9 & 0 & 1 \\ 13 & -1 & 0 & 1 & -2 & 0 & 10 \\ 9 & 10 & 0 & 0 & 7 & 1 & 7 \\ 5 & 1 & 0 & 4 & 1 & 7 & 28 \end{bmatrix}.$$

(a) What is the optimal solution of the minimization problem?

Solution Since there are three variables in the minimization problem, there must be three slack variables in the dual maximization problem, namely, x_4, x_5, and x_6. Looking at the bottom of the columns corresponding to these slack variables, we see that the solution of the minimization problem is

$$y_1 = 4, \qquad y_2 = 1, \qquad y_3 = 7, \qquad \text{with a minimal value of } w = 28.$$

(b) What is the optimal solution of the dual maximization problem?

Solution Since x_4, x_5, and x_6 are slack variables by part (a), the variables in the dual problem are x_1, x_2, and x_3. Read the solution from the final tableau, as in Sections 7.4 and 7.5:

$$x_1 = 0, \qquad x_2 = 0 \qquad x_3 = 1, \qquad \text{with a maximal value of } z = 28. \; ▲$$

Further Uses of the Dual

The dual is useful not only in solving minimization problems but also in seeing how small changes in one variable will affect the value of the objective function. For example, suppose an animal breeder needs at least 6 units per day of nutrient A and at least 3 units of nutrient B and that the breeder can choose between two different feeds: feed 1 and feed 2. Find the minimum cost for the breeder if each bag of feed 1 costs \$3 and provides 1 unit of nutrient A and 2 units of B, while each bag of feed 2 costs \$2 and provides 3 units of nutrient A and 1 of B.

If y_1 represents the number of bags of feed 1 and y_2 represents the number of bags of feed 2, the given information leads to

$$\text{Minimize} \quad w = 3y_1 + 2y_2$$
$$\text{subject to} \quad y_1 + 3y_2 \geq 6$$
$$2y_1 + y_2 \geq 3$$
$$y_1 \geq 0, y_2 \geq 0.$$

This minimization linear programming problem is the one we solved in Example 6 of this section. In that example, we formed the dual and reached the following final tableau:

$$
\begin{array}{cccc}
x_1 & x_2 & x_3 & x_4 \\
\end{array}
$$
$$
\left[
\begin{array}{cccc|c}
0 & 1 & \frac{3}{5} & -\frac{1}{5} & \frac{7}{5} \\
1 & 0 & -\frac{1}{5} & \frac{2}{5} & \frac{1}{5} \\
\hline
0 & 0 & \frac{3}{5} & \frac{9}{5} & \frac{27}{5} \\
\end{array}
\right].
$$

This final tableau shows that the breeder will obtain minimum feed costs by using 3/5 bag of feed 1 and 9/5 bags of feed 2 per day, for a daily cost of $27/5 = 5.40$ dollars.

Now look at the data from the feed problem shown in this table.

	UNITS OF NUTRIENT (PER BAG)		Cost per Bag
	A	B	
Feed 1	1	2	\$3
Feed 2	3	1	\$2
Minimum Nutrient Needed	6	3	

If x_1 and x_2 are the cost per unit of nutrients A and B, the constraints of the dual problem can be stated as follows (see page 417).

$$\text{Cost of feed 1:} \quad x_1 + 2x_2 \leq 3$$
$$\text{Cost of feed 2:} \quad 3x_1 + x_2 \leq 2$$

The solution of the dual problem, which maximizes nutrients, also can be read from the final tableau above:

$$x_1 = \frac{1}{5} = .20 \quad \text{and} \quad x_2 = \frac{7}{5} = 1.40.$$

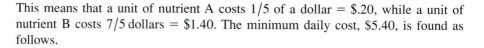

⟨6⟩

The final tableau of the dual of the problem about filing cabinets in Example 2, Section 7.3, and Example 6, Section 7.4 is

$$
\begin{array}{cccc}
x_1 & x_2 & x_3 & x_4 \\
\begin{bmatrix}
0 & 1 & \frac{1}{2} & -\frac{1}{4} & \bigg| & 1 \\
1 & 0 & -\frac{1}{20} & -\frac{1}{80} & \bigg| & \frac{1}{20} \\
0 & 0 & 8 & 3 & \bigg| & 100
\end{bmatrix}
\end{array}
$$

What are the shadow values of the cost and the floor space?

Answers:
$\frac{1}{20}$, 1

This means that a unit of nutrient A costs $1/5$ of a dollar = \$.20, while a unit of nutrient B costs $7/5$ dollars = \$1.40. The minimum daily cost, \$5.40, is found as follows.

$$
\begin{aligned}
(\$.20 \text{ per unit of A}) \times (6 \text{ units of A}) &= \$1.20 \\
+ \ (\$1.40 \text{ per unit of B}) \times (3 \text{ units of B}) &= \$4.20 \\
\hline
\text{Minimum daily cost} &= \$5.40
\end{aligned}
$$

The numbers .20 and 1.40 are called the **shadow costs** of the nutrients. These two numbers from the dual, \$.20 and \$1.40, also allow the breeder to estimate feed costs for "small" changes in nutrient requirements. For example, an increase of 1 unit in the requirement for each nutrient would produce total cost as follows.

\$5.40	6 units of A, 3 of B
.20	1 extra unit of A
1.40	1 extra unit of B
\$7.00	Total cost per day ⟨6⟩

◆7.6 Exercises

Find the transpose of each matrix. (See Example 1.)

1.
$$
\begin{bmatrix}
3 & -4 & 5 \\
1 & 10 & 7 \\
0 & 3 & 6
\end{bmatrix}
$$

2.
$$
\begin{bmatrix}
3 & -5 & 9 & 4 \\
1 & 6 & -7 & 0 \\
4 & 18 & 11 & 9
\end{bmatrix}
$$

3.
$$
\begin{bmatrix}
3 & 0 & 14 & -5 & 3 \\
4 & 17 & 8 & -6 & 1
\end{bmatrix}
$$

4.
$$
\begin{bmatrix}
15 & -6 & -2 \\
13 & -1 & 11 \\
10 & 12 & -3 \\
24 & 1 & 0
\end{bmatrix}
$$

State the dual problem for each of the following, but do not solve it. (See Examples 2 and 3.)

5. Minimize $w = 3y_1 + 5y_2$
subject to $3y_1 + y_2 \geq 4$
$-y_1 + 2y_2 \geq 6$
$y_1 \geq 0, y_2 \geq 0.$

6. Minimize $w = 4y_1 + 7y_2$
subject to $y_1 + y_2 \geq 17$
$3y_1 + 6y_2 \geq 21$
$2y_1 + 4y_2 \geq 19$
$y_1 \geq 0, y_2 \geq 0.$

7. Minimize $w = 2y_1 + 8y_2$
subject to $y_1 + 7y_2 \geq 18$
$4y_1 + y_2 \geq 15$
$5y_1 + 3y_2 \geq 20$
$y_1 \geq 0, y_2 \geq 0.$

8. Minimize $w = 5y_1 + y_2 + 3y_3$
subject to $7y_1 + 6y_2 + 8y_3 \geq 18$
$4y_1 + 5y_2 + 10y_3 \geq 20$
$y_1 \geq 0, y_2 \geq 0, y_3 \geq 0.$

9. Minimize $w = y_1 + 2y_2 + 6y_3$
subject to $3y_1 + 4y_2 + 6y_3 \geq 8$
$y_1 + 5y_2 + 2y_3 \geq 12$
$y_1 \geq 0, y_2 \geq 0, y_3 \geq 0.$

10. Minimize $w = 4y_1 + 3y_2 + y_3$
subject to $y_1 + 2y_2 + 3y_3 \geq 115$
$2y_1 + y_2 + 8y_3 \geq 200$
$y_1 \qquad - y_3 \geq 50$
$y_1 \geq 0, y_2 \geq 0, y_3 \geq 0.$

11. Minimize $w = 8y_1 + 9y_2 + 3y_3$
subject to $y_1 + y_2 + y_3 \geq 5$
$y_1 + y_2 \qquad \geq 4$
$2y_1 + y_2 + 3y_3 \geq 15$
$y_1 \geq 0, y_2 \geq 0, y_3 \geq 0.$

12. Minimize $w = y_1 + 2y_2 + y_3 + 5y_4$
subject to $y_1 + y_2 + y_3 + y_4 \geq 50$
$3y_1 + y_2 + 2y_3 + y_4 \geq 100$
$y_1 \geq 0, y_2 \geq 0, y_3 \geq 0, y_4 \geq 0.$

Use duality to solve the following problems. (See Examples 5 and 6.)

13. Minimize $w = 2y_1 + y_2 + 3y_3$
subject to $y_1 + y_2 + y_3 \geq 100$
$2y_1 + y_2 \quad\quad \geq 50$
$y_1 \geq 0, y_2 \geq 0, y_3 \geq 0.$

14. Minimize $w = 2y_1 + 4y_2$
subject to $4y_1 + 2y_2 \geq 10$
$4y_1 + y_2 \geq 8$
$2y_1 + y_2 \geq 12$
$y_1 \geq 0, y_2 \geq 0.$

15. Minimize $w = 3y_1 + y_2 + 4y_3$
subject to $2y_1 + y_2 + y_3 \geq 6$
$y_1 + 2y_2 + y_3 \geq 8$
$2y_1 + y_2 + 2y_3 \geq 12$
$y_1 \geq 0, y_2 \geq 0, y_3 \geq 0.$

16. Minimize $w = y_1 + y_2 + 3y_3$
subject to $2y_1 + 6y_2 + y_3 \geq 8$
$y_1 + 2y_2 + 4y_3 \geq 12$
$y_1 \geq 0, y_2 \geq 0, y_3 \geq 0.$

17. Minimize $w = 6y_1 + 4y_2 + 2y_3$
subject to $2y_1 + 2y_2 + y_3 \geq 2$
$y_1 + 3y_2 + 2y_3 \geq 3$
$y_1 + y_2 + 2y_3 \geq 4$
$y_1 \geq 0, y_2 \geq 0, y_3 \geq 0.$

18. Minimize $w = 12y_1 + 10y_2 + 7y_3$
subject to $2y_1 + y_2 + y_3 \geq 7$
$y_1 + 2y_2 + y_3 \geq 4$
$y_1 \geq 0, y_2 \geq 0, y_3 \geq 0.$

19. Minimize $w = 20y_1 + 12y_2 + 40y_3$
subject to $y_1 + y_2 + 5y_3 \geq 20$
$2y_1 + y_2 + y_3 \geq 30$
$y_1 \geq 0, y_2 \geq 0, y_3 \geq 0.$

20. Minimize $w = 4y_1 + 5y_2$
subject to $10y_1 + 5y_2 \geq 100$
$20y_1 + 10y_2 \geq 150$
$y_1 \geq 0, y_2 \geq 0.$

21. Minimize $w = 4y_1 + 2y_2 + y_3$
subject to $y_1 + y_2 + y_3 \geq 4$
$3y_1 + y_2 + 3y_3 \geq 6$
$y_1 + y_2 + 3y_3 \geq 5$
$y_1 \geq 0, y_2 \geq 0, y_3 \geq 0.$

22. Minimize $w = 3y_1 + 2y_2$
subject to $2y_1 + 3y_2 \geq 60$
$y_1 + 4y_2 \geq 40$
$y_1 \geq 0, y_2 \geq 0.$

23. Health   Glenn Russell, who is dieting, requires two food supplements: I and II. He can get these supplements from two different products—A and B—as shown in the following table.

Product	Supplement (Grams per Serving)	
	I	II
A	4	2
B	2	5

Glenn's physician has recommended that he include at least 20 grams of supplement I and 18 grams of supplement II in his diet. If product A costs 24¢ per serving and product B costs 40¢ per serving, how can he satisfy these requirements most economically?

24. Business   An animal food must provide at least 54 units of vitamins and 60 calories per serving. One gram of soybean meal provides at least 2.5 units of vitamins and 5 calories. One gram of meat by-products provides at least 4.5 units of vitamins and 3 calories. One gram of grain provides at least 5 units of vitamins and 10 calories. If a gram of soybean meal costs 8¢, a gram of meat by-products 9¢, and a gram of grain 10¢, what mixture of these three ingredients will provide the required vitamins and calories at minimum cost?

25. Business   A brewery produces regular beer and a lower carbohydrate "light" beer. Steady customers of the brewery buy 12 units of regular beer and 10 units of light beer monthly. While setting up the brewery to produce the beers, the management decides to produce extra beer, beyond the need to satisfy the customers. The cost per unit of regular beer is $36,000, and the cost per unit of light beer is $48,000. Every unit of regular beer brings in $100,000 in revenue, while every unit of light beer brings in $300,000. The brewery wants at least $7,000,000 in revenue. At least 20 additional units of beer can be sold. How much of each type of beer should be made so as to minimize total production costs?

26. Business   Joan McKee has a part-time job conducting public-opinion interviews. She has found that a political interview takes 45 minutes and a market interview takes 55 minutes. To allow more time for her full-time job, she needs to minimize the time she spends doing interviews. Unfortunately, to keep her part-time job, she must complete at least 8 interviews each week. Also, she must earn at least $60 per week at this job, at which she earns $8 for each political interview and $10 for each market interview. Finally, to stay in good standing with her supervisor, she must earn at least 40 bonus points per week; she receives 6 bonus points for each political interview and 5 points for each market interview. How many of each interview should she do each week to minimize the time spent?

27. You are given the following linear programming problem (P):*

$$\text{Minimize} \quad z = x_1 + 2x_2$$
$$\text{subject to} \quad -2x_1 + x_2 \geq 1$$
$$x_1 - 2x_2 \geq 1$$
$$x_1 \geq 0, x_2 \geq 0.$$

The dual of (P) is (D). Which of the following statements is true?
(a) (P) has no feasible solution and the objective function of (D) is unbounded.
(b) (D) has no feasible solution and the objective function of (P) is unbounded.
(c) The objective functions of both (P) and (D) are unbounded.
(d) Both (P) and (D) have optimal solutions.
(e) Neither (P) nor (D) has a feasible solution.

28. **Business** Refer to the end of this section in the text, on minimizing the daily cost of feeds.
(a) Find a combination of feeds that will cost $7.00 and give 7 units of A and 4 units of B.
(b) Use the dual variables to predict the daily cost of feed if the requirements change to 5 units of A and 4 units of B. Find a combination of feeds to meet these requirements at the predicted price.

29. **Business** A small toy-manufacturing firm has 200 squares of felt, 600 ounces of stuffing, and 90 feet of

trim available to make two types of toys: a small bear and a monkey. The bear requires 1 square of felt and 4 ounces of stuffing. The monkey requires 2 squares of felt, 3 ounces of stuffing, and 1 foot of trim. The firm makes $1 profit on each bear and $1.50 profit on each monkey. The linear program to maximize profit is

$$\text{Maximize} \quad z = x_1 + 1.5x_2$$
$$\text{subject to} \quad x_1 + 2x_2 \leq 200$$
$$4x_1 + 3x_2 \leq 600$$
$$x_2 \leq 90$$
$$x_1 \geq 0, x_2 \geq 0.$$

The final simplex tableau is

$$\begin{bmatrix} 1 & 0 & -.6 & .4 & 0 & 120 \\ 0 & 0 & -8 & .2 & 1 & 50 \\ 0 & 1 & .8 & -.2 & 0 & 40 \\ 0 & 0 & .6 & .1 & 0 & 180 \end{bmatrix}.$$

(a) What is the corresponding dual problem?
(b) What is the optimal solution to the dual problem?
(c) Use the shadow values to estimate the profit the firm will make if its supply of felt increases to 210 squares.
(d) How much profit will the firm make if its supply of stuffing is cut to 590 ounces and its supply of trim is cut to 80 feet?

30. Refer to Example 1 in Section 7.5.
(a) Give the dual problem.
(b) Use the shadow values to estimate the farmer's profit if land is cut to 90 acres but capital increases to $21,000.
(c) Suppose the farmer has 110 acres but only $19,000. Find the optimum profit and the planting strategy that will produce this profit.

*Problem 2 from "November 1989 Course 130 Examination Operations Research" of the *Education and Examination Committee of the Society of Actuaries.* Reprinted by permission of the Society of Actuaries.

7.7 The Simplex Method: Nonstandard Problems

So far, the simplex method has been used to solve problems in which the variables are nonnegative and all the other constraints are of one type (either all $\leq$ or all $\geq$). Now we extend the simplex method to linear programming problems with nonnegative variables and mixed constraints ($\leq$, $=$, and $\geq$).

The solution method to be used here requires that all inequality constraints be written so that the constant on the right side is nonnegative. For instance, the inequality

$$4x_1 + 5x_2 - 12x_3 \leq -30$$

can be replaced by the equivalent one obtained by multiplying both sides by -1 and reversing the direction of the inequality sign:

$$-4x_1 - 5x_2 + 12x_3 \geq 30.$$

Maximization with ≤ and ≥ Constraints

As is always the case when the simplex method is involved, each inequality constraint must be written as an equation. Constraints involving $\leq$ are converted to equations by adding a nonnegative slack variable, as in Section 7.4. Similarly, constraints involving $\geq$ are converted to equations by *subtracting* a nonnegative **surplus variable.** For example, the inequality $2x_1 - x_2 + 5x_3 \geq 12$ is written as

$$2x_1 - x_2 + 5x_3 - x_4 = 12,$$

where $x_4 \geq 0$. The surplus variable x_4 represents the amount by which $2x_1 - x_2 + 5x_3$ exceeds 12.

▼ **EXAMPLE 1**   Restate the following problem in terms of equations, and write its initial simplex tableau:

$$\begin{aligned}\text{Maximize} \quad & z = 4x_1 + 10x_2 + 6x_3 \\ \text{subject to} \quad & x_1 + 4x_2 + 4x_3 \geq 8 \\ & x_1 + 3x_2 + 2x_3 \leq 6 \\ & 3x_1 + 4x_2 + 8x_3 \leq 22 \\ & x_1 \geq 0, x_2 \geq 0, x_3 \geq 0. \end{aligned}$$

Solution In order to write the constraints as equations, subtract a surplus variable from the $\geq$ constraint and add a slack variable to each $\leq$ constraint. So the problem becomes

$$\begin{aligned}\text{Maximize} \quad & z = 4x_1 + 10x_2 + 6x_3 \\ \text{subject to} \quad & x_1 + 4x_2 + 4x_3 - x_4 \qquad\qquad = 8 \\ & x_1 + 3x_2 + 2x_3 \qquad + x_5 \qquad = 6 \\ & 3x_1 + 4x_2 + 8x_3 \qquad\qquad + x_6 = 22 \\ & x_1 \geq 0, x_2 \geq 0, x_3 \geq 0, x_4 \geq 0, x_5 \geq 0, x_6 \geq 0. \end{aligned}$$

Write the objective function as $z - 4x_1 - 10x_2 - 6x_3 = 0$ and use the coefficients of the four equations to write the initial simplex tableau (omitting the z column):

$$\begin{array}{cccccc} x_1 & x_2 & x_3 & x_4 & x_5 & x_6 \\ \left[\begin{array}{cccccc|c} 1 & 4 & 4 & -1 & 0 & 0 & 8 \\ 1 & 3 & 2 & 0 & 1 & 0 & 6 \\ 3 & 4 & 8 & 0 & 0 & 1 & 22 \\ \hline -4 & -10 & -6 & 0 & 0 & 0 & 0 \end{array}\right] \end{array}$$ ▲ ◇ ①

The tableau in Example 1 resembles those which have appeared previously, and similar terminology is used. The variables whose columns have one entry ± 1 and the rest 0 will be called **basic variables;** the other variables are nonbasic. A solution obtained by setting the nonbasic variables equal to 0 and solving for the basic variables (by looking at the constants in the right-hand column) will be called a **basic solution.** A basic solution that is feasible is called a **basic feasible solution.** In the tableau of Example 1, for instance, the basic variables are x_4, x_5, and x_6, and the basic solution is

$$x_1 = 0, \qquad x_2 = 0, \qquad x_3 = 0, \qquad x_4 = -8, \qquad x_5 = 6, \qquad x_6 = 22.$$

---

①

(a) Restate this problem in terms of equations:

$$\begin{aligned}\text{Maximize} \quad & z = 3x_1 - 2x_2 \\ \text{subject to} \quad & 2x_1 + 3x_2 \leq 8 \\ & 6x_1 - 2x_2 \geq 3 \\ & x_1 + 4x_2 \geq 1 \\ & x_1 \geq 0, x_2 \geq 0. \end{aligned}$$

(b) Write the initial simplex tableau.

Answers:

(a) Maximize $z = 3x_1 - 2x_2$
subject to
$$\begin{aligned} & 2x_1 + 3x_2 + x_3 = 8 \\ & 6x_1 - 2x_2 - x_4 = 3 \\ & x_1 + 4x_2 - x_5 = 1 \\ & x_1 \geq 0, x_2 \geq 0, x_3 \geq 0, \\ & x_4 \geq 0, x_5 \geq 0. \end{aligned}$$

(b)
$$\begin{array}{ccccc} x_1 & x_2 & x_3 & x_4 & x_5 \\ \left[\begin{array}{ccccc|c} 2 & 3 & 1 & 0 & 0 & 8 \\ 6 & -2 & 0 & -1 & 0 & 3 \\ 1 & 4 & 0 & 0 & -1 & 1 \\ \hline -3 & 2 & 0 & 0 & 0 & 0 \end{array}\right] \end{array}$$

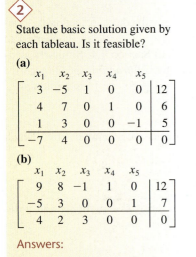

State the basic solution given by each tableau. Is it feasible?

(a)

$$\begin{bmatrix} x_1 & x_2 & x_3 & x_4 & x_5 & \\ 3 & -5 & 1 & 0 & 0 & 12 \\ 4 & 7 & 0 & 1 & 0 & 6 \\ 1 & 3 & 0 & 0 & -1 & 5 \\ \hline -7 & 4 & 0 & 0 & 0 & 0 \end{bmatrix}$$

(b)

$$\begin{bmatrix} x_1 & x_2 & x_3 & x_4 & x_5 & \\ 9 & 8 & -1 & 1 & 0 & 12 \\ -5 & 3 & 0 & 0 & 1 & 7 \\ \hline 4 & 2 & 3 & 0 & 0 & 0 \end{bmatrix}$$

Answers:

(a) $x_1 = 0$, $x_2 = 0$, $x_3 = 12$, $x_4 = 6$, $x_5 = -5$; no.

(b) $x_1 = 0$, $x_2 = 0$, $x_3 = 0$, $x_4 = 12$, $x_5 = 7$; yes.

However, because one variable is negative, this solution is not feasible.

The solution method for problems such as the one in Example 1 consists of two stages. **Stage I** consists of finding a basic *feasible* solution that can be used as the starting point for the simplex method. (This stage is unnecessary in a standard maximization problem, because the solution given by the initial tableau is always feasible.) There are many systematic ways of finding a feasible solution, all of which depend on the fact that row operations (such as pivoting) produce a tableau that represents a system with the same solutions as the original one. One such technique is explained in the next example. Since the immediate goal is to find a feasible solution, not necessarily an optimal one, the procedures for choosing pivots differ from those in the ordinary simplex method.

▼ **EXAMPLE 2** Find a basic feasible solution for the problem in Example 1, whose initial tableau is

$$\begin{bmatrix} x_1 & x_2 & x_3 & x_4 & x_5 & x_6 & \\ 1 & 4 & 4 & -1 & 0 & 0 & 8 \\ 1 & 3 & 2 & 0 & 1 & 0 & 6 \\ 3 & 4 & 8 & 0 & 0 & 1 & 22 \\ \hline -4 & -10 & -6 & 0 & 0 & 0 & 0 \end{bmatrix}.$$

Solution In the basic solution given by this tableau, x_4 has a negative value. The only nonzero entry in its column is the -1 in row one. Choose any *positive* entry in row one except the entry on the far right. The column that the chosen entry is in will be the pivot column. We choose the first positive entry in row one: the 1 in column one. The pivot row is determined in the usual way by considering quotients (the constant at the right end of the row, divided by the positive entry in the pivot column) in each row except the objective row:

$$8/1 = 8, \qquad 6/1 = 6, \qquad 22/3 = 7\frac{1}{3}.$$

The smallest quotient is 6, so the pivot is the 1 in row two, column one. Pivoting in the usual way leads to the tableau

$$\begin{bmatrix} x_1 & x_2 & x_3 & x_4 & x_5 & x_6 & \\ 0 & 1 & 2 & -1 & -1 & 0 & 2 \\ 1 & 3 & 2 & 0 & 1 & 0 & 6 \\ 0 & -5 & 2 & 0 & -3 & 1 & 4 \\ \hline 0 & 2 & 2 & 0 & 4 & 0 & 24 \end{bmatrix} \begin{array}{l} -R_2 + R_1 \\ \\ -3R_2 + R_3 \\ 4R_2 + R_4 \end{array}$$

and the basic solution

$$x_1 = 6, \qquad x_2 = 0, \qquad x_3 = 0, \qquad x_4 = -2, \qquad x_5 = 0, \qquad x_6 = 4.$$

Since the basic variable x_4 is negative, this solution is not feasible. So we repeat the pivoting process. The x_4 column has a -1 in row one, so we choose a positive entry in that row, namely, the 1 in row one, column two. This choice makes column two the pivot column. The pivot row is determined by the quotients $2/1 = 2$ and $6/3 = 2$ (negative entries in the pivot column and the entry in the objective row are not used).

Since there is a tie, we can choose either row one or row two. We choose row one and use the 1 in row one, column 2, as the pivot. Pivoting produces this tableau

$$
\begin{array}{c}
\begin{array}{cccccc} x_1 & x_2 & x_3 & x_4 & x_5 & x_6 \end{array} \\
\left[\begin{array}{cccccc|c}
0 & 1 & 2 & -1 & -1 & 0 & 2 \\
1 & 0 & -4 & 3 & 4 & 0 & 0 \\
0 & 0 & 12 & -5 & -8 & 1 & 14 \\
\hline
0 & 0 & -2 & 2 & 6 & 0 & 20
\end{array}\right]
\begin{array}{l}
\\ -3R_1 + R_2 \\ 5R_1 + R_3 \\ -2R_1 + R_4
\end{array}
\end{array}
$$

and the basic *feasible* solution

$$ x_1 = 0, \qquad x_2 = 2, \qquad x_3 = 0, \qquad x_4 = 0, \qquad x_5 = 0, \qquad x_6 = 14. \ \blacktriangle $$

Once a basic feasible solution has been found, Stage I is ended. The procedures used in Stage I are summarized here.*

Finding a Basic Feasible Solution

1. If any basic variable has a negative value, locate the -1 in that variable's column and note the row it is in.

2. In the row determined in Step 1, choose a positive entry (other than the one at the far right) and note the column it is in. This is the pivot column.

3. Use the positive entries in the pivot column (except in the objective row) to form quotients and select the pivot.

4. Pivot as usual, which results in the pivot column's having one entry 1 and the rest 0's.

5. Repeat Steps 1–4 until every basic variable is nonnegative, so that the basic solution given by the tableau is feasible. If it ever becomes impossible to continue, then the problem has no feasible solution.

⬦ **3**

The initial tableau of a maximization problem is shown. Use column one as the pivot column for carrying out Stage I, and state the basic feasible solution that results.

$$
\begin{array}{c}
\begin{array}{cccc} x_1 & x_2 & x_3 & x_4 \end{array} \\
\left[\begin{array}{cccc|c}
1 & 3 & 1 & 0 & 70 \\
2 & 4 & 0 & -1 & 50 \\
\hline
-8 & -10 & 0 & 0 & 0
\end{array}\right]
\end{array}
$$

Answer:

$$
\begin{array}{c}
\begin{array}{cccc} x_1 & x_2 & x_3 & x_4 \end{array} \\
\left[\begin{array}{cccc|c}
0 & 1 & 1 & \frac{1}{2} & 45 \\
1 & 2 & 0 & -\frac{1}{2} & 25 \\
\hline
0 & 6 & 0 & -4 & 200
\end{array}\right]
\end{array}
$$

$x_1 = 25, x_2 = 0, x_3 = 45,$ $x_4 = 0.$

One way to make the required choices systematically is to choose the first possibility in each case (going from the top for rows or from the left for columns). However, any choice meeting the required conditions may be used. For maximum efficiency, it is usually best to choose the pivot column in Step 2, so that the pivot is in the same row chosen in Step 1, if this is possible. ⬦**3**

In **Stage II,** the simplex method is applied as usual to the tableau that produced the basic feasible solution in Stage I. Just as in Section 7.4, each round of pivoting replaces the basic feasible solution of one tableau with the basic feasible solution of a new tableau in such a way that the value of the objective function is increased, until an optimal value is obtained (or it becomes clear that no optimal solution exists).

---

*Except in rare cases that do not occur in this book, this method either eventually produces a basic feasible solution or shows that one does not exist. The *two-phase method* using artificial variables, which is discussed in more advanced texts, works in all cases and often is more efficient.

▼ **EXAMPLE 3** Solve the linear programming problem in Example 1.

Solution A basic feasible solution for this problem was found in Example 2 by using the tableau shown below. However, this solution is not maximal, because there is a negative indicator in the objective row. So we use the simplex method: the most negative indicator determines the pivot column, and the usual quotients determine that the number 2 in row one, column three, is the pivot.

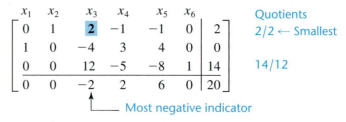

Pivoting leads to the final tableau.

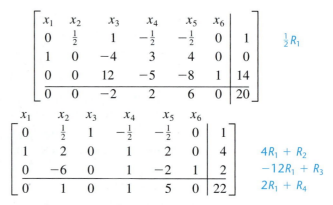

Therefore, the maximum value of z occurs when $x_1 = 4$, $x_2 = 0$, and $x_3 = 1$, in which case $z = 22$. ▲ ◇4◇

◇4◇

Complete Stage II and find an optimal solution for side problem 3 on page 425. What is the optimal value of the objective function z?

Answer:

The optimal value $z = 560$ occurs when $x_1 = 70$, $x_2 = 0$, $x_3 = 0$, and $x_4 = 90$.

Minimization Problems

The two-stage method for maximization problems illustrated in Examples 1–3 also provides a means of solving minimization problems. To see why, consider this simple fact: when a number t gets smaller, $-t$ gets larger, and vice versa. For instance, if t goes from 6 down to -8, then $-t$ goes from -6 up to 8. Thus, if w is the objective function of a linear programming problem, the feasible solution that produces the minimum value of w also produces the maximum value of $-w$, and vice versa. Therefore, to solve a minimization problem with objective function w, we need only solve the maximization problem with the same constraints and objective function $z = -w$.

▼ **EXAMPLE 4**

$$\text{Minimize} \quad w = 2y_1 + y_2 - y_3$$
$$\text{subject to} \quad -y_1 - y_2 + y_3 \le -4$$
$$y_1 + 3y_2 + 3y_3 \ge 6$$
$$y_1 \ge 0, y_2 \ge 0, y_3 \ge 0.$$

Solution Make the constant in the first constraint positive by multiplying both sides by -1. Then solve this maximization problem:

$$\text{Maximize} \quad z = -w = -2y_1 - y_2 + y_3$$
$$\text{subject to} \quad y_1 + y_2 - y_3 \geq 4$$
$$y_1 + 3y_2 + 3y_3 \geq 6$$
$$y_1 \geq 0, y_2 \geq 0, y_3 \geq 0.$$

Convert the constraints to equations by subtracting surplus variables, and set up the first tableau.

$$
\begin{array}{ccccc}
y_1 & y_2 & y_3 & y_4 & y_5 \\
\end{array}
$$
$$
\left[
\begin{array}{ccccc|c}
1 & 1 & -1 & -1 & 0 & 4 \\
1 & 3 & 3 & 0 & -1 & 6 \\
\hline
2 & 1 & -1 & 0 & 0 & 0 \\
\end{array}
\right]
$$

The basic solution given by this tableau, namely, $y_1 = 0$, $y_2 = 0$, $y_3 = 0$, $y_4 = -4$, $y_5 = -6$, is not feasible, so the procedures of Stage I must be used to find a basic feasible solution. In the column of the negative basic variable y_4, there is a -1 in row one; we choose the first positive entry in that row, so that column one will be the pivot column. The quotients $4/1 = 4$ and $6/1 = 6$ show that the pivot is the 1 in row one, column one. Pivoting produces this tableau:

$$
\begin{array}{ccccc}
y_1 & y_2 & y_3 & y_4 & y_5 \\
\end{array}
$$
$$
\left[
\begin{array}{ccccc|c}
1 & 1 & -1 & -1 & 0 & 4 \\
0 & 2 & 4 & 1 & -1 & 2 \\
\hline
0 & -1 & 1 & 2 & 0 & -8 \\
\end{array}
\right]
\quad
\begin{array}{l}
-R_1 + R_2 \\
-2R_1 + R_3 \\
\end{array}
$$

The basic solution $y_1 = 4$, $y_2 = 0$, $y_3 = 0$, $y_4 = 0$, $y_5 = -2$ is not feasible because y_5 is negative, so we repeat the process. We choose the first positive entry in row two (the row containing the -1 in the y_5 column), which is in column two, so that column two is the pivot column. The relevant quotients are $4/1 = 4$ and $2/2 = 1$, so the pivot is the 2 in row two, column two. Pivoting produces a new tableau.

$$
\begin{array}{ccccc}
y_1 & y_2 & y_3 & y_4 & y_5 \\
\end{array}
$$
$$
\left[
\begin{array}{ccccc|c}
1 & 1 & -1 & -1 & 0 & 4 \\
0 & 1 & 2 & \frac{1}{2} & -\frac{1}{2} & 1 \\
\hline
0 & -1 & 1 & 2 & 0 & -8 \\
\end{array}
\right]
\quad
\begin{array}{l}
\frac{1}{2}R_2 \\
\end{array}
$$

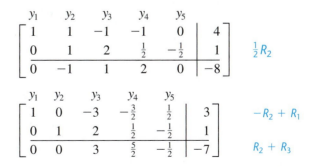

$$
\begin{array}{ccccc}
y_1 & y_2 & y_3 & y_4 & y_5 \\
\end{array}
$$
$$
\left[
\begin{array}{ccccc|c}
1 & 0 & -3 & -\frac{3}{2} & \frac{1}{2} & 3 \\
0 & 1 & 2 & \frac{1}{2} & -\frac{1}{2} & 1 \\
\hline
0 & 0 & 3 & \frac{5}{2} & -\frac{1}{2} & -7 \\
\end{array}
\right]
\quad
\begin{array}{l}
-R_2 + R_1 \\
\\
R_2 + R_3 \\
\end{array}
$$

The basic solution $y_1 = 3$, $y_2 = 1$, $y_3 = 0$, $y_4 = 0$, $y_5 = 0$ is feasible, so Stage I is complete. However, this solution is not optimal, because the objective row contains the negative indicator $-1/2$ in column five. According to the simplex method, column five is the next pivot column. The only positive ratio $3/\frac{1}{2} = 6$ is in row one, so the pivot is $1/2$ in row one, column five. Pivoting produces the final tableau.

$$
\begin{array}{ccccc}
y_1 & y_2 & y_3 & y_4 & y_5 \\
\left[\begin{array}{ccccc|c}
2 & 0 & -6 & -3 & 1 & 6 \\
0 & 1 & 2 & \frac{1}{2} & -\frac{1}{2} & 1 \\
\hline
0 & 0 & 3 & \frac{5}{2} & -\frac{1}{2} & -7
\end{array}\right] & & & & & 2R_1
\end{array}
$$

$$
\begin{array}{ccccc}
y_1 & y_2 & y_3 & y_4 & y_5 \\
\left[\begin{array}{ccccc|c}
2 & 0 & -6 & -3 & 1 & 6 \\
1 & 1 & -1 & -1 & 0 & 4 \\
\hline
1 & 0 & 0 & 1 & 0 & -4
\end{array}\right] & & & & & \begin{array}{c}\frac{1}{2}R_1 + R_2 \\ \frac{1}{2}R_1 + R_3\end{array}
\end{array}
$$

Since there are no negative indicators, the solution given by this tableau ($y_1 = 0$, $y_2 = 4$, $y_3 = 0$, $y_4 = 0$, $y_5 = 6$) is optimal. The maximum value of $z = -w$ is -4. Therefore, the minimum value of the original objective function w is $-(-4) = 4$, which occurs when $y_1 = 0$, $y_2 = 4$, $y_3 = 0$. ▲ ⑤

⑤
Minimize $w = 2y_1 + 3y_2$
subject to
$$y_1 + y_2 \geq 10$$
$$2y_1 + y_2 \geq 16$$
$$y_1 \geq 0, y_2 \geq 0.$$

Answer:

$y_1 = 10, y_2 = 0; w = 20$

Equation Constraints

Recall that, for any real numbers a and b,

$$a = b \qquad \text{exactly when } a \geq b \text{ and simultaneously } a \leq b.$$

Thus, an equation such as $y_1 + 3y_2 + 3y_3 = 6$ is equivalent to this pair of inequalities:

$$y_1 + 3y_2 + 3y_3 \geq 6$$
$$y_1 + 3y_2 + 3y_3 \leq 6.$$

In a linear programming problem, each equation constraint should be replaced in this way by a pair of inequality constraints. Then the problem can be solved by the two-stage method.

▼ **EXAMPLE 5**

$$\text{Minimize} \quad w = 2y_1 + y_2 - y_3$$
$$\text{subject to} \quad -y_1 - y_2 + y_3 \leq -4$$
$$y_1 + 3y_2 + 3y_3 = 6$$
$$y_1 \geq 0, y_2 \geq 0, y_3 \geq 0.$$

Solution Multiply the first inequality by -1 and replace the equation by an equivalent pair of inequalities, as just explained, to obtain this problem:

$$\text{Maximize} \quad z = -w = -2y_1 - y_2 + y_3$$
$$\text{subject to} \quad y_1 + y_2 - y_3 \geq 4$$
$$y_1 + 3y_2 + 3y_3 \geq 6$$
$$y_1 + 3y_2 + 3y_3 \leq 6$$
$$y_1 \geq 0, y_2 \geq 0, y_3 \geq 0.$$

Convert the constraints to equations by subtracting surplus variables y_4 and y_5 from the first two inequalities and adding a slack variable y_6 to the third. Then the first tableau is

$$
\begin{array}{cccccc}
y_1 & y_2 & y_3 & y_4 & y_5 & y_6 \\
\end{array}
$$
$$
\left[\begin{array}{cccccc|c}
1 & 1 & -1 & -1 & 0 & 0 & 4 \\
1 & 3 & 3 & 0 & -1 & 0 & 6 \\
1 & 3 & 3 & 0 & 0 & 1 & 6 \\
\hline
2 & 1 & -1 & 0 & 0 & 0 & 0
\end{array}\right].
$$

The basic solution given by this tableau is $y_1 = 0$, $y_2 = 0$, $y_3 = 0$, $y_4 = -4$, $y_5 = -6$, $y_6 = 6$, which is not feasible. So we carry out the procedures of Stage I.

$$
\begin{array}{cccccc}
y_1 & y_2 & y_3 & y_4 & y_5 & y_6 \\
\end{array}
$$
$$
\left[\begin{array}{cccccc|c}
1 & 1 & -1 & -1 & 0 & 0 & 4 \\
0 & 2 & 4 & 1 & -1 & 0 & 2 \\
0 & 2 & 4 & 1 & 0 & 1 & 2 \\
\hline
0 & -1 & 1 & 2 & 0 & 0 & -8
\end{array}\right]
$$

$$
\begin{array}{cccccc}
y_1 & y_2 & y_3 & y_4 & y_5 & y_6 \\
\end{array}
$$
$$
\left[\begin{array}{cccccc|c}
1 & 0 & -3 & -1.5 & .5 & 0 & 3 \\
0 & 1 & 2 & .5 & -.5 & 0 & 1 \\
0 & 0 & 0 & 0 & 1 & 1 & 0 \\
\hline
0 & 0 & 3 & 2.5 & -.5 & 0 & 7
\end{array}\right]
$$

This tableau gives the basic feasible solution $y_1 = 3$, $y_2 = 1$, $y_3 = 0$, $y_4 = 0$, $y_5 = 0$, $y_6 = 0$, so Stage I is complete. Now apply the simplex method. One round of pivoting produces the final tableau.

$$
\begin{array}{cccccc}
y_1 & y_2 & y_3 & y_4 & y_5 & y_6 \\
\end{array}
$$
$$
\left[\begin{array}{cccccc|c}
1 & 0 & -3 & -1.5 & 0 & -.5 & 3 \\
0 & 1 & 2 & .5 & 0 & .5 & 1 \\
0 & 0 & 0 & 0 & 1 & 1 & 0 \\
\hline
0 & 0 & 3 & 2.5 & 0 & .5 & -7
\end{array}\right]
$$

Therefore, the minimum value of $w = -z$ is $w = -(-7) = 7$, which occurs when $y_1 = 3$, $y_2 = 1$, and $y_3 = 0$. ▲

You may have noticed that Example 5 is just Example 4 with the last inequality constraint replaced by an equation constraint. Note, however, that the optimal solutions are different in the two examples. The minimal value of w found in Example 4 is smaller than the one found in Example 5, but does not satisfy the equation constraint in Example 5.

The two-stage method used in Examples 1–5 is summarized here.

Solving Nonstandard Problems

1. Replace each equation constraint by an equivalent pair of inequality constraints.

continued

2. If necessary, write each constraint with a positive constant.

3. Convert a minimization problem to a maximization problem by letting $z = -w$.

4. Add slack variables and subtract surplus variables as needed to convert the constraints into equations.

5. Write the initial simplex tableau.

6. Find a basic feasible solution for the problem if such a solution exists (Stage I).

7. When a basic feasible solution is found, use the simplex method to solve the problem (Stage II).

NOTE It may happen that the tableau which gives the basic feasible solution in Stage I has no negative indicators in its last row. In this case, the solution found is already optimal and Stage II is not necessary.

Applications

Many real-world applications of linear programming involve mixed constraints. Since they typically include a large number of variables and constraints, technology is normally required to solve such problems.

▼ **EXAMPLE 6**   A college textbook publisher has received orders from two colleges: C_1 and C_2. C_1 needs 500 books and C_2 needs 1000. The publisher can supply the books from either of two warehouses. Warehouse W_1 has 900 books available and warehouse W_2 has 700. The costs to ship a book from each warehouse to each college are as follows.

		To	
		C_1	C_2
From	W_1	$1.20	$1.80
	W_2	$2.10	$1.50

How many books should be sent from each warehouse to each college to minimize the shipping costs?

Solution   To begin, let

$$y_1 = \text{the number of books shipped from } W_1 \text{ to } C_1;$$
$$y_2 = \text{the number of books shipped from } W_2 \text{ to } C_1;$$
$$y_3 = \text{the number of books shipped from } W_1 \text{ to } C_2;$$
$$y_4 = \text{the number of books shipped from } W_2 \text{ to } C_2.$$

C_1 needs 500 books, so $y_1 + y_2 = 500$, which is equivalent to this pair of inequalities:

$$y_1 + y_2 \geq 500$$
$$y_1 + y_2 \leq 500.$$

Similarly, $y_3 + y_4 = 1000$, which is equivalent to

$$y_3 + y_4 \geq 1000$$
$$y_3 + y_4 \leq 1000.$$

Since W_1 has 900 books available and W_2 has 700 available,

$$y_1 + y_3 \leq 900 \qquad \text{and} \qquad y_2 + y_4 \leq 700.$$

The company wants to minimize shipping costs, so the objective function is

$$w = 1.20y_1 + 2.10y_2 + 1.80y_3 + 1.50y_4.$$

Now write the problem as a system of linear equations, adding slack or surplus variables as needed, and let $z = -w$.

$$
\begin{aligned}
y_1 + y_2 \phantom{{}+{}} & - y_5 \phantom{{}+ y_6} && = 500 \\
y_1 + y_2 + &   y_6 && = 500 \\
& y_3 + y_4 - y_7 && = 1000 \\
& y_3 + y_4 + y_8 && = 1000 \\
y_1 + & \phantom{y_3 +{}} y_3 + y_9 && = 900 \\
\phantom{y_1 +{}} y_2 \phantom{{}+{}} & \phantom{y_3 +{}} y_4 + y_{10} && = 700 \\
1.20y_1 + 2.10y_2 + 1.80y_3 + 1.50y_4 & + z && = 0
\end{aligned}
$$

Set up the initial simplex tableau.

y_1	y_2	y_3	y_4	y_5	y_6	y_7	y_8	y_9	y_{10}	
1	1	0	0	−1	0	0	0	0	0	500
1	1	0	0	0	1	0	0	0	0	500
0	0	1	1	0	0	−1	0	0	0	1000
0	0	1	1	0	0	0	1	0	0	1000
1	0	1	0	0	0	0	0	1	0	900
0	1	0	1	0	0	0	0	0	1	700
1.20	2.10	1.80	1.50	0	0	0	0	0	0	0

The basic solution here is not feasible, because $y_5 = -500$ and $y_7 = -1000$. Stages I and II could be done by hand here, but because of the large size of the matrix, it's more efficient to use technology, such as the program in the Graphing Calculator Appendix. Stage I takes four rounds of pivoting and produces the feasible solution in Figure 7.28.

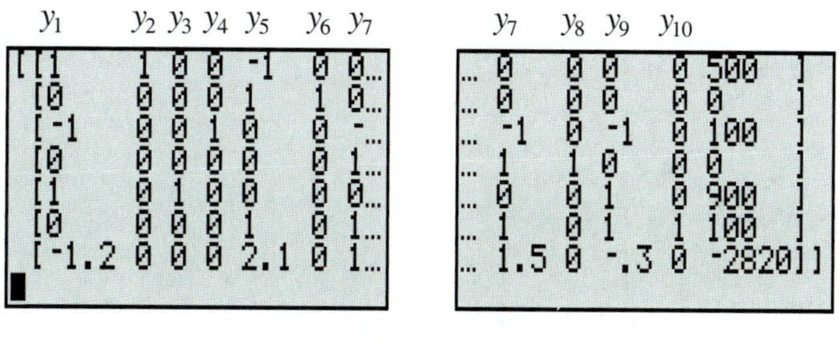

FIGURE 7.28

Because of the small size of a calculator screen, you must scroll to the right to see the entire matrix. Now Stage II begins. Two rounds of pivoting produce the final tableau (Figure 7.29).

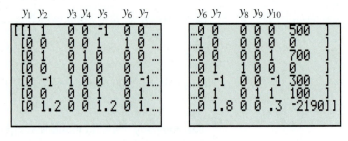

FIGURE 7.29

The optimal solution is $y_1 = 500$, $y_2 = 0$, $y_3 = 300$, $y_4 = 700$, which results in a minimum shipping cost of $2190. (Remember that the optimal value for the original minimization problem is the negative of the optimal value of the associated maximization problem.) ▲

◆7.7 Exercises

In Exercises 1–4, (a) restate the problem in terms of equations by introducing slack and surplus variables and (b) write the initial simplex tableau. (See Example 1.)

1. Maximize $z = 5x_1 + 2x_2 - x_3$
   subject to $2x_1 + 3x_2 + 5x_3 \geq 8$
        $4x_1 - x_2 + 3x_3 \leq 7$
        $x_1 \geq 0, x_2 \geq 0, x_3 \geq 0.$

2. Maximize $z = x_1 + 4x_2 + 6x_3$
   subject to $5x_1 + 8x_2 - 5x_3 \leq 10$
        $6x_1 + 2x_2 + 3x_3 \geq 7$
        $x_1 \geq 0, x_2 \geq 0, x_3 \geq 0.$

3. Maximize $z = 2x_1 - 3x_2 + 4x_3$
   subject to $x_1 + x_2 + x_3 \leq 100$
        $x_1 + x_2 + x_3 \geq 75$
        $x_1 + x_2 \geq 27$
        $x_1 \geq 0, x_2 \geq 0, x_3 \geq 0.$

4. Maximize $z = -x_1 + 5x_2 + x_3$
   subject to $2x_1 + x_3 \leq 40$
        $x_1 + x_2 \geq 18$
        $x_1 + x_3 = 20$
        $x_1 \geq 0, x_2 \geq 0, x_3 \geq 0.$

Convert Exercises 5–8 into maximization problems with positive constants on the right side of each constraint and write the initial simplex tableau. (See Examples 4 and 5.)

5. Minimize $w = 2y_1 + 5y_2 - 3y_3$
   subject to $y_1 + 2y_2 + 3y_3 \geq 115$
        $2y_1 + y_2 + y_3 \leq 200$
        $y_1 + y_3 \geq 50$
        $y_1 \geq 0, y_2 \geq 0, y_3 \geq 0.$

6. Minimize $w = 7y_1 + 6y_2 + y_3$
   subject to $y_1 + y_2 + y_3 \geq 5$
        $-y_1 + y_2 \leq -4$
        $2y_1 + y_2 + 3y_3 \geq 15$
        $y_1 \geq 0, y_2 \geq 0, y_3 \geq 0.$

7. Minimize $w = y_1 - 4y_2 + 2y_3$
   subject to $-7y_1 + 6y_2 - 8y_3 \leq -18$
        $4y_1 + 5y_2 + 10y_3 \geq 20$
        $y_1 \geq 0, y_2 \geq 0, y_3 \geq 0.$

8. Minimize $w = y_1 + 2y_2 + y_3 + 5y_4$
   subject to $-y_1 + y_2 + y_3 + y_4 \leq -50$
        $3y_1 + y_2 + 2y_3 + y_4 = 100$
        $y_1 \geq 0, y_2 \geq 0, y_3 \geq 0, y_4 \geq 0.$

Use the two-stage method to solve Exercises 9–20. (See Examples 1–5.)

9. Maximize $z = 12x_1 + 10x_2$
   subject to $x_1 + 2x_2 \geq 24$
        $x_1 + x_2 \leq 40$
        $x_1 \geq 0, x_2 \geq 0.$

10. Find $x_1 \geq 0$, $x_2 \geq 0$, and $x_3 \geq 0$ such that

$x_1 + x_2 + x_3 \leq 150$

$x_1 + x_2 + x_3 \geq 100$

and $z = 2x_1 + 5x_2 + 3x_3$ is maximized.

11. Find $x_1 \geq 0$, $x_2 \geq 0$, and $x_3 \geq 0$ such that

$x_1 + x_2 + 2x_3 \leq 38$

$2x_1 + x_2 + x_3 \geq 24$

and $z = 3x_1 + 2x_2 + 2x_3$ is maximized.

12. Maximize $z = 6x_1 + 8x_2$
subject to $3x_1 + 12x_2 \geq 48$
$2x_1 + 4x_2 \leq 60$
$x_1 \geq 0, x_2 \geq 0.$

13. Find $x_1 \geq 0$ and $x_2 \geq 0$ such that

$x_1 + 2x_2 \leq 18$

$x_1 + 3x_2 \geq 12$

$2x_1 + 2x_2 \leq 30$

and $z = 5x_1 + 10x_2$ is maximized.

14. Find $y_1 \geq 0, y_2 \geq 0$ such that

$10y_1 + 5y_2 \geq 100$

$20y_1 + 10y_2 \geq 160$

and $w = 4y_1 + 5y_2$ is minimized.

15. Minimize $w = 3y_1 + 2y_2$
subject to $2y_1 + 3y_2 \geq 60$
$y_1 + 4y_2 \geq 40$
$y_1 \geq 0, y_2 \geq 0.$

16. Minimize $w = 3y_1 + 4y_2$
subject to $y_1 + 2y_2 \geq 10$
$y_1 + y_2 \geq 8$
$2y_1 + y_2 \leq 22$
$y_1 \geq 0, y_2 \geq 0.$

17. Maximize $z = 3x_1 + 2x_2$
subject to $x_1 + x_2 = 50$
$4x_1 + 2x_2 \geq 120$
$5x_1 + 2x_2 \leq 200$
with $x_1 \geq 0, x_2 \geq 0.$

18. Maximize $z = 10x_1 + 9x_2$
subject to $x_1 + x_2 = 30$
$x_1 + x_2 \geq 25$
$2x_1 + x_2 \leq 40$
with $x_1 \geq 0, x_2 \geq 0.$

19. Minimize $w = 32y_1 + 40y_2$
subject to $20y_1 + 10y_2 = 200$
$25y_1 + 40y_2 \leq 500$
$18y_1 + 24y_2 \geq 300$
with $y_1 \geq 0, y_2 \geq 0.$

20. Minimize $w = 15y_1 + 12y_2$
subject to $y_1 + 2y_2 \leq 12$
$3y_1 + y_2 \geq 18$
$y_1 + y_2 = 10$
with $y_1 \geq 0, y_2 \geq 0.$

In Exercises 21–24, set up the initial simplex tableau, but do not solve the problem. (See Example 6.)

21. **Business**   A company is developing a new additive for gasoline. The additive is a mixture of three liquid ingredients: I, II, and III. For proper performance, the total amount of additive must be at least 10 ounces per barrel of gasoline. However, for safety reasons, the amount of

additive should not exceed 15 ounces per barrel of gasoline. At least 1/4 ounce of ingredient I must be used for every ounce of ingredient II, and at least 1 ounce of ingredient III must be used for every ounce of ingredient I. If the costs of I, II, and III are $.30, $.09, and $.27 per ounce, respectively, find the mixture of the three ingredients that produces the minimum cost of the additive. What is the minimum cost?

22. **Business**   A popular soft drink called Sugarlo, which is advertised as having a sugar content of no more than 10%, is blended from five ingredients, each of which has some sugar content. Water may also be added to dilute the mixture. The sugar content of the ingredients and their costs per gallon are given in the table.

	Ingredient					
	1	2	3	4	5	Water
Sugar content (%)	.28	.19	.43	.57	.22	0
Cost ($/gal.)	.48	.32	.53	.28	.43	.04

At least .01 of the content of Sugarlo must come from ingredient 3 or 4, .01 must come from ingredient 2 or 5, and .01 from ingredient 1 or 4. How much of each ingredient should be used in preparing at least 15,000 gallons of Sugarlo to minimize the cost? What is the minimum cost?

23. **Business**   The manufacturer of a popular personal computer has orders from two dealers. Dealer D_1 wants 32 computers, and dealer D_2 wants 20 computers. The manufacturer can fill the orders from either of two warehouses, W_1 or W_2. W_1 has 25 of the computers on hand, and W_2 has 30. The costs (in dollars) to ship one computer to each dealer from each warehouse are as follows.

		To	
		D_1	D_2
From	W_1	$14	$22
	W_2	$12	$10

How should the orders be filled to minimize shipping costs? What is the minimum cost?

24. **Natural Science**   Mark, who is ill, takes vitamin pills. Each day, he must have at least 16 units of vitamin A, 5 units of vitamin B_1, and 20 units of vitamin C. He can choose between pill 1, which costs 10¢ and contains 8 units of A, 1 of B_1, and 2 of C, and pill 2, which costs 20¢ and contains 2 units of A, 1 of B_1, and 7 of C. How many of each pill should he buy in order to minimize his cost?

Use the two-stage method to solve Exercises 23–30. (See Examples 5 and 6.)

25. Transportation Southwestern Oil supplies two distributors in the Northwest from two outlets: S_1 and S_2. Distributor D_1 needs at least 3000 barrels of oil, and distributor D_2 needs at least 5000 barrels. The two outlets can each furnish up to 5000 barrels of oil. The costs per barrel to ship the oil are given in the table.

		Distributors	
		D_1	**D_2**
Outlets	S_1	$30	$20
	S_2	$25	$22

There is also a shipping tax per barrel as given in the table below. Southwestern Oil is determined to spend no more than $40,000 on shipping tax.

	D_1	**D_2**
S_1	$2	$6
S_2	$5	$4

How should the oil be supplied to minimize shipping costs?

26. Transportation Change Exercise 25 so that the two outlets each furnish exactly 5000 barrels of oil, with everything else the same. Solve the problem as in Example 5.

27. Business Topgrade Turf lawn seed mixture contains three types of seeds: bluegrass, rye, and Bermuda. The costs per pound of the three types of seed are 12¢, 15¢, and 5¢, respectively. In each batch, there must be at least 20% bluegrass seed and the amount of Bermuda must be no more than two-thirds the amount of rye. To fill current orders, the company must make at least 5000 pounds of the mixture. How much of each kind of seed should be used to minimize cost?

28. Business Change Exercise 27 so that the company must make exactly 5000 pounds of the mixture. Solve the problem as in Example 5.

29. Finance A bank has set aside a maximum of $25 million for commercial and home loans. Every million dollars in commercial loans requires 2 lengthy application

forms, while every million dollars in home loans requires 3 lengthy application forms. The bank cannot process more than 72 application forms at this time. The bank's policy is to loan at least four times as much for home loans as for commercial loans. Because of prior commitments, at least $10 million will be used for these two types of loans. The bank earns 12% on home loans and 10% on commercial loans. What amount of money should be allotted for each type of loan to maximize the interest income?

30. Finance Virginia Keleske has decided to invest a $100,000 inheritance in government securities that earn 7% per year, municipal bonds that earn 6% per year, and mutual funds that earn an average of 10% per year. She will spend at least $40,000 on government securities, and she wants at least half the inheritance to go to bonds and mutual funds. Government securities have an initial fee of 2%, municipal bonds an initial fee of 1%, and mutual funds an initial fee of 3%. Virginia has $2400 available to pay initial fees. How much money should go into each type of investment to maximize the interest yet meet the constraints? What is the maximum interest she can earn?

31. Business A brewery produces regular beer and a lower carbohydrate "light" beer. Steady customers of the brewery buy 12 units of regular beer and 10 units of light beer. While setting up the brewery to produce the beers, the management decides to produce extra beer, beyond that needed to satisfy the steady customers. The cost per unit of regular beer is $36,000, and the cost per unit of light beer is $48,000. The number of units of light beer should not exceed twice the number of units of regular beer. At least 20 additional units of beer can be sold. How much of each type of beer should be made so as to minimize total production costs?

32. Business The chemistry department at a local college decides to stock at least 800 small test tubes and 500 large test tubes. It wants to buy at least 1500 test tubes to take advantage of a special price. Since the small tubes are broken twice as often as the large, the department will order at least twice as many small tubes as large. If the small test tubes cost 15¢ each and the large ones, made of a cheaper glass, cost 12¢ each, how many of each size should the department order to minimize cost?

Business *Use technology to solve the following exercises, whose initial tableaus were set up in Exercises 21–23.*

33. Exercise 21   **34.** Exercise 22

35. Exercise 23

Chapter *7* Summary

Key Terms and Symbols

7.1 linear inequality
graphs of linear inequalities
boundary
half plane
system of inequalities
region of feasible solutions
   (feasible region)

7.2 linear programming
objective function
constraints
corner point
bounded feasible region
unbounded feasible region
corner point theorem

7.3 applications of linear
   programming

7.4 standard maximum form
slack variable
simplex tableau
indicator
pivot and pivoting
final simplex tableau
basic variables
nonbasic variables
basic feasible solution

7.6 transpose of a matrix
dual
theorem of duality
shadow costs

7.7 surplus variable
basic variables
basic solution
basic feasible solution
Stage I
Stage II

Key Concepts

Graphing a Linear Inequality

Graph the boundary line as a solid line if the inequality includes "or equal," a dashed line otherwise. Shade the half plane for which the inequality is true. The graph of a system of inequalities, called the **region of feasible solutions,** includes all points that satisfy all the inequalities of the system at the same time.

Solving Linear Programming Problems

Graphically: Determine the objective function and all necessary constraints. Graph the region of feasible solutions. The maximum or minimum value will occur at one or more of the corner points of this region.

Simplex Method: Determine the objective function and all necessary constraints. Convert each constraint into an equation by adding slack variables. Set up the initial simplex tableau. Locate the most negative indicator. Form the quotients to determine the pivot. Use row operations to change the pivot to 1 and all other numbers in that column to 0. If the indicators are all positive or 0, this is the final tableau. If not, choose a new pivot and repeat the process until no indicators are negative. Read the solution from the final tableau. The optimum value of the objective function is the number in the lower right corner of the final tableau. For problems with **mixed constraints,** replace each equation constraint by a pair of inequality constraints. Then add slack variables and subtract surplus variables as needed to convert each constraint into an equation. In Stage I, use row operations to transform the matrix until the solution is feasible. In Stage II, use the simplex method as just described. For **minimization** problems, let the objective function be w and set $-w = z$. Then proceed as with mixed constraints.

Solving Minimization Problems with Duals

Find the dual maximization problem. Solve the dual with the simplex method. The minimum value of the objective function w is the maximum value of the dual objective function z. The optimal solution is found in the entries in the bottom row of the columns corresponding to the slack variables.

Chapter 7 Review Exercises

Graph each of the following linear inequalities.

1. $y \leq 3x + 2$ **2.** $2x - y \geq 6$

3. $3x + 4y \geq 12$ **4.** $y \leq 4$

Graph the solution of each of the following systems of inequalities.

5. $x + y \leq 6$ **6.** $4x + y \geq 8$
   $2x - y \geq 3$ $2x - 3y \leq 6$

7. $2 \leq x \leq 5$ **8.** $x + 2y \leq 4$
   $1 \leq y \leq 6$ $2x - 3y \leq 6$
   $x - y \leq 3$ $x \geq 0$
                                        $y \geq 0$

Set up a system of inequalities for each of the following problems; then graph the region of feasible solutions.

9. Business A bakery makes both cakes and cookies. Each batch of cakes requires 2 hours in the oven and 3 hours in the decorating room. Each batch of cookies needs $1\frac{1}{2}$ hours in the oven and 2/3 of an hour in the decorating room. The oven is available no more than 15 hours a day, while the decorating room can be used no more than 13 hours a day.

10. Business A company makes two kinds of pizza: special and basic. The special has toppings of cheese, tomatoes, and vegetables. Basic has just cheese and tomatoes. The company sells at least 6 units a day of the special pizza and 4 units a day of the basic. The cost of the vegetables (including tomatoes) is $2 per unit for special and $1 per unit for basic. No more than $32 per day can be spent on vegetables (including tomatoes). The cheese used for the special is $5 per unit, and the cheese for the basic is $4 per unit. The company can spend no more than $100 per day on cheese.

Use the given regions to find the maximum and minimum values of the objective function $z = 5x + y$.

11.

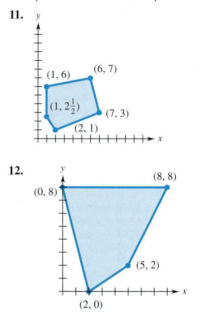

12.

Use the graphical method to solve Exercises 13–16.

13. Maximize $z = 2x + 4y$
     subject to $2x + 7y \leq 14$
                   $2x + 3y \leq 10$
                   $x \geq 0, y \geq 0.$

14. Find $x \geq 0$ and $y \geq 0$ such that

$$8x + 9y \geq 72$$
$$6x + 8y \geq 72$$

and $w = 10x + 2y$ is minimized.

15. Find $x \geq 0$ and $y \geq 0$ such that

$$x + y \leq 50$$
$$2x + y \geq 20$$
$$x + 2y \geq 30$$

and $w = 3x + 7y$ is minimized.

16. Maximize $z = -5x + 2y$
     subject to $3x + 2y \leq 12$
                   $5x + y \geq 5$
                   $x \geq 0, y \geq 0.$

17. Business How many batches of cakes and cookies should the bakery of Exercise 9 make in order to maximize profits if cookies produce a profit of $20 per batch and cakes produce a profit of $30 per batch?

18. Business How many units of each kind of pizza should the company of Exercise 10 make in order to maximize revenue if special pizza sells for $20 per unit and basic for $18 per unit?

For Exercises 19–22, (a) select appropriate variables, (b) write the objective function, (c) write the constraints as inequalities.

19. Business Roberta Hernandez sells three items—A, B, and C—in her gift shop. Each unit of A costs her $2 to buy, $1 to sell, and $2 to deliver. For each unit of B, the costs are $3, $2, and $2, respectively, and for each unit of C, the costs are $6, $2, and $4, respectively. The profit on A is $4, on B it is $3, and on C $3. How many of each item should she order to maximize her profit if she can spend $1200 to buy, $800 to sell, and $500 to deliver?

20. Business An investor is considering three types of investment: a high-risk venture into oil leases with a potential return of 15%, a medium-risk investment in bonds with a 9% return, and a relatively safe stock investment with a 5% return. He has $50,000 to invest. Because of the risk, he will limit his investment in oil leases and bonds to 30% and his investment in oil leases and stock to 50%. How much should he invest in each to maximize his return, assuming investment returns are as expected?

21. Business The Aged Wood Winery makes two white wines—Fruity and Crystal—from two kinds of grapes and sugar. The wines require the following amounts of each ingredient per gallon and produce a profit per gallon as shown in the table.

	Grape A (bushels)	Grape B (bushels)	Sugar (pounds)	Profit (dollars)
Fruity	2	2	2	12
Crystal	1	3	1	15

The winery has available 110 bushels of grape A, 125 bushels of grape B, and 90 pounds of sugar. How much of each wine should be made to maximize profit?

22. Business A company makes three sizes of plastic bags: 5 gallon, 10 gallon, and 20 gallon. The production time in hours for cutting, sealing, and packaging a unit of each size is as follows.

Size	Cutting	Sealing	Packaging
5 gallon	1	1	2
10 gallon	1.1	1.2	3
20 gallon	1.5	1.3	4

There are at most 8 hours available each day for each of the three operations. If the profit per unit is $1 for 5-gallon

bags, $.90 for 10-gallon bags, and $.95 for 20-gallon bags, how many of each size should be made per day to maximize the profit?

23. When is it necessary to use the simplex method rather than the graphical method?

24. What types of problems can be solved with the use of slack variables and surplus variables?

25. What kind of problem can be solved with the method of duals?

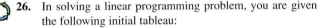 **26.** In solving a linear programming problem, you are given the following initial tableau:

$$\begin{bmatrix} 4 & 2 & 3 & 1 & 0 & 0 & | & 9 \\ 5 & 4 & 1 & 0 & 1 & 0 & | & 10 \\ -6 & -7 & -5 & 0 & 0 & 1 & | & 0 \end{bmatrix}.$$

  (a) What is the problem being solved?
  (b) If the 1 in row 1, column 4, were a -1 rather than a 1, how would it change your answer to part (a)?
  (c) After several steps of the simplex algorithm, the following tableau results:

$$\begin{bmatrix} 3 & 0 & 1 & 2 & -1 & 0 & | & 8 \\ 11 & 1 & 0 & -1 & 3 & 0 & | & 21 \\ 47 & 0 & 0 & 13 & 11 & 1 & | & 227 \end{bmatrix}.$$

What is the solution? (List only the values of the original variables and the objective function. Do not include slack or surplus variables.)
  (d) What is the dual of the problem you found in part (a)?
  (e) What is the solution of the dual you found in part (d)? (Do not perform any steps of the simplex algorithm; just examine the tableau given in part (c).)

For each of the following problems, (a) add slack variables and (b) set up the initial simplex tableau.

27. Maximize $z = 2x_1 + 9x_2$
   subject to $3x_1 + 5x_2 \le 47$
        $x_1 + x_2 \le 25$
        $5x_1 + 2x_2 \le 35$
        $2x_1 + x_2 \le 30$
        $x_1 \ge 0, x_2 \ge 0.$

28. Maximize $z = 15x_1 + 12x_2$
   subject to $2x_1 + 5x_2 \le 50$
        $x_1 + 3x_2 \le 25$
        $4x_1 + x_2 \le 18$
        $x_1 + x_2 \le 12$
        $x_1 \ge 0, x_2 \ge 0.$

29. Maximize $z = 5x_1 + 6x_2 + 3x_3$
   subject to $x_1 + x_2 + x_3 \le 100$
        $2x_1 + 3x_2 \le 500$
        $x_1 + 2x_3 \le 350$
        $x_1 \ge 0, x_2 \ge 0, x_3 \ge 0.$

30. Maximize $z = x_1 + 8x_2 + 2x_3$
subject to $x_1 + x_2 + x_3 \leq 90$
$2x_1 + 5x_2 + x_3 \leq 120$
$x_1 + 3x_2 \leq 80$
$x_1 \geq 0, x_2 \geq 0, x_3 \geq 0.$

For each of the following, use the simplex method to solve the maximization linear programming problems with initial tableaus as given.

31.
$$\begin{array}{ccccc} x_1 & x_2 & x_3 & x_4 & x_5 \\ \end{array}$$
$$\begin{bmatrix} 1 & 2 & 3 & 1 & 0 & | & 28 \\ 2 & 4 & 8 & 0 & 1 & | & 32 \\ -5 & -2 & -3 & 0 & 0 & | & 0 \end{bmatrix}$$

32.
$$\begin{array}{cccc} x_1 & x_2 & x_3 & x_4 \\ \end{array}$$
$$\begin{bmatrix} 2 & 1 & 1 & 0 & | & 10 \\ 9 & 3 & 0 & 1 & | & 15 \\ -2 & -3 & 0 & 0 & | & 0 \end{bmatrix}$$

33.
$$\begin{array}{cccccc} x_1 & x_2 & x_3 & x_4 & x_5 & x_6 \\ \end{array}$$
$$\begin{bmatrix} 1 & 2 & 2 & 1 & 0 & 0 & | & 50 \\ 4 & 24 & 0 & 0 & 1 & 0 & | & 20 \\ 1 & 0 & 2 & 0 & 0 & 1 & | & 15 \\ -5 & -3 & -2 & 0 & 0 & 0 & | & 0 \end{bmatrix}$$

34.
$$\begin{array}{ccccc} x_1 & x_2 & x_3 & x_4 & x_5 \\ \end{array}$$
$$\begin{bmatrix} 1 & -2 & 1 & 0 & 0 & | & 38 \\ 1 & -1 & 0 & 1 & 0 & | & 12 \\ 2 & 1 & 0 & 0 & 1 & | & 30 \\ -1 & -2 & 0 & 0 & 0 & | & 0 \end{bmatrix}$$

Convert the following problems into maximization problems without using duals.

35. Minimize $w = 18y_1 + 10y_2$
subject to $y_1 + y_2 \geq 17$
$5y_1 + 8y_2 \geq 42$
$y_1 \geq 0, y_2 \geq 0.$

36. Minimize $w = 12y_1 + 20y_2 - 8y_3$
subject to $y_1 + y_2 + 2y_3 \geq 48$
$y_1 + y_2 \geq 12$
$y_3 \geq 10$
$3y_1 + y_3 \geq 30$
$y_1 \geq 0, y_2 \geq 0, y_3 \geq 0.$

37. Minimize $w = 6y_1 - 3y_2 + 4y_3$
subject to $2y_1 + y_2 + y_3 \geq 112$
$y_1 + y_2 + y_3 \geq 80$
$y_1 + y_2 \geq 45$
$y_1 \geq 0, y_2 \geq 0, y_3 \geq 0.$

Use the simplex method to solve the following mixed constraint problems.

38. Maximize $z = 2x_1 + 4x_2$
subject to $3x_1 + 2x_2 \leq 12$
$5x_1 + x_2 \geq 5$
$x_1 \geq 0, x_2 \geq 0.$

39. Minimize $w = 4y_1 - 8y_2$
subject to $y_1 + y_2 \leq 50$
$2y_1 - 4y_2 \geq 20$
$y_1 - y_2 \leq 22$
$y_1 \geq 0, y_2 \geq 0.$

The following tableaus are the final tableaus of minimization problems solved by letting $w = -z$. Give the solution and the minimum value of the objective function for each problem.

40.
$$\begin{bmatrix} 0 & 1 & 0 & 2 & 5 & 0 & | & 17 \\ 0 & 0 & 1 & 3 & 1 & 1 & | & 25 \\ 1 & 0 & 0 & 4 & 2 & \frac{1}{2} & | & 8 \\ 0 & 0 & 0 & 2 & 5 & 0 & | & -427 \end{bmatrix}$$

41.
$$\begin{bmatrix} 0 & 0 & 2 & 1 & 0 & 6 & 6 & | & 92 \\ 1 & 0 & 3 & 0 & 0 & 0 & 2 & | & 47 \\ 0 & 1 & 0 & 0 & 0 & 1 & 0 & | & 68 \\ 0 & 0 & 4 & 0 & 1 & 0 & 3 & | & 35 \\ 0 & 0 & 5 & 0 & 0 & 2 & 9 & | & -1957 \end{bmatrix}$$

The tableaus in Exercises 42–44 are the final tableaus of minimization problems solved by the method of duals. State the solution and the minimum value of the objective function for each problem.

42.
$$\begin{bmatrix} 1 & 0 & 0 & 3 & 1 & 2 & | & 12 \\ 0 & 0 & 1 & 4 & 5 & 3 & | & 5 \\ 0 & 1 & 0 & -2 & 7 & -6 & | & 8 \\ 0 & 0 & 0 & 5 & 7 & 3 & | & 172 \end{bmatrix}$$

43.
$$\begin{bmatrix} 0 & 0 & 1 & 6 & 3 & 1 & | & 2 \\ 1 & 0 & 0 & 4 & -2 & 2 & | & 8 \\ 0 & 1 & 0 & 10 & 7 & 0 & | & 12 \\ 0 & 0 & 0 & 9 & 5 & 8 & | & 62 \end{bmatrix}$$

44.
$$\begin{bmatrix} 1 & 0 & 7 & -1 & | & 100 \\ 0 & 1 & 1 & 3 & | & 27 \\ 0 & 0 & 7 & 2 & | & 640 \end{bmatrix}$$

Business *Solve the following maximization problems, which were begun in Exercises 19–22.*

45. Exercise 19

46. Exercise 20

47. Exercise 21

48. Exercise 22

Business *Solve the following minimization problems.*

49. Cauchy Canners produces canned corn, beans, and carrots. Demand for vegetables requires the company to produce at least 1000 cases per month. Based on past sales, it should produce at least twice as many cases of corn as of beans and at least 340 cases of carrots. It costs $10 to produce a case of corn, $15 to produce a case of beans, and $25 to produce a case of carrots. How many cases of each vegetable should be produced to minimize costs? What is the minimum cost?

50. A contractor builds boathouses in two basic models: the Atlantic and the Pacific. Each Atlantic model requires 1000 feet of framing lumber, 3000 cubic feet of concrete, and $2000 for advertising. Each Pacific model requires 2000 feet of framing lumber, 3000 cubic feet of concrete, and $3000 for advertising. Contracts call for using at least 8000 feet of framing lumber, 18,000 cubic feet of concrete, and $15,000 worth of advertising. If the total spent on each Atlantic model is $3000 and the total spent on each Pacific model is $4000, how many of each model should be built to minimize costs?

Business *Solve these mixed constraint problems.*

51. Brand X Cannery produces canned whole tomatoes and tomato sauce. This season, the company has available 3,000,000 kilograms of tomatoes for these two products. To meet the demands of regular customers, it must produce at least 80,000 kilograms of sauce and 800,000 kilograms of whole tomatoes. The cost per kilogram is $4 to produce canned whole tomatoes and $3.25 to produce tomato sauce. Labor agreements require that at least 110,000 person-hours be used. Each 1-kilogram can of sauce requires 3 minutes for one worker to produce, and each 1-kilogram can of whole tomatoes requires 6 minutes for one worker. How many kilograms of tomatoes should Brand X use for each product to minimize cost? (For simplicity, assume that the production of y_1 kilograms of canned whole tomatoes and y_2 kilograms of tomato sauce requires $y_1 + y_2$ kilograms of tomatoes.)

52. A steel company produces two types of alloys. A run of type I requires 3000 pounds of molybdenum and 2000 tons of iron ore pellets, as well as $2000 in advertising. A run of type II requires 3000 pounds of molybdenum and 1000 tons of iron ore pellets, as well as $3000 in advertising. Total costs are $15,000 on a run of type I and $6000 on a run of type II. The company has on hand 18,000 pounds of molybdenum and 7000 tons of iron ore pellets and wants to use all of it. It plans to spend at least $14,000 on advertising. How much of each type should be produced to minimize cost? What is the minimum cost?

Case 7

Cooking with Linear Programming

Constructing a nutritious recipe can be a difficult task. The recipe must produce food that tastes good, and it must also balance the nutrients that each ingredient brings to the dish. This balancing of nutrients is very important in several diet plans that are currently popular. Many of these plans restrict the intake of certain nutrients (usually fat) while allowing for large amounts of other nutrients (protein and carbohydrates are popular choices). The number of calories in the dish is also often minimized. Linear programming can be used to help create recipes that balance nutrients.

In order to develop solutions to this type of problem, we will need to have nutritional data for the ingredients in our recipes. This data can be found in the U.S. Department of Agriculture's USDA Nutrient Database for Standard Reference, available at www.na1.udsa.gov/fnic/foodcomp/data/sr13.html. This database contains the nutrient levels for hundreds of basic foods. The nutrient levels are given per 100 grams of food. Unfortunately, grams are not often used in recipes; instead, kitchen measures like cups, tablespoons, and fractions of vegetables are used. Table 1 shows the conversion factors from grams to more familiar kitchen units and also gives serving sizes for various food.

Table 1   Serving Sizes of Various Food

Food	Serving Size
Beef	6 oz = 170 g
Egg	1 egg = 61 g
Feta Cheese	1/4 cup = 38 g
Lettuce	1/2 cup = 28 g
Milk	1 cup = 244 g
Oil	1 Tbsp = 13.5 g
Onion	1 onion = 110 g
Salad Dressing	1 cup = 250 g
Soy Sauce	1 Tbsp = 18 g
Spinach	1 cup = 180 g
Tomato	1 tomato = 123 g

Consider creating a recipe for a spinach omelet from eggs, milk, vegetable oil, and spinach. The nutrients of interest will be protein, fat, and carbohydrates. Calories will also be monitored. The amounts of the nutrients and calories for these ingredients are given in Table 2.

Table 2   Nutritional Values per 100 grams of Food

Nutrient (units)	Eggs	Milk	Oil	Spinach
Calories (kcal)	152	61.44	884	23
Protein (g)	10.33	3.29	0	2.9
Fat (g)	11.44	3.34	100	.26
Carbohydrates (g)	1.04	4.66	0	3.75

Let x_1 be the number of 100-gram units of eggs to use in the recipe, x_2 be the number of 100-gram units of milk, x_3 be the number of 100-gram units of oil, and x_4 be the number of 100-gram units of spinach. We will want to minimize the amount of calories in the dish while providing at least 4 grams of protein, 15 grams of carbohydrates, and 20 grams of fat. The cooking technique specifies that at least 1/8 of a cup of milk (30.5 grams) must be used in the recipe. We should thus minimize the objective function (using 100-gram units of food):

$$z = 152x_1 + 61.44x_2 + 884x_3 + 23x_4$$

subject to

$$10.33x_1 + 3.29x_2 + 0x_3 + 2.90x_4 \geq 4$$
$$11.44x_1 + 3.34x_2 + 100x_3 + .26x_4 \geq 20$$
$$1.04x_1 + 4.66x_2 + 0x_3 + 3.75x_4 \geq 15$$
$$0x_1 + 1x_2 + 0x_3 + 0x_4 \geq .305.$$

Of course, all variables are subject to nonnegativity constraints:

$$x_1 \geq 0, \qquad x_2 \geq 0, \qquad x_3 \geq 0, \qquad x_4 \geq 0.$$

Using a graphing calculator or a computer with linear programming software, we get the following solution:

$$x_1 = 1.2600, \qquad x_2 = .3050, \qquad x_3 = .0448, \qquad x_4 = .338.$$

This recipe produces an omelet with 257.63 calories. The amounts of each ingredient are 126 grams of eggs, 30.5 grams of milk, 4.48 grams of oil, and 33.8 grams of spinach. Converting to kitchen units using Table 1, we find the recipe to be approximately 2 eggs, 1/8 cup milk, 1 teaspoon oil, and 1/4 cup spinach.

◆ # Exercises

1. Consider preparing a high-carbohydrate Greek salad using feta cheese, lettuce, salad dressing, and tomato. The amount of carbohydrates in the salad should be maximized. In addition, the salad should have less than 260 calories, over 210 milligrams of calcium, and over 6 grams of protein. The salad should also weigh less than 400 grams and be dressed with at least 2 tablespoons (1/8 cup) of salad dressing. The amounts of the nutrients and calories for these ingredients are given in Table 3.

Table 3   Nutritional Values per 100 g of Food

Nutrient (units)	Feta Cheese	Lettuce	Salad Dressing	Tomato
Calories (kcal)	263	14	448.8	21
Calcium (mg)	492.5	36	0	5
Protein (g)	10.33	1.62	0	.85
Carbohydrates (g)	4.09	2.37	2.5	4.64

Use linear programming to find the number of 100-gram units of each ingredient in such a Greek salad, and convert to kitchen units by using Table 1.

2. Consider preparing a stir-fry using beef, oil, onion, and soy sauce. A low-calorie stir-fry is desired, which also contains less than 10 grams of carbohydrates, more than 50 grams of protein, and more than 3.5 grams of vitamin C. In order for the wok to function correctly, at least one teaspoon (or 4.5 grams) of oil must be used in the recipe. The amounts of the nutrients and calories for these ingredients are given in Table 4.

Table 4   Nutritional Values per 100 g of Foodstuff

Nutrient (units)	Beef	Oil	Onion	Soy Sauce
Calories (kcal)	215	884	38	60
Protein (g)	26	0	1.16	10.51
Carbohydrates (g)	0	1	8.63	5.57
Vitamin C (g)	0	0	6.4	0

Use linear programming to find the number of 100-gram units of each ingredient to be used in the stir-fry, and convert to kitchen units by using Table 1.

Chapter 8

Sets and Probability

We often use the relative frequency of an event in a survey to estimate unknown probabilities. For example, we can estimate the number of times per week Americans attend religious services. See Exercise 31 on page 468. Other applications of probability are in the health and social sciences, as well as in business. Examples include estimating the probability of having a low-birth-weight infant, earning particular annual incomes, or being selected for a jury. See Exercises 55 and 56 on page 478.

Federal officials cannot predict exactly how traffic deaths are affected by the trends toward fewer drunken drivers and the increased use of seat belts. Economists cannot tell exactly how stricter federal regulations on bank loans affect the U.S. economy. The number of traffic deaths and the growth of the economy are subject to many factors that cannot be predicted precisely.

Probability theory enables us to deal with uncertainty. The basic concepts of probability are discussed in this chapter, and applications of probability are discussed in the next chapter. Sets and set operations are the basic tools for the study of probability, so we begin with them.

8.1 Sets

Think of a set as a well-defined collection of objects. A set of coins might include one of each type of coin now put out by the U.S. government. Another set might be made up of all the students in your English class. By contrast, a collection of young adults does not constitute a set unless the designation "young adult" is clearly defined. For example, this set might be defined as those aged 18–29.

In mathematics, sets are often made up of numbers. The set consisting of the numbers 3, 4, and 5 is written

$$\{3, 4, 5\},$$

where **set braces,** { }, are used to enclose the numbers belonging to the set. The numbers, 3, 4, and 5 are called the **elements** or **members** of this set. To show that 4 is an element of the set $\{3, 4, 5\}$, we use the symbol $\in$ and write

$$4 \in \{3, 4, 5\},$$

read "4 is an element of the set containing 3, 4, and 5."

Also, $5 \in \{3, 4, 5\}$. Place a slash through the symbol $\in$ to show that 8 is *not* an element of this set.

$$8 \notin \{3, 4, 5\}$$

This is read "8 is not an element of the set $\{3, 4, 5\}$."

Sets are often named with capital letters, so that if

$$B = \{5, 6, 7\},$$

then, for example, $6 \in B$ and $10 \notin B$. ◁ 1 ▷

Sometimes a set has no elements. Some examples are the set of female presidents of the United States in the period 1788–2008, the set of counting numbers less than 1, and the set of men more than 10 feet tall. A set with no elements is called the **empty set.** The symbol $\emptyset$ is used to represent the empty set.

> **CAUTION** Be careful to distinguish between the symbols 0, $\emptyset$, and $\{0\}$. The symbol 0 represents a *number;* $\emptyset$ represents a *set* with no elements; and $\{0\}$ represents a *set* with one element, the number 0. Do not confuse the empty set symbol $\emptyset$ with the zero $\emptyset$ on a computer screen or printout.

Two sets are **equal** if they contain exactly the same elements. The sets $\{5, 6, 7\}$, $\{7, 6, 5\}$, and $\{6, 5, 7\}$ all contain exactly the same elements and are equal. In symbols,

◇ 1 ◇

Write *true* or *false*.

(a) $9 \in \{8, 4, -3, -9, 6\}$

(b) $4 \notin \{3, 9, 7\}$

(c) If $M = \{0, 1, 2, 3, 4\}$, then $0 \in M$.

Answers:

(a) False

(b) True

(c) True

$$\{5, 6, 7\} = \{7, 6, 5\} = \{6, 5, 7\}.$$

This means that the ordering of the elements in a set is unimportant. Sets that do not contain exactly the same elements are *not equal*. For example, the sets $\{5, 6, 7\}$ and $\{5, 6, 7, 8\}$ do not contain exactly the same elements and are not equal. We show this by writing

$$\{5, 6, 7\} \neq \{5, 6, 7, 8\}.$$

Sometimes we describe a set by a common property of its elements rather than by a list of its elements. This common property can be expressed with **set-builder notation;** for example,

$$\{x | x \text{ has property } P\}$$

(read "the set of all elements x such that x has property P") represents the set of all elements x having some property P.

▼ **EXAMPLE 1** List the elements belonging to each of the following sets.

(a) $\{x | x \text{ is a natural number less than 5}\}$

Solution The natural numbers less than 5 make up the set $\{1, 2, 3, 4\}$.

(b) $\{x | x \text{ is a state that borders Florida}\} = \{\text{Alabama, Georgia}\}$ ▲ ②

The **universal set** in a particular discussion is a set that contains all of the objects being discussed. In grade-school arithmetic, for example, the set of whole numbers might be the universal set, whereas in a college calculus class the universal set might be the set of all real numbers. When it is necessary to consider the universal set being used, it will be clearly specified or easily understood from the context of the problem.

Sometimes, every element of one set also belongs to another set. For example, if

$$A = \{3, 4, 5, 6\}$$

and

$$B = \{2, 3, 4, 5, 6, 7, 8\},$$

then every element of A is also an element of B. This is an example of the following definition.

A set A is a **subset** of a set B (written $A \subseteq B$) provided that every element of A is also an element of B.

▼ **EXAMPLE 2** Decide whether $M \subseteq N$.

(a) M is the set of all small businesses with fewer than 20 employees. N is the set of all businesses.

Solution Each business with fewer than 20 employees is also a business, so $M \subseteq N$.

②

List the elements in the following sets.

(a) $\{x | x \text{ is a counting number more than 5 and less than 8}\}$

(b) $\{x | x \text{ is an integer,} -3 < x \leq 1\}$

Answers:

(a) $\{6, 7\}$

(b) $\{-2, -1, 0, 1\}$

(b) *M* is the set of all fourth-grade students in a school at the end of the school year, and *N* is the set of all nine-year-old students in the school at the end of the school year.

Solution By the end of the school year, some fourth-grade students are ten years old, so there are elements in *M* that are not in *N*. Thus, *M* is not a subset of *N*, written $M \not\subseteq N$. ▲

Every set *A* is a subset of itself, because the statement "every element of *A* is also an element of *A*" is always true. It is also true that the empty set is a subset of every set.*

For any set *A*,

$$\emptyset \subseteq A \quad \text{and} \quad A \subseteq A.$$

A set *A* is said to be a **proper subset** of a set *B* (written $A \subset B$) if every element of *A* is an element of *B*, but *B* contains at least one element that is not a member of *A*.

▼ **EXAMPLE 3** Decide whether $E \subset F$.

(a) $E = \{2, 4, 6, 8\}$ and $F = \{1, 2, 3, 4, 5, 6, 7, 8, 9, 10\}$

Solution Since each element of *E* is an element of *F* and *F* contains several elements not in *E*, $E \subset F$.

(b) *E* is the set of registered voters in Texas. *F* is the set of adults aged 18 years or older.

Solution To register to vote, one must be at least 18 years old. Not all adults at least 18 years old, however, are registered. Thus, every element of *E* is contained in *F* and *F* contains elements not in *E*. Therefore, $E \subset F$.

(c) *E* is the set of diet soda drinks. *F* is the set of diet soda drinks sweetened with Nutrasweet®.

Solution Some diet soda drinks are sweetened with the sugar substitute Splenda®. In this case, *E* is not a proper subset of *F* (written $E \not\subset F$) nor is it a subset ($E \not\subseteq F$). ▲ ③

▼ **EXAMPLE 4** List all possible subsets for each of the following sets.

(a) $\{7, 8\}$

Solution There are 4 subsets of $\{7, 8\}$:

$$\emptyset, \quad \{7\}, \quad \{8\}, \quad \{7, 8\}.$$

(b) $\{a, b, c\}$

Solution There are 8 subsets of $\{a, b, c\}$:

$$\emptyset, \quad \{a\}, \quad \{b\}, \quad \{c\}, \quad \{a, b\}, \quad \{a, c\}, \quad \{b, c\}, \quad \{a, b, c\}.$$

③

Write *true* or *false*.

(a) $\{3, 4, 5\} \subseteq \{2, 3, 4, 6\}$

(b) $\{x | x$ is an automobile$\} \subseteq \{x | x$ is a motor vehicle$\}$

(c) $\{3, 6, 9, 10\} \subset \{3, 9, 11, 13\}$

(d) $\{x | x$ is a Great Lake of the United States$\} \subset \{$Ontario, Erie, Michigan, Superior, Huron$\}$

Answers:

(a) False

(b) True

(c) False

(d) False

---

*This fact is not intuitively obvious to most people. If you wish, you can think of it as a convention that we agree to adopt in order to simplify the statements of several results later.

A good way to find the subsets of $\{7, 8\}$ and the subsets of $\{a, b, c\}$ in Example 4 is to use a **tree diagram**—a systematic way of listing all the subsets of a given set. Figures 8.1(a) and (b) show tree diagrams for finding the subsets of $\{7, 8\}$ and $\{a, b, c\}$. ▲ ④

⟨4⟩

List all subsets of $\{w, x, y, z\}$.

Answer:

$\emptyset, \{w\}, \{x\}, \{y\}, \{z\}, \{w, x\},$
$\{w, y\}, \{w, z\}, \{x, y\}, \{x, z\},$
$\{y, z\}, \{w, x, y\}, \{w, x, z\},$
$\{w, y, z\}, \{x, y, z\}, \{w, x, y, z\}$

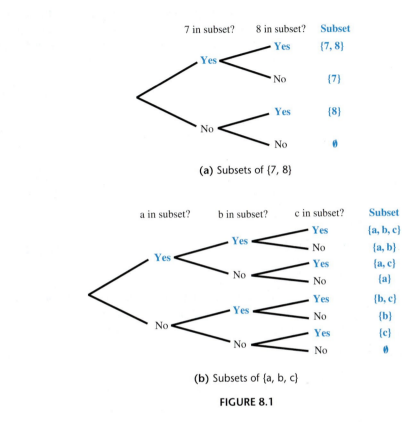

(a) Subsets of $\{7, 8\}$

(b) Subsets of $\{a, b, c\}$

FIGURE 8.1

By using the fact that there are two possibilities for each element (either it is in the subset or it is not), we have found that a set with 2 elements has 4 $(= 2^2)$ subsets and a set with 3 elements has 8 $(= 2^3)$ subsets. Similar arguments work for any finite set and lead to the following conclusion.

A set of n distinct elements has 2^n subsets.

▼ **EXAMPLE 5**   Find the number of subsets for each of the following sets.

(a) $\{3, 4, 5, 6, 7\}$

Solution Since this set has 5 elements, it has 2^5, or 32, subsets.

(b) $\{x \mid x \text{ is a day of the week}\}$

Solution This set has 7 elements and therefore has $2^7 = 128$ subsets.

5

Find the number of subsets for each of the following sets.

(a)
$\{x | x$ is a season of the year$\}$

(b)
$\{-6, -5, -4, -3, -2, -1, 0\}$

(c) $\{6\}$

Answers:

(a) 16

(b) 128

(c) 2

6

Refer to sets $A, B, C,$ and U in the diagram.

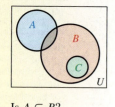

(a) Is $A \subseteq B$?

(b) Is $C \subseteq B$?

(c) Is $C \subseteq U$?

(d) Is $\emptyset \subseteq A$?

Answers:

(a) No

(b) Yes

(c) Yes

(d) Yes

(c) $\emptyset$

Solution Since the empty set has 0 elements, it has $2^0 = 1$ subset, $\emptyset$ itself. ▲

Venn diagrams are sometimes used to illustrate relationships among sets. The Venn diagram in Figure 8.2 shows a set A, which is a subset of a set B, because A is entirely in B. (The areas of the regions are not meant to be proportional to the sizes of the corresponding sets.) The rectangle represents the universal set U.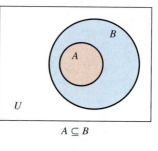

$A \subseteq B$

FIGURE 8.2

Some sets have infinitely many elements. We often use the notation ... to indicate such sets. One example of an infinite set is the set of natural numbers $\{1, 2, 3, 4, \dots\}$. Another infinite set is the set of integers $\{\dots, -3, -2, -1, 0, 1, 2, 3, \dots\}$.

Operations on Sets

Given a set A and a universal set U, the set of all elements of U that do *not* belong to A is called the **complement** of set A. For example, if A is the set of all the female students in your class and U is the set of all students in the class, then the complement of A would be the set of all male students in the class. The complement of set A is written A' (read "A-prime"). The Venn diagram of Figure 8.3 shows a set B. Its complment, B', is shown in color.

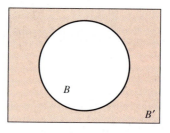

FIGURE 8.3

Some textbooks use $\overline{A}$ to denote the complement of A. This notation conveys the same meaning as A'.

▼ **EXAMPLE 6** Let $U = \{1, 2, 3, 4, 5, 6, 7\}$, $A = \{1, 3, 5, 7\}$, and $B = \{3, 4, 6\}$. Find the following sets.

(a) A'

Solution Set A' contains the elements of U that are not in A.

$$A' = \{2, 4, 6\}$$

(b) $B' = \{1, 2, 5, 7\}$

(c) $\emptyset' = U$ and $U' = \emptyset$ ▲ ⟨7⟩

Given two sets A and B, the set of all elements belonging to *both* set A and set B is called the **intersection** of the two sets, written $A \cap B$. For example, the elements that belong to both $A = \{1, 2, 4, 5, 7\}$ and $B = \{2, 4, 5, 7, 9, 11\}$ are 2, 4, 5, and 7, so

$$\begin{aligned} A \text{ and } B = A \cap B \\ = \{1, 2, 4, 5, 7\} \cap \{2, 4, 5, 7, 9, 11\} \\ = \{2, 4, 5, 7\}. \end{aligned}$$

The Venn diagram of Figure 8.4 shows two sets A and B, with their intersection, $A \cap B$, shown in color.

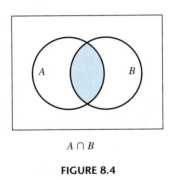

$A \cap B$

FIGURE 8.4

⟨7⟩

Let $U = \{a, b, c, d, e, f, g\}$, with $K = \{c, d, f, g\}$ and $R = \{a, c, d, e, g\}$. Find

(a) K';

(b) R'.

Answers:

(a) $\{a, b, e\}$

(b) $\{b, f\}$

⟨8⟩

Find the following.

(a) $\{1, 2, 3, 4\} \cap \{3, 5, 7, 9\}$

(b) Suppose set K is the set of all blue-eyed blondes in a class and J is the set of all blue-eyed brunettes in the class. Let $P = \{x \mid x$ is a brown-eyed redhead$\}$. If the class has only blondes or brunettes, find $K \cap P$.

Answers:

(a) $\{3\}$

(b) $\emptyset$

▼ **EXAMPLE 7** **(a)** $\{9, 15, 25, 36\} \cap \{15, 20, 25, 30, 35\} = \{15, 25\}$
The elements 15 and 25 are the only ones belonging to both sets.

(b) $\{x \mid x$ is a teen-ager$\} \cap \{x \mid x$ is a senior citizen$\}$ is an empty set. ▲ ⟨8⟩

Two sets that have no elements in common are called **disjoint sets.** For example, there are no elements common to both $\{50, 51, 54\}$ and $\{52, 53, 55, 56\}$, so these two sets are disjoint, and

$$\{50, 51, 54\} \cap \{52, 53, 55, 56\} = \emptyset.$$

The result of this example can be generalized as follows.

For any sets A and B,

if A and B are disjoint sets, then $A \cap B = \emptyset$.

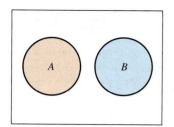

A and B are disjoint sets.

FIGURE 8.5

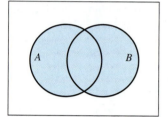

$A \cup B$

FIGURE 8.6

⑨

Find the following.

(a) $\{a, b, c\} \cup \{a, c, e\}$

(b) Describe $K \cup J$ in words for the sets given in side problem 8(b).

Answers:

(a) $\{a, b, c, e\}$

(b) All members of the class who have blue eyes

Figure 8.5 is a Venn diagram of disjoint sets.

The set of all elements belonging to set A or to set B, or to both sets, is called the **union** of the two sets, written $A \cup B$. For example, for sets A = $\{1, 3, 5\}$ and B = $\{3, 5, 7, 9\}$,

$$A \text{ or } B = A \cup B$$
$$= \{1, 3, 5\} \cup \{3, 5, 7, 9\}$$
$$= \{1, 3, 5, 7, 9\}.$$

The Venn diagram of Figure 8.6 shows two sets A and B, with their union, $A \cup B$, shown in color.

▼ **EXAMPLE 8** **(a)** Find the union of $\{1, 2, 5, 9, 14\}$ and $\{1, 3, 4, 8\}$. Begin by listing the elements of the first set, $\{1, 2, 5, 9, 14\}$. Then include any elements from the second set *that are not already listed*. Doing this gives

$$\{1, 2, 5, 9, 14\} \cup \{1, 3, 4, 8\} = \{1, 2, 3, 4, 5, 8, 9, 14\}.$$

(b) $\{$terriers, spaniels, chows, dalmatians$\} \cup \{$spaniels, collies, bulldogs$\} = \{$terriers, spaniels, chows, dalmatians, collies, bulldogs$\}$ ▲ ⑨

Finding the complement of a set, the intersection of two sets, or the union of two sets is an example of a *set operation*. The set operations are summarized next.

Operations on Sets

Let A and B be any sets, with U the universal set. Then

the **complement** of A, written A', is

$$A' = \{x | x \notin A \text{ and } x \in U\};$$

the **intersection** of A and B is

$$A \cap B = \{x | x \in A \text{ and } x \in B\};$$

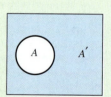

the **union** of A and B is

$$A \cup B = \{x | x \in A \text{ or } x \in B \text{ or both}\}.$$

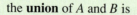

CAUTION As shown in the preceding definitions, an element is in the intersection of sets A and B if it is in *both* A and B at the same time, but an element is in the union of sets A and B if it is in either set A or set B or in both sets A and B.

▼ **EXAMPLE 9** The following table gives the 52-week high and low prices, the last price, and the change from the day before for five stocks on a recent day:*

Stock	High	Low	Last	Change
Allstate	59.02	43.44	59.02	+ .53
General Motors	48.27	24.67	31.24	−.15
Goodyear	16.09	8.46	14.34	−.07
UPS	89.11	66.65	73.64	−.01
Wendy's	45.89	31.74	44.59	−.40

Let the universal set U consist of the five stocks listed in the table. Let A contain all stocks with a high price greater than $50, B all stocks with a last price between $20 and $45, and C all stocks with a negative change. Find the following.

(a) B'

> **Solution** Set B consists of General Motors and Wendy's. Set B' contains all the listed stocks that are not in set B, so
>
> $$B' = \{\text{Allstate, Goodyear, UPS}\}.$$

(b) $A \cap C$

> **Solution** Set A consists of Allstate and UPS, and set C consists of General Motors, Goodyear, UPS, and Wendy's. Hence,
>
> $$A \cap C = \{\text{UPS}\}.$$

(c) $A \cup B$

> **Solution** $A \cup B = \{\text{Allstate, UPS, General Motors, Wendy's}\}$ ▲ ⟨10⟩

*Cleveland Plain Dealer, June 3, 2005.

⟨10⟩

Find all stocks

(a) with a last price between $20 and $45 and with a negative price change.

(b) with a last price between $20 and $45 or with a negative price change.

Answers:

(a) {General Motors, Wendy's}

(b) {General Motors, Goodyear, UPS, Wendy's}

◆8.1 Exercises

Write true or false for each statement.

1. $3 \in \{2, 5, 7, 9, 10\}$

2. $6 \in \{-2, 6, 9, 5\}$

3. $9 \notin \{2, 1, 5, 8\}$

4. $3 \notin \{7, 6, 5, 4\}$

5. $\{2, 5, 8, 9\} = \{2, 5, 9, 8\}$

6. $\{3, 7, 12, 14\} = \{3, 7, 12, 14, 0\}$

7. {all whole numbers greater than 7 and less than 10} = {8, 9}

8. {all counting numbers not greater than 3} = {0, 1, 2}

9. $\{x \mid x$ is an odd integer, $6 \le x \le 18\} = \{7, 9, 11, 15, 17\}$

10. $\{x \mid x$ is a vowel$\} = \{a, e, i, o, u\}$

11. The elements of a set may be sets themselves, as in $\{1, \{1, 3\}, \{2\}, 4\}$. Explain why the set $\{\emptyset\}$ is not the same set as $\{0\}$.

12. What is set-builder notation? Give an example.

Let $A = \{-3, 0, 3\}$, $B = \{-2, -1, 0, 1, 2\}$, $C = \{-3, -1\}$, $D = \{0\}$, $E = \{-2\}$, *and* $U = \{-3, -2, -1, 0, 1, 2, 3\}$. *Insert* $\subseteq$ *or* $\nsubseteq$ *to make the following statements true. (See Example 2.)*

13. A _____ U

14. E _____ A

15. A _____ E

16. B _____ C

17. $\emptyset$ _____ A

18. $\{0, 2\}$ _____ D

19. D _____ B

20. A _____ C

Find the number of subsets for each set. (See Example 5.)

21. $\{A, B, C\}$

22. {red, yellow, blue, black, white}

23. $\{x \mid x$ is an integer strictly between 0 and 8$\}$

24. $\{x \mid x$ is a whole number less than 4$\}$

Find the complement of each set. (See Example 6.)

25. The set in Exercise 23 if U is the set of all integers.

26. The set in Exercise 24 if U is the set of all whole numbers.

27. Describe the intersection and union of sets. How do they differ?

Insert $\cap$ *or* $\cup$ *to make each statement true. (See Examples 7 and 8.)*

28. $\{5, 7, 9, 19\}$ _____ $\{7, 9, 11, 15\} = \{7, 9\}$

29. $\{8, 11, 15\}$ _____ $\{8, 11, 19, 20\} = \{8, 11\}$

30. $\{2, 1, 7\}$ _____ $\{1, 5, 9\} = \{1\}$

31. $\{6, 12, 14, 16\}$ _____ $\{6, 14, 19\} = \{6, 14\}$

32. $\{3, 5, 9, 10\}$ _____ $\emptyset = \emptyset$

33. $\{3, 5, 9, 10\}$ _____ $\emptyset = \{3, 5, 9, 10\}$

34. $\{1, 2, 4\}$ _____ $\{1, 2, 4\} = \{1, 2, 4\}$

35. $\{1, 2, 4\}$ _____ $\{1, 2\} = \{1, 2, 4\}$

36. Is it possible for two nonempty sets to have the same intersection and union? If so, give an example.

Let $U = \{1, 2, 3, 4, 5, 7, 9, 11, 15\}$, $X = \{2, 3, 4, 5, 6\}$, $Y = \{3, 5, 7, 9\}$, *and* $Z = \{2, 4, 11, 15\}$.

List the members of each of the following sets, using set braces. (See Examples 6–8.)

37. $X \cap Y$

38. $X \cup Y$

39. X'

40. Y'

41. $X' \cap Y'$

42. $X' \cap Z$

43. $X \cup (Y \cap Z)$

44. $Y \cap (X \cup Z)$

Let $U = $ {all students in this school}, $M = $ {all students taking this course}, $N = $ {all students taking accounting}, $P = $ {all students taking philosophy}.

Describe each of the following sets in words.

45. M'

46. $M \cup N$

47. $N \cap P$

48. $N' \cap P'$

49. Refer to the sets listed in the directions for Exercises 13–20. Which pairs of sets are disjoint?

50. Refer to the sets listed in the directions for Exercises 37–44. Which pairs of sets are disjoint?

Refer to Example 9 in the text. Describe each of the sets in Exercises 51–54 in words; then list the elements of each set.

51. C'

52. $A \cap B$

53. $(A \cup B)'$

54. $(B \cap C)'$

Business *An electronics store classifies credit applicants by sex, marital status, and employment status. Let the universal set be the set of all applicants, M be the set of male applicants, S be the set of single applicants, and E be the set of employed applicants. Describe the following sets in words.*

55. $M \cap E$

56. $M' \cap S$

57. $M' \cup S'$

Business *The U.S. advertising volume (in millions of dollars) spent by specific media in 2002 and 2003 is shown in the table.**

Medium	2002	2003
Newspapers	44,102	44,939
Magazines	10,995	11,435
Broadcast Television	42,068	41,932
Cable Television	16,297	18,814
Radio	18,887	19,100
Direct Mail	46,067	48,370
Internet	4,883	5,650

List the elements of each set.

58. The set of all media that collected more than $40,000 million in both 2002 and 2003.

*Newspaper Association of America, www.naa.org.

59. The set of all media that collected less than $11,000 million in 2002 or 2003 (or both years).

60. The dollar amounts spent for magazine ads in each year shown.

Business *The top seven cable television providers as of December 2004 are listed here.* Use this information for Exercises 61–66.*

Rank	Cable Provider	Subscribers
1	Comcast Cable Communications	21,548,000
2	Time Warner Cable	10,884,000
3	Cox Communications	6,287,400
4	Charter Communications	5,992,000
5	Adelphia Communications	5,211,600
6	Cablevision Systems Corporation	2,963,000
7	Bright House Networks	2,167,000

List the elements of the following sets.

61. *F*, the set of cable providers with more than 7 million subscribers.

*National Cable Television Association, www.ncta.com.

62. *G*, the set of cable providers with between 3 and 6 million subscribers.

63. *H*, the set of cable providers with over 2.9 millions subscribers.

64. $F \cup G$ **65.** $H \cap F$ **66.** G'

Health *The table shows some symptoms of an underactive thyroid and an overactive thyroid:*

Underactive Thyroid	Overactive Thyroid
Sleepiness, s	Insomnia, i
Dry hands, d	Moist hands, m
Intolerance of cold, c	Intolerance of heat, h
Goiter, g	Goiter, g

Let U be the smallest possible set that includes all the symptoms listed, N be the set of symptoms for an underactive thyroid, and O be the set of symptoms for an overactive thyroid. Use the lowercase letters in the table to list the elements of each set.

67. O' **68.** N'

69. $N \cap O$ **70.** $N \cup O$

8.2 Applications of Venn Diagrams

We used Venn diagrams in the last section to illustrate set union and intersection. The rectangular region in a Venn diagram represents the universal set U. Including only a single set A inside the universal set, as in Figure 8.7, divides U into two nonoverlapping regions. Region 1 represents A', those elements outside set A, while region 2 represents those elements belonging to set A. (The numbering of these regions is arbitrary.)

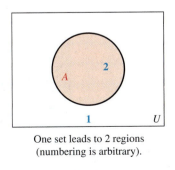

One set leads to 2 regions
(numbering is arbitrary).

FIGURE 8.7

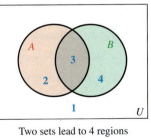

Two sets lead to 4 regions
(numbering is arbitrary).

FIGURE 8.8

The Venn diagram of Figure 8.8 shows two sets inside U. These two sets divide the universal set into four nonoverlapping regions. As labeled in Figure 8.8,

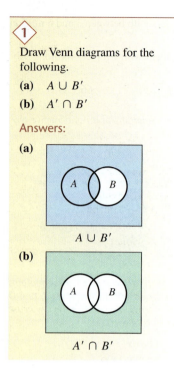
region 1 includes those elements outside both set A and set B. Region 2 includes those elements belonging to A and not to B. Region 3 includes those elements belonging to both A and B. Which elements belong to region 4? (Again, the numbering is arbitrary.)

▼ **EXAMPLE 1** Draw a Venn diagram similar to Figure 8.8, and shade the regions representing the following sets.

(a) $A' \cap B$

Solution Set A' contains all the elements outside set A. As labeled in Figure 8.8, A' is represented by regions 1 and 4. Set B is represented by the elements in regions 3 and 4. The intersection of sets A' and B, the set $A' \cap B$, is given by the region common to regions 1 and 4 and regions 3 and 4. The result, region 4, is shaded in Figure 8.9.

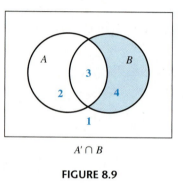

$A' \cap B$

FIGURE 8.9

(b) $A' \cup B'$

Solution Again, set A' is represented by regions 1 and 4 and set B' by regions 1 and 2. To find $A' \cup B'$, identify the region that represents the set of all elements in A', B', or both. The result, which is shaded in Figure 8.10, includes regions 1, 2, and 4. ▲ ①

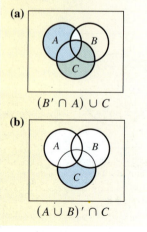
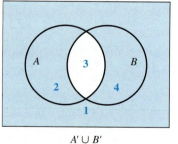

$A' \cup B'$

FIGURE 8.10

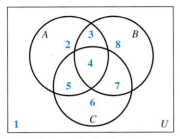

Three sets lead to 8 regions.

FIGURE 8.11

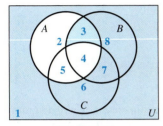

$A' \cup (B \cap C')$

FIGURE 8.12

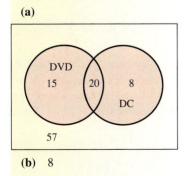

⟨3⟩

(a) Place numbers in the regions on a Venn diagram if the data on the 100 households showed

    35 DVD players;

    28 DCs;

    20 with both.

(b) How many have a DC but not a DVD player?

Answers:

(a)

(b) 8

Venn diagrams also can be drawn with three sets inside U. These three sets divide the universal set into eight nonoverlapping regions that can be numbered (arbitrarily) as in Figure 8.11.

▼ **EXAMPLE 2** Shade $A' \cup (B \cap C')$ in a Venn diagram.

Solution First find $B \cap C'$. See Figure 8.12. Set B is represented by regions 3, 4, 7, and 8, and set C' by regions 1, 2, 3, and 8. The overlap of these regions, regions 3 and 8, represents the set $B \cap C'$. Set A' is represented by regions 1, 6, 7, and 8. The union of regions 3 and 8 and regions 1, 6, 7, and 8 contains regions 1, 3, 6, 7, and 8, which are shaded in Figure 8.12. ▲ ⟨2⟩

Venn diagrams can be used to solve problems that result from surveying groups of people. As an example, suppose a researcher collecting data on 100 households finds that

    29 have a DVD player (DVD)

    21 have a digital camera (DC)

    15 have both.

The researcher wants to answer the following questions.

    (a) How many do not have a digital camera?

    (b) How many have neither a DVD player nor a digital camera?

    (c) How many have a DVD player but not a digital camera?

A Venn diagram like the one in Figure 8.13 will help sort out the information. In Figure 8.13(a), we put the number 15 in the region common to both a digital camera and a DVD player, because 15 households have both. Of the 29 with a DVD player, $29 - 15 = 14$ have no digital camera, so in Figure 8.13(b) we put 14 in the region for a DVD player but no digital camera. Similarly, $21 - 15 = 6$ households have a digital camera but not a DVD player, so we put 6 in that region. Finally, the diagram shows that $100 - 6 - 15 - 14 = 65$ households have neither a digital camera nor a DVD player. Now we can answer the questions.

    (a) $65 + 14 = 79$

    (b) 65 have neither.

    (c) 14 have a DVD player but not a digital camera. ⟨3⟩

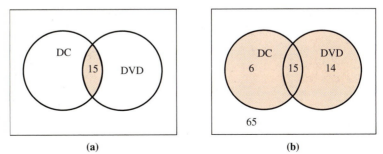

(a)   (b)

FIGURE 8.13

▼ **EXAMPLE 3** A group of 60 freshman business students at a large university was surveyed, with the following results.

19 of the students read *Business Week;*

18 read the *Wall Street Journal;*

50 read *Fortune;*

13 read *Business Week* and the *Journal;*

11 read the *Journal* and *Fortune;*

13 read *Business Week* and *Fortune;*

9 read all three magazines.

Use the preceding data to answer the following questions.

(a) How many students read none of the publications?

(b) How many read only *Fortune?*

(c) How many read *Business Week* and the *Journal,* but not *Fortune?*

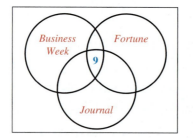

FIGURE 8.14(a)

Solution Once again, use a Venn diagram to represent the data. Since 9 students read all three publications, begin by placing 9 in the area in Figure 8.14(a) that belongs to all three regions.

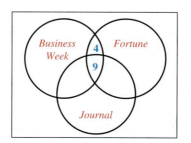

FIGURE 8.14(b)

Of the 13 students who read *Business Week* and *Fortune,* 9 also read the *Journal.* Therefore, only 13 − 9 = 4 students read just *Business Week* and *Fortune.* So place a 4 in the region common only to *Business Week* and *Fortune* readers, as in Figure 8.14(b).

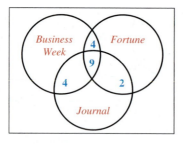

FIGURE 8.14(c)

In the same way, place a 4 in the region of Figure 8.14(c) common only to *Business Week* and the *Journal,* and 2 in the region common only to *Fortune* and the *Journal.*

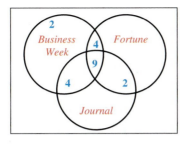

FIGURE 8.14(d)

The data shows that 19 students read *Business Week*. However, $4 + 9 + 4 = 17$ readers have already been placed in the *Business Week* region. The balance of this region in Figure 8.14(d) will contain only $19 - 17 = 2$ students. These 2 students read *Business Week* only—not *Fortune* and not the *Journal*.

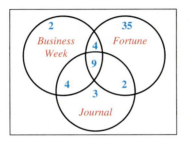

FIGURE 8.14(e)

In the same way, 3 students read only the *Journal* and 35 read only *Fortune*, as shown in Figure 8.14(e).

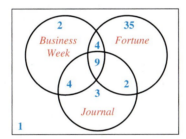

FIGURE 8.14(f)

A total of $2 + 4 + 3 + 4 + 9 + 2 + 35 = 59$ students are placed in the various regions of Figure 8.14(e). Since 60 students were surveyed, $60 - 59 = 1$ student reads none of the three publications and 1 is placed outside the other regions in Figure 8.14(f).

Figure 8.14(f) can now be used to answer the questions asked at the beginning of this example.

    (a) Only 1 student reads none of the publications.
    (b) There are 35 students who read only *Fortune*.
    (c) The overlap of the regions representing *Business Week* and the *Journal* shows that 4 students read *Business Week* and the *Journal*, but not *Fortune*. ▲ ◇4

◇4

In Example 3, how many students read exactly

(a) 1 of the publications?

(b) 2 of the publications?

Answers:

(a) 40

(b) 10

▼ **EXAMPLE 4**   Jeff Friedman is a section chief for an electric utility company. The employees in his section cut down tall trees, climb poles, and splice wire. Friedman reported the following information to the management of the utility:

Of the 100 employees in my section,

45 can cut tall trees;

50 can climb poles;

57 can splice wire;

28 can cut trees and climb poles;

20 can climb poles and splice wire;

25 can cut trees and splice wire;

11 can do all three;

  9 can't do any of the three (management trainees).

The data supplied by Friedman leads to the numbers shown in Figure 8.15. Add the numbers from all the regions to get the total number of Friedman's employees.

$$9 + 3 + 14 + 23 + 11 + 9 + 17 + 13 = 99$$

Friedman claimed to have 100 employees, but his data indicates only 99. The management decided that Friedman didn't qualify as a section chief and reassigned him as a night-shift meter reader in Guam. (Moral: He should have taken this course.) ▲

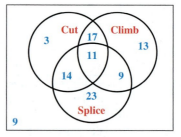

FIGURE 8.15

NOTE In all the preceding examples, we started in the innermost region with the intersection of the categories. This is usually the best way to begin solving problems of this type.

We use the symbol **_n(A)_** to denote the *number* of elements in _A_. For instance, if $A = \{w, x, y, z\}$, then $n(A) = 4$. The following useful fact is proved below.

Addition Rule for Counting

$$n(A \cup B) = n(A) + n(B) - n(A \cap B)$$

For example, if $A = \{r, s, t, u, v\}$ and $B = \{r, t, w\}$, then $A \cap B = \{r, t\}$, so that $n(A) = 5$, $n(B) = 3$, and $n(A \cap B) = 2$. By the formula in the box, $n(A \cup B) = 5 + 3 - 2 = 6$, which is certainly true, since $A \cup B = \{r, s, t, u, v, w\}$.

Here is a proof of the statement in the box: let x be the number of elements in A that are not in B, y be the number of elements in $A \cap B$, and z be the number of elements in B that are not in A, as indicated in Figure 8.16. That diagram shows that

---

5

In Example 4, suppose 46 employees can cut tall trees. Then how many

(a) can only cut tall trees?

(b) can cut trees or climb poles?

(c) can cut trees or climb poles or splice wire?

Answers:

(a) 4

(b) 68

(c) 91

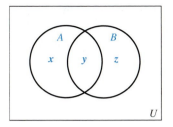

FIGURE 8.16

$n(A \cup B) = x + y + z$. It also shows that $n(A) = x + y$ and $n(B) = y + z$, so that

$$n(A) + n(B) - n(A \cap B) = (x + y) + (z + y) - y$$
$$= x + y + z$$
$$= n(A \cup B).$$

▼ **EXAMPLE 5**   A group of 10 students meets to plan a school function. All are majoring in accounting or economics or both. Five of the students are economics majors and 7 are majors in accounting. How many major in both subjects?

Solution Let A represent the set of accounting majors and B represent the set of economics majors. Use the union rule, with $n(A) = 5$, $n(B) = 7$, and $n(A \cup B) = 10$. We must find $n(A \cap B)$.
$$n(A \cup B) = n(A) + n(B) - \boldsymbol{n(A \cap B)}$$
$$10 = 5 + 7 - \boldsymbol{n(A \cap B)},$$

so

$$\boldsymbol{n(A \cap B)} = 5 + 7 - 10 = 2. \quad \blacktriangle \; \langle 6 \rangle$$

⟨6⟩

If $n(A) = 10$, $n(B) = 7$, and $n(A \cap B) = 3$, find $n(A \cup B)$.

Answer:

14

▼ **EXAMPLE 6**   On May 20–22, 2005, the Gallup Organization conducted a national poll that examined American's reading habits.* One of the questions asked was, *"During the past year, about how many books, either hardcover or paperback, did you read either all or part of the way through?"* The following table gives results consistent with Gallup's results for 100 people in 2005 and compares these results with those of a similar poll taken December 5–8, 2002.

		A None	B 1–5	C 6–10	D 11–50	E 51 +	F No Answer	Total
G	2005	16	38	14	25	6	1	100
H	2002	18	31	15	27	8	1	100
	Total	34	69	29	52	14	2	200

Let G denote the set of people interviewed in 2005 and H the set of people interviewed in 2002. Let A be the set of people who read no books, B the set of people who read 1–5 books, C the set of people who read 6–10 books, D the set of people who read 11–50 books, and E the set of people who read 51 or more books. Finally, let F be the set of people who gave no answer. Find the number of people in each of the following sets.

(a) $G \cap B$

Solution The set $G \cap B$ consists of all those who indicated that they read 1–5 books in the last year *and* were interviewed in 2005. From the table, we see there were 38 such people.

(b) $G \cup B$

*The Gallup Organization, June 3, 2005.

Solution The set $G \cup B$ consists of all those who indicated that they read 1–5 books in the last year *or* they were interviewed in 2005. We include all 100 interviewed in 2005, plus the 31 who read 1–5 books and were polled in 2002, for a total of 131. Alternatively, we could use the formula $n(G \cup B) = n(G) + n(B) - n(G \cap B) = 100 + 69 - 38 = 131$.

(c) $(A \cup C) \cap H'$

Solution Begin with the set $A \cup C$, which contains everyone who read none or 6–10 books. This consists of the four categories with 16, 18, 14, and 15 people. Of this set, take those who did *not* get interviewed in 2002, for a total of $16 + 14 = 30$ people. This is the number of people who read none or 6 to 10 books and who did not get interviewed in 2002. ▲ ⬦7

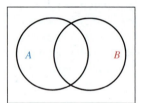

⬦7

In Example 6 find the number of people in each set.

(a) $H \cup D$

(b) $(H \cap E) \cup B'$

Answers:

(a) 125

(b) 131

◆8.2 Exercises

Sketch a Venn diagram like the one shown, and use shading to show each of the following sets. (See Example 1.)

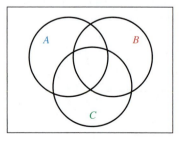

1. $A \cap B'$ **2.** $A \cup B'$ **3.** $B' \cup A'$ **4.** $A' \cap B'$

5. $B' \cup (A \cap B')$ **6.** $(A \cap B) \cup A'$

7. U' **8.** $\emptyset'$

9. Three sets divide the universal set into at most _____ regions.

10. What does the notation $n(A)$ represent?

Sketch a Venn diagram like the one shown, and use shading to show each of the following sets. (See Example 2.)

11. $(A \cap C') \cup B$ **12.** $A \cap (B \cup C')$

13. $A' \cap (B \cap C)$ **14.** $(A' \cap B') \cap C$

15. $(A \cap B') \cup C$ **16.** $(A \cap B') \cap C$

Use Venn diagrams to answer the following questions. (See Examples 2 and 4.)

17. **Social Science** In 2003, the percent of children under 18 years of age who lived with both parents was 68.4, the percent of children under 18 years of age who lived only with their father was 4.6, and the percent of children under 18 years of age who lived with neither parent was 4.1.* What percent of children under age 18 lived with their mother only?

18. **Social Science** In 2003, there were 5,243,000 children under age 6 living with their mother (and not their father). Of these children, 3,983,000 children were living with the mother as head of the household, 898,000 were living with a grandparent as head of the household, and 213,000 were living with another relative as head of the household.* How many children living with their mother had a nonrelative as head of the household?

19. **Social Science** A survey of people attending a Lunar New Year celebration in Chinatown yielded the following results:

120 were women;

150 spoke Cantonese;

170 lit firecrackers;

108 of the men spoke Cantonese;

100 of the men did not light firecrackers;

18 of the non-Cantonese-speaking women lit firecrackers;

---

*U.S. Census Bureau, *America's Families and Living Arrangements*, March 2003.

78 non-Cantonese-speaking men did not light firecrackers;

30 of the women who spoke Cantonese lit firecrackers.

(a) How many attended the celebration?
(b) How many of those who attended did not speak Cantonese?
(c) How many women did not light firecrackers?
(d) How many of those who lit firecrackers were Cantonese-speaking men?

20. **Business** Jeff Friedman (see Example 4) was again reassigned, this time to the home economics department of the electric utility. He interviewed 140 people in a suburban shopping center to find out some of their cooking habits. He obtained the following results. Should he be reassigned yet one more time?

58 use microwave ovens;

63 use electric ranges;

58 use gas ranges;

19 use microwave ovens and electric ranges;

17 use microwave ovens and gas ranges;

 4 use both gas and electric ranges;

 1 uses all three cooking devices;

 2 cook only with solar energy.

21. **Natural Science** After a genetics experiment, the number of pea plants having certain characteristics was tallied, with the following results:

22 were tall;

25 had green peas;

39 had smooth peas;

 9 were tall and had green peas;

17 were tall and had smooth peas;

20 had green peas and smooth peas;

 6 had all three characteristics;

 4 had none of the characteristics.

(a) Find the total number of plants counted.
(b) How many plants were tall and had peas that were neither smooth nor green?
(c) How many plants were not tall, but had peas that were smooth and green?

22. **Health** Human blood can contain either no antigens, the A antigen, the B antigen, or both the A and B antigens. A third antigen, called the Rh antigen, is important in human reproduction and again may or may not be present in an individual. Blood is called type A positive if the individual has the A and Rh antigens, but not the B antigen. A person having only the A and B antigens is said to have type AB-negative blood. A person having only the Rh antigen has type O-positive blood. Other blood types are defined in a similar manner. Identify the blood type of the individuals in regions (a)–(g) of the Venn diagram.

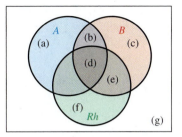

23. **Natural Science** Use the diagram from Exercise 22. In a certain hospital, the following data was recorded.

25 patients had the A antigen;

17 had the A and B antigens;

27 had the B antigen;

22 had the B and Rh antigens;

30 had the Rh antigen;

12 had none of the antigens;

16 had the A and Rh antigens;

15 had all three antigens.

How many patients
(a) were represented?
(b) had exactly one antigen?
(c) had exactly two antigens?
(d) had O-positive blood?
(e) had AB-positive blood?
(f) had B-negative blood?
(g) had O-negative blood?
(h) had A-positive blood?

24. **Social Science** At a powwow in Arizona, Native Americans from all over the Southwest came to participate in the ceremonies. A coordinator of the powwow took a survey and found that

15 families brought food, costumes, and crafts;

25 families brought food and crafts;

42 families brought food;

20 families brought costumes and food;

 6 families brought costumes and crafts, but not food;

 4 families brought crafts, but neither food nor costumes;

10 families brought none of the three items;

18 families brought costumes but not crafts.

(a) How many families were surveyed?
(b) How many families brought costumes?
(c) How many families brought crafts, but not costumes?

(d) How many families did not bring crafts?

(e) How many families brought food or costumes?

25. **Social Science** The following table lists the cross-classification of marital status and sex for 500 adults selected at random from the 2002 General Social Survey (GSS).* Using the letters given in the table, find the number of respondents in each set.

	Male (*M*)	Female (*F*)
Married (*A*)	100	128
Widowed (*B*)	6	32
Divorced (*C*)	36	52
Separated (*D*)	7	11
Never Married (*E*)	63	65

(a) $A \cap M$

(b) $C \cap (F \cup M)$

(c) $D \cup F$

(d) $B' \cap E'$

26. **Social Science** The number of female military personnel on active duty in March 2001 is given in the table.[†]

	Army (*A*)	Air Force (*B*)	Navy (*C*)	Marines (*D*)
Officers (*O*)	10,671	11,662	7,761	951
Enlisted (*E*)	61,871	54,537	43,010	9,387

Use this information and the letters given to find the number of female military personnel in each of the following sets.

(a) $A \cup B$

(b) $E \cup (C \cup D)$

(c) $O' \cap B'$

27. **Social Science** For U.S. people living in families in 2003 (in thousands), the poverty status and age cross-classification is given in the table.[‡]

	In Poverty (*A*)	Not in Poverty (*B*)
Under 18 (*C*)	12,389	59,742
18–64 (*D*)	11,947	131,624
65 and Over (*E*)	1,348	21,853

Using the letters given in the table, find the number of people in each set. (See Example 6.)

(a) $A \cap E$

(b) $E \cup B$

(c) $D \cup (B \cap E)$

(d) $E' \cap (A \cup C)$

(e) $C' \cup A$

(f) $D' \cap A'$

28. **Health** The table shows the numbers for five types of cancer classified as "in situ" cancer (cancer confined to the point of origin) and invasive cancer (cancer that has spread to other tissues) for subjects enrolled in the California Teachers Study.*

Type of Cancer	In Situ (*A*)	Invasive (*B*)
Breast (*C*)	507	2,596
Endometrial (*D*)	783	54
Cervix (*E*)	433	98
Lung (*F*)	0	135
Melanoma (*G*)	165	409

Using the letters given in the table, find the number of women in each set. (See Example 6.)

(a) $B \cap G$

(b) $A \cup C'$

(c) $E' \cap (A \cup C)$

(d) $F \cap B'$

29. Restate the union rule in words.

Use Venn diagrams to answer the following questions. (See Example 5.)

30. If $n(A) = 5$, $n(B) = 8$, and $n(A \cap B) = 4$, what is $n(A \cup B)$?

31. If $n(A) = 12$, $n(B) = 27$, and $n(A \cup B) = 30$, what is $n(A \cap B)$?

32. Suppose $n(B) = 7$, $n(A \cap B) = 3$, and $n(A \cup B) = 20$. What is $n(A)$?

33. Suppose $n(A \cap B) = 5$, $n(A \cup B) = 35$, and $n(A) = 13$. What is $n(B)$?

Draw a Venn diagram and use the given information to fill in the number of elements for each region.

34. $n(U) = 38, n(A) = 16, n(A \cap B) = 12, n(B') = 20$

35. $n(A) = 26, n(B) = 10, n(A \cup B) = 30, n(A') = 17$

*The National Data Program for Social Sciences, National Opinion Research Center, University of Chicago, www.norc.uchicago.edu/projects/gensoc.asp.

†U.S. Department of Defense, Defense Manpower Data Center, unpublished data, March 30, 2001. Available at www.infoplease.com/ipa/A0004600.html.

‡U.S. Census Bureau, Current Population Survey, 2004 Annual Social and Economic Supplement, www.census.gov.

*Arti Parikh-Patel, Mark Allen, William E. Wright, and California Teachers Study Steering Committee, "Validation of Self-reported Cancers in the California Teachers Study," *American Journal of Epidemiology*, 2003; 157:539–545.

36. $n(A \cup B) = 17,$ $n(A \cap B) = 3,$ $n(A) = 8,$
$n(A' \cup B') = 21$

37. $n(A') = 28,$ $n(B) = 25,$ $n(A' \cup B') = 45,$
$n(A \cap B) = 12$

38. $n(A) = 28,\ n(B) = 34,\ n(C) = 25,\ n(A \cap B) = 14,$
$n(B \cap C) = 15,\ n(A \cap C) = 11,\ n(A \cap B \cap C) = 9,$
$n(U) = 59$

39. $n(A) = 54,$ $n(A \cap B) = 22,$ $n(A \cup B) = 85,$
$n(A \cap B \cap C) = 4,\ n(A \cap C) = 15,\ n(B \cap C) = 16,$
$n(C) = 44, n(B') = 63$

In Exercises 40–43, show that the statements are true by drawing Venn diagrams and shading the regions representing the sets on each side of the equals signs.*

40. $(A \cup B)' = A' \cap B'$

41. $(A \cap B)' = A' \cup B'$

42. $A \cap (B \cup C) = (A \cap B) \cup (A \cap C)$

43. $A \cup (B \cap C) = (A \cup B) \cap (A \cup C)$

44. Explain in words the statement about sets in question 40.

45. Explain in words the statement about sets in question 41.

46. Explain in words the statement about sets in question 42.

47. Explain in words the statement about sets in question 43.

*The statements in Exercises 40 and 41 are known as De Morgan's laws. They are named for the English mathematician Augustus De Morgan (1806–71).

8.3   Introduction to Probability

If you go to a supermarket and buy 5 pounds of peaches at $1.26 per pound, you can easily find the *exact* price of your purchase: $6.30. On the other hand, the produce manager of the market is faced with the problem of ordering peaches. The manager may have a good estimate of the number of pounds of peaches that will be sold during the day, but it is impossible to predict the *exact* amount. The number of pounds that customers will purchase during a day is *random:* the quantity cannot be predicted exactly. A great many problems that come up in applications of mathematics involve random phenomena—those for which exact prediction is impossible. The best that we can do is determine the *probability* of the possible outcomes.

Sample Spaces

In probability, an **experiment** is an activity or occurrence with an observable result. Each repetition of an experiment is called a **trial.** The possible results of each trial are called **outcomes.** The set of all possible outcomes for an experiment is the **sample space** for that experiment. A sample space for the experiment of tossing a coin is made up of the outcomes heads (*h*) and tails (*t*). If *S* represents this sample space, then

$$S = \{h, t\}.$$

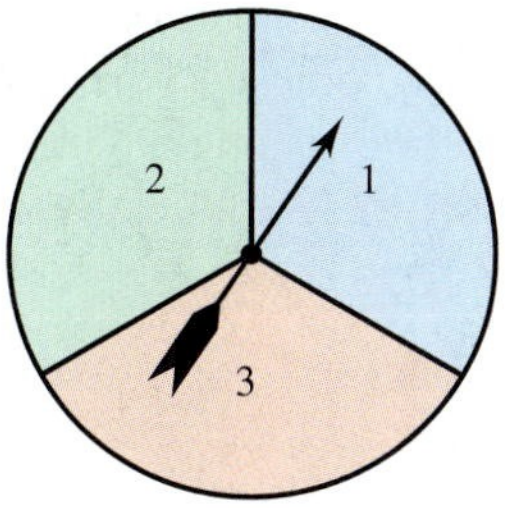

FIGURE 8.17

▼ **EXAMPLE 1**   Give the sample space for each experiment.

(a) A spinner like the one in Figure 8.17 is spun.

  Solution The three outcomes are 1, 2, or 3, so the sample space is

$$\{1, 2, 3\}.$$

(b) For the purposes of a public opinion poll, respondents are classified as young, middle aged, or senior and as male or female.

Solution A sample space for this poll could be written as a set of ordered pairs:

$$\{(\text{young, male}), (\text{young, female}), (\text{middle aged, male}),$$

$$(\text{middle aged, female}), (\text{senior, male}), (\text{senior, female})\}$$

(c) An experiment consists of studying the numbers of boys and girls in families with exactly 3 children. Let *b* represent *boy* and *g* represent *girl*.

Solution A three-child family can have 3 boys, written *bbb*, 3 girls, *ggg*, or various combinations, such as *bgg*. A sample space with four outcomes (not equally likely) is

$$S_1 = \{3 \text{ boys, 2 boys and 1 girl, 1 boy and 2 girls, 3 girls}\}.$$

Notice that a family with 3 boys or 3 girls can occur in just one way, but a family of two boys and 1 girl or 1 boy and 2 girls can occur in more than one way. If the order of the births is considered, so that, for example, *bgg* is different from *gbg* or *ggb*, another sample space is

$$S_2 = \{bbb, bbg, bgb, gbb, bgg, gbg, ggb, ggg\}.$$

The second sample space, S_2, has equally likely outcomes if we assume that having a boy and having a girl are equally likely. This assumption, while not quite true, is approximately true, so we will use it throughout this book. The outcomes in S_1 are not equally likely, since there is more than one way to get a family with 2 boys and 1 girl (*bbg, bgb,* or *gbb*) or a family with 2 girls and 1 boy (*ggb, gbg,* or *bgg*), but only one way to get 3 boys (*bbb*) or 3 girls (*ggg*). ▲

CAUTION An experiment may have more than one sample space, as shown in Example 1(c). The most convenient sample spaces have equally likely outcomes, but many situations do not have such a sample space. For example, a sample space with outcomes that are not equally likely consists of a driver having a car crash within the year or not having a car crash within the year. ◁1▷

Events

An **event** is a subset of a sample space. If the sample space for tossing a coin is $S = \{h, t\}$, then one event is $E = \{h\}$, which represents the outcome "heads."

An ordinary die is a cube whose six different faces show the following numbers of dots: 1, 2, 3, 4, 5, and 6. If the die is fair (not "loaded" to favor certain faces over others), then any one of the faces is equally likely to come up when the die is rolled. The sample space for the experiment of rolling a single fair die is $S = \{1, 2, 3, 4, 5, 6\}$. Some possible events are as follows:

The die shows an even number: $E_1 = \{2, 4, 6\}$.

The die shows a 1: $E_2 = \{1\}$.

The die shows a number less than 5: $E_3 = \{1, 2, 3, 4\}$.

The die shows a multiple of 3: $E_4 = \{3, 6\}$.

◁1▷
(a) Write an equally likely sample space for the experiment of tossing 2 fair coins.

(b) Write a sample space for the experiment of tossing 2 fair coins if we are interested only in the number of heads. Are the outcomes in this sample space equally likely?

Answers:

(a) {hh, ht, th, tt}

(b) {0, 1, 2}; no

▼ **EXAMPLE 2**   For the sample space S_2 in Example 1(c), write the following events.

(a) Event H: the family has exactly 2 girls.

Solution Families with three children can have exactly 2 girls with either bgg, gbg, or ggb, so that event H is

$$H = \{bgg, gbg, ggb\}.$$

(b) Event K: the three children are the same sex.

Solution Two outcomes satisfy this condition: all boys and all girls.

$$K = \{bbb, ggg\}$$

(c) Event J: the family has three girls.

Solution Only ggg satisfies this condition, so

$$J = \{ggg\}. \quad ▲ \; ⟨2⟩$$

In Example 2(c), event J had only one possible outcome: ggg. Such an event, with only one possible outcome, is a **simple event.** If an event E equals the sample space S, then E is a **certain event.** If event $E = \emptyset$, then E is an **impossible event.**

▼ **EXAMPLE 3**   Suppose a fair die is rolled. Then the sample space is $\{1, 2, 3, 4, 5, 6\}$.

(a) The event "the die shows a 4," $\{4\}$, has only one possible outcome. It is a simple event.

(b) The event "the number showing is less than 10" equals the sample space $S = \{1, 2, 3, 4, 5, 6\}$. This event is a certain event; if a die is rolled, the number showing (either 1, 2, 3, 4, 5, or 6) must be less than 10.

(c) The event "the die shows a 7" is the empty set, $\emptyset$; this is an impossible event. ▲ ⟨3⟩

Since events are sets, we can use set operations to find unions, intersections, and complements of events. Here is a summary of the set operations for events.

⟨2⟩
Suppose a die is tossed. Write the following events.

(a) The number showing is less than 3.

(b) The number showing is 5.

(c) The number showing is 8.

Answers:

(a) $\{1, 2\}$

(b) $\{5\}$

(c) $\emptyset$

⟨3⟩
Which of the events listed in side problem 2 is

(a) simple?

(b) certain?

(c) impossible?

Answers:

(a) Part (b)

(b) None

(c) Part (c)

Set Operations for Events

Let E and F be events for a sample space S. Then

$E \cap F$ occurs when both E **and** F occur;

$E \cup F$ occurs when E **or** F **or both** occur;

E' occurs when E does **not** occur.

▼ **EXAMPLE 4** A study of workers earning the minimum wage grouped such workers into various categories, which can be interpreted as events when a worker is selected at random. Consider the following events:

E: worker is under 20;

F: worker is white;

G: worker is female.

Describe each of the following events in words.

(a) E'

Solution E' is the event that the worker is 20 or older.

(b) $F \cap G'$

Solution $F \cap G'$ is the event that the worker is white and not a female—that is, the worker is a white male.

(c) $E \cup G$

Solution $E \cup G$ is the event that the worker is under 20 or is female. Note that this event includes all workers under 20, both male and female, and all female workers of any age. ▲ ⟨4⟩

Two events that cannot both occur at the same time, such as getting both a head and a tail on the same toss of a coin, are called **disjoint events.** (Disjoint events are sometimes referred to as *mutually exclusive events*.)

⟨4⟩

Give the following events for the experiment of rolling a fair die if $E = \{1, 3\}$ and $F = \{2, 3, 4, 5\}$.

(a) $E \cap F$

(b) $E \cup F$

(c) E'

Answers:

(a) $\{3\}$

(b) $\{1, 2, 3, 4, 5\}$

(c) $\{2, 4, 5, 6\}$

Disjoint Events

Events E and F are disjoint events if $E \cap F = \emptyset$.

For any event E, E and E' are disjoint events.

⟨5⟩

In Example 5, let $F = \{2, 4, 6\}$, $K = \{1, 3, 5\}$, and G remain the same. Are the following events disjoint?

(a) F and K

(b) F and G

Answers:

(a) Yes

(b) No

▼ **EXAMPLE 5** Let $S = \{1, 2, 3, 4, 5, 6\}$, the sample space for tossing a die. Let $E = \{4, 5, 6\}$ and let $G = \{1, 2\}$. Then E and G are disjoint events, because they have no outcomes in common; $E \cap G = \emptyset$. See Figure 8.18. ▲ ⟨5⟩

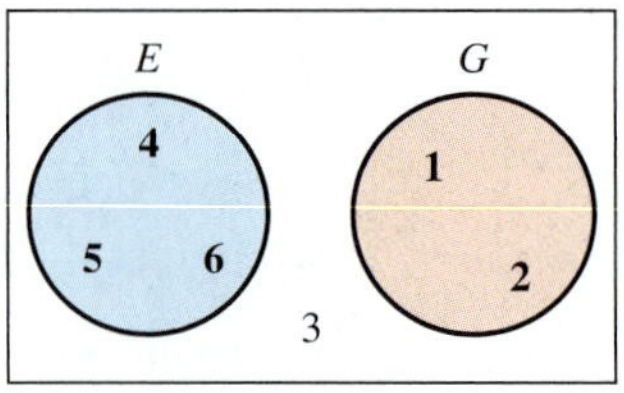

FIGURE 8.18

Probability

For sample spaces with *equally likely* outcomes, the probability of an event is defined as follows.

Basic Probability Principle

Let S be a sample space of equally likely outcomes, and let event E be a subset of S. Then the **probability that event E occurs** is

$$P(E) = \frac{n(E)}{n(S)}.$$

By this definition, the **probability of an event** is a number that indicates the relative likelihood of the event.

CAUTION The basic probability principle applies only when the outcomes are equally likely.

▼ **EXAMPLE 6**   Suppose a single fair die is rolled. Use the sample space $S = \{1, 2, 3, 4, 5, 6\}$, and give the probability of each of the following events.

(a) E: the die shows an even number.

Solution Here, $E = \{2, 4, 6\}$, a set with three elements. Because S contains six elements,

$$P(E) = \frac{3}{6} = \frac{1}{2}.$$

(b) F: the die shows a number less than 10.

Solution Event F is a certain event, with

$$F = \{1, 2, 3, 4, 5, 6\},$$

so that

$$P(F) = \frac{6}{6} = 1.$$

(c) G: the die shows an 8.

Solution This event is impossible, so

$$P(G) = \frac{0}{6} = 0. \; \blacktriangle \; \langle 6 \rangle$$

⬦ **6**

A fair die is rolled. Find the probability of rolling

(a) an odd number;

(b) 2, 4, 5, or 6;

(c) a number greater than 5;

(d) the number 7.

Answers:

(a) 1/2

(b) 2/3

(c) 1/6

(d) 0

A standard deck of 52 cards has four suits: hearts (♥), clubs (♣), diamonds (♦), and spades (♠), with 13 cards in each suit. The hearts and diamonds are red, and the spades and clubs are black. Each suit has an ace (A), a king (K), a queen (Q), a jack (J), and cards numbered from 2 to 10. The jack, queen, and king are called face cards and for many purposes can be thought of as having values 11, 12, and 13, respectively. The ace can be thought of as the low card (value 1) or the high card (value 14). See Figure 8.19. We will refer to this standard deck of cards often in our discussion of probability.

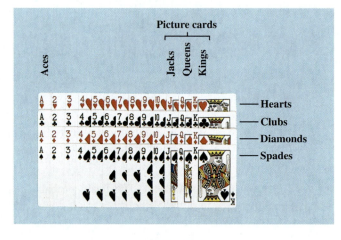

FIGURE 8.19

▼ **EXAMPLE 7** If a single card is drawn at random from a standard, well-shuffled, 52-card deck, find the probability of each of the following events.

(a) Drawing an ace

Solution There are 4 aces in the deck. The event "drawing an ace" is

{heart ace, diamond ace, club ace, spade ace}.

Therefore,

$$P(\text{ace}) = \frac{4}{52} = \frac{1}{13}.$$

(b) Drawing a face card

Solution Since there are 12 face cards,

$$P(\text{face card}) = \frac{12}{52} = \frac{3}{13}.$$

(c) Drawing a spade

Solution The deck contains 13 spades, so

$$P(\text{spade}) = \frac{13}{52} = \frac{1}{4}.$$

(d) Drawing a spade or a heart

Solution Besides the 13 spades, the deck contains 13 hearts, so

$$P(\text{spade or heart}) = \frac{26}{52} = \frac{1}{2}. \; \blacktriangle \; \langle 7 \rangle$$

In the preceding examples, the probability of each event was a number between 0 and 1, inclusive. The same thing is true in general. Any event E is a subset of the sample space S, so $0 \le n(E) \le n(S)$. Since $P(E) = n(E)/n(S)$, it follows that $0 \le P(E) \le 1$.

For any event E,

$$0 \le P(E) \le 1.$$

▼ **EXAMPLE 8**   The table gives the number of years of service of senators in the 109th Congress of the United States of America, which convened on January 3, 2005.*

Years of Service	Number of Senators
0–9	50
10–19	28
20–29	15
30–39	4
40 or more	3

Find the probability that a randomly selected senator of the 109th Congress had served 20–29 years when Congress convened.

Solution This probability is found by dividing the number of senators who served 20–29 years by the total number of senators. Thus,

$$P(20\text{–}29 \text{ years}) = \frac{15}{100} = .15. \; \blacktriangle \; \langle 8 \rangle$$

*Data compiled from www.infoplease.com.

7
A single playing card is drawn at random from an ordinary 52-card deck. Find the probability of drawing

(a) a queen;

(b) a diamond;

(c) a red card.

Answers:

(a) 1/13

(b) 1/4

(c) 1/2

8
Find the probability that a randomly selected senator of the 109th Congress had served 30 or more years when Congress convened.

Answer:

.07

◆8.3 Exercises

1. What is meant by a "fair" coin or die?

2. What is the sample space for an experiment?

Write sample spaces for the experiments in Exercises 3–8. (See Example 1.)

3. A month of the year is chosen for a wedding.

4. A day in April is selected for a bicycle race.

5. A student is asked how many points she earned on a recent 80-point test.

6. A person is asked the number of hours (to the nearest hour) he watched television yesterday.

7. The management of an oil company must decide whether to go ahead with a new oil shale plant or to cancel it.

8. A coin is tossed and a die is rolled.

9. Define an event.

10. What is a simple event?

For the experiments in Exercises 11–16, write out an equally likely sample space, and then write the indicated events in set notation. (See Examples 2 and 3.)

11. A marble is drawn at random from a bowl containing 3 yellow, 4 white, and 8 blue marbles.
   (a) A yellow marble is drawn.
   (b) A blue marble is drawn.
   (c) A white marble is drawn.
   (d) A black marble is drawn.

12. Six people live in a dorm suite. Two are to be selected to go to the campus café to pick up a pizza. Of course, no one wants to go, so the six names (Connie, Casey, Lindsey, Jackie, Taisa, and Lisa) are placed in a hat. After the hat is shaken, two names are selected.
   (a) Both names selected begin with the letter "L".
   (b) The first name selected begins with "L" and the second name selected begins with "J" or "C".
   (c) The first name selected begins with the letter "C".

13. An unprepared student takes a three-question true–false quiz in which he flips a coin to guess the answers. If the coin is heads, he guesses true, and if the coin is tails, he guesses false.
   (a) The student guesses true twice and guesses false once.
   (b) The student guesses all false.
   (c) The student guesses true once and guesses false twice.

14. In deciding what color and style to paint a room, Greg has narrowed his choices to three colors—forest sage, evergreen whisper, and opaque emerald—and two styles—rag painting and colorwash.
   (a) Greg picks a combination with colorwash.
   (b) Greg picks a combination with opaque emerald or rag painting.

15. Tami goes shopping and sees three kinds of shoes: flats, 2″ heels, and 3″ heels. They come in 2 shades of beige (light and dark) and black.
   (a) The shoe selected has a heel and is black.
   (b) The shoe selected has no heel and is beige.
   (c) The shoe selected has a heel and is beige.

16. From 5 employees—Strutz, Martin, Hampton, Williams, and Ewing—2 are selected to attend a conference.
   (a) Hampton is selected.
   (b) Strutz and Martin are not both selected.
   (c) Both Williams and Ewing are selected.

A single fair die is rolled. Find the probabilities of the following events. (See Example 6.)

17. Getting a 5

18. Getting a number less than 4

19. Getting a number greater than 4

20. Getting a 2 or a 5   **21.** Getting a multiple of 3

22. Getting any number except 3

A card is drawn from a well-shuffled deck of 52 cards. Find the probability of drawing each of the following. (See Example 7.)

23. An 8   **24.** A red card

25. A red 10   **26.** A club

27. The 8 of diamonds   **28.** A 7 or a king

29. A red 6 or a black 9   **30.** A black face card

31. **Social Science** Respondents for the 2002 General Social Survey (GSS) indicated the following percentages pertaining to attendance at religious services.*

Attendance	Percent
Never	18.6
Less than Once a Year	7.0
Once a Year	13.9
Several Times a Year	12.9
Once a Month	6.8
2–3 Times a Month	9.3
Nearly Every Week	6.5
Every Week	16.5
More than Once a Week	7.7
Don't Know/No Answer	.8
Total	100.0

Find the probability that a randomly chosen respondent would indicate the following religious service attendance.
   (a) Nearly Every Week
   (b) Once a Year
   (c) Every Week
   (d) Less than Every Week
   (e) Every Week or More

32. **Finance** As of October 31, 2004, the Janus Mercury fund invested in equities throughout the world as follows:†

Region	Percent
Canada	3.50%
Europe	12.30%
South Korea	3.00%
Cayman Islands	0.30%
United States	80.90%

Find the probability that a randomly selected equity would be from each of the following regions.
   (a) Europe
   (b) The United States
   (c) Cayman Islands

*The National Data Program for Social Sciences, National Opinion Research Center, University of Chicago, www.norc.uchicago.edu/projects/gensoc.asp.

†www.janus.com.

33. **Health**   For a medical experiment, people are classified as to whether they smoke, have a family history of heart disease, or are overweight. Define events E, F, and G as follows.

    E: person smokes

    F: person has a family history of heart disease

    G: person is overweight

    Describe each of the following events in words.
    (a) G'
    (b) $F \cap G$
    (c) $E \cup G'$

34. **Health**   Refer to Exercise 33. Describe each of the following events in words.
    (a) $E \cup F$
    (b) $E' \cap F$
    (c) $F' \cup G'$

35. **Social Sciences**   The population of the United States by race in 2000 and the projected population by race for the year 2025 are given in the table (in thousands).* Find the probability of a randomly selected person being each of the following.
    (a) Hispanic in 2000
    (b) Hispanic in 2025
    (c) African-American in 2000
    (d) African-American in 2025

Race	2000	2025
White	196,700	209,900
Hispanic	32,500	56,900
African-American	33,500	44,700
Asian-American	10,600	24,000
Other	2,100	2,800

*"Projections of the Total Resident Population by 5-Year Age Groups, Race, and Hispanic Origin, 1999 to 2000," from U.S. Census Bureau, *Current Population Reports.*

8.4 Basic Concepts of Probability

We determine the probability of more complex events in this section.

To find the probability of the union of two sets E and F in a sample space, we use the union rule for counting given in Section 8.2.

$$n(E \cup F) = n(E) + n(F) - n(E \cap F).$$

Dividing both sides by $n(S)$ yields

$$\frac{n(E \cup F)}{n(S)} = \frac{n(E)}{n(S)} + \frac{n(F)}{n(S)} - \frac{n(E \cap F)}{n(S)}$$
$$P(E \cup F) = P(E) + P(F) - P(E \cap F).$$

This discussion is summarized in the next rule.

Addition Rule for Probability

For any events E and F from a sample space S,

$$P(E \cup F) = P(E) + P(F) - P(E \cap F).$$

In words, we have

$$P(E \text{ or } F) = P(E) + P(F) - P(E \text{ and } F).$$

(Although the addition rule applies to any events E and F from any sample space, the derivation we have given is valid only for sample spaces with equally likely simple events.)

▼ **EXAMPLE 1** If a single card is drawn from an ordinary deck of cards, find the probability that it will be red or a face card.

Solution Let R represent the event "red card" and F the event "face card." There are 26 red cards in the deck, so $P(R) = 26/52$. There are 12 face cards in the deck, so $P(F) = 12/52$. Since there are 6 red face cards in the deck, $P(R \cap F) = 6/52$. By the addition rule, the probability that the card will be red or a face card is

$$P(R \cup F) = P(R) + P(F) - P(R \cap F)$$
$$= \frac{26}{52} + \frac{12}{52} - \frac{6}{52} = \frac{32}{52} = \frac{8}{13}.$$ ▲ ◇ ①

CAUTION Recall from Section 8.1 that the word "or" always indicates use of the addition rule.

①

A single card is drawn from an ordinary deck. Find the probability that it is black or a 9.

Answer:

7/13

▼ **EXAMPLE 2** Suppose two fair dice (plural of *die*) are rolled. Find each of the following probabilities.

(a) The first die shows a 2 or the sum of the results is 6 or 7.

Solution The sample space for the throw of two dice is shown in Figure 8.20, where 1-1 represents the event "the first die shows a 1 and the second die shows a 1," 1-2 represents "the first die shows a 1 and the second die shows a 2," and so on. Let A represent the event "the first die shows a 2" and B represent the event "the sum of the results is 6 or 7." These events are indicated in color in Figure 8.20. From the diagram, event A has 6 elements, B has 11 elements, and the sample space has 36 elements. Thus,

$$P(A) = \frac{6}{36}, \quad P(B) = \frac{11}{36}, \quad \text{and} \quad P(A \cap B) = \frac{2}{36}.$$

By the addition rule,

$$P(A \cup B) = P(A) + P(B) - P(A \cap B),$$
$$P(A \cup B) = \frac{6}{36} + \frac{11}{36} - \frac{2}{36} = \frac{15}{36} = \frac{5}{12}.$$

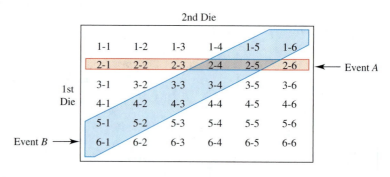

FIGURE 8.20

(b) The sum is 11 or the second die shows a 5.

Solution $P(\text{sum is 11}) = 2/36$, $P(\text{second die shows 5}) = 6/36$, and $P(\text{sum is 11 and second die shows 5}) = 1/36$, so

$$P(\text{sum is 11 or second die shows 5}) = \frac{2}{36} + \frac{6}{36} - \frac{1}{36} = \frac{7}{36}. \; \blacktriangle$$

2

②

In the experiment of Example 2, find the following probabilities.

(a) The sum is 5 or the second die shows a 3.

(b) Both dice show the same number, or the sum is at least 11.

Answers:

(a) 1/4

(b) 2/9

CAUTION You may wonder why we did not use $S = \{2,3,4,5, \ldots ,12\}$ as the sample space in Example 2. Remember, we prefer to use a sample space with equally likely outcomes. The outcomes in set S are not equally likely: a sum of 2 can occur in just one way, a sum of 3 in two ways, a sum of 4 in three ways, and so on, as shown in Figure 8.20.

If events E and F are disjoint, then $E \cap F = \emptyset$ by definition; hence; $P(E \cap F) = 0$. Applying the addition rule yields this useful fact.

Addition Rule for Disjoint Events

For disjoint events E and F,

$$P(E \cup F) = P(E) + P(F).$$

▼ **EXAMPLE 3** Assume that the probability of a couple having a boy is the same as the probability of having a girl. If the couple has 3 children, find the probability that at least 2 of them are girls.

Solution The event of having at least 2 girls is the union of the disjoint events $E =$ "the family has exactly 2 girls" and $F =$ "the family has exactly 3 girls." Using the equally likely sample space

$$\{ggg, ggb, gbg, bgg, gbb, bgb, bbg, bbb\},$$

we see that $P(2 \text{ girls}) = 3/8$ and $P(3 \text{ girls}) = 1/8$. Therefore,

$$P(\text{at least 2 girls}) = P(2 \text{ girls}) + P(3 \text{ girls})$$

$$= \frac{3}{8} + \frac{1}{8} = \frac{1}{2}. \; \blacktriangle$$

③

In Example 3, find the probability of no more than 2 girls.

Answer:

7/8

By definition of E', for any event E from a sample space S,

$$E \cup E' = S \quad \text{and} \quad E \cap E' = \emptyset.$$

Because $E \cap E' = \emptyset$, events E and E' are disjoint, so that

$$P(E \cup E') = P(E) + P(E').$$

However, $E \cup E' = S$, the sample space, and $P(S) = 1$. Thus,

$$P(E \cup E') = P(E) + P(E') = 1.$$

Rearranging these terms gives the following useful rule.

Complement Rule

For any event E,

$$P(E') = 1 - P(E) \quad \text{and} \quad P(E) = 1 - P(E').$$

▼ **EXAMPLE 4** If a fair die is rolled, what is the probability that any number but 5 will come up?

Solution If E is the event that 5 comes up, then E' is the event that any number but 5 comes up. $P(E) = 1/6$, so we have $P(E') = 1 - 1/6 = 5/6$. ▲ ④

▼ **EXAMPLE 5** If two fair dice are rolled, find the probability that the sum of the numbers showing is greater than 3.

Solution To calculate this probability directly, we must find each of the probabilities that the sum is 4, 5, 6, 7, 8, 9, 10, 11, and 12 and then add them. It is much simpler to first find the probability of the complement, the event that the sum is less than or equal to 3.

$$P(\text{sum} \le 3) = P(\text{sum is } 2) + P(\text{sum is } 3)$$
$$= \frac{1}{36} + \frac{2}{36} = \frac{3}{36} = \frac{1}{12}$$

Now use the fact that $P(E) = 1 - P(E')$ to get

$$P(\text{sum} > 3) = 1 - P(\text{sum} \le 3) = 1 - \frac{1}{12} = \frac{11}{12}. \quad ▲ ⑤$$

Odds

Sometimes probability statements are given in terms of **odds:** a comparison of $P(E)$ with $P(E')$. For example, suppose $P(E) = \frac{4}{5}$. Then $P(E') = 1 - \frac{4}{5} = \frac{1}{5}$. These probabilities predict that E will occur 4 out of 5 times and E' will occur 1 out of 5 times. Then we say that the **odds in favor** of E are 4 to 1, or 4:1.

> ## Odds
>
> The **odds in favor** of an event E are defined as the ratio of $P(E)$ to $P(E')$, or
>
> $$\frac{P(E)}{P(E')}, \qquad P(E') \ne 0.$$

▼ **EXAMPLE 6** Suppose the weather forecaster says that the probability of rain tomorrow is $1/3$. Find the odds in favor of rain tomorrow.

Solution Let E be the event "rain tomorrow." Then E' is the event "no rain tomorrow." Since $P(E) = 1/3$, $P(E') = 2/3$. By the definition of odds, the odds in favor of rain are

$$\frac{1/3}{2/3} = \frac{1}{2}, \quad \text{written 1 to 2} \quad \text{or} \quad 1:2.$$

On the other hand, the odds that it will *not* rain, or the odds *against* rain, are

$$\frac{2/3}{1/3} = \frac{2}{1}, \quad \text{written 2 to 1.}$$

Sidebar

④

(a) Let $P(K) = 2/3$. Find $P(K')$.

(b) If $P(X') = 3/4$, find $P(X)$.

Answers:

(a) 1/3

(b) 1/4

⑤

In Example 5, find the probability that the sum of the numbers rolled is at least 5.

Answer:

5/6

If the odds in favor of an event are, say, 3 to 5, then the probability of the event is 3/8, while the probability of the complement of the event is 5/8. (Odds of 3 to 5 indicate 3 outcomes in favor of the event out of a total of 8 outcomes.) This example suggests the following generalization:

If the odds favoring event E are m to n, then

$$P(E) = \frac{m}{m+n} \quad \text{and} \quad P(E') = \frac{n}{m+n}.$$

▼ **EXAMPLE 7** Often, weather forecasters give probability in terms of percent. Suppose the weather forecaster says that there is a 40% chance that it will snow tomorrow. Find the odds of snow tomorrow.

Solution In this case, we can let E be the event "snow tomorrow." Then E' is the event "no snow tomorrow." Now, we have $P(E) = .4 = 4/10$, $P(E') = .6 = 6/10$. By the definition of odds in favor of snow are

$$\frac{4/10}{6/10} = \frac{4}{6} = \frac{2}{3}, \text{ written 2 to 3 or 2:3.}$$

It is important to put the final fraction into lowest terms to communicate the odds. ▲ ⑥

▼ **EXAMPLE 8** The odds that a particular bid will be the low bid are 4 to 5.

(a) Find the probability that the bid will be the low bid.

Solution Odds of 4 to 5 show 4 favorable chances out of $4 + 5 = 9$ chances altogether.

$$P(\text{bid will be low bid}) = \frac{4}{4+5} = \frac{4}{9}$$

(b) Find the odds against that bid being the low bid.

Solution There is a 5/9 chance that the bid will not be the low bid, so the odds against a low bid are

$$\frac{P(\text{bid will not be low})}{P(\text{bid will be low})} = \frac{5/9}{4/9} = \frac{5}{4}$$

or 5:4. ▲ ⑦

Relative Frequency Probability

In many real-life problems, it is not possible to establish exact probabilities for events. Instead, useful estimates are often found by drawing on past experience. This approach is called **relative frequency probability.** We calculate our estimate of the probability by determining the percentage of the responses with the characteristic of interest. Estimates based on relative frequency probability are sometimes called empirical probabilities. The next example shows one approach to finding such relative frequency probabilities.

⑥
Suppose $P(E) = 9/10$. Find the odds

(a) in favor of E;

(b) against E.
Suppose the chance of snow in 80%. Find the odds

(c) in the favor of snow;

(d) against snow

Answers:

(a) 9 to 1

(b) 1 to 9

(c) 4 to 1

(d) 1 to 4

⑦
If the odds in favor of event E are 1 to 5, find

(a) $P(E)$;

(b) $P(E')$.

Answers:

(a) 1/6

(b) 5/6

▼ **EXAMPLE 9**   The table lists the number of siblings indicated by respondents in the 2002 General Social Survey.*

Number of Siblings	Frequency
0	140
1	505
2	583
3	457
4	314
5	224
6	157
7	115
8	77
9	49
10 or more	137
Total	2758

(a) Find the relative frequency probability of having 0, 1, 2, . . . , 10 or more siblings by first finding the total number of respondents and then dividing the number in each category by the total.

Solution Verify that the number of responses in each category sum to 2758. The probability that a respondent has zero siblings, for example, is $P(0) = 140/2758 \approx .051$. Similarly, we could divide each count by 2758, with the following results (rounded to three decimal places):

Number of Siblings	Probabilities
0	.051
1	.183
2	.211
3	.166
4	.114
5	.081
6	.057
7	.042
8	.028
9	.018
10 or More	.050

The numbers in this table actually sum to 1.001. In theory, they should total 1.000, but this does not always occur with rounded numbers.

When conducting a study such as the General Social Survey, we use the probability estimates from the sample to make estimates for the population of the United States. So we estimate that the probability that a randomly chosen American has no siblings is .051.

*The National Data Program for Social Sciences, National Opinion Research Center, University of Chicago, www.norc.uchicago.edu/projects/gensoc.asp.

(b) Find the probability that a randomly chosen American has one or two siblings.

Solution The categories in the table are disjoint events. Thus, to find the probability that a randomly chosen American has one or two siblings, we use the addition rule to calculate

$$P(1 \text{ or } 2) = .183 + .211 = .394.$$

We could get this same result by summing the counts for 1 and 2 siblings and then dividing by 2758. ▲ ⑧

A table of probabilities, as in Example 9, sets up a probability distribution; that is, for each possible outcome of an experiment, a number, called the probability of that outcome, is assigned. This assignment may be done in any reasonable way (on a relative frequency basis, as in Example 9, or by theoretical reasoning, as in Section 8.3), provided that it satisfies the following conditions.

⑧
Let E be the event of having more than 4 siblings. Find $P(E)$.

Answer:

.276

Properties of Probability

Let S be a sample space consisting of n distinct outcomes $s_1, s_2, \ldots, s_n$. An acceptable probability assignment consists of assigning, to each outcome s_i, a number p_i (the probability of s_i) according to these rules.

1. The probability of each outcome is a number between 0 and 1.

$$0 \le p_1 \le 1, \quad 0 \le p_2 \le 1, \ldots, \quad 0 \le p_n \le 1$$

2. The sum of the probabilities of all possible outcomes is 1.

$$p_1 + p_2 + p_3 + \cdots + p_n = 1$$

▼ **EXAMPLE 10**   Again using the 2002 General Social Survey,* let L indicate the event that the respondent had a "liberal" political tendency, and let M indicate that the respondent believes marijuana use should be legal. From the survey, we have the following estimates:

$$P(L) = .27, \quad P(M) = .37, \quad P(L \cap M) = .15.$$

(a) Find the probability that a respondent does not have a liberal tendency and does not support legalizing the use of marijuana.

Solution Place the given information on a Venn diagram, starting with .15 in the intersection of the regions L and M. (See Figure 8.21.) As stated earlier, event L has probability .27. Since .15 has already been placed inside the intersection of L and M,

$$.27 - .15 = .12$$

goes inside region L, but outside the intersection of L and M. In the same way,

$$.37 - .15 = .22$$

goes inside the region for M and outside the overlap.

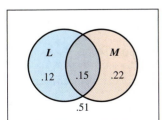

FIGURE 8.21

*The National Data Program for Social Sciences, National Opinion Research Center, University of Chicago, www.norc.uchicago.edu/projects/gensoc.asp.

Using regions L and M, we see that the event we want is $L' \cap M'$. From the Venn diagram in Figure 8.21, the labeled regions have a total probability of

$$.12 + .15 + .22 = .49.$$

Since the entire region of the Venn diagram must have probability 1, the region outside L and M, or $L' \cap M'$, has probability

$$1 - .49 = .51.$$

The probability that a respondent does not have a liberal tendency and does not support marijuana legalization is .51.

(b) Find the probability that the respondent does not have a liberal tendency *or* does not support marijuana legalization.

Solution The corresponding region, $L' \cup M'$, has probability

$$.51 + .21 + .22 = .85. \; \blacktriangle \; \langle 9 \rangle$$

9

Find the probability that a respondent in Example 10 is

(a) not supportive of the legalization of marijuana;

(b) not a liberal and a supporter of the legalization of marijuana.

Answers:

(a) .63

(b) .22

◆ 8.4 Exercises

1. Define disjoint events in your own words.

Decide whether the events in Exercises 2–7 are disjoint.

2. Owning a SUV and owning a Hummer

3. Wearing a hat and wearing glasses

4. Being married and being under 30 years old

5. Being a doctor and being under 5 years old

6. Being male and being a nurse

7. Being female and being a pilot

Two dice are rolled. Find the probabilities of rolling the given sums. (See Examples 2, 4, and 5.)

8. **(a)** 2
   (b) 4
   (c) 5
   (d) 6

9. **(a)** 8
   (b) 9
   (c) 10
   (d) 13

10. **(a)** 9 or more
   (b) Less than 7
   (c) Between 5 and 8 (exclusive)

11. **(a)** Not more than 5
   (b) Not less than 8
   (c) Between 3 and 7 (exclusive)

Tami goes shopping and sees three kinds of shoes: flats, 2" heels, and 3" heels. The shoes come in 2 shades of beige (light and dark)

and black. If each option has an equal chance of being selected, find the probabilities of the following events:

12. The shoes have a heel.

13. The shoes are black.

14. The shoes have a heel and are beige.

One card is drawn from an ordinary deck of 52 cards. Find the probabilities of drawing the following cards. (See Example 1).

15. **(a)** A 4 or a queen
   (b) A 3 or a spade
   (c) A black card or a 9

16. **(a)** A 10 or a jack
   (b) A red card or a 5
   (c) A 8 or a black 7
   (d) A heart or a black card
   (e) A face card or a club

17. **(a)** Less than a 4 (count aces as ones)
   (b) A club or a 4
   (c) A red card or king
   (d) A club or a queen
   (e) A black card or a face card

Ms. Elliott invites 10 relatives to a party: her mother, 3 aunts, 2 uncles, 2 sisters, 1 male cousin, and 1 female cousin. If the chances of any 1 guest arriving first are equally likely, find the probabilities that the first guest to arrive is as follows.

18. **(a)** A sister or an aunt
   (b) A sister or a cousin
   (c) A sister or her mother

19. (a) An aunt or a cousin
   (b) A male or an uncle
   (c) A female or a cousin

Six people live in a dorm suite. Two are to be selected to go to the campus café to pick up a pizza. Of course, no one wants to go, so the six names (Connie, Casey, Lindsey, Jackie, Tasia, and Lisa) are placed in a hat. After they are mixed, two names are selected (without replacement). Find the probabilities in Exercises 20 and 21.

20. Both names selected begin with the letter "L".

21. The first name selected begins with "L" and the second name selected begins with "J" or "C".

Use Venn diagrams to work Exercises 22 and 23.

22. Suppose $P(E) = .30$, $P(F) = .51$, and $P(E \cap F) = .19$. Find each of the following.
   (a) $P(E \cup F)$ **(b)** $P(E' \cap F)$
   (c) $P(E \cap F')$ **(d)** $P(E' \cup F')$

23. Let $P(Z) = .40$, $P(Y) = .30$, and $P(Z \cup Y) = .58$. Find each of the following probabilities.
   (a) $P(Z' \cap Y')$ **(b)** $P(Z' \cup Y')$
   (c) $P(Z' \cup Y)$ **(d)** $P(Z \cap Y')$

24. Define what is meant by odds.

A single fair die is rolled. Find the odds in favor of getting the results in Exercises 25–28. (See Examples 6 and 7.)

25. 2   **26.** 2, 3, 4

27. 2, 3, 5, or 6   **28.** Some number more than 5

29. A marble is drawn from a box containing 3 yellow, 4 white, and 8 blue marbles. Find the odds in favor of drawing the following.
   (a) A yellow marble   **(b)** A blue marble
   (c) A white marble

30. Find the odds of *not* drawing a white marble in Exercise 29.

31. Two dice are rolled. Find the odds of rolling a 7 or an 11.

32. In the "Ask Marilyn" column of *Parade* magazine, a reader wrote about the following game: You and I each roll a die. If your die is higher than mine, you win. Otherwise, I win. The reader thought that the probability that each player wins is 1/2. Is this correct? If not, what is the probability that each player wins?*

33. On page 134 of Roger Staubach's autobiography, *First Down, Lifetime to Go*, Staubach makes the following statement regarding his experience in Vietnam:

> Odds against a direct hit are very low but when your life is in danger, you don't worry too much about the odds.

Is this wording consistent with our definition of odds, for and against? How could it have been said so as to be technically correct?

---

*Parade Magazine, November 6, 1994, p. 10. Reprinted by permission of the William Morris Agency, Inc., on behalf of the author. Copyright © 1994 by Marilyn vos Savant.

34. The table gives the odds that a particular event will occur.* Convert each odds to the probability that the event will occur.

Event	Odds for the Event
A resident of the former East Germany wants the Berlin Wall back.	1:7
A Russian scientist says he or she would consider working for North Korea.	1:6
An American would say that driving a fuel-efficient car is an act of patriotism.	33:17
A traffic fatality from anywhere in the world occurs in China.	1:4

Which of Exercises 35–43 are examples of relative frequency probability?

35. The probability of heads on 5 consecutive tosses of a coin

36. The probability that a freshman entering college will graduate with a degree

37. The probability that a person is allergic to penicillin

38. The probability of drawing an ace from a standard deck of 52 cards

39. The probability that a person will get lung cancer from smoking cigarettes

40. A weather forecast that predicts a 70% chance of rain tomorrow

41. A gambler's claim that $P(\text{even}) = 1/2$ on a roll of a fair die

42. A surgeon's prediction that a patient has a 90% chance of full recovery

43. A first-year student's claim that he will have a date with a senior by the semester's end.

An experiment is conducted for which the sample space is $S = \{s_1, s_2, s_3, s_4, s_5\}$. Which of the probability assignments in Exercises 44–49 is possible for this experiment? If an assignment is not possible, tell why.

44.
Outcomes	s_1	s_2	s_3	s_4	s_5
Probabilities	.09	.32	.21	.25	.13

45.
Outcomes	s_1	s_2	s_3	s_4	s_5
Probabilities	.92	.03	0	.02	.03

46.
Outcomes	s_1	s_2	s_3	s_4	s_5
Probabilities	1/3	1/4	1/6	1/8	1/10

---

*Harper's Index for May 2005, www.harpers.org.

47.

Outcomes	s_1	s_2	s_3	s_4	s_5
Probabilities	1/5	1/3	1/4	1/5	1/10

48.

Outcomes	s_1	s_2	s_3	s_4	s_5
Probabilities	.64	−.08	.30	.12	.02

49.

Outcomes	s_1	s_2	s_3	s_4	s_5
Probabilities	.05	.35	.5	.2	−.3

*One way to solve a probability problem is to repeat the experiment many times, keeping track of the results. Then the probability can be approximated by using the basic definition of the probability of an event E: $P(E) = n(E)/n(S)$, where E occurs $n(E)$ times out of $n(S)$ trials of an experiment. This is called the **Monte Carlo method** of finding probabilities. If physically repeating the experiment is too tedious, it may be simulated with the use of a random-number generator, available on most computers and scientific or graphing calculators. To simulate a coin toss or the roll of a die on a graphing calculator, change the setting to fixed decimal mode with 0 digits displayed. To simulate multiple tosses of a coin, press RAND (or RANDOM or RND#) in the PROB submenu of the MATH (or OPTN) menu, and then press ENTER repeatedly. Interpret 0 as a head and 1 as a tail. To simulate multiple rolls of a die, press RAND × 6 + .5, and then press ENTER repeatedly.*

50. Suppose two dice are rolled. Use the Monte Carlo method with at least 50 repetitions to approximate the following probabilities. Compare with the results of Exercise 11.
   (a) P(the sum is not more than 5)
   (b) P(the sum is not less than 8)

51. Suppose two dice are rolled. Use the Monte Carlo method with at least 50 repetitions to approximate the following probabilities. Compare with the results of Exercise 10.
   (a) P(the sum is 9 or more)
   (b) P(the sum is less than 7)

52. Suppose three dice are rolled. Use the Monte Carlo method with at least 100 repetitions to approximate the following probabilities.
   (a) P(the sum is 5 or less)
   (b) P(neither a 1 nor a 6 is rolled)

53. Suppose a coin is tossed 5 times. Use the Monte Carlo method with at least 50 repetitions to approximate the following probabilities.
   (a) P(exactly 4 heads)   **(b)** P(2 heads and 3 tails)

54. **Business**   Suppose that 8% of a certain batch of calculators have a defective case and that 11% have defective batteries. Also, 3% have both a defective case and defective batteries. A calculator is selected from the batch at random. Find the probability that the calculator has a good case and good batteries.

55. **Business**   The table shows the probabilities of a family's household income in 2002.* Find the probabilities that a family's household incomes are the following.

(a) $35,000 or more   **(b)** Less than $75,000
(c) $25,000–74,999   **(d)** $34,999 or less

Income (Dollars)	Probability
Less than $15,000	.161
$15,000–24,999	.132
$25,000–34,999	.123
$35,000–49,999	.151
$50,000–74,999	.083
$75,000–99,999	.110
$100,000 or More	.140

56. **Health**   The table gives the probabilities of a live infant birth weight for the state of Georgia in 2003.* Find the probabilities that new infants' weights are the following.
   (a) 500–2499 grams
   (b) Less than 2500 grams
   (c) 2500 grams or more
   (d) Between 1500 and 3999 grams

Birthweight (grams)	Probability
Less than 500	.002
500–1499	.015
1500–2499	.072
2500–3999	.833
4000 or more	.078

Business   The table shows the probability that a customer of a department store will make a purchase in the indicated range.

Amount Spent	Probability
Below $10	.07
$10–$24.99	.18
$25–$49.99	.21
$50–$74.99	.16
$75–$99.99	.11
$100–$199.99	.09
$200–$349.99	.07
$350–$499.99	.08
$500 or more	.03

Statistical Abstract of the United States, 2004–2005.

*Georgia Office of Health Information and Policy, *Vital Statistics Reports—Natality, 2003*; http://health.state.ga.us.

Find the probabilities that a customer makes a purchase in the following ranges.

57. (a) Less than $25
   (b) More than $24.99
   (c) $50 to $199.99

58. (a) Less than $350
   (b) $75 or more
   (c) $200 or more

59. **Business**   The probability that a company will make a profit this year is .74. Find the odds against the company making a profit.

60. **Natural Science**   A study on body types gave the following results: 45% were short, 25% were short and overweight, and 24% were tall and not overweight. Find the probabilities that a person is the following.
   (a) Overweight
   (b) Short, but not overweight
   (c) Tall and overweight

61. **Natural Science**   Color blindness is an inherited characteristic that is more common in males than in females. If M represents male and C represents red–green color blindness, we use the relative frequencies of the incidences of males and red–green color blindness as probabilities to get

$P(C) = .039$, $P(M \cap C) = .035$, $P(M \cup C) = .495$.*
   Find the following probabilities.
   (a) $P(C')$
   (b) $P(M)$
   (c) $P(M')$
   (d) $P(M' \cap C')$
   (e) $P(C \cap M')$
   (f) $P(C \cup M')$

62. **Natural Science**   Gregor Mendel, an Austrian monk, was the first to use probability in the study of genetics. In an effort to understand the mechanism of character transmittal from one generation to the next in plants, he counted the number of occurrences of various characteristics. Mendel found that the flower color in certain pea plants obeyed this scheme:

   Pure red crossed with pure white produces red.

   From its parents, the red offspring received genes for both red (R) and white (W), but in this case red is *dominant* and white *recessive*, so the offspring exhibits the color red. However, the offspring still caries both genes, and when two such offspring are crossed, several things can happen in the third generation. The table in the next col-

umn, which is called a *Punnett square*, shows the equally likely outcomes.

		Second Parent	
		R	*W*
FIRST PARENT	*R*	*RR*	*RW*
	W	*WR*	*WW*

Use the fact that red is dominant over white to find each of the following. Assume that there are an equal number of red and white genes in the population.
   (a) P(a flower is red)
   (b) P(a flower is white)

63. **Natural Science**   Mendel found no dominance in snapdragons, with one red gene and one white gene producing pink-flowered offspring. These second-generation pinks, however, still carry one red and one white gene, and when they are crossed, the next generation still yields the Punnett square in Exercise 62. Find each of the following probabilities.
   (a) P(red)
   (b) P(pink)
   (c) P(white)
   (Mendel verified these probability ratios experimentally and did the same for many characteristics other than flower color. His work, published in 1866, was not recognized until 1890.)

64. **Natural Science**   Researchers studied bird electrocution in Northeast Spain.* In their study, there were a total of 50 bird electrocutions, with the following probabilities related to the type of bird.

Type of Bird	Probability
Raptors	.48
Crows	.22
Owls	.18
Other	.12

Find the probability of the following events.
   (a) An owl or crow death
   (b) Not a raptor death

---

*The probabilities of a person being male or female are from *The World Almanac and Book of Facts*, 2002. The probabilities of a male and female being color blind are from *Parsons' Diseases of the Eye* (18th ed.), by Stephen J. H. Miller (Churchill Livingston, 1990), p. 269. This reference gives a range of 3 to 4% for the probability of gross color blindness in men; we used the midpoint of that range.

---

*Albert Tintó, Joan Real, Santi Mañosa, "A Classification Method of Power Lines to Prevent Forest Fires Caused by Bird Electrocution," on-line, http://www.ctfc.es/confeinfor/articles/posters/PAPER%20TINT%C3%93.pdf.

Business *Exercise 65 and 66 deal with a May 2005 report from the Bureau of Labor Statistics on the employment status of adults age 20 or older. All numbers are in thousands.*

65. Business There were a total of 76,439 males in the workforce, 73,100 males who were employed, and 135,564 employed (both males and females).* With the entire number of males and females being 141,918, find the probabilities that a person is
   (a) Male and not employed
   (b) Female and not employed
   (c) Not employed
   (d) Female and employed
   (*Hint:* Draw two circles (denoting male and employed, respectively) that overlap in the middle)

66. Business There were a total of 7,614 black males in the work-force, 6,914 black males who were employed, and 68,250 males employed (both whites and blacks).* With the total number of white and black males together being 71,460, find the probabilities that a person is
   (a) White and not employed
   (b) Black and employed
   (c) Not employed
   (d) White and employed
   (*Hint:* Draw two circles (denoting black and employed respectively) that overlap in the middle.)

67. Social Science There were a total of 199,850 female military personnel on active duty in March 2001 in various ranks and military branches, as listed in the table:[†]
   (a) Convert the numbers in the table to probabilities.

	Army (A)	Air Force (B)	Navy (C)	Marines (D)
Officers (O)	10,671	11,662	7,761	951
Enlisted (E)	61,871	54,537	43,010	9,387

   (b) Find the probability that a randomly selected woman is enlisted in the Army.
   (c) Find the probability that a randomly selected woman is an officer in the Navy or Marine Corps.
   (d) Find $P(A \cup B)$.
   (e) Find $P(E \cup (C \cup D))$.

68. Social Science In the 2002 General Social Survey, respondents were asked their political leaning and whether they favored the legalization of marijuana.* Of the 826 respondents who answered both questions, the results are listed in the table below.

	Liberal	Moderate	Conservative
In favor of Legalization	120	102	78
Not in favor of Legalization	98	222	206

   (a) Find the probability that a randomly chosen respondent is a liberal and favors legalization.
   (b) Find the probability that a randomly chosen respondent is conservative and favors legalization.
   (c) Find the probability that a randomly chosen respondent is moderate and favors legalization.

---

*Bureau of Labor Statistics, www.bls.gov.

†U.S. Department of Defense, Defense Manpower Data Center, unpublished data, March 30, 2001. Available at www.infoplease.com/ipa/A0004600.html.

*The National Data Program for Social Services, National Opinion Research Center, University of Chicago, www.norc.uchicago.edu/projects/gensoc.asp.

8.5 Conditional Probability and Independent Events

The training manager for a large brokerage firm has noticed that some of the company's brokers use its research advice, while other brokers tend to go with their own beliefs about which stocks will go up. To see whether the research department performs better than the beliefs of the brokers, the manager conducted a survey of 100 brokers, with the results shown in the following table:

	Picked a Stock That Went Up	Didn't Pick a Stock That Went Up	Totals
Used Research	30	15	45
Didn't Use Research	30	25	55
Totals	60	40	100

Letting A be the event "picked a stock that went up" and letting B be the event "used research," $P(A)$, $P(A')$, $P(B)$, and $P(B')$ can be found. For example, the chart shows that a total of 60 brokers picked stocks that went up, so $P(A) = 60/100 = .6$.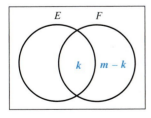

Use the data in the table to find

(a) $P(B)$,

(b) $P(A')$,

(c) $P(B')$.

Answers:

(a) .45

(b) .4

(c) .55

Suppose we want to find the probability that a broker using research will pick a stock that goes up. From the table, of the 45 brokers who use research, 30 picked stocks that went up, so

$$P(\text{broker who uses research picks stocks that go up}) = \frac{30}{45} \approx .667.$$

This is a number different from the probability that a broker picks a stock that goes up, .6, because *we have additional information* (the broker uses research) *that has reduced the sample space.* In other words, we found the probability that a broker picks a stock that goes up, A, given the additional information that the broker uses research, B. This is called the *conditional probability* of event A, given that event B has occurred, written $P(A|B)$. ($P(A|B)$ may also be read as "the probability of A given B.")

In the preceding example,

$$P(A|B) = \frac{30}{45}.$$

If we divide the numerator and denominator by 100 (the size of the sample space), this can be written as

$$P(A|B) = \frac{\dfrac{30}{100}}{\dfrac{45}{100}} = \frac{P(A \cap B)}{P(B)},$$

where $P(A \cap B)$ represents, as usual, the probability that both A and B will occur.

To generalize this result, assume that E and F are two events for a particular experiment. Assume also that the sample space S for the experiment has n possible equally likely outcomes. Suppose event F has m elements and $E \cap F$ has k elements ($k \le m$). Then, using the fundamental principle of probability yields

$$P(F) = \frac{m}{n} \quad \text{and} \quad P(E \cap F) = \frac{k}{n}.$$

$E \qquad F$

$k \quad m - k$

Event F has a total of m elements.

FIGURE 8.22

We now want to find $P(E|F)$: the probability that E occurs, given that F has occurred. Since we assume that F has occurred, we reduce the sample space to F: we look only at the m elements inside F. (See Figure 8.22.) Of these m elements, there are k elements where E also occurs, because $E \cap F$ has k elements. This makes

$$P(E|F) = \frac{k}{m}.$$

Divide numerator and denominator by n to get

$$P(E|F) = \frac{k/n}{m/n} = \frac{P(E \cap F)}{P(F)}.$$

The last result motivates the following definition of conditional probability.

The **conditional probability** of an event E, given event F, written $P(E|F)$, is

$$P(E|F) = \frac{P(E \cap F)}{P(F)} = \frac{P(E \text{ and } F)}{P(F)}, P(F) \neq 0.$$

This definition tells us that, for equally likely outcomes, conditional probability is found by *reducing the sample space to event F* and then finding the number of outcomes in F that are also in event E. Thus,

$$P(E|F) = \frac{n(E \cap F)}{n(F)}.$$

Although the definition of conditional probability was motivated by an example with equally likely outcomes, it is valid in all cases. For an intuitive explanation, think of the formula as giving the probability that both E and F occur, compared with the entire probability of F occuring.

▼ **EXAMPLE 1**   Use the information given in the table on the previous page to find the following probabilities.

(a) $P(B|A)$

Solution This represents the probability that the broker used research, given that the broker picked a stock that went up. Reduce the sample space to A. Then find $n(A \cap B)$ and $n(A)$.

$$P(B|A) = \frac{n(A \cap B)}{n(A)} = \frac{30}{60} = \frac{1}{2}$$

If a broker picked a stock that went up, then the probability is $1/2$ that the broker used research.

(b) $P(A'|B)$

Solution In words, this is the probability that a broker picked a stock that did not go up, even though he or she used research.

$$P(A'|B) = \frac{n(A' \cap B)}{n(B)} = \frac{15}{45} = \frac{1}{3}$$

(c) $P(B'|A')$

Solution Here, we want the probability that a broker who picked a stock that did not go up did not use research.

$$P(B'|A') = \frac{n(B' \cap A')}{n(A')} = \frac{25}{40} = \frac{5}{8} \; ▲ \; ⟨2⟩$$

Venn diagrams can be used to illustrate problems in conditional probability. A Venn diagram for Example 1, in which the probabilities are used to indicate the number in the set defined by each region, is shown in Figure 8.23. In the diagram, $P(B|A)$ is found by *reducing the sample space to just set A.* Then $P(B|A)$ is the ratio of the number in that part of set B which is also in A to the number in set A, or $.3/.6 = .5$.

⟨2⟩

The table shows the results of a survey of a buffalo herd.

	Males	Females	Totals
Adults	500	1300	1800
Calves	520	500	1020
Totals	1020	1800	2820

Let M represent "male" and A represent "adult." Find each of the following.

(a) $P(M|A)$
(b) $P(M'|A)$
(c) $P(A|M')$
(d) $P(A'|M)$
(e) State the probability in part (d) in words.

Answers:

(a) 5/18
(b) 13/18
(c) 13/18
(d) 26/51
(e) The probability that a buffalo is a calf, given that it is a male

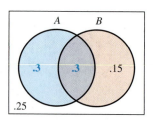

FIGURE 8.23

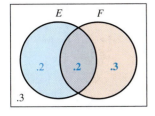

FIGURE 8.24

▼ **EXAMPLE 2**   Given $P(E) = .4, P(F) = .5$, and $P(E \cup F) = .7$, find $P(E|F)$.

Solution Find $P(E \cap F)$ first. Then use a Venn diagram to find $P(E|F)$. By the addition rule,

$$P(E \cup F) = P(E) + P(F) - P(E \cap F)$$
$$.7 = .4 + .5 - P(E \cap F)$$
$$P(E \cap F) = .2.$$

Now use the probabilities to indicate the number in each region of the Venn diagram in Figure 8.24. $P(E|F)$ is the ratio of the probability of that part of E which is in F to the probability of F, or

$$P(E|F) = \frac{P(E \cap F)}{P(F)} = \frac{.2}{.5} = \frac{2}{5} = .4. \quad \blacktriangle \; \langle 3 \rangle$$

3

Find $P(F|E)$ if $P(E) = .3$, $P(F) = .4$, and $P(E \cup F) = .6$.

Answer:

1/3

▼ **EXAMPLE 3**   Two fair coins were tossed, and it is known that at least one was a head. Find the probability that both were heads.

Solution The sample space has four equally likely outcomes: $S = \{hh, ht, th, tt\}$. Define two events:

$$E_1 = \text{at least 1 head} = \{hh, ht, th\}$$

and

$$E_2 = 2 \text{ heads} = \{hh\}.$$

Because there are four equally likely outcomes, $P(E_1) = 3/4$. Also, $P(E_1 \cap E_2) = 1/4$. We want the probability that both were heads, given that at least one was a head; that is, we want to find $P(E_2|E_1)$. Because of the condition that at least one coin was a head, the reduced sample space is

$$\{hh, ht, th\}.$$

Since only one outcome in this reduced sample space is 2 heads,

$$P(E_2|E_1) = \frac{1}{3}.$$

Alternatively, use the definition given earlier.

4

In Example 3, find the probability that exactly one coin showed a head, given that at least one was a head.

Answer:

2/3

$$P(E_2|E_1) = \frac{P(E_2 \cap E_1)}{P(E_1)} = \frac{1/4}{3/4} = \frac{1}{3} \quad \blacktriangle \; \langle 4 \rangle$$

It is important not to confuse $P(A|B)$ with $P(B|A)$. For example, in a criminal trial, a prosecutor may point out to the jury that the probability of the defendant's DNA profile matching that of a sample taken at the scene of the crime, given that the defendant is innocent, is very small. What the jury must decide, however, is the probability that the defendant is innocent, given that the defendant's DNA profile matches the sample. Confusing the two is an error sometimes called "the prosecutor's fallacy," and the 1990 conviction of a rape suspect in England was overturned by a panel of judges who ordered a retrial, because the fallacy made the original trial unfair.* This mistake is often called "confusion of the inverse."

In the next section, we will see how to compute $P(A|B)$ when we know $P(B|A)$.

Product Rule

If $P(E) \neq 0$ and $P(F) \neq 0$, then the definition of conditional probability shows that

$$P(E|F) = \frac{P(E \cap F)}{P(F)} \quad \text{and} \quad P(F|E) = \frac{P(F \cap E)}{P(E)}.$$

Using the fact that $P(E \cap F) = P(F \cap E)$, and solving each of these equations for $P(E \cap F)$, we obtain the following rule.

Product Rule of Probability

If E and F are events, then $P(E \cap F)$ may be found by either of these formulas:

$$P(E \cap F) = P(F) \cdot P(E|F) \quad \text{or} \quad P(E \cap F) = P(E) \cdot P(F|E).$$

The **product rule** gives a method for finding the probability that events E and F both occur. Here is a simple way to remember the ordering of E and F in the probability rule:

$$P(E \cap F) = P(F) \cdot P(E|F) \quad \text{or} \quad P(E \cap F) = P(E) \cdot P(F|E).$$

▼ **EXAMPLE 4**   The 2001–2002 National Health and Nutrition Examination Survey (NHANES) estimates that, of adults age 20 or older, approximately 2/3 are between the ages of 20 and 59 and approximately 1/3 are age 60 or older.* The survey also found that approximately 70% of adults age 20 to 59 had at least 12 alcoholic drinks in the last year.

Let A and Y represent the events "drank at least 12 alcoholic drinks in the last year" and "age 20 to 59," respectively. We want to find $P(A \cap Y)$. By the product rule,

$$P(A \cap Y) = P(Y) \cdot P(A|Y).$$

From the given information, $P(Y) = 2/3 = .67$, and the probability that someone age 20 to 59 drank at least 12 alcoholic drinks in the last year is $P(A|Y) = .7$. Then

$$P(A \cap Y) = .67(.70) = .47. \; \blacktriangle \; \langle 5 \rangle$$

⟨5⟩
In a litter of puppies, 3 were female and 4 were male. Half the males were black. Find the probability that a puppy chosen at random from the litter would be a black male.

Answer:

2/7

---

*David Pringle, "Who's the DNA Fingerprinting Pointing At?" *New Scientist*, January 29, 1994, pp. 51–52.

In Section 8.1, we used a tree diagram to find the number of subsets of a given set. By including the probabilities for each branch of a tree diagram, we convert it to a **probability tree.** The next examples show how a probability tree is used with the product rule to find the probability of a sequence of events.

▼ **EXAMPLE 5**   A company needs to hire a new director of advertising. It has decided to try to hire either person A or person B, both of whom are assistant advertising directors for its major competitor. To decide between A and B, the company does research on the campaigns managed by A or B (none are managed by both) and finds that A is in charge of twice as many advertising campaigns as B. Also, A's campaigns have satisfactory results three out of four times, while B's campaigns have satisfactory results only two out of five times. Suppose one of the competitor's advertising campaigns (managed by A or B) is selected randomly.

We can represent this situation schematically as follows. Let A denote the event "Person A does the job" and B the event "Person B does the job." Let S be the event "satisfactory results" and U the event "unsatisfactory results." Then the given information can be summarized in the probability tree in Figure 8.25. Since A does twice as many jobs as B, $P(A) = 2/3$ and $P(B) = 1/3$, as noted on the first-stage branches of the tree. When A does a job, the probability of satisfactory results is $3/4$ and of unsatisfactory results $1/4$, as noted on the second-stage branches. Similarly, the probabilities when B does the job are noted on the remaining second-stage branches. The composite branches labeled 1–4 represent the four disjoint possibilities for the running and outcome of the campaign.

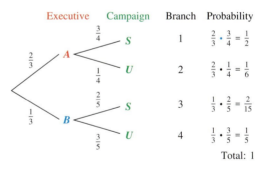

FIGURE 8.25

(a) Find the probability that A is in charge of a campaign that produces satisfactory results.

Solution   We are asked to find $P(A \cap S)$. We know that when A does the job, the probability of success is $3/4$; that is, $P(S|A) = 3/4$. Hence, by the product rule,

$$P(A \cap S) = P(A) \cdot P(S|A) = \frac{2}{3} \cdot \frac{3}{4} = \frac{1}{2}.$$

*2001–2002 National Health and Examination Survey, Centers for Disease Control, http://www.cdc.gov/nchs/nhanes.htm.

The event $(A \cap S)$ is represented by branch 1 of the tree, and as we have just seen, its probability is the product of the probabilities that make up that branch.

(b) Find the probability that B runs a campaign that produces satisfactory results.

Solution We must find $P(B \cap S)$. This event is represented by branch 3 of the tree, and as before, its probability is the product of the probabilities of the pieces of that branch:

$$P(B \cap S) = P(B) \cdot P(S|B) = \frac{1}{3} \cdot \frac{2}{5} = \frac{2}{15}.$$

(c) What is the probability that the selected campaign is satisfactory?

Solution The event S is the union of the disjoint events $A \cap S$ and $B \cap S$, which are represented by branches 1 and 3 of the tree diagram. By the addition rule,

$$P(S) = P(A \cap S) + P(B \cap S) = \frac{1}{2} + \frac{2}{15} = \frac{19}{30}.$$

Thus, the probability of an event that appears on several branches is the sum of the probabilities of each of these branches.

(d) What is the probability that the selected campaign is unsatisfactory?

Solution $P(U)$ can be read from branches 2 and 4 of the tree.

$$P(U) = \frac{1}{6} + \frac{1}{5} = \frac{11}{30}$$

Alternatively, because U is the complement of S,

$$P(U) = 1 - P(S) = 1 - \frac{19}{30} = \frac{11}{30}.$$

(e) Find the probability that either A runs the campaign or the results are satisfactory (or possibly both).

Solution Event A combines branches 1 and 2, while event S combines branches 1 and 3, so use branches 1, 2, and 3.

$$P(A \cup S) = \frac{1}{2} + \frac{1}{6} + \frac{2}{15} = \frac{4}{5} \quad \blacktriangle \; \langle 6 \rangle$$

6

Find each of the following probabilities for Example 5.

(a) $P(U|A)$

(b) $P(U|B)$

Answers:

(a) $1/4$

(b) $3/5$

▼ **EXAMPLE 6** Suppose 6 potential jurors remain in a jury pool and 2 are to be selected to sit on the jury for the trial. The races of the 6 potential jurors are 1 Hispanic, 3 Caucasian, and 2 African-American. If we select one juror at a time, find the probability that one Caucasian and one African-American are drawn.

Solution A probability tree showing the various possible outcomes is given in Figure 8.26. In this diagram, C represents the event "selecting a Caucasian juror" and A represents "selecting an African-American juror." On the first draw, P(C on the 1st) = $3/6 = 1/2$ because three of the six jurors are Caucasian. On the second draw, P(A on the 2nd | C on the 1st) = 2/5. One Caucasian juror has been removed, leaving 5, of which 2 are African-American.

We want to find the probability of selecting exactly one Caucasian and exactly one African-American. Two events satisfy this condition: selecting a Caucasian first and then selecting an African-American (branch 2 of the tree) and drawing an African-American juror first and then selecting a Caucasian juror (branch 4). For branch 2,

⟨7⟩
Find the probability of selecting an African-American juror and then a Caucasian juror

Answer:

1/5

$$P(C \text{ on 1st}) \cdot P(A \text{ on 2nd}|C \text{ on 1st}) = \frac{1}{2} \cdot \frac{2}{5} = \frac{1}{5} \quad ⟨7⟩$$

For branch 4, on which the African-American juror is selected first,

$$P(A \text{ first}) \cdot P(C \text{ second}|A \text{ first}) = \frac{1}{3} \cdot \frac{3}{5} = \frac{1}{5}.$$

Since these two events are disjoint, the final probability is the sum of the two probabilities.

⟨8⟩
In Example 6, find the probability of selecting a Caucasian juror and a Hispanic juror.

Answer:

1/10

$$P(\text{one } C, \text{one } A) = P(C \text{ on 1st}) \cdot P(A \text{ on 2nd}|C \text{ on 1st})$$
$$+ P(A \text{ on 1st}) \cdot P(C \text{ on 2nd}|A \text{ on 1st}) = \frac{2}{5} \quad ▲ \quad ⟨8⟩$$

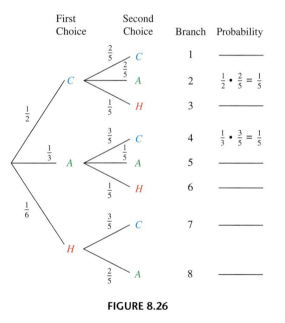

FIGURE 8.26

The product rule is often used in dealing with *stochastic processes*, which are mathematical models that evolve over time in a probabilistic manner. For example, selecting different jurors is such a process, in which the probabilities change with each successive selection. (Particular stochastic processes are studied further in Section 9.5.)

▼ **EXAMPLE 7**   Two cards are drawn without replacement from an ordinary deck (52 cards). Find the probability that the first card is a heart and the second card is red.

Solution   Start with the probability tree of Figure 8.27. (You may wish to refer to the deck of cards shown in Figure 8.19.) On the first draw, since there are 13 hearts in the 52 cards, the probability of drawing a heart first is $13/52 = 1/4$. On the second draw, since a (red) heart has been drawn already, there are 25 red cards in the remaining 51 cards. Thus, the probability of drawing a red card on the second draw, given that the first is a heart, is $25/51$. By the product rule of probability,

$$P(\text{heart on 1st and red on 2nd})$$
$$= P(\text{heart on 1st}) \cdot P(\text{red on 2nd}|\text{heart on 1st})$$
$$= \frac{1}{4} \cdot \frac{25}{51} = \frac{25}{204} \approx .1225. \; \blacktriangle \; \text{⑨}$$

▼ **EXAMPLE 8**   Three cards are drawn, without replacement, from an ordinary deck. Find the probability that exactly 2 of the cards are red.

Solution   Here we need a probability tree with three stages, as shown in Figure 8.28. The three branches indicated with arrows produce exactly 2 red cards from the draws. Multiply the probabilities along each of these branches and then add.

$$P(\text{exactly 2 red cards}) = \frac{26}{52} \cdot \frac{25}{51} \cdot \frac{26}{50} + \frac{26}{52} \cdot \frac{26}{51} \cdot \frac{25}{50} + \frac{26}{52} \cdot \frac{26}{51} \cdot \frac{25}{50}$$
$$= \frac{50,700}{132,600} = \frac{13}{34} \approx .382 \; \blacktriangle \; \text{⑩}$$

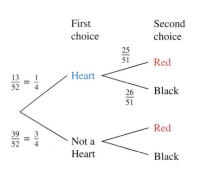

FIGURE 8.27

⑨ Find the probability of drawing a heart on the first draw and a black card on the second if two cards are drawn without replacement.

Answer:

$13/102 \approx .1275$

⑩ Use the tree in Example 8 to find the probability that exactly one of the cards is red.

Answer:

$13/34 \approx .382$

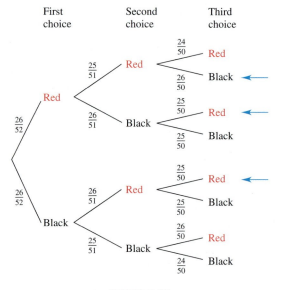

FIGURE 8.28

Independent Events

Suppose in Example 7 that we draw the two cards *with* replacement rather than without replacement. (That is, we put the first card back, shuffle them, then draw the second card.) If the first card is a heart, then the probability of drawing a red card on the second draw is 26/52, rather than 25/51, because there are still 52 cards in the deck, 26 of them red. In this case, $P(\text{red second}|\text{heart first})$ is the same as $P(\text{red second})$. The value of the second card is not affected by the value of the first card. We say that the event that the second card is red is *independent* of the event that the first card is a heart, since knowledge of the first card does not influence what happens to the second card. On the other hand, when we draw *without* replacement, the events that the first card is a heart and the second is red are *dependent* events. The fact that the first card is a heart means that there is one less red card in the deck, influencing the probability that the second card is red.

As another example, consider tossing a fair coin twice. If the first toss shows heads, the probability that the next toss is heads is still 1/2. Coin tosses are independent events, since the outcome of one toss does not influence the outcome of the next toss. Similarly, rolls of a fair die are independent events. On the other hand, the events "the milk is old" and "the milk is sour" are dependent events: if the milk is old, there is an increased chance that it is sour. Also, in the example at the beginning of this section, the events A (broker picked a stock that went up) and B (broker used research) are dependent events, because information about the use of research affects the probability of picking stocks that go up. That is, $P(A|B)$ is different from $P(A)$.

If events E and F are independent, then the knowledge that E has occurred gives no (probability) information about the occurrence or nonoccurrence of event F. That is, $P(F)$ is exactly the same as $P(F|E)$, or

$$P(F|E) = P(F).$$

This, in fact, is the formal definition of independent events.

E and F are **independent events** if

$$P(F|E) = P(F) \quad \text{or} \quad P(E|F) = P(E).$$

If the events are not independent, they are **dependent events.**

When E and F are independent events, $P(F|E) = P(F)$ and the product rule becomes

$$P(E \cap F) = P(E) \cdot P(F|E) = P(E) \cdot P(F).$$

Conversely, if this equation holds, it follows that $P(F) = P(F|E)$. Consequently, we have the following useful fact.

Product Rule for Independent Events

E and F are independent events if and only if

$$P(E \cap F) = P(E) \cdot P(F).$$

▼ **EXAMPLE 9** A calculator requires a keystroke assembly and a logic circuit. Assume that 99% of the keystroke assemblies are satisfactory and 97% of the logic circuits are satisfactory. Find the probability that a finished calculator will be satisfactory.

Solution If the failure of a keystroke assembly and the failure of a logic circuit are independent events, then

$$P(\text{satisfactory calculator})$$
$$= P(\text{satisfactory keystroke assembly}) \cdot P(\text{satisfactory logic circuit})$$
$$= (.99)(.97) \approx .96. \; \blacktriangle \; ⟨11⟩$$

⟨11⟩

Find the probability of getting 4 successive heads on 4 tosses of a fair coin.

Answer:

1/16

CAUTION It is common for students to confuse the ideas of *disjoint* events and *independent* events. Events E and F are disjoint if $E \cap F = \emptyset$. For example, if a family has exactly one child, the only possible outcomes are $B = \{\text{boy}\}$ and $G = \{\text{girl}\}$. These two events are disjoint. However, the events are *not* independent, since $P(G|B) = 0$ (if a family with only one child has a boy, the probability that it has a girl is then 0). Since $P(G|B) \neq P(G)$, the events are not independent. Of all the families with exactly *two* children, the events $G_1 = \{\text{first child is a girl}\}$ and $G_2 = \{\text{second child is a girl}\}$ are independent, because $P(G_2|G_1)$ equals $P(G_2)$. However, G_1 and G_2 are not disjoint, since $G_1 \cap G_2 = \{\text{both children are girls}\} \neq \emptyset$.

To show that two events E and F are independent, we can show that $P(F|E) = P(F)$ or that $P(E|F) = P(E)$ or that $P(E \cap F) = P(E) \cdot P(F)$. Another way is to observe that knowledge of one outcome does not influence the probability of the other outcome, as we did for coin tosses.

NOTE In some cases, it may not be apparent from the physical description of the problem whether two events are independent or not. For example, it is not obvious whether the event that a baseball player gets a hit tomorrow is independent of the event that he got a hit today. In such cases, it is necessary to use the definition and calculate whether $P(F|E) = P(F)$ or, equivalently, whether $P(E \cap F) = P(E) \cdot P(F)$.

▼ **EXAMPLE 10** On a typical January day in Manhattan, the probability of snow is .10, the probability of a traffic jam is .80, and the probability of snow or a traffic jam (or both) is .82. Are the event "it snows" and the event "a traffic jam occurs" independent?

⟨12⟩

In the U.S. population, the probability of being Hispanic is .11, the probability of living in California is .12, and the probability of being a Hispanic living in California is .04. Are the events being Hispanic and living in California independent?

Answer:

No

Solution Let S represent the event "it snows" and T represent the event "a traffic jam occurs." We must determine whether

$$P(T|S) = P(T) \quad \text{or} \quad P(S|T) = P(S).$$

We know that $P(S) = .10$, $P(T) = .8$, and $P(S \cup T) = .82$. We can use the addition rule (or a Venn diagram) to find $P(S \cap T) = .08$, $P(T|S) = .8$, and $P(S|T) = .1$. Since

$$P(T|S) = P(T) = .8 \quad \text{and} \quad P(S|T) = P(S) = .1,$$

the events "it snows" and "a traffic jam occurs" are independent. ▲ ⟨12⟩

Although we showed that $P(T|S) = P(T)$ and $P(S|T) = P(S)$ in Example 10, only one of these results is needed to establish independence. It is also important to note that independence of events does not necessarily follow our intuition; it is established from the mathematical definition of independence.

◆ 8.5 Exercises

If a single fair die is rolled, find the probability of rolling the following. (See Examples 1 and 2.)

1. 3, given that the number rolled was odd

2. 5, given that the number rolled was even

3. An odd number, given that the number rolled was 3

If two fair dice are rolled (recall the 36-outcome sample space), find the probability of rolling the following.

4. A sum of 8, given that the sum was greater than 7

5. A sum of 6, given that the roll was a "double" (two identical numbers)

6. A double, given that the sum was 9

If two cards are drawn without replacement from an ordinary deck (see Example 7), find the following probabilities.

7. The second is a heart, given that the first is a heart

8. The second is black, given that the first is a spade

9. A jack and a 10 are drawn.

10. An ace and a 4 are drawn.

11. In your own words, explain how to find the conditional probability $P(E|F)$.

12. Your friend asks you to explain how the product rule for independent events differs from the product rule for dependent events. How would you respond?

13. Another friend asks you to explain how to tell whether two events are dependent or independent. How would you reply? (Use your own words.)

Decide whether the two events listed are independent.

14. S is the event that it snows tomorrow and L is the event that the instructor is late for class.

15. R is the event that four semesters of theology are required to graduate from a certain college and A is that the college is a religiously affiliated school.

16. R is the event that it rains in the Amazon jungle and H is the event that an instructor in New York City writes a difficult exam.

17. T is the event that Tom Cruise's next movie grosses over $200 million and R is the event that the Republicans have a majority in Congress in 2008.

18. A student reasons that the probability in Example 3 of both coins being heads is just the probability that the other coin is a head—that is, $1/2$. Explain why this reasoning is wrong.

19. In a two-child family, if we assume that the probabilities of a male child and a female child are each .5, are the events *each child is the same sex* and *at most one child is male* independent? Are they independent for a three-child family?

20. Let A and B be independent events with $P(A) = \dfrac{1}{4}$ and $P(B) = \dfrac{1}{5}$. Find $P(A \cap B)$ and $P(A \cup B)$.

Business   *Use a probability tree or Venn diagram in Exercises 21–24. (See Examples 2 and 5–9.) A shop that produces custom kitchen cabinets has two employees: Sitlington and Čapek. 95% of Čapek's work is satisfactory and 10% of Sitlington's work is unsatisfactory. 60% of the shop's cabinets are made by Čapek (the rest by Sitlington). Find the following probabilities.*

21. An unsatisfactory cabinet was made by Čapek. (*Hint:* Consider which event came first.)

22. A finished cabinet is unsatisfactory.

23. **Social Science**   In Pakistan, 50% of marriages are consanguineous (between first cousins or people even more closely related). A recent study in Pakistan showed that 16% of children from unrelated marriages died by age 10, while 21% of children from consanguineous marriages died by age 10. Find the following probabilities.
(a) A child survives.
(b) A child from a consanguineous marriage survives.

24. **Business**   According to the Bureau of Labor Statistics, 41% of women age 16 years or older were not in the labor force in May 2005. Also, 93% of the women were age 20 or older. In addition, 97% of the women age 16 or older either were not in the labor force or were age 20 or older. Find the probability that
(a) a woman is not in the labor force if she is age 20 years or older.
(b) a woman is age 20 years or older, given that she is not in the labor force.

25. **Health**   The following table, based on data from the Centers for Disease Control, gives the estimated number of people with HIV/AIDS for men and women in 32 states with confidential, name-based reporting of HIV infection.*

Method	Male	Female	Total
Homosexual Contact (HC)	54,424	0	54,424
Intravenous Drug Use (IDU)	12,998	6,723	19,721
HC and IDU	5,022	0	5,022
Heterosexual Contact	15,413	26,882	42,295
Other	1,009	987	1,996
Total	88,866	34,592	123,458

(a) Given that the person living with HIV/AIDS is male, find the probability that he contracted the disease via heterosexual contact.

(b) Given that the person living with HIV/AIDS is female, find the probability that she contracted the disease via intravenous drug use.

(c) Find the probability that a person living with HIV/AIDS is female.

(d) Given that the person living with HIV/AIDS contracted the disease via heterosexual contact, find the probability that the person is female.

Health *The table shows the numbers for five types of cancer classified as "in situ" cancer (cancer confined to the point of origin) and invasive cancer (cancer that has spread to other tissues) for subjects enrolled in the California Teachers Study.†*

Type of Cancer	In Situ (A)	Invasive (B)
Breast (C)	507	2,596
Endometrial (D)	783	54
Cervix (E)	433	98
Lung (F)	0	135
Melanoma (G)	165	409

For a randomly selected teacher in the study, find the following probabilities.

26. $P(B \cap G)$

27. $P(A \cup C')$

28. $P(E' \cap (A \cup C))$

29. $P(F \cap B')$

Natural Science *In a letter to the journal Nature, Robert A. J. Matthews gives the following table of outcomes of forecast and*

weather over 1000 1-hour walks, based on the United Kingdom's Meteorological Office's 83% accuracy in 24-hour forecasts:*

	Rain	No Rain	Sum
Forecast of Rain	66	156	222
Forecast of No Rain	14	764	778
Sum	80	920	1000

30. Verify that the probability that the forecast called for rain, given that there was rain, is indeed 83%. Also, verify that the probability that the forecast called for no rain, given that there was no rain, is 83%.

31. Calculate the probability that there was rain, given that the forecast called for rain.

32. Calculate the probability that there was no rain, given that the forecast called for no rain.

33. Observe that your answer to part (c) is higher than 83% and that your answer to part (b) is much lower. Discuss which figure best describes the accuracy of the weather forecast in recommending whether you should carry an umbrella.

Natural Science *The following table shows frequencies for red–green color blindness, where M represents male and C represents color blind:*

	M	M'	Totals
C	.042	.007	.049
C'	.485	.466	.951
Totals	.527	.473	1.000

Use the table to find the following probabilities.

34. $P(M)$ 35. $P(C)$

36. $P(M \cap C)$ 37. $P(M \cup C)$

38. $P(M|C)$ 39. $P(M'|C)$

40. Are the events C and M dependent? Recall that two events E and F are dependent if $P(E|F) \neq P(E)$. (See Example 10.)

41. Are the events M' and C dependent?

42. **Natural Science** A scientist wishes to determine whether there is any dependence between color blindness (C) and deafness (D). Given the probabilities listed in the table, what should his findings be? (See Example 10.)

	D	D'	Totals
C	.0004	.0796	.0800
C'	.0046	.9154	.9200
Totals	.0050	.9950	1.0000

---

*Centers for Disease Control and Prevention, as reported in *USA Today,* December 1, 2004.

†Arti Parikh-Patel, Mark Allen, William E. Wright, and California Teachers Study Steering Committee, "Validation of Self-reported Cancers in the California Teachers Study," *American Journal of Epidemiology,* 2003; 157:539–545.

*Robert A. J. Matthews, *Nature* 382, August 29, 1996: p 3.

Social Science *The Motor Vehicle Department has found that the probability of a person passing the test for a driver's license on the first try is .75. The probability that an individual who fails on the first test will pass on the second try is .80, and the probability that an individual who fails the first and second tests will pass the third time is .70. Find the probability that an individual*

43. fails both the first and second tests;

44. will fail three times in a row;

45. will require at least two tries to pass the test.

On May 20–22, 2005, the Gallup Organization conducted a national poll that examined Americans' reading habits.* One of the questions asked was "During the past year, about how many books, either hardcover or paperback, did you read either all or part of the way through?" The following table gives results consistent with Gallup's results for 100 people in 2005 and compares them with results from a similar poll taken December 5–8, 2002:

	A None	B 1–5	C 6–10	D 11–50	E 51 +	F No Answer	Total
2005	16	38	14	25	6	1	100
2002	18	31	15	27	8	1	100
Total	34	69	29	52	14	2	200

Let G denote the set of people interviewed in 2005 and H the set of people interviewed in 2002. Let A be the set of people who read no books, B the set of people who read 1–5 books, C the set of people who read 6–10 books, D the set of people who read 11–50 books, and E the set of people who read 51 or more books. Finally, let F be the set of people who gave no answer. Find the probability that a person chosen at random

46. in 2005 read 1–5 books.

47. in 2002 read 51 or more books.

48. in 2005 read no more than 5 books.

49. in 2005 read between 1 and 10 books.

Business *The number (in thousands) of automobiles, buses and trucks registered in the United States for selected years is shown in the following table:†*

Year	Autos	Busses	Trucks
1999	132,432	729	83,148
2000	133,621	746	87,108
2001	137,633	750	92,045
2002	135,921	761	92,939

Among vehicles registered in 1999–2002, find the probability that

50. a certain vehicle was an automobile.

*www.gallup.com.

†Source: U.S. Federal Highway Administration, *Highway Statistics,* annual. See www.fhwa.dot.gov/policy/ohpi/hss/hsspubs.htm.

51. a certain vehicle was registered in 2002.

52. in 2002, a randomly selected vehicle was a truck.

53. a bus was registered in 1999.

Suppose the probability that the first record by a singing group will be a hit is .32. If the first record is a hit, so are all the group's subsequent records. If the first record is not a hit, the probability of the group's second record and all subsequent ones being hits is .16. If the first two records are not hits, the probability that the third is a hit is .08. The probability of a hit continues to decrease by half with each successive nonhit record. Find the probability that

54. the group will have at least one hit in its first four records.

55. the group will have exactly one hit in its first three records.

56. the group will have a hit in its first six records if the first three are not hits.

Work the following problems on independent events. (See Examples 9 and 10.)

57. **Business** Corporations such as banks, where a computer is essential to day-to-day operations, often have a second, backup computer in case of failure by the main computer. Suppose that there is a .003 chance that the main computer will fail in a given period and a .005 chance that the backup computer will fail while the main computer is being repaired. Suppose these failures represent independent events, and find the fraction of the time the corporation can assume that it will have computer service. How realistic is our assumption of independence?

58. **Business** According to a press release from Continental Airlines, the airline had an "on-time" performance rating of 84% in May 2005. Use this information, and assume that the event that a given flight takes place on time is independent of the event that another flight is on time to answer the following questions.
   (a) Elisabeta Gueyara plans to visit her company's branch offices; her journey requires 3 separate flights on Continental Airlines. What is the probability that all of these flights will be on time?
   (b) How reasonable do you believe it is to suppose the independence of being on time from flight to flight?

59. **Social Science** The probability that a key component of a space rocket will fail is .03.
   (a) How many such components must be used as backups to ensure that the probability that at least one of the components will work is .999999?
   (b) Is it reasonable to assume independence here?

60. **Natural Science** A medical experiment showed that the probability that a new medicine is effective is .75, the probability that a patient will have a certain side effect is .4, and the probability that both events occur is .3. Decide whether these events are dependent or independent.

61. **Social Science** A teacher has found that the probability that a student studies for a test is .6, the probability that a student gets a good grade on a test is .7, and the probability that both occur is .52. Are these events independent?

62.   Health   Refer to Exercises 26–29. Are the events of having breast cancer and having invasive cancer independent?

8.6 Bayes' Formula

Suppose the probability that a person gets lung cancer, given that the person smokes a pack or more of cigarettes daily, is known. For a research project, it might be necessary to know the probability that a person smokes a pack or more of cigarettes daily, given that the person has lung cancer. More generally, if $P(E|F)$ is known for two events E and F, can $P(F|E)$ be found? The answer is yes, we can find $P(F|E)$ by using the formula to be developed in this section. To develop this formula, we can use a probability tree to find $P(F|E)$. Since $P(E|F)$ is known, the first outcome is either F or F'. Then, for each of these outcomes, either E or E' occurs, as shown in Figure 8.29.

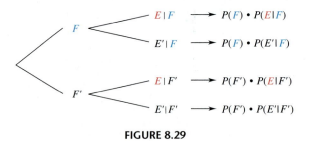

FIGURE 8.29

The four cases have the probabilities shown on the right. By the definition of conditional probability and the product rule,

$$P(E) = P(F \cap E) + P(F' \cap E),$$

$$P(F \cap E) = P(F) \cdot P(E|F), \quad \text{and} \quad P(F' \cap E) = P(F') \cdot P(E|F').$$

By substitution,

$$P(E) = P(F) \cdot P(E|F) + P(F') \cdot P(E|F')$$

and

$$P(F|E) = \frac{P(F \cap E)}{P(E)} = \frac{P(F) \cdot P(E|F)}{P(F) \cdot P(E|F) + P(F') \cdot P(E|F')}.$$

We have proved a special case of Bayes' formula, which is generalized later in this section. ①

① Use the special case of Bayes' formula to find $P(F|E)$ if $P(F) = .2$, $P(E|F) = .1$, and $P(E|F') = .3$.
   (*Hint:* $P(F') = 1 - P(F)$.)

Answer:

   $1/13 \approx .077$

Bayes' Formula (Special Case)

$$P(F|E) = \frac{P(F) \cdot P(E|F)}{P(F) \cdot P(E|F) + P(F') \cdot P(E|F')}.$$

▼ **EXAMPLE 1**   For a fixed length of time, the probability of worker error on a certain production line is .1, the probability that an accident will occur when there is a worker error is .3, and the probability that an accident will occur when there is no worker error is .2. Find the probability of a worker error if there is an accident.

Solution   Let E represent the event of an accident, and let F represent the event of a worker error. From the given information,

$$P(F) = .1, \quad P(E|F) = .3, \quad \text{and} \quad P(E|F') = .2.$$

These probabilities are shown on the probability tree in Figure 8.30.

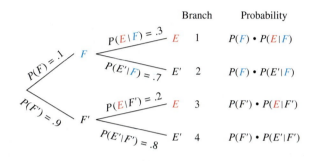

FIGURE 8.30

We find $P(F|E)$ by dividing the probability that both E and F occur, given by branch 1, by the probability that E occurs, given by the sum of branches 1 and 3.

$$P(F|E) = \frac{P(F) \cdot P(E|F)}{P(F) \cdot P(E|F) + P(F') \cdot P(E|F')}$$

$$= \frac{(.1)(.3)}{(.1)(.3) + (.9)(.2)} \approx .143 \; \blacktriangle \; \langle 2 \rangle$$

⟨2⟩

In Example 1, find $P(F'|E)$.

Answer:

$6/7 \approx .857$

The special case of Bayes' formula can be generalized to more than two possibilities with the probability tree of Figure 8.31. This diagram shows the paths that can produce an event E. We assume that events $F_1, F_2, \ldots, F_n$ are pairwise disjoint events (that is, events which, taken two at a time, are disjoint) whose union is the sample space and that E is an event that has occurred. See Figure 8.32 on the next page.

FIGURE 8.31

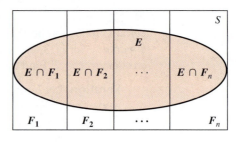

FIGURE 8.32

The probability $P(F_i|E)$, where $1 \leq i \leq n$, can be found by dividing the probability of the branch containing $P(E|F_i)$ by the sum of the probabilities of all the branches producing event E.

Bayes' Formula

$$P(F_i|E) = \frac{P(F_i) \cdot P(E|F_i)}{P(F_1) \cdot P(E|F_1) + \cdots + P(F_n) \cdot P(E|F_n)}.$$

This result is known as **Bayes' formula,** after the Reverend Thomas Bayes (1702–61), whose paper on probability was published about two hundred and forty years ago.

The statement of Bayes' formula can be daunting. Actually, it is easier to remember the formula by thinking of the probability tree that produced it. Go through the following steps.

Using Bayes' Formula

Step 1 Start a probability tree with branches representing events $F_1, F_2, \ldots, F_n$. Label each branch with its corresponding probability.

Step 2 From the end of each of these branches, draw a branch for event E. Label this branch with the probability of getting to it, or $P(E|F_i)$.

Step 3 There are now n different paths that result in event E. Next to each path, put its probability: the product of the probabilities that the first branch occurs, $P(F_i)$— and that the second branch occurs, $P(E|F_i)$: that is, $P(F_i) \cdot P(E|F_i)$.

Step 4 $P(F_i|E)$ is found by dividing the probability of the branch for F_i by the sum of the probabilities of all the branches producing event E.

▼ **EXAMPLE 2**   The 2002 General Social Survey of women who are age 18 or older indicated that 87% of married women have one or more children, 40% of never-married women have one or more children, and 88% of women who are divorced, separated, or widowed have one or more children.* The survey also indicted that 43% of women age 18 or older were currently married, 24% had never been married, and 33% were divorced, separated, or widowed (labeled "other"). Find the probability that a woman who has one or more children is married.

Solution   Let E represent the event "having one or more children," with F_1 representing "married women," F_2 representing "never-married women," and F_3 "other."

$$P(F_1) = .43 \quad P(E|F_1) = .87$$
$$P(F_2) = .24 \quad P(E|F_2) = .40$$
$$P(F_3) = .33 \quad P(E|F_3) = .88$$

We need to find $P(F_1|E)$, the probability that a woman is married, given that she has one or more children. First, draw a probability tree using the given information, as in Figure 8.33. The steps leading to event E are shown.

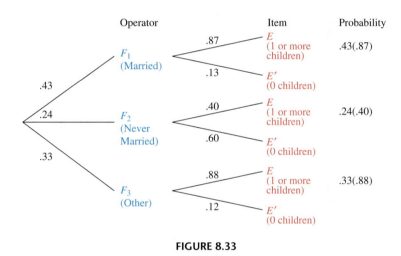

FIGURE 8.33

Find $P(F_1|E)$, using the top branch of the tree shown in Figure 8.33: divide the probability of this branch by the sum of the probabilities of all the branches leading to E.

$$P(F_1|E) = \frac{.43(.87)}{.43(.87) + .24(.40) + .33(.88)} = \frac{.3741}{.7605} \approx .4919 \; ▲ \; ⬥3⬥$$

⬥3⬥
In Example 2, find

(a) $P(F_2|E)$;

(b) $P(F_3|E)$.

Answers:

(a) .1262

(b) .3819

▼ **EXAMPLE 3**   A manufacturer buys items from six different suppliers. The fraction of the total number of items obtained from each supplier, along with the probability that an item purchased from that supplier is defective, is shown in the following table:

*The National Data Program for Social Sciences, National Opinion Research Center, University of Chicago, www.norc.uchicago.edu/projects/gensoc.asp.

Find the probability that a defective item came from supplier 5.

Supplier	Fraction of Total Supplied	Probability of Defective
1	.05	.04
2	.12	.02
3	.16	.07
4	.23	.01
5	.35	.03
6	.09	.05

Solution Let F_1 be the event that an item came from supplier 1, with F_2, F_3, F_4, F_5, and F_6 defined in a similar manner. Let E be the event that an item is defective. We want to find $P(F_5|E)$. Use the probabilities in the table to prepare a probability tree. By Bayes' formula,

$$P(F_5|E)$$

$$= \frac{(.35)(.03)}{(.05)(.04)+(.12)(.02)+(.16)(.07)+(.23)(.01)+(.35)(.03)+(.09)(.05)}$$

$$= \frac{.0105}{.0329} \approx .319.$$

There is about a 32% chance that a defective item came from supplier 5. Even though supplier 5 has only 3% defectives, his probability of being "guilty" is relatively high, about 32%, because of the large fraction of items supplied by 5. ▲ ◇④

④

In Example 3, find the probability that the defective item came from

(a) supplier 3;

(b) supplier 6.

Answers:

(a) .340

(b) .137

◆8.6 Exercises

For two events M and N, $P(M) = .4$, $P(N|M) = .3$, and $P(N|M') = .4$. Find each of the following. (See Example 1.)

1. $P(M|N)$

2. $P(M'|N)$

For disjoint events R_1, R_2, and R_3, $P(R_1) = .05$, $P(R_2) = .6$, and $P(R_3) = .35$. In addition, $P(Q|R_1) = .40$, $P(Q|R_2) = .30$, and $P(Q|R_3) = .60$. Find each of the following. (See Examples 2 and 3.)

3. $P(R_1|Q)$

4. $P(R_2|Q)$

5. $P(R_3|Q)$

6. $P(R_1'|Q)$

Suppose three jars have the following contents: 2 black balls and 1 white ball in the first; 1 black ball and 2 white balls in the second; 1 black ball and 1 white ball in the third. If the probability of selecting one of the three jars is 1/2, 1/3, and 1/6, respectively, find the probability that if a white ball is drawn, it came from the

7. second jar;

8. third jar.

Social Science According to estimates from the 2002 General Social Survey, 7.4% of adults living in the United States categorized themselves as Hispanic. The probability that a Hispanic was living outside the United States at age 16 is .265. The probability that a non-Hispanic was living outside the United States at age 16 is .055.* Find the following probabilities.

9. A person who was living outside the United States at age 16 is Hispanic.

10. A person who was living in the United States at age 16 is not Hispanic.

Social Science A federal study showed that 68% of all vehicle crashes in 2001 produced property damage only (as opposed to injuries or fatalities). Of the property-damage-only crashes, 11.0%

---

*The National Data Program for Social Sciences, National Opinion Research Center, University of Chicago, www.norc.uchicago.edu/projects/gensoc.asp.

occurred during rain and 3.2% occurred during snow or sleet. For the injury–fatality crashes, 10.5% occurred during rain and 2.2% occurred during snow or sleet.*

11. Find the probability that a randomly chosen crash that occurred during snow or sleet resulted in property damage only.

12. Find the probability that a randomly chosen crash that occurred during rain conditions was not property damage only.

Business   In March 2005, Qantas Airlines and its partners made up 51% of the flights in Australia. Qantas flights were on time 85% of the time. The other airlines, taken together, were on time 86% of the time.[†] Give the probabilities that a flight which is not on time is from.

13. Qantas/Qantas Partners   **14.** Any non-Qantas airline

Business   In 2001, 26.2% of Americans were college graduates. The probability that a college graduate used direct deposit with his or her financial institution was .78. For non–college graduates, the probability was .62.[‡] Suppose a person used direct deposit. Give the probabilities that a person

15. is a college graduate   **16.** is not a college graduate

Social Science   According to 2002 data from the Census Bureau, 20.3% of households are headed by someone age 65 or older. Of these households, the probability that the household income is less than $15,000 is .307. Of households headed by someone less than 65 years old, the probability that the household income is less than $15,000 is .225.[§] Find the probability that

17. a household with income less than $15,000 is headed by someone 65 years old or older.

18. a household with income $15,000 or higher is headed by someone younger than age 65.

19. **Management**   The following information pertains to three shipping terminals operated by Krag Corp.:[||]

Terminal	Percentage of Cargo Handled	Percent Error
Land	50	2
Air	40	4
Sea	10	14

Krag's internal auditor randomly selects one set of shipping documents, ascertaining that the set selected contains an error. Which of the following gives the probability that the error occurred in the Land Terminal?
(a) .02
(b) .10
(c) .25
(d) .50

Health   In a test for toxemia, a disease that affects pregnant women, the woman lies on her left side and then rolls over on her back. The test is considered positive if there is a 20-mm rise in her blood pressure within 1 minute. The results have produced the following probabilities, where T represents having toxemia at some time during the pregnancy and N represents a negative test:

$$P(T'|N) = .90 \quad and \quad P(T|N') = .75.$$

Assume that $P(N') = .11$, and find each of the following.

20. $P(N|T)$

21. $P(N'|T)$

Health   In 2002, 25.6% of Americans were 18 years old or younger. Of these, the probability that the person went the entire year without health insurance was .196. For adults 18 years or older, the probability of going the entire year without health insurance was .165.[§] Find the probability that

22. a person without health insurance is 18 years old or younger.

23. a person with health insurance is over 18 years old.

24. **Health**   The probability that a person with certain symptoms has hepatitis is .8. The blood test used to confirm this diagnosis gives positive results for 90% of people with the disease and 5% of those without the disease. What is the probability that an individual who has the symptoms and who reacts positively to the test actually has hepatitis?

25. **Health**   The sensitivity of a medical test is defined as the probability that the test will be positive, given that a person has a disease, written $P(T^+|D^+)$. The specificity of a test is defined as the probability that the test will be negative, given that the person does not have the disease, written $P(T^-|D^-)$. For example, the sensitivity and specificity for breast cancer during a clinical breast examination by a trained expert are approximately .54 and .94, respectively.*

(a) If 2% of U.S. women have breast cancer,[†] find the probability that a woman who tests positive during a clinical breast examination actually has breast cancer.

(b) Given that a woman tests negative during a clinical breast examination, find the probability that she does not have breast cancer.

*U.S. Department of Transportation, National Highway Traffic Safety Administration, *Traffic Safety Facts 2001*.

†Australian Government, Department of Transportation and Regional Services, www.btre.gov.au.

‡*Statistical Abstract of the United States, 2004–2005*.

§*Statistical Abstract of the United States, 2004–2005*.

||Uniform CPA Examination, November 1989.

*Mary B. Barton, Russell Harris, and Suzanne Fletcher, "Does This Patient Have Breast Cancer? The Screening Clinical Breast Examinations: Should It Be Done? How?" *JAMA*, 282 no. 13 (October 6, 1999): 1270–1280.

†*The World Almanac and Book of Facts 2000*.

(c) How many false positives could be expected for every 1000 clinical breast examinations? (A false positive refers to a person who does not have the disease, but tests positive.)

26. Health   A test for the virus that causes AIDS, developed by Octopus Diagnostics Research of Hantsport, Nova Scotia, shows the presence or absence of HIV in a drop of blood in two minutes, compared with five days for a conventional test.* Preliminary results indicate a false positive rate (an indication that the HIV virus is present when it is not) of less than 2% and a false negative rate (a failure to detect the presence of the HIV virus) of up to 5%. Assume for this exercise that these rates are exactly 2% and 5%. According to estimates, there were 956,000 people in North America with the HIV virus[†] out of a population of 425.3 million in 2003.[‡] Suppose a resident of North American is chosen at random and given this test. If the result is positive, what is the probability that the person actually has the HIV virus?

27. Social Science   A recent study by the Harvard School of Public Health reported that 78.9% of college students living in a fraternity or sorority house are binge drinkers. For students living in a regular dormitory, the rate of binge drinking is 44.5%, and for students living off campus, the rate is 43.7%.[§] Suppose that 10% of U.S. students live in a fraternity or sorority house, 20% live in regular dormitories, and 70% live off campus.

(a) What is the probability that a randomly selected student is a binge drinker?

(b) If a randomly selected student is a binge drinker, what is the probability that he or she lives in a fraternity or sorority house?

28. Finance   A bank finds that the relationship between mortgage defaults and the size of the down payment is given by the table.

Down Payment (%)	5%	10%	20%	25%
Number of Mortgages with This Down Payment	1260	700	560	280
Probability of Default	.05	.03	.02	.01

If a default occurs, what is the probability that it is on a mortgage with a 5% down payment? (See Examples 2 and 3.)

Business   *The table shows the percent of U.S. households in each region and the percent per region using central air-conditioning.*

	Northeast	Midwest	South	West
Region	19.4	23.7	35.2	21.5
Uses Central AC	22.1	51.3	69.4	26.9

Find the probability that a household using central air-conditioning is

29. in the South

30. in the Midwest

Social Science   *The table gives the proportions of people over age 20 in the U.S. population, and the proportions of those that live alone, in a recent year.* Use this table for Exercises 31–35.*

Age	Proportion in Population Age 20 or Higher	Proportion Living Alone
20–34	.293	.088
35–54	.412	.107
55–74	.212	.191
75 and older	.083	.393

31. Find the probability that a randomly selected person age 20 or older who lives alone is between the ages of 55 and 74.

32. Find the probability that a randomly selected person age 20 or older who lives alone is age 75 or older.

33. Find the probability of not living alone for any adult age 20 or higher.

34. Find the probability of not living alone for any adult age 20 to 34.

35. Find the probability of not living alone for any adult age 75 or older.

Macleans's, February 17, 1997.

[†]AVERT International AIDS Charity, www.avert.org.

[‡]CIA World Factbook, March 2005, www.nationmaster.com.

[§]Wechsler, H., Lee, J., Kuo, M., Lee, H., "College Binge Drinking in the 1990s: A Continuing Problem, Results of the Harvard School of Public Health 1999 College Alcohol Study," online at http://www.hsph.harvard.edu/cas/rpt2000/CAS2000rpt.shtml.

Statistical Abstract of the United States, 2004–2005.

Chapter 8 Summary

Key Terms and Symbols

{}	set braces
$\in$	is an element of
$\notin$	is not an element of
$\varnothing$	empty set
$\subseteq$	is a subset of
$\not\subseteq$	is not a subset of
$\subset$	is a proper subset of
A'	complement of set A
$\cap$	set intersection
$\cup$	set union
$P(E)$	probability of event E
$P(F|E)$	probability of F, given that E has occurred

8.1
set
element (member)
empty set
set-builder notation
universal set
subset
set operations
tree diagram
Venn diagram
complement
intersection
disjoint sets
union

8.2 addition rule for counting

8.3 experiment
trial
outcome
sample space

event
simple event
certain event
impossible event
disjoint events
basic probability principle

8.4 addition rule for probability
complement rule
odds
relative frequency probability
probability distribution

8.5 conditional probability
product rule of probability
probability tree
independent events
dependent events

8.6 Bayes' formula

Key Concepts

Sets

Set A is a **subset** of set B if every element of A is also an element of B.

A set of n elements has 2^n subsets.

Let A and B be any sets with universal set U.

The **complement** of A is $A' = \{x|x \notin A \text{ and } x \in U\}$.
The **intersection** of A and B is $A \cap B = \{x|x \in A \text{ and } x \in B\}$.
The **union** of A and B is $A \cup B = \{x|x \in A \text{ or } x \in B \text{ or both}\}$.

$n(A \cup B) = n(A) + n(B) - n(A \cap B)$, where $n(X)$ is the number of elements in set X

Probability Summary

Basic Probability Principle   Let S be a sample space of equally likely outcomes, and let event E be a subset of S. Then the probability that event E occurs is

$$P(E) = \frac{n(E)}{n(S)}.$$

Addition Rule   For any events E and F from a sample space S,

$$P(E \cup F) = P(E) + P(F) - P(E \cap F).$$

For disjoint events E and F,

$$P(E \cup F) = P(E) + P(F).$$

Complement Rule $P(E) = 1 - P(E')$ and $P(E') = 1 - P(E)$

Odds   The odds in favor of event E are $\dfrac{P(E)}{P(E')}$, $P(E') \neq 0$.

Properties of Probability
1. For any event E in sample space S, $0 \leq P(E) \leq 1$.
2. The sum of the probabilities of all possible distinct outcomes is 1.

Conditional Probability   The conditional probability of event E, given that event F has occurred, is

$$P(E|F) = \frac{P(E \cap F)}{P(F)}, \quad \text{where} \quad P(F) \neq 0.$$

For equally likely outcomes, conditional probability is found by reducing the sample space to event F; then

$$P(E|F) = \frac{n(E \cap F)}{n(F)}.$$

Product Rule of Probability   If E and F are events, then $P(E \cap F)$ may be found by either of these formulas:

$$P(E \cap F) = P(F) \cdot P(E|F) \quad \text{or} \quad P(E \cap F) = P(E) \cdot P(F|E).$$

If E and F are independent events, then $P(E \cap F) = P(E) \cdot P(F)$.

Bayes' Formula $P(F_i|E) = \dfrac{P(F_i) \cdot P(E|F_i)}{P(F_1) \cdot P(E|F_1) + P(F_2) \cdot P(E|F_2) + \cdots + P(F_n) \cdot P(E|F_n)}$

Chapter 8 Review Exercises

Write true or false for each of the following.

1. $9 \in \{8, 4, -3, -9, 6\}$

2. $4 \in \{3, 9, 7\}$

3. $2 \notin \{0, 1, 2, 3, 4\}$

4. $0 \notin \{0, 1, 2, 3, 4\}$

5. $\{3, 4, 5\} \subseteq \{2, 3, 4, 5, 6\}$

6. $\{1, 2, 5, 8\} \subseteq \{1, 2, 5, 10, 11\}$

7. $\{1, 5, 9\} \subset \{1, 5, 6, 9, 10\}$

8. $0 \subseteq \emptyset$

List the elements in the following sets.

9. $\{x | x \text{ is a national holiday}\}$

10. $\{x | x \text{ is an integer, } -3 \leq x < 1\}$

11. $\{\text{all counting numbers less than } 5\}$

12. $\{x | x \text{ is a leap year between 1989 and 2006}\}$

Let $U = \{$Vitamins $A, B_1, B_2, B_3, B_6, B_{12}, C, D, E\}$, $M = \{A, C, D, E\}$, and $N = \{A, B_1, B_2, C, E\}$. Find the following.

13. M'

14. N'

15. $M \cap N$

16. $M \cup N$

17. $M \cup N'$

18. $M' \cap N$

*Let $U = \{$all students in a class$\}$,
$A = \{$all students majoring in business$\}$,
$B = \{$all students with a GPA $< 3.0\}$,
$C = \{$all students with brown eyes$\}$, and
$D = \{$all students younger than 25$\}$. Describe the following sets in words.*

19. $A \cap C$

20. $B \cap D$

21. $A \cup D$

22. $A' \cap D$

23. $B' \cap C'$

Draw a Venn diagram and shade the given set.

24. $B \cup A'$

25. $A' \cap B$

26. $A' \cap (B' \cap C)$

27. $(A \cup B)' \cap C$

Social Science   *A survey of a group of military personnel revealed that the respondents could be categorized as follows:*

20 officers
27 minorities
19 women
5 women officers
8 minority women
10 minority officers
3 women minority officers
6 Caucasian male enlisted personnel

28. How many were interviewed?

29. How many were enlisted minority women?

30. How many were male minority officers?

Write sample spaces for the following.

31. A die is rolled and the number of points showing is noted.

32. A card is drawn from a deck containing only 4 aces.

33. A color is selected from the set {red, blue, green}, and then a number is chosen from the set {10, 20, 30}.

A jar contains 5 discs labeled 2, 4, 6, 8, 10, and another jar contains 2 blue and 3 yellow balls. One disc is drawn and then a ball is drawn. Give the following.

34. The sample space

35. Event F, the ball is blue.

36. Event E, the disc shows a number greater than 5.

37. Are the outcomes in this sample space equally likely?

Business   *A company sells computers and copiers. Let E be the event "a customer buys a computer," and let F be the event "a customer buys a copier." In Exercises 38 and 39, write each of the following, using ∩, ∪, or ' as necessary.*

38. A customer buys neither a computer nor a copier.

39. A customer buys at least one computer or copier.

40. A student gives the answer to a probability problem as 6/5. Explain why this answer must be incorrect.

41. Describe what is meant by disjoint sets, and give an example.

42. Describe what is meant by mutually exclusive events, and give an example.

43. How are disjoint sets and mutually exclusive events related?

A single card is drawn from an ordinary deck of playing cards. Find the probability of drawing each of the following.

44. A red queen

45. A face card

46. A red card or a face card

47. A black card, given that it is a 10

48. A jack, given that it is a face card

49. A face card, given that it is a king

Find the odds in favor of drawing the following.

50. A club

51. A red queen

52. A black face card or a 9

Business   *A sample shipment of five electric motors is chosen at random. The probability of exactly 0, 1, 2, 3, 4, or 5 motors being defective is given in the following table:*

Number defective	0	1	2	3	4	5
Probability	.31	.25	.18	.12	.08	.06

Find the probability that

53. no more than 3 are defective.

54. at least 3 are defective.

Health   *The partial table shows the four possible (equally likely) combinations when both parents are carriers of the sickle-cell anemia trait. Each carrier parent has normal cells (N) and trait cells (T).*

		2nd Parent	
		N_2	T_2
1st Parent	N_1		N_1T_2
	T_1		

55. Complete the table.

56. If the disease occurs only when two trait cells combine, find the probability that a child born to these parents will have sickle-cell anemia.

57. The child will carry the trait, but not have the disease, if a normal cell combines with a trait cell. Find this probability.

58. Find the probability that the child neither is a carrier nor has the disease.

Find the probabilities for the following sums when two fair dice are rolled.

59. 8

60. No more than 4

61. At least 9

62. Odd and greater than 8

63. 2, given that the sum is less than 4

64. 7, given that at least one die shows a 4

Suppose $P(E) = .62$, $P(F) = .45$, and $P(E \cap F) = .28$. Find each of the following probabilities.

65. $P(E \cup F)$

66. $P(E \cap F')$

67. $P(E' \cup F)$

68. $P(E' \cap F')$

69. For the events E and F, $P(E) = .2$, $P(E|F) = .3$, and $P(F) = .4$. Find each of the following.
  (a) $P(E'|F)$
  (b) $P(E|F')$

70. Define independent events and give an example.

71. Are independent events always disjoint? Are they ever disjoint? Give examples.

Business *Of the appliance repair shops listed in the phone book, 80% are competent and 20% are not. A competent shop can repair an appliance correctly 95% of the time; an incompetent shop can repair an appliance correctly 60% of the time. Suppose an appliance was repaired correctly. Find the probability that it was repaired by*

72. a competent shop;

73. an incompetent shop.

Suppose an appliance was repaired incorrectly. Find the probability that it was repaired by

74. a competent shop;

75. an incompetent shop.

76. **Business** A manufacturer buys items from four different suppliers. The fraction of the total number of items that is obtained from each supplier, along with the probability that an item purchased from that supplier is defective, is shown in the following table.

Supplier	Fraction of Total Supplied	Probability of Defective
1	.17	.04
2	.39	.02
3	.35	.07
4	.09	.03

(a) Find the probability that a defective item came from supplier 4.

(b) Find the probability that a defective item came from supplier 2.

77. **Social Science** The following tables list the number of passengers who were on the Titanic and the number of passengers who survived, according to class of ticket.*

	CHILDREN		WOMEN	
	On	Survived	On	Survived
First Class	6	6	144	140
Second Class	24	24	165	76
Third Class	79	27	93	80
Total	109	57	402	296

	MEN		TOTALS	
	On	Survived	On	Survived
First Class	175	57	325	203
Second Class	168	14	357	114
Third Class	462	75	634	182
Total	805	146	1316	499

Use this information to determine the following probabilities. (Round answers to two decimal places.)
  (a) What is the probability that a randomly selected passenger was in second class?
  (b) What is the overall probability of surviving?
  (c) What is the probability of a first-class passenger surviving?
  (d) What is the probability of a child who was also in third class surviving?
  (e) Given that the survivor is from first class, what is the probability that she was a woman?
  (f) Given that a male has survived, what is the probability that he was in third class?
  (g) Are the events third-class survival and male survival independent events? What does this imply?

78. **Social Science** The table gives partial results of the 2002 General Social Survey. It shows the results of answering the question "Are federal income taxes too high, about right, too low, or don't know?" categorized by sex.[†]

---

*Sandra L. Takis, "Titanic: A Statistical Exploration," *Mathematics Teacher* 92, no. 8, (November 1999): p. 660–664. Reprinted with permission. ©1999 by the National Council of Teachers of Mathematics. All rights reserved.

†The National Data Program for Social Sciences, National Opinion Research Center, University of Chicago, www.norc.uchicago.edu/projects/gensoc.asp.

	Too High	About Right	Too Low	Don't Know	Total
Male	289		6	10	497
Female		153	3		
Total	546	345		24	

(a) Complete the table.
(b) How many were surveyed?
(c) How many men think taxes are about right?
(d) How many women think taxes are too high?
(e) How many women are in the survey?
(f) How many who think taxes are too high are male?
(g) Rewrite the event stated in part (f), using the expression "given that."
(h) Find the probability of the outcome in parts (f) and (g).
(i) Find the probability that a woman thinks that taxes are about right.

Additional Probability Review Exercises

Use these exercises for practice deciding which rule, principle, or formula to apply.

1. Suppose $P(E) = .4$, $P(F) = .22$, and $P(E \cup F) = .52$. Find
   (a) $P(E \cup F')$
   (b) $P(E \cap F')$
   (c) $P(E' \cup F)$

2. A jar contains 2 white, 3 orange, 5 yellow, and 8 black marbles. If a marble is drawn at random, find the probability that it is
   (a) White
   (b) Orange
   (c) Not black
   (d) Orange or yellow

3. **Finance** American Century Equity Growth Fund has invested in the following sectors.*

Sector	Percent
Financials	19.8
Consumer Discretionary	18.1
Health Care	17.0
Information Technology	13.4
Energy	9.1
Other	22.6

*American Century Equity Growth Fund as of March 31, 2005, www.americancentury.com.

Find the probability that an investment selected at random from this fund is
(a) in health care or energy
(b) in financials or consumer discretionary
(c) not in energy

4. **Finance** American Century Global Growth Fund has invested in the following geographical regions.*

Region	Percent
Asia	13.0
Europe	36.8
North America	46.4
Other	3.8

Find the probability that an investment selected at random from this fund is
(a) not in Europe
(b) in Europe or North America
(c) not in either Asia or Europe

5. A single fair die is rolled. Find the following probabilities if the die shows
   (a) a 2, given that the number was odd;
   (b) a 4, given that the number was even;
   (c) an even number, given that the number was 6.

6. **Finance** In 2002, the United States Department of Education spent the following dollar amounts on elementary and secondary education programs (dollars in millions—for example, 9,398.3 = $9,398,300,000).†

Program	Amount
Grants for the disadvantaged	9,398.3
School Improvement Programs	4,700.5
Indian education	113.0
Special education	6,924.6
Vocational and adult education	1,995.0
Education reform: Goals 2000	1,792.7

Find the probability that funds for a particular project
(a) came from special education
(b) came from special education, or vocational and adult education
(c) did not come from education reform: Goals 2000

*American Century Global Growth Fund as of February 28, 2005, www.americancentury.com.

†*Statistical Abstract of the United States, 2004–2005.*

Social Science *The population of the United States (in thousands) by age groups in 2000 and the projected population by age groups for the year 2020 are given in the table.**

Age Group (Years)	2000	2020
0–4	19,218	22,932
5–19	61,331	65,955
20–44	104,075	108,632
45–64	62,440	83,653
65–84	30,794	47,363
85 and older	4,267	7,269

Find the probability that a randomly selected person is

7. Age 45–64 in 2000

8. Younger than 65 in 2000

9. Younger than 5 in 2020

10. 65 or older in 2020

11. **Social Science** In one area, 4% of the population drives luxury cars. However, 17% of the certified public accountants (CPAs) drive luxury cars. Are the events "person drives a luxury car" and "person is a CPA" independent?

12. Suppose $P(E) = .05$, $P(F) = .1$, and $P(E \cap F) = .02$. Find
    (a) $P(E' \cap F)$
    (b) $P(E' \cup F')$
    (c) $P(E \cap F')$.

13. One orange and four red slips of paper are placed in a box. Two red and three orange slips are placed in a second box. A box is chosen at random, and a slip of paper is selected from it. The probability of choosing the first box is 3/8. If the selected slip of paper is orange, what is the probability that it came from the first box?

14. Find the probability that the slip of paper in Exercise 13 came from the second box, given that it is red.

15. **Business** A manufacturing firm finds that 70% of its new hires turn out to be good workers and 30% poor workers. All current workers are given a reasoning test. Of the good workers, 80% pass it; 40% of the poor workers pass it. Assume that these figures will hold true in the future. If the company makes the test part of its hiring procedure and hires only people who meet the previous requirements and pass the test, what percent of the new hires will turn out to be good workers?

Social Science *The 2002 General Social Survey estimates the number of women working full time at 61.4%. Of these women,* 52.4% are married. It also estimates the percentage of women working part time as 7.8%, with 33.3% of these women married. The percentage of women not working full or part time, but who are married, is 49.2%.* Find the probability that a woman is

16. married;

17. married and works full time;

18. not married and works part time;

19. not married and not working full or part time.

Business *On a given weekend in the fall, a tire company can buy television advertising time for a college football game, a baseball game, or a professional football game. If the company sponsors the college game, there is a 70% chance of a high rating. There is a 50% chance if the company sponsors a baseball game and a 60% chance if it sponsors a professional football game. The probability of the company sponsoring these various games is .5, .2, and .3, respectively. Suppose the company does get a high rating; find the probability that it sponsored*

20. a college game;

21. a professional football game.

22. As reported by the Associated Press, Arizona was the state with the highest seat-belt use in 2004. The odds that a driver was using a seat belt were 19:1.[†] What is the probability that a driver does *not* use a seat belt?

23. The same article as in Exercise 22 stated that Tennessee had odds of seat-belt use of 17:8 in 2003. What is the probability that a driver was using a seat belt?[†]

24. If a marble is drawn from a bag containing 2 yellow, 5 red, and 3 blue marbles, what are the probabilities of the following?
    (a) The marble is red.
    (b) The marble is either yellow or blue.
    (c) The marble is yellow or red.
    (d) The marble is green.

25. The law firm of Alam, Bartolini, Chinn, Dickinson, and Ellsberg has two senior partners: Alam and Bartolini. Two of the attorneys are to be selected to attend a conference. Assuming that all are equally likely to be selected, find the following probabilities.
    (a) Chinn is selected.
    (b) Ellsberg is not selected.
    (c) Alam and Dickinson are selected.
    (d) At least one senior partner is selected.

---

*The National Data Program for Social Sciences, National Opinion Research Center, University of Chicago, www.norc.uchicago.edu/projects/gensoc.asp.

†Exercises 22 and 23 are based on *Associated Press* article in *The Detroit News,* November 23, 2004.

Statistical Abstract of the United States, 2004–2005.

Case 8

Medical Diagnosis

When a patient is examined, information, typically incomplete, is obtained about his or her state of health. Probability theory provides a mathematical model appropriate for this situation, as well as a procedure for quantitatively interpreting such partial information in order to arrive at a reasonable diagnosis.*

To do the latter, list the states of health that can be distinguished in such a way that the patient can be in one and only one state at the time of the examination. Each state of health H is associated with a number $P(H)$ between 0 and 1 such that the sum of all these numbers is 1. This number $P(H)$ represents the probability, before the exam, that a patient is in the state of health H, and $P(H)$ may be chosen subjectively from medical experience, using any information available prior to the examination. The probability may be most conveniently established from clinical records; that is, a mean probability is established for patients in general, although the number would vary from patient to patient.

For example, limiting the discussion to the condition of a patient's heart, suppose there are exactly 3 states of health, with probabilities as follows:

	State of Health H	P(H)
H_1	patient has a normal heart	.8
H_2	patient has minor heart irregularities	.15
H_3	patient has a severe heart condition	.05

Having selected $P(H)$, we can process the information from the examination. First, the results of the examination must be classified. The examination itself consists of observing the state of a number of characteristics of the patient. Let us assume that the examination for a heart condition consists of a stethoscope examination and a cardiogram. The outcome of such an examination, C, might be one of the following:

C_1 = stethoscope shows normal heart and cardiogram shows normal heart;

C_2 = stethoscope shows normal heart and cardiogram shows minor irregularities;

and so on.

*From "Probabilistic Medical Diagnosis," Roger Wright, *Some Mathematical Models in Biology*, Robert M. Thrall, ed. (The University of Michigan, 1967), by permission of Robert M. Thrall.

It remains to assess, for each state of health H, the conditional probability $P(C|H)$ of each examination outcome C, using only the knowledge that a patient is in a given state of health. (This knowledge may be based on the medical knowledge and clinical experience of the doctor.) The conditional probabilities $P(C|H)$ will not vary from patient to patient, so that they may be built into a diagnostic system, although they should be reviewed periodically.

Suppose the result of the examination is C_1. Let us assume the following probabilities:

$$P(C_1|H_1) = .9$$
$$P(C_1|H_2) = .4$$
$$P(C_1|H_3) = .1.$$

Now, for a given patient, the appropriate probability associated with each state of health H, after examination, is $P(H|C)$, where C is the outcome of the examination. This probability can be calculated by using Bayes' formula. For example, to find $P(H_1|C_1)$—that is, the probability that the patient has a normal heart, given that the examination showed a normal stethoscope examination and a normal cardiogram—we use Bayes' formula as follows:

$$P(H_1|C_1)$$
$$= \frac{P(C_1|H_1)P(H_1)}{P(C_1|H_1)P(H_1) + P(C_1|H_2)P(H_2) + P(C_1|H_3)P(H_3)}$$
$$= \frac{(.9)(.8)}{(.9)(.8) + (.4)(.15) + (.1)(.05)} \approx .92.$$

Hence, the probability is about .92 that the patient has a normal heart on the basis of the examination results. This means that in 8 out of 100 patients, some abnormality will be present and not be detected by the stethoscope or the cardiogram.

◆ Exercises

1. Find $P(H_2|C_1)$.
2. Assuming the following probabilities, find $P(H_1|C_2)$:

   $$P(C_2|H_1) = .2, \quad P(C_2|H_2) = .8, \quad P(C_2|H_3) = .3.$$

3. Assuming the probabilities of Exercise 2, find $P(H_3|C_2)$.

Counting, Probability Distributions, and Further Topics in Probability

Chapter **9**

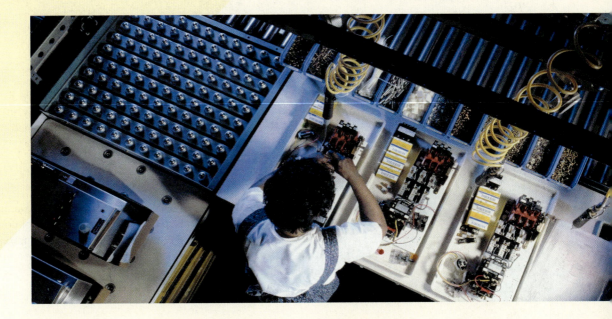

Probability has applications to quality control in manufacturing, and to business decision-making. It plays a role in testing new medications, in evaluating DNA evidence in criminal trials, and in a host of other situations. See Exercises 1 and 2 on page 540, Exercise 41 on page 519, and Exercises 42 and 43 on page 520. Sophisticated counting techniques are often necessary for determining the probabilities used in these applications. See Exercise 33–35 on page 532.

Probability distributions enable us to compute the "average value" or "expected outcome" when an experiment or process is repeated a number of times. These distributions are introduced in Section 9.1, and later are used in Sections 9.4 and 9.6. The other focus of this chapter is the development of effective ways to count the possible outcomes of an experiment without actually listing them all (which can be *very* tedious when large numbers are involved). Theses counting techniques are introduced in Section 9.2 and are used to find probabilities throughout the rest of the chapter.

9.1 Probability Distributions and Expected Value

In this section, we shall see that the *expected value* of a probability distribution is a type of average. Probability distributions were introduced briefly in Section 8.4. Now we take a more complete look at them. A probability distribution depends on the idea of a *random variable*, so we begin with that.

Random Variables

One of the questions asked in the National Health and Nutrition Examination Study (NHANES) had to do with respondents' daily hours of TV, video, or computer use.* The answer to that question, which we will label x, is one of the numbers 0 through 6 (corresponding to the numbers of hours of use). Since the value of x is random, x is called a random variable.

Random Variable

A **random variable** is a function that assigns a real number to each outcome of an experiment.

Table 1 gives the outcomes of the study in regard to TV, video, or computer use, together with the probability $P(x)$ of each outcome x.

Table 1

x	0	1	2	3	4	5	6
$P(x)$	.11	.15	.25	.17	.11	.19	.02

A table that lists all the outcomes with the corresponding probabilities is called a **probability distribution.** The sum of the probabilities in a probability distribution must always equal 1. (The sum in some distributions may vary slightly from 1 because of rounding.)

*National Health and Nutrition Examination Study, www.cdc.gov/nchs/nhanes.htm.

Instead of writing the probability distribution as a table, we could write the same information as a set of ordered pairs:

$$\{(0, .11), (1, .15), (2, .25), (3, .17), (4, .11), (5, .19), (6, .02)\}.$$

There is just one probability for each value of the random variable.

The information in a probability distribution is often displayed graphically as a special kind of bar graph called a **histogram.** The bars of a histogram all have the same width, usually 1. The heights of the bars are determined by the probabilities. A histogram for the data in Table 1 is given in Figure 9.1. A histogram shows important characteristics of a distribution that may not be readily apparent in tabular form, such as the relative sizes of the probabilities and any symmetry in the distribution.

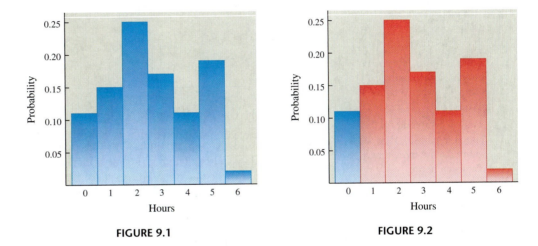

FIGURE 9.1   **FIGURE 9.2**

The area of the bar above $x = 0$ in Figure 9.1 is the product of 1 and .11, or $1 \cdot .11 = .11$. Since each bar has a width of 1, its area is equal to the probability which corresponds to its x value. The probability that a particular value will occur is thus given by the area of the appropriate bar of the graph. For example, the probability that one or more hours are spent watching TV or a video or using the computer is the sum of the areas for $x = 1$, $x = 2$, $x = 3$, $x = 4$, $x = 5$, and $x = 6$. This area, shown in red in Figure 9.2, corresponds to .89 of the total area, since

$$P(x \geq 1) = P(x = 1) + P(x = 2) + P(x = 3) + P(x = 4)$$
$$+ P(x = 5) + P(x = 6)$$
$$= .15 + .25 + .17 + .11 + .19 + .02$$
$$= .89.$$

▼ **EXAMPLE 1**

(a) Give the probability distribution for the number of heads showing when two coins are tossed.

Solution Let x represent the random variable "number of heads." Then x can take on the value 0, 1, or 2. Now find the probability of each outcome. When two

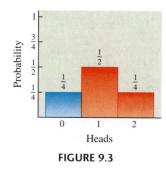

FIGURE 9.3

coins are tossed, the sample space is {TT, TH, HT, HH}. So the probability of getting one head is $2/4 = 1/2$. Similar analysis of the other cases produces Table 2.

Table 2			
x	0	1	2
$P(x)$	1/4	1/2	1/4

(b) Draw a histogram for the distribution in Table 2. Find the probability that at least one coin comes up heads.

Solution   The histogram is shown in Figure 9.3. The portion in red represents

$$P(x \geq 1) = P(x = 1) + P(x = 2)$$

$$= \frac{3}{4}. \; \blacktriangle \; \langle 1 \rangle$$

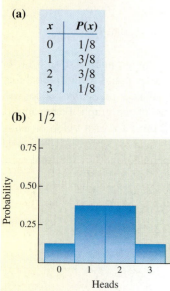

⟨1⟩

(a)   Give the probability distribution for the number of heads showing when 3 coins are tossed.

(b)   Draw a histogram for the distribution in part (a). Find the probability that no more than 1 coin comes up heads.

Answers:

(a)

x	$P(x)$
0	1/8
1	3/8
2	3/8
3	1/8

(b)   1/2

TECHNOLOGY TIP   Virtually all graphing calculators can produce histograms. The procedures differ on various calculators, but you usually are required to enter the outcomes in one list and the corresponding frequencies in a second list. For specific details, check your instruction manual under "statistics graphs" or "statistical plotting." To get the histogram in Figure 9.3 with a TI-84+ calculator, we entered the outcomes 0, 1, and 2 in the first list and entered the probabilities .25, .5, and .25 in a second list. Two versions of the histogram are shown in Figure 9.4. They differ slightly because different viewing windows were used. With some calculators, the probabilities must be entered as integers, so make the entries in the second list 1, 2, and 1 (corresponding to 1/4, 2/4, and 1/4 respectively), and use a window with $0 \leq y \leq 4$.

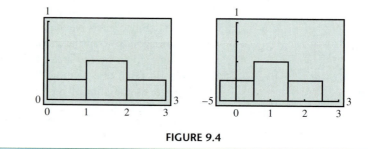

FIGURE 9.4

Expected Value

In working with probability distributions, it is useful to have a concept of the typical or average value that the random variable takes on. In Example 1, for instance, it seems reasonable that, on the average, one head shows when two coins are tossed. This does not tell what will happen the next time we toss two coins; we may get two heads, or we may get none. If we tossed two coins many times, however, we would expect that, in the long run, we would average about one head for each toss of two coins.

A way to solve such problems in general is to imagine flipping two coins 4 times. Based on the probability distribution in Example 1, we would expect that 1 of the 4 times we would get 0 heads, 2 of the 4 times we would get 1 head, and 1 of the 4 times we would get 2 heads. The total number of heads we would get, then, is

$$0 \cdot 1 + 1 \cdot 2 + 2 \cdot 1 = 4.$$

The expected number of heads per toss is found by dividing the total number of heads by the total number of tosses, or

$$\frac{0 \cdot 1 + 1 \cdot 2 + 2 \cdot 1}{4} = 0 \cdot \frac{1}{4} + 1 \cdot \frac{1}{2} + 2 \cdot \frac{1}{4} = 1.$$

Notice that the expected number of heads turns out to be the sum of the three values of the random variable x, multiplied by their corresponding probabilities. We can use this idea to define the *expected value* of a random variable as follows.

Expected Value

Suppose that the random variable x can take on the n values $x_1, x_2, x_3, \ldots, x_n$. Suppose also that the probabilities that these values occur are, respectively, $p_1, p_2, p_3, \ldots, p_n$. Then the **expected value** of the random variable is

$$E(x) = x_1 p_1 + x_2 p_2 + x_3 p_3 + \cdots + x_n p_n.$$

▼ **EXAMPLE 2** In the example with the TV, video, and computer usage on page 510, find the expected number of hours per day of viewing.

Solution Multiply each outcome in Table 1 by its probability, and sum the products.

$$E(x) = 0 \cdot .11 + 1 \cdot .15 + 2 \cdot .25 + 3 \cdot .17 + 4 \cdot .11 + 5 \cdot .19 + 6 \cdot .02$$
$$= 2.67$$

On the average, a respondent of the survey will indicate 2.67 hours of TV, video, and computer viewing. ▲ ◇2

⬧2

Find the expected value of the number of heads showing when four coins are tossed.

Answer:

2

Physically, the expected value of a probability distribution represents a balance point. If we think of the histogram in Figure 9.1 as a series of weights with magnitudes represented by the heights of the bars, then the system would balance if supported at the point corresponding to the expected value.

▼ **EXAMPLE 3** Suppose a local symphony decides to raise money by raffling off a microwave oven worth $400, a dinner for two worth $80, and 2 books worth $20 each. A total of 2000 tickets are sold at $1 each. Find the expected value of winning for a person who buys 1 ticket in the raffle.

Solution Here the random variable represents the possible amounts of net winnings, where net winnings = amount won − cost of ticket. The net winnings of the person winning the oven are $400 (amount won) − $1 (cost of ticket) = $399. The net winnings for each losing ticket are $0 − $1 = −$1.

The net winnings of the various prizes, as well as their respective probabilities, are shown in Table 3. The probability of winning $19 is 2/2000, because there are 2 prizes worth $20. (We have not reduced the fractions in order to keep all the denominators equal.) Because there are 4 winning tickets, there are 1996 losing tickets, so the probability of winning $-\$1$ is 1996/2000.

Table 3

x	$399	$79	$19	$-\$1$
$P(x)$	1/2000	1/2000	2/2000	1996/2000

The expected winnings for a person buying 1 ticket are

$$399\left(\frac{1}{2000}\right) + 79\left(\frac{1}{2000}\right) + 19\left(\frac{2}{2000}\right) + (-1)\left(\frac{1996}{2000}\right) = -\frac{1480}{2000}$$
$$= -.74.$$

On the average, a person buying 1 ticket in the raffle will lose $.74, or 74¢.

It is not possible to lose 74¢ in this raffle: either you lose $1, or you win a prize worth $400, $80, or $20, minus the $1 you paid to play. But if you bought tickets in many such raffles over a long time, you would lose 74¢ per ticket, on the average. It is important to note that the expected value of a random variable may be a number that can never occur in any one trial of the experiment. ▲ ⟨3⟩

NOTE An alternative way to compute expected value in this and other examples is to calculate the expected amount won and then subtract the cost of the ticket afterward. The amount won is either $400 (with probability 1/2000), $80 (with probability 1/2000), $20 (with probability 2/2000), or $0 (with probability 1996/2000). The expected winnings for a person buying one ticket are then

$$400\left(\frac{1}{2000}\right) + 80\left(\frac{1}{2000}\right) + 20\left(\frac{2}{2000}\right) + 0\left(\frac{1996}{2000}\right) - 1 = -\frac{1480}{2000}$$
$$= -.74.$$

▼ **EXAMPLE 4** Each day Lynette and Tanisha toss a coin to see who buys coffee (at $1.25 a cup). One tosses, while the other calls the outcome. If the person who calls the outcome is correct, the other buys the coffee; otherwise the caller pays. Find Lynette's expected winnings.

Solution Assume that an honest coin is used, that Tanisha tosses the coin, and that Lynette calls the outcome. The possible results and corresponding probabilities are shown in the following table.

Possible Results				
Result of Toss	Heads	Heads	Tails	Tails
Call	Heads	Tails	Heads	Tails
Caller Wins?	Yes	No	No	Yes
Probability	1/4	1/4	1/4	1/4

⟨3⟩
Suppose you buy 1 of 10,000 tickets at $1 each in a lottery where the prize is $5,000. What are your expected net winnings? What does this answer mean?

Answer:
$-\$.50$. On the average you lose $.50 per ticket purchased.

Lynette wins a $1.25 cup of coffee whenever the results and calls match, and she loses $1.25 when there is no match. Her expected winnings are

$$1.25\left(\frac{1}{4}\right) + (-1.25)\left(\frac{1}{4}\right) + (-1.25)\left(\frac{1}{4}\right) + 1.25\left(\frac{1}{4}\right) + = 0.$$

On the average, over the long run, Lynette breaks even. ▲ ◈4

A game with an expected value of 0 (such as the one in Example 4) is called a **fair game.** Casinos do not offer fair games. If they did, they would win (on the average) $0 and have a hard time paying the help! Casino games have expected winnings for the house that vary from 1.5 cents per dollar to 60 cents per dollar. Exercises 17–20 at the end of the section ask you to find the expected winnings for certain games of chance.

The idea of expected value can be very useful in decision making, as shown by the next example.

▼ **EXAMPLE 5** At age 50, suppose you receive a letter from Mutual of Mauritania Insurance Company. According to the letter, you must tell the company immediately which of the following two options you will choose: take $50,000 at age 60 (if you are alive, $0 otherwise), or $65,000 at age 70 (again, if you are alive, $0 otherwise). Based *only* on the idea of expected value, which should you choose?

Solution Life insurance companies have constructed elaborate tables showing the probability of a person living a given number of years into the future. From a recent such table, the probability of living from age 50 to age 60 is .88, while the probability of living from age 50 to 70 is .64. The expected values of the two options are as follows.

First Option: $(50,000)(.88) + (0)(.12) = 44,000$
Second Option: $(65,000)(.64) + (0)(.36) = 41,600$

Strictly on the basis of expected values, choose the first option. ▲ ◈5

▼ **EXAMPLE 6** Table 4 gives the probability distribution for the number of children of respondents to the 2002 General Social Survey,* for those with 7 or fewer children.

Table 4

x	0	1	2	3	4	5	6	7
$P(x)$	.293	.172	.241	.176	.068	.027	.015	.008

Find the expected value for the number of children.

---

*The National Data Program for Social Sciences, National Opinion Research Center, University of Chicago, Chicago, Illinois.

Solution Using the formula for the expected value, we have

$$E(x) = 0(.293) + 1(.172) + 2(.241) + 3(.176)$$
$$+ 4(.068) + 5(.027) + 6(.015) + 7(.008)$$
$$= 1.735.$$

For those respondents with 7 or fewer children, the number of children, on average, is 1.74. ▲

▼ **EXAMPLE 7** Suppose a family has 3 children.

(a) Find the probability distribution for the number of girls.

Solution Notice that, in the sample space S of all 3-child families, there are eight equally likely outcomes: $S = \{ggg, ggb, gbg, gbb, bgg, bgb, bbg, bbb\}$. One of the outcomes has 0 girls (or $P(0) = 1/8$), three have 1 girl (or $P(1) = 3/8$), three have 2 girls (or $P(2) = 3/8$), and one has 3 girls (or $P(3) = 1/8$). The respective probabilities are given in Table 5.

Table 5

x	0	1	2	3
$P(x)$	1/8	3/8	3/8	1/8

(b) Use the distribution from part (a) to find the expected number of girls in a 3-child family.

Solution Using the formula for expected value, we have

$$\text{Expected number of girls} = 0\left(\frac{1}{8}\right) + 1\left(\frac{3}{8}\right) + 2\left(\frac{3}{8}\right) + 3\left(\frac{1}{8}\right)$$
$$= \frac{12}{8} = 1.5.$$

On average, a 3-child family will have 1.5 girls. This result agrees with our intuition that, on the average, half the children born will be girls. ▲ ⬦6

⬦6

A recent estimate of the probabilty that an adult works full time is 51%. Suppose 5 people are chosen at random. The table gives the probability distribution for the number that work full time. Find $E(x)$,

x	$P(x)$
0	.028
1	.147
2	.306
3	.318
4	.166
5	.035

Answer:

$E(x) = 2.552$

◆9.1 Exercises

For each of the experiments described, let x determine a random variable and use your knowledge of probability to prepare a probability distribution. (Hint: Use a tree diagram.)

1. Four children are born, and the number of boys is noted. (Assume an equal chance of a boy or a girl for each birth.)

2. Two dice are rolled, and the total number of points is recorded.

3. Three cards are drawn from a deck. The number of queens is counted.

4. Two names are drawn from a hat signifying who should go pick up pizza. Three of the names are on the swim team and two are not. The number of swimmers selected is counted.

Draw a histogram for each of the following, and shade the region that gives the indicated probability. (See Example 1.)

5. Exercise 1; $P(x \le 2)$

6. Exercise 2; $P(x \ge 11)$

7. Exercise 3; P(at least one queen)

8. Exercise 4; P(fewer than two swimmers)

Find the expected value for each random variable. (See Example 2.)

9.

x	1	3	5	7
$P(x)$	.2	.4	.3	.1

10.

y	0	15	30	40
$P(y)$	.1	.3	.3	.2

11.

z	0	2	4	8	16
$P(z)$	.14	.22	.36	.18	.10

12.

x	5	1	15	20	25
$P(x)$	.31	.30	.29	.06	.04

Find the expected values for the random variables x having the probability functions graphed.

13.

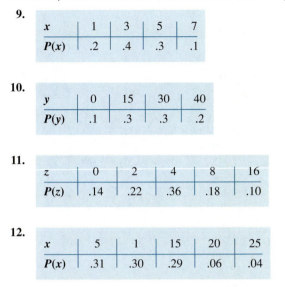

14.

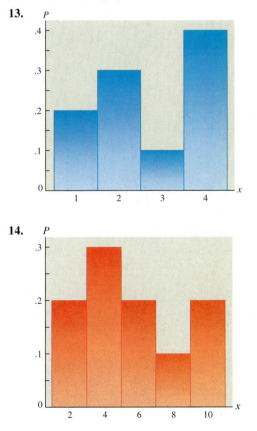

15.

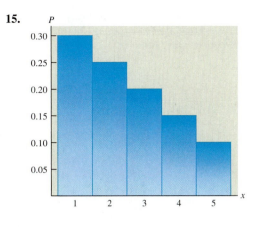

16.

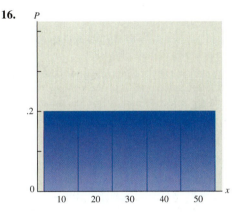

Find the expected winnings for the games of chance described in Exercises 17–20.

17. In one form of roulette, you bet $1 on "even." If 1 of the 18 even numbers comes up, you get your dollar back, plus another one. If 1 of the 20 noneven (18 odd, 0, and 00) numbers comes up, you lose your dollar.

18. Repeat Exercise 17 if there are only 19 noneven numbers (no 00).

19. *Numbers* is a game in which you bet $1 on any three-digit number from 000 to 999. If your number comes up, you get $500.

20. In one form of the game Keno, the house has a pot containing 80 balls, each marked with a different number from 1 to 80. You buy a ticket for $1 and mark one of the 80 numbers on it. The house then selects 20 numbers at random. If your number is among the 20, you get $3.20 (for a net winning of $2.20).

21. **Business** An online gambling site offers a first prize of $50,000 and two second prizes of $10,000 each for registered users when they place a bet. A random bet will be selected over a 24-hour period. Two million bets are received in the contest. Find the expected winnings if you can place one registered bet of $1 in the given period.

Business *A contest at a fast-food restaurant offered the following cash prizes and probabilities of winning on one visit when one buys a large order of French fries for $1.09.*

Prize	Probability
$100,000	1/8,504,860
$50,000	1/302,500
$10,000	1/282,735
$1000	1/153,560
$100	1/104,560
$25	1/9,540

22. Find the expected winnings if the player buys one large order of French fries.

23. Find the expected winnings if the player buys 25 large orders of French fries in multiple visits.

24. According to the website of Mars, the makers of M&M's Plain Chocolate Candies, 16% of the candies produced are green.* If we select 4 candies from a bag at random and record the number of green candies, the probability distribution is as follows:

x	0	1	2	3	4
P(x)	.4979	.3793	.1084	.0138	.0007

Find the expected value for the number of green candies.

25. **Health** A survey asked respondents how many days they had eaten food prepared at a restaurant in the previous week. The probability distribution is an follows:

x	0	1	2	3	4	5	6	7
P(x)	.15	.25	.18	.15	.13	.09	.03	.02

Find the expected value.

For Exercises 26–30, determine whether the probability distributions are valid or not. If not, explain why.

26.

x	0	2	4	6	8	10	11
P(x)	.01	.02	.03	.04	.05	.06	.07

27.

x	5	10	15	20	25	30	35
P(x)	.01	.09	.25	.45	.05	.20	−.05

28.

x	−2	−1	0	1	2	3	4
P(x)	.05	.10	.75	.02	.03	.04	.01

29.

x	−10	−5	0	5	10
P(x)	.50	.10	−.20	.30	.30

30.

x	1	3	5	7	9	11
P(x)	.01	.02	.03	.04	.05	.85

For Exercises 31–35, fill in the missing value(s) to make a valid probability distribution.

31.

x	5	10	15	20	25	30
P(x)	.01	.09	.25	.45	.05	

32.

x	−3	−2	−1	0	1	2	3
P(x)		.15	.15	.15	.15	.15	.15

33.

x	10	20	30	40
P(x)	.20		.25	.30

34.

x	−50	−40	−30	−20	−10	0	10
P(x)	.05	.25	.10	.10	.05		

35.

x	1	2	3	6	12	24	48
P(x)	.10	.10	.20	.25	.05		

36. **Business** A customer service specialist at a home improvement store handles returns. During the month of July, the store sold a great many air-conditioning units. The following is the probability distribution for the daily number of returns of air-conditioning units sold in July:

x	0	1	2	3	4	5
$P(x)$	.55	.31	.08	.04	.01	.01

Find the expected number of returns per day.

37. **Finance** An insurance company has written 100 policies of $15,000, 250 of $10,000, and 500 of $5000 for people age 20. If experience shows that the probability that a person will die in the next year at age 20 is .0007, how much can the company expect to pay out during the year the policies were written?

38. **Social Science** According to a recent study, 70% of children have television sets in their bedrooms.* If 6 children are selected at random, the probability distribution for the number of children with televisions in their bedrooms is as follows.

x	0	1	2	3	4	5	6
$P(x)$	.001	.010	.059	.186	.324	.302	.118

Find the expected number of children with televisions in their room in a random sample of 6 children.

39. **Social Science** According to the center for Immigration Studies, 23% of children born in the United States in 2002 had a foreign-born mother.[†] If 5 newly born infants are selected at random, the probability distribution for the number with foreign-born mothers is as follows.

x	0	1	2	3	4	5
$P(x)$	.2707	.4043	.2415	.0721	.0108	.0006

Find the expected number of foreign-born mothers when 5 children are selected at random.

40. **Natural Science** Poaching is the biggest killer of gorillas, but new research shows that approximately 25% of mountain gorilla deaths in Rwanda and Uganda were due to illnesses such as influenza and other viruses.[‡] Suppose 4 gorillas are selected at random from these regions. Then the probability distribution for the number dying from influenza or another virus is as follows:

x	0	1	2	3	4
$P(x)$	.3164	.4219	.2109	.0469	.0039

Find the expected number of gorilla deaths due to influenza or other viruses in a sample of 4 gorillas.

41. **Business** Levi Strauss and Company uses expected value to help its salespeople rate their accounts.* For each account, a salesperson estimates potential additional volume and the probability of getting it. The product of these figures gives the expected value of the potential, which is added to the existing volume. The totals are then classified as A, B, or C as follows: $40,000 or below, class C; above $40,000 up to and including $55,000, class B; above $55,000, class A. Complete the chart at the bottom for one salesperson.

Account Number	Existing Volume	Potential Additional Volume	Probability of Additional Volume	Expected Value of Potential	Existing Volume + Expected Value of Potential	Class
1	$15,000	$10,000	.25	$2,500	$17,500	C
2	40,000	0	—	—	40,000	C
3	20,000	10,000	.20			
4	50,000	10,000	.10			
5	5,000	50,000	.50			
6	0	100,000	.60			
7	30,000	20,000	.80			

*New York Times, July 5, 2005.

[†]Reuters, July 6, 2005.

[‡]Associated Press, July 5, 2005.

*This example was supplied by James McDonald, Levi Strauss and Company, San Francisco.

42. According to Len Pasquarelli, in the first 10 games of the 2004 professional football season in the United States, two-point conversions were successful 51.2% of the time.[*] We can compare this rate with the historical success rate of extra-point kicks of 94%.[†]

(a) Calculate the expected value of each strategy.

(b) Over the long run, which strategy will maximize the number of points scored?

(c) From this information, should a team always use only one strategy? Explain.

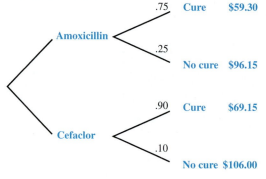

43. **Natural Science** Otitis media, or middle-ear infection, is initially treated with an antibiotic. Researchers have compared two antibiotics—amoxicillin and cefaclor—for their cost-effectiveness. Amoxicillin is inexpensive, safe, and effective. Cefaclor is also safe. However, it is considerably more expensive and is generally more effective. Use the given tree diagram (in which costs are estimated as the total cost of medication, an office visit, an ear check, and hours of lost work) to figure out the following.[‡]

(a) Find the expected cost of using each antibiotic to treat a middle-ear infection.

(b) To minimize the total expected cost, which antibiotic should be chosen?

*Len Pasquarelli, "Teams More Successful Going for Two," November 18, 2004, www.espn.com.

†David Leonhardt, "In Football, 6 + 2 Often Equals 6," *New York Times*, Sunday, January 16, 2000, p. 4-2.

‡Jeffrey Weiss and Shoshana Melman, "Cost Effectiveness in the Choice of Antibiotics for the Initial Treatment of Otitis Media in Children: A Decision Analysis Approach." *Journal of Pediatric Infectious Disease*, vol. 7, no. 1 (1998): 23–26.

44. **Physical Science** One of the few methods that can be used in an attempt to cut the severity of a hurricane is to *seed* the storm. In this process, silver iodide crystals are dropped into the storm. Unfortunately, silver iodide crystals sometimes cause the storm to *increase* its speed. Wind speeds may also increase or decrease even with no seeding. Use the given tree diagram to figure out the following.[*]

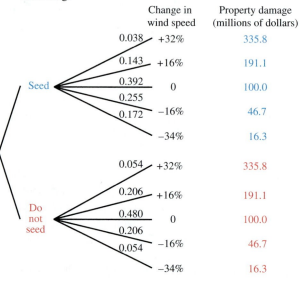

(a) Find the expected amount of damage under each of the options, "seed" and "do not seed."

(b) To minimize total expected damage, what option should be chosen?

45. At the end of the Wimbledon Championships in 2005, the top two male seeds on grass (Roger Federer and Andy Roddick) played in the finals. The prize money for the winner was £630,000 (British pounds sterling) and the prize money for the runner-up was £315,000. Find the expected winnings for Andy Roddick if

(a) we assume that both players were of equal ability;

(b) we use the players' prior head-to-head match record, whereby Roddick has only a 1/9 probability of winning.

46. Bryan Miller has 2 cats and a dog. Each pet has a 35% probability of climbing into the chair in which Bryan is sitting, independently of how many pets are already in the chair with Bryan.

(a) Find the probability distribution for the number of pets in the chair with Bryan. (*Hint:* List the sample space.)

(b) Use the probability distribution in part (a) to find the expected number of pets in the chair with Bryan.

*Data from "The Decision to Seed Hurricanes," by R. A. Howard from *Science*, Vol. 176, pp. 1191–1202, Copyright 1972 by the American Association for the Advancement of Science.

9.2 The Multiplication Principle, Permutations, and Combinations

We begin with a simple example. If there are three roads from town A to town B and 2 roads from town B to town C, in how many ways can someone travel from A to C by way of B? We can solve this simple problem with the help of Figure 9.5, which lists all the possible ways to go from A to C.

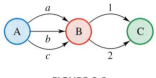

FIGURE 9.5

The possible ways to go from A through B to C are a1, a2, b1, b2, c1, and c2. So there are 6 possible ways. Note that 6 is the product of $3 \cdot 2$, 3 is the number of ways to go from A to B, and 2 is the number of ways to go from B to C.

Another way to solve this problem is to use a tree diagram, as shown in Figure 9.6. This diagram shows that, for each of the 3 roads from A, there are 2 different routes leading from B to C, making $3 \cdot 2 = 6$ different ways.

This example is an illustration of the *multiplication principle*.

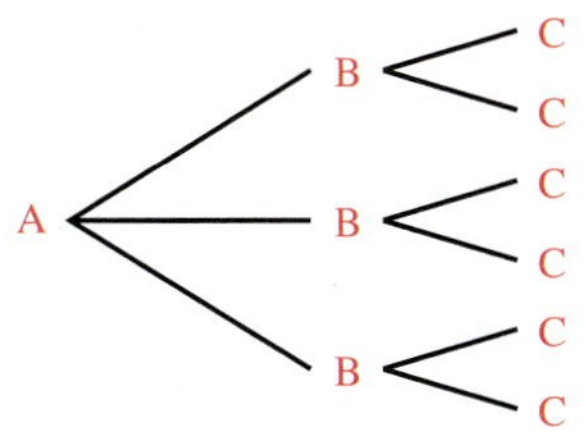

FIGURE 9.6

Multiplication Principle

Suppose n choices must be made, with

$$m_1 \text{ ways to make choice } 1,$$

and for each of these,

$$m_2 \text{ ways to make choice } 2,$$

and so on, with

$$m_n \text{ ways to make choice } n.$$

Then there are

$$m_1 \cdot m_2 \cdots \cdots m_n$$

different ways to make the entire sequence of choices.

▼ **EXAMPLE 1** Suppose Angela has 9 skirts, 8 blouses, and 13 different pairs of shoes. If she is willing to wear any combination, how many different skirt–blouse–shoe choices does she have?

Solution By the multiplication principle, there are 9 skirt choices, 8 blouse choices, and 13 shoe choices, for a total of $9 \cdot 8 \cdot 13 = 936$ skirt–blouse–shoe outfits. ▲

▼ **EXAMPLE 2**   One day in 2005, Amazon.com listed 179 items (books, audio-tapes, etc.) by Sue Grafton, 257 items by Dean Koontz, and 1098 items by Stephen King. How many ways could you buy three items, one by each author?

Solution A tree (or other) diagram would be far too complicated to use here, but the multiplication principle easily answers the question. There are

$$179 \cdot 257 \cdot 1098 = 50,511,294$$

ways. ▲

▼ **EXAMPLE 3**   A combination lock can be set to open to any 3-letter sequence.

(a) How many sequences are possible?

Solution Since there are 26 letters of the alphabet, there are 26 choices for each of the 3 letters, and, by the multiplication principle, $26 \cdot 26 \cdot 26 = 17,576$ different sequences.

(b) How many sequences are possible if no letter is repeated?

Solution There are 26 choices for the first letter. It cannot be used again, so there are 25 choices for the second letter and then 24 choices for the third letter. Consequently, the number of such sequences is $26 \cdot 25 \cdot 24 = 15,600$. ▲ ①

Factorial Notation

The use of the multiplication principle often leads to products such as $5 \cdot 4 \cdot 3 \cdot 2 \cdot 1$, the product of all the natural numbers from 5 down to 1. If n is a natural number, the symbol $n!$ (read "*n factorial*") denotes the product of all the natural numbers from n down to 1. The factorial is an algebraic shorthand. For example, instead of writing $5 \cdot 4 \cdot 3 \cdot 2 \cdot 1$, we simply write 5! If $n = 1$, this formula is understood to give $1! = 1$.

> ### *n*-Factorial
>
> For any natural number n,
>
> $$n! = n(n-1)(n-2) \ldots (3)(2)(1).$$
>
> By definition, $0! = 1$.

Note that $6! = 6 \cdot 5 \cdot 4 \cdot 3 \cdot 2 \cdot 1 = 6 \cdot (5 \cdot 4 \cdot 3 \cdot 2 \cdot 1) = 6 \cdot 5!$. Similarly, the definition of $n!$ shows that

$$n! = n \cdot (n-1)!$$

One reason that 0! is defined to be 1 is to make the preceding formula valid when $n = 1$. For when $n = 1$, we have $1! = 1$ and $1 \cdot (1-1)! = 1 \cdot 0! = 1 \cdot 1 = 1$, so $n! = n \cdot (n-1)!$. ②

---

①

(a) In how many ways can 6 business tycoons line up their golf carts at the country club?

(b) How many ways can 4 pupils be seated in a row with 4 seats?

Answers:

(a) 720

(b) 24

---

②

Evaluate:

(a) 4!

(b) 6!

(c) 1!

(d) 6!/4!

Answers:

(a) 24

(b) 720

(c) 1

(d) 30

Almost all calculators have an $n!$ key. A calculator with a 10-digit display and scientific notation capability will usually give the exact value of $n!$ for $n \leq 13$ and approximate values of $n!$ for $14 \leq n \leq 69$. The value of $70!$ is approximately $1.198 \cdot 10^{100}$, which is too large for most calculators. So how would you simplify $\dfrac{100!}{98!}$? Depending on the type of calculator, there may be an overflow problem. The next two examples show how to avoid this problem.

TECHNOLOGY TIP   The factorial key on a graphing calculator is usually located in the PRB or PROB submenu of the MATH or OPTN menu.

▼ **EXAMPLE 4**   Evaluate $\dfrac{100!}{98!}$.

Solution We use the fact that $n! = n \cdot (n-1)!$ several times

$$\frac{100!}{98!} = \frac{100 \cdot 99!}{98!} = \frac{100 \cdot 99 \cdot 98!}{98!} = 100 \cdot 99 = 9900. \; ▲$$

▼ **EXAMPLE 5**   Evaluate $\dfrac{5!}{2!\,3!}$.

Solution $\dfrac{5!}{2!\,3!} = \dfrac{5 \cdot 4!}{2!\,3!} = \dfrac{5 \cdot 4 \cdot 3!}{2!\,3!} = \dfrac{5 \cdot 4}{2 \cdot 1} = 10. \; ▲$

▼ **EXAMPLE 6**   Morse code uses a sequence of dots and dashes to represent letters and words. How many sequences are possible with at most 3 symbols?

Solution "At most 3" means "1 or 2 or 3." Each symbol may be either a dot or a dash. Thus, the following numbers of sequences are possible in each case.

Number of Symbols	Number of Sequences
1	2
2	$2 \cdot 2 = 4$
3	$2 \cdot 2 \cdot 2 = 8$

Altogether, $2 + 4 + 8 = 14$ different sequences of at most 3 symbols are possible. Because there are 26 letters in the alphabet, some letters must be represented by sequences of 4 symbols in Morse code. ▲ ③

⟨3⟩

How many sequences are possible with at most 4 symbols?

Answer:

30

Permutations

A **permutation** of a set of elements is an ordering of the elements. For instance, there are six permutations (orderings) of the letters A, B, C, namely,

ABC, ACB, BAC, BCA, CAB, CBA,

as you can easily verify. As this listing shows, order counts when determining the number of permutations of a set of elements. By saying "order counts," we mean that the event ABC is indeed distinct from CBA or any other ordering of the three letters. We can use the multiplication principle to determine the number of possible permutations of any set.

▼ **EXAMPLE 7**   How many batting orders are possible for a 9-person baseball team?

Solution There are 9 possible choices for the first batter, 8 possible choices for the second batter, 7 for the third batter, and so on, down to the eighth batter (2 possible choices) and the ninth batter (1 possibility). So the total number of batting orders is

$$9 \cdot 8 \cdot 7 \cdot 6 \cdot 5 \cdot 4 \cdot 3 \cdot 2 \cdot 1 = 362,880.$$

In other words, the number of permutations of a 9-person set is 9!. ▲

The argument in Example 7 applies to any set that leads to the following conclusion.

The number of permutations of an n element set is $n!$.

Sometimes we want to order only some of the elements in a set, rather than all of them.

▼ **EXAMPLE 8**   A teacher has 5 books and wants to display 3 of them side by side on her desk. How many arrangements of 3 books are possible?

Solution The teacher has 5 ways to fill the first space, 4 ways to fill the second space, and 3 ways to fill the third space. Because she wants only 3 books on the desk, there are only 3 spaces to fill, giving $5 \cdot 4 \cdot 3 = 60$ possible arrangements. ▲ ◇4◇

In Example 8, we say that the possible arrangements are *the permutations of 5 things taken 3 at a time*, and we denote the number of such permutations by $_5P_3$. In other words, $_5P_3 = 60$. More generally, on ordering of r elements from a set of n elements is called a **permutation of n things taken r at a time,** and the number of such permutations is denoted $_nP_r$.* To see how to compute this number, look at the answer in Example 8, which can be expressed like this:

$$_5P_3 = 5 \cdot 4 \cdot 3 = 5 \cdot 4 \cdot 3 \cdot \frac{2 \cdot 1}{2 \cdot 1} = \frac{5 \cdot 4 \cdot 3 \cdot 2 \cdot 1}{2 \cdot 1} = \frac{5!}{2!} = \frac{5!}{(5-3)!}.$$

A similar analysis in the general case lead to the following useful fact.

Permutations

If $_nP_r$ (where $r \le n$) is the number of permutations of n elements taken r at a time, then

$$_nP_r = \frac{n!}{(n-r)!}.$$

◇4◇

In how many ways can 3 of 7 items be arranged?

Answer:

$7 \cdot 6 \cdot 5 = 210$

_____

*Another notation that is sometimes used is $P(n, r)$.

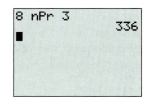

FIGURE 9.7

<hr>

5

Find the number of permutations of

(a) 5 things taken 2 at a time;

(b) 9 things taken 3 at a time.

Find each of the following.

(c) $_3P_1$

(d) $_7P_3$

(e) $_{12}P_2$

Answers:

(a) 20

(b) 504

(c) 3

(d) 210

(e) 132

<hr>

6

A collection of 3 paintings by one artist and 2 by another is to be displayed. In how many ways can the paintings be shown

(a) in a row?

(b) if the works of the artists are to be alternated?

(c) if one painting by each artist is displayed?

Answers:

(a) 120

(b) 12

(c) 6

<hr>

To find $_nP_r$, we can either use the preceding rule or apply the multiplication principle directly, as the next example shows.

▼ **EXAMPLE 9**   Early in 2004, 8 candidates sought the Democratic nomination for president. In how many ways could voters rank their first, second, and third choices?

Solution This is the same as finding the number of permutations of 8 elements taken 3 at a time. Since there are 3 choices to be made, the multiplication principle gives $_8P_3 = 8 \cdot 7 \cdot 6 = 336$. Alternatively, by the formula for $_nP_r$,

$$_8P_3 = \frac{8!}{(8-3)!} = \frac{8!}{5!} = \frac{8 \cdot 7 \cdot 6 \cdot 5 \cdot 4 \cdot 3 \cdot 2 \cdot 1}{5 \cdot 4 \cdot 3 \cdot 2 \cdot 1} = 8 \cdot 7 \cdot 6 = 336.$$

Figure 9.7 shows this result on a TI-84+ graphing calculator. ▲ **5**

▼ **EXAMPLE 10**   A televised talk show will include 4 women and 3 men as panelists.

(a) In how many ways can the panelists be seated in a row of 7 chairs?

Solution Find $_7P_7$, the total number of ways to seat 7 panelists in 7 chairs.

$$_7P_7 = \frac{7!}{(7-7)!} = \frac{7!}{0!} = \frac{7!}{1} = 7 \cdot 6 \cdot 5 \cdot 4 \cdot 3 \cdot 2 \cdot 1 = 5040$$

There are 5040 ways to seat the 7 panelists.

(b) In how many ways can the panelists be seated if the men and women are to be alternated?

Solution Use the multiplication principle. In order to alternate men and women, a woman must be seated in the first chair (since there are 4 women and only 3 men), any of the men next, and so on. Thus, there are 4 ways to fill the first seat, 3 ways to fill the second seat, 3 ways to fill the third seat (with any of the 3 remaining women), and so on.

$$4 \cdot 3 \cdot 3 \cdot 2 \cdot 2 \cdot 1 \cdot 1 = 144$$

There are 144 ways to seat the panelists.

(c) In how many ways can the panelists be seated if the men must sit together and the women must sit together?

Solution Use the multiplication principle. We first must decide how to arrange the two groups (men and women). There are 2! ways of doing this. Next, there are 4! ways of arranging the women and 3! ways of arranging the men, for a total of

$$2! \, 4! \, 3! = 2 \cdot 24 \cdot 6 = 288$$

ways. ▲ **6**

Combinations

In Example 8, we found that there are 60 ways a teacher can arrange 3 of 5 different books on a desk. That is, there are 60 permutations of 5 things taken 3 at a time. Suppose now that the teacher does not wish to arrange the books on her desk, but rather wishes to choose, at random, any 3 of the 5 books to give to a book sale to raise money for her school. In how many ways can she do this?

At first glance, we might say 60 again, but that is incorrect. The number 60 counts all possible *arrangements* of 3 books chosen from 5. However, the following arrangements would all lead to the same set of 3 books being given to the book sale.

mystery–biography–textbook	biography–textbook–mystery
mystery–textbook–biography	textbook–biography–mystery
biography–mystery–textbook	textbook–mystery–biography

The list shows 6 different *arrangements* of 3 books, but only one *subset* of 3 books. A subset of items selected *without regard to order* is called a **combination.** The number of combinations of 5 things taken 3 at a time is written $_5C_3$. Since they are subsets, combinations are *not ordered.*

To evaluate $_5C_3$ start with the $5 \cdot 4 \cdot 3$ *permutations* of 5 things taken 3 at a time. Combinations are unordered; therefore, we find the number of combinations by dividing the number of permutations by the number of ways each group of 3 can be ordered—that is, by 3!.

$$_5C_3 = \frac{5 \cdot 4 \cdot 3}{3!} = \frac{5 \cdot 4 \cdot 3}{3 \cdot 2 \cdot 1} = 10$$

There are 10 ways that the teacher can choose 3 books at random for the book sale.

Generalizing this discussion gives the formula for the number of combinations of n elements taken r at a time, written $_nC_r$.* In general, a set of r elements can be ordered in $r!$ ways, so we divide $_nP_r$ by $r!$ to get $_nC_r$.

$$_nC_r = \frac{_nP_r}{r!}$$
$$= _nP_r \frac{1}{r!}$$
$$= \frac{n!}{(n-r)!} \cdot \frac{1}{r!} \qquad \text{Definition of } _nP_r$$
$$= \frac{n!}{(n-r)!r!}$$

This last form is the most useful for setting up the calculation. ◇7

7

Evaluate $\dfrac{_nP_r}{r!}$ for the following values.

(a) $n = 6, r = 2$

(b) $n = 8, r = 4$

(c) $n = 7, r = 0$

Answers:

(a) 15

(b) 70

(c) 1

Combinations

The number of combinations of n elements taken r at a time, where $r \le n$, is

$$_nC_r = \frac{n!}{(n-r)!r!}.$$

---

*Another notation that is sometimes used in place of $_nC_r$ is $\binom{n}{r}$.

▼ **EXAMPLE 11**   From a group of 10 students, a committee is to be chosen to meet with the dean. How many different 3-person committees are possible?

Solution A committee is not ordered, so we compute

$$_{10}C_3 = \frac{10!}{(10-3)!\,3!} = \frac{10!}{7!\,3!} = \frac{10 \cdot 9 \cdot 8 \cdot 7!}{7!\,3!} = \frac{10 \cdot 9 \cdot 8}{3 \cdot 2 \cdot 1} = 120. ▲$$

| TECHNOLOGY TIP   The key for obtaining $_nC_r$ on a graphing calculator is located in the same menu as the key for obtaining $_nP_r$.

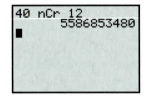

```
40 nCr 12
        5586853480
■
```

FIGURE 9.8

▼ **EXAMPLE 12**   In how many ways can a 12-person jury be chosen from a pool of 40 people?

Solution Since the order in which the jurors are chosen doesn't matter, we use combinations. The number of combinations of 40 things taken 12 at a time is

$$_{40}C_{12} = \frac{40!}{(40-12)!\,12!} = \frac{40!}{28!\,12!}.$$

Using a calculator to compute this number (Figure 9.8), we see that there are 5,586,853,480 possible ways to choose a jury. ▲

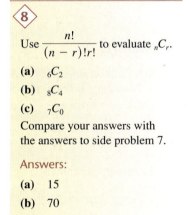

⟨8⟩

Use $\dfrac{n!}{(n-r)!\,r!}$ to evaluate $_nC_r$.

(a) $_6C_2$

(b) $_8C_4$

(c) $_7C_0$

Compare your answers with the answers to side problem 7.

Answers:

(a) 15

(b) 70

(c) 1

▼ **EXAMPLE 13**   Three managers are to be selected from a group of 30 to work on a special project.

(a) In how many different ways can the managers be selected?

Solution Here we wish to know the number of 3-element combinations that can be formed from a set of 30 elements. (We want combinations, not permutations, since order within the group of 3 does not matter.)

$$_{30}C_3 = \frac{30!}{27!\,3!} = 4060$$

There are 4060 ways to select the project group.

(b) In how many ways can the group of 3 be selected if a certain manager must work on the project?

Solution Since 1 manager has already been selected for the project, the problem is reduced to selecting 2 more from the remaining 29 managers.

$$_{29}C_2 = \frac{29!}{27!\,2!} = 406$$

In this case, the project group can be selected in 406 ways.

(c) In how many ways can a nonempty group of at most 3 managers be selected from these 30 managers?

Solution The group is to be nonempty; therefore, "at most 3" means "1 or 2 or 3." Find the number of ways for each case.

Case	Number of Ways
1	$_{30}C_1 = \dfrac{30!}{29!\,1!} = \dfrac{30 \cdot 29!}{29!\,1!} = 30$
2	$_{30}C_2 = \dfrac{30!}{28!\,2!} = \dfrac{30 \cdot 29 \cdot 28!}{28! \cdot 2 \cdot 1} = 435$
3	$_{30}C_3 = \dfrac{30!}{27!\,3!} = \dfrac{30 \cdot 29 \cdot 28 \cdot 27!}{27! \cdot 3 \cdot 2 \cdot 1} = 4060$

The total number of ways to select at most 3 managers will be the sum

$$30 + 435 + 4060 = 4525. \ \blacktriangle \ \langle 9 \rangle$$

⟨**9**⟩

Five orchids from a collection of 20 are to be selected for a flower show.

(a) In how many ways can this be done?

(b) In how many different ways can the group of 5 be selected if 2 particular orchids must be included?

(c) In how many ways can at least 1 and at most 5 orchids be selected? (*Hint:* Use a calculator.)

Answers:

(a) 15,504

(b) 816

(c) 21,699

Choosing a Method

The formulas for permutations and combinations given in this section will be very useful in solving probability problems in later sections. Any difficulty in using these formulas usually comes from being unable to differentiate among them. Both permutations and combinations give the number of ways to choose r objects from a set of n objects. The differences between permutations and combinations are outlined in the following summary.

Permutations	**Combinations**
Different orderings or arrangements of the r objects are different permutations.	Each choice or subset of r objects gives 1 combination. Order within the r objects does not matter.
$$_nP_r = \dfrac{n!}{(n-r)!}$$	$$_nC_r = \dfrac{n!}{(n-r)!\,r!}$$
Clue words: arrangement, schedule, order	Clue words: group, committee, set, sample
Order matters!	Order does not matter!

In the following examples, concentrate on recognizing which of the formulas should be applied.

▼ **EXAMPLE 14** For each of the following problems, tell whether permutations or combinations should be used to solve the problem.

(a) How many 4-digit code numbers are possible if no digits are repeated?

Solution Since changing the order of the 4 digits results in a different code, we use permutations.

(b) A sample of 3 lightbulbs is randomly selected from a batch of 15 bulbs. How many different samples are possible?

Solution The order in which the 3 lightbulbs are selected is not important. The sample is unchanged if the bulbs are rearranged, so combinations should be used.

(c) In a basketball conference with 8 teams, how many games must be played so that each team plays every other team exactly once?

Solution The selection of 2 teams for a game is an *unordered* subset of 2 from the set of 8 teams. Use combinations again.

(d) In how many ways can 4 patients be assigned to 6 hospital rooms so that each patient has a private room?

Solution The room assignments are an *ordered* selection of 4 rooms from the 6 rooms. Exchanging the rooms of any 2 patients within a selection of 4 rooms gives a different assignment, so permutations should be used. ▲ ⟨10⟩

⟨10⟩

Solve the problems in Example 14.

Answers:

(a) 5040

(b) 455

(c) 28

(d) 360

▼ **EXAMPLE 15** A manager must select 4 employees for promotion. 12 employees are eligible.

(a) In how many ways can the 4 be chosen?

Solution Because there is no reason to consider the order in which the 4 are selected, we use combinations.

$$_{12}C_4 = \frac{12!}{4! \, 8!} = 495$$

(b) In how many ways can 4 employees be chosen (from 12) to be placed in 4 different jobs?

Solution In this case, once a group of 4 is selected, its members can be assigned in many different ways (or arrangements) to the 4 jobs. Therefore, this problem requires permutations.

$$_{12}P_4 = \frac{12!}{8!} = 11{,}880 \; ▲ \; ⟨11⟩$$

⟨11⟩

A mailman has special-delivery mail for 7 customers.

(a) In how many ways can he arrange his schedule to deliver to all 7?

(b) In how many ways can he schedule deliveries if he can deliver to only 4 of the 7?

Answers:

(a) 5040

(b) 840

▼ **EXAMPLE 16** Powerball is a lottery game played in 28 states (plus the District of Columbia and the U.S. Virgin Islands). For a $1 ticket, a player selects five different numbers from 1 to 53 and one powerball number from 1 to 42 (which may be the same as one of the first five chosen). A match of all six numbers wins the jackpot. How many different selections are possible?

Solution The order in which the first five numbers are chosen doesn't matter. So we use combinations to find the number of combinations of 53 things taken 5 at a time—that is, $_{53}C_5$. There are 42 ways to choose one powerball number from 1 to 42. So, by the multiplication principle, the number of different selections is

$$_{53}C_5 \cdot 42 = \frac{53!}{(53-5)! \, 5!} \cdot 42 = \frac{53! \cdot 42}{48! \, 5!} = 120{,}526{,}770,$$

⟨12⟩

Under earlier Powerball rules you had to choose five different numbers from 1 to 45 and then choose one powerball number from 1 to 45. Under those rules, how many different selections were possible?

Answer:

54,979,155

as shown in two ways on a graphing calculator in Figure 9.9 on the next page. ▲ ⟨12⟩

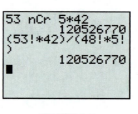

FIGURE 9.9

The following problems involve a standard deck of 52 playing cards, shown in Figure 8.19 in Chapter 8.

▼ **EXAMPLE 17** Five cards are dealt from a standard 52-card deck.

(a) How many hands have all face cards?

Solution The face cards are the king, queen, and jack of each suit. There are 4 suits, so there are 12 face cards. The order of the 5 cards is not important, so use combinations to get

$$_{12}C_5 = \frac{12!}{5!\,7!} = 792.$$

(b) How many hands have exactly 2 hearts?

Solution There are 13 hearts in the deck, so the 2 hearts will be selected from those 13 cards. The other 3 cards must come from the remaining 39 cards that are not hearts. Use combinations and the multiplication principle to get

$$_{13}C_2 \cdot _{39}C_3 = 78 \cdot 9139 = 712,842.$$

(c) How many 5-card hands have all cards of a single suit?

Solution The arrangement of the 5 cards is not important, so use combinations. The total number of ways that 5 cards of a particular suit of 13 cards can be dealt is $_{13}C_5$.

Since there are 4 different suits, by the multiplication principle, there are

$$4 \cdot _{13}C_5 = 4 \cdot 1287 = 5148$$

ways to deal 5 cards of the same suit. ▲

⟨13⟩

In how many ways can 5 red cards be dealt?

Answer:

65,780

As Examples 16 and 17 show, often both combinations and the multiplication principle must be used in the same problem.

▼ **EXAMPLE 18** To illustrate the differences between permutations and combinations in another way, suppose 2 cans of soup are to be selected from 4 cans on a shelf: noodle (N), bean (B), mushroom (M), and tomato (T). As shown in Figure 9.10(a), there are 12 ways to select 2 cans from the 4 cans if the order matters (if noodle first and bean second is considered different from bean and then noodle, for example). However, if order is unimportant, then there are 6 ways to choose 2 cans of soup from the 4, as illustrated in Figure 9.10(b). ▲

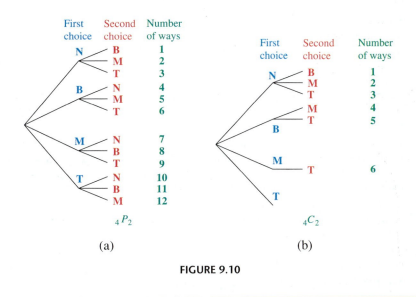

FIGURE 9.10

> **CAUTION** It should be stressed that not all counting problems lend themselves to either permutations or combinations.

◆9.2 Exercises

Evaluate the following factorials, permutations, and combinations.

1. $_4P_2$ **2.** $3!$ **3.** $_8C_5$
4. $7!$ **5.** $_8P_1$ **6.** $_7C_2$
7. $4!$ **8.** $_4P_4$ **9.** $_9C_6$
10. $_{10}C_8$ **11.** $_{13}P_2$ **12.** $_{10}P_3$

Use a calculator to find values for Exercises 13–20.

13. $_{25}P_5$ **14.** $_{40}P_5$
15. $_{14}P_5$ **16.** $_{17}P_8$
17. $_{18}C_5$ **18.** $_{32}C_9$
19. $_{25}C_{16}$ **20.** $_{50}C_8$

21. Some students find it puzzling that $0! = 1$, and they think that $0!$ should equal 0. If this were true, what would be the value of $_4P_4$ according to the permutations formula?

22. If you already knew the value of $8!$, how could you find the value of $9!$ quickly?

Use the multiplication principle to solve the following problems. (See Examples 1–6.)

23. **Social Science** An ancient Chinese philosophical work known as the *I Ching* (*Book of Changes*) is often used as an oracle from which people can seek and obtain advice. The philosophy describes the duality of the universe in terms of two primary forces: *yin* (passive, dark, receptive) and *yang* (active, light, creative). See the figure. The yin energy is represented by a broken line (– –) and the yang by a solid line (—). These lines are written on top of one another in groups of three, known as *trigrams*. For example, the trigram ☱ is called *Tui*, the Joyous, and has the image of a lake.

(a) How many trigrams are there altogether?
(b) The trigrams are grouped together, one on top of the other, in pairs known as hexagrams. Each hexagram represents one aspect of the *I Ching* philosophy. How many hexagrams are there?

24. **Business** How many different heating–cooling units are possible if a home owner has 3 choices for the efficiency rating of the furnace, 3 options for the fan speed, and 6 options for the air condenser?

25. An auto manufacturer produces 6 models, each available in 8 different colors, with 4 different upholstery fabrics and 3 interior colors. How many varieties of the auto are available?

26. **Business**   How many different 4-letter radio station call letters can be made
    (a) if the first letter must be K or W and no letter may be repeated?
    (b) if repeats are allowed (but the first letter is K or W)?
    (c) How many of the 4-letter call letters (starting with K or W) with no repeats end in R?

27. **Social Science**   A social security number has 9 digits. How many social security numbers are possible? The U.S. population in 2005 was approximately 296 million. Was it possible for every U.S. resident to have a unique social security number? (Assume no restrictions.)

28. **Social Science**   The United States Postal Service currently uses 5-digit zip codes in most areas. How many zip codes are possible if there are no restrictions on the digits used? How many would be possible if the first number could not be 0?

29. **Social Science**   The Postal Service is encouraging the use of 9-digit zip codes in some areas, adding 4 digits after the usual 5-digit code. How many such zip codes are possible with no restrictions?

30. **Social Science**   For many years, the state of California used 3 letters followed by 3 digits on its automobile license plates.
    (a) How many different license plates are possible with this arrangement?
    (b) When the state ran out of new numbers, the order was reversed to 3 digits followed by 3 letters. How many new license plate numbers were then possible?
    (c) Several years ago, the numbers described in part (b) were also used up. The state then issued plates with 1 letter followed by 3 digits and then 3 letters. How many new license plate numbers will this provide?

31. **Social Science**   A polka band from Cheektowaga, New York, will play 6 traditional and 4 original compositions at a local concert. In how many ways can the musicians arrange the program if
    (a) they begin with a traditional piece?
    (b) they play an original piece last?

32. **Business**   A real-estate agent has the names of seven potential clients interested in selling their homes.
    (a) In how many ways can she arrange her schedule if she calls on all 7?
    (b) In how many ways can she arrange her schedule if she can call on only 5 of the 7?

Social Science   *The United States is rapidly running out of telephone numbers. In large cities, telephone companies have introduced new area codes as numbers are used up.*

33. (a) Until recently, all area codes had a 0 or a 1 as the middle digit and the first digit could not be 0 or 1. How many area codes were possible with this arrangement? How many telephone numbers does the current

7-digit sequence permit per area code? (The 3-digit sequence that follows the area code cannot start with 0 or 1. Assume that there are no other restrictions.)
    (b) The actual number of area codes under the previous system was 152. Explain the discrepancy between this number and your answer to part (a).

34. The shortage of area codes under the previous system was avoided by removing the restriction on the second digit. How many area codes are available under this new system?

35. A problem with the plan in Exercise 34 was that the second digit in the area code had been used to tell phone company equipment that a long-distance call was being made. To avoid changing all equipment, an alternative plan proposed a 4-digit area code and restricted the first and second digits as before. How many area codes would this plan have provided?

36. Still another alternative solution is to increase the local dialing sequence to 8 digits instead of 7. How many additional numbers would this plan create? (Assume the same restrictions.)

37. Define permutation in your own words.

Use permutations to solve each of the following problems. (See Examples 7–10.)

38. A baseball team has 15 players. How many 9-player batting orders are possible?

39. In a game of musical chairs, 10 children will sit in 9 chairs arranged in a row. (One child will be left out.) In how many ways can the 9 children find seats?

40. **Business**   From a cooler with 8 cans of a different kinds of soda, 3 are selected for 3 people. In how many ways can this be done?

41. **Social Science**   In an election with 4 candidates for one office and 3 candidates for another office, how many different ballot orders may be used?

42. **Business**   From a pool of 8 new hires, 4 are selected to be assigned to 4 managers. In how many ways can they be selected?

43. The student activity club at the college has 32 members. In how many different ways can the club select a president, a vice president, a treasurer, and a secretary?

44. A suitcase contains 6 distinct pairs of socks and 4 distinct pairs of pants. If a traveler randomly picks 2 pairs of socks and then 3 pairs of pants, how many ways can this be done?

45. In a club with 17 members, how many ways can the club elect a president and a treasurer?

Use combinations to solve each of the following problems. (See Examples 11–13.)

46. **Business**   Four items are to be randomly selected from the first 25 items on an assembly line in order to deter-

mine the defect rate. How many different samples of 4 items can be chosen?

47. **Social Science**   A group of 4 students is to be selected from a group of 10 students to take part in a class in cell biology.
    (a) In how many ways can this be done?
    (b) In how many ways can the group that will *not* take part be chosen?

48. **Natural Science**   From a group of 15 smokers and 21 nonsmokers, a researcher wants to randomly select 7 smokers and 6 nonsmokers for a study. In how many ways can the study group be selected?

49. Five cards are drawn from an ordinary deck. In how many ways is it possible to draw
    (a) only 7's;
    (b) only 2's, 3's or 4's;
    (c) no 2's, 3's or 4's;
    (d) exactly 2 kings or queens;
    (e) 2 hearts and 3 diamonds.

50. **Business**   A salesman has 8 accounts in a certain city.
    (a) In how many ways can he select 3 accounts to call on?
    (b) In how many ways can he select at least 6 of the 8 accounts to use in preparing a report?

51. Explain the difference between a permutation and a combination.

52. Padlocks with digit dials are often referred to as "combination locks." According to the mathematical definition of combination, is this an accurate description? Explain.

Exercises 53–70 are mixed problems that may require permutations, combinations, or the multiplication principle. (See Examples 14–17.)

53. Use a tree diagram to find the number of ways 2 letters can be chosen from the set {P, Q, R} if order is important and
    (a) if repetition is allowed.
    (b) if no repeats are allowed.
    (c) Find the number of combinations of 3 elements taken 2 at a time. Does this answer differ from that to part (a) or (b)?

54. Repeat Exercise 53, using the set {P, Q, R, S} and 4 in place of 3 in part (c).

55. **Social Science**   A legislative committee consists of 6 Republicans and 4 Democrats. A delegation of 3 is to be selected to visit Iran.
    (a) How many different delegations are possible?
    (b) How many delegations would have all Republicans?
    (c) How many delegations would have 2 Republicans and 1 Democrat?
    (d) How many delegations would include at least 1 Democrat?

56. **Natural Science**   In an experiment on plant hardiness, a researcher gathers 6 wheat plants, 5 barley plants, and 3 rye plants. She wishes to select 4 plants at random.

    (a) In how many ways can this be done?
    (b) In how many ways can this be done if 2 wheat plants must be included?

57. Baskin-Robbins® keeps 21 flavors of ice cream in the company's permanent collection.
    (a) How many different double-scoop cones can be made?
    (b) How many different triple-scoop cones can be made?

58. A concert to raise money for an economics prize is to consist of 8 works: 3 overtures, 3 sonatas, and 2 piano concertos.
    (a) In how many ways can the program be arranged?
    (b) In how many ways can the program be arranged if an overture must come first?

59. A lottery game requires that you pick 6 different numbers from 1 to 99. If you pick all 6 winning numbers, you win $4 million.
    (a) How many ways are there to choose 6 numbers if order is not important?
    (b) How many ways are there to choose 6 numbers if order matters?

60. In Exercise 59, if you pick 5 of the 6 numbers correctly, you win $5,000. In how many ways can you pick exactly 5 of the 6 winning numbers without regard to order?

61. The game of Sets* consists of a special deck of cards. Each card has on it either one, two, or three shapes. The shapes on each card are all the same color, either green, purple, or red. The shapes on each card are the same style, either solid, shaded, or outline. There are three possible shapes: squiggle, diamond, and oval, and only one type of shape appears on a card. The deck consists of all possible combinations of shape, color, style, and number. How many cards are in a deck?

62. **Health**   Nine drugs have been found to be effective in the treatment of a certain disease. It is believed that the sequence in which the drugs are administered is important in the effectiveness of the treatment. In how many orders can 4 of the 9 drugs be administered?

63. **Natural Science**   A biologist is attempting to classify 52,000 species of insects by assigning 3 initials to each species. Is it possible to classify all the species in this way? If not, how many initials should be used?

64. One play in a state lottery consists of choosing 6 numbers from 1 to 44. If your 6 numbers are drawn (in any order), you win the jackpot.
    (a) How many possible ways are there to draw the 6 numbers?
    (b) If you get 2 plays for a dollar, how much would it cost to guarantee that 1 of your choices would be drawn?

---

*Copyright © Marsha J. Falco.

(c) Assuming that you work alone and can fill out a betting ticket (for 2 plays) every second, and assuming that the lotto drawing will take place 3 days from now, can you place enough bets to guarantee that 1 of your choices will be drawn?

65. A cooler contains 5 cans of Pepsi®, 1 can of Diet Coke®, and 3 cans of 7UP®; you pick 3 cans at random. How many samples are possible in which the soda cans are

(a) only Pepsi   **(b)** only Diet Coke

(c) only 7UP   **(d)** 2 Pepsi, 1 Diet Coke

(e) 2 Pepsi, 1 7UP   **(f)** 2 7UP, 1 Pepsi

(g) 2 Diet Coke, 1 7UP

66. A class has 10 male students and 12 female students. How many ways can the class select a committee of four people to petition the teacher not to make the final exam cumulative if the committee has to have 2 males and 2 females?

67. A gardener has 4 flower beds to plant. He has 4 colors of impatients, 3 colors of begonias, and 5 kinds of daylilies to choose from. Treating each plant–color combination as a selection, and assuming that each flower bed is planted with only one selection, how many ways can he plant the 4 flower beds if

(a) order matters.

(b) order does not matter.

68. Suppose a pizza shop offers 5 choices of cheese and 8 toppings. If the order of the cheeses and toppings does not matter, how many different pizza selections are possible when choosing two cheeses and 3 toppings?

69. In the game of bingo, each card has 5 columns. Column 1 has spaces for 5 numbers, chosen from 1 to 15. Column 2 similarly has 5 numbers, chosen from 16 to 30. Column 3 has a free space in the middle, plus 4 numbers chosen from 31 to 45. The 5 numbers in columns 4 and 5 are chosen from 46 to 60 and from 61 to 75, respectively. The numbers in each card can be in any order. How many different bingo cards are there?

70. A television commercial for Little Caesars® pizza announced that, with the purchase of two pizzas, one could receive free any combination of up to five toppings on each pizza. The commercial shows a young child waiting in line at one of the company's stores who calculates that there are 1,048,576 possibilities for the toppings on the two pizzas.* Verify the child's calculation. Use the fact

that Little Caesars has 11 toppings to choose from. Assume that the order of the two pizzas matters; that is, if the first pizza has combination 1 and the second pizza has combination 2, that arrangement is different from combination 2 on the first pizza and combination 1 on the second.

If the n objects in a permutations problem are not all distinguishable—that is, if there are n_1 of type 1, n_2 of type 2, and so on for r different types, then the number of distinguishable permutations is

$$\frac{n!}{n_1!\, n_2! \cdots n_r!}.$$

Example *In how many ways can you arrange the letters in the word Mississippi?*

This word contains 1 m, 4 i's, 4 s's, and 2 p's. To use the formula, let $n = 11$, $n_1 = 1$, $n_2 = 4$, $n_3 = 4$, and $n_4 = 2$ to get

$$\frac{11!}{1!\,4!\,4!\,2!} = 34{,}650$$

arrangements. The letters in a word with 11 different letters can be arranged in $11! = 39{,}916{,}800$ ways.

71. Find the number of distinguishable permutations of the letters in each of the following words.

(a) martini   **(b)** nunnery   **(c)** grinding

72. A printer has 5 W's, 4 X's, 3 Y's, and 2 Z's. How many different "words" are possible that use all these letters? (A "word" does not have to have any meaning here.)

73. Shirley is a shelf stocker at the local grocery store. She has 4 varieties of Stouffer's® frozen dinners, 3 varieties of Lean Cuisine® frozen dinners, and 5 varieties of Weight-Watchers® frozen dinners. In how many distinguishable ways can she stock the shelves if

(a) the dinners can be arranged in any order?

(b) dinners from the same company are considered alike and have to be shelved together?

(c) dinners from the same company are considered alike, but do not have to be shelved together?

74. A child has a set of different-shaped plastic objects. There are 2 pyramids, 5 cubes, and 6 spheres. In how many ways can she arrange them in a row

(a) if they are all different colors?

(b) if the same shapes must be grouped?

(c) In how many distinguishable ways can they be arranged in a row if objects of the same shape are also the same color, but need not be grouped?

---

*Joseph F. Heiser, "Pascal and Gauss meet Little Caesars," *Mathematics Teacher*, 87 (September 1994): 389. In a letter to *The Mathematics Teacher*, Heiser argued that the two combinations should be counted as the same, so the child has actually overcounted. In that case there would be 524,800 possibilities.

9.3   Applications of Counting

Many of the probability problems involving *dependent* events that were solved with probability trees in Chapter 8 can also be solved by using counting principles—that is, permutations and combinations. Permutations and combinations are especially helpful when the number of choices is large. The use of counting rules to solve probability problems depends on the basic probability principle introduced in Section 8.3 and repeated here.

If event E is a subset of sample space S, then the probability that event E occurs, written $P(E)$, is

$$P(E) = \frac{n(E)}{n(S)}.$$

It is also helpful to keep in mind that, in probability statements,

"and" corresponds to multiplication

and

"or" corresponds to addition.

▼ **EXAMPLE 1**   From a potential jury pool with 1 Hispanic, 3 Caucasian, and 2 African-American members, 2 jurors are selected one at a time without replacement. Find the probability that 1 Caucasian and 1 African-American are selected.

Solution   In Example 6 of Section 8.5, it was necessary to consider the order in which the jurors were selected. With combinations, it is not necessary: Simply count the number of ways in which 1 Caucasian and 1 African-American juror can be selected. The Caucasian can be selected in $_3C_1$ ways, and the African-American juror can be selected in $_2C_1$ ways. By the multiplication principle, both results can occur in

$$_3C_1 \cdot {_2C_1} = 3 \cdot 2 = 6 \text{ ways,}$$

giving the numerator of the probability fraction. For the denominator, 2 jurors are selected from a total of 6 candidates. This can occur in $_6C_2 = 15$ ways. The required probability is

$$P(1 \text{ Caucasian and 1 African American}) = \frac{_3C_1 \cdot {_2C_1}}{_6C_2} = \frac{3 \cdot 2}{15} = \frac{6}{15} = \frac{2}{5} = .40.$$

This agrees with the answer found earlier. ▲

▼ **EXAMPLE 2**   From a baseball team of 15 players, 4 are to be selected to present a list of grievances to the coach.
(a) In how many ways can this be done?

Solution   Four players from a group of 15 can be selected in $_{15}C_4$ ways. (Use combinations, since the order in which the group of 4 is selected is unimportant.)

$$_{15}C_4 = \frac{15!}{4! \, 11!} = \frac{15(14)(13)(12)}{4(3)(2)(1)} = 1365$$

There are at 1365 ways to choose 4 players from 15.

(b) One of the players is Michael Branson. Find the probability that Branson will be among the 4 selected.

Solution   The probability that Branson will be selected is the number of ways the chosen group includes him, divided by the total number of ways the group of 4 can be chosen. If Branson must be one of the 4 selected, the problem reduces to finding the number of ways the additional 3 players can be chosen. There are 3 chosen from 14 players; this can be done in

$$_{14}C_3 = \frac{14!}{3!\,11!} = 364$$

ways. The number of ways 4 players can be selected from 15 is

$$n = {}_{15}C_4 = 1365.$$

The probability that Branson will be one of the 4 chosen is

$$P(\text{Branson is chosen}) = \frac{364}{1365} \approx .267.$$

(c) Find the probability that Branson will not be selected.

Solution   The probability that he will not be chosen is $1 - .267 = .733.$ ▲ ◇1◇

1 ◇

The ski club has 8 women and 7 men. What is the probability that if the club elects 3 officers, all 3 of them will be women?

Answer:

.431

▼ **EXAMPLE 3**   In shipping diesel engines abroad, it is common to pack 12 engines in 1 container that is then loaded on a railcar and sent to a port. Suppose that a company has received complaints from its customers that many of the engines arrive in nonworking condition. To help solve this problem, the company decides to make a spot check of containers after loading. The company will test 3 engines from a container at random; if any of the 3 is nonworking, the container will not be shipped until each engine in it is checked. Suppose a given container has 2 non-working engines. Find the probability that the container will not be shipped.

Solution   The container will not be shipped if the sample of 3 engines contains 1 or 2 defective engines. Letting $P(1 \text{ defective})$ represent the probability of exactly 1 defective engine in the sample, we obtain

$$P(\text{not shipping}) = P(1 \text{ defective}) + P(2 \text{ defectives}).$$

There are $_{12}C_3$ ways to choose the 3 engines for testing:

$$_{12}C_3 = \frac{12!}{3!\,9!} = \frac{12(11)(10)}{3(2)(1)} = 220.$$

2 ◇

Calculate $_2C_1 \cdot {}_{10}C_2$.

Answer:

90

There are $_2C_1$ ways of choosing 1 defective engine from the 2 in the container, and for each of these ways, there are $_{10}C_2$ ways of choosing 2 good engines from among the 10 in the container. By the multiplication principle, there are

$$_2C_1 \cdot {}_{10}C_2$$

ways of choosing a sample of 3 engines containing 1 that is defective. ◇2◇

Using the result from side problem 2 yields

$$P(1 \text{ defective}) = \frac{90}{220}.$$

There are $_2C_2$ ways of choosing 2 defective engines from the 2 defective engines in the container and $_{10}C_1$ ways of choosing 1 good engine from among the 10 good engines, giving

$$_2C_2 \cdot {}_{10}C_1$$

ways of choosing a sample of 3 engines containing 2 defective ones. ◁③▷

Using the result from side problem 3 gives

$$P(2 \text{ defectives}) = \frac{10}{220}$$

and

$$P(\text{not shipping}) = P(1 \text{ defective}) + P(2 \text{ defectives})$$
$$= \frac{90}{220} + \frac{10}{220} = \frac{100}{220} = \frac{5}{11} \approx .455.$$

Notice that the probability is $1 - .455 = .545$ that the container *will* be shipped, even though it has 2 defective engines. The management must decide whether this probability is acceptable; if not, it may be necessary to test more than 3 engines from a container. ▲

Instead of finding the sum $P(1 \text{ defective}) + P(2 \text{ defectives})$, the result in Example 3 could be obtained by calculating $1 - P(\text{no defectives})$.

$$P(\text{not shipping}) = 1 - P(\text{no defectives in sample})$$
$$= 1 - \frac{{}_2C_0 \cdot {}_{10}C_3}{{}_{12}C_3} = 1 - \frac{1(120)}{220}$$
$$= 1 - \frac{120}{220} = \frac{100}{220} \approx .455 \text{ ◁④▷}$$

▼ EXAMPLE 4 In a common form of 5-card draw poker, a hand of 5 cards is dealt to each player from a deck of 52 cards. There are a total of

$$_{52}C_5 = \frac{52!}{5! \, 47!} = 2{,}598{,}960$$

such hands possible. Find the probability of being dealt each of the following:

(a) Heart flush hands (5 hearts)

Solution There are 13 hearts in a deck; there are

$$_{13}C_5 = \frac{13!}{5! \, 8!} = \frac{13(12)(11)(10)(9)}{5(4)(3)(2)(1)} = 1287$$

---

③

Calculate $_2C_2 \cdot {}_{10}C_1$.

Answer:

10

---

④

In Example 3, if a sample of 2 engines is tested, what is the probability that the container will not be shipped?

Answer:

.318

different hands containing only hearts. The probability of a heart flush is

$$P(\text{heart flush}) = \frac{1287}{2{,}598{,}960} \approx .000495.$$

(b) A flush of any suit (5 cards, all from 1 suit)

Solution There are 4 suits to a deck, so

$$P(\text{flush}) = 4 \cdot P(\text{heart flush}) = 4(.000495) \approx .00198.$$

(c) A full house of aces and eights (3 aces and 2 eights)

Solution There are $_4C_3$ ways to choose 3 aces from among the 4 in the deck and $_4C_2$ ways to choose 2 eights.

$$P(3 \text{ aces}, 2 \text{ eights}) = \frac{_4C_3 \cdot _4C_2}{_{52}C_5} \approx .00000923$$

(d) Any full house (3 cards of one value, 2 of another)

Solution There are 13 values in a deck, so there are **13** choices for the first value mentioned, leaving **12** choices for the second value. (Order *is* important here, since a full house of aces and eights, for example, is not the same as a full house of eights and aces.)

$$P(\text{full house}) = \frac{13 \cdot _4C_3 \cdot 12 \cdot _4C_2}{_{52}C_5} \approx .00144 \; \blacktriangle \; \langle 5 \rangle$$

⟨5⟩
Find the probability of being dealt a poker hand (5 cards) with 4 kings.

Answer:

.00001847

▼ EXAMPLE 5 A cooler contains 8 different kinds of soda, among which 3 cans are Pepsi®, Classic Coke®, and Sprite®. What is the probability, when picking at random, of selecting the 3 cans in the particular order listed in the previous sentence?

Solution Use permutations to find the number of arrangements in the sample, because order matters.

$$n = _8P_3 = 8(7)(6) = 336$$

Since each can is different, there is only 1 way to choose Pepsi, Classic Coke, and Sprite in that order, so the probability is

$$\frac{1}{336} = .0030. \; \blacktriangle \; \langle 6 \rangle$$

▼ EXAMPLE 6 Suppose a group of 5 people is in a room. Find the probability that at least 2 of the people have the same birthday.

Solution "Same birthday" refers to the month and the day, not necessarily the same year. Also, ignore leap years, and assume that each day in the year is equally likely as a birthday. First find the probability that *no 2 people* among 5 people have the same birthday. There are 365 different birthdays possible for the first of the 5 people, 364 for the second (so that the people have different birthdays), 363 for the third, and so on. The number of ways the 5 people can have different birthdays is thus the number of permutations of 365 things (days) taken 5 at a time, or

$$_{365}P_5 = 365 \cdot 364 \cdot 363 \cdot 362 \cdot 361.$$

⟨6⟩
Martha, Leonard, Calvin, and Sheila will be handling the officer duties of president, vice president, treasurer, and secretary.

(a) If the offices are assigned randomly, what is the probability that Calvin is the president?

(b) If the offices are assigned randomly, what is the probability that Sheila is president, Martha is vice president, Calvin is treasurer, and Leonard is secretary?

Answers:

(a) 1/4

(b) 1/24

The number of ways that the 5 people can have the same or different birthdays is

$$365 \cdot 365 \cdot 365 \cdot 365 \cdot 365 = (365)^5.$$

Finally, the *probability* that none of the 5 people have the same birthday is

$$\frac{{}_{365}P_5}{(365)^5} = \frac{365 \cdot 364 \cdot 363 \cdot 362 \cdot 361}{365 \cdot 365 \cdot 365 \cdot 365 \cdot 365} \approx .973.$$

The probability that at least 2 of the 5 people *do* have the same birthday is $1 - .973 = .027.$ ▲

Example 6 can be extended for more than 5 people. In general, the probability that no 2 people among n people have the same birthday is

$$\frac{{}_{365}P_n}{(365)^n}.$$

The probability that at least 2 of the n people *do* have the same birthday is

$$1 - \frac{{}_{365}P_n}{(365)^n}. \quad ⟨7⟩$$

The following table shows this probability for various values of n:

Number of People, n	Probability That at Least 2 Have the Same Birthday
5	.027
10	.117
15	.253
20	.411
22	.476
23	.507
25	.569
30	.706
35	.814
40	.891
50	.970
365	1

The probability that 2 people among 23 have the same birthday is .507, a little more than half. Many people are surprised at this result; somehow it seems that a larger number of people should be required. ⟨8⟩

Using a graphing calculator, we can graph the probability formula in the previous example as a function of n, but the graphing calculator must be set to evaluate the function at integer points. Figure 9.11 was produced on a TI-84+ by letting $Y_1 = 1 - (365 \text{ nPr } X)/365^X$ on the interval $0 \le x \le 47$. (This domain ensures integer values for x.) Notice that the graph does not extend past $x = 39$. This is because $P(365, n)$ and 365^n are too large for the calculator when $n \ge 40$.

⟨7⟩

Evaluate $1 - \dfrac{{}_{365}P_n}{(365)^n}$ for

(a) $n = 3$;

(b) $n = 6$.

Answers:

(a)   .008

(b)   .040

⟨8⟩

Set up (do not calculate) the probability that at least 2 of the 9 members of the Supreme Court have the same birthday.

Answer:

$1 - {}_{365}P_9/365^9$

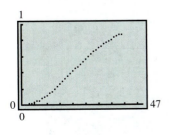

FIGURE 9.11

◆9.3 Exercises

Business *Refer to Example 3. The management feels that the probability of .545 that a container will be shipped even though it contains 2 defectives is too high. Management decides to increase the sample size chosen. Find the probability that a container will be shipped even though it contains 2 defectives if the sample size is increased to*

1. 4.   **2.** 5.

Business *A shipment of 8 computers contains 3 with defects. Find the probability that with a sample of the following size, drawn from the 8, will not contain a defective computer. (See Example 3.)*

3. 1   **4.** 2

5. 3   **6.** 5

A radio station runs a promotion at an auto show with a money box with 10 $100 tickets, 12 $50 tickets, and 20 $25 tickets. The box contains an additional 200 "dummy" tickets with no value. Three tickets are randomly drawn. Find the following probabilities (See Examples 1 and 2).

7. All $100 tickets   **8.** All $50 tickets

9. Exactly two $25 tickets and no other money winners

10. One ticket of each money amount

11. No tickets with money

12. At least one money ticket

Two cards are drawn at random from an ordinary deck of 52 cards. (See Example 4.)

13. How many 2-card hands are possible?

Find the probability that the 2-card hand in Exercise 13 contains the following.

14. 2 kings   **15.** No deuces (2's)

16. 2 face cards   **17.** Different suits

18. At least 1 black card   **19.** No more than 1 diamond

20. Discuss the relative merits of using probability trees versus combinations to solve probability problems. When would each approach be most appropriate?

21. Several examples in this section used the rule $P(E') = 1 - P(E)$. Explain the advantage (especially in Example 6) of using this rule.

A bridge hand consists of 13 cards from a deck of 52. Set up the probabilities that a bridge hand includes each of the following.

22. 4 face cards   **23.** 2 10's and 3 jacks

24. 8 cards of one suit and 5 of another

25. In Exercise 59 in Section 9.2, we found the number of ways to pick 6 different numbers from 1 to 99 in a state lottery.

(a) Assuming that order is unimportant, what is the probability of picking all 6 numbers correctly to win the big prize?

(b) What is the probability if order matters?

26. In Exercise 25 (a), what is the probability of picking exactly 5 of the 6 numbers correctly?

27. Example 16 in Section 9.2 shows that the probability of winning the Powerball lottery is 1/120,526,770. If Juanita and Michelle each play Powerball on one particular evening, what is the probability that both will select the winning numbers if they make their selections independently of each other?

28. **Business** A cellular phone manufacturer randomly selects 5 of every 100 phones from the assembly line and tests them. If at least 4 of the 5 pass inspection, the batch of 100 is considered acceptable. Find the probability that a batch is considered acceptable if it contains the following.
(a) 2 defective phones
(b) No defective phones
(c) 3 defective phones

29. **Social Science** Of the 14 members of President George W. Bush's first cabinet, 3 were women. Suppose the president randomly selected 4 advisors from the cabinet for a meeting. Find the probability that the group of 4 would be composed as follows.
(a) 3 women and 1 man
(b) All men
(c) At least one woman

30. **Business** A car dealer has 6 red, 10 gray, and 7 blue cars in stock. Ten cars are randomly chosen to be displayed in front of the dealership. Find the probability that
(a) 4 are red and the others are blue.
(b) 3 are red, 3 are blue, and 4 are gray.
(c) exactly 5 are gray and none are blue.
(d) all 10 are gray.

31. Social Science On the first day of school, the teacher of a multicultural first-grade class found that in her class of 25 students, 10 spoke only English, 6 spoke only Spanish, 4 spoke only Russian, 3 spoke only Vietnamese, and 2 spoke only Hmong. The children were assigned randomly to groups of 5. Find the probability that a group included the following.

(a) 2 English-speaking and 3 Russian-speaking children

(b) All English-speaking children

(c) No English-speaking children

(d) At least two children who spoke Vietnamese or Hmong

For Exercises 32–34, refer to Example 6 in this section.

32. Set up the probability that at least 2 of the 42 men who have served as president of the United States have had the same birthday.*

33. Set up the probability that at least 2 of the 100 U.S. senators have the same birthday.

34. What is the probability that at least 2 of the 50 U.S. governors have the same birthday?

One version of the New York State lottery game Quick Draw has players selecting 4 numbers at random from the numbers 1–80. The state picks 20 winning numbers. If the player's 4 numbers are selected by the state, the player wins $55.[†]

35. What is the probability of winning?

36. If the state picks 3 of the player's numbers, the player wins $5. What is the probability of winning $5?

37. What is the probability of having none of your 4 numbers selected by the state?

38. During the 1988 college football season, the Big Eight Conference ended the season in a "perfect progression," as shown in the following table.[‡]

Won	Lost	Team
7	0	Nebraska (NU)
6	1	Oklahoma (OU)
5	2	Oklahoma State (OSU)
4	3	Colorado (CU)
3	4	Iowa State (ISU)
2	5	Missouri (MU)
1	6	Kansas (KU)
0	7	Kansas State (KSU)

Someone wondered what the probability of such an outcome might be.

(a) Assuming no ties and assuming that each team had an equally likely probability of winning each game, find the probability of the perfect progression shown in the table.

(b) Under the same assumptions, find a general expression for the probability of a perfect progression in an n-team league.

39. Use a computer or a graphing calculator and the Monte Carlo method with $n = 50$ to estimate the probabilities of the following hands at poker. (See the directions for Exercises 50–53 on page 000.) Assume that aces are either high or low. Since each hand has 5 cards, you will need $50 \cdot 5 = 250$ random numbers to "look at" 50 hands. Compare these experimental results with the theoretical results.

(a) A pair of aces

(b) Any two cards of the same value

(c) Three of a kind

40. Use a computer or a graphing calculator and the Monte Carlo method with $n = 20$ to estimate the probabilities of the following 13-card bridge hands. Since each hand has 13 cards, you will need $20 \cdot 13 = 260$ random numbers to "look at" 20 hands.

(a) No aces

(b) 2 kings and 2 aces

(c) No cards of any one suit—that is, only 3 suits represented

---

*In fact, James Polk and Warren Harding were both born on November 2. Although George W. Bush is the 43rd president, the 22nd and 24th presidents were the same man: Grover Cleveland.

†www.nylottery.org/index.php.

‡From Richard Madsen. "On the Probability of a Perfect Progression." *American Statistics* 45, no. 3 (August 1991): 214.

9.4 Binomial Probability

In Section 9.1, we learned about probability distributions where we listed each outcome and its associated probability. After learning in Sections 9.2 and 9.3 how to count the number of possible outcomes, we are now ready to understand a special probability distribution known as the *binomial distribution*. This distribution occurs when the same experiment is repeated many times and each repetition is independent of previous ones. One outcome is designated a success and any other outcome is considered a failure. For example, you might want to find the probability of rolling 8 twos in 12 rolls of a die (rolling two is success, rolling anything else is a failure). The individual trials (rolling the die once) are called **Bernoulli trials** or **Bernoulli processes,** and are named after the Swiss mathematician Jakob Bernoulli (1654–1705).

If the probability of a success in a single trial is p, then the probability of failure is $1 - p$ (by the Complement Rule). When Bernoulli trials are repeated a fixed number of times, the resulting distribution of outcomes is called a **binomial distribution** or a **binomial experiment.** A binomial experiment must satisfy the following conditions.

Binomial Experiment

1. The same experiment is repeated a fixed number of times.
2. There are only two possible outcomes: success and failure.
3. The probability of success for each trial is constant.
4. The repeated trials are independent.

The basic facts about binomial experiment are illustrated by a recent poll conducted by Charlton Research Corp. that was commissioned by Research America and *Parade* magazine. It found that approximately 60% of Americans indicated "favor" when asked, "Do you favor or oppose medical research using embryonic stem cells?"* If we sample 5 Americans at random and use .60 as the estimate of the probability of answering "favor," we will generate a binomial experiment because we will repeat the sampling a fixed number of times (5), we have only two outcomes of interest (answering "favor" or answering "oppose"), the probability of success is constant with $p = 3/5$, and "at random" will guarantee that the trials are independent.

To calculate the probability that all 5 randomly chosen Americans will answer "favor" to the given question, we use the product rule for independent events of Section 8.5 (F represents answering "favor") and $P(F) = 3/5$ to obtain

$$P(FFFFF) = P(F) \cdot P(F) \cdot P(F) \cdot P(F) \cdot P(F) = \left(\frac{3}{5}\right)^5 = .078.$$

Determining the probability that 4 out of 5 Americans chose answer "favor" is slightly more complicated. We can write out the possible outcomes. The person not answering "favor" could be the first, second, third, fourth, or fifth person chosen. Letting O represent those who respond with "oppose," we have the following picture of the possible outcomes:

$$O\,F\,F\,F\,F$$
$$F\,O\,F\,F\,F$$
$$F\,F\,O\,F\,F$$
$$F\,F\,F\,O\,F$$
$$F\,F\,F\,F\,O$$

*Survey conducted June 4–9, 2005. Results available at www.researchamerica.org.

Using combinations, we find that the total number of ways in which 4 successes (and 1 failure) can occur is $_5C_4 = 5$ as shown above. The probability of each of these 5 outcomes is

$$P(F) \cdot P(F) \cdot P(F) \cdot P(F) \cdot P(O) = \left(\frac{3}{5}\right)^4\left(\frac{2}{5}\right) \approx .052.$$

Since the 5 outcomes represent disjoint alternative events, we add the 5 probabilities, or multiply by $_5C_4 = 5$.

$$P(4F\text{'s out of }5\text{ trials}) = {_5C_4}\left(\frac{3}{5}\right)^4\left(\frac{2}{5}\right)^{5-4} = 5\left(\frac{3}{5}\right)^4\left(\frac{2}{5}\right)^1 \approx .259$$

The probability of obtaining exactly 3 F's and 2 O's can be computed in the same way. The probability of any one way of achieving 3 F's and 2 O's will be

$$\left(\frac{3}{5}\right)^3\left(\frac{2}{5}\right)^2.$$

Again, the desired outcome can occur in more than one way. Using combinations, we find that the number of ways in which 3 F's (and 2 O's) can occur is $_5C_3 = 10$.

$$P(3F\text{'s and }2O\text{'s}) = {_5C_3}\left(\frac{3}{5}\right)^3\left(\frac{2}{5}\right)^{5-3} = 10\left(\frac{3}{5}\right)^3\left(\frac{2}{5}\right)^2 \approx .346 \;\; \diamondsuit \; \text{(1)}$$

With the probabilities just generated and the answers to side problem 1, we can write the probability distribution for the number of Americans who answer "favor" when 5 are selected at random:

When the outcomes and their associated probabilities are written in this form, it is very easy to calculate answers to questions such as

x	0	1	2	3	4	5
$P(x)$	.010	.077	.230	.346	.259	.078

$$P(3\text{ or more }F\text{'s}) = .346 + .259 + .078 = .683$$

and

$$P(1\text{ or fewer }F\text{'s}) = .077 + .010 = .087.$$

This example illustrates the following fact.

---

(1)

Find the probability of obtaining

(a) exactly 2 Americans answering "favor";

(b) exactly 1 American answering "favor";

(c) no Americans answering "favor".

Answers:

(a) $_5C_2\left(\frac{3}{5}\right)^2\left(\frac{2}{5}\right)^3 \approx .230$

(b) $_5C_1\left(\frac{3}{5}\right)^1\left(\frac{2}{5}\right)^4 \approx .077$

(c) $_5C_0\left(\frac{3}{5}\right)^0\left(\frac{2}{5}\right)^5 \approx .010$

Binomial Probability

If p is the probability of success in a single trial of a binomial experiment, the probability of x successes and $n - x$ failures in n independent repeated trials of the experiment is

$$_nC_x p^x(1 - p)^{n-x}.$$

TECHNOLOGY TIP
On the TI-84+ calculator, use "binompdf(*n,p,x*)" in the DISTR menu to compute the probability of exactly *x* successes in *n* trials (where *p* is the probability of success in a single trial). Use "binomcdf(*n,p,x*)" to compute the probability of at most *x* successes in *n* trials. Figure 9.12 shows the probability of exactly 3 successes in 5 trials and the probability of at most 3 successes in 5 trials, with the probability of success 3/5 for each case.

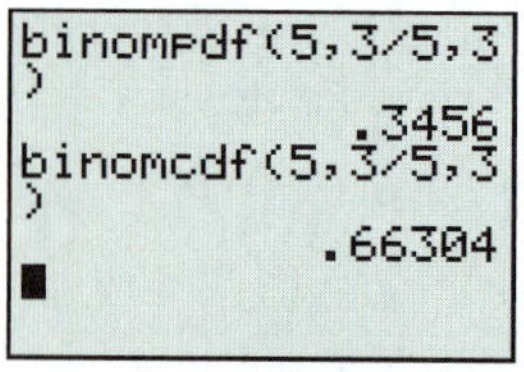

```
binompdf(5,3/5,3
)
             .3456
binomcdf(5,3/5,3
)
            .66304
■
```

FIGURE 9.12

2

According to Harper's Index for December, 2004,* the percentage of poor Americans living in the suburbs in 2003 was 39%. If 4 Americans living in the suburbs are selected at random, find the probability that exactly

(a) 1 of the 4 is poor;

(b) all 4 are poor.

Answers:

(a) about .354

(b) about .023

*www.harpers.org.

▼ **EXAMPLE 1** On July 1, 2005, Supreme Court Justice Sandra Day O'Connor announced that she would retire from the court. On July 6, 2005, Gallup reported the results of a poll which indicated that 30% of Americans wanted the new justice appointed to be more liberal. Suppose that a random sample of 6 people is chosen. Suppose also that all responses are independent of each other. Find the probability of the following.

(a) Exactly 4 of the 6 people prefer a more liberal justice.

Solution We can think of the 6 responses as independent trials, and a success occurs if a person prefers a liberal. (Note that "liberal" is labeled a success because the context of the questions refers to liberal. Labeling liberal a success is not a value judgment.) This is a binomial experiment with $p = .3$, $n = 6$, and $x = 4$. By the Binomial Probability rule,

$$P(\text{exactly } 4) = {_6}C_4(.3)^4(.7)^{6-4}$$
$$= 15(.3)^4(.7)^2$$
$$= 15(.0081)(.49)$$
$$= .059535.$$

(b) None of the 6 people prefers a liberal.

Solution Let $x = 0$.

$$P(\text{exactly } 0) = {_6}C_0(.3)^0(.7)^{6-0}$$
$$= 1(1)(.7)^6$$
$$\approx .1176 \; ▲ \; 2$$

▼ **EXAMPLE 2** In Example 7 of Section 9.1, we calculated the probability distribution for the number of girls in a family with 3 children by listing all the possibilities. Here we do the same problem using combinations and binomial probability.

(a) Find the probability distribution for the number of girls.

Solution Let $x =$ the number of girls in three births. According to the Binomial Probability rule, the probability of exactly one girl being born is

$$P(x = 1) = {_3}C_1\left(\frac{1}{2}\right)^1\left(\frac{1}{2}\right)^2 = 3\left(\frac{1}{2}\right)^3 = \frac{3}{8}.$$

The other probabilities in this distribution are found similarly, as shown here.

x	0	1	2	3
$P(x)$	${_3}C_0\left(\frac{1}{2}\right)^0\left(\frac{1}{2}\right)^3 = \frac{1}{8}$	${_3}C_1\left(\frac{1}{2}\right)^1\left(\frac{1}{2}\right)^2 = \frac{3}{8}$	${_3}C_2\left(\frac{1}{2}\right)^2\left(\frac{1}{2}\right)^1 = \frac{3}{8}$	${_3}C_3\left(\frac{1}{2}\right)^3\left(\frac{1}{2}\right)^0 = \frac{1}{8}$

(b) Find the expected number of girls in a 3-child family.

Solution For a binomial distribution, we can use the following method (which is presented here with a "plausibility argument," but not a full proof). Because 50% of births are girls, it is reasonable to expect that 50% of a sample of children will be girls. Since 50% of 3 is $3(.50) = 1.5$, we conclude that the expected number of girls is 1.5—the same answer obtained by using the expected value formula in Example 7 of Section 9.1. ▲ ⟨3⟩

The expected value in Example 2(b) was the product of the number of births and the probability of a single birth being a girl; that is, the product of the number of trials and the probability of success in a single trial. The same conclusion holds in the general case.

⟨3⟩

Find the probability of getting 2 fours in 8 tosses of a die.

Answer:

about .2605

Expected Value for a Binomial Distribution

When an experiment meets the four conditions of a binomial experiment with n fixed trials and constant probability of success p, the expected value is

$$E(x) = np.$$

▼ **EXAMPLE 3** *Men's Health* magazine reported the results of a study showing one-third of otherwise healthy 18- to 29-year-olds in Boston had significantly low vitamin D levels by the end of the winter.* Suppose a researcher randomly selects 15 males between the ages of 18 and 29 in the Boston area.

(a) Find the probability that 6 of the males have low vitamin D levels at the end of the winter.

Solution The experiment, selecting a young male, is repeated 15 times. If having low vitamin D levels is labeled a success, then not having low vitamin D levels is labeled a failure. The probability of success is $1/3$ and the probability of failure is $2/3$. Since the selection was done at random, researchers consider such trials to be independent. Thus, the probability that 6 young males in the sample of 15 have low vitamin D is

$$_{15}C_6\left(\frac{1}{3}\right)^6\left(\frac{2}{3}\right)^9 \approx .179.$$

(b) Find the probability that at most 3 of the 15 males in the sample of 15 have low vitamin D levels.

Solution "At most 3" means 0 or 1 or 2 or 3. We must find the probability for each case and then use the addition rule for disjoint events.

$$P(0 \text{ low vitamin D males}) = _{15}C_0\left(\frac{1}{3}\right)^0\left(\frac{2}{3}\right)^{15} \approx .0023$$

$$P(1 \text{ low vitamin D males}) = _{15}C_1\left(\frac{1}{3}\right)^1\left(\frac{2}{3}\right)^{14} \approx .0171$$

*Posted on www.menshealth.com, 8/20/2004.

$$P(2 \text{ low vitamin D males}) = {}_{15}C_2\left(\frac{1}{3}\right)^2\left(\frac{2}{3}\right)^{13} \approx .0599$$

$$P(3 \text{ low vitamin D males}) = {}_{15}C_3\left(\frac{1}{3}\right)^3\left(\frac{2}{3}\right)^{12} \approx .1299$$

By the addition rule for disjoint events,

$$P(\text{at most 3 males with low vitamin D levels}) \approx .0023 + .0171 + .0599 + .1299$$

$$\approx .2092.$$

(c) Find the expected number of males in a sample of 15 that will have low vitamin D levels.

Solution Using the formula because this is a binomial experiment, we obtain $E(x) = np = 15(1/3) = 5$. In repeated samples of size 15, the average number of males with low vitamin D levels is 5.0. ▲ ⟨4⟩

▼ **EXAMPLE 4** A survey in China, as reported by the *Los Angeles Times*, found that 66% of young people feel themselves to be under heavy pressure.* If a sample of 8 young people in China is selected, what is the probability that at least 1 will report being under heavy pressure?

Solution We can treat this as a binomial experiment, letting $n = 8$, $p = .66$, and x represent the number selected from 8 who are under heavy pressure. At least 1 of 8 means 1 or 2 or 3, up to 8. It will be simpler here to find the probability of selecting no young people feeling heavy pressure—that is, $P(x = 0)$—and then find the probability of at least 1 feeling heavy pressure, which is the number $1 - P(x = 0)$.

$$P(x = 0) = {}_8C_0(.66)^0(.34)^8 \approx .00018$$

$$P(x \geq 1) = 1 - P(x = 0) = 1 - .00018 \approx .99982 \; \blacktriangle \; ⟨5⟩$$

▼ **EXAMPLE 5** If each member of a 9-person jury acts independently of the other members and makes the correct determination of guilt or innocence with probability .65, find the probability that the majority of jurors will reach a correct verdict.†

Solution Since the jurors in this particular situation act independently, we can treat the experiment as a binomial one. Thus, the probability that the majority of the jurors will reach the correct verdict is given by

$$P(\text{at least 5}) = {}_9C_5(.65)^5(.35)^4 + {}_9C_6(.65)^6(.35)^3 + {}_9C_7(.65)^7(.35)^2$$

$$+ {}_9C_8(.65)^8(.35)^1 + {}_9C_9(.65)^9$$

$$\approx .2194 + .2716 + .2162 + .1004 + .0207$$

$$= .8283. \; \blacktriangle$$

⟨4⟩

Five percent of the clay pots fired in a certain way are defective. Assume independence, and find the probability of getting each result in a sample of 12 pots.

(a) 2 defective pots

(b) At most 2 defective pots

(c) What is the expected number of defective pots?

Answer:

(a) ${}_{12}C_2(.05)^2(.95)^{10} \approx .0988$

(b) .980

(c) .6

⟨5⟩

In Example 4, find the probability that

(a) at least 6 of the people selected are under heavy pressure

(b) at most 5 of the people selected are under heavy pressure

(c) What is the expected value?

Answers:

(a) about .452

(b) about .548

(c) 5.28

Los Angeles Times, July 11, 2005.

†Bernard Grofman, "A Preliminary Model of Jury Decision Making as a Function of Jury Size, Effective Jury Decision Rule, and Mean Juror Judgmental Competence," *Frontiers in Economics* (1979), pp. 98–110.

TECHNOLOGY TIP   Some spread sheets provide binomial probabilities. In Microsoft Excel, for example, the command "=BINOMDIST (5, 9, .65, 0)" gives .21939, which is the probability for $x = 5$ in Example 5. Alternatively, the command "=BINOMDIST (4, 9, .65, 1)" gives .17172 as the probability that 4 or fewer jurors will make the correct decision. Subtract .17172 from 1 to get .82828 as the probability that the majority of the jurors will make the correct decision. This value agrees with the value found in Example 5.

◆9.4 Exercises

In Exercises 1–39, see Examples 1–5.

Social Science   *In 2003, the percentage of children under 18 years of age who lived with both parents was approximately 70%.* Find the probabilities that the following number of persons selected at random from 10 children under 18 years of age in 2003 lived with both parents.*

1. Exactly 6
2. Exactly 5
3. None
4. All
5. At least 1
6. At most 4

Social Science   *The aforementioned study also found that approximately 5% of children under 18 years of age lived with their father only. Find the probabilities that the following number of persons selected at random from 10 children under 18 years of age in 2003 lived with their father only.*

7. Exactly 2
8. Exactly 1
9. None
10. All
11. At least 1
12. At most 2

A coin is tossed 5 times. Find the probability of getting

13. All heads;
14. Exactly 3 heads;
15. No more than 3 heads;
16. At least 3 heads.
17. How do you identify a probability problem that involves a binomial experiment?
18. Why do combinations occur in the binomial probability formula?

Natural Science   *According to the Alzheimer's Disease Education & Referral Center, 3% of Americans age 65 to 74 have Alzheimer's.[†] A group of 12 people age 65 to 74 are selected at random. Find the following probabilities.*

19. 3 people have the disease.
20. No one has the disease.
21. At most 2 person has the disease.
22. 10 people do not have the disease.

23. If 200 people aged 65 to 74 were selected, what is the expected number having the disease?

Health   *The survey discussed on page 542 of this section also asked the question, "How closely have you been following the issue of stem cell research?" Approximately 12% of the respondents answered "not at all." If 15 Americans are selected at random and asked the same question, find the following probabilities.*

24. None say "not at all"
25. One says "not at all"
26. 13 do not say "not at all"
27. At most 4 say "not at all"
28. What is the expected number who will answer "not at all"?

Natural Science   *The probability that a birth will result in twins is .027.* Assuming independence (perhaps not a valid assumption), what are the probabilities that, out of 100 births in a hospital, there will be the following numbers of sets of twins?*

29. Exactly 2 sets of twins
30. At most 2 sets of twins

Social Science   *According to the website Answers.com, 10–13% of Americans are left-handed. Assume that the percentage is 11%. If we select 9 people at random, find the probability that the number who are left-handed is:*

31. Exactly 2
32. At least 2
33. None
34. At most 3
35. In a class of 35 students, how many left-handed students should the instructor expect?

Social Science   *Respondents to the 2002 General Social Survey (GSS) indicated that approximately 8% of Americans attend church services more than once a week.[†] If 16 Americans are chosen at random, find the following probabilities.*

36. Exactly 2 attend church services more than once a week
37. At most 3 attend church services more than once a week

---

*U.S Census Bureau, *America's Families and Living Arrangments, March 2003.*

†www.alzheimers.org/generalinfo.htm.

**The World Almanac and Book of Facts,* 2001, p. 873.

†The National Data Program for Social Sciences, National Opinion Research Center, University of Cihcago, Chicago, IL.

Health *According to the Discovery Health Channel website, "More than two-thirds of all sexually transmitted diseases occur in people younger than 25."* Assume that the true probability is .66 that a person with a sexually transmitted disease is younger than 25.*

38. If we select 10 people at random with a sexually transmitted disease, what is the probability that at least 8 are under age 25?

39. If we select 500 at random with a sexually transmitted disease, what is the expected number that will be younger than 25?

40. Why can't we answer the question, "If we select 10 people at random, what is the probability that exactly 8 will have a sexually transmitted disease?"

41. Social Science In the "Numbers" section of *Time* magazine, it was reported that 15.2% of low-birth-weight babies graduate from high school by age 19. It was also reported that 57.5% of their normal-birth-weight siblings graduated from high school.[†]

  (a) If 40 low-birth-weight babies were tracked through high school, what is the probability that fewer than 15 will graduate from high school by age 19?

  (b) What are some of the factors that may contribute to the wide difference in high school success between these siblings? Do you believe that low birth weight is the primary cause of the difference? What other information do you need to better answer these questions?

42. Natural Science In a placebo-controlled trial of Adderall XR®, a medication for Attention Deficit and Hyperactivity Disorder (ADHD), 22% of users of the drug reported loss of appetite. Only 2% of the patients taking the placebo reported loss of appetite.[‡]

  (a) If 100 patients who are taking Adderall XR are selected at random, what is the probability that 15 or more will experience loss of appetite?

  (b) If 100 patients who are taking the placebo are selected at random, what is the probability that 15 or more will experience loss of appetite?

  (c) Do you believe Adderall XR causes a loss of appetite? Why?

43. Natural Science The use of DNA has become an integral part of many court cases. When DNA is extracted from cells and body fluids, genetic information is represented by bands of information, which look similar to a bar code at a grocery store. It is generally accepted that, in unrelated people, the probability of a particular band matching is 1 in 4.*

  (a) If 5 bands are compared in unrelated people, what is the probability that all 5 of the bands match? (Express your answer in terms of "1 chance in ?".)

  (b) If 20 bands are compared in unrelated people, what is the probability that all 20 of the bands match? (Express your answer in terms of "1 chance in ?".)

  (c) If 20 bands are compared in unrelated people, what is the probability that 16 or more bands match? (Express your answer in terms of "1 chance in ?".)

  (d) If you were deciding paternity and there were 16 matches out of 20 bands compared, would you believe that the person being tested was the father? Explain.

44. Social Science In England, a woman was found guilty of smothering her two infant children. Much of the Crown's case against her was based on the testimony from a pediatrician who indicated that the chances of 2 crib deaths occurring in both siblings was only about 1 in 73 million. This number was calculated by assuming that the probability of a single crib death is 1 in 8500 and the probability of 2 crib deaths is 1 in 8500^2 (i.e., binomial).[†] Why is the use of binomial probability not correct in this situation?

---

*health.discovery.com/centers.teen.relationships.sexstds.html.

[†]"Numbers," *Time*, July 17, 2000, p. 21.

[‡]Advertisement in *Newsweek*, June 13, 2005, for Adderall XR®, marketed by Shire US, Inc.

*"Genetic Fingerprinting Worksheet." Centre for Innovation in Mathematics Teaching, http://www.ex.ac.uk/cimt/resource/fgrprnts.htm.

[†]Stephen J. Watkins, "Conviction by Mathematical Error?" *British Medical Journal* 320. no. 7226 (January 1, 2000):2–3.

9.5 Markov Chains

In Section 8.5, we touched on **stochastic processes**—mathematical models that evolve over time in a probabilistic manner. In the current section, we study a special kind of stochastic process called a **Markov chain,** in which the outcome of an experiment depends only on the outcome of the previous experiment. In other words, the next state of the system depends only on the present state, not on preceding states. Such experiments are common enough in applications to make their study worthwhile. Markov chains are named after the Russian mathematician A. A. Markov (1856–1922), who started the theory of stochastic processes. To see how Markov chains work, we look at an example.

▼ **EXAMPLE 1** A small town has only two dry cleaners: Johnson and NorthClean. Johnson's manager hopes to increase the firm's market share by an extensive advertising campaign. After the campaign, a market research firm finds that there is a probability of .8 that a Johnson customer will bring his next batch of dirty items to Johnson and a .35 chance that a NorthClean customer will switch to Johnson for his next batch. Assume that the probability that a customer comes to a given cleaner depends only on where the last load of clothes was taken. If there is an .8 chance that a Johnson customer will return to Johnson, then there must be a $1 - .8 = .2$ chance that the customer will switch to NorthClean. In the same way, there is a $1 - .35 = .65$ chance that a NorthClean customer will return to NorthClean. If an individual bringing a load to Johnson is said to be in state 1 and an individual bringing a load to NorthClean is said to be in state 2, then these probabilities of change from one cleaner to the other are as shown in the following table:

		SECOND LOAD	
	State	**1**	**2**
FIRST LOAD	**1**	.8	.2
	2	.35	.65

The information from the table can be written in other forms. Figure 9.13 is a **transition diagram** that shows the two states and the probabilities of going from one to another.

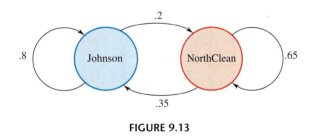

FIGURE 9.13

In a **transition matrix,** the states are indicated at the side and top, as follows.

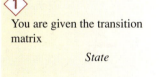

You are given the transition matrix

$$\text{State} \quad \begin{array}{c} \\ \text{State} \begin{array}{c} 1 \\ 2 \end{array} \end{array} \begin{array}{cc} 1 & 2 \\ \begin{bmatrix} .3 & .7 \\ .1 & .9 \end{bmatrix} \end{array}$$

(a) What is the probability of changing from state 1 to state 2?

(b) What does the number .1 represent?

(c) Draw a transition diagram for this information.

Answers:

(a) .7

(b) The probability of changing from state 2 to state 1

(c)
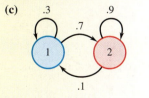

Second Load
Johnson NorthClean
First Load Johnson NorthClean $\begin{bmatrix} .8 & .2 \\ .35 & .65 \end{bmatrix}$ ▲ ◇①

A **transition matrix** has the following features:

1. It is square, since all possible states must be used both as rows and as columns.
2. All entries are between 0 and 1, inclusive, because all entries represent probabilities.
3. The sum of the entries in any row must be 1, because the numbers in the row give the probability of changing from the state at the left to one of the states indicated across the top.

▼ **EXAMPLE 2** Suppose that when the new promotional campaign began, Johnson had 40% of the market and NorthClean had 60%. Use the probability tree in Figure 9.14 to find how these proportions would change after another week of advertising.

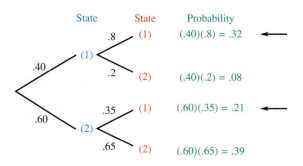

State 1: taking cleaning to Johnson
State 2: taking cleaning to NorthClean

FIGURE 9.14

Solution Add the numbers indicated with arrows to find the proportion of people taking their cleaning to Johnson after 1 week.

$$.32 + .21 = .53$$

Similarly, the proportion taking their cleaning to NorthClean is

$$.08 + .39 = .47.$$

The initial distribution of 40% and 60% becomes 53% and 47%, respectively, after 1 week. ▲

These distributions can be written as the *probability vectors*

$$[.40 \quad .60] \quad \text{and} \quad [.53 \quad .47].$$

A **probability vector** is a one-row matrix with nonnegative entries and with the sum of the entries equal to 1.

The results from the probability tree of Figure 9.14 are exactly the same as the result of multiplying the initial probability vector by the transition matrix. (Multiplication of matrices was discussed in Section 6.4.)

$$[.4 \quad .6]\begin{bmatrix} .8 & .2 \\ .35 & .65 \end{bmatrix} = [.53 \quad .47]$$

If v denotes the original probability vector $[.4 \quad .6]$ and P denotes the transition matrix, then the market share vector after 1 week is $vP = [.53 \quad .47]$. To find the market share after 2 weeks, multiply the vector $vP = [.53 \quad .47]$ by P; this amounts to finding vP^2. ②

Problem 2 at the side shows that after 2 weeks the market share vector is $vP^2 = [.59 \quad .41]$. To get the market share after 3 weeks, multiply this vector by P; that is, find vP^3. Do not use the rounded answer from Problem 2. ③

Continuing this process gives each cleaner's share of the market after additional weeks.

Weeks after Start	Johnson	NorthClean	
0	.4	.6	v
1	.53	.47	vP^1
2	.59	.41	vP^2
3	.61	.39	vP^3
4	.63	.37	vP^4
12	.64	.36	vP^{12}
13	.64	.36	vP^{13}

The results seem to approach the probability vector $[.64 \quad .36]$.

What happens if the initial probability vector is different from $[.4 \quad .6]$? Suppose $[.75 \quad .25]$ is used; then the same powers of the transition matrix as before give the following results.

Week after Start	Johnson	NorthClean	
0	.75	.25	v
1	.69	.31	vP^1
2	.66	.34	vP^2
3	.65	.35	vP^3
4	.64	.36	vP^4
5	.64	.36	vP^5
6	.64	.36	vP^6

The results again seem to be approaching the numbers in the probability vector $[.64 \quad .36]$, the same numbers approached with the initial probability vector $[.4 \quad .6]$. In either case, the long-range trend is for a market share of about 64% for Johnson and 36% for NorthClean. The example suggests that this long-range

② Find the product

$$[.53 \quad .47]\begin{bmatrix} .8 & .2 \\ .35 & .65 \end{bmatrix}$$

Answer:

$[.59 \quad .41]$ (rounded)

③ Find each cleaner's market share after 3 weeks.

Answer:

$[.61 \quad .39]$ (rounded)

trend does not depend on the initial distribution of market shares. This means that if the initial market share for Johnson was less than 64%, the advertising campaign has paid off in terms of a greater long-range market share. If the initial share was more than 64%, the campaign did not pay off.

Regular Transition Matrices

One of the many applications of Markov chains is in finding the long-range predictions. It is not possible to make long-range predictions with all transition matrices, but for a large set of transition matrices, long-range predictions *are* possible. Such predictions are always possible with **regular transition matrices.** A transition matrix is **regular** if some power of the matrix contains all positive entries. A Markov chain is a **regular Markov chain** if its transition matrix is regular.

▼ **EXAMPLE 3**   Decide whether the following transition matrices are regular.

(a) $A = \begin{bmatrix} .3 & .2 & .5 \\ .0 & .4 & .6 \\ .2 & .8 & 0 \end{bmatrix}$

Solution Square A.

$$A^2 = \begin{bmatrix} .19 & .54 & .27 \\ .12 & .64 & .24 \\ .06 & .36 & .58 \end{bmatrix}$$

Since all entries in A^2 are positive, matrix A is regular.

(b) $B = \begin{bmatrix} .2 & 0 & 8 \\ 0 & 1 & 0 \\ 0 & 0 & 1 \end{bmatrix}$

Solution Find various powers of B.

$$B^2 = \begin{bmatrix} .04 & 0 & .96 \\ 0 & 1 & 0 \\ 0 & 0 & 1 \end{bmatrix} \quad B^3 = \begin{bmatrix} .008 & 0 & .992 \\ 0 & 1 & 0 \\ 0 & 0 & 1 \end{bmatrix} \quad B^4 = \begin{bmatrix} .0016 & 0 & .9984 \\ 0 & 1 & 0 \\ 0 & 0 & 1 \end{bmatrix}$$

Notice that all of the powers of B shown here have zeros in the same locations. Thus, further powers of B will still give the same zero entries, so that no power of matrix B contains all positive entries. For this reason, B is not regular. ▲

4
Decide whether the following transition matrices are regular.

(a) $\begin{bmatrix} 0 & 1 \\ 1 & 0 \end{bmatrix}$

(b) $\begin{bmatrix} .25 & .75 \\ 1 & 0 \end{bmatrix}$

Answers:

(a) No

(b) Yes

NOTE If a transition matrix P has some zero entries, and P^2 does as well, you may wonder how far you must compute P^k to be certain that the matrix is not regular. The answer is that if zeros occur in the identical places in both P^k and P^{k+1} for any k, they will appear in those places for all higher powers of P, so P is not regular.

Suppose that v is any probability vector. It can be shown that, for a regular Markov chain with a transition matrix P, there exists a single vector V that does not depend on v, such that $v \cdot P^n$ gets closer and closer to V as n gets larger and larger.

Equilibrium Vector of a Markov Chain

If a Markov chain with transition matrix P is regular, then there is a unique vector v such that, for any probability vector v and for large values of n,

$$v \cdot P^n \approx V.$$

Vector V is called the equilibrium vector, or the fixed vector, of the Markov chain.

In the example with Johnson Cleaners, the equilibrium vector V is approximately $[.64 \quad .36]$. Vector V can be determined by finding P^n for larger and larger values of n and then looking for a vector that the product $v \cdot P^n$ approaches. Such a strategy can be very tedious, however, and is prone to error. To find a better way, start with the fact that, for a large value of n,

$$v \cdot P^n \approx V,$$

as mentioned in the proceeding box. From this result, $v \cdot P^n \cdot P \approx V \cdot P$, so that

$$v \cdot P^n \cdot P = v \cdot P^{n+1} \approx VP.$$

Since $v \cdot P^n \approx V$ for large values of n, it is also true that $v \cdot P^{n+1} \approx V$ for large values of n. (The product $v \cdot P^n$ approaches V, so that $v \cdot P^{n+1}$ must also approach V.) Thus, $v \cdot P^{n+1} \approx V$ and $v \cdot P^{n+1} \approx VP$, which suggests that

$$VP = V.$$

If a Markov chain with transition matrix P is regular, then the equilibrium vector V satisties

$$VP = V.$$

The equilibrium vector V can be found by solving a system of linear equations, as shown in the remaining examples.

▼ **EXAMPLE 4**   Find the long-range trend for the Markov chain in Examples 1 and 2, with transition matrix

$$P = \begin{bmatrix} .8 & .2 \\ .35 & .65 \end{bmatrix}.$$

Solution This matrix is regular, since all entries are positive. Let P represent this transition matrix and let V be the probability vector $[v_1 \quad v_2]$. We want to find V such that

$$VP = V,$$

or

$$[v_1 \quad v_2]\begin{bmatrix} .8 & .2 \\ .35 & .65 \end{bmatrix} = [v_1 \quad v_2].$$

Multiply on the left to get

$$[.8v_1 + .35v_2 \quad .2v_1 + .65v_2] = [v_1 \quad v_2].$$

Set corresponding entries from the two matrices equal to obtain

$$.8v_1 + .35v_2 = v_1 \quad .2v_1 + .65v_2 = v_2.$$

Simplify each of these equations.

$$-.2v_1 + .35v_2 = 0 \quad .2v_1 - .35v_2 = 0$$

These last two equations are really the same. (The equations in the system obtained from $VP = V$ are always dependent.) To find the values of v_1 and $v_2,$ recall that $V = [v_1 \quad v_2]$ is a probability vector, so that

$$v_1 + v_2 = 1.$$

Find v_1 and v_2 by solving the system

$$-.2v_1 + .35v_2 = 0$$
$$v_1 + \quad v_2 = 1.$$

From the second equation, $v_1 = 1 - v_2.$ Substitute for v_1 in the first equation:

$$-.2(1 - v_2) + .35v_2 = 0. \quad \langle 5 \rangle$$

⟨5⟩

Solve the equation for v_2.
Round to the nearest thousandth.

Answer:

.364

Since $v_2 = .364$ (from problem 5 at the side) and $v_1 = 1 - v_2,$ it follows that $v_1 = 1 - .364 = .636,$ and the equilibrium vector is $[.636 \quad .364] \approx [.64 \quad .36].$ ▲

▼ **EXAMPLE 5** The probability that a complex assembly line works correctly depends on whether the line worked correctly the last time it was used. The various probabilities are as given in the following transition matrix

	Works Properly Now	Does Not
Worked Properly Before	.84	.16
Did Not	.72	.28

Find the long-range probability that the assembly line will work properly.

Solution Begin by finding the equilibrium vector $[v_1 \quad v_2],$ where

$$[v_1 \quad v_2]\begin{bmatrix} .84 & .16 \\ .72 & .28 \end{bmatrix} = [v_1 \quad v_2].$$

Multiplying on the left and setting corresponding entries equal gives the equations

$$.84v_1 + .72v_2 = v_1 \quad \text{and} \quad .16v_1 + .28v_2 = v_2,$$

or

$$-.16v_1 + .72v_2 = 0 \quad \text{and} \quad .16v_1 - .72v_2 = 0.$$

Substitute $v_1 = 1 - v_2$ in the first of these equations to get

$$-.16(1 - v_2) + .72v_2 = 0$$
$$-.16 + .16v_2 + .72v_2 = 0$$
$$-.16 + .88v_2 = 0$$
$$.88v_2 = .16$$
$$v_2 = \frac{2}{11},$$

and $v_1 = 1 - \dfrac{2}{11} = \dfrac{9}{11}$. The equilibrium vector is $\begin{bmatrix} 9/11 & 2/11 \end{bmatrix}$. In the long run, the company can expect the assembly line to run properly $\dfrac{9}{11} \approx 82\%$ of the time. ▲ ⟨6⟩

⟨6⟩

In Example 5, suppose the company modifies the line so that the transition matrix becomes

$$\begin{bmatrix} .95 & .05 \\ .8 & .2 \end{bmatrix}$$

Find the long-range probability that the assembly line will work properly.

Answer:

$16/17 \approx 94\%$

▼ **EXAMPLE 6** Find the equilibrium vector for the transition matrix

$$K = \begin{bmatrix} .2 & .6 & .2 \\ .1 & .1 & .8 \\ .3 & .3 & .4 \end{bmatrix}.$$

Solution Matrix K has all positive entries and thus is regular. For this reason, an equilibrium vector V must exist such that $VK = V$. Let $V = \begin{bmatrix} v_1 & v_2 & v_3 \end{bmatrix}$.

Then

$$\begin{bmatrix} v_1 & v_2 & v_3 \end{bmatrix} \begin{bmatrix} .2 & .6 & .2 \\ .1 & .1 & .8 \\ .3 & .3 & .4 \end{bmatrix} = \begin{bmatrix} v_1 & v_2 & v_3 \end{bmatrix}.$$

Use matrix multiplication on the left.

$$\begin{bmatrix} .2v_1 + .1v_2 + .3v_3 & .6v_1 + .1v_2 + .3v_3 & .2v_1 + .8v_2 + .4v_3 \end{bmatrix} = \begin{bmatrix} v_1 & v_2 & v_3 \end{bmatrix}$$

Set corresponding entries equal.

$$.2v_1 + .1v_2 + .3v_3 = v_1$$
$$.6v_1 + .1v_2 + .3v_3 = v_2$$
$$.2v_1 + .8v_2 + .4v_3 = v_3$$

Simplify these equations.

$$-.8v_1 + .1v_2 + .3v_3 = 0$$
$$6v_1 - .9v_2 + .3v_3 = 0$$
$$.2v_1 + .8v_2 - .6v_3 = 0$$

Since V is a probability vector,

$$v_1 + v_2 + v_3 = 1.$$

This gives a system of four equations in three unknowns:

$$v_1 + v_2 + v_3 = 1.$$
$$-.8v_1 + .1v_2 + .3v_3 = 0$$
$$.6v_1 - .9v_2 + .3v_3 = 0$$
$$.2v_1 + .8v_2 - .6v_3 = 0.$$

The system can be solved with the Gauss-Jordan method set forth in Section 6.2. Start with the augmented matrix

$$\begin{bmatrix} 1 & 1 & 1 & | & 1 \\ -.8 & .1 & .3 & | & 0 \\ .6 & -.9 & .3 & | & 0 \\ .2 & .8 & -.6 & | & 0 \end{bmatrix}$$

The solution of this system is $v_1 = 5/23$, $v_2 = 7/23$, $v_3 = 11/23$, and

$$V = \begin{bmatrix} \dfrac{5}{23} & \dfrac{7}{23} & \dfrac{11}{23} \end{bmatrix} \approx [.22 \quad .30 \quad .48]. \quad \blacktriangle \quad \diamond$$

In Example 4, we found that $[.64 \quad .36]$ was the equilibrium vector for the regular transition matrix.

$$P = \begin{bmatrix} .8 & .2 \\ .35 & .65 \end{bmatrix}.$$

Observe what happens when you take powers of the matrix P. (Displayed entries have been rounded for easy reading, but the full decimals were used in the calculations.)

$$P^2 = \begin{bmatrix} .71 & .29 \\ .51 & .49 \end{bmatrix} \quad P^3 = \begin{bmatrix} .67 & .33 \\ .58 & .42 \end{bmatrix} \quad P^4 = \begin{bmatrix} .65 & .35 \\ .61 & .39 \end{bmatrix}$$

$$P^5 = \begin{bmatrix} .64 & .36 \\ .62 & .38 \end{bmatrix} \quad P^6 = \begin{bmatrix} .64 & .36 \\ .63 & .37 \end{bmatrix} \quad P^{10} = \begin{bmatrix} .64 & .36 \\ .64 & .36 \end{bmatrix}$$

As these results suggest, higher and higher powers of the transition matrix P approach a matrix having all rows identical: rows that have as entries the entries of the equilibrium vector V.

If you have the technology to compute matrix powers easily (such as a graphing calculator), you can approximate the equilibrium vector by taking higher and higher powers of the transition matrix until all its rows are identical. Figure 9.15 shows part of this process for the transition matrix

$$B = \begin{bmatrix} .84 & .16 \\ .72 & .28 \end{bmatrix}$$

of Example 5.

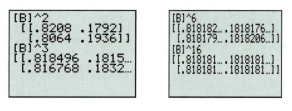

FIGURE 9.15

Figure 9.15 indicates that the equilibrium vector is $[.818 \quad .182] = [9/11 \quad 2/11]$, which is what was found algebraically in Example 5.

The results of this section can be summarized as follows.

⟨7⟩

Find the equilibrium vector for the transition matrix

$$P = \begin{bmatrix} .3 & .7 \\ .5 & .5 \end{bmatrix}.$$

Answer:

$[5/12 \quad 7/12]$

Properties of Regular Markov Chains

Suppose a regular Markov chain has a transition matrix P.

1. As n gets larger and larger, the product $v \cdot P^n$ approaches a unique vector V for any initial probability vector v. Vector V is called the *equilibrium vector*, or *fixed vector*.
2. Vector V has the property that $VP = V$.
3. To find V, solve a system of equations obtained from the matrix equation $VP = V$ and from the fact that the sum of the entries of V is 1.
4. The powers P^n come closer and closer to a matrix whose rows are made up of the entries of the equilibrium vector V.

◆9.5 Exercises

Decide which of the following could be a probability vector.

1. $\begin{bmatrix} \frac{1}{3} & \frac{2}{3} \end{bmatrix}$

2. $\begin{bmatrix} \frac{5}{8} & \frac{3}{8} \end{bmatrix}$

3. $\begin{bmatrix} 0 & 1 \end{bmatrix}$

4. $\begin{bmatrix} .4 & .2 & .1 \end{bmatrix}$

5. $\begin{bmatrix} .3 & -.1 & .6 \end{bmatrix}$

6. $\begin{bmatrix} \frac{1}{4} & \frac{1}{8} & \frac{5}{8} \end{bmatrix}$

Decide which of the following could be a transition matrix. Sketch a transition diagram for any transition matrices.

7. $\begin{bmatrix} .7 & .2 \\ .5 & .5 \end{bmatrix}$

8. $\begin{bmatrix} \frac{1}{4} & \frac{3}{4} \\ 0 & 1 \end{bmatrix}$

9. $\begin{bmatrix} \frac{4}{9} & \frac{1}{3} \\ \frac{1}{5} & \frac{7}{10} \end{bmatrix}$

10. $\begin{bmatrix} 0 & 1 & 0 \\ .3 & .3 & .3 \\ 1 & 0 & 0 \end{bmatrix}$

11. $\begin{bmatrix} \frac{1}{2} & \frac{1}{4} & 1 \\ \frac{2}{3} & 0 & \frac{1}{3} \\ \frac{1}{3} & 1 & 0 \end{bmatrix}$

12. $\begin{bmatrix} .2 & .3 & .5 \\ .0 & 0 & 1 \\ .1 & .9 & 0 \end{bmatrix}$

In Exercises 13–15, write any transition diagrams as transition matrices.

13.

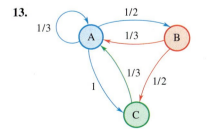

14.

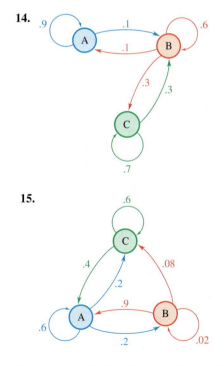

15.

Decide whether the following transition matrices are regular. (See Example 3.)

16. $\begin{bmatrix} 1 & 0 \\ .6 & .4 \end{bmatrix}$

17. $\begin{bmatrix} .2 & .8 \\ .9 & .1 \end{bmatrix}$

18. $\begin{bmatrix} .3 & .5 & .2 \\ 1 & 0 & 0 \\ .5 & .1 & .4 \end{bmatrix}$ **19.** $\begin{bmatrix} 0 & 1 & 0 \\ .3 & .3 & .4 \\ 1 & 0 & 0 \end{bmatrix}$

20. $\begin{bmatrix} .25 & .40 & .30 & .05 \\ .18 & .23 & .59 & 0 \\ 0 & .15 & .36 & .49 \\ .28 & .32 & .24 & .16 \end{bmatrix}$ **21.** $\begin{bmatrix} .10 & .70 & 0 & .20 \\ 0 & .36 & .39 & .25 \\ 0 & 0 & 1 & 0 \\ .54 & 0 & .42 & .04 \end{bmatrix}$

Find the equilibrium vector for each of the following transition matrices. (See Examples 4 and 5.)

22. $\begin{bmatrix} .3 & .7 \\ .4 & .6 \end{bmatrix}$ **23.** $\begin{bmatrix} .66 & .34 \\ .12 & .88 \end{bmatrix}$

24. $\begin{bmatrix} \frac{5}{8} & \frac{3}{8} \\ \frac{7}{9} & \frac{2}{9} \end{bmatrix}$ **25.** $\begin{bmatrix} \frac{2}{3} & \frac{1}{3} \\ \frac{1}{8} & \frac{7}{8} \end{bmatrix}$

26. $\begin{bmatrix} .25 & .35 & .4 \\ .1 & .3 & .6 \\ .55 & .4 & .05 \end{bmatrix}$ **27.** $\begin{bmatrix} .16 & .28 & .56 \\ .43 & .12 & .45 \\ .86 & .05 & .09 \end{bmatrix}$

28. $\begin{bmatrix} .1 & .1 & .8 \\ .4 & .4 & .2 \\ .1 & .2 & .7 \end{bmatrix}$ **29.** $\begin{bmatrix} .44 & .31 & .25 \\ .80 & .11 & .09 \\ .26 & .31 & .43 \end{bmatrix}$

For each of the following transition matrices, use a graphing calculator or computer to find the first five powers of the matrix. Then find the probability that state 2 changes to state 4 after 5 repetitions of the experiment.

30. $\begin{bmatrix} .1 & .2 & .2 & .3 & .2 \\ .2 & .1 & .1 & .2 & .4 \\ .2 & .1 & .4 & .2 & .1 \\ .3 & .1 & .1 & .2 & .3 \\ .1 & .3 & .1 & .1 & .4 \end{bmatrix}$ **31.** $\begin{bmatrix} .3 & .2 & .3 & .1 & .1 \\ .4 & .2 & .1 & .2 & .1 \\ .1 & .3 & .2 & .2 & .2 \\ .2 & .1 & .3 & .2 & .2 \\ .1 & .1 & .4 & .2 & .2 \end{bmatrix}$

32. Social Science The chart shows the percentage of the poor, middle class, and affluent that move into another class. The first graph shows the figures for 1967–1979, the second those for 1980–1991.* For each period, assume that the number of people who move directly from poor to affluent or from affluent to poor is essentially 0.

(a) Find the long-range percentage of poor, middle-class, and affluent people if the 1967–1979 trends were to continue.

The poor are more likely to stay poor and the affluent are more likely to stay affluent.

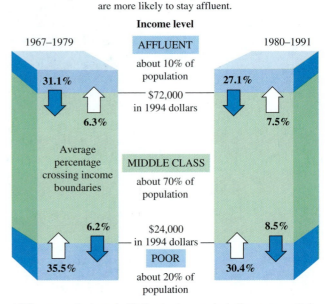

All figures are for household after-tax income, including wages, salaries, and some Government assistance programs like food stamps.

Sources: Greg Duncan, Northwestern University; Timothy Smeeding, Syracuse University

(b) Find the long-range percentage of poor, middle-class, and affluent people if the 1980–1991 trends were to continue.[†]

33. Natural Science Markov chains can be utilized in research into earthquakes. Researchers in Italy give the

	2.5	2.6	2.7	2.8
2.5	3/7	1/7	2/7	1/7
2.6	1/2	0	1/4	1/4
2.7	1/3	1/3	0	1/3
2.8	1/4	1/2	0	1/4

following example of a transition matrix in which the rows are magnitudes of an earthquake and the columns are magnitudes of the next earthquake in the sequence.* Thus, the probability of a 2.5-magnitude earthquake being followed by a 2.8-magnitude earthquake is $1/7$. If these trends were to persist, find the long-range trend for the probabilities of each magnitude for the subsequent earthquake.

New York Times, June 4, 1995, p. E4.

†*Sources:* Greg Duncan, Northwestern University; Timothy Smeeding, Syracuse University.

*Michele Lovallo, Vincenzo Lapenna, Luciano Telesca, "Transition matrix analysis of earthquake magnitude sequences," *Chaos, Solitons, and Fractals* 24 (2005) 33–43.

34. Business The probability that a complex assembly line works correctly depends on whether the line worked correctly the last time it was used. There is a .91 chance that the line will work correctly if it worked correctly the time before and a .68 chance that it will work correctly if it did *not* work correctly the time before. Set up a tran-sition matrix with this information, and find the long- run probability that the line will work correctly. (See Example 5.)

35. Business Suppose something unplanned occurred to the assembly line of Exercise 34, so that the transition matrix becomes

$$
\begin{array}{c c}
 & \begin{array}{c c} \text{Works} & \text{Doesn't Work} \end{array} \\
\begin{array}{c} \text{Works} \\ \text{Doesn't Work} \end{array} &
\begin{bmatrix} .81 & .19 \\ .77 & .23 \end{bmatrix}
\end{array}
$$

Find the new long-run probability that the line will work properly.

36. Natural Science In Exercises 62 and 63 of Section 8.4, we discussed the effect on flower color of cross-pollinating pea plants. As shown there, since the gene for red is dominant and the gene for white is recessive, 75% of the pea plants have red flowers and 25% have white flowers, because plants with 1 red and 1 white gene appear red. If a red-flowered plant is crossed with a red-flowered plant known to have 1 red and 1 white gene, then 75% of the offspring will be red and 25% will be white. Crossing a red-flowered plant that has 1 red and 1 white gene with a white-flowered plant produces 50% red-flowered offspring and 50% white-flowered offspring.
(a) Write a transition matrix using this information.
(b) Write a probability vector for the initial distribution of colors.
(c) Find the distribution of colors after 4 generations.
(d) Find the long-range distribution of colors.

37. Natural Science Snapdragons with 1 red gene and 1 white gene produce pink-flowered offspring. If a red snapdragon is crossed with a pink snapdragon, the probabilities that the offspring will be red, pink, or white are $1/2$, $1/2$, and 0, respectively. If 2 pink snapdragons are crossed, the probabilities of red, pink, or white offspring are $1/4$, $1/2$, and $1/4$, respectively. For a cross between a white and a pink snapdragon, the corresponding probabilities are 0, $1/2$, and $1/2$. Set up a transition matrix and find the long-range prediction for the fraction of red, pink, and white snapdragons.

38. Professors at Cleveland State University have 3 basic options to eat lunch: the cafeteria, a coffee shop, or a pizza place. We can treat this situation as a simple random walk. Every day, a professor decides to pick a place to eat at random. If she ate at the cafeteria yesterday, she is equally likely to eat at the coffee shop or pizza place today. If she ate at the coffee shop (her favorite) yesterday, she is equally likely to eat at any of the 3 places

	Cafeteria	Coffee Shop	Pizza Place
Cafeteria	?	?	?
Coffee Shop	?	?	?
Pizza Place	?	?	?

today, and if she ate at the pizza place yesterday, she is equally likely to eat at the cafeteria or coffee shop today.
(a) Set up the transition matrix for this situation.
(b) Find the long-range trend for the fraction of the time she eats at each location.

39. Social Science The probability that a homeowner will become a renter in five years is .03. The probability that a renter will become a homeowner in five years is .1. Suppose the proportions in the population are 64% homeowners (O), 35.5% renters (R), and .5% homeless (H), with the following transition matrix. Assume that these figures continue to apply.

$$
\begin{array}{c c c c}
 & \text{O} & \text{R} & \text{H} \\
\begin{array}{c} \text{O} \\ \text{R} \\ \text{H} \end{array} &
\begin{bmatrix} .94 & .06 & 0 \\ .12 & .879 & .001 \\ 0 & .32 & .68 \end{bmatrix}
\end{array}
$$

Find the long-range probabilities for the three categories.

40. Business An insurance company classifies its drivers into three groups: G_0 (no accidents), G_1 (one accident), and G_2 (more than one accident). The probability that a driver in G_0 will stay in G_0 after 1 year is .85, that he will become a G_1 is .10, and that he will become a G_2 is .05. A driver in G_1 cannot move to G_0. (This insurance company has a long memory!) There is an .80 probability that a G_1 driver will stay in G_1 and a .20 probability that he will become a G_2. A driver in G_2 must stay in G_2.
(a) Write a transition matrix using this information.

Suppose that the company accepts 50,000 new policy-holders, all of whom are in group G_0. Find the number in each group
(b) after 1 year;   (c) after 2 years;
(d) after 3 years;   (e) after 4 years.
(f) Find the equilibrium vector here. Interpret your result.

41. Business The difficulty with the mathematical model of Exercise 40 is that no "grace period" is provided; there should be a certain probability of moving from G_1 or G_2 back to G_0 (say, after 4 years with no accidents). A new system with this feature might produce the following tran-sition matrix.

$$
\begin{bmatrix} .85 & .10 & .05 \\ .15 & .75 & .10 \\ .10 & .30 & .60 \end{bmatrix}
$$

Suppose that when this new policy is adopted, the company has 50,000 policyholders in group G_0. Find the number of these in each group

(a) after 1 year;   (b) after 2 years;
(c) after 3 years.
(d) Find the equilibrium vector here. Interpret your result.

	Will Switch to		
	Company A	Company B	Company C
Company A	.91	.07	.02
Now has Company B	.03	.87	.10
Company C	.14	.04	.82

42. Suppose research on 3 major cell phone companies revealed the following transition matrix for the probability that a person with one cell phone carrier switches to another.

The current share of the market is $[.26, .36, .38]$ for Companies A, B, and C, respectively. Find the share of the market held by each company after

(a) 1 year;   (b) 2 years;   (c) 3 years.
(d) What is the long-range prediction?

43. The results of cricket matches between England and Australia have been found to be modeled by a Markov chain.* The probability that England wins, loses, or draws is based on the result of the previous game, with the following transition matrix.

	Wins	Loses	Draws
Wins	.443	.364	.193
Loses	.277	.436	.287
Draws	.266	.304	.430

(a) Compute the transition matrix for the game after the next one, based on the result of the last game.
(b) Use your answer from part (a) to find the probability that, if England won the last game, England will win the game after the next one.
(c) Use your answer from part (a) to find the probability that, if Australia won the last game, England will win the game after the next one.

44. **Social Science**   At one liberal arts college, students are classified as humanities majors, science majors, or undecided. There is a 23% chance that a humanities major will change to a science major from one year to the next and a 40% chance that a humanities major will change to undecided. A science major will change to humanities with

probability .12 and to undecided with probability .38. An undecided will switch to humanities or science with probabilities of .45 and .28, respectively. Find the long-range prediction for the fraction of students in each of these three majors.

45. **Business**   In the queuing chain, we assume that people are queuing up to be served by, say, a bank teller. For simplicity, let us assume that once two people are in line, no one else can enter the line. Let us further assume that one person is served every minute, as long as someone is in line. Assume further that, in any minute, there is a probability of .4 that no one enters the line, a probability of .3 that exactly one person enters the line, and a probability of .3 that exactly two people enter the line, assuming that there is room. If there is not enough room for two people, then the probability that one person enters the line is .5. Let the state be given by the number of people in line.

(a) Give the transition matrix for the number of people in line.

	0	1	2
0	?	?	?
1	?	?	?
2	?	?	?

(b) Find the transition matrix for a 2-minute period.
(c) Use your result from part (b) to find the probability that a queue with no one in line has two people in line 2 minutes later.

Use a graphing calculator or computer for Exercises 46 and 47.

46. **Business**   A company with a new training program classified each employee in one of four states: s_1, never in the program; s_2, currently in the program; s_3, discharged; s_4, completed the program. The transition matrix for this company is as follows.

	s_1	s_2	s_3	s_4
s_1	.4	.2	.05	.35
s_2	0	.45	.05	.5
s_3	0	0	1	0
s_4	0	0	0	1

(a) What percentage of employees who had never been in the program (state s_1) completed the program (state s_4) after the program had been offered five times?
(b) If the initial percentage of employees in each state was $[.5 \quad .5 \quad 0 \quad 0]$, find the corresponding percentages after the program had been offered four times.

47. **Business**   Find the long-range prediction for the percentage of employees in each state for the company training program from Exercise 46.

*From Derek Colwell, Brian Jones, and Jack Gillett, "A Markov Chain in Cricket," *Mathematical Gazette*, June 1991.

9.6 Decision Making

John F. Kennedy once remarked that he had assumed that, as president, it would be difficult to choose between distinct, opposite alternatives when a decision needed to be made. Actually, he found that such decisions were easy to make; the hard decisions came when he was faced with choices that were not as clear cut. Most decisions fall into this last category—decisions that must be made under conditions of uncertainty. In Section 9.1, we saw how to use expected values to help make a decision. Those ideas are extended in this section, where we consider decision making in the face of uncertainty. We begin with an example.

▼ **EXAMPLE 1**   Freezing temperatures are endangering the orange crop in central California. A farmer can protect his crop by burning smudge pots; the heat from the pots keeps the oranges from freezing. However, burning the pots; is expensive, costing $4000. The farmer knows that if he burns smudge pots, he will be able to sell his crop for a net profit (after the costs of the pots are deducted) of $10,000, provided that the freeze does develop and wipes out other orange crops in California. If he does nothing, he will either lose $2000 already invested in the crop if it does freeze or make a profit of $9600 if it does not freeze. (If it does not freeze, there will be a large supply of oranges, and thus his profit will be lower than if there were a small supply.) What should the farmer do?

Solution   He should begin by carefully defining the problem. First, he must decide on the **states of nature**—the possible alternatives over which he has no control. Here there are two: freezing temperatures and no freezing temperatures. Next, the farmer should list the things he can control—his actions or **strategies.** He has two possible strategies: to use smudge pots or not. The consequences of each action under each state of nature, called **payoffs,** are summarized in a **payoff matrix,** as follows. The payoffs in this case are the profits for each possible combination of events.

		States of Nature	
		Freeze	No Freeze
Strategies of Farmer	Use Smudge Pots	$10,000	$5600
	Do Not Use Pots	−$2000	$9600

To get the $5600 entry in the payoff matrix, use the profit if there is no freeze, namely, $9600, and subtract the $4000 cost of using the pots. ◇ 1

Once the farmer makes the payoff matrix, what then? The farmer might be an optimist (some might call him a gambler); in this case, he might assume that the best will happen and go for the biggest number of the matrix ($10,000). For that profit, he must adopt the strategy "use smudge pots."

On the other hand, if the farmer is a pessimist, he would want to minimize the worst thing that could happen. If he uses smudge pots, the worst thing that could happen to him would be a profit of $5600, which will result if there is no freeze. If he does not use smudge pots, he might face a loss of $2000. To minimize the worst, he once again should adopt the strategy "use smudge pots."

Suppose the farmer decides that he is neither an optimist nor a pessimist but would like further information before choosing a strategy. For example, he might

◇ 1
Explain how each of the following payoffs in the matrix were obtained.

(a)   −$2000

(b) $10,000

Answers:

(a)   If it freezes and smudge pots are not used, the farmer's profit is −$2000 for labor costs.

(b)   If it freezes and smudge pots are used, the farmer makes a profit of $10,000.

call the weather forecaster and ask for the probability of a freeze. Suppose the forecaster says that this probability is only .1. What should the farmer do? He should recall the discussion of expected value and work out the expected profit for each of his two possible strategies. If the probability of a freeze is .1, then the probability that there is no freeze is .9. This information leads to the following expected values.

If smudge pots are used: $\quad 10{,}000(.1) + 5600(.9) = 6040$

If no smudge pots are used: $\quad -2000(.1) + 9600(.9) = 8440$

Here the maximum expected profit, \$8440, is obtained if smudge pots are not used. ▲ ⟨2⟩

As the example shows, the farmer's beliefs about the probabilities of a freeze affect his choice of strategies.

⟨2⟩

What should the farmer do if the probability of a freeze is .6? What is his expected profit?

Answer:

Use smudge pots; \$8240

▼ **EXAMPLE 2**   An owner of several greeting-card stores must decide in July about the type of displays to emphasize for Sweetest Day in October. He has three possible choices: emphasize chocolates, emphasize collectible gifts, or emphasize gifts that can be engraved. His success is dependent on the state of the economy in October. If the economy is strong, he will do well with the collectable gifts, while in a weak economy, the chocolates do very well. In a mixed economy, the gifts that can be engraved will do well. He first prepares a payoff matrix for all three possibilities. The numbers in the matrix represent his profits in thousands of dollars.

		States of Nature		
		Weak Economy	Mixed	Strong Economy
Strategies	Chocolates	85	30	75
	Collectibles	45	45	110
	Engraved	60	95	85

(a) What would an optimist do?

Solution If the owner is an optimist, he should aim for the biggest number on the matrix, 110 (representing \$110,000 in profit). His strategy in this case would be to display collectibles.

(b) How would a pessimist react?

Solution A pessimist wants to find the best of the worst of all bad things that can happen. If he displays collectibles, the worst that can happen is a profit of \$45,000. For displaying engravable items, the worst is a profit of \$60,000 and for displaying chocolates, the worst is a profit of \$30,000. His strategy here is to use the engravable items.

(c) Suppose the owner reads in a business magazine that leading experts believe that there is a 50% chance of a weak economy in October, a 20% chance of a mixed economy, and a 30% chance of a strong economy. How might he use this information?

3

Suppose the owner reads another article, which gives the following predictions: a 35% chance of a weak economy, a 25% chance of an in-between economy, and a 40% chance of a strong economy. What is the best strategy now? What is the expected profit?

Answer:

Engravable; $78,750

Solution The owner can now find his expected profit for each possible strategy.

Chocolates	$85(.5) + 30(.20) + .75(.30) = 71$
Collectibles	$45(.5) + 45(.20) + 110(.30) = 64.5$
Engraved	$60(.5) + 95(.20) + 85(.30) = 74.5$

Here the best strategy is to display gifts that can be engraved; the expected profit is 74.5, or $74,500. ▲ ⟨3⟩

9.6 Exercises

1. **Business** A developer has $100,000 to invest in land. He has a choice of two parcels (at the same price): one on the highway and one on the coast. With both parcels, his ultimate profit depends on whether he faces light opposition from environmental groups or heavy opposition. He estimates that the payoff matrix is as follows. (The numbers represent his profit.)

Opposition

	Light	Heavy
Highway	$70,000	$30,000
Coast	$150,000	−$40,000

What should the developer do if he is
(a) an optimist?   (b) a pessimist?
(c) Suppose the probability of heavy opposition is .8. What is his best strategy? What is the expected profit?
(d) What is the best strategy if the probability of heavy opposition is only .4?

2. **Business** Mount Union College has sold out all tickets for a jazz concert to be held in the stadium. If it rains, the show will have to be moved to the gym, which has a much smaller seating capacity. The dean must decide in advance whether to set up the seats and the stage in the gym or in the stadium, or both, just in case. The following payoff matrix shows the net profit in each case.

States of Nature

		Rain	No Rain
	Set up in Stadium	−$1550	$1500
Strategies	Set up in Gym	$1000	$1000
	Set up Both	$750	$1400

What strategy should the dean choose if she is

(a) an optimist?   (b) a pessimist?
(c) If the weather forecaster predicts rain with a probability of .6, what strategy should she choose to maximize the expected profit? What is the maximum expected profit?

3. **Business** An analyst must decide what fraction of the automobile tires produced at a particular manufacturing plant are defective. She has already decided that there are three possibilities for the fraction of defective items: .02, .09, and .16. She may recommend two courses of action: upgrade the equipment at the plant or make no upgrades. The following payoff matrix represents the *costs* to the company in each case, in hundreds of dollars.

Defectives

		.02	.09	.16
Strategies	Upgrade	130	130	130
	No Upgrade	28	180	450

What strategy should the analyst recommend if she is
(a) an optimist?   (b) a pessimist?
(c) Suppose the analyst is able to estimate probabilities for the three states of nature as follows.

Fraction of Defectives	Probability
.02	.70
.09	.20
.16	.10

Which strategy should she recommend? Find the expected cost to the company if that strategy is chosen.

4. **Business** The research department of the Allied Manufacturing Company has developed a new process that it believes will result in an improved product. Management must decide whether or not to go ahead and market the new product. The new product may be better than the old one, or it may not be better. If the new product is better and the company decides to market it, sales should increase by $50,000. If it is not better and the old product is replaced with the new product on the market, the company will lose $25,000 to competitors. If management decides not to market the new product, the company will lose $40,000 if it is better and will lose research costs of $10,000 if it is not.

   (a) Prepare a payoff matrix.

   (b) If management believes that the probability that the new product is better is .4, find the expected profits under each strategy and determine the best action.

5. **Business** A businessman is planning to ship a used machine to his plant in Nigeria. He would like to use it there for the next 4 years. He must decide whether to overhaul the machine before sending it. The cost of overhaul is $2600. If the machine fails when it is in operation in Nigeria, it will cost him $6000 in lost production and repairs. He estimates that the probability that it will fail is .3 if he does not overhaul it and .1 if he does overhaul it. Neglect the possibility that the machine might fail more than once in the 4 years.

   (a) Prepare a payoff matrix.

   (b) What should the businessman do to minimize his expected costs?

6. **Business** A contractor prepares to bid on a job. If all goes well, his bid should be $25,000, which will cover his costs plus his usual profit margin of $4000. However, if a threatened labor strike actually occurs, his bid should be $35,000 to give him the same profit. If there is a strike and he bids $25,000, he will lose $5500. If his bid is too high, he may lose the job entirely, while if it is too low, he may lose money.

   (a) Prepare a payoff matrix.

   (b) If the contractor believes that the probability of a strike is .6, how much should he bid?

7. **Natural Science** A community is considering an anti-smoking campaign.* The city council will choose one of three possible strategies: a campaign for everyone over age 10 in the community, a campaign for youths only, or no campaign at all. The two states of nature are a true cause-effect relationship between smoking and cancer and no cause-effect relationship. The costs to the community (including loss of life and productivity) in each case are as shown at the top of the next column.

	States of Nature	
Strategies	Cause–Effect Relationship	No Cause–Effect Relationship
Campaign for All	$100,000	$800,000
Campaign for Youth	$2,820,000	$20,000
No Campaign	$3,100,100	$0

What action should the city council choose if it is

   (a) optimistic?

   (b) pessimistic?

   (c) If the director of public health estimates that the probability of a true cause–effect relationship is .8, which strategy should the city council choose?

8. **Business** An investor has $20,000 to invest in stocks. She has two possible strategies: buy conservative blue-chip stocks or buy highly speculative stocks. There are two states of nature: the market goes up and the market goes down. The following payoff matrix shows the net amounts she will have under the various circumstances.

	Market Up	Market Down
Buy Blue Chip	$25,000	$18,000
Buy Speculative	$30,000	$11,000

What should the investor do if she is

   (a) an optimist?   (b) a pessimist?

   (c) Suppose there is a .7 probability of the market going up. What is the best strategy? What is the expected profit?

   (d) What is the best strategy if the probability of a market rise is .2?

Sometimes the numbers (or payoffs) in a payoff matrix do not represent money (profits or costs, for example). Instead, they may represent utility. A utility is a number that measures the satisfaction (or lack of it) that results from a certain action. Utility numbers must be assigned by each individual, depending on how he or she feels about a situation. For example, one person might assign a utility of +20 for a week's vacation in San Francisco, with −6 being assigned if the vacation were moved to Sacramento. Work the following problems in the same way as the preceding ones.

9. **Social Science** A politician must plan her reelection strategy. She can emphasize jobs or she can emphasize the environment. The voters can be concerned about jobs or about the environment. Following is a payoff matrix showing the utility of each possible outcome.

		Voters	
		Jobs	Environment
Candidate	Jobs	+40	−10
	Environment	−12	+30

The political analysts feel that there is a .35 chance that the voters will emphasize jobs. What strategy should the candidate adopt? What is its expected utility?

10. In an accounting class, the instructor permits the students to bring a calculator or a reference book (but not both) to an examination. The examination itself can emphasize either numbers or definitions. In trying to decide which aid to take to an examination, a student first decides on the utilities shown in the following payoff matrix.

Exam Emphasizes

		Numbers	Definition
Student Chooses	Calculator	+50	0
	Book	+15	+35

(a) What strategy should the student choose if the probability that the examination will emphasize numbers is .6? What is the expected utility in this case?

(b) Suppose the probability that the examination emphasizes numbers is .4. What strategy should be chosen by the student?

Chapter 9 Summary

Key Terms and Symbols

9.1 random variable
probability distribution
histogram
expected value
fair game

9.2 *n! n* factorial
multiplication principle
permutations
combinations

9.4 Bernoulli trials (processes)
binomial experiment
binomial probability

9.5 stochastic processes
Markov chain
state
transition diagram
transition matrix
probability vector

regular transition matrix
regular Markov chain
equilibrium vector (fixed vector)

9.6 states of nature
strategies
payoffs
payoff matrix

Key Concepts

Expected Value of a Probability Distribution: For a random variable x with values $x_1, x_2, \ldots, x_n$ and probabilities $p_1, p_2, \ldots, p_n$, the expected value is

$$E(x) = x_1 p_1 + x_2 p_2 + \cdots + x_n p_n.$$

Multiplication Principle: If there are m_1 ways to make a first choice, m_2 ways to make a second choice, and so on, where none of the choices depend on any of the others, then there are $m_1 m_2 \cdots m_n$ different ways to make the entire sequence of choices.

The number of **permutations** of n elements taken r at a time is $_nP_r = \dfrac{n!}{(n-r)!}$.

The number of **combinations** of n elements taken r at a time is

$$_nC_r = \frac{n!}{(n-r)!r!}.$$

Binomial experiments have the following characteristics: (1) the same experiment is repeated several times; (2) there are only *two* outcomes, labeled success and failure; (3) the probability of success is the same for each trial; and (4) the trials are independent. Then if the probability of success in a single trial is p, the probability of x successes in n trials is

$$_nC_x p^x (1-p)^{n-x}.$$

Markov Chains: A **transition matrix** must be square, with all entries between 0 and 1 inclusive, and the sum of the entries in any row must be 1. A Markov chain is *regular* if some power of its transition matrix P contains all positive entries. The long-range probabilities for a regular Markov chain are given by the **equilibrium** or **fixed vector** V, where, for any initial probability vector v, the products vP^n approach V as n gets larger and larger and $VP = V$. To find V, solve the system of equations formed by $VP = V$ and the fact that the sum of the entries of V is 1.

Decision Making: A **payoff matrix,** which includes all available strategies and states of nature, is used in decision making to define the problem and the possible solutions. The expected value of each strategy can help to determine the best course of action.

Chapter 9 Review Exercises

In Exercises 1–3, using the given probability distribution, (a) sketch its histogram and (b) find the expected value.

1.

x	0	1	2	3
$P(x)$	.22	.54	.16	.08

2.

x	−3	−2	−1	0	1	2	3
$P(x)$	.15	.20	.25	.18	.12	.06	.04

3.

x	−10	0	10
$P(x)$	.333	.333	.333

In Exercises 4–7, (a) give the probability distribution and (b) find the expected value.

4. In one form of roulette, you bet $1 on "red." If 1 of the 18 red numbers comes up, you get your dollar back, plus another one. If one of the 20 nonred (18 black and 2 green) numbers comes up, you lose your dollar.

5. A fair coin is tossed 4 times and the number of tails is recorded.

6. **Business** A grocery store has 10 bouquets of flowers for sale, 3 of which are red rose displays. Two bouquets are selected at random, and the number of rose bouquets is noted.

7. **Social Science** In a small class of 10 students, 3 did not do their homework. The professor selects 3 members of the class to present solutions to homework problems on the board and records how many of those selected did not do their homework.

Solve the following problems.

8. Suppose someone offers to pay you $100 if you draw 3 cards from a standard deck of 52 cards and all the cards are hearts. What should you pay for the chance to win if it is a fair game?

9. You pay $2 to play a game of "Over/Under," in which you will roll two dice and note the sum of the results. You can bet that the sum will be less than 7 (under), exactly 7, or greater than 7 (over). If you bet "under" and you win, you get your $2 back, plus $2 more. If you bet 7 and you win, you get your $2 back, plus $4, and if you bet "over" and win, you get your $2 back, plus $2 more. What are the expected winnings for each type of bet?

10. A lottery has a first prize of $10,000, two second prizes of $1000 each, and two $100 third prizes. Ten thousand tickets are sold, at $2 each. Find the expected winnings of a person buying 1 ticket.

11. Find the expected number of girls in a family of 5 children.

12. Three cards are drawn from a standard deck of 52 cards.
(a) What is the expected number of kings?
(b) What is the expected number of diamonds?

13. In how many ways can 5 shuttle vans line up at the airport?

14. How many variations in first-, second-, and third-place finishes are possible in a 100-yard dash with 7 runners?

15. In how many ways can a sample of 3 monitors be taken from a batch of 12 monitors?

16. If 4 of the 12 monitors in Exercise 15 are broken, in how many ways can the sample of 3 include the following?
(a) 1 broken monitor
(b) No broken monitors
(c) At least 1 broken monitor

17. In how many ways can 6 students from a class of 30 be arranged in the first row of seats? (There are 6 seats in the first row.)

18. In how many ways can the six students in Exercise 17 be arranged in a row if a certain student must be first?

19. In how many ways can the 6 students in Exercise 17 be arranged if half the students are science majors and the other half are business majors and if
    (a) like majors must be together?
    (b) science and business majors are alternated?

20. Explain under what circumstances a permutation should be used in a probability problem and under what circumstances a combination should be used.

21. Discuss under what circumstances the binomial probability formula should be used in a probability problem.

Suppose 2 cards are drawn without replacement from an ordinary deck of 52 cards. Find the probabilities of the following results.

22. Both cards are black.   23. Both cards are hearts.

24. Exactly 1 is a face card.   25. At most 1 is an ace.

An ice cream stand contains 4 custard flavors, 6 ice-cream flavors, and 2 frozen yogurt selections. Three customers come to the window. If each customer's selection is random, find the probability that the selections include

26. All ice cream   27. All custard

28. At least one frozen yogurt

29. One custard, one ice cream, and one frozen yogurt

30. At most one ice cream

31. In this exercise we study the connection between sets (from Chapter 8) and combinations.
    (a) Given a set with n elements, what is the number of subsets of size 0? of size 1? of size 2? of size n?
    (b) Using your answer from part (a), give an expression for the total number of subsets of a set with n elements.
    (c) Using your answer from part (b) and a result from Chapter 8, explain why the following equation must be true:

    $$_nC_0 + {_nC_1} + {_nC_2} + \cdots + {_nC_n} = 2.$$

    (d) Verify the equation in part (c) for $n = 4$ and $n = 5$.

Business *A bottle capping machine has an error rate of .01. A random sample of 20 bottles is selected. Find the following probabilities.*

32. Exactly 4 bottles are improperly capped.

33. No more than 3 bottles are improperly capped.

The July 18, 2005, issue of Newsweek magazine reported that 25% of U.S. adult Internet users have stopped using file-sharing software for fear of infecting their PCs with spyware. Suppose 6 U.S. adult Internet users are selected at random and the number who have stopped using file-sharing software is noted.

34. Give the resulting probability distribution.

35. What is the expected number of those who stopped using the file-sharing software.

36. In Exercise 32 of Section 8.3, we learned that 81% of equities in the Janus Mercury fund were invested in the United States. If 4 equities are selected at random, find the probability that
    (a) All 4 are invested in the United States.
    (b) At least 2 are invested in the United States.
    (c) At most 1 is invested in the United States.

37. In Exercise 34 of Section 8.4, we learned that the odds that a traffic fatality from anywhere in the world occurs in China is 1:4.
    (a) What is the probability that a traffic fatality occurred in China?
    (b) Suppose 8 traffic fatalities are selected at random from around the world. Give the probability distribution for the number of fatalities that occurred in China.
    (c) What is the expected number of fatalities that occurred in China?

Decide whether each of the following is a regular transition matrix.

38. $\begin{bmatrix} 0 & 1 \\ .77 & .23 \end{bmatrix}$

39. $\begin{bmatrix} -.2 & .4 \\ .3 & .7 \end{bmatrix}$

40. $\begin{bmatrix} .21 & .15 & .64 \\ .50 & .12 & .38 \\ 1 & 0 & 0 \end{bmatrix}$

41. $\begin{bmatrix} .22 & 0 & .78 \\ .40 & .33 & .27 \\ 0 & .61 & .39 \end{bmatrix}$

42. **Business** Using e-mail for professional correspondence has become a major component of a worker's day. A study classified e-mail use into 3 categories for an office day: no use, light use (1–60 minutes), and heavy use (more than 60 minutes). Researchers observed a pool of 100 office workers over a month and developed the following transition matrix of probabilities from day to day.

		Current Day	
	No Use	Light Use	Heavy Use
Previous Day No Use	.35	.15	.50
Light Use	.30	.35	.35
Heavy Use	.15	.30	.55

Suppose the initial distribution for the three states is $[.2, .4, .4]$. Find the distribution after
    (a) 1 day;
    (b) 2 days.
    (c) What is the long-range prediction for the distribution of e-mail use?

43. **Business** A major brokerage firm that invests in Europe, North America, and Asia has examined the investment records for a particular international stock mutual fund over several years. The analyst constructed the following transition matrix for the probability of switching the location of an equity from year to year.

Current Year

		North America	Asia
	Europe	America	

		Europe	North America	Asia
Previous Year	Europe	.80	.14	.06
	North America	.04	.85	.11
	Asia	.03	.13	.84

If the initial investment vector is 15% in Europe, 60% in North America, and 25% in China,
- (a) find the percentages in Europe, North America, and China after 1 year;
- (b) find the percentages in Europe, North America, and China after 3 years;
- (c) find the long-range percentages in Europe, North America, and China.

44. Social Science A candidate for city council can come out in favor of a new factory, be opposed to it, or waffle on the issue. The change in votes for the candidate depends on what her opponent does, with payoffs as shown.

Opponent

		Favors	Waffles	Opposes
Candidate	Favors	0	−1000	−4000
	Waffles	1000	0	−500
	Opposes	5000	2000	0

- (a) What should the candidate do if she is an optimist?
- (b) What should she do if she is a pessimist?
- (c) Suppose the candidate's campaign manager feels that there is a 40% chance that the opponent will favor the plant and a 35% chance that he will waffle. What strategy should the candidate adopt? What is the expected change in the number of votes?
- (d) The opponent conducts a new poll that shows strong opposition to the new factory. This changes the probability that he will favor the factory to 0 and the probability that he will waffle to .7. What strategy should our candidate adopt? What is the expected change in the number of votes now?

45. Social Science When teaching, an instructor can adopt a strategy using either active learning or lecturing to help students learn best. A class often reacts very differently to these two strategies. A class can prefer lecturing or active learning. A department chair constructs the following payoff of the average point gain (out of 500 possible points) on the final exam after studying many classes that use active learning and many that use lecturing and polling students as to their preference.

Students in Class prefer

		Lecture	Active Learning
Instructor uses	Lecture	50	−80
	Active Learning	−30	100

- (a) If the department chair uses the preceding information to decide how to teach herself, what should she do if she is an optimist?
- (b) What if she is a pessimist?
- (c) If the polling data shows that there is a 75% chance that a class will prefer the lecture format, what strategy should she adopt? What is the expected payoff?
- (d) If the chair finds out that her next class has had more experience with active learning, so that there is now a 60% chance that the class will prefer active learning, what strategy should she adopt? What is the expected payoff?

Exercises 46 and 47 are taken from actuarial examinations given by the Society of Actuaries.*

46. Business A company is considering the introduction of a new product that is believed to have probability .5 of being successful and probability .5 of being unsuccessful. Successful products pass quality control 80% of the time. Unsuccessful products pass quality control 25% of the time. If the product is successful, the net profit to the company will be $40 million; if unsuccessful, the net loss will be $15 million. Determine the expected net profit if the product passes quality control.
- (a) $23 million (b) $24 million
- (c) $25 million (d) $26 million
- (e) $27 million

47. Business A merchant buys boxes of fruit from a grower and sells them. Each box of fruit is either Good or Bad. A Good box contains 80% excellent fruit and will earn $200 profit on the retail market. A Bad box contains 30% excellent fruit and will produce a loss of $1000. The a priori probability of receiving a Good box of fruit is .9. Before the merchant decides to put the box on the market, he can sample one piece of fruit to test whether it is excellent. Based on that sample, he has the option of rejecting the box without paying for it. Determine the expected value of the right to sample. (*Hint:* If the merchant samples the fruit, what are the probabilities of accepting a Good box, accepting a Bad box, and not accepting the box? What are these probabilities if he does not sample the fruit?)
- (a) 0 (b) $16 (c) $34
- (d) $72 (e) $80

48. Business The March 1982 issue of *Mathematics Teacher* included "Overbooking Airline Flights," an article by Joe Dan Austin. In this article, Austin developed a model for the expected income for an airline flight. With appropriate assumptions, the probability that exactly x of n people with reservations show up at the airport to buy a

*Problem from "Course 130 Examination, Operations Research," of the Education and Examination Committee of the Society of Actuaries. Reprinted by permission of the Society of Actuaries.

ticket is given by the binomial probability formula. Assume the following: 6 reservations have been accepted for 3 seats, $p = .6$ is the probability that a person with a reservation will show up, a ticket costs \$100, and the airline must pay \$100 to anyone with a reservation who does not get a ticket. Complete the following table:

Number Who Show Up (x)	0	1	2	3	4	5	6
Airline's Income							
$P(x)$							

(a) Use the table to find $E(I)$, the expected airline income from the 3 seats.

(b) Find $E(I)$ for $n = 3, n = 4$, and $n = 5$. Compare these answers with $E(I)$ for $n = 6$. For these values of n, how many reservations should the airline book for the 3 seats in order to maximize the expected revenue?

Case 9

Optimal Inventory for a Service Truck

For many different items, it is difficult or impossible to take the item to a central repair facility when service is required. Washing machines, large television sets, office copiers, and computers are only a few examples of such items. Service for items of this type is commonly performed by sending a repair person to the item, with the person driving to the location in a truck containing various parts that might be required in repairing the item. Ideally, the truck should contain all the parts that might be required. However, most parts would be needed only infrequently, so that inventory costs for the parts would be high.

An optimum policy for deciding on which parts to stock on a truck would require that the probability of not being able to repair an item without a trip back to the warehouse for needed parts be as low as possible, consistent with minimum inventory costs. An analysis similar to the one that follows was developed at the Xerox Corporation.*

*Reprinted by permission of Stephen Smith, John Chambers, and Eli Shlifer. "Optimal Inventories Based on Job Completion Rate for Repairs Requiring Multiple Items." *Management Sciences*, Vol. 26, No. 8, August 1980, copyright © 1980 by The Institute of Management Science.

To set up a mathematical model for deciding on the optimum truck-stocking policy, let us assume that a broken machine might require one of 5 different parts (we could assume any number of different parts—we use 5 to simplify the notation). Suppose also that the probability that a particular machine requires part 1 is p_1; that it requires part 2 is p_2; and so on. Assume also that failures of different part types are independent, and that at most one part of each type is used on a given job.

Suppose that, on the average, a repair person makes N service calls per time period. If the repair person is unable to make a repair because at least one of the parts is unavailable, there is a penalty cost, L, corresponding to wasted time for the repair person, an extra trip to the parts depot, customer unhappiness, and so on. For each of the parts carried on the truck, an average inventory cost is incurred. Let H_i be the average inventory cost for part i, where $1 \le i \le 5$.

Let M_1 represent a policy of carrying only part 1 on the repair truck, M_{24} represent a policy of carrying only parts 2 and 4, with M_{12345} and M_0 representing policies of carrying all parts and no parts, respectively.

For policy M_{35}, carrying parts 3 and 5 only, the expected cost per time period per repair person, written $C(M_{35})$, is
$$C(M_{35}) =$$
$$(H_3 + H_5) + NL[1 - (1 - p_1)(1 - p_2)(1 - p_4)].$$

(The expression in brackets represents the probability of needing at least one of the parts not carried, 1, 2, or 4 here.) As further examples,

$$C(M_{125}) = (H_1 + H_2 + H_5) + NL[1 - (1 - p_3)(1 - p_4)],$$

while

$$C(M_{12345}) = (H_1 + H_2 + H_3 + H_4 + H_5) + NL[1-1]$$
$$= (H_1 + H_2 + H_3 + H_4 + H_5),$$

and

$$C(M_0) =$$
$$NL[1 - (1 - p_1)(1 - p_2)(1 - p_3)(1 - p_4)(1 - p_5)].$$

To find the best policy, evaluate $C(M_0), C(M_1), \ldots,$ $C(M_{12345})$, and choose the smallest result. (A general solution method is in the *Management Science* paper.)

▼ **EXAMPLE 1** Suppose that, for a particular item, only 3 parts might need to be replaced. By studying past records of failures of the item and finding necessary inventory costs, suppose that the following values have been found:

p_1	p_2	p_3
.09	.24	.17

H_1	H_2	H_3
$15	$40	$9

Suppose $N = 3$ and L is $54. Then, as an example,

$$C(M_1) = H_1 + NL[1 - (1 - p_2)(1 - p_3)]$$
$$= 15 + 3(54)[1 - (1 - .24)(1 - .17)]$$
$$= 15 + 3(54)[1 - (.76)(.83)]$$
$$\approx 15 + 59.81$$
$$= 74.81.$$

Thus, if policy M_1 is followed (carrying only part 1 on the truck), the expected cost per repair person per period is $74.81. Also,

$$C(M_{23}) = H_2 + H_3 + NL[1 - (1 - p_1)]$$
$$= 40 + 9 + 3(54)(.09)$$
$$= 63.58,$$

so that M_{23} is a better policy than M_1. By finding the expected values for all other possible policies (see the exercises), the optimum policy may be chosen. ▲

◆ **Exercises**

1. Refer to the example and find each of the following.
   (a) $C(M_0)$
   (b) $C(M_2)$
   (c) $C(M_3)$
   (d) $C(M_{12})$
   (e) $C(M_{13})$
   (f) $C(M_{123})$
2. Which policy leads to the lowest expected cost?
3. In Example 1, $p_1 + p_2 + p_3 = .09 + .24 + .17 = .50$. Why is it not necessary that the probabilities add up to 1?
4. Suppose an item to be repaired might need one of n different parts. How many different policies would then need to be evaluated?

Chapter 10 Introduction to Statistics

10.1 Frequency Distributions and Measures of Central Tendency

10.2 Measures of Variation

10.3 Normal Distributions

10.4 Normal Approximation to the Binomial Distribution

Case 10 Statistics in the Law: The Castaneda Decision

Statistics has applications to almost every aspect of modern life. The digital age is creating a wealth of data that needs to be summarized, visualized, and analyzed whether we want to know executives' earnings per year, movies' earnings at the box office on a given weekend, or the sales within sectors of the home and leisure industry. See Exercises 35 and 36 on page 584, and Exercises 53–57 on page 609.

Statistics is the science that deals with the collection and summarization of data. Methods of statistical analysis make it possible to draw conclusions about a population on the basis of data from a sample of the population. Statistical models have become increasingly useful in manufacturing, government, agriculture, medicine, and the social sciences and in all types of research. An Indianapolis race-car team is using statistics to improve its performance by gathering data on each run around the track. The team samples data 300 times a second and uses computers to process the data. In this chapter, we give a brief introduction to some of the key topics from statistical methodology.

10.1 Frequency Distributions and Measures of Central Tendency

Researchers often wish to learn characteristics or traits about a specific **population** of individuals, objects, or units. The traits of interest are called **variables,** and it is these that we measure or label. Often, however, a population of interest is very large or constantly changing, so measuring each unit is impossible. Thus, researchers are forced to collect data on a subset of the population of interest, called a **sample.**

Sampling is a complex topic, but the universal aim of all sampling methods is to obtain a sample that "represents" the population of interest. One common way of obtaining a representative sample is to perform simple random sampling, in which every unit of the population has an equal chance to be selected to be in the sample. Suppose we wanted to study the height of students enrolled in a class. To obtain a random sample, we could place the names of everyone in class in a hat, mix the names, and draw ten names blindly. We would then record the height (the variable of interest) for each student selected.

A simple random sample can be difficult to obtain in real life. For example, suppose you want to take a random sample of voters in your congressional district to see which candidate they prefer in the next election. If you do a telephone survey, you have a representative sample of people who are at home to answer the telephone, but those who work a lot of hours and are rarely home to answer the phone, those who have an unlisted number, those who cannot afford a telephone, and those who refuse to answer telephone surveys are underrepresented. Such people may have an opinion different from those of the people you interview.

A famous example of an inaccurate poll was made by the *Literary Digest* in 1936. Their survey indicated that Alfred Landon would win the presidential election; in fact, Franklin Roosevelt won with 62% of the popular vote. The *Digest's* major error was mailing their surveys to a sample of those listed in telephone directories. During the depression, many poor people did not have telephones, and the poor voted overwhelmingly for Roosevelt. Modern pollsters use sophisticated techniques to ensure that their sample is as representative as possible.

Once a sample has been collected and all data of interest are recorded, the data must be organized so that conclusions may be more easily drawn. With numeric responses, one method of organization is to group the data into intervals, usually of equal size.

▼ **EXAMPLE 1**   A survey asked a random sample of 30 business executives for their recommendations as to the number of college units in management that a business major should obtain. The results are shown below. Identify the population and the variable, group the data into intervals, and find the frequency of each interval.

3	25	22	16	0	9	14	8	34	21
15	12	9	3	8	15	20	12	28	19
17	16	23	19	12	14	29	13	24	18

Solution The population of interest is all business executives, and the variable of interest is the number of college units in management an undergraduate business major should take prior to graduation. The highest number in the list is 34 and the lowest is 0; one convenient way to group the data is in intervals of size 5, starting with 0–4 and ending with 30–34. This grouping gives an interval for each number in the list and results in seven equal intervals of a convenient size. Too many intervals of smaller size would not simplify the data enough, while too few intervals of larger size would conceal information that the data might provide. A rule of thumb is to use from 6 to 15 intervals.

First tally the number of college units falling into each interval. Then total the tallies in each interval, as in the following table. This table is an example of a **grouped frequency distribution.**

College Units	Tally	Frequency
0–4	|||	3
5–9	||||	4
10–14	⊮⊮		6
15–19	⊮⊮ |||	8
20–24	⊮⊮	5
25–29	|||	3
30–34			1
		Total: 30

▲ ◇ ①

The frequency distribution in Example 1 shows information about the data that might not have been noticed before. For example, the interval with the largest number of units is 15–19, and 19 executives (more than half) recommended between 9 and 25 units. Also, the frequency in each interval increases rather evenly (up to 8) and then decreases at about the same pace. However, some information has been lost; for example, we no longer know how many executives recommended 12 units.

Picturing Data

The information in a grouped frequency distribution can be displayed graphically with a **histogram,** which is similar to a bar graph. In a histogram, the number of observations in each interval determines the height of each bar, and the size of each interval determines the width of each bar. If equally sized intervals are used, all the bars have the same width.

A **frequency polygon** is another form of graph that illustrates a grouped frequency distribution. The polygon is formed by joining consecutive midpoints of the tops of the histogram bars with straight-line segments. Sometimes the midpoints of the first and last bars are joined to endpoints on the horizontal axis where the next midpoint would appear. (See Figure 10.1 on the next page.)

▼ **EXAMPLE 2** A grouped frequency distribution of suggested college units was found in Example 1. Draw a histogram and a frequency polygon for this distribution.

◇ **1**

An accounting firm selected 24 complex tax returns prepared by a certain tax preparer. The number of errors per return were as follows:

8 12 0 6 10 8 0 14

8 12 14 16 4 14 7 11

9 12 7 15 11 21 22 19

Prepare a grouped frequency distribution for this data. Use intervals 0–4, 5–9, and so on.

Answer:

Interval	Frequency
0–4	3
5–9	7
10–14	9
15–19	3
20–24	2
	Total: 24

Solution First, draw a histogram, shown in blue in Figure 10.1. To get a frequency polygon, connect consecutive midpoints of the tops of the bars. The frequency polygon is shown in red. ▲ ⟨2⟩

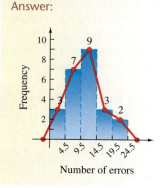

⟨2⟩

Make a histogram and a frequency polygon for the distribution found in side problem 1.

Answer:

Number of errors

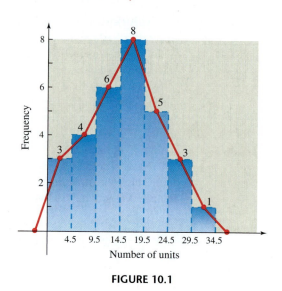

FIGURE 10.1

TECHNOLOGY TIP As noted in Section 9.1, most graphing calculators can display histograms. Many will also display frequency polygons (which are usually labeled LINE or xyLINE in calculator menus). When dealing with grouped frequency distributions, however, certain adjustments must be made on a calculator.

1. *A calculator list of outcomes must consist of single numbers, not intervals.* The table in Example 1, for instance, cannot be entered as shown. To convert the first column of the table for calculator use, choose one number in each interval—say, 2 in the interval 0–4, 7 in the interval 5–9, 12 in the interval 10–14, etc. Then use 2, 7, 12, . . . as the list of outcomes to be entered into the calculator. The frequency list (the last column of the table) remains the same.

2. *The histogram bar width affects the shape of the graph.* If you use a bar width of 4 in Example 1, the calculator may produce a histogram with gaps in it. To avoid this, use the interval $0 \le x < 5$ in place of $0 \le x \le 4$, and similarly for the other intervals, and make 5 the bar width.

Following this procedure, we obtain the calculator-generated histogram and frequency polygon in Figure 10.2 for the data from Example 1. Note that the width of each histogram bar is 5. Some calculators cannot display both the histogram and the frequency polygon on the same screen, as is done here.

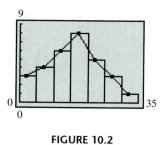

FIGURE 10.2

Stem-and-leaf plots allow us to organize the data into a distribution without the disadvantage of losing the original information. In a **stem-and-leaf plot,** we separate the digits in each data point into two parts consisting of the first one or two digits (the stem) and the remaining digit (the leaf). We also provide a key to show the reader the units of the data that was recorded.

▼ **EXAMPLE 3** Construct a stem-and-leaf plot for the data in Example 1.

Solution Since the data are two-digit numbers, we use the first digit for the stems: 0, 1, 2, and 3. The second digits provide the leaves. For example, as shown below, the first stem 0 has leaves 0, 3, 3 corresponding to the entries 3 and 0 in the first row of the data list and 3 in the second row. Here, each stem corresponds to an interval in the frequency table. The stems and leaves are separated by a vertical line.

Stem	Leaves
0	033
0	8899
1	222344
1	55667899
2	01234
2	589
3	4

Units: 3|4 = 34 credits

If we turn the page on its side, the distribution looks like a histogram but still retains each of the original values. We used each stem digit twice, because, as with a histogram, using too few intervals conceals useful information about the shape of the distribution. ▲ ③

▼ **EXAMPLE 4** List the original data for the following stem-and-leaf plot of resting pulses taken on the first day of class for 36 students.

Stem	Leaves
4	8
5	278
6	034455688888
7	02222478
8	2269
9	00002289

Units: 9|0 = 90 beats per minute

The first stem and its leaf correspond to the data point 48 beats per minute. Similarly, the rest of the data are 52, 57, 58, 60, 63, 64, 64, 65, 65, 66, 68, 68, 68, 68, 68, 70, 72, 72, 72, 72, 74, 77, 78, 82, 82, 86, 89, 90, 90, 90, 90, 92, 92, 98, and 99 beats per minute. ▲ ④

NOTE The remainder of this section deals with topics that are generally referred to as "measures of central tendency." Computing these various measures is greatly simplified by the statistical capabilities of most scientific and graphing calculators.

③
Make a stem-and-leaf plot for the data in Example 1, using one stem each for 0, 1, 2, and 3.

Answer:

Stem	Leaves
0	0338899
1	22234455667899
2	01234589
3	4

Units: 3|4 = 34 credits

④
List the original data for the following heights of students.

Stem	Leaves
5	9
6	00012233334444
6	55567777799
7	0111134
7	558

Units: 7|5 = 75 inches

Answer:

59, 60, 60, 60, 61, 62, 62, 63, 63, 63, 63, 64, 64, 64, 64, 65, 65, 65, 66, 67, 67, 67, 67, 67, 69, 69, 70, 71, 71, 71, 71, 73, 74, 75, 75, 78

Calculators vary considerably in how data is entered, so read your instruction manual to learn how to enter lists of data and the corresponding frequencies. On scientific calculators with statistical capabilities, there are keys for finding most of the measures of central tendency discussed below. On graphing calculators, most or all of these measures can be obtained with a single keystroke. (Look for *one-variable statistics,* which is often labeled 1-VAR, in the STAT menu or its CALC submenu.)

Mean

Three measures of central tendency, or "averages," are used with frequency distributions: the mean, the median, and the mode. The most important of these is the mean, which is similar to the expected value of a probability distribution. The **arithmetic mean** (the mean) of a set of numbers is the sum of the numbers, divided by the total number of numbers. We write the sum of n numbers $x_1, x_2, x_3, \ldots, x_n$ in a compact way with **summation notation,** also called **sigma notation.** With the Greek letter Σ (sigma), the sum

$$x_1 + x_2 + x_3 + \cdots + x_n$$

is written

$$x_1 + x_2 + x_3 + \cdots + x_n = \sum_{i=1}^{n} x_i.$$

In statistics, $\sum_{i=1}^{n} x_i$ is often abbreviated as just Σx. The symbol $\bar{x}$ (read "x-bar") is used to represent the mean of a sample.

Mean

The mean of the n numbers $x_1, x_2, x_3, \ldots, x_n$ is

$$\bar{x} = \frac{x_1 + x_2 + \cdots + x_n}{n} = \frac{\Sigma x}{n}.$$

▼ **EXAMPLE 5**   The number of businesses filing for bankruptcy for the years 2000–2004 are given in the following table.* Find the mean number of business bankruptcies filed annually during this period.

Year	Petition Filed
2000	36,065
2001	38,490
2002	39,091
2003	36,183
2004	34,817

*www.uscourts.gov/Press_Releases/fy04bk.pdf

Solution Let $x_1 = 36{,}065$, $x_2 = 38{,}490$, and so on. Here, $n = 5$, since there are 5 numbers in the list.

$$\bar{x} = \frac{36{,}065 + 38{,}490 + 39{,}091 + 36{,}183 + 34{,}817}{5} \approx 36{,}929.$$

The mean number of business bankruptcy petitions filed during the given years is 36,929. ▲ ⟨5⟩

The mean of data that has been arranged into a frequency distribution is found in a similar way. For example, suppose the following quiz score data is collected.

Value	Frequency
84	2
87	4
88	7
93	4
99	3
Total:	20

The value 84 appears twice, 87 four times, and so on. To find the mean, first add 84 two times, 87 four times, and so on; or get the same result faster by multiplying 84 by 2, 87 by 4, and so on, and then adding the results. Dividing the sum by 20, the total of the frequencies, gives the mean.

$$\begin{aligned}
\bar{x} &= \frac{(84 \cdot 2) + (87 \cdot 4) + (88 \cdot 7) + (93 \cdot 4) + (99 \cdot 3)}{20} \\
&= \frac{168 + 348 + 616 + 372 + 297}{20} \\
&= \frac{1801}{20} \\
\bar{x} &= 90.05
\end{aligned}$$

Verify that your calculator gives the same result.

▼ **EXAMPLE 6** Find the mean for the lengths, in inches, of Northern Pike caught in a local lake during June. The data is shown in the following frequency distribution.

Value	Frequency	Value × Frequency
30	6	$30 \cdot 6 = 180$
32	9	$32 \cdot 9 = 288$
33	7	$33 \cdot 7 = 231$
37	12	$37 \cdot 12 = 444$
42	6	$42 \cdot 6 = 252$
Total:	40	Total: 1395

⟨5⟩
Find the mean dollar purchases of eight students selected at random at the campus bookstore during the first week of classes.

$250.56	$567.32
$45.29	$321.56
$120.22	$561.04
$321.07	$226.90

Answer:

$301.75

TECHNOLOGY TIP The mean of the five numbers in Example 5 is easily found by using the $\bar{x}$ key on a scientific calculator or the one-variable statistics key on a graphing calculator. A graphing calculator will also display additional information, which will be discussed in the next section.

Solution The value 30 appears six times, 32 nine times, and so on. To find the mean, first multiply 30 times 6, 32 times 9, and so on, to get the new column "Value × Frequency," that has been added to the frequency distribution. Adding the products from this column gives a total of 1395. The total from the frequency column is 40. The mean length is

$$\overline{x} = \frac{1395}{40} = 34.875. \quad \blacktriangle \quad$$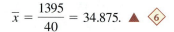

The mean of grouped data is found in a similar way. For grouped data, intervals are used, rather than single values. To calculate the mean, it is assumed that all of the values in a given interval are located at the midpoint of the interval. The letter x is used to represent the midpoints and f represents the frequencies, as shown in the next example.

▼ **EXAMPLE 7** The grouped frequency distribution for the 30 business executives described in Example 1 is listed below. Find the mean from the grouped frequency distribution.

Interval	Midpoint, x	Frequency, f	Product, xf
0–4	2	3	6
5–9	7	4	28
10–14	12	6	72
15–19	17	8	136
20–24	22	5	110
25–29	27	3	81
30–34	32	1	32
		Total: 30	Total: 465

Solution A column for the midpoint of each interval has been added. The numbers in this column are found by adding the endpoints of each interval and dividing by 2. For the interval 0–4, the midpoint is $(0 + 4)/2 = 2$. The numbers in the product column on the right are found by multiplying each frequency by its corresponding midpoint. Finally, we divide the total of the product column by the total of the frequency column to get

$$\overline{x} = \frac{465}{30} = 15.5.$$

Notice that this mean is slightly different from the earlier mean of 15.93. The reason for this difference is that we have acted as if each piece of data is at the midpoint, which is not true here and is not true in most cases. Information is always lost when the data is grouped. It is more accurate to use the original data, rather than the grouped frequency, when calculating the mean, but the original data might not be available. Furthermore, the mean based upon the grouped data is typically not too different from the mean based upon the original data, and there may be situations in which the extra accuracy is not worth the extra effort. ▲ ⟨7⟩

⟨6⟩

Find $\overline{x}$ for the following frequency distribution for the variable of years of schooling for a sample of construction workers.

Years	Frequency
7	2
9	3
11	6
13	4
15	4
16	1

Answer:

$\overline{x} = 11.75$

⟨7⟩

Find the mean of the following grouped frequency distribution for the number of classes completed thus far in the college careers of a random sample of 52 students.

Classes	Frequency
0–5	6
6–10	10
11–20	12
21–30	15
31–40	9

Answer:

18.90

NOTE 1. The midpoint of an interval in a grouped frequency distribution may be a value that none of the data assumes. For example, if we grouped the data for the 30 business executives into the intervals 0–5, 6–11, 12–17, 18–23, 24–29, and 30–35, the midpoints would be 2.5, 8.5, 14.5, 20.5, 26.5, and 32.5, even though all the data are whole numbers.

  2. If we used different intervals in Example 7, the mean would come out to be a slightly different number. Verify that with the intervals 0–5, 6–11, 12–17, 18–23, 24–29, and 30–35, the mean in Example 7 is 16.1.

  The formula for the mean of a grouped frequency distribution is as follows.

Mean of a Grouped Distribution

The mean of a distribution in which x represents the midpoints, f denotes the frequencies, and $n = \Sigma f$, is

$$\overline{x} = \frac{\Sigma(xf)}{n}.$$

The mean of a random sample is a random variable, and for this reason it is sometimes called the **sample mean**. The sample mean is a random variable because it assigns a number to the experiment of taking a random sample. If a different random sample were taken, the mean would probably have a different value, with some values more probable than others. If another set of 30 business executives were selected in Example 1, the mean number of college units in management recommended for a business major might be 13.22 or 17.69. It is unlikely that the mean would be as small as 1.21 or as large as 32.75, although these values are remotely possible.

  We saw in Section 9.1 how to calculate the expected value of a random variable when we know its probability distribution. The expected value is sometimes called the **population mean,** denoted by the Greek letter μ. In other words,

$$E(x) = \mu.$$

Furthermore, it can be shown that the expected value of $\overline{x}$ is also equal to μ; that is,

$$E(\overline{x}) = \mu.$$

For instance, consider again the 30 business executives in Example 1. We found that $\overline{x} = 15.93$, but the value of μ, the average for all possible business executives, is unknown. If a good estimate of μ were needed, the best guess (based on this data) is 15.93. Although $E(x) = \mu$ and $E(\overline{x}) = \mu$, $E(x)$ and $E(\overline{x})$ are not always the same value. $E(x)$ is the population mean, while $E(\overline{x})$ is the expected value of all possible sample means.

Median

Asked by a reporter to give the average height of the players on his team, the Little League coach lined up his 15 players by increasing height. He picked out the player in the middle and pronounced this player to be of average height. This kind of average, called the **median,** is defined as the middle entry in a set of data arranged in either increasing or decreasing order. If there is an even number of entries, the median is defined to be the mean of the two middle entries. The following table shows how to find the median for two sets of data: {8, 7, 4, 3, 1} and {2, 3, 4, 7, 9, 12}.

Odd Number of Entries		Even Number of Entries
8		2
7		3
Median = 4		4 ⎫ $Median = \dfrac{4+7}{2} = 5.5$
3		7 ⎭
1		9
		12

NOTE As shown in the table, when there is an even number of entries, the median is not always equal to one of the data entries.

▼ **EXAMPLE 8**   Find the median hours worked per week

(a) for a sample of 7 male students whose work hours were

$$0, 7, 10, \mathbf{20,}\ 22, 25, 30.$$

Solution The median is the middle number, in this case 20 hours per week. (Note that the numbers are already arranged in numerical order.) In this list, three numbers are smaller than 20 and three are larger.

(b) for a sample of 11 female students whose work hours were

$$20, 0, 20, 30, 35, 30, 20, 23, 16, 38, 25.$$

Solution First arrange the numbers in numerical order, from smallest to largest or vice versa.

$$0, 16, 20, 20, 20, \mathbf{23,}\ 25, 30, 30, 35, 38$$

The middle number can now be determined; the median is 23 hours per week.

(c) For a sample of 10 students of either gender whose work hours were

$$25, 18, 25, 20, 16, 12, 10, 0, 35, 32.$$

Solution Write the numbers in numerical order.

$$0, 10, 12, 16, \mathbf{18, 20,}\ 25, 25, 32, 35$$

There are six 10 numbers here; the median is the mean of the two middle numbers, or

$$\text{Median} = \frac{18 + 20}{2} = 19.$$

The median is 19 hours per week. ▲

⟨8⟩

Find the median for the following heights in inches.

(a) 60, 72, 64, 75, 72, 65, 68, 70

(b) 73, 58, 77, 66, 69, 69, 66, 68, 67

Answers:

(a) 69 inches

(b) 68 inches

CAUTION Remember, the data must be arranged in numerical order before you locate the median. ⟨8⟩

Both the mean and the median are examples of a **statistic,** which is simply a number that gives information about a sample. In some situations, the median gives a truer representative or typical element of the data than the mean does. For example, suppose that in an office there are 10 salespersons, 4 secretaries, the sales manager, and Ms. Daly, who owns the business. Their annual salaries are as follows: support staff, $30,000 each; salespersons, $50,000 each; manager, $70,000; and owner, $400,000. The mean salary is

$$\bar{x} = \frac{(30,000)4 + (50,000)10 + 70,000 + 400,000}{16} = \$68,125.$$

However, since 14 people earn less than $68,125 and only 2 earn more, the mean does not seem very representative. The median salary is found by ranking the salaries by size: $30,000, $30,000, $30,000, $30,000, $50,000, $50,000, . . . , $400,000. There are 16 salaries (an even number) in the list, so the mean of the 8th and 9th entries will give the value of the median. The 8th and 9th entries are both $50,000, so the median is $50,000. In this example, the median is more representative of the distribution than the mean is.

The preceding example shows that the mean is quite sensitive to extreme values ($70,000 and $400,000). Thus, the median is often reported for data sets with extreme values. For example, the U.S. government frequently reports the median of housing selling prices and annual household incomes.

TECHNOLOGY TIP Many graphing calculators (including most TI-models, HP-39+, and Casio 9850) display the median when doing one-variable statistics. You may have to scroll down to a second screen to find it.

Mode

Sue's scores on ten class quizzes include one 7, two 8's, six 9's, and one 10. She claims that her average grade on quizzes is 9, because most of her scores are 9's. This kind of "average," found by selecting the most frequent entry, is called the **mode.**

⟨9⟩

Find the mode for each of the following data sets.

(a) highway miles per gallon of an automobile: 25, 28, 32, 19, 15, 25, 30, 25

(b) price paid for last haircut or styling: $11, $35, $35, $10, $0, $12, $0, $35, $38, $42, $0, $25

(c) Class enrollment in six sections of calculus: 30, 35, 26, 28, 29, 19

Answers:

(a) 25 miles per gallon

(b) $0 and $35

(c) No mode

▼ **EXAMPLE 9** Find the mode for the following data.

(a) ages of retirement: 55, 60, 63, 63, 70, 55, 60, 65, 68, 65, 65, 71, 65, 65

Solution The number 65 occurs more often than any other, so it is the mode. It is sometimes convenient, but is not necessary to place the numbers in numerical order when looking for the mode.

(b) total cholesterol score: 180, 200, 220, 260, 220, 205, 255, 240, 190, 300, 240

Solution Both 220 and 240 occur twice. This list has *two* modes, so it is bimodal.

(c) prices of new cars: $25,789, $43,231, $33,456, $19,432, $22,971, $29,876

Solution No number occurs more than once. This list has no mode. ▲ ⟨9⟩

The mode has the advantages of being easily found and not being influenced by data that are extreme values. It is often used in samples where the data to be "averaged" are not numerical. A major disadvantage of the mode is that there may be more than one, in case of ties, or there may be no mode at all when all entries occur with the same frequency.

The mean is the most commonly used measure of central tendency. Its advantages are that it is easy to compute, it takes all the data into consideration, and it is reliable—that is, repeated samples are likely to give similar means. A disadvantage of the mean is that it is influenced by extreme values, as illustrated in the salary example.

The median can be easy to compute and is influenced very little by extremes. A disadvantage of the median is the need to rank the data in order; this can be tedious when the number of items is large.

Vegetable	Years
Beans	3
Cabbage	4
Carrots	1
Cauliflower	4
Corn	2
Cucumbers	5
Melons	4
Peppers	2
Pumpkin	4
Tomatoes	3

▼ **EXAMPLE 10** Seeds that are dried, placed in an airtight container, and stored in a cool, dry place remain ready to be planted for a long time. The table in the margin gives the amount of time that each type of seed can be stored and still remain viable for planting.* Find the mean, median, and mode of the information in the table.

Solution The mean amount of time that the seeds can be stored is

$$\bar{x} = \frac{3 + 4 + 1 + 4 + 2 + 5 + 4 + 2 + 4 + 3}{10} = 3.2 \text{ years.}$$

After the numbers are arranged in order from smallest to largest, the middle number, or median, is found; it is 3.5.

The number 4 occurs more often than any other, so it is the mode. ▲

*The Handy Science Answer Book, Second Edition, The Carnegie Library of Pittsburgh, p. 247.

◆10.1 Exercises

The data for Exercises 1–4 consists of a random sample of 50 births taken from the 2002 North Carolina Birth Registry.* For each variable, (a) group the data as indicated; (b) prepare a frequency distribution with columns for intervals and frequencies; (c) construct a histogram; (d) construct a frequency polygon. (See Examples 1 and 2).

1. The variable is the age, in years, of the mother giving birth. Use 9 intervals, starting with 15–17 (inclusive).

32	26	22	17	16	16	29	24	30	23
21	21	32	37	23	26	29	26	31	37
31	16	28	37	18	18	34	31	20	19
35	23	22	33	27	20	41	22	24	18
20	24	30	24	29	25	29	24	25	28

*Data compiled by author.

2. The variable is weeks of gestation of the infant. Use 11 intervals, starting with 25–26 (inclusive).

40	34	42	42	41	41	43	40	39	40
41	38	38	37	43	40	45	39	44	25
38	38	36	40	36	41	37	40	38	25
38	40	40	38	40	38	40	41	36	36
39	38	39	41	39	34	40	38	36	35

3. The variable is weight (in pounds) gained by the mother during pregnancy. Use 8 intervals, starting with 0–9 (inclusive).

38	7	10	27	35	38	50	25	45	20
15	25	20	29	71	15	18	26	25	25
25	28	23	45	35	51	35	45	32	11
48	40	15	33	14	18	29	39	41	10
43	0	33	30	31	16	19	19	26	55

4. The variable is weight (in grams) of the infant at birth. Use 8 intervals, starting with 500–999 (inclusive).

3147 3572 1559 3544 2183 3119 3799 3147 3232 709

2977 3487 3799 3459 3204 3912 4054 3289 3175 624

3119 3147 2750 2807 3005 3572 3430 3629 3657 2778

3487 3345 3487 3856 3090 3884 3459 3657 3600 3572

4111 2665 3317 3260 3430 3090 3827 3175 3572 2977

Construct a stem-and-leaf plot for the data in the indicated exercise. (See Example 3.)

5. Exercise 1 (Use Stems 1, 1, 2, 2, 3, 3, 4)

6. Exercise 3

7. How does a frequency polygon differ from a histogram?

8. Discuss the advantages and disadvantages of the mean as a measure of central tendency.

Find the mean for each data set. Round to the nearest tenth. (See Example 5.)

9. Secretarial salaries (U.S. dollars):

$21,900, $22,850, $24,930, $29,710, $28,340, $40,000

10. Starting teaching salaries (U.S. dollars):

$38,400, $39,720, $28,458, $29,679, $33,679

11. Earthquakes on the Richter scale:

3.5, 4.2, 5.8, 6.3, 7.1, 2.8, 3.7, 4.2, 4.2, 5.7

12. Body temperature of self classified "healthy" students (degrees Fahrenheit):

96.8, 94.1, 99.2, 97.4, 98.4, 99.9, 98.7, 98.6

13. Length of foot (inches) for adult men:

9.2, 10.4, 13.5, 8.7, 9.7

Find the mean for each distribution. Round to the nearest tenth. (See Examples 5, 6, and 7.)

14. Scores on a quiz with a scale from 0 to 10.

Value	Frequency
7	4
8	6
9	7
10	11

15. Age (years) of student in an introductory accounting class.

Value	Frequency
19	3
20	5
21	25
22	8
23	2
24	1
28	1

16. Commuting distance (miles) to the university.

Value	Frequency
0	15
1	12
2	8
5	6
10	5
17	2
20	1
25	1

17. Estimated miles per gallon of automobiles.

Value	Frequency
9	5
11	10
15	12
17	9
20	6
28	1

18–22. *Find the median of the data in Exercises 9–13.*

Find the mode or modes for each of the following lists of numbers. (See Example 9.)

23. Ages (years) of children in a day-care facility:

1, 2, 2, 1, 2, 2, 1, 1, 2, 2, 3, 4, 2, 3, 4, 2, 3, 2, 3

24. Ages (years) in the intensive care unit at a local hospital:

68, 64, 23, 68, 70, 72, 72, 68

25. Heights (inches) of students in a statistics class:

62, 65, 71, 74, 71, 76, 71, 63, 59, 65, 65, 64, 72, 71, 77, 63, 65

26. Minutes of pain relief from acetaminophen after childbirth:

60, 240, 270, 180, 240, 210, 240, 300, 330, 360, 240, 120

27. Artistic impression scores for Gold Medalist Sarah Hughes in the 2002 Winter Olympics:*

5.7, 5.7, 5.8, 5.6, 5.8, 5.8, 5.8, 5.8, 5.8

28. Length of foot (inches) for adult men:

9.2, 10.4, 13.5, 8.7, 9.7

29. When is the median the most appropriate measure of central tendency?

30. Under what circumstances would the mode be an appropriate measure of central tendency?

For grouped data, the modal class is the interval containing the most data values. Give the mean and modal class for each of the following collections of grouped data. (See Example 7.)

31. The distribution in Exercise 3.

32. The distribution in Exercise 4.

33. To predict the outcome of the next congressional election, you take a survey of your friends. Is this a random sample of the voters in your congressional district? Explain why or why not.

Work each problem. (See Example 10.)

34. **Social Science**   The table shows the number of nations participating in the winter Olympic games from 1972 to 2002.[†]

Year	Nations Participating
1972	35
1976	37
1980	37
1984	49
1988	57
1992	64
1994	67
1998	72
2002	77

Find the following statistics for the data.
(a) Mean
(b) Median
(c) Mode

35. **Business**   The total compensation (pay plus bonus, in thousands of dollars) for the ten highest-paid CEOs in 2004 is given in the following table:[‡]

Person, Company	Total Compensation
Reuben Mark, Colgate–Palmolive	147,970
George David, United Technologies	70,527
Richard S. Fuld, Jr., Lehman Bros. Holding	67,682
Henry R. Silverman, Cendant	60,023
Dwight C. Schar, NVR	58,105
Lawrence J. Ellison, Oracle	40,589
Richard M. Kovacevich, Wells Fargo	37,842
Howard Solomon, Forest Labs	36,089
James E. Cayne, Bear Stearns Cos.	33,925
Todd S. Nelson, Apollo-Education Group	32,812

(a) Find the mean compensation for this group of people.
(b) Find the median compensation for the group of people.
(c) What might account for the difference between these values?
(d) Write the salary for Silverman.

36. **Business**   The top ten box-office receipts (in millions of dollars) for the weekend of August 19–21, 2005, at the movies are given here.*

Movie	Weekend Receipts (millions)
The 40-Year-Old Virgin	21.4
Red Eye	16.2
Four Brothers	12.5
Wedding Crashers	8.0
The Skeleton Key	7.7
Marche de l'empereur, la	6.5
The Dukes of Hazard	6.0
Valiant	5.9
Charlie and the Chocolate Factory	4.4
Sky High	4.0

(a) Find the mean weekend receipts in dollars for this group.
(b) Find the median weekend receipts in dollars for the group.

37. **Natural Science**   The number of recognized blood types varies by species, as indicated in the following table.[†] Find the mean, median, and mode of this data.

---

*http://www.icecalc.com/events/owg2002/results/SEG004.HTM

†http://en.wikipedia.org/wiki/Winter_olympics#2002_Winter_Olympic

‡www.forbes.com

*www.imdb.com

†*The Handy Science Answer Book,* Carnegie Library of Pittsburgh, Pennsylvania, p. 264.

Animal	Number of Blood Types
Pig	16
Cow	12
Chicken	11
Horse	9
Human	8
Sheep	7
Dog	7
Rhesus Monkey	6
Mink	5
Rabbit	5
Mouse	4
Rat	4
Cat	2

Month	Maximum	Minimum
January	39	16
February	39	18
March	44	21
April	50	26
May	60	32
June	69	37
July	79	43
August	78	42
September	70	37
October	51	31
November	47	24
December	40	20

38. **Social Science**   The number of weddings in each month of 1996 is given in the following table.*

Month	Number (in thousands)
January	110
February	155
March	118
April	172
May	241
June	242
July	235
August	239
September	225
October	231
November	171
December	184

(a) Calculate the mean and median for this data.
(b) Which month is closest to the mean?

Physical Science   *The table gives the average monthly temperatures in degrees Fahrenheit for a certain area.*

Find the mean and median for each of the following.

39.   The maximum temperature

40.   The minimum temperature

Business   *U.S. wheat prices and production figures for a recent decade are as follows.**

Year	Price ($ per bushel)	Production (millions of bushels)
1994	3.45	2321
1995	4.55	2183
1996	4.30	2285
1997	3.38	2482
1998	2.65	2550
1999	2.48	2290
2000	2.62	2232
2001	2.78	1957
2002	3.56	1619
2003	3.40	2337

Find the mean and median for each of the following.

41.   Price per bushel of wheat

42.   Wheat production

---

*The Amazing Almanac, Blackbirch Press, Inc., p. 245.

*The World Almanac and Book of Facts 2005, pp. 138–141.

43. Social Science The total household income for full-time African-American workers making under $100,000 in 2001 is given in the table.* (See Example 7.)

Income Range	Midpoint Salary	Frequency (in thousands)
Under $5,000	$2500	905
$5,000–$9,999	$7500	1451
$10,000–$14,999	$12,500	1158
$15,000–$24,999	$20,000	2197
$25,000–$34,999	$30,000	1904
$35,000–$49,999	$42,500	1984
$50,000–$74,999	$62,500	2051
$75,000–$99,999	$87,500	865

Use this table to estimate the mean household income for full-time African-American workers in 2001.

44. Social Science The total household income for full-time Caucasian workers making under $100,000 in 2001 is given in the table.* (See Example 7.)

Income Range	Midpoint Salary	Frequency (in thousands)
Under $5,000	$2500	2176
$5,000–$9,999	$7500	4715
$10,000–$14,999	$12,500	6076
$15,000–$24,999	$20,000	11,789
$25,000–$34,999	$30,000	11,063
$35,000–$49,999	$42,500	14,056
$50,000–$74,999	$62,500	17,048
$75,000–$99,999	$87,500	13,421

(a) Use this table to estimate the mean household income for full-time Caucasian-American workers in 2001.

(b) Compare your estimate from part (a) with the estimate found in Exercise 43. Does this comparison provide some evidence that full-time Caucasian-American workers have higher household earnings than full-time African-American workers?

45. Social Science The histogram below shows estimates of the percent of the U.S. population in each age group in the year 2000.* What percent of the population is estimated to be in each of the following age groups?

(a) 10–19   (b) 60–69

(c) What age range has the largest percent of the population?

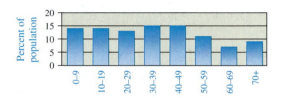

46. Social Science The histogram below shows estimates of the percent of the U.S. population in each age group in the year 2025.* What percent of the population is estimated to be in each of the following age groups then?

(a) 20–29   (b) 70+

(c) What age group will have the smallest percent of the population?

(d) Compare the histogram in Exercise 45 with the histogram presented here. What seems to be true of the U.S. population?

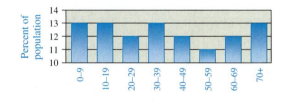

*www.infoplease.com/ipa/A0104552.html

*U.S. Census Bureau, January 13, 2000.

10.2 Measures of Variation

The mean, median, and mode are measures of central tendency for a list of numbers, but tell nothing about the *spread* of the numbers in the list. For example, look at the following data sets of number of times per week three people ate meals at restaurants.

Jill:	3	5	6	3	3
Miguel:	4	4	4	4	4
Sharille:	10	1	0	0	9

Each of these three data sets has a mean of 4, but the amount of dispersion or variation within the lists is different. This difference may reflect different dining patterns over time. Thus, in addition to a measure of central tendency, another kind of measure is needed that describes how much the numbers vary.

The largest number of restaurant meals for Jill is 6, while the smallest is 3, a difference of 3. For Miguel, the difference is 0; for Sharille, it is 10. The difference between the largest and smallest number in a sample is called the **range,** one example of a measure of variation. The range is 3 for Jill, 0 for Miguel, and 10 for Sharille. The range has the advantage of being very easy to compute and gives a rough estimate of the variation among the data in the sample. However, it depends only on the two extremes and tells nothing about how the other data are distributed between the extremes.

TECHNOLOGY TIP   Many graphing calculators show the largest and smallest numbers in a list when displaying one-variable statistics, usually on the second screen of the display.

▼ **EXAMPLE 1**   Find the range for each data set for the following small samples of people.

(a) Years of education: 12, 12, 16, 8, 10, 12, 15, 13, 12, 16, 16, 19

Solution The highest number here is 19; the lowest is 8. The range is the difference of these numbers, or

$$19 - 8 = 11.$$

(b) Miles driven to a vacation location: 74, 112, 59, 88, 200, 73, 92, 175

Solution Range $= 200 - 59 = 141$ ▲ ⟨1⟩

To find a useful measure of variation, we begin by finding the **deviations from the mean**—the differences found by subtracting the mean from each number in a distribution.

⟨1⟩

Find the range for the this sample of miles traveled to a vacation location: 159, 283, 490, 75, 90, 120.

Answer:

415

▼ **EXAMPLE 2**   Find the deviations from the mean for the following sample of ages.

$$32, \quad 41, \quad 47, \quad 53, \quad 57$$

Solution Adding these numbers and dividing by 5 gives a mean of 46 years. To find the deviations from the mean, subtract 46 from each number in the sample. For example, the first deviation from the mean is $32 - 46 = -14$; the last is $57 - 46 = 11$ years.

Age	Deviation from Mean
32	−14
41	−5
47	1
53	7
57	11

⟨2⟩

Find the deviations from the mean for the sample of miles traveled to a vacation location:

135, 60, 50, 425, 380

Answer:

Mean is 210; deviations are −75, −150, −160, 215, 170.

To check your work, find the sum of the deviations. It should always equal 0. (The answer is always 0 because the positive and negative deviations cancel each other.) ▲ ⟨2⟩

To find a measure of variation, we might be tempted to use the mean of the deviations. However, as just mentioned, this number is always 0, no matter how widely the data are dispersed. To avoid the problem of the positive and negative deviations averaging 0, we could take absolute values and find $\Sigma|x - \bar{x}|$, then divide it by n to get the *mean deviation*. However, statisticians generally prefer to square each deviation to get nonnegative numbers, then take the square root of the mean of the squared variations to preserve the units of the original data (such as inches, pounds). (Using squares instead of absolute values allows us to take advantage of some algebraic properties that make other important statistical methods much easier.) The squared deviations for the data in Example 2 are shown in the table.

Number	Deviation from Mean	Square of Deviation
32	−14	196
41	−5	25
47	1	1
53	7	49
57	11	121

In this case, the mean of the squared deviations is

$$\frac{196 + 25 + 1 + 49 + 121}{5} = \frac{392}{5} = 78.4.$$

This number is called the **population variance,** because the sum was divided by $n = 5$, the number of items in the original list.

Since the deviations from the mean must add up to 0, if we know any 4 of the 5 deviations, the 5th will be determined. That is, only $n - 1$ of the deviations are free to vary, so we really have only $n - 1$ independent pieces of information, or *degrees of freedom.* Using $n - 1$ as the divisor in the formula for the mean gives

$$\frac{196 + 25 + 1 + 49 + 121}{5 - 1} = \frac{392}{4} = 98.$$

This number, 98, is called the **sample variance** of the distribution and is denoted s^2, because it is found by averaging a list of squares. In this case, the population and sample variances differ by quite a bit. But when n is relatively large, as is the case in real-life applications, the difference between them is rather small.

Sample Variance

The variance of a sample of n numbers $x_1, x_2, x_3, \ldots, x_n$, with mean $\bar{x}$, is

$$s^2 = \frac{\Sigma(x - \bar{x})^2}{n - 1}.$$

When computing the sample variance by hand, it is often convenient to use the following shortcut formula, which can be derived algebraically from the definition in the preceding box.

$$s^2 = \frac{\Sigma x^2 - n\bar{x}^2}{n - 1}$$

To find the sample variance, we square the deviations from the mean, so the variance is in squared units. To return to the same units as the data, we use the *square root* of the variance, called the **sample standard deviation,** denoted s.

Sample Standard Deviation

The standard deviation of a sample of n numbers $x_1, x_2, x_3, \ldots, x_n$, with mean $\bar{x}$, is

$$s = \sqrt{\frac{\Sigma(x - \bar{x})^2}{n - 1}}.$$

NOTE The **population standard deviation** is

$$\sigma = \sqrt{\frac{\Sigma(x - \bar{x})^2}{n}},$$

where n is the population size.

TECHNOLOGY TIP   When a graphing calculator computes one-variable statistics for a list of data, it usually displays the following information (not necessarily in this order and sometimes on two screens) and, possibly, other information as well.

Information	Notation
Number of data entries	n or $N\Sigma$
Mean	$\bar{x}$ or mean Σ
Sum of all data entries	Σx or TOT Σ
Sum of the squares of all data entries	Σx^2
Sample standard deviation	Sx or sx or $x\sigma_{n-1}$ or SSDEV
Population standard deviation	σx or $x\sigma_n$ or PSDEV
Largest/smallest data entries	maxX/minX or MAXΣ/MINΣ
Median	Med or MEDIAN

NOTE In the rest of this section, we shall deal exclusively with the sample variance and the sample standard deviation. So whenever standard deviation is mentioned, it means "sample standard deviation," not population standard deviation.

As its name indicates, the standard deviation is the most commonly used measure of variation. The standard deviation is a measure of the variation from the mean. The size of the standard deviation indicates how spread out the data are from the mean.

▼ **EXAMPLE 3**   Find the standard deviation for this sample of the lengths (in minutes) of eight consecutive cell phone conversations by one person.

$$2, \quad 8, \quad 3, \quad 2, \quad 6, \quad 11, \quad 31, \quad 9$$

Work by hand, using the shortcut variance formula at the top of page 589.

Solution   Arrange the work in columns, as shown in the table in the margin. Now use the first column to find the mean.

Time	Square of the Time
2	4
8	64
3	9
2	4
6	36
11	121
31	961
9	81
72	1280

$$\bar{x} = \frac{\sum x}{8} = \frac{72}{8} = 9 \text{ minutes}$$

The total of the second column gives $\sum x^2 = 1280$. The variance is

$$s^2 = \frac{\sum x^2 - n\bar{x}^2}{n - 1}$$

$$= \frac{1280 - 8(9)^2}{8 - 1}$$

$$= 90.3 \text{ (rounded)},$$

and the standard deviation is

$$s \approx \sqrt{90.3} \approx 9.5 \text{ minutes.} \quad \blacktriangle \quad \langle 3 \rangle$$

3

Find the standard deviation for the sample of miles traveled to a vacation location:

135, 60, 50, 425, 380

Answer:

179.5 miles

TECHNOLOGY TIP   The screens in Figure 10.3 show two ways to find variance and standard deviation on a TI-84+ calculator: with the LIST menu and with the STAT menu. The data points are first entered in a list, here L_5. See your instruction book for details.

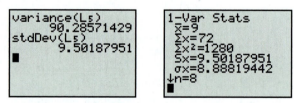

FIGURE 10.3

With a spreadsheet, the data are entered in cells A1 through A8. Then, in cell A9, type "=VAR (A1..A8)" and press Enter. The standard deviation can be calculated either by taking the square root of cell A9 or by typing "=STDEV (A1..A8)" in cell A10 and pressing Enter.

CAUTION   We must be careful to divide by $n - 1$, not n, when calculating the standard deviation of a sample. Many calculators are equipped with statistical keys that compute the variance and standard deviation. Some of these calculators use $n - 1$

and others use n for these computations; some may have keys for both. Check your calculator's instruction book before using a statistical calculator for the exercises.

One way to interpret the standard deviation uses the fact that, for many populations, most of the data is within three standard deviations of the mean. (See Section 10.3.) This implies that, in Example 3, most of the population data from which this sample is taken are between

$$\overline{x} - 3s = 9 - 3(9.5) = -19.5$$

and

$$\overline{x} + 3s = 9 + 3(9.5) = 37.5.$$

For Example 3, the preceding calculations imply that most phone conversations are less than 37.5 minutes long. This approach of determining whether sample observations are beyond 3 standard deviations of the mean is often employed in conducting quality control in many industries.

For data in a grouped frequency distribution, a slightly different formula for the standard deviation is used.

Standard Deviation for a Grouped Distribution

The standard deviation for a sample distribution with mean $\overline{x}$, where x is an interval midpoint with frequency f and $n = \Sigma f$, is

$$s = \sqrt{\frac{\Sigma f x^2 - n\overline{x}^2}{n - 1}}.$$

The formula indicates that the product $f x^2$ is to be found for each interval. Then all the products are summed, n times the square of the mean is subtracted, and the difference is divided by one less than the total frequency—that is, by $n - 1$. The square root of this result is s, the standard deviation. The standard deviation found by this formula may (probably will) differ somewhat from the standard deviation found from the original data.

CAUTION In calculating the standard deviation for either a grouped or an ungrouped distribution, using a rounded value for the mean or variance may produce an inaccurate value.

▼ **EXAMPLE 4**   Find s for the grouped data of Example 7 of Section 10.1.

Solution Begin by including columns for x^2 (where x is the midpoint of the interval) and $f x^2$.

Classes	x	x^2	f	fx^2
0–4	2	4	3	12
5–9	7	49	4	196
10–14	12	144	6	864
15–19	17	289	8	2312
20–24	22	484	5	2420
25–29	27	729	3	2187
30–34	32	1024	1	1024
			Total: 30	Total: 9015

4

Find the standard deviation for the following grouped frequency distribution of the number of classes completed thus far in the college careers of a random sample of 52 students.

Classes	Frequency
0–5	6
6–10	10
11–20	12
21–30	15
31–40	9

Answer:

10.92 classes

Recall from Example 7 of Section 10.1 that $\bar{x} = 15.5$. Use the formula for the standard deviation with $n = 30$ to find s.

$$s = \sqrt{\frac{\Sigma fx^2 - n\bar{x}^2}{n-1}}$$
$$= \sqrt{\frac{9015 - 30(15.5)^2}{30-1}}$$
$$\approx 7.89 \; \blacktriangle \; \text{(4)}$$

NOTE A calculator is almost a necessity for finding a standard deviation. With a nongraphing calculator, a good procedure to follow is to first calculate $\bar{x}$. Then, for each x, square that number, and multiply the result by the appropriate frequency. If your calculator has a key that accumulates a sum, use it to accumulate the total in the last column of the table. With a graphing calculator, simply enter the midpoints and the frequencies, and then ask for the 1-variable statistics.

◆10.2 Exercises

1. How are the variance and the standard deviation related?

2. Why can't we use the sum of the deviations from the mean as a measure of dispersion of a distribution?

Find the range and standard deviation for each of the following samples. (See Examples 1 and 3.)

3. Age (years) of women giving birth: 32, 21, 31, 35, 20

4. Age (years) of women giving birth: 26, 21, 16, 23, 24

5. Weeks of gestation for a mother giving birth: 40, 25, 25, 36, 35

6. Weeks of gestation for a mother giving birth: 39, 44, 38, 36, 36

7. Weight (pounds) gained by the mother during pregnancy: 38, 15, 25, 48, 43, 7, 25, 28

8. Weight (pounds) gained by the mother during pregnancy: 40, 0, 10, 20, 23, 15, 33

9. Weight (grams) of the infant at birth: 3147, 3572, 1559, 3544, 2183, 3119, 3799, 3147, 3232

10. Weight (grams) of the infant at birth: 2977, 3487, 3799, 3459, 3204, 3912, 4054, 3289

Find the standard deviation for the grouped data in Exercises 11 and 12. (See Example 4.)

11. Number of credits for a sample of college students

College Credits	Frequency
0–24	4
25–49	3
50–74	6
75–99	3
100–124	5
125–149	9

12. Scores on a calculus exam

Scores	Frequency
30–39	1
40–49	6
50–59	13
60–69	22
70–79	17
80–89	13
90–99	8

13. Natural Science Twenty-five laboratory rats used in an experiment to test the food value of a new product made the following weight gains in grams.

5.25	5.03	4.90	4.97	5.03
5.12	5.08	5.15	5.20	4.95
4.90	5.00	5.13	5.18	5.18
5.22	5.04	5.09	5.10	5.11
5.23	5.22	5.19	4.99	4.93

Find the mean gain and the standard deviation of the gains.

14. Business An assembly-line machine turns out washers with the following thicknesses (in millimeters).

1.20	1.01	1.25	2.20	2.58	2.19	1.29	1.15
2.05	1.46	1.90	2.03	2.13	1.86	1.65	2.27
1.64	2.19	2.25	2.08	1.96	1.83	1.17	2.24

Find the mean and standard deviation of these thicknesses.

An application of standard deviation is given by Chebyshev's theorem. (P. L. Chebyshev was a Russian mathematician who lived from 1821 to 1894.) This theorem, which applies to any distribution of numerical data, states,

For any distribution of numerical data, at least $1 - 1/k^2$ of the numbers lie within k standard deviations of the mean.

Example *For any distribution, at least*

$$1 - \frac{1}{3^2} = 1 - \frac{1}{9} = \frac{8}{9}$$

of the numbers lie within 3 standard deviations of the mean. Find the fraction of all the numbers of a data set lying within the following numbers of standard deviations from the mean.

15. 2   **16.** 4   **17.** 1.5

In a certain distribution of numbers, the mean is 50 with a standard deviation of 6. Use Chebyshev's theorem to tell what percent of the numbers are

18. between 32 and 68;

19. between 26 and 74;

20. less than 38 or more than 62;

21. less than 32 or more than 68;

22. less than 26 or more than 74.

23. Health A sample of 8 male and 8 female first-year college students had their heights measured in centimeters. The data appear in the following table.

Males	182	178	179	182	173	170	167	171
Females	174	162	157	172	164	162	168	163

(a) Compute the mean and standard deviation for each sample. Compare the means and standard deviations for the two groups of students, and then answer the following questions.
(b) Which gender has less variability in height?
(c) Which gender has the higher average height? Is this a surprise?

24. Business The weekly wages of the six employees of Harold's Hardware Store are $300, $320, $380, $420, $500, and $2000.
(a) Find the mean and standard deviation of this distribution.
(b) How many of the employees earn within one standard deviation of the mean? How many earn within two standard deviations of the mean?

25. Social Science The number of unemployed workers in the United States in recent years (in millions) is as follows.*

Year	Number Unemployed
1997	6.7
1998	6.2
1999	5.9
2000	5.7
2001	6.8
2002	8.4
2003	8.8

Use sample statistics to find
(a) the mean number unemployed (in millions) during the given period. Which year has unemployment closest to the mean?
(b) the standard deviation for the data.

The World Almanac and Book of Facts 2000, p. 145.

(c) how many of the years unemployment was within 1 standard deviation of the mean.

(d) how many of the years unemployment was within 3 standard deviations of the mean.

26. **Life Science** A medical laboratory tested 21 samples of human blood for acidity on the pH scale, with the following results.

7.1	7.5	7.3	7.4	7.6	7.2	7.3
7.4	7.5	7.3	7.2	7.4	7.3	7.5
7.5	7.4	7.4	7.1	7.3	7.4	7.4

(a) Find the mean and standard deviation.

(b) What percent of the data is within 2 standard deviations of the mean?

27. **Life Science** The number of recognized blood types between species is given in the following table.* In Exercise 37 of the previous section, the mean was found to be 7.38.

Animal	Number of Blood Types
Pig	16
Cow	12
Chicken	11
Horse	9
Human	8
Sheep	7
Dog	7
Rhesus Monkey	6
Mink	5
Rabbit	5
Mouse	4
Rat	4
Cat	2

(a) Find the variance and the standard deviation of these data.

(b) How many of these animals have blood types that are within 1 standard deviation of the mean?

28. **Social Science** In 2003, 13 state governors earned at least $125,000 annually (excluding their expense allowances). The salaries of these officials are given in the following table in thousands of dollars and are rounded to the nearest $1000.[†]

State	Salary
California	175
Georgia	127
Illinois	152
Maryland	135
Massachusetts	135
Michigan	172
New Jersey	157
New York	179
Ohio	131
Pennsylvania	138
Vermont	126
Washington	139
Wyoming	150

(a) Find the mean salary of these governors. Which state has the governor with the salary closest to the mean?

(b) Find the standard deviation for the data.

(c) What percent of these governors have salaries within 1 standard deviation of the mean?

(d) What percent of these governors have salaries within 3 standard deviations of the mean?

29. **Health** The amounts of time that it takes for various slow-growing tumors to double in size are listed in the following table.*

Type of Cancer	Doubling Time (days)
Breast cancer	84
Rectal cancer	91
Synovioma	128
Skin cancer	131
Lip cancer	143
Testicular cancer	153
Esophageal cancer	164

(a) Find the mean and standard deviation of these data.

(b) How many of these cancers have doubling times that are within 2 standard deviations of the mean?

(c) If a person had a nonspecified tumor that was doubling every 200 days, discuss whether this particular tumor was growing at a rate that would be expected.

*The Handy Science Answer Book, Carnegie Library of Pittsburgh, Pennsylvania, p. 264.

†www.infoplease.com/ipa/A0108309.html

*Vincent Collins, R. Kenneth Lodffer, and Harold Tivey, "Observations on Growth Rates of Human Tumors," American Journal of Roentgen, 76, no. 5 (November 1956): 988–1000.

30. Business   The Quaker Oats Company conducted a survey to determine whether a proposed premium, to be included in the firm's cereal, was appealing enough to generate new sales.* Four cities were used as test markets, where the cereal was distributed with the premium, and four cities as control markets, where the cereal was distributed without the premium. The eight cities were chosen on the basis of their similarity in terms of population, per capita income, and total cereal purchase volume. The results were as follows.

		Percent Change in Average Market Shares per Month
Test Cities	1	+18
	2	+15
	3	+7
	4	+10
Control Cities	1	+1
	2	−8
	3	−5
	4	0

(a) Find the mean of the change in market share for the four test cities.

(b) Find the mean of the change in market share for the four control cities.

(c) Find the standard deviation of the change in market share for the test cities.

(d) Find the standard deviation of the change in market share for the control cities.

(e) Find the difference between the mean of part (a) and the mean of part (b). This represents the estimate of the percent change in sales due to the premium.

(f) The two standard deviations from part (c) and part (d) were used to calculate an "error" of ± 7.95 for the estimate in part (e). With this amount of error, what is the smallest and largest estimate of the increase in sales? (*Hint:* Use the answer to part (e).)

On the basis of the results of Exercise 30, the company decided to mass-produce the premium and distribute it nationally.

31. Business   The following table gives 10 samples of three measurements, made during a production run.

SAMPLE NUMBER									
1	2	3	4	5	6	7	8	9	10
2	3	−2	−3	−1	3	0	−1	2	0
−2	−1	0	1	2	2	1	2	3	0
1	4	1	2	4	2	2	3	2	2

(a) Find the mean $\bar{x}$ for each sample of three measurements.

(b) Find the standard deviation s for each sample of three measurements.

(c) Find the mean $\bar{\bar{x}}$ of the sample means.

(d) Find the mean $\bar{s}$ of the sample standard deviations.

(e) The upper and lower control limits of the sample means here are $\bar{\bar{x}} \pm 1.954\bar{s}$. Find these limits. If any of the measurements are outside these limits, the process is out of control. Decide whether this production process is out of control.

✏ **32.** Discuss what the standard deviation tells us about a distribution.

Social Science   *The reading scores of a second-grade class given individualized instruction are shown. Also shown are the reading scores of a second-grade class given traditional instruction in the same school.*

Scores	Individualized Instruction	Traditional Instruction
50–59	2	5
60–69	4	8
70–79	7	8
80–89	9	7
90–99	8	6

33. Find the mean and standard deviation for the individualized instruction scores.

34. Find the mean and standard deviation for the traditional instruction scores.

✏ **35.** Discuss a possible interpretation of the differences in the means and the standard deviations in Exercises 33 and 34.

36. Business   Refer to Section 10.1. Exercise 35, which gives total pay for CEOs.

(a) Calculate the standard deviation of the data.

(b) What percent of the CEOs listed have compensation more than 2 standard deviations from the mean?

✏ (c) What does your answer to part (b) suggest?

*This example was supplied by Jeffery S. Berman Senior Analyst, Marketing Information, Quaker Oats Company.

10.3 Normal Distributions

Suppose a bank is interested in improving its services to customers. The manager decides to begin by finding the amount of time tellers spend on each transaction, rounded to the nearest minute. The times for 75 different transactions are recorded, with the results shown in the following table. The frequencies listed in the second column are divided by 75 to find the empirical probabilities.

Time	Frequency	Probability
1	3	$3/75 = .04$
2	5	$5/75 \approx .07$
3	9	$9/75 = .12$
4	12	$12/75 = .16$
5	15	$15/75 = .20$
6	11	$11/75 \approx .15$
7	10	$10/75 \approx .13$
8	6	$6/75 = .08$
9	3	$3/75 = .04$
10	1	$1/75 \approx .01$

Figure 10.4(a) shows a histogram and frequency polygon for the data. The heights of the bars are the empirical probabilities, rather than the frequencies. The transaction times are given to the nearest minute. Theoretically at least, they could have been timed to the nearest tenth of a minute, or hundredth of a minute, or even more precisely. In each case, a histogram and frequency polygon could be drawn. If the times are measured with smaller and smaller units, there are more bars in the histogram and the frequency polygon begins to look more and more like the curve in Figure 10.4(b) instead of a polygon. Actually, it is possible for the transaction times to take on any real number value greater than 0. A distribution in which the outcomes can take any real number value within some interval is a **continuous distribution.** The graph of a continuous distribution is a curve.

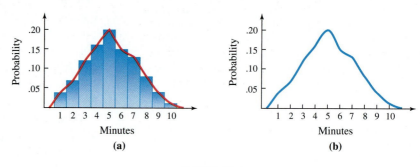

FIGURE 10.4

The distribution of heights (in inches) of college women is another example of a continuous distribution, since these heights include infinitely many possible measurements, such as 53, 58.5, 66.3, 72.666, and so on. Figure 10.5 shows the continuous distribution of heights of college women. Here, the most frequent heights occur near the center of the interval displayed.

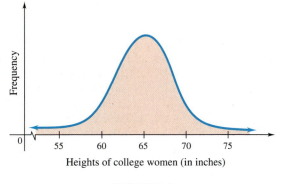

FIGURE 10.5

Another continuous curve, which approximates the distribution of yearly incomes in the United States, is given in Figure 10.6. The graph shows that the most frequent incomes are grouped near the low end of the interval. This kind of distribution, where the peak is not at the center, is called **skewed.**

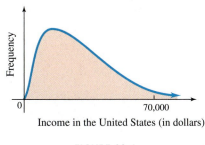

FIGURE 10.6

Normal Distributions

Many natural and social phenomena produce continuous probability distributions whose graphs are approximated very well by bell-shaped curves, such as those shown in Figure 10.7. Such distributions are called **normal distributions** and their graphs are called **normal curves.** Examples of distributions that are approximately normal are the heights of college women and the errors made in filling 1-pound cereal boxes. We use the Greek letters μ (mu) to denote the mean and σ (sigma) to denote the standard deviation of a normal distribution.

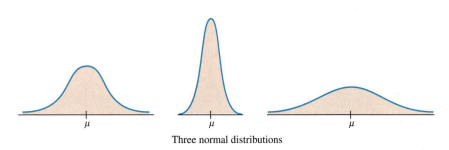

Three normal distributions

FIGURE 10.7

There are many normal distributions. Some of the corresponding normal curves are tall and thin and others short and wide, as shown in Figure 10.7. But every normal curve has the following properties:

1. Its peak occurs directly above the mean μ
2. The curve is symmetric about the vertical line through the mean (that is, if you fold the graph along this line, the left half of the graph will fit exactly on the right half).
3. The curve never touches the x-axis—it extends indefinitely in both directions.
4. The area under the curve (and above the horizontal axis) is 1. (As can be shown with calculus, this is a consequence of the fact that the sum of the probabilities in any distribution is 1.)

A normal distribution is completely determined by its mean μ and standard deviation σ.* A small standard deviation leads to a tall, narrow curve like the one in the center of Figure 10.7, because most of the data are close to the mean. A large standard deviation means the data are very spread out, producing a flat, wide curve like the one on the right in Figure 10.7.

Since the area under a normal curve is 1, parts of this area can be used to determine certain probabilities. For instance, Figure 10.8(a) is the probability distribution of the annual rainfall in a certain region. The probability that the annual rainfall will be between 25 and 35 inches is the area under the curve from 25 to 35. The general case, shown in Figure 10.8(b), can be stated as follows.

The area of the shaded region under the normal curve from a to b is the probability that an observed data value will be between a and b.

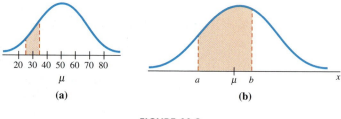

FIGURE 10.8

To use normal curves effectively, we must be able to calculate areas under portions of them. These calculations have already been done for the normal curve with mean $\mu = 0$ and standard deviation $\sigma = 1$ (which is called the **standard normal curve**) and are available in Table 2 at the back of the book. The examples below

---

*As shown in more advanced courses, its graph is the graph of the function

$$f(x) = \frac{1}{\sigma\sqrt{2\pi}}\, e^{-(x-\mu)^2/(2\sigma^2)},$$

where $e \approx 2.71828$ is the real number introduced in Section 4.1.

demonstrate how to use Table 2 to find such areas. Later we shall see how the standard normal curve may be used to find areas under any normal curve.

The horizontal axis of the standard normal curve is usually labeled z. Since the standard deviation of the standard normal curve is 1, the numbers along the horizontal axis (the z-values) measure the number of standard deviations above or below the mean $z = 0$.

TECHNOLOGY TIP   Some graphing calculators (such as the TI-84+ and Casio 9850) have the ability to graph a normal distribution, given its mean and standard deviation, and to find areas under the curve between two x-values. For an area under the curve, some calculators will give the corresponding z-value. For details, see your instruction book. (Look for "distribution" or "probability distribution.") A calculator-generated graph of the standard normal curve is shown in Figure 10.9.

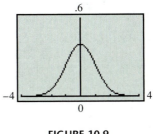

FIGURE 10.9

▼ **EXAMPLE 1**   Find the following areas under the standard normal curve.

(a) The area between $z = 0$ and $z = 1$, the shaded region in Figure 10.10

Solution Find the entry 1 in the z-column of Table 2. The entry next to it in the A-column is .3413, which means that the area between $z = 0$ and $z = 1$ is .3413. Since the total area under the curve is 1, the shaded area in Figure 10.10 is 34.13% of the total area under the normal curve.

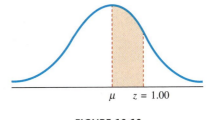

FIGURE 10.10

(b) The area between $z = -2.43$ and $z = 0$

Solution Table 2 lists only positive values of z. But the normal curve is symmetric around the mean $z = 0$, so the area between $z = 0$ and $z = -2.43$ is the same as the area between $z = 0$ and $z = 2.43$. Find 2.43 in the z-column of Table 2. The entry next to it in the A-column shows that the area is .4925. Hence, the shaded area in Figure 10.11 on the next page is 49.25% of the total area under the curve. ▲ ⟨1⟩ ⟨2⟩

⟨1⟩

Find the percent of the area between the mean and

(a) $z = 1.51$;

(b) $z = -2.04$.

(c) Find the percent of the area in the shaded region.

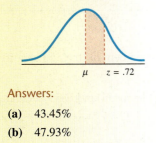

Answers:

(a) 43.45%

(b) 47.93%

(c) 26.42%

◇2

If your calculator can graph probability distributions and find areas, use it to find the areas requested in Example 1.

Answers:

(a) 34.13%

(b) 49.25%

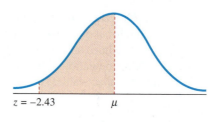

$z = -2.43$ μ

FIGURE 10.11

TECHNOLOGY TIP Because of convenience and accuracy, graphing calculators and computers have made normal curve tables less important. Figure 10.12 shows how part (b) of Example 1 can be done on a TI-84+ calculator using a command from the DISTR menu. The second result in the calculator screen gives the area between $-\infty$ and $z = -2.43$; the entry $-1E99$ represents $-1 \cdot 10^{99}$, which is used to approximate $-\infty$.

```
normalcdf(-2.43,
0,0,1)
        .4924505896
normalcdf(-1E99,
-2.43,0,1)
        .007549411
■
```

FIGURE 10.12

Many statistical software packages are widely used today. All of these packages are set up in a way that is similar to a spreadsheet, and they all can be used to generate normal curve values. In addition, most spreadsheets can perform a wide range of statistical calculations.

▼ **EXAMPLE 2** Use technology or Table 2 to find the percent of the total area for the following areas under the standard normal curve.

(a) The area between .88 standard deviations *below* the mean and 2.35 standard deviations *above* the mean (that is, between $z = -.88$ and $z = 2.35$)

Solution First, draw a sketch showing the desired area, as in Figure 10.13. From Table 2, the area between the mean and .88 standard deviations below the mean is .3106. Also, the area from the mean to 2.35 standard deviations above the mean is .4906. As the figure shows, the total desired area can be found by *adding* these numbers.

$$
\begin{array}{r}
.3106 \\
+ .4906 \\
\hline
.8012
\end{array}
$$

The shaded area in Figure 10.13 represents 80.12% of the total area under the normal curve.

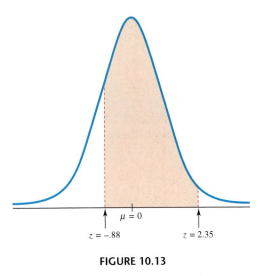

FIGURE 10.13

(b) The area between .58 standard deviation above the mean and 1.94 standard deviations above the mean

Solution Figure 10.14 shows the desired area. The area between the mean and .58 standard deviation above the mean is .2190. The area between the mean and 1.94 standard deviations above the mean is .4738. As the figure shows, the desired area is found by *subtracting* one area from the other.

$$
\begin{array}{r}
.4738 \\
-.2190 \\
\hline
.2548
\end{array}
$$

The shaded area of Figure 10.14 represents 25.48% of the total area under the normal curve.

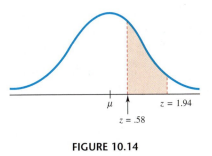

FIGURE 10.14

(c) The area to the right of 2.09 standard deviations above the mean

Solution The total area under a normal curve is 1. Thus, the total area to the right of the mean is 1/2, or .5000. From Table 2, the area from the mean to

Find the following standard normal curve areas as percentages of the total area.

(a) Between .31 standard deviations below the mean and 1.01 standard deviations above the mean

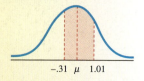

−.31 μ 1.01

(b) Between .38 standard deviations and 1.98 standard deviations below the mean

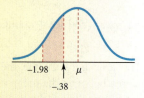

−1.98 μ
−.38

(c) To the right of 1.49 standard deviations above the mean

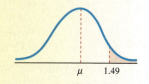

μ 1.49

(d) What percent of the area is within 1 standard deviation of the mean? within 2 standard deviations of the mean? within 3 standard deviations of the mean? What can you conclude from the last answer?

Answers:

(a) 46.55%

(b) 32.82%

(c) 6.81%

(d) 68.3%, 95.47%, 99.7%
    Almost all the data lies within 3 standard deviations of the mean.

2.09 standard deviations above the mean is .4817. The area to the right of 2.09 standard deviations is found by subtracting .4817 from .5000.

$$
\begin{array}{r}
.5000 \\
-.4817 \\
\hline
.0183
\end{array}
$$

A total of 1.83% of the total area is to the right of 2.09 standard deviations above the mean. Figure 10.15 (which is not to scale) shows the desired area. ▲ ③

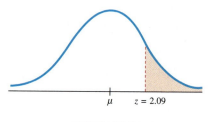

μ $z = 2.09$

FIGURE 10.15

The key to finding areas under *any* normal curve is to express each number x on the horizontal axis in terms of standard deviations above or below the mean. The **z-score** for x is the number of standard deviations that x lies from the mean (positive if x is above the mean, negative if x is below the mean).

▼ **EXAMPLE 3**   If a normal distribution has mean 60 and standard deviation 5, find the following z-scores.

(a) The z-score for $x = 65$

Solution Since 65 is 5 units above 60 and the standard deviation is 5, 65 is 1 standard deviation above the mean. So its z-score is 1.

(b) The z-score for $x = 52.5$

Solution The z-score is -1.5, because 52.5 is 7.5 units below the mean (since $52.5 - 60 = -7.5$) and 7.5 is 1.5 standard deviations (since $7.5/5 = 1.5$). ▲ ④

In Example 3(b) we found the z-score by taking the difference between 52.5 and the mean and dividing this difference by the standard deviation. The same procedure works in the general case.

If a normal distribution has mean μ and standard deviation σ, then the z-score for the number x is

$$
z = \frac{x - \mu}{\sigma}.
$$

The importance of z-scores is the following fact, whose proof is omitted.

4

Find each z-score using the information in Example 3.

(a) $x = 36$

(b) $x = 55$

Answers:

(a) -4.8

(b) -1

Area under a Normal Curve

The area under a normal curve between $x = a$ and $x = b$ is the same as the area under the standard normal curve between the z-score for a and the z-score for b.

Therefore, by converting to z-scores and using a graphing calculator or Table 2 for the standard normal curve, we can find areas under any normal curve. Since these areas are probabilities (as explained earlier), we can now handle a variety of applications.

Graphing calculators, computer programs, and CAS programs (such as DERIVE) can be used to find areas under the normal curve and, hence, probabilities. The equation of the standard normal curve, with $\mu = 0$ and $\sigma = 1$, is

$$f(x) = (1/\sqrt{2\pi})e^{-x^2/2}.$$

A good approximation of the area under this curve (and above $y = 0$) can be found by using the x-interval $[-4, 4]$. See Chapter 13 for more information on finding such areas.

▼ **EXAMPLE 4**   Dixie Office Supplies finds that its sales force drives an average of 1200 miles per month per person, with a standard deviation of 150 miles. Assume that the number of miles driven by a salesperson is closely approximated by a normal distribution.

(a) Find the probability that a salesperson drives between 1200 and 1600 miles per month.

Solution   Here, $\mu = 1200$ and $\sigma = 150$, and we must find the area under the normal distribution curve between $x = 1200$ and $x = 1600$. We begin by finding the z-score for $x = 1200$.

$$z = \frac{x - \mu}{\sigma} = \frac{1200 - 1200}{150} = \frac{0}{150} = 0$$

The z-score for $x = 1600$ is

$$z = \frac{x - \mu}{\sigma} = \frac{1600 - 1200}{150} = \frac{400}{150} = 2.67.^*$$

So the area under the curve from $x = 1200$ to $x = 1600$ is the same as the area under the standard normal curve from $z = 0$ to $z = 2.67$, as indicated in Figure 10.16 on the next page. A graphing calculator or Table 2 shows that this area is .4962. Therefore, the probability that a salesperson drives between 1200 and 1600 miles per month is .4962.

---

*All z-scores here are rounded to two decimal places.

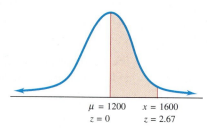

FIGURE 10.16

(b) Find the probability that a salesperson drives between 1000 and 1500 miles per month.

Solution As shown in Figure 10.17, z-scores for both $x = 1000$ and $x = 1500$ are needed.

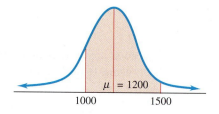

FIGURE 10.17

For $x = 1000$,

$$z = \frac{1000 - 1200}{150}$$

$$= \frac{-200}{150}$$

$$= -1.33.$$

For $x = 1500$,

$$z = \frac{1500 - 1200}{150}$$

$$= \frac{300}{150}$$

$$= 2.00.$$

From the table, $z = 1.33$ leads to an area of .4082, while $z = 2.00$ corresponds to .4773. A total of .4082 + .4773 = .8855, or 88.55%, of all drivers travel between 1000 and 1500 miles per month. From this calculation, the probability that a driver travels between 1000 and 1500 miles per month is .8855. ▲ ⑤

▼ **EXAMPLE 5** With data from the 2001–2002 National Health and Nutritional Examination Survey (NHANES), we can use 187 (mg/dL) as an estimate of the mean total cholesterol level for all Americans and 43 (mg/dL) as an estimate of the standard deviation.* Assuming total cholesterol levels to be normally distributed, what is the probability that an American chosen at random has a cholesterol level higher than 250? If 200 Americans are chosen at random, how many can we expect to have total cholesterol higher than 250?

Solution Here, $\mu = 187$ and $\sigma = 43$. The probability that a randomly chosen American has cholesterol higher than 250 is the area under the normal curve to the right of $x = 250$. The z-score for $x = 250$ is

⑤

The heights of female sophomore college students at one school have $\mu = 172$ centimeters, with $\sigma = 10$ centimeters. Find the probability that the height of such a student is

(a) between 172 cm and 185 cm;

(b) between 160 cm and 180 cm;

(c) less than 165 cm.

Answers:

(a) .4032

(b) .6730

(c) .2420

*Values calculated by author from data available at www.cdc.gov/nchs/nhanes.htm

$$z = \frac{x - \mu}{\sigma} = \frac{250 - 187}{43} = \frac{63}{43} = 1.47.$$

From Table 2, we see that the area to the right of 1.47 is $.5 - .4292 = .0708$, which is 7.08% of the total area under the curve. Therefore, the probability of a randomly chosen American having cholesterol higher than 250 is .0708.

With 7.08% of Americans having total cholesterol higher than 250, selecting 200 Americans at random yields

$$7.08\% \text{ of } 200 = .0708 \cdot 200 = 14.16.$$

Approximately 14 of these Americans can be expected to have total cholesterol higher than 250. ▲

NOTE Notice in Example 5 that $P(z \geq 1.47) = P(z > 1.47.)$. The area under the curve is the same whether we include the endpoint or not. Notice also that $P(z = 1.47) = 0$, because no area is included.

CAUTION When calculating the normal probability, it is wise to draw a normal curve with the mean and the z-scores every time. This will avoid confusion as to whether you should add or subtract probabilities.

As mentioned earlier, z-scores are standard deviations, so $z = 1$ corresponds to 1 standard deviation above the mean, and so on. As found in side problem 3(d) of this section, 68.3% of the area under a normal curve lies within 1 standard deviation of the mean. Also, 95.47% lies within 2 standard deviations of the mean, and 99.7% lies within 3 standard deviations of the mean. These results, summarized in Figure 10.18, can be used to get quick estimates when you work with normal curves.

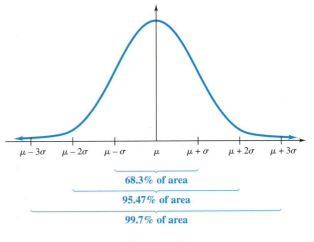

FIGURE 10.18

Boxplots

The normal curve is useful because you can easily read various characteristics of the data from the picture. Boxplots are another graphical means of presenting key characteristics of a data set. The idea is to arrange the data in increasing order and choose three numbers Q_1, Q_2, and Q_3 that divide it into four equal parts, as indicated schematically here.

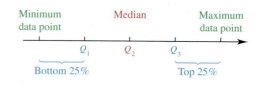

The number Q_1 is called the **first quartile,** the median Q_2 is called the **second quartile,** and Q_3 is called the **third quartile.** The minimum, Q_1, Q_2, Q_3, and the maximum, are often called the five-number summary, and they are used to construct a boxplot, as illustrated in the following examples.

▼ **EXAMPLE 6**   The Britelite company conducted tests on the life of its lightbulbs and those of competitor (Brand X), with the following results for samples of 10 bulbs of each brand.

	Hours of Use (in 100s)									
Britelite	20	22	22	25	26	27	27	28	30	35
Brand X	15	18	19	23	25	25	28	30	34	38

Construct a boxplot for the Britelite data.

Solution   The data is already ranked.

$$20, 22, 22, 25, \textbf{26, 27}, 27, 28, 30, 35.$$

The minimum score is 20 and the maximum score is 35. Since there is an even number of scores, the median score Q_2 is 26.5 (halfway between the two center entries). To find Q_1, which separates the bottom 25% of the data from the rest, we first calculate

25% of n (rounded up to the nearest integer),

where n is the number of data. Here $n = 10$ and $.25(10) = 2.5$, which rounds to 3. Now count to the *third* score to get $Q_1 = 22$.

Similarly, since Q_3 separates the bottom 75% of the data from the rest, we calculate

75% of n (rounded up to the nearest integer).

When $n = 10$, then $.75(10) = 7.5$, which rounds to 8. Count to the *eighth* score, getting $Q_3 = 28$.

The five key numbers for constructing the boxplot are

$$\text{Minimum,} \quad Q_1, \quad Q_2, \quad Q_3, \quad \text{Maximum}$$
$$20, \quad 22, \quad 26.5, \quad 28, \quad 35$$

Draw a horizontal line parallel to the number axis, from the minimum to the maximum score. Around this line, construct a box whose ends are at Q_1 and Q_3, and mark the location of Q_2 by a vertical line in the box, as shown in Figure 10.19. ▲

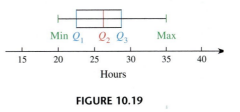

FIGURE 10.19

Because they don't show the detail of a histogram or a stem-and-leaf plot, box-plots are not as useful for analyzing a single data set. However, they can be useful for comparing two sets of data, as long as the same scale is used.

▼ **EXAMPLE 7** Construct a boxplot for the Brand X data shown in Example 6, and compare it with the one for the Britelite data.

Solution The data is again ranked from low to high:

$$15, 18, 19, 23, 25, 25, 28, 30, 34, 38.$$

The minimum is 15, the maximum is 38, and the median is $Q_2 = 25$. Again, $n = 10$ and

$$25\% \text{ of } n = .25(10) = 2.5, \text{ which rounds up to } 3,$$

so that $Q_1 = 19$ (the *third* data score). Similarly,

$$75\% \text{ of } n = .75(10) = 7.5, \text{ which rounds up to } 8,$$

so that $Q_3 = 30$ (the *eighth* data score). The minimum, the maximum, and the three quartiles (namely, 15, 19, 25, 30, and 38) are used to get the boxplot in Figure 10.20.

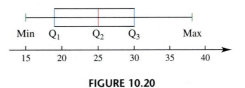

FIGURE 10.20

6

Construct a box plot for the number of blood types data given in Exercise 27 of Section 10.2.

Answer:

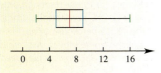

Comparing the graphs in Figures 10.19 and 10.20 shows that the range of hours in use is less for Brand X bulbs than for Britelite bulbs, the width of the boxplot indicates where the middle 50% of the distribution lies, and the median indicates how that 50% is distributed. The boxplot for Britelite bulbs is narrower than that for Brand X, showing that Britelites are more consistent in number of hours of use, and the higher placement of the median indicates that they tend to last longer than brand X. ▲ 6

| TECHNOLOGY TIP   Many graphing calculators can graph boxplots for single-variable data. The procedure is similar to plotting data with the STAT PLOT menu.

◆10.3 Exercises

1. The peak in a normal curve occurs directly above _____.

2. The total area under a normal curve (above the horizontal axis) is _____.

3. How are z-scores found for normal distributions with $\mu \neq 0$ or $\sigma \neq 1$?

4. How is the standard normal curve used to find probabilities for normal distributions?

Find the percentage of the area under a normal curve between the mean and the following number of standard deviations from the mean. (See Example 2.)

5. 1.75

6. .26

7. −.43

8. −2.4

Find the percentage of the total area under the standard normal curve between the following z-scores. (See Examples 1 and 2.)

9. $z = 1.41$ and $z = 2.83$

10. $z = .64$ and $z = 2.11$

11. $z = -2.48$ and $z = -.05$

12. $z = -1.63$ and $z = -1.08$

13. $z = -3.05$ and $z = 1.36$

14. $z = -2.91$ and $z = -.51$

Find a z-score satisfying each of the following conditions. (Hint: Use Table 2 backward or a graphing calculator.)

15. 5% of the total area is to the right of z.

16. 1% of the total area is to the left of z.

17. 15% of the total area is to the left of z.

18. 25% of the total area is to the right of z.

19. For any normal distribution, what is the value of $P(x \leq \mu)$? $P(x \geq \mu)$?

20. Using Chebyshev's theorem and the normal distribution, compare the probability that a number will lie within 2 standard deviations of the mean of a probability distribution. (See Exercises 15–22 of Section 10.2.) Explain what you observe.

21. Repeat Exercise 20, using 3 standard deviations.

Assume the following distributions are all normal. (See Example 4.)

22. **Business**   According to the label, a regular can of Campbell's™ soup holds an average of 305 grams with a standard deviation of 4.2 grams. What is the probability that a can will be sold that holds more than 306 grams?

23. **Business**   A jar of Adams Old Fashioned Peanut Butter contains 453 grams with a standard deviation of 10.1 grams. Find the probability that one of these jars contains less than 450 grams.

24. **Business**   A General Electric soft white three-way bulb has an average life of 1200 hours with a standard devia-

tion of 50 hours. Find the probability that the life of one of these bulbs will be between 1150 and 1300 hours.

25. **Business**   A 100-watt lightbulb has an average brightness of 1640 lumens with a standard deviation of 62 lumens. What is the probability that a 100-watt bulb will have a brightness between 1600 and 1700 lumens?

26. **Social Science**   The scores on a standardized test in a suburban high school have a mean of 80 with a standard deviation of 12. What is the probability that a student will have a score less than 60?

27. **Health**   Using the data from the same study as Example 5, we find that the average HDL cholesterol level is 51.6 mg/dL with a standard deviation of 14.3 mg/dL. Find the probability that an individual will have an HDL cholesterol level greater than 60 mg/dL?

28. **Business**   The production of cars per day at an assembly plant has mean 120.5 and standard deviation 6.2. Find the probability that fewer than 100 cars are produced on a random day.

29. **Business**   Starting salaries for accounting majors have mean $41,000 with standard deviation $3,200. What is the probability an individual will start at a salary above $50,000?

30. **Social Science**   The driving distance to work for residents of a certain community has mean 21 miles and standard deviation 3.6 miles. What is the probability that an individual drives between 10 and 20 miles to work?

Business   *A certain type of lightbulb has an average life of 500 hours with a standard deviation of 100 hours. The length of life of the bulb can be closely approximated by a normal curve. An amusement park buys and installs 10,000 such bulbs. Find the total number that can be expected to last for each of the following periods of time. (See Example 5.)*

31. At least 500 hours

32. Less than 500 hours

33. Between 650 and 780 hours

34. Between 290 and 540 hours

35. Less than 740 hours

36. More than 300 hours

37. Find the shortest and longest lengths of life for the middle 80% of the bulbs.

Social Science   *New studies by Federal Highway Administration traffic engineers suggest that speed limits on many thoroughfares are set arbitrarily and often are artificially low. According to traffic engineers, the ideal limit should be the "85th-percentile speed," the speed at or below which 85 percent of the traffic moves. Assuming that speeds are normally distributed, find the 85th-percentile speed for roads with the following conditions.*

38. The mean speed is 55 mph with a standard deviation of 10 mph.

39. The mean speed is 40 mph with a standard deviation of 5 mph.

Social Science *One professor uses the following system for assigning letter grades in a course.*

Grade	Total Points
A	Greater than $\mu + \frac{3}{2}\sigma$
B	$\mu + \frac{1}{2}\sigma$ to $\mu + \frac{3}{2}\sigma$
C	$\mu - \frac{1}{2}\sigma$ to $\mu + \frac{1}{2}\sigma$
D	$\mu - \frac{3}{2}\sigma$ to $\mu - \frac{1}{2}\sigma$
F	Below $\mu - \frac{3}{2}\sigma$

What percentage of the students receive the following grades?

40. A **41.** B **42.** C

43. Do you think the system in Exercises 40–42 would be more likely to be fair in a large freshman class in psychology or in a graduate seminar of five students? Why?

Health *In nutrition, the recommended daily allowance of vitamins is a number set by the government as a guide to an individual's daily vitamin intake. Actually, vitamin needs vary drastically from person to person, but the needs are very closely approximated by a normal curve. To calculate the recommended daily allowance, the government first finds the average need for vitamins among people in the population and then the standard deviation. The recommended daily allowance is defined as the mean plus 2.5 times the standard deviation.*

44. What percentage of the population will receive adequate amounts of vitamins under this plan?

Find the recommended daily allowance for the following vitamins.

45. Mean = 550 units, standard deviation = 46 units

46. Mean = 1700 units, standard deviation = 120 units

47. Mean = 155 units, standard deviation = 14 units

48. Mean = 1080 units, standard deviation = 86 units

Social Science *The mean performance score of a large group of fifth-grade students on a math achievement test is 88. The scores are known to be normally distributed. What percentage of the students had scores as follows?*

49. More than 1 standard deviation above the mean

50. More than 2 standard deviations above the mean

Social Science *Studies have shown that women are charged an average of $500 more than men for cars.* Assume a normal distribution of overcharges with a mean of $500 and a standard deviation of $65. Find the probability of a woman's paying the following additional amounts for a car.*

51. Less than $400

52. At least $700

53. Between $350 and $600

Business *The table gives total sales estimates for selected market sectors from the 2003 Annual Retail Survey for 2002 and 2003.† Estimates are in billions of dollars. Use this information for Exercises 54–58.*

Sector	2002	2003
Books and magazines	4.1	4.2
Computer software	4.4	3.9
Electronics and appliances	4.8	6.0
Food, beer, and wine	2.1	2.3
Furniture and home furnishings	7.5	8.3
Music and videos	4.2	3.8
Office equipment and supplies	6.5	7.0
Sporting goods	2.8	2.6
Toys, hobby goods, and games	3.7	3.9
Nonmerchandise receipts	4.9	5.8

54. What is the five-number summary for 2002 sales?

55. What is the five-number summary for 2003 sales?

56. Construct a boxplot for 2002 sales.

57. Construct a boxplot for 2003 sales.

58. Are the distributions similar for the two years? Is this unexpected?

**"From repair shops to cleaners, women pay more," by Bob Dart, as appeared in The Chicago Tribune, May 27, 1993. Reprinted by permission of the author.

†2003 E-Commerce Multi-Sector Data Tables, www.census.gov/eos/www/ebusiness614.htm

Health *The table shows the percentages of births classified as low birth weight (less than 2500 grams) for 15 selected U.S. states for 1990 and 2002.**

Sector	1990	2002
Alabama	8.4	9.9
Arizona	8.2	8.6
Colorado	8.0	8.9
Florida	7.4	8.4
Idaho	7.1	6.1
Iowa	5.4	6.6
Louisiana	9.2	10.4
Maine	5.1	6.3
Massachusetts	5.9	7.5
Nevada	5.3	7.2
New Jersey	7.0	8.0
North Carolina	8.0	9.0
Oklahoma	6.6	8.0
South Dakota	5.1	7.2
West Virginia	7.1	9.0

*Statistical Abstract of the United States, 2004–2005, Section 81, www.census.gov.

59. What is the five-number summary for the 1990 percentages?

60. What is the five-number summary for the 2002 percentages?

61. Construct a boxplot for the 1990 percentages.

62. Construct a boxplot for the 2002 percentages.

63. Does it appear that, in general, the percentage of low-birth-weight children has increased over the 12 years covered in the table? Why?

10.4 Normal Approximation to the Binomial Distribution

As we saw in Section 9.4, many practical experiments have only two possible outcomes, sometimes referred to as success or failure. Such experiments are called Bernoulli trials or Bernoulli processes. Examples of Bernoulli trials include flipping a coin (with heads being a success, for instance, and tails a failure) or testing a computer chip coming off the assembly line to see whether it is defective. A binomial experiment consists of repeated independent Bernoulli trials, such as flipping a coin 10 times or taking a random sample of 20 computer chips from the assembly line. In Section 9.4, we found the probability distribution for several binomial experiments, such as sampling five people with bachelor's degrees in education and counting how many are women. The probability distribution for a binomial experiment is known as a **binomial distribution.**

As another example, it is reported that 29% of drivers on Interstate 10 in Texas exceed the 70 mph speed limit.* Suppose a state trooper wants to verify this statistic and records the speed of 10 randomly selected drivers. The trooper finds that 4 out of 10, or 40%, exceed the speed limit. How likely is this result if the 29% figure is accurate? We can answer that question with the binomial probability formula

$$ {}_nC_x \cdot p^x \cdot (1 - p)^{n-x}, $$

*Time, May 13, 1996, p. 34.

where n is the size of the sample (10 in this case), x is the number of speeders (4 in this example), and p is the probability that a driver is a speeder (.29). We obtain

$$P(x = 4) = {}_{10}C_4 \cdot .29^4 \cdot (1 - .29)^6$$
$$= 210(.007073)(.1281) \approx .1903.$$

The probability is almost 20%, so this result is not unusual.

Suppose that the state trooper takes a larger random sample, say, of 100 drivers. What is the probability that 40 or more drivers speed if the 29% figure is accurate? Calculating $P(x = 40) + P(x = 41) + \cdots + P(x = 100)$ is a formidable task. One solution is provided by graphing calculators or computers. On the TI-83, for example, we can first calculate the probability that 39 or fewer drivers exceed the speed limit using the DISTR menu command binomcdf (100, .29, 39). Subtracting the answer from 1 gives a probability of .0119. But this high-tech method fails as n becomes larger; the command binomcdf (1000, .29, 300) gives an error message. On the other hand, there is a low-tech method that works regardless of the size of n. It has further interest because it connects two different distributions: the normal and the binomial. The normal distribution is continuous, since the random variable can take on any real number. The binomial distribution is *discrete,* because the random variable can take on only integer values between 0 and n. Nevertheless, the normal distribution can be used to give a good approximation to binomial probability.

In order to use the normal approximation, we first need to know the mean and standard deviation of the binomial distribution. Recall from Section 9.4 that, for the binomial distribution, $E(x) = np$. In Section 10.1, we referred to $E(x)$ as μ, and that notation will be used here. It is shown in more advanced courses in statistics that the standard deviation of the binomial distribution is given by $\sigma = \sqrt{np(1 - p)}$.

Mean and Standard Deviation for the Binomial Distribution

For the binomial distribution, the mean and standard deviation are respectively given by

$$\mu = np \qquad \text{and} \qquad \sigma = \sqrt{np(1 - p)},$$

where n is the number of trials and p is the probability of success on a single trial. ◇1

◇1

Find μ and σ for a binomial distribution having $n = 120$ and $p = 1/6$.

Answer:

$\mu = 20; \sigma = 4.08$

▼ **EXAMPLE 1** Suppose a fair coin is flipped 15 times.

(a) Find the mean and standard deviation for the number of heads. With $n = 15$ and $p = 1/2$, the mean is

$$\mu = np = 15\left(\frac{1}{2}\right) = 7.5.$$

The standard deviation is

$$\sigma = \sqrt{np(1-p)} = \sqrt{15\left(\frac{1}{2}\right)\left(1 - \frac{1}{2}\right)}$$

$$= \sqrt{15\left(\frac{1}{2}\right)\left(\frac{1}{2}\right)} = \sqrt{3.75} \approx 1.94.$$

We expect, on average, to get 7.5 heads out of 15 tosses. Most of the time, the number of heads will be within three standard deviations of the mean, or between $7.5 - 3(1.94) = 1.68$ and $7.5 + 3(1.94) = 13.32$.

(b) Find the probability distribution for the number of heads and draw a histogram of the probabilities.

The probability distribution is found by putting $n = 15$ and $p = 1/2$ into the formula for binomial probability. For example, the probability of getting 9 heads is given by

$$P(x = 9) = {}_{15}C_9\left(\frac{1}{2}\right)^9\left(1 - \frac{1}{2}\right)^6 \approx .15274.$$

Probabilities for the other values of x between 0 and 15, as well as a histogram of the probabilities, are shown in Figure 10.21.

x	$P(x)$
0	.00003
1	.00046
2	.00320
3	.01389
4	.04166
5	.09164
6	.15274
7	.19638
8	.19638
.9	.15274
10	.09164
11	.04166
12	.01389
13	.00320
14	.00046
15	.00003

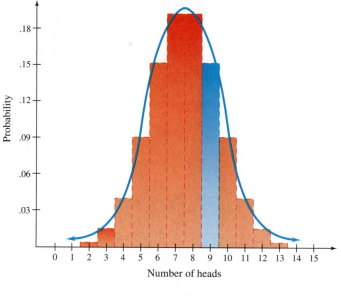

FIGURE 10.21

In Figure 10.21, we have superimposed the normal curve with $\mu = 7.5$ and $\sigma = 1.94$ over the histogram of the distribution. Notice how well the normal distribution fits the binomial distribution. This approximation was first discovered in 1718 by Abraham De Moivre (1667–1754) for the case $p = 1/2$. The result was generalized by the French mathematician Pierre-Simon Laplace (1749–1827) in a book published in 1812. As n becomes larger and larger, a histogram for the binomial distribution looks more and more like a normal curve. Figures 10.22 (a) and (b), on the next page, show histograms of the binomial distribution with $p = .3$, using $n = 8$ and $n = 50$, respectively.

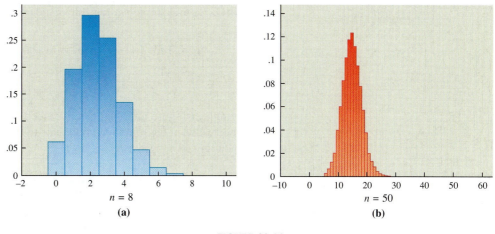

FIGURE 10.22

The probability of getting exactly 9 heads in the 15 tosses, or .15274, is the same as the area of the blue bar in Figure 10.21. As the graph suggests, the blue area is approximately equal to the area under the normal curve from $x = 8.5$ to $x = 9.5$. The normal curve is higher than the top of the bar in the left half but lower in the right half.

To find the area under the normal curve from $x = 8.5$ to $x = 9.5$, first find z-scores, as in the previous section. Use the mean and the standard deviation for the distribution, which we have already calculated, to get z-scores for $x = 8.5$ and $x = 9.5$.

For $x = 8.5$,
$$z = \frac{8.5 - 7.5}{1.94}$$
$$= \frac{1.00}{1.94}$$
$$z \approx .52.$$

For $x = 9.5$,
$$z = \frac{9.5 - 7.5}{1.94}$$
$$= \frac{2.00}{1.94}$$
$$z \approx 1.03.$$

From Table 2, $z = .52$ gives an area of .1985 and $z = 1.03$ gives .3485. The difference between these two numbers is the desired result.

$$.3485 - .1985 = .1500$$

This answer (.1500) is not far from the more accurate answer of .15274 found earlier. ▲

CAUTION The normal curve approximation to a binomial distribution is quite accurate, *provided that n is large and p is not close to 0 or 1.* As a rule of thumb, the normal curve approximation can be used as long as both np and $n(1 - p)$ are at least 5.

▼ **EXAMPLE 2** In August 2004, The Gallup Organization conducted a poll that asked, "If you were free to do either, would you prefer to have a job outside the home, or would you prefer to stay at home and take care of the house and family?" They found that 54% of respondents answered, "Outside the home."* Suppose we select 100 respondents at random.

(a) Use the normal distribution to approximate the probability that at least 65 respondents would answer "Outside the home."

*Gallup Organization, August 8–August 11, 2005.

2

Use the normal distribution to find the probability of getting exactly the following number of heads in 15 tosses of a coin.

(a) 7

(b) 10

Answers:

(a) .1985

(b) .0909

Solution First find the mean and the standard deviation, using $n = 100$ and $p = .54$.

$$\mu = 100(.54) \qquad \sigma = \sqrt{100(.54)(1.54)}$$
$$= 54 \qquad\qquad = \sqrt{100(.54)(.46)}$$
$$\approx 4.98$$

As the graph in Figure 10.23 shows, we need to find the area to the right of $x = 64.5$ (since we want 65 or more respondents to say "Outside the home"). The z-score corresponding to $x = 64.5$ is

$$z = \frac{64.5 - 54}{4.98} \approx 2.11.$$

From the table, $z = 2.11$ leads to an area of .4826, so

$$P(z > 2.11) = .5 - .4826 = .0174.$$

This is an extremely low probability. If a survey researcher obtained 65 respondents who chose "Outside the home" in a sample of 100, she would suspect that either her sample is not truly random or the 54% figure is too low.

(b) Find the probability of finding between 55 and 62 respondents who chose "Outside the home" in a random sample of 100.

Solution As Figure 10.24 shows, we need to find the area between $x = 54.5$ and $x = 61.5$.

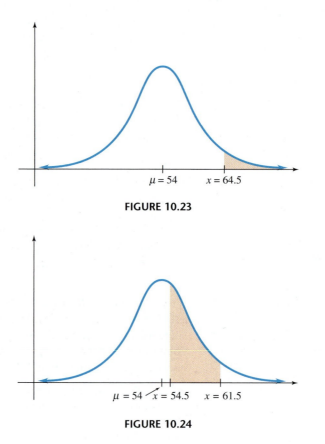

FIGURE 10.23

FIGURE 10.24

If $x = 54.5$, then $z = \dfrac{54.5 - 54}{4.98} \approx .10$.

If $x = 61.5$, then $z = \dfrac{61.5 - 54}{4.98} \approx 1.51$.

Use the table to find that $z = .10$ gives an area of .0398 and $z = 1.51$ yields .4345. The final answer is the difference of these numbers.

$$P(.10 \leq z \leq 1.51) = .4345 - .0398 = .3947$$

The probability of finding between 55 and 62 respondents who chose "Outside the home" is approximately .3947. ▲ ③

---

③

About 9% of the transistors produced by a certain factory are defective. Find the approximate probability that, in a sample of 200, the following numbers of transistors will be defective. (*Hint:* $\sigma = 4.05$)

(a) Exactly 11

(b) 16 or fewer

(c) More than 14

Answers:

(a) .0226

(b) .3557

(c) .8051

◆10.4 Exercises

1. What must be known to find the mean and standard deviation of a binomial distribution?

2. What is the rule of thumb for using the normal distribution to approximate a binomial distribution?

Suppose 16 coins are tossed. Find the probability of getting each of the following results by using (a) the binomial probability formula, and (b) the normal curve approximation. (See Examples 1 and 2.)

3. Exactly 8 heads

4. Exactly 9 heads

5. More than 13 tails

6. Fewer than 6 tails

For the remaining exercises in this section, use the normal curve approximation to the binomial distribution.

Suppose 1000 coins are tossed. Find the probability of getting each of the following results.

7. Exactly 500 heads

8. Exactly 510 heads

9. 475 or more heads

10. Fewer than 460 tails

A die is tossed 120 times. Find the probability of getting each of the following results.

11. Exactly twenty 5's

12. Exactly twenty-four 4's

13. More than seventeen 3's

14. Fewer than twenty-one 6's

15. A reader asked Mr. Statistics (a feature in *Fortune* magazine) about the game of 26 once played in the bars of Chicago.* The player chooses a number between 1 and 6 and then rolls a cup full of 10 dice 13 times. Out of the 130 numbers rolled, if the number chosen appears at least 26 times, the player wins. Calculate the probability of winning.

16. **Social Science** According to the National Center for Education Statistics, in 2001, 56% of the graduates with bachelor's degrees were women.* Suppose five holders of bachelor's degrees are picked at random.
    (a) Find the probability that three of the five are women.
    (b) Find the probability that at least four of the five are women.

17. **Natural Science** For certain bird species, with appropriate assumptions, the number of nests escaping predation has a binomial distribution.[†] Suppose the probability of success (that is, a nest escaping predation) is .3. Find the probability that at least half of 26 nests escape predation.

18. **Natural Science** Let us assume, under certain appropriate assumptions, the probability of a young animal eating x units of food is binomially distributed, with n equal to the maximum number of food units the animal can acquire and p equal to the probability per time unit that an animal eats a unit of food. Suppose $n = 120$ and $p = .6$.

---

*Daniel Seligman and Patty De Llosa, "Ask Mr. Statistics," *Fortune*, May 1, 1995, p. 141.

*National Center for Education Statistics, http://nces.ed.gov/das/epubs/2005169/gender_1.asp

†From G. deJong, *American Naturalist*, vol. 110.

(a) Find the probability that an animal consumes 80 units of food.

(b) Suppose the animal must consume at least 70 units of food to survive. What is the probability that this happens?

19. **Social Science** According to the Census Bureau, in 2002, 11.4% of U.S. residents lived in poverty.* Suppose a random sample of 600 residents is selected. What is the probability that over 80 residents live in poverty?

20. **Social Science** In Exercise 19, what is the probability that fewer than 70 residents live in poverty?

21. **Health** *HealthDay* reported that researchers estimate that, in 2004, 24.5% of American adults were obese.[†] A random sample of 40 adults is selected. Find the probability that 15 or fewer are obese.

22. **Health** In Exercise 21, what is the probability that more than 10 adults selected are obese?

23. **Natural Science** A flu vaccine has a probability of 80% of preventing a person who is inoculated from getting the flu. A county health office inoculates 134 people. Find the probabilities of the following.
    (a) Exactly 12 of the people inoculated get the flu.
    (b) No more than 12 of the people inoculated get the flu.
    (c) None of the people inoculated get the flu.

24. **Natural Science** The probability that a male will be color blind is .042. Find the probabilities that, in a group of 53 men, the following will be true.
    (a) Exactly 6 are color blind.
    (b) No more than 6 are color blind.
    (c) At least 1 is color blind.

25. **Business** The probability that a certain machine turns out a defective item is .05. Find the probabilities that, in a run of 75 items, the following results are obtained.
    (a) Exactly 7 defectives
    (b) No defectives
    (c) At least 1 defective

26. **Business** A company is taking a survey to find out whether people like its product. Their last survey indicated that 70% of the population likes the product. On the basis of that survey, of a sample of 58 people, find the probabilities of the following.
    (a) All 58 like the product.
    (b) From 28 to 30 (inclusive) like the product.

27. **Natural Science** The blood types B– and AB– are the rarest of the eight human blood types, representing 1.5% and .6% of the population, respectively.[‡]

(a) If the blood types of a random sample of 1000 blood donors are recorded, what is the probability that 10 or more of the samples are AB–?

(b) If the blood types of a random sample of 1000 blood donors are recorded, what is the probability that 20 to 40, inclusive, of the samples are B–?

(c) If a particular city had a blood drive in which 500 people gave blood and 3% of the donations were B–, would we have reason to believe that this town has a higher-than-normal number of donors who are B–? (*Hint:* Calculate the probability of 15 or more donors being B– for a random sample of 500, and then consider the probability obtained.)

28. **Social Science** According to the National Center for Statistics and Analysis, in 2003, there were 43,220 traffic deaths. Of those, 3,592 were motorcycle rider deaths. If 800 traffic death reports from 2003 are selected at random, find the probability that between 50 and 70 (inclusive) are motorcycle rider deaths.

29. **Social Science** In one state, 55% of the voters expect to vote for Bill Poole. Suppose 1400 people are asked the name of the person for whom they expect to vote. Find the probability that at least 750 people will say that they expect to vote for Poole.

30. **Social Science** A poll of 2000 teenagers found that 1 in 8 reported carrying a weapon for protection.* In a typical high school with 1200 students, what is the probability that more than 125 students, but fewer than 175, carry a weapon?

31. In the 1989 U.S. Open, four golfers each made a hole in one on the same par-3 hole on the same day. *Sports Illustrated* writer R. Reilly stated the probability of a hole in one for a given golf pro on a given par-3 hole to be 1/3709.[†]
    (a) For a specific par-3 hole, use the binomial distribution to find the probability that 4 or more of the 156 golf pros in the tournament field shoot a hole in one.[‡]
    (b) For a specific par-3 hole, use the normal approximation to the binomial distribution to find the probability that 4 or more of the 156 golf pros in the tournament field shoot a hole in one. Why must we be very cautious when using this approximation for this application?
    (c) If the probability of a hole in one remains constant and is 1/3709 for any par-3 hole, find the probability that, in 20,000 attempts by golf pros, there will be 4 or more holes in one. Discuss whether this assumption is reasonable.

*Statistical Abstract of the United States. 2004–2005, No. 688, www.census.gov.

†*HealthDay,* August 23, 2005, www.healthday.com

‡National Center for Statistics and Analysis, http://www-nrd.nhtsa.dot.gov/pdf/nrd-30/NCSA/RNotes/2004/809-734/

*New York Times, January 12, 1996, p. A6.

†R. Reilly, "King of the Hill," *Sports Illustrated,* June 1989, pp. 20–25.

‡Bonnie Litwiller and David Duncan, "The Probability of a Hole in One," *School Science and Mathematics* 91, no. 1, (January 1991): 30.

Chapter **10** Summary

Key Terms and Symbols

10.1 Σ summation (sigma) notation
$\bar{x}$ sample mean
μ population mean
random sample
grouped frequency distribution
frequency polygon
(arithmetic) mean
median
statistic
mode

10.2 s^2 sample variance

s sample standard deviation
σ population standard deviation
range
deviations from the mean
variance
standard deviation

10.3 μ mean of a continuous
distribution
σ standard deviation of a
normal distribution
continuous distribution

skewed distribution
normal distributions
normal curves
standard normal curve
z-score
boxplot
quartile

10.4 binomial distribution

Key Concepts

To organize the data from a sample, we use a **grouped frequency distribution**—a set of intervals with their corresponding frequencies. The same information can be displayed with a histogram—a type of bar graph with a bar for each interval. Each bar has width 1 and height equal to the probability of the corresponding interval. A **stem-and-leaf plot** presents the individual data in a similar form, so it can be viewed as a bar graph as well. Another way to display this information is with a **frequency polygon,** which is formed by connecting the midpoints of consecutive bars of the histogram with straight-line segments.

The **mean** $\bar{x}$ of a frequency distribution is the expected value.

For n numbers $x_1, x_2, \ldots, x_n$,

$$\bar{x} = \frac{\Sigma x}{n}.$$

For a grouped distribution,

$$\bar{x} = \frac{\Sigma(xf)}{n}.$$

The **median** is the middle entry in a set of data arranged in either increasing or decreasing order.

The **mode** is the most frequent entry in a set of numbers.

The **range** of a distribution is the difference between the largest and smallest numbers in the distribution.

The **sample standard deviation** s is the square root of the sample **variance.**

For n numbers,

$$s = \sqrt{\frac{\Sigma x^2 - n\bar{x}^2}{n - 1}}.$$

For a grouped distribution,

$$s = \sqrt{\frac{\Sigma f x^2 - n\bar{x}^2}{n - 1}}.$$

A **normal distribution** is a continuous distribution with the following properties: The highest frequency is at the mean; the graph is symmetric about a vertical line through the mean; the total area under the curve, above the x-axis, is 1. If a normal distribution has mean μ and standard deviation σ, then the z-score for the number x is $z = \dfrac{x - \mu}{\sigma}$.

A **boxplot** organizes a list of data using the minimum and maximum values, the median, and the first and third quartiles to give a visual overview of the distribution.

Area under a Normal Curve The area under a normal curve between $x = a$ and $x = b$ gives the probability that an observed data value will be between a and b.

The **binomial distribution** is a distribution with the following properties: For n independent repeated trials, in which the probability of success in a single trial is p, the probability of x successes is $_nC_x p^x(1 - p)^{n-x}$. The mean is $\mu = np$ and the standard deviation is

$$\sigma = \sqrt{np(1 - p)}.$$

Chapter 10 Review Exercises

1. Discuss some reasons for organizing data into a grouped frequency distribution.

2. What is the rule of thumb for an appropriate interval in a grouped frequency distribution?

In Exercises 3 and 4, (a) write a frequency distribution; (b) draw a histogram; and (c) draw a frequency polygon. (d) In Exercise 4, also construct a stem-and-leaf plot.

3. The following numbers give the sales in dollars for the lunch hour at a local hamburger store for the last 20 Fridays. (Use intervals of 450–474, 475–499, and so on.)

480 451 501 478 512 473 509 515 458 566
516 535 492 558 488 547 461 475 492 471

4. The number of units carried in one semester by the students in a business mathematics class was as follows. (Use intervals of 9–10, 11–12, 13–14, 15–16.)

10 9 16 12 13 15 13 16 15 11 13
12 12 15 12 14 10 12 14 15 15 13

Find the mean for each of the following.

5. The data in Exercise 3.

6. The data in Exercise 4.

7. The table gives the frequency counts for 44 first-year college students' waist circumference in cm.

Interval	Frequency
60–69	10
70–79	24
80–89	6
90–99	3
100–109	1

8. The table gives the frequency counts for 44 first-year college students' caloric intake on a random day.

Interval	Frequency
0–999	1
1000–1999	12
2000–2999	14
3000–3999	11
4000–4999	5
5000–5999	1

9. What do the mean, median, and mode of a distribution have in common? How do they differ? Describe each in a sentence or two.

Find the median and the mode (or modes) for each of the following.

10. Ages (years) of senior citizens tested for low calcium levels:

78, 72, 72, 73, 73, 73, 65, 68, 89, 84, 71, 80

11. Ages (years) of senior citizens tested for low calcium levels:

68, 80, 76, 66, 72, 73, 72, 74, 72, 71, 67, 77, 70

The modal class is the interval containing the most data values. Find the modal class for the distributions of

12. Exercise 7.

13. Exercise 8.

14. What is meant by the range of a distribution?

15. How are the variance and the standard deviation of a distribution related? What is measured by the standard deviation?

Find the range and standard deviation for each of the following distributions.

16. Number of days in a month of sunshine for a village.

14, 17, 18, 19, 30

17. Ages of drivers entering a fast-food restaurant.

26, 43, 17, 20, 25, 37, 54, 28, 20, 19

Find the standard deviation for the following.

18. Exercise 7.

19. Exercise 8.

✏ **20.** Describe the characteristics of a normal distribution.

✏ **21.** What is meant by a skewed distribution?

Find the following areas under the standard normal curve.

22. Between $z = 0$ and $z = 1.35$

23. To the left of $z = .38$

24. Between $z = -1.88$ and $z = 2.41$

25. Between $z = 1.53$ and $z = 2.82$

26. Find a z-score such that 8% of the area under the curve is to the right of z.

✏ **27.** Why is the normal distribution not a good approximation of a binomial distribution that has a value of p close to 0 or 1?

Business   *The table gives the probability distribution of buyers of jogging or running shoes, by age, at a shoe store.*

Age	Probability
10–19	.12
20–29	.25
30–39	.20
40–49	.22
50–59	.09
60–69	.10
70 and over	.02

28. (a) Draw a histogram for this distribution.
   (b) From the shape of the histogram, is this a normal distribution?

29. As shown in the distribution in Exercise 28, 12% of buyers of jogging or running shoes are age 10–19. Assume a binomial distribution with $n = 6$ and $p = .12$.
   (a) Find the probability that 2 of 6 randomly selected purchasers of jogging or running shoes are age 10–19.
   (b) Find the mean and standard deviation of the binomial distribution. Interpret your results.

30. Finance   The annual returns of two stocks for three years are given in the following table.

	2003	2004	2005
Stock I	11%	−1%	14%
Stock II	9%	5%	10%

(a) Find the mean and standard deviation for each stock over the three-year period.
(b) If you are looking for security with an 8% return, which of the two stocks would you choose, using your answers from part (a) to decide?

31. Natural Science   The weight gains of two groups of 10 rats fed on two different experimental diets were as follows.

Weight Gains									
Diet A	1	0	3	7	1	1	5	4	1	4
Diet B	2	1	1	2	3	2	1	0	1	0

Compute the mean and standard deviation for each group, and compare them to answer the following questions.
   (a) Which diet produced the greatest mean gain?
   (b) Which diet produced the most consistent gain?

32. Natural Science   Refer to Exercise 31.
   (a) Construct boxplots for each diet data.
✏ **(b)** Use the boxplots to compare weight gains for the two diets.

33. Business   The following are total sales (in billions of dollars) for the sporting goods industry for the years 1997–2003.*

Year	1997	1998	1999	2000	2001	2002	2003
Sales	67.3	69.8	71.2	74.4	74.3	77.9	79.8

(a) Compute the mean and standard deviation of the sales.
(b) How many standard deviations from the mean is the highest sales? The smallest sales?

Business   *The following are 1990 sales and 2003 projected sales (in millions of dollars) of various items by the sporting goods industry.*

Industry	1990 Sales	2003 Sales
Archery	265	285
Baseball and softball	217	315
Billiards and pool	192	559
Camping	1072	1458
Exercise equipment	1824	4553
Fishing tackle	1910	2028
Golf	2514	3372
Hunting and firearms	2202	2495
In-line skating and wheel sports	150	469
Optics	438	850
Skin diving and scuba	294	337
Skiing, alpine	475	539
Tennis	333	401

Statistical Abstract of the United States, 2004–2005, Table No. 1247, www.census.gov.

34. **(a)** Find the mean and standard deviation for these sales in 1990 and 2003.
(b) Which industry is closest to the mean in 1990? In 2003?

35. Find the 5-number-summary for these sales in 1990 and 2003.

36. **(a)** Construct boxplots for these sales.
(b) Does it appear that sales for the industry as a whole have increased in the 13 years covered by the table. Explain.

37. **Social Science** On standard IQ tests, the mean is 100, with a standard deviation of 15. The results are very close to fitting a normal curve. Suppose an IQ test is given to a very large group of people. Find the percentage of people whose IQ score is
(a) more than 130;
(b) less than 85;
(c) between 85 and 115.

38. **Business** A machine that fills quart orange juice cartons is set to fill them with 32.1 oz. If the actual contents of the cartons vary normally, with a standard deviation of .1 oz, what percentage of the cartons contains less than a quart (32 oz)?

39. **Natural Science** An area infested with fruit flies is to be sprayed with a chemical that is known to be 98% effective for each application. A sample of 100 flies is checked. Assume a binomial distribution and use the normal distribution to approximate the following probabilities.
(a) 95% of the flies are killed in one application.
(b) At least 95% of the flies are killed in one application.
(c) At least 90% of the flies are killed in one application.
(d) All the flies are killed in one application. Compare your answer with the probability found by using the binomial distribution.

Case 10

Statistics in the Law: The *Castañeda* Decision

Statistical evidence is now routinely presented in both criminal and civil cases. In this application, we'll look at a famous case that established use of the binomial distribution and measurement by standard deviation as an accepted procedure.*

Defendants who are convicted in criminal cases sometimes appeal their conviction on the grounds that the jury which indicted or convicted them was drawn from a pool of jurors that does not represent the population of the district in which they live. These appeals almost always cite the Supreme Court's decision in *Castañeda v. Partida* [430 U.S. 482], a case that dealt with the selection of grand juries in the state of Texas. The decision summarizes the facts this way:

> After respondent, a Mexican-American, had been convicted of a crime in a Texas District Court and had exhausted his state remedies on his claim of discrimination in the selection of the grand jury that had indicted him, he filed a habeas corpus petition in the Federal District Court, alleging a denial of due process and equal protection under the Fourteenth Amendment, because of gross underrepresentation of Mexican-Americans on the county grand juries.

*The Castañeda case and many other interesting applications of statistics in law are discussed in Michael O. Finkelstein and Bruce Levin, *Statistics for Lawyers*, New York, Springer-Verlag, 1990. U.S. Supreme Court decisions are online at http://www.findlaw.com/casecode/supreme.html, and most states now have important state court decisions online.

The case went to the Appeals Court, which noted that "the county population was 79% Mexican-American, but, over an 11-year period, only 39% of those summoned for grand jury service were Mexican-American," and concluded that together with other testimony about the selection process, "the proof offered by respondent was sufficient to demonstrate a prima facie case of intentional discrimination in grand jury selection. . . ."

The state appealed to the Supreme Court, and the Supreme Court needed to decide whether the underrepresentation of Mexican-Americans on grand juries was indeed too extreme to be an effect of chance. To do so, they invoked the binomial distribution. Here is the argument:

> Given that 79.1% of the population is Mexican-American, the expected number of Mexican-Americans among the 870 persons summoned to serve as grand jurors over the 11-year period is approximately 688. The observed number is 339. Of course, in any given drawing some fluctuation from the expected number is predicted. The important point, however, is that the statistical model shows that the results of a random drawing are likely to fall in the vicinity of the expected value. . . .
>
> The measure of the predicted fluctuations from the expected value is the standard deviation, defined for the binomial distribution as the square root of the product of the total number in the sample (here 870) times the probability of selecting a Mexican-American (.791) times the probability of selecting a non-Mexican-American (.209). . . . Thus, in this case the standard deviation is approximately 12. As a general rule for such large samples, if the difference between the expected value and the observed number is greater than two or three standard deviations, then the hypothesis that the jury drawing was random would be suspect to a social scientist. The 11-year data here reflect a difference between the expected and observed number of Mexican-Americans of approximately 29 standard deviations. A detailed calculation reveals that the likelihood that such a substantial departure from the expected value would occur by chance is less than 1 in 10^{140}.

The Court decided that the statistical evidence supported the conclusion that jurors were not randomly selected, and that it was up to the state to show that its selection process did not discriminate against Mexican-Americans. The Court concluded,

> The proof offered by respondent was sufficient to demonstrate a prima facie case of discrimination in grand jury

selection. Since the State failed to rebut the presumption of purposeful discrimination by competent testimony, despite two opportunities to do so, we affirm the Court of Appeals' holding of a denial of equal protection of the law in the grand jury selection process in respondent's case.

◆ Exercises

1. Check the Court's calculation of 29 standard deviations as the difference between the expected number of Mexican-Americans and the number actually chosen.

2. Where do you think the Court's figure of 1 in 10^{140} came from?

3. The *Castañeda* decison also presents data from a $2\frac{1}{2}$-year period during which the state district judge supervised the selection process. During this period, 220 persons were called to serve as grand jurors, and only 100 of these were Mexican-American.

   (a) Considering the 220 jurors as a random selection from a large population, what is the expected number of Mexican-Americans, given the 79.1% population figure?

   (b) If we model the drawing of jurors as a sequence of 220 independent Bernoulli trials, what is the standard deviation of the number of Mexican-Americans?

   (c) About how many standard deviations is the actual number of Mexican-Americans drawn (100) from the expected number that you calculated in (a)?

   (d) What does the normal distribution table at the back of the book tell you about this result?

4. The following information is from a case brought by Hy-Vee stores before the Iowa Supreme Court, appealing a ruling by the Iowa Civil Rights Commission in favor of a female employee of one of their grocery stores.

   > In 1985, there were 112 managerial positions in the ten Hy-Vee stores located in Cedar Rapids. Only 6 of these managers were women. During that same year there were 294 employees; 206 were men and 88 were women.

   (a) How far from the expected number of women in management was the actual number, assuming that gender had nothing to do with promotion? Measure the difference in standard deviations.

   (b) Does this look like evidence of purposeful discrimination?

Appendix A Graphing Calculators

This is just a summary of the contents of Appendix A. The entire appendix is available at our Website: www.aw-bc.com/MWA9. You can print as much or as little of it as you want.

Part 1: Using Your Calculator

This is a brief introduction to the features of a graphing calculator that are relevant to the topics in this text, including the following.

Basics: keyboard, edit/replay, scientific notation
Function Graphing: trace, zoom, maximum/minimum finder, intersection finder
Function Evaluation: tables, functional notation, eval key
Equation Solving: root finder, polynomial solver, equation solver
Linear Regression and Other Regression Procedures
Matrices: matrix editor, matrix arithmetic, row operations

Part 2: Calculator Programs

The following programs are available for TI and Casio 9850 (they work on some other Casio models as well). TI users may download them directly if they have the appropriate link hardware and software. All programs may also be printed and then entered into your calculator by hand.

1. Fraction Conversion for Casio
2. Quadratic Formula for TI-82/83
3. Table Maker for TI-85
4. Present and Future Value of an Annuity
5. Loan Payment
6. Loan Balance after n Payments
7. Amortization Table
8. RREF Program for TI-82 and Casio 9850
9. Simplex Method
10. Two-Stage Method
11. Rectangle Approximation of $\int_a^b f(x)\ dx$ (using left endpoints)

Programs 1–3 and 8 are built in on most other models. Programs 7 and 9–11 are not built in on any calculators. Programs 4–6 are included in the TI-83 Financial menu (and in similar menus that can be downloaded from TI for the TI-86 and TI-89), although some students may find the versions here easier to use.

Appendix B Tables

Table 1 Formulas from Geometry

CIRCLE
Area: $A = \pi r^2$
Circumference: $C = 2\pi r$

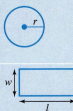

RECTANGLE
Area: $A = lw$
Perimeter: $P = 2l + 2w$

PARALLELOGRAM
Area: $A = bh$
Perimeter: $P = 2a + 2b$

TRIANGLE
Area: $A = \dfrac{1}{2}bh$

SPHERE
Volume: $V = \dfrac{4}{3}\pi r^3$

Surface area: $A = 4\pi r^2$

RECTANGULAR BOX
Volume: $V = lwh$
Surface area: $A = 2lh + 2wh + 2lw$

CIRCULAR CYLINDER
Volume: $V = \pi r^2 h$
Surface area: $A = 2\pi r^2 + 2\pi rh$

TRIANGULAR CYLINDER
Volume: $V = \dfrac{1}{2}bhl$

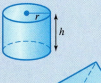

CONE
Volume: $V = \dfrac{1}{3}\pi r^2 h$

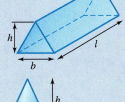

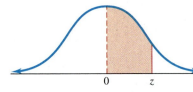

Table 2   Areas under the Normal Curve
The column under A gives the proportion of the area under the entire curve that is between $z = 0$ and a positive value of z.

z	A	z	A	z	A	z	A
.00	.0000	.48	.1844	.96	.3315	1.44	.4251
.01	.0040	.49	.1879	.97	.3340	1.45	.4265
.02	.0080	.50	.1915	.98	.3365	1.46	.4279
.03	.0120	.51	.1950	.99	.3389	1.47	.4292
.04	.0160	.52	.1985	1.00	.3413	1.48	.4306
.05	.0199	.53	.2019	1.01	.3438	1.49	.4319
.06	.0239	.54	.2054	1.02	.3461	1.50	.4332
.07	.0279	.55	.2088	1.03	.3485	1.51	.4345
.08	.0319	.56	.2123	1.04	.3508	1.52	.4357
.09	.0359	.57	.2157	1.05	.3531	1.53	.4370
.10	.0398	.58	.2190	1.06	.3554	1.54	.4382
.11	.0438	.59	.2224	1.07	.3577	1.55	.4394
.12	.0478	.60	.2258	1.08	.3599	1.56	.4406
.13	.0517	.61	.2291	1.09	.3621	1.57	.4418
.14	.0557	.62	.2324	1.10	.3643	1.58	.4430
.15	.0596	.63	.2357	1.11	.3665	1.59	.4441
.16	.0636	.64	.2389	1.12	.3686	1.60	.4452
.17	.0675	.65	.2422	1.13	.3708	1.61	.4463
.18	.0714	.66	.2454	1.14	.3729	1.62	.4474
.19	.0754	.67	.2486	1.15	.3749	1.63	.4485
.20	.0793	.68	.2518	1.16	.3770	1.64	.4495
.21	.0832	.69	.2549	1.17	.3790	1.65	.4505
.22	.0871	.70	.2580	1.18	.3810	1.66	.4515
.23	.0910	.71	.2612	1.19	.3830	1.67	.4525
.24	.0948	.72	.2642	1.20	.3849	1.68	.4535
.25	.0987	.73	.2673	1.21	.3869	1.69	.4545
.26	.1026	.74	.2704	1.22	.3888	1.70	.4554
.27	.1064	.75	.2734	1.23	.3907	1.71	.4564
.28	.1103	.76	.2764	1.24	.3925	1.72	.4573
.29	.1141	.77	.2794	1.25	.3944	1.73	.4582
.30	.1179	.78	.2823	1.26	.3962	1.74	.4591
.31	.1217	.79	.2852	1.27	.3980	1.75	.4599
.32	.1255	.80	.2881	1.28	.3997	1.76	.4608
.33	.1293	.81	.2910	1.29	.4015	1.77	.4616
.34	.1331	.82	.2939	1.30	.4032	1.78	.4625
.35	.1368	.83	.2967	1.31	.4049	1.79	.4633
.36	.1406	.84	.2996	1.32	.4066	1.80	.4641
.37	.1443	.85	.3023	1.33	.4082	1.81	.4649
.38	.1480	.86	.3051	1.34	.4099	1.82	.4656
.39	.1517	.87	.3079	1.35	.4115	1.83	.4664
.40	.1554	.88	.3106	1.36	.4131	1.84	.4671
.41	.1591	.89	.3133	1.37	.4147	1.85	.4678
.42	.1628	.90	.3159	1.38	.4162	1.86	.4686
.43	.1664	.91	.3186	1.39	.4177	1.87	.4693
.44	.1700	.92	.3212	1.40	.4192	1.88	.4700
.45	.1736	.93	.3238	1.41	.4207	1.89	.4706
.46	.1772	.94	.3264	1.42	.4222	1.90	.4713
.47	.1808	.95	.3289	1.43	.4236	1.91	.4719

Table 2   *(continued)*

z	A	z	A	z	A	z	A
1.92	.4726	2.42	.4922	2.92	.4983	3.42	.4997
1.93	.4732	2.43	.4925	2.93	.4983	3.43	.4997
1.94	.4738	2.44	.4927	2.94	.4984	3.44	.4997
1.95	.4744	2.45	.4929	2.95	.4984	3.45	.4997
1.96	.4750	2.46	.4931	2.96	.4985	3.46	.4997
1.97	.4756	2.47	.4932	2.97	.4985	3.47	.4997
1.98	.4762	2.48	.4934	2.98	.4986	3.48	.4998
1.99	.4767	2.49	.4936	2.99	.4986	3.49	.4998
2.00	.4773	2.50	.4938	3.00	.4987	3.50	.4998
2.01	.4778	2.51	.4940	3.01	.4987	3.51	.4998
2.02	.4783	2.52	.4941	3.02	.4987	3.52	.4998
2.03	.4788	2.53	.4943	3.03	.4988	3.53	.4998
2.04	.4793	2.54	.4945	3.04	.4988	3.54	.4998
2.05	.4798	2.55	.4946	3.05	.4989	3.55	.4998
2.06	.4803	2.56	.4948	3.06	.4989	3.56	.4998
2.07	.4808	2.57	.4949	3.07	.4989	3.57	.4998
2.08	.4812	2.58	.4951	3.08	.4990	3.58	.4998
2.09	.4817	2.59	.4952	3.09	.4990	3.59	.4998
2.10	.4821	2.60	.4953	3.10	.4990	3.60	.4998
2.11	.4826	2.61	.4955	3.11	.4991	3.61	.4999
2.12	.4830	2.62	.4956	3.12	.4991	3.62	.4999
2.13	.4834	2.63	.4957	3.13	.4991	3.63	.4999
2.14	.4838	2.64	.4959	3.14	.4992	3.64	.4999
2.15	.4842	2.65	.4960	3.15	.4992	3.65	.4999
2.16	.4846	2.66	.4961	3.16	.4992	3.66	.4999
2.17	.4850	2.67	.4962	3.17	.4992	3.67	.4999
2.18	.4854	2.68	.4963	3.18	.4993	3.68	.4999
2.19	.4857	2.69	.4964	3.19	.4993	3.69	.4999
2.20	.4861	2.70	.4965	3.20	.4993	3.70	.4999
2.21	.4865	2.71	.4966	3.21	.4993	3.71	.4999
2.22	.4868	2.72	.4967	3.22	.4994	3.72	.4999
2.23	.4871	2.73	.4968	3.23	.4994	3.73	.4999
2.24	.4875	2.74	.4969	3.24	.4994	3.74	.4999
2.25	.4878	2.75	.4970	3.25	.4994	3.75	.4999
2.26	.4881	2.76	.4971	3.26	.4994	3.76	.4999
2.27	.4884	2.77	.4972	3.27	.4995	3.77	.4999
2.28	.4887	2.78	.4973	3.28	.4995	3.78	.4999
2.29	.4890	2.79	.4974	3.29	.4995	3.79	.4999
2.30	.4893	2.80	.4974	3.30	.4995	3.80	.4999
2.31	.4896	2.81	.4975	3.31	.4995	3.81	.4999
2.32	.4898	2.82	.4976	3.32	.4996	3.82	.4999
2.33	.4901	2.83	.4977	3.33	.4996	3.83	.4999
2.34	.4904	2.84	.4977	3.34	.4996	3.84	.4999
2.35	.4906	2.85	.4978	3.35	.4996	3.85	.4999
2.36	.4909	2.86	.4979	3.36	.4996	3.86	.4999
2.37	.4911	2.87	.4980	3.37	.4996	3.87	.5000
2.38	.4913	2.88	.4980	3.38	.4996	3.88	.5000
2.39	.4916	2.89	.4981	3.39	.4997	3.89	.5000
2.40	.4918	2.90	.4981	3.40	.4997		
2.41	.4920	2.91	.4982	3.41	.4997		

Answers to Selected Exercises

Chapter 1

Section 1.1 (Page 9)

1. True   **3.** Answers vary with the calculator, but 2,508,429,787/798,458,000 is the best.   **5.** Distributive property   **7.** Commutative property of addition   **9.** Answers vary.   **11.** -54 **13.** $-1/2$ **15.** 18%   **17.** .75%   **19.** -12 **21.** 4   **23.** -4 **25.** -1 **27.** $\frac{2040}{523}, \frac{189}{37}, \sqrt{27}, \frac{4587}{691},$ 6.375, $\sqrt{47}$ **29.** $12 < 18.5$ **31.** $x \geq 5.7$ **33.** $z \leq 7.5$

35.

37.

39.

41.

43.

45. 4   **47.** 6   **49.** 0   **51.** (a) 17.6 (b) no   **53.** (a) 23.2 (b) yes   **55.** 45°   **57.** 24°   **59.** 4   **61.** -19 **63.** $=$ **65.** $=$ **67.** $=$ **69.** $>$ **71.** $7 - a$ **73.–75.** Writing   **77.** 2000, 2003, 2004   **79.** Let x be the percentage of international students from India. Then $x > 13\%$.   **81.** Let x be the rank of UNC. Then $x > 25$.

Section 1.2 (Page 18)

1. 1,973,822.685   **3.** 289.0991339   **5.** Answers vary.   **7.** 4^5 **9.** $(-6)^7$ **11.** $(5u)^{28}$ **13.** degree 4; coefficients: 6.2, $-5, 4, -3$, 3.7; constant term 3.7   **15.** 3   **17.** $-x^3 + x^2 - 13x$ **19.** $-6y^2 + 3y + 6$ **21.** $-6x^2 + 4x - 4$ **23.** $-18m^3 - 54m^2 + 9m$ **25.** $12z^3 + 14z^2 - 7z + 5$ **27.** $12k^2 + 16k - 3$ **29.** $6y^2 + 13y + 5$ **31.** $18k^2 - 7kq - q^2$ **33.** $4.34m^2 + 5.68m - 4.42$ **35.** $-k + 3$ **37.** $R = 5000x$; $C = 150,000 + 2250x$; $P = 2750x - 150,000$ **39.** $R = 7500x$; $C = -3x^2 + 3480x + 249,675$; $P = 3x^2 + 4020x - 249,675$ **41.** (a) 102,000,000 (b) 101,550,000   **43.** (a) 157,600,000 (b) 154,680,000   **45.** 199,350,000   **47.** 235,560,000   **49.** .866   **51.** .505   **53.** (a) approximately 60,501,067 cu ft (b) The shape becomes a rectangular box with a square base, with volume b^2h. (c) yes   **55.** (a) 0, 1, 2, 3, or no degree (if one is the negative of the other) (b) 0, 1, 2, 3, or no degree (if they are equal) (c) 6

Section 1.3 (Page 25)

1. $12x(x - 2)$ **3.** $r(r^2 - 5r + 1)$ **5.** $6z(z^2 - 2z + 3)$ **7.** $(2y - 1)^2(14y - 4) = 2(2y - 1)^2(7y - 2)$ **9.** $(x + 5)^4(x^2 + 10x + 28)$ **11.** $(x + 1)(x + 4)$ **13.** $(x + 3)(x + 4)$ **15.** $(z + 4)(z + 6)$ **17.** $(2x - 1)(x - 4)$ **19.** $(3p - 4)(5p - 1)$ **21.** $(2z - 5)(2z - 3)$ **23.** $(2x + 1)(3x - 4)$ **25.** $(5y - 2)(2y + 5)$ **27.** $(2x - 1)(3x + 4)$ **29.** $(3a + 5)(a - 1)$ **31.** $(x + 9)(x - 9)$ **33.** $(3p - 2)^2$ **35.** $(r - 2t)(r + 5t)$ **37.** $(m - 4n)^2$ **39.** $(2p + 3)(2p - 3)$ **41.** $3(x - 4z)^2$ **43.** cannot be factored   **45.** $(-x + 4)(x - 3)$ or $(x - 4)(-x + 3)$ **47.** $(3a + 5)(a - 6)$ **49.** $(7m + 2n)(3m + n)$ **51.** $(4y - x)(5y + 11x)$ **53.** $(y - 7z)(y + 3z)$ **55.** $(11x + 8)(11x - 8)$ **57.** $m^3(m - 1)^2(m^2 + m + 1)^2(5 + 3m^2 - 3m^5)$ **59.** $(a - 6)(a^2 + 6a + 36)$ **61.** $(2r - 3s)(4r^2 + 6rs + 9s^2)$ **63.** $(4m + 5)(16m^2 - 20m + 25)$ **65.** $(10y - z)(100y^2 + 10yz + z^2)$ **67.** $(x^2 + 3)(x^2 + 2)$ **69.** $b^2(b + 1)(b - 1)$ **71.** $(x + 2)(x - 2)(x^2 + 3)$ **73.** $(4a^2 + 9b^2)(2a + 3b)(2a - 3b)$ **75.** $x^2(x^2 + 2)(x^4 - 2x^2 + 4)$ **77.** Answers vary.   **79.** Answers vary.

Section 1.4 (Page 31)

1. $\frac{x}{7}$ **3.** $\frac{5}{7p}$ **5.** $\frac{5}{4}$ **7.** $\frac{4}{w + 6}$ **9.** $\frac{y - 4}{3y^2}$ **11.** $\frac{m - 2}{m + 3}$ **13.** $\frac{x + 5}{x + 1}$ **15.** $\frac{2p}{7}$ **17.** $\frac{9a^3}{4}$ **19.** $\frac{5}{4c}$ **21.** $\frac{3}{4}$ **23.** $\frac{3}{10}$ **25.** $\frac{2(a + 4)}{a - 3}$ **27.** $\frac{k + 2}{k + 3}$ **29.** Answers vary.   **31.** $\frac{4}{15z}$ **33.** $\frac{4}{3}$ **35.** $\frac{20 + x}{5x}$ **37.** $\frac{3m - 2}{m(m - 1)}$ **39.** $\frac{17}{3(a - 1)}$ **41.** $\frac{33}{20(k - 2)}$ **43.** $\frac{7x - 1}{(x - 3)(x - 1)(x + 2)}$ **45.** $\frac{y^2}{(y + 4)(y + 3)(y + 2)}$ **47.** $\frac{k^2 - 13k}{(2k - 1)(k + 2)(k - 3)}$ **49.** $\frac{x + 1}{x - 1}$ **51.** $\frac{-1}{x(x + h)}$ **53.** (a) $\frac{x - 5}{x^2 - 10x}$ (b) 32.7 sec; 8 sec; 4.5 sec   **55.** (a) $2.4 million (b) No

Section 1.5 (Page 43)

1. 49 **3.** $16c^2$ **5.** $32/x^5$ **7.** $108u^{12}$ **9.** $1/6$

11. $1/32$ **13.** $-1/2401$ **15.** $-1/y^3$ **17.** 343 **19.** $9/16$

21. b^3/a **23.** 7 **25.** 1.55 **27.** 9 **29.** -16

31. $81/16$ **33.** $1/4^5$ **35.** 64 **37.** 2 **39.** 8^5

41. $\dfrac{1}{9^{32/15}}$ **43.** z^3 **45.** $\dfrac{p}{4}$ **47.** $\dfrac{q^5}{r^3}$ **49.** $\dfrac{8}{25p^7}$

51. $2^{5/6}p^{3/2}$ **53.** $2p + 5p^{5/3}$ **55.** $\dfrac{1}{3y^{2/3}}$ **57.** $\dfrac{a^{1/2}}{49b^{5/2}}$

59. $x^{7/6} - x^{11/6}$ **61.** $x - y$ **63.** (f) **65.** (h) **67.** (g)

69. (c) **71.** 5 **73.** 5 **75.** -2 **77.** 9 **79.** $\sqrt{77}$

81. 7 **83.** $16\sqrt{3}$ **85.** $13\sqrt{3}$ **87.** -7

89. $\sqrt{15} - 4\sqrt{3} + 4\sqrt{5} - 16$ **91.** $-3 - 3\sqrt{2}$

93. $3 + \sqrt{2}$ **95.** $\dfrac{1}{7 + 4\sqrt{3}}$ **97. (a)** 14 **(b)** 85 **(c)** 58.0

99. 179.61 **101.** 188.68 **103.** About 11,397 **105.** About 14,328 **107.** About 15,092 **109. (a)** About 3.996 million; about 4.629 million; about 4.991 million **(b)** About 2012; no, not in the next 40 years.

Section 1.6 (Page 53)

1. 4 **3.** 7 **5.** $-\dfrac{10}{9}$ **7.** 4 **9.** $\dfrac{40}{7}$ **11.** $\dfrac{26}{3}$ **13.** $-\dfrac{12}{5}$

15. $-\dfrac{59}{6}$ **17.** $-\dfrac{9}{4}$ **19.** $x = .72$ **21.** $r \approx -13.26$

23. $\dfrac{b - 5a}{2}$ **25.** $x = \dfrac{3b}{a + 5}$ **27.** $V = \dfrac{k}{P}$ **29.** $g = \dfrac{V - V_0}{t}$

31. $B = \dfrac{2A}{h} - b$ or $B = \dfrac{2A - bh}{h}$ **33.** $-2, 3$ **35.** $-8, 2$

37. $\dfrac{5}{2}, \dfrac{7}{2}$ **39.** $-7, -\dfrac{3}{7}$ **41.** 14 **43.** 64.4 **45.** 2006

47. 2011 **49.** 1995 **51.** 2006 **53.** 1984 **55.** 2000

57. 2000 **59.** 2007 **61.** \$205.41 **63. (a)** .0352

(b) about .015, or 15% **(c)** approximately one case **65.** \$8000

67. \$5000 **69.** about 838 mi **71.** $\dfrac{400}{3}$ L **73.** 142 mi

75. 70 mph **77.** 10 cm **79.** 5 cm

Section 1.7 (Page 63)

1. $-4, 14$ **3.** $0, -6$ **5.** $0, 3$ **7.** $-7, -8$ **9.** $\dfrac{1}{2}, 3$

11. $-\dfrac{1}{2}, \dfrac{1}{3}$ **13.** $\dfrac{5}{2}, 4$ **15.** $-5, -2$ **17.** $\dfrac{4}{3}, -\dfrac{4}{3}$

19. $0, 1$ **21.** $2 \pm \sqrt{7}$ **23.** $\dfrac{1 \pm 2\sqrt{5}}{4}$

25. $\dfrac{-7 \pm \sqrt{41}}{4}$; $-.1492, -3.3508$

27. $\dfrac{-1 \pm \sqrt{5}}{4}$; $.3090, -.8090$ **29.** $\dfrac{-5 \pm \sqrt{65}}{10}$; $.3062, -1.3062$

31. No real number solutions **33.** $-\dfrac{5}{2}, 1$ **35.** no real number

solutions **37.** $-5, \dfrac{3}{2}$ **39.** 1 **41.** 2

43. $x \approx .4701$ or 1.8240 **45.** $x \approx -1.0376$ or $.6720$

47. (a) 1.5 sec **(b)** about 3.35 sec **49. (a)** about 32 and 62 **(b)** about 17 and 77 **51. (a)** 2003 **(b)** 2007 **(c)** mid-2010 **53. (a)** $x + 20$ **(b)** northbound: $5x$; eastbound: $5(x + 20)$ or $5x + 100$ **(c)** $(5x)^2 + (5x + 100)^2 = 300^2$ **(d)** about 31.23 mph and 51.23 mph **55. (a)** $150 - x$ **(b)** $x(150 - x) = 5000$ **(c)** length 100 m; width 50 m **57.** 9 ft by 12 ft **59. (a)** 2 sec

(b) $\dfrac{1}{2}$ sec or $\dfrac{7}{2}$ sec **(c)** It reaches thegiven height twice: once on the way up and once on the way down

61. $t = \pm\dfrac{\sqrt{2Sg}}{g}$ **63.** $h = \dfrac{\pm d^2\sqrt{kL}}{L}$

65. $R = \dfrac{-2Pr + E^2 \pm E\sqrt{E^2 - 4Pr}}{2P}$

67. (a) $x^2 - 2x = 15$ **(b)** $x = 5$ or $x = -3$ **(c)** $z = \pm\sqrt{5}$

69. $\pm\dfrac{\sqrt{6}}{2}$ **71.** $\pm\sqrt{\dfrac{3 + \sqrt{13}}{2}}$

Chapter 1 Review (Page 66)

1. $0, 6$ **3.** $-12, -6, -\dfrac{9}{10}, -\sqrt{4}, 0, \dfrac{1}{8}, 6$ **5.** Commutative property of multiplication **7.** Distributive property **9.** $x \geq 9$
11. $-7, -3, -2, 0, \pi, 8$ **13.** $-|3 - (-2)|, -|-2|, |6 - 4|, |8 + 1|$ **15.** -2 **17.** -1
19.

-3

21.

-2

23. -18 **25.** $-\dfrac{3}{4}$ **27.** $4x^4 - 4x^2 + 11x$

29. $2q^4 + 14q^3 - 4q^2$ **31.** $12z^2 - 2z - 4$ **33.** $16k^2 - 9h^2$
35. $36x^2 + 36xy + 9y^2$ **37.** $k(2h^2 - 4h + 5)$
39. $a^2(3a + 1)(a + 4)$ **41.** $(2y - 1)(5y - 3)$
43. $(2a - 5)^2$ **45.** $(12p + 13q)(12p - 13q)$

47. $(2y - 1)(4y^2 + 2y + 1)$ **49.** $\dfrac{7x^2}{4}$ **51.** 4

53. $\dfrac{(y - 5)(4y - 5)}{7y}$ **55.** $\dfrac{(m - 1)^2}{3(m + 1)}$ **57.** $\dfrac{1}{6z}$ **59.** $\dfrac{64}{35q}$

61. $\dfrac{1}{6^3}$ or $\dfrac{1}{216}$ **63.** -1 **65.** $\dfrac{5^2}{6^2}$ or $\dfrac{25}{36}$ **67.** 4^3 **69.** $\dfrac{1}{9}$

71. 9^3 **73.** $\dfrac{3}{4}$ **75.** 25 **77.** $\dfrac{1}{3^5}$ **79.** $\dfrac{1}{5^{2/3}}$ **81.** $3^{7/2}a^{5/2}$

83. 3 **85.** $3\sqrt{6}$ **87.** $3pq\sqrt[3]{2q^2}$ **89.** $\dfrac{n\sqrt{30m}}{6m}$

91. $-8\sqrt{3}$ **93.** 5 **95.** $\dfrac{-\sqrt{2} + \sqrt{6}}{2}$ **97.** \$27,000,000

99. \$36,482,872.69 **101.** 1 **103.** -2 **105.** No solution

107. $x = \dfrac{1}{6a - 1}$ **109.** $x = \dfrac{c - 3}{3a - ac - 2}$ **111.** $14, -4$

113. $-38, 42$ **115. (a)** 2010 **(b)** 2021 **117. (a)** 2004 **(b)** 2007 **119.** \$60,000 at 8%; \$40,000 at 5% **121.** 2

123. 1   **125.** 0   **127.** $\dfrac{-1 \pm \sqrt{7}}{2}$ **129.** $3, -\dfrac{5}{2}$

131. $-2 \pm \sqrt{5}$ **133.** $\dfrac{-1 \pm \sqrt{3}}{3}$ **135.** $\dfrac{1}{2}, \dfrac{1}{6}$ **137.** $-\dfrac{8}{3}, 2$

139. $\dfrac{\pm\sqrt{3}}{3}$ **141.** $\dfrac{\pm\sqrt{3}}{3}$ **143.** $r = \dfrac{-Rp \pm E\sqrt{Rp}}{p}$

145. $s = \dfrac{a \pm \sqrt{a^2 + 4K}}{2}$ **147. (a)** $171{,}000$ **(b)** 2008

149. 50 m by 225 m or 100 m by 112.5 m

Case 1 (Page 70)

1. $700 + 85x$ **3.** The \$700 refrigerator costs \$300 more.

Chapter 2

Section 2.1 (Page 79)

1. IV, II, I, III   **3.** yes   **5.** no

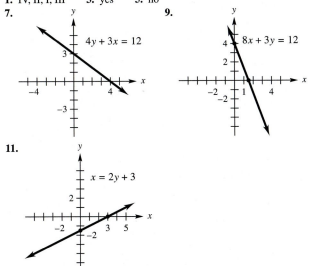

7.

9.

11.

13. x-intercepts $-1, 3.5$; y-intercept 1   **15.** x-intercepts $-2, 0, 2$; y-intercept 0   **17.** x-intercept 4; y-intercept 3   **19.** x-intercept 12; y-intercept -8 **21.** x-intercepts $3, -3$; y-intercept -9 **23.** x-intercepts $-5, 4$; y-intercept -20 **25.** no x-intercept; y-intercept 7

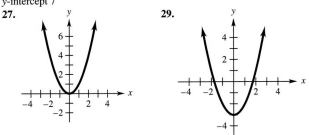

27.

29.

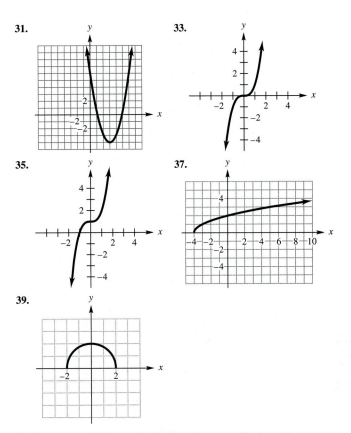

31.

33.

35.

37.

39.

41. Fargo, about 2:00 P.M.; Seattle, about 5 P.M.   **43.** from 11 A.M. until 6 P.M.   **45. (a)** about \$1,250,000   **(b)** about \$1,750,000   **(c)** about \$4,250,000   **47. (a)** about 500,000   **(b)** about \$1,000,000   **(c)** about \$1,5000,000   **49.** about 12 sec   **51. (a)** year 5, about \$750; year 15, about \$600; year 25, about \$300   **(b)** during the 22nd year.   **53.** From about mid-2003 through 2008   **55.** About 14% in mid-2004   **57.** Men, about \$695; women, about \$520   **59.** About \$635 in 1994.   **61.** Men, about \$10; women, about \$140

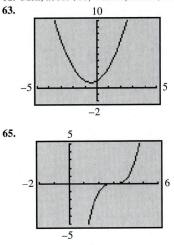

63.

65.

67.

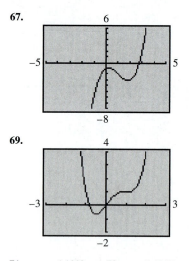

69.

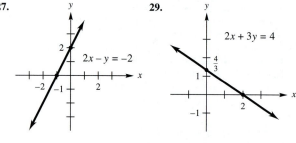

71. $x \approx -1.1038$ **73.** $x \approx 2.1017$ **75.** $x \approx -1.7521$
77. $r \approx 4.6580$ in.   **79.** 2004   **81.** About 2.7 doctors per 1000 residents in mid 2006.

Section 2.2 (Page 92)

1. $-\dfrac{3}{2}$ **3.** -2 **5.** $-\dfrac{5}{2}$ **7.** not defined   **9.** $y = 4x + 5$

11. $y = -2.3x + 1.5$ **13.** $y = -\dfrac{3}{4}x + 4$ **15.** $m = 2; b = -9$

17. $m = 3; b = -2$ **19.** $m = \dfrac{2}{3}; b = -\dfrac{16}{9}$ **21.** $m = \dfrac{2}{3}; b = 0$

23. $m = 1; b = 5$ **25. (a)** C   **(b)** B   **(c)** B   **(d)** D

27.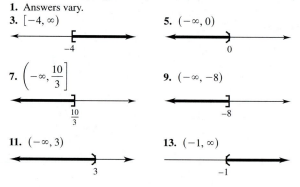

29.

31.

33. perpendicular   **35.** parallel   **37.** neither

39. (a) $\dfrac{2}{5}, \dfrac{9}{8}, -\dfrac{5}{2}$ **(b)** yes

41. $3y = -2x + 4$ or $y = -\dfrac{2}{3}x + \dfrac{4}{3}$ **43.** $y = 4x + 6$

45. $y = 2$ **47.** $x = 6$ **49.** $y = 2x + 3$ **51.** $2y = 7x - 3$

53. $y = 5x$ **55.** $x = 6$ **57.** $y = 2x - 2$ **59.** $y = x - 6$
61. $y = -x + 2$ **63.** \$1248.20   **65.** \$5178.57
67. (a) 3,215,000,000   **(b)** 3,487,500,000   **(c)** 2015
69. (a) About \$10,181,000,000   **(b)** 2011   **71. (a)** $(0, 60), (7, 65)$

(b) $y = \dfrac{5}{7}x + 60$ **(c)** About 62.14 million   **(d)** In 2011

73. (a) $y = .1913x + 1.1$ **(b)** About 7,412,900
(c) In late 2002 $(x \approx 25.7)$ **75. (a)** $T = .03t + 15$
(b) Approximately 2103   **77. (a)** The slope indicates that, *on average,* the race is being run about .01786 second faster each year (although the year-to-year changes may vary significantly from this average). The negative slope indicates that times were generally decreasing during the 68-year period shown.   **(b)** 12.99 minutes
(c) No games were held during World War II (1939–1945).

Section 2.3 (Page 103)

1. (a) $y = \dfrac{5}{9}(x - 32)$ **(b)** 10°C and 23.89°C   **3.** 463.89°C

5. 186.7; 202.7   **7.** \$204.5 billion   **9.** \$246.33 billion
11. (a) For $y = .5x + 1.5$: 0, $-.5$, 0, $-.5$, 1; sum 0. For $y = x$: 1, 0, 0, -1, 0; sum 0   **(b)** For $y = .5x + 1.5$: 1.5. For $y = x$: 2.   **(c)** $y = .5 + 1.5$ **13.** No   **15. (a)** Two-point model: $y = -7.5952x + 559$; regression model: $y = -7.7135x + 563.22$
(b) Two-point model: 194.43 per 100,000; regression model: 192.97 per 100,000   **17. (a)** Two-point model: $y = 15x + 2750$; regression line model: $y = 14.9x + 2822$ **(b)** Two-point model: 5000, 6950, 9050; regression line model: 5055, 6992, 9078   **(c)** Two-point model: 6275, 6500; regression line model: 6321.5, 6500
19. (a) $(4, 1.1), (5, 1.3), (6, 1.8), (7, 2.3), (8, 2.5), (9, 3.1), (10, 3.9),$
$(11, 3.8), (12, 4.0), (13, 4.4)$ **(b)** $y = .3879x - .477$ **(c)** About \$6.5 billion   **21. (a)** $y = .03119x + .5635$ **(b)** About .938 billion; about 1.06 billion   **23. (a)** $y = 11.1143x + 340.81$; 396,000; 508,000; 619,000   **(b)** 2037; 2060
25. (a) $y = 34.9125x + 277.75$ **(b)** Fairly well; the coefficient of correlation is near 1.   **(c)** About \$941.09 billion   **(d)** In 2016

Section 2.4 (Page 113)

1. Answers vary.
3. $[-4, \infty)$

5. $(-\infty, 0)$

7. $\left(-\infty, \dfrac{10}{3}\right]$

9. $(-\infty, -8)$

11. $(-\infty, 3)$

13. $(-1, \infty)$

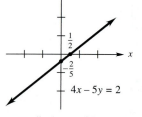

15. $(-\infty, 1]$

17. $\left(\dfrac{1}{5}, \infty\right)$

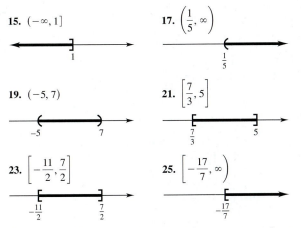

19. $(-5, 7)$

21. $\left[\dfrac{7}{3}, 5\right]$

23. $\left[-\dfrac{11}{2}, \dfrac{7}{2}\right]$

25. $\left[-\dfrac{17}{7}, \infty\right)$

27. $x \geq 2$ **29.** $-3 < x \leq 5$ **31. (a)** Let x represent the number of mg per L of lead in the water. **(b)** $.038 \leq x \leq .042$
   (c) yes   **33. (a)** $0 < x \leq 7300$; $7300 < x \leq 29{,}700$;
$29{,}700 < x \leq 71{,}950$; $71{,}950 < x \leq 150{,}150$; $150{,}150 < x \leq 326{,}450$;
$x > 326{,}450$ **(b)** $0 < T \leq 730$; $730 < T \leq 4090$;
$4090 < T \leq 14{,}652.50$; $14{,}652.50 < T \leq 36{,}548.50$;
$36{,}548.50 < T \leq 94{,}727.50$; $T > 94{,}727.50$

35. $(-2, 2)$

37. no solution

39. $(-3, -2)$

41. $\left(-\infty, -\dfrac{5}{3}\right]$ or $[1, \infty)$

43. $\left(-\dfrac{3}{2}, \dfrac{13}{10}\right)$

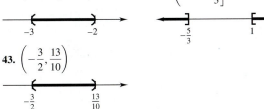

45. $29 \leq T \leq 69$ **47.** $43 \leq T \leq 81$
49. (a) $25.33 \leq R_L \leq 28.17$; $36.58 \leq R_E \leq 40.92$
(b) $5699.25 \leq T_L \leq 6338.25$ $8230.5 \leq T_E \leq 9207$
51. 2005–2010   **53.** Ages 20 through 35   **55.** 2 to 6 miles
57. $x \geq 400$ **59.** $x \geq 50$ **61.** Impossible to break even
63. $|x - 2| \leq 3$ **65.** $|z - 12| \geq 2$

Section 2.5 (Page 121)

1. $\left[-4, \dfrac{3}{2}\right]$

3. $(-\infty, -3)$ or $(-1, \infty)$

5. $\left[-2, \dfrac{1}{4}\right]$

7. $(-\infty, -1)$ or $\left(\dfrac{1}{4}, \infty\right)$

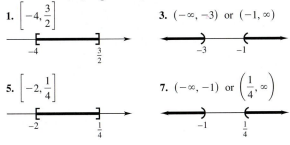

9. $[-6, 6]$

11. $(-\infty, 0)$ or $(16, \infty)$

13. $[-3, 0]$ or $[3, \infty)$ **15.** $[-6, -1]$ or $[4, \infty)$
17. $(-\infty, -3)$ or $(-1, 3)$ **19.** $\left(-\infty, -\dfrac{1}{2}\right)$ or $\left(0, \dfrac{4}{3}\right)$
21. no   **23.** $(.1565, 2.5565)$
25. $[-2.2635, .7556]$ or $[3.5079, \infty)$ **27.** $(.5, .8393)$
29. $(-\infty, 1)$ or $[4, \infty)$ **31.** $\left(\dfrac{7}{2}, 5\right)$ **33.** $(-\infty, 2)$ or $(5, \infty)$
35. $(-\infty, -1)$ **37.** $(-\infty, -2)$ or $(0, 3)$ **39.** $[-1, .5]$
41. $\left(0, \dfrac{5}{3}\right)$ or $(10, \infty)$ **43.** $(100, 150)$ **45.** 2005–present
47. From mid-2014 $(x = 14.53)$ onward.

Chapter 2 Review (Page 123)
1. $(-2, 3)$, $(0, -5)$, $(3, -2)$, $(4, 3)$
3.   **5.**

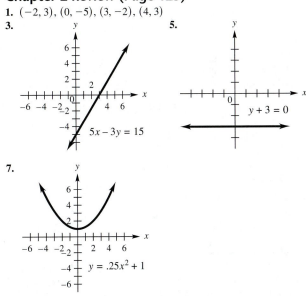

7.

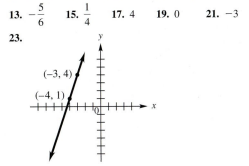

9. (a) about 11:30 A.M. to about 7:30 P.M.   **(b)** from midnight until about 5 P.M. and after about 10:30 P.M.   **11.** Answers vary.
13. $-\dfrac{5}{6}$ **15.** $\dfrac{1}{4}$ **17.** 4   **19.** 0   **21.** -3
23.

25. $3y = 2x - 13$ **27.** $4y = -5x + 17$ **29.** $x = -1$
31. $3y = 5x + 15$

33. (a) $y = -\dfrac{7}{30}x + 52.3$ **(b)** Negative; smoking is decreasing
(c) 21.1%   **35. (a)** $y = .47x + 9.24$;   **(b)** $y = .4785x + 9.188$
(c) Two-point model: \$13.94; regression model: \$13.97; one model
was low by 6 cents and the other by 3 cents   **(d)** Two-point model:
\$17.70; regression model: \$17.80
37. (a) $y = 606.5386x + 14,863.20$ **(b)** $r \approx .98$; yes   **(c)** About
\$25,781

39. $\left(\dfrac{3}{8}, \infty\right)$ **41.** $\left(-\infty, \dfrac{1}{4}\right]$ **43.** $\left[-\dfrac{1}{2}, 2\right]$ **45.** $[-8, 8]$

47. $(-\infty, 2]$ or $[5, \infty)$ **49.** $\left[-\dfrac{9}{5}, 1\right]$ **51. (d)**

53. (a) $y = 23x + 415$ **(b)** until 1998   **(c)** after 2003

55. $(-3, 2)$ **57.** $(-\infty, -5]$ or $\left[\dfrac{3}{2}, \infty\right)$

59. $(-\infty, -5]$ or $[-2, 3]$ **61.** $[-2, 0)$ **63.** $\left(-1, \dfrac{3}{2}\right)$
65. $[-19, -5)$ or $(2, \infty)$ **67.** 1950–1992

Case 2 (Page 126)

3. The poor prediction isn't surprising, since we were extrapolating far
beyond the range of the original data.   **5.** It's not clear that any
simple smooth function will fit these data—there seems to be a break
in the pattern between 1970 and 1980. This will make is difficult to
predict the life expectancy for females born in 2010.   **7.** They used
a regression equation of some kind to predict this value!

Chapter 3

Section 3.1 (Page 136)

1. function   **3.** function   **5.** not a function   **7.** function
9. $(-\infty, \infty)$ **11.** $(-\infty, \infty)$ **13.** $(-\infty, 0]$ **15.** all real
numbers except 2   **17.** all real numbers except 2 and -2
19. all real numbers x such that $x > -4$ and $x \neq 3$.   **21.** $(-\infty, \infty)$
23. (a) 8   **(b)** 8   **(c)** 8   **(d)** 8   **25. (a)** 48   **(b)** 6   **(c)** 25.38
(d) 28.42   **27. (a)** $\sqrt{7}$ **(b)** 0   **(c)** $\sqrt{5.7}$ **(d)** not defined
29. (a) 12   **(b)** 23   **(c)** 12.91   **(d)** 49.41
31. (a) $\dfrac{\sqrt{3}}{15}$ **(b)** not defined   **(c)** $\dfrac{\sqrt{1.7}}{6.29}$ **(d)** not defined
33. (a) 13   **(b)** 9   **(c)** 6.5   **(d)** 24.01

35. (a) $6 - p$ **(b)** $6 + r$ **(c)** $3 - m$
37. (a) $\sqrt{4 - p}$ $(p \leq 4)$ **(b)** $\sqrt{4 + r}$ $(r \geq -4)$ **(c)** $\sqrt{1 - m}$
$(m \leq 1)$
39. (a) $p^3 + 1$ **(b)** $-r^3 + 1$ **(c)** $m^3 + 9m^2 + 27m + 28$
41. (a) $\dfrac{3}{p - 1}$ $(p \neq 1)$ **(b)** $\dfrac{3}{-r - 1}$ $(r \neq -1)$ **(c)** $\dfrac{3}{m + 2}$
$(m \neq -2)$ **43.** 2   **45.** $2x + h$

47.

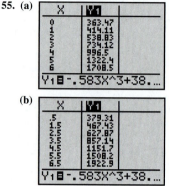

49. (a) \$142.50   **(b)** \$1247.50   **(c)** \$322.80   **51. (a)** about
89%   **(b)** about 47%   **53. (a)** 76,699; 110,570   **(b)** 157,300
55. (a)

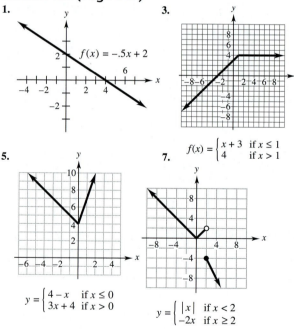

57. $f(t) = 2000 - 475t$
59. (a) $C(x) = 125x + 36,000$ **(b)** \$185

Section 3.2 (Page 147)

1. $f(x) = -.5x + 2$

3.

5. $y = \begin{cases} 4 - x & \text{if } x \leq 0 \\ 3x + 4 & \text{if } x > 0 \end{cases}$

7. $f(x) = \begin{cases} x + 3 & \text{if } x \leq 1 \\ 4 & \text{if } x > 1 \end{cases}$

$y = \begin{cases} |x| & \text{if } x < 2 \\ -2x & \text{if } x \geq 2 \end{cases}$

9.

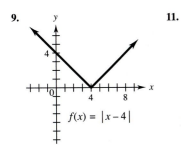

$f(x) = |x - 4|$

11.

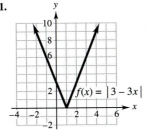

$f(x) = |3 - 3x|$

25.

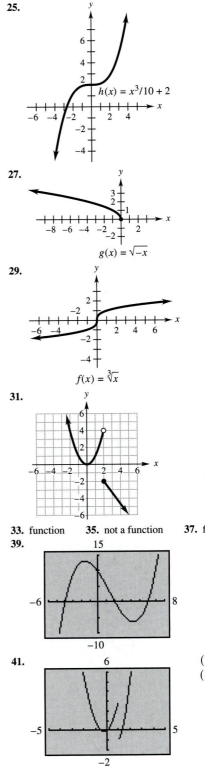

$h(x) = x^3/10 + 2$

13.

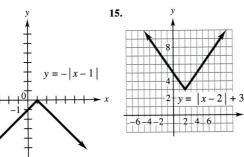

$y = -|x - 1|$

15.

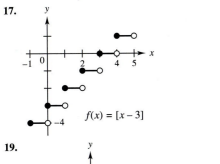

$y = |x - 2| + 3$

27.

$g(x) = \sqrt{-x}$

29.

$f(x) = \sqrt[3]{x}$

17.

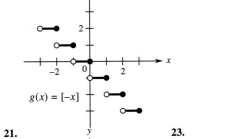

$f(x) = [x - 3]$

31.

19.

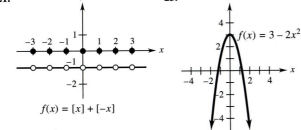

$g(x) = [-x]$

33. function   **35.** not a function   **37.** function

39.

21.

$f(x) = [x] + [-x]$

23.

$f(x) = 3 - 2x^2$

41.

$(1, -1)$ is on the graph;
$(1, 3)$ is not on the graph

43. $x = -4, 2, 6$ **45.** peak at $(.5078, .3938)$; valleys at $(-1.9826, -4.2009)$ and $(3.7248, -8.7035)$
47. (a)

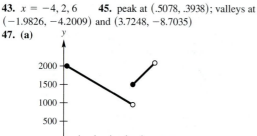

(b) Adjusted for inflation, the maximum yearly IRA contribution fell from $2000 to $1000 during this period.
49. (a)

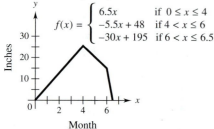

$$f(x) = \begin{cases} 6.5x & \text{if } 0 \le x \le 4 \\ -5.5x + 48 & \text{if } 4 < x \le 6 \\ -30x + 195 & \text{if } 6 < x \le 6.5 \end{cases}$$

(b) at the beginning of February; 26 in.   **(c)** begins in early October; ends in mid-April

51. (a) $f(x) = \begin{cases} 1.3x + 15.1 & \text{if } 0 \le x \le 25 \\ 8.9x - 175 & \text{if } x > 25 \end{cases}$

(b)

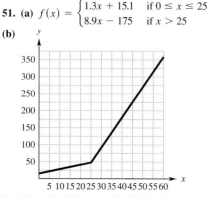

(c) 270   **(d)** 350.1   **53. (a)** No. The graph for all items is always above the x-axis, which means that the percent change was positive every year; hence, the CPI was always increasing.   **(b)** From early 1997 to late 1998 and from early 2001 to early 2002.   **(c)** From 1990 to early 1997; from late 1998 to early 2001; and after early 2002. The percent change was positive during these periods, which means that the CPI was increasing.
55. (a)   **(b)** $y = .044x - 86.7$

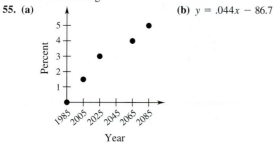

(c) $f(x) = .044x - 86.7$ **(d)** 4.16; yes; yes   **57. (a)** 33; 37
(b) The figure has vertical line segments, which can't be part of the graph of a function. (Why?) To make the figure into the graph of f, delete the vertical line segments; then, for each horizontal segment of the graph, put a closed dot on the left end and an open circle on the right end (as in Figure 3.7).   **59. (a)** $29 **(b)** $29 **(c)** $33
(d) $33
(e)

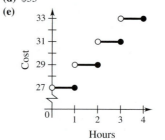

61. There are many correct answers including

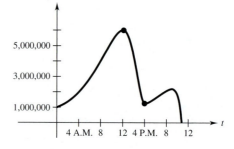

Section 3.3 (Page 159)
1. Let $C(x)$ be the cost of renting a saw for x for hours; $C(x) = 12 + x$.   **3.** Let $P(x)$ be the cost (in cents) of parking for x half hours; $P(x) = 200 + 50x$ **5.** $C(x) = 36x + 200$
7. $C(x) = 120x + 3800$ **9.** $48; $15.60; $13.80   **11.** $55.50; $11.40; $8.46; $8.97   **13. (a)** $f(x) = -1360x + 15,350$
(b) $7190   **(c)** It is decreasing in value at a rate of $1360 per year.
15. (a) $f(x) = -11,875x + 120,000$ **(b)** $[0, 8]$ **(c)** $48,750
17. (a) $80,000   **(b)** $42.50   **(c)** $122,500; $1,440,000
(d) $122.50; $45   **19. (a)** $C(x) = .097x + 1.32$ **(b)** $98.32
(c) $98.417   **(d)** $.097, or 9.7¢   **(e)** $.097, or 9.7¢
21. $R(x) = 2.0337x + 1,845,000$ **23. (a)** $C(x) = 10x + 500$
(b) $R(x) = 35x$ **(c)** $P(x) = 25x - 500$ **(d)** $2000
25. (a) $C(x) = 18x + 250$ **(b)** $R(x) = 28x$
(c) $P(x) = 10x - 250$ **(d)** $750
27. (a) $C(x) = 12.5x + 18,000$ **(b)** $R(x) = 25x$
(c) $P(x) = 12.5x - 18,000$ **(d)** $-$16,750 (loss)
29. $(3, -1)$ **31.** $\left(-\dfrac{11}{4}, -\dfrac{61}{4}\right)$
33. (a) 200,000 policies $(x = 200)$

(b)

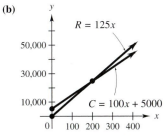

(c) revenue: $12,500; cost: $15,000   **35. (a)** $c = .126x + 1.5$
(b) $2.382 million   **(c)** about 17.857 units   **37.** Break-even point
is about 467 units; do not produce the item.   **39.** Break-even point
is about 1037 units; produce the item.   **41.** The compensation of
about $20/hour was the same in late 2000.   **43. (a)** Net imports:
$f(x) = .3077x + 7.2$; domestic supply: $g(x) = -.1x + 9.7$
(b)

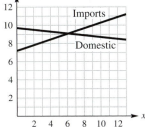

(c) $(6.1320, 9.0868)$; this means that net imports were the same as
domestic supply in early 1996, at about 9,086,800 barrels per day each.
45. $140   **47.** 10 items   **49. (a)** $16   **(b)** $11   **(c)** $6
(d) 8 units   **(e)** 4 units   **(f)** 0 units

(g)   **(h)** 0 units   **(i)** $\dfrac{40}{3}$ units

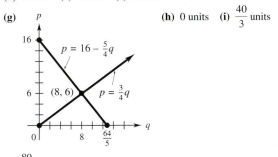

(j) $\dfrac{80}{3}$ units   **(k)** See part (g).   **(l)** 8 units   **(m)** $6

51. (a)   **(b)** 125 units   **(c)** 50¢

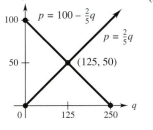

(d) $[0, 125)$ **53.** Total cost increases when more items are made
(because it includes the cost of all previously made items), so the
graph cannot move downward. No; the average cost can decrease as
more items are made, so its graph can move downward.

Section 3.4 (Page 168)

1. upward   **3.** downward   **5.** upward   **7.** $(4, 5)$, downward
9. $(-6, -7)$, upward   **11.** $(3.5, -9)$, upward   **13.** I

15. K   **17.** J   **19.** F   **21.** $f(x) = \dfrac{1}{4}(x - 1)^2 + 2$

23. $f(x) = (x + 1)^2 - 2$ **25.** $f(x) = 3x^2$ **27.** $(4, -12)$

29. $(-3, -21)$ **31.** $\left(\dfrac{1}{2}, \dfrac{1}{4}\right)$ **33.** $\left(\dfrac{2}{3}, \dfrac{19}{3}\right)$

35. x-intercepts 1, 3; y-intercept 9
37. x-intercepts -1, -3; y-intercept 6

39. $(-2, 0)$, $x = -2$ **41.** $(1, -3)$, $x = 1$

$f(x) = (x + 2)^2$ $f(x) = (x - 1)^2 - 3$

43. $(2, 2)$, $x = 2$ **45.** $(1, 3)$, $x = 1$

$f(x) = x^2 - 4x + 6$
$f(x) = (x - 2)^2 + 2$

$f(x) = 2x^2 - 4x + 5$
$f(x) = 2(x - 1)^2 + 3$

47. about 47   **49. (a)** 100 ft   **(b)** 5
51. (a)–(d)

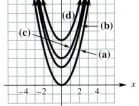

(e) The graph of $f(x) = x^2 + c$ is the graph of $k(x) = x^2$
shifted c units upward.

53. (a)–(d)

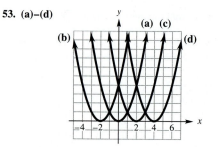

(e) The graph of $f(x) = (x + c)^2$ is the graph of $k(x) = x^2$ shifted c units to the left. The graph of $f(x) = (x - c)^2$ is the graph of $k(x) = x^2$ shifted c units to the right.

Section 3.5 (Page 176)

1. (a) $30; $9; $105   **(b)**

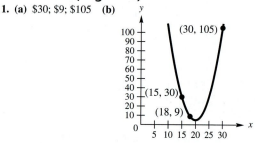

(c) $(20, 5)$ **(d)** 20 boxes per day; $5 per box
3. (a) 10 milliseconds   **(b)** 40 responses per millisecond
5. (a) 27 cases   **(b)** Answers vary.   **(c)** 15 cases   **7. (a)** about
12 books   **(b)** 10 books   **(c)** about 7 books   **(d)** 0 books
(e) 5 books   **(f)** about 7 books   **(g)** 10 books   **(h)** about 12 books
(i)

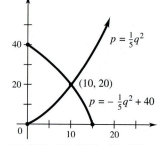

9. (a) $640   **(b)** $515   **(c)** $140
(d)   **(e)** 800 units   **(f)** $320

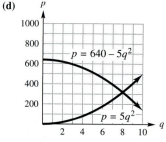

11. equilibrium quantity, 80; equilibrium price, $3600   **13.** equilibrium quantity, 30; equilibrium price, $1500   **15.** 20   **17.** 10
19. (a) $R(x) = (100 - x)(200 + 4x)$
$= 20,000 + 200x - 4x^2$
(b)

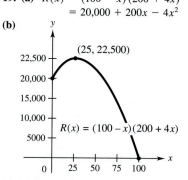

(c) 25 seats   **(d)** $22,500   **21.** 13 weeks; $96.10/hog
23. (a) $g(x) = .0625x^2 + 2.1$ **(b)** $4.35 billion; $6.1 billion (The
answer given by quadratic regression provides a better model and
suggests less spending; see Exercise 27.)
25. (a) $f(x) = 75.5(x - 7)^2 + 3820$ **(b)** $8652; $14,692 (The
answer given by quadratic regression
provides a better model and lower costs; see Exercise 29.)
27. $g(x) = .02875x^2 + .1457x + 2.0771$
estimates approximate spending of $3.98 billion in 2006 and
$5.07 billion in 2008, somewhat lower than the corresponding numbers in Exercise 23.

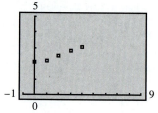

29. The model is $f(x) = 58.667x^2 - 725.7x + 6040.17$.

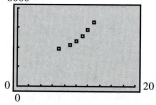

(a) Since r^2 is very close to 1, the model is a good one.   **(b)** About
$8355; about $13,4 31—lower than in Exercise 25.
31. 80 ft by 160 ft   **33. (a)** 11.3 and 88.7   **(b)** 50   **(c)** $3000
(d) $x < 11.3$ or $x > 88.7$ **(e)** $11.3 < x < 88.7$

Section 3.6 (Page 186)

1.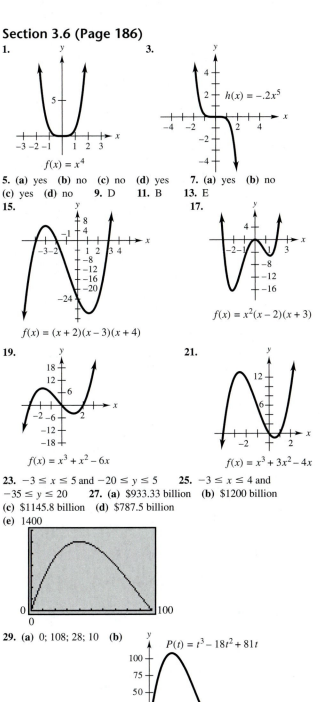

$f(x) = x^4$

3.

$h(x) = -.2x^5$

5. (a) yes **(b)** no **(c)** no **(d)** yes   **7. (a)** yes   **(b)** no
(c) yes **(d)** no   **9.** D   **11.** B   **13.** E
15.

$f(x) = (x + 2)(x - 3)(x + 4)$

17.

$f(x) = x^2(x - 2)(x + 3)$

19.

$f(x) = x^3 + x^2 - 6x$

21.

$f(x) = x^3 + 3x^2 - 4x$

23. $-3 \le x \le 5$ and $-20 \le y \le 5$ **25.** $-3 \le x \le 4$ and
$-35 \le y \le 20$ **27. (a)** \$933.33 billion   **(b)** \$1200 billion
(c) \$1145.8 billion   **(d)** \$787.5 billion
(e)

29. (a) 0; 108; 28; 10   **(b)**

$P(t) = t^3 - 18t^2 + 81t$

(c) increasing for years 0 to 3 and from the 9th year on; decreasing for
the years 3 to 9   **31. (a)** 1,357,500,000; 1,452,300,000
(b) When x is large, the graph must resemble the graph of
$y = -.00096x^3$, which drops down forever at the far right. (See the
chart on page 180.) This would mean that China's population would
become 0 at some point.
33. (a)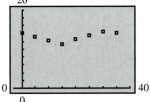

(b) $f(x) = -.000035606x^4 + .0021x^3 - .02714x^2 - .12059x + 14.2996$ **(c)** It fits reasonably well.

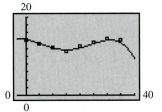

(d) about 13.7 million; about 14.2 million   **(e)** Late 1989

Section 3.7 (Page 196)

1. $x = -5, y = 0$ **3.** $x = -\dfrac{5}{2}, y = 0$

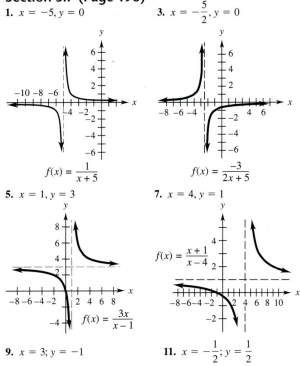

$f(x) = \dfrac{1}{x + 5}$ $f(x) = \dfrac{-3}{2x + 5}$

5. $x = 1, y = 3$ **7.** $x = 4, y = 1$

$f(x) = \dfrac{3x}{x - 1}$ $f(x) = \dfrac{x + 1}{x - 4}$

9. $x = 3; y = -1$ **11.** $x = -\dfrac{1}{2}; y = \dfrac{1}{2}$

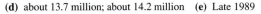

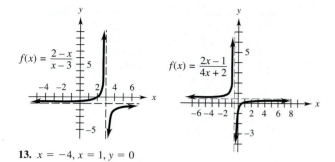

$$f(x) = \frac{2-x}{x-3}$$

$$f(x) = \frac{2x-1}{4x+2}$$

13. $x = -4, x = 1, y = 0$

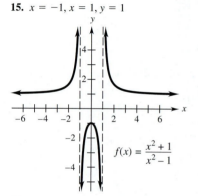

$$h(x) = \frac{x+1}{x^2+3x-4}$$

15. $x = -1, x = 1, y = 1$

$$f(x) = \frac{x^2+1}{x^2-1}$$

17. $x = -2; x = 1$

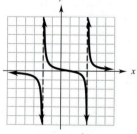

$$f(x) = \frac{x-1}{x^2-x-6}$$

19. $x = -1; x = 5$ **21. (a)** \$4300   **(b)** \$10,033.33
(c) \$17,200   **(d)** \$38,700   **(e)** \$81,700   **(f)** \$210,700
(g) \$425,700   **(h)** no

(i)

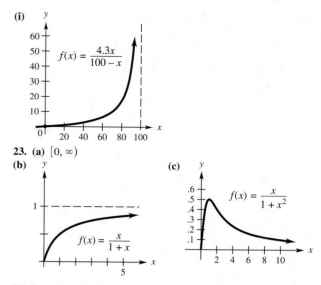

$$f(x) = \frac{4.3x}{100-x}$$

23. (a) $[0, \infty)$

(b)

$$f(x) = \frac{x}{1+x}$$

(c)

$$f(x) = \frac{x}{1+x^2}$$

(d) Increasing b makes the next generation smaller when this
generation is larger.   **25. (a)** 6 min   **(b)** 1.5 min
(c) .6 min   **(d)** $A = 0$ **(e)**

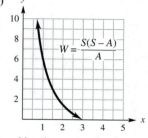

$$W = \frac{S(S-A)}{A}$$

(f) W becomes negative. The waiting time approaches 0 as A
approaches 3. The formula does not apply for $A > 3$ because there will
be no waiting if people arrive more than 3 min apart.
27. 30,000 reds; 10,000 blues

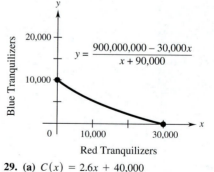

Blue Tranquilizers

$$y = \frac{900,000,000 - 30,000x}{x + 90,000}$$

Red Tranquilizers

29. (a) $C(x) = 2.6x + 40,000$

(b) $\overline{C}(x) = \dfrac{2.6x + 40,000}{x} = 2.6 + \dfrac{40,000}{x}$

(c) $y = 2.6$; the average cost may get close to, but will never equal,
\$2.60.   **31.** about 73.9

33. (a)

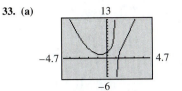

(b) They appear almost identical, because the parabola is an asymptote of the graph.

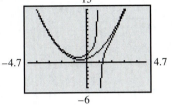

Chapter 3 Review (Page 199)

1. not a function   **3.** function   **5.** not a function   **7. (a)** 23
(b) -9 **(c)** $4p - 1$ **(d)** $4r + 3$ **9. (a)** -28 **(b)** -12
(c) $-p^2 + 2p - 4$ **(d)** $-r^2 - 3$ **(b)** 6   **(c)** $8 - p - p^2$
(d) $-r^2 - 3r + 6$ **11. (a)** -13 **(b)** 3   **(c)** $-k^2 - 4k$
(d) $-9m^2 + 12m$ **(e)** $-k^2 + 14k - 45$ **(f)** $12 - 5p$

13.

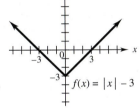

$f(x) = |x| - 3$

15.

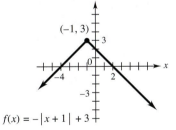

$f(x) = -|x + 1| + 3$

17.

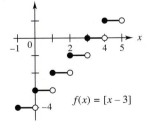

$f(x) = [x - 3]$

19.

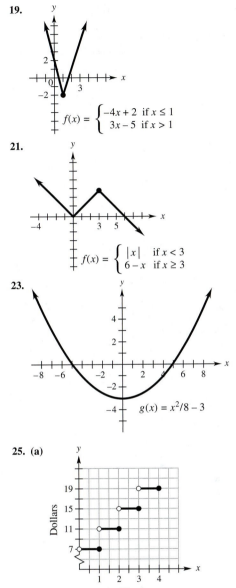

$f(x) = \begin{cases} -4x + 2 & \text{if } x \le 1 \\ 3x - 5 & \text{if } x > 1 \end{cases}$

21.

$f(x) = \begin{cases} |x| & \text{if } x < 3 \\ 6 - x & \text{if } x \ge 3 \end{cases}$

23.

$g(x) = x^2/8 - 3$

25. (a)

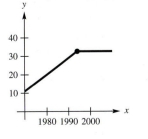

Dollars

Hours

(b) domain: $(0, \infty)$; range: $\{7, 11, 15, 19, \dots\}$ **(c)** 3 days
27. These births appear to be leveling off.

1980 1990 2000

29. (a) $C(x) = 30x + 60$ (b) $30   (c) $30.60
31. (a) $C(x) = 30x + 85$ (b) $30   (c) $30.85
33. (a) $18,000   (b) $R(x) = 28x$ (c) 4500 cartridges
(d) $126,000   **35.** Equilibrium quantity is 36 million subscribers
at a price of $12.95 per month.   **37.** upward; $(2, 6)$
39. downward; $(-1, 8)$

41.

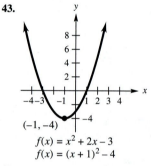

$f(x) = x^2 - 4$

43.

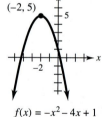

$f(x) = x^2 + 2x - 3$
$f(x) = (x + 1)^2 - 4$

45.

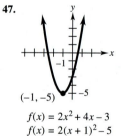

$f(x) = -x^2 - 4x + 1$
$f(x) = -(x + 2)^2 + 5$

47.

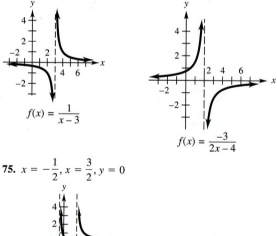

$f(x) = 2x^2 + 4x - 3$
$f(x) = 2(x + 1)^2 - 5$

49. minimum value; -11 **51.** maximum value; 7   **53.** 4
months after she began   **55.** 125 units
$f(x) = .00405(x - 5)^2 + .38$

57. (a)
about $2.9113 trillion   (b) About $2.5225 trillion;

59.

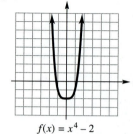

$f(x) = x^4 - 2$

61.

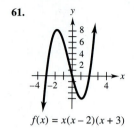

$f(x) = x(x - 2)(x + 3)$

63.

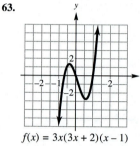

$f(x) = 3x(3x + 2)(x - 1)$

65.

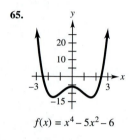

$f(x) = x^4 - 5x^2 - 6$

67. about 313,152; about $690.72 per thousand
69. (a) $R(x) = 23x$; $P(x) = .000006x^3 - .07x^2 + 21x - 1200$
(b) about 76.54, which means that 77 must be made to earn a profit
(you can't make .54 of a rack)   (c) 230   (d) 153; about $395.86
71. $x = 3, y = 0$ **73.** $x = 2, y = 0$

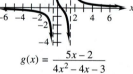

$f(x) = \dfrac{1}{x - 3}$

$f(x) = \dfrac{-3}{2x - 4}$

75. $x = -\dfrac{1}{2}, x = \dfrac{3}{2}, y = 0$

$g(x) = \dfrac{5x - 2}{4x^2 - 4x - 3}$

77. (a) about $10.83 **(b)** about $4.64 **(c)** about $3.61 **(d)** about $2.71 **(e)**

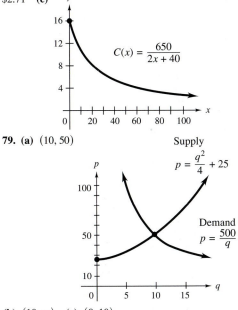

$$C(x) = \frac{650}{2x + 40}$$

79. (a) $(10, 50)$

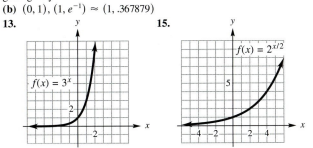

Supply

$$p = \frac{q^2}{4} + 25$$

Demand

$$p = \frac{500}{q}$$

(b) $(10, \infty)$ **(c)** $(0, 10)$

Case 3 (Page 204)

1. $f(x) = -\frac{15}{64}x^2 + 15$ **3.** $h(x) = \sqrt{64 - x^2} + 7$; 7 ft

Chapter 4

Section 4.1 (Page 212)

1. exponential   **3.** quadratic   **5.** exponential   **7. (a)** The graph is entirely above the x-axis and falls from left to right, crossing the y-axis at 1 and then getting very close to the x-axis.   **(b)** $(0, 1)$, $(1, .6)$ **9. (a)** The graph is entirely above the x-axis

and rises from left to right, less steeply than the graph of $f(x) = 2^x$.
(b) $(0, 1)$, $(1, 2^{.5}) = (1, \sqrt{2})$ **11. (a)** The graph is entirely above the x-axis and falls from left to right, crossing the y-axis at 1 and then getting very close to the x-axis.
(b) $(0, 1)$, $(1, e^{-1}) \approx (1, .367879)$

13.

$$f(x) = 3^x$$

15.

$$f(x) = 2^{x/2}$$

17.

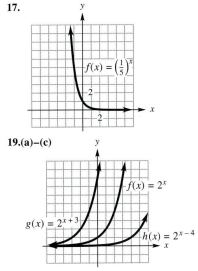

$$f(x) = \left(\frac{1}{5}\right)^x$$

19.(a)–(c)

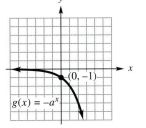

$$g(x) = 2^{x+3}$$ $$f(x) = 2^x$$ $$h(x) = 2^{x-4}$$

21. 2.3   **23.** .75   **25.** .31   **27. (a)** $a > 1$ **(b)** domain: $(-\infty, \infty)$; range; $(0, \infty)$
(c)

$(0, -1)$

$$g(x) = -a^x$$

(d) domain: $(-\infty, \infty)$; range; $(-\infty, 0)$
(e)

$(0, 1)$

$$h(x) = a^{-x}$$

(f) domain: $(-\infty, \infty)$; range; $(0, \infty)$ **29. (a)** 3   **(b)** $\frac{1}{3}$ **(c)** 9
(d) 1

31.

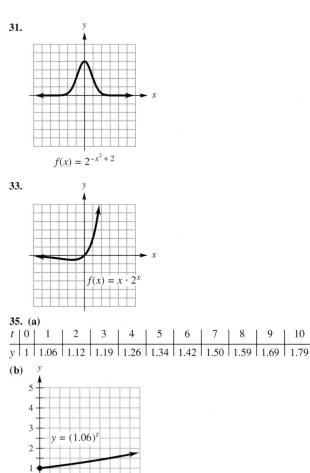

$$f(x) = 2^{-x^2+2}$$

33.

$$f(x) = x \cdot 2^x$$

35. (a)

t	0	1	2	3	4	5	6	7	8	9	10
y	1	1.06	1.12	1.19	1.26	1.34	1.42	1.50	1.59	1.69	1.79

(b)

$y = (1.06)^t$

37. (a) about \$141,892 **(b)** about \$64.10 **39. (a)** about 95,648,000 **(b)** about 164,290,000 **(c)** No; the model estimates about 580,490,000 accounts in 2010, when the U.S. population is projected to be about 309,000,000 **41. (a)** about .97 kg **(b)** about .75 kg **(c)** about .65 kg **(d)** about 24,360 years **43. (a)** about 6.2 billion **(b)** about 6.5 billion **(c)** about 8 billion **(d)** answers vary **45.** \$19,420.26 **47.** Answers vary. **49. (a)** about 5951, 5711, and 5618 **(b)** In 2014

Section 4.2 (Page 219)

1. (a) \$752.27 **(b)** \$707.39 **(c)** \$432.45 **(d)** \$298.98 **(e)** Answers vary. **3. (a)** about \$1084.55 **(b)** about \$544.97 **5. (a)** $f(t) = 32.5(1.022362^t)$ **(b)** about 40,500,000; about 70,500,000 **(c)** 2023 **7. (a)** $f(t) = 2(1.2917^t)$, where $f(t)$ is in billions of dollars **(b)** about \$9,289,700,000 in 2003 and \$15,500,000,000 in 2005. **9. (a)** two-point: $f(t) = .928(.97098599^t)$; regression: $g(t) = .90383(.9709574^t)$, **(b)** two-point: .597; .500; .471; regression: .581, .487, .459 **(c)** two-point: 2013; regression: 2012 **11. (a)** two-point:

$f(t) = 492.7(.97783^t)$; regression: $g(t) = 502.67(.977755^t)$ **(b)** two-point: 281.3, 224.8; regression: 286.4, 228.7 **(c)** 2041 **13. (a)** about 6 items **(b)** about 23 items **(c)** 25 items **(b)** about 42 **(c)** about 56 **(d)** about 59 **15.** $-4.98°C$ **17. (a)** .13 **(b)** .23 **(c)** about 2 weeks **19. (a)** about 71,295,000; about 73,170,000 **(b)** 100

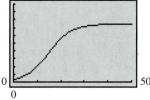

(c) In 2012 **(d)** no **21. (a)** about \$5,756,700,000,000; about \$7,308,000,000,000; about \$8,414,800,000,000 **(b)** 35

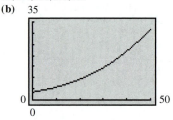

(c) In 2011 **(d)** yes, but not in your lifetime

Section 4.3 (Page 230)

1. a^y **3.** It is missing the value that equals b^y. If that value is x, the expression should read $y = \log_b x$. **5.** $10^5 = 100,000$

7. $9^2 = 81$ **9.** $\log 75 = 1.8751$ **11.** $\log_3\left(\dfrac{1}{9}\right) = -2$

13. 3 **15.** 2 **17.** 3 **19.** -2 **21.** $\dfrac{1}{2}$ **23.** 8.77

25. 1.724 **27.** $-.794$ **29.** Because $a^0 = 1$ for every valid base a **31.** $\log 16$ **33.** $\ln 5$

35. $\log\left(\dfrac{u^2 w^3}{v^6}\right)$ **37.** $\ln\left(\dfrac{(x+1)^2}{x+2}\right)$ **39.** $\dfrac{1}{2}\ln 6 + 2\ln m + \ln n$

41. $\dfrac{1}{2}\log x - \dfrac{5}{2}\log z$ **43.** $2u + 5v$ **45.** $3u - 2v$

47. 3.26296 **49.** 2.5151 **51.** Many correct answers, including $b = 1, c = 2$.

53.

55.

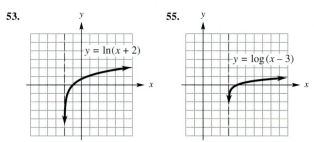

$y = \ln(x+2)$

$y = \log(x-3)$

57. Answers vary.   **59.**

61. (a) 17.67 yr   (b) 9.01 yr   (c) 4.19 yr   (d) 2.25 yr

63. (a) about 21,511; about 28,608; about 32,760

(b)

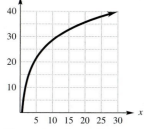

(c) The number of restaurants is increasing at a slower rate as time goes on. What feature of the graph tells you this?   **65.** (a) about 20.2%; about 21.8%; about 23.3%; about 24.4%

(b)

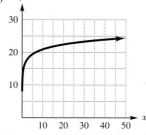

(c) It appears to be leveling off at a bit below 25%.   **67.** 1.5887

Section 4.4 (Page 239)

1. 9   **3.** no solution   **5.** 11   **7.** $\dfrac{11}{6}$ **9.** $\dfrac{5}{9}$ **11.** 10

13. 5.2378   **15.** 5   **17.** $\dfrac{4 + b}{4}$ **19.** $\dfrac{10^{2-b} - 5}{6}$

21. Answers vary.   **23.** 3   **25.** $-\dfrac{5}{6}$ **27.** -2

29. -2 **31.** 2.3219   **33.** 2.710   **35.** -1.825 **37.** .973

39. $-.123$ **41.** $\dfrac{\log d + 3}{4}$ **43.** $\dfrac{\ln b + 1}{2}$ **45.** 3

47. no solution   **49.** $-4, 4$ **51.** 1, 10   **53.** 1   **55.** 4, -4

57. ±2.0789 **59.** 1.386   **61.** Answers vary.

63. (a) 11.1; 9.4   (b) in 2019   **65.** 14.2 hr   **67.** (a) 25 g

(b) about 4.95 yr   **69.** about 3689 yr old   **71.** (a) approximately $3,981,072i_0$ (b) approximately $3,162,278i_0$ (c) about 1.26 times stronger.   **73.** (a) 21   (b) 100   (c) 105   (d) 120

(e) 140   **75.** (a) 27.5%   (b) $130.14

77. (a)

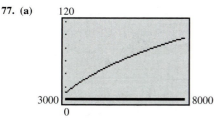

(b) about 4717 ft

79. (a) $f(x) = -4964.2 + 6284 \ln x$ (b) 14,460   (c) 2010

Chapter 4 Review (Page 243)

1. (c)   **3.** (d)   **5.** $0 < a < 1$ **7.** all positive real numbers

9.   **11.**

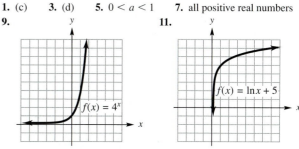

13. (a) about 15.24%; about 40.89%   (b) in 2009; no—the model seems unlikely to be accurate after 2008.   **15.** $\log 55 = 1.7404$

17. $\ln 45 = 3.8067$ **19.** $10^4 = 10,000$ **21.** $e^{4.3957} = 81.1$

23. 5   **25.** 8.9   **27.** $\dfrac{1}{2}$ **29.** $\log 12x^5$ **31.** $\log\left(\dfrac{b^3}{c^2}\right)$

33. 3   **35.** 8   **37.** 6   **39.** 2   **41.** -1 **43.** 2

45. 1.416   **47.** -2.807 **49.** -3.305 **51.** .747

53. 28.463   **55.** (a) $15 million   (b) $15.6 million

(c) $16.4 million   **57.** (a) 10 g   (b) about 140 days   (c) about 243 days   **59.** 81.25°C   **61.** (a) $f(x) = 272.6(.70795^x)$

(b) $g(x) = 350.29(.7116^x)$ (c) 2-point: about 12,177,000; regression: about 16,389,000   (d) 2-point: 2010; regression: 2011

63. (a) $f(x) = 27.8 + 16.80345 \ln x$

(b) $g(x) = 25.83 + 16.04 \ln x$ (c) two-point: about 75,408,000; regression: about 71,275,000   (d) two-point: 2013; regression: 2020

Case 4 (Page 248)

1. about 23.6, 47.7, and 58.4; these estimates are a bit low.

Chapter 5

Section 5.1 (Page 254)

1. time and interest rate   **3.** $133   **5.** $217.48   **7.** $86.26

9. $158.82   **11.** $315.45   **13.** $3231.18   **15.** Answers vary.

17. $46,451.61   **19.** $28,048.80   **21.** $8898.75

23. $48,178.45   **25.** $r \approx 7.34\%$ **27.** $r \approx 11.75\%$

29. $27,894.30   **31.** $3056.25   **33.** 5.0%   **35.** 11.4%

37. $1750.04   **39.** $5825.24   **41.** $3773; 13.58%

43. $7278.68; no   **45.** (a) $372.77   (b) $26.94   (c) No, he will still owe $166.45.

Section 5.2 (Page 265)

1. Answers vary.   **3.** interest rate and number of compounding periods   **5.** Answers vary.   **7.** $1368.57   **9.** $1204.75

11. $9020.99   **13.** $8793.87   **15.** $2307.95   **17.** $1968.48

19. 4.50%   **21.** 5.25%   **23.** $28,187.42   **25.** $45,552.97

27. $83,002.92   **29.** 4.04%   **31.** 6.09%   **33.** 5.268%

35. $8954.58   **37.** $3384.27   **39.** $11,572.58

41. $1000 now   **43.** $30,611.30   **45.** about $1.946 million
47. Flagstar, 4.45%; Principal, 4.46%; Principal   **49.** 3.20%,
3.69%, 3.93%, 4.17%, 4.65%   **51.** $7522.50
53. (a) $16,659.95 **(b)** $21,472.67   **55.** $1000 now
57. 17.7 years   **59.** about 11.9 years   **61.** 10%
63. (a) $16,288.95 **(b)** $16,436.19 **(c)** $16,470.09
(d) $16,486.65   **65. (a)**

Section 5.3 (Page 276

1. Answers vary.   **3. (a)** $a_1 = 1169$; $r = .916$ **(b)** $a_{10} = 531$;
$a_{20} = 221$. This means that a person who is 10 years from retirement
should have savings of 531% of his or her annual salary; a person
20 years from retirement should have savings of 221% of his or her
annual salary.   **5.** 108   **7.** 3   **9.** 2315.25   **11.** 45
13. 6.24   **15.** 594.048   **17.** 15.91713   **19.** 21.82453
21. 48.88637   **23.** $119,625.61   **25.** $23,242.87
27. $72,482.38   **29.** $777.93   **31.** $2165.29   **33.** Answers
vary.   **35.** $6398.20   **37.** $1539.06   **39.** $275.39
41. $6603.39   **43.** $234,295.32   **45.** $26,671.23
47. $3928.88   **49.** $144,872.14   **51.** $4168.30
53. (a) $137,895.79 **(b)** $132,318.77 **(c)** $5577.02
55. $130,159.72   **57.** $284,527.35   **59. (a)** $256.08
(b) $247.81   **61.** $626.75   **63.** 7.397%   **65. (a)** $1200
(b) $3511.58
(c)

Payment Number	Amount of Deposit	Interest Earned	Total
1	$3511.58	$0.00	$3511.58
2	3511.58	105.35	7128.51
3	3511.58	213.86	10,853.95
4	3511.58	325.62	14,691.15
5	3511.58	440.73	18,643.46
6	3511.58	559.30	22,714.34

Section 5.4 (Page 285)

1. (c)   **3.** 10.379658   **5.** 12.65930   **7.** 14.13126
9. Answers vary.   **11.** $8693.71   **13.** $1,566,346.66
15. $221,358.80   **17.** $111,183.87   **19.** Answers vary.
21. $5181.53   **23.** $11,727.32   **25.** $252.53   **27.** $1149.60
29. $781.69   **31.** $86.24   **33.** $13.02   **35.** $48,677.34
37. (a) $158 **(b)** $1584   **39.** $19,900,932.64
41. (a) $1,800,000 **(b)** $29,671,601.18   **43.** $414.18;
$14,701.60   **45. (a)** $2717.36 **(b)** 2   **47. (a)** $1151.22
(b) about $152,320.58 **(c)** about $2549.78 **(d)** about 21.22 years
49. $320.03   **51.** $127,831.45
53.

Payment Number	Amount of Payment	Interest for Period	Portion to Principal	Principal at End of Period
0	–	–	–	$72,000.00
1	$9247.24	$2160.00	$7087.24	64,912.76
2	9247.24	1947.38	7299.85	57,612.91
3	9247.24	1728.39	7518.85	50,094.06
4	9247.24	1502.82	7744.42	42,349.64

55.

Payment Number	Amount of Payment	Interest for Period	Portion to Principal	Principal at End of Period
0	–	–	–	$20,000.00
1	$2404.83	$700.00	$1704.83	18,295.17
2	2404.83	640.33	1764.50	16,530.67
3	2404.83	578.57	1826.26	14,704.42
4	2404.83	514.65	1890.18	12,814.24

Chapter 5 Review (Page 289)

1. $292.08   **3.** $62.05   **5.** Answers vary.   **7.** $78,742.54
9. $742.03   **11.** Answers vary.   **13.** $73,402.52; $15,593.18
15. $5282.19; $604.96   **17.** $12,857.07   **19.** $1935.77
21. 16, 8, 4, 2   **23.** -32 **25.** 5500   **27.** Answers vary.
29. $162,753.15   **31.** $22,643.29   **33.** Answers vary.
35. $2619.29   **37.** $916.12   **39.** $31,921.91
41. $14,222.42   **43.** $3581.11   **45.** $392.70   **47.** $896.06
49. $2696.12   **51.** $10,550.54   **53.** $8369.51
55. (a) about 22.3716% **(b)** about 5.8434%   **57.** about 3.014%
59. $24,818.76; $2418.76   **61.** $32.49   **63.** $5596.62
65. $3560.61   **67.** (d)   **69.** $5927.56

Case 5 (Page 294)

1.

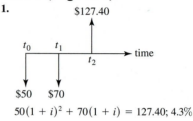

$50(1 + i)^2 + 70(1 + i) = 127.40$; 4.3%

3. (a)

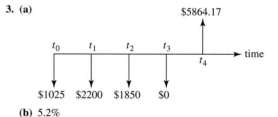

(b) 5.2%

5. (a)

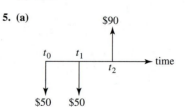

$50(1 + i)^2 + 50(1 + i) - 90 = 0$; $-.068$, -2.932
(b) $-.068$ is reasonable, -2.932 is not.

Chapter 6

Section 6.1 (Page 309)

1. yes   **3.** $(-1, -4)$ **5.** $\left(\dfrac{2}{7}, -\dfrac{11}{7}\right)$

7. $\left(\dfrac{5}{2}, -\dfrac{5}{2}\right)$ **9.** $\left(\dfrac{11}{5}, -\dfrac{7}{5}\right)$ **11.** $(28, 22)$ **13.** $(2, -1)$

15. no solution   **17.** $(4y + 1, y)$ for any real number y

19. no solution   **21.** (a)   **23.** $(5, 10)$ **25.** (a) $r = 175{,}000$
and $b = 375{,}000$ (b) Answers vary.   **27.** Never   **29.** 300 of
Boeing; 100 of GE   **31.** 5 of model 201; 8 of model 301

33.
$$\begin{aligned} x \quad\quad - 3z &= 2 \\ 2x - 4y + 5z &= 1 \\ 5x - 8y + 7z &= 6 \\ 3x - 4y + 2z &= 3 \end{aligned}$$

35.
$$\begin{aligned} 3x \quad\quad + z + 2w + 18v &= 0 \\ -4x + y \quad\quad - w - 24v &= 0 \\ 7x - y + z + 3w + 42v &= 0 \\ 4x \quad\quad + z + 2w + 24v &= 0 \end{aligned}$$

37.
$$\begin{aligned} x + y + 2z + 3w &= 1 \\ - y - z - 2w &= -1 \\ 3x + y + 4z + 5w &= 2 \end{aligned}$$

39.
$$\begin{aligned} x + 12y - 3z + 4w &= 10 \\ 2y + 3z + w &= 4 \\ - z &= -7 \\ 6y - 2z - 3w &= 0 \end{aligned}$$

41. $(-68, 13, -6, 3)$ **43.** $\left(20, -9, \dfrac{15}{2}, 3\right)$

45. $\begin{bmatrix} 2 & 1 & 1 & \bigm| & 3 \\ 3 & -4 & 2 & \bigm| & -5 \\ 1 & 1 & 1 & \bigm| & 2 \end{bmatrix}$

47.
$$\begin{aligned} 2x + 3y + 8z &= 20 \\ x + 4y + 6z &= 12 \\ 3y + 5z &= 10 \end{aligned}$$

49. $\begin{bmatrix} 1 & 2 & 3 & \bigm| & -1 \\ 2 & 0 & 7 & \bigm| & -4 \\ 6 & 5 & 4 & \bigm| & 6 \end{bmatrix}$ **51.** $\begin{bmatrix} -4 & -3 & 1 & -1 & \bigm| & 2 \\ 0 & -4 & 7 & -2 & \bigm| & 10 \\ 0 & -2 & 9 & 4 & \bigm| & 5 \end{bmatrix}$

53. dependent   **55.** inconsistent   **57.** independent

59. $\left(\dfrac{3}{2}, \dfrac{3}{2}, -\dfrac{3}{2}\right)$ **61.** $(-14, -6, 2)$ **63.** $(100, 50, 50)$

65. $(1, 2, -1)$ **67.** $(2, 0, 3)$ **69.** no solution   **71.** $(0, 2, 4)$

73. $\left(\dfrac{1 - z}{2}, \dfrac{11 - z}{2}, z\right)$ for any real number z

75. $\left(z, \dfrac{1}{2} - 2z, z\right)$ for any real number z

77. (a) $y = x + 1$ (b) $y = 3x + 4$ (c) $\left(-\dfrac{3}{2}, -\dfrac{1}{2}\right)$

79.

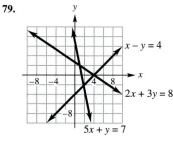

There is no point that is on
all three lines.

81. \$20,000 in AAA; \$10,000 in B; none in A

83. 40 lb corn chips; $46\dfrac{2}{3}$ lb nuts; $13\dfrac{1}{3}$ lb pretzels   **85.** 400 box
seats; 5400 grandstand seats; 1200 bleacher seats   **87.** (a) $b + c$
(b) .35, .45, .2; A is tumorous, B is bone, and C is healthy.
89. (a) $x_2 + x_3 = 700$
$\qquad\quad x_3 + x_4 = 600$
(b) $(1000 - x_4, 100 + x_4, 600 - x_4, x_4)$ (c) 600; 0
(d) x_1: 1000; 400
$\quad\; x_2$: 700; 100
$\quad\; x_3$: 600; 0
(e) Answers vary.

Section 6.2 (Page 322)

1. $\left(\dfrac{3}{2}, 7, -3, 0\right)$

3. $(2 - w, -3 - 2w, 5, w)$ for any real number w
5. $(-3, 4, 0)$ **7.** $(1, 0, -1)$ **9.** $(-14, -6, 2)$

11. $\left(\dfrac{-9z + 5}{23}, \dfrac{10z - 3}{23}, z\right)$ for any real number z

13. $(-1, 23, 16)$

15. $\left(2 - \dfrac{4}{3}z, -1 + \dfrac{1}{3}z, z\right)$ for any real number z

17. no solution   **19.** $(-7, 5)$ **21.** $(-1, 2)$
23. $(-3, z - 17, z)$ for any real number z
25. $(-1, 1, -3, -2)$ **27.** no solution

29. $\left(\dfrac{1}{2}, \dfrac{1}{3}, -\dfrac{1}{4}\right)$ **31.** Garcia 20 hr;
Wong 15 hr   **33.** 40 lb pretzels, 20 lb dried fruit, 80 lb nuts
35. (a) .056057, 1.06657   (b) 228 ft   **37.** \$15,000 in the mutual
fund, \$30,000 in bonds, and \$25,000 in the food franchise
39. 6 cups of Roasted Chicken Rotini, 9 cups of Hearty Chicken,
and 2 cups of Chunky Chicken Noodle; serving size 1.7 cups
41. (a) \$12,000 at 6%, \$7000 at 7%, and \$6000 at 10%
(b) \$10,000 at 6%, \$15,000 at 7%, and \$5000 at 10%
(c) \$20,000 at 6%, \$10,000 at 7%, and \$10,000 at 10%
43. (a) $y = .01x^2 - .3x + 4.24$ (b) 15 platters; \$1.99
45. (a) $f(x) = .006463x^2 + .159634x + 10.4629$
(b) about \$12.4 trillion; about \$14.3 trillion   (c) in 2036
47. (a) $C = -.0000108S^2 + .034896S + 22.9$
(b) about 864.7 knots

Section 6.3 (Page 331)

1. 2×3; $\begin{bmatrix} -7 & 8 & -4 \\ 0 & -13 & -9 \end{bmatrix}$

3. 3×3; square matrix; $\begin{bmatrix} 3 & 0 & -11 \\ -1 & -\frac{1}{4} & 7 \\ -5 & 3 & -9 \end{bmatrix}$

5. 2×1; column matrix; $\begin{bmatrix} -7 \\ -11 \end{bmatrix}$ row matrix; column matrix; $[5]$

7. B is a 5×3 zero matrix.

9. $\begin{bmatrix} 9 & 14 & 0 & 4 \\ 1 & -1 & 2 & -4 \end{bmatrix}$ **11.** $\begin{bmatrix} 3 & 13 & 2 \\ 3 & 1 & 8 \end{bmatrix}$

13. $\begin{bmatrix} 3 & -7 & 7 \\ 6 & 4 & -8 \end{bmatrix}$ **15.** $\begin{bmatrix} -4 & 0 \\ 10 & 6 \end{bmatrix}$ **17.** $\begin{bmatrix} 0 & -8 \\ -16 & 24 \end{bmatrix}$

19. $\begin{bmatrix} 8 & 10 \\ 0 & -42 \end{bmatrix}$ **21.** $\begin{bmatrix} 4 & -\frac{7}{2} \\ 4 & \frac{21}{2} \end{bmatrix}$

23. $X + T = \begin{bmatrix} x & y \\ z & w \end{bmatrix} + \begin{bmatrix} r & s \\ t & u \end{bmatrix} = \begin{bmatrix} x+r & y+s \\ z+t & w+u \end{bmatrix}$ a 2×2 matrix

25. $X + (T + P) = \begin{bmatrix} x+(r+m) & y+(s+n) \\ z+(t+p) & w+(u+q) \end{bmatrix}$

$\qquad = \begin{bmatrix} (x+r)+m & (y+s)+n \\ (z+t)+p & (w+u)+q \end{bmatrix}$

$\qquad = (X+T)+P$

27. $P + O = \begin{bmatrix} m+0 & n+0 \\ p+0 & q+0 \end{bmatrix} = \begin{bmatrix} m & n \\ p & q \end{bmatrix} = P$

29. Several possible correct answers, including

	basketball	hockey	football	baseball
percent of no shows	16	16	20	18
lost revenue per fan($)	18.20	18.25	19	15.40
lost annual revenue (millions $)	22.7	35.8	51.9	96.3

31. Several possible answers, including

	1998	2000	2005
Heart	4121	4143	3140
Lung	3171	3614	3601
Liver	12,070	15,359	17,376
Kidney	38,270	45,273	62,130

33. Several possible answers, including

	1995	2000	2001	2002
Ages 15–24	2.9	2.6	2.5	2.5
Ages 45–54	111	94.2	92.9	90.7
Ages 65–74	799.9	666.6	635.1	615.9

35. (a) $A = \begin{bmatrix} 2.61 & 4.39 & 6.29 & 9.08 \\ 1.63 & 2.77 & 4.61 & 6.92 \\ .92 & .75 & .62 & .54 \end{bmatrix}$

(b) $B = \begin{bmatrix} 1.38 & 1.72 & 1.94 & 3.31 \\ 1.26 & 1.48 & 2.82 & 2.28 \\ .41 & .33 & .27 & .40 \end{bmatrix}$

(c) $\begin{bmatrix} 1.23 & 2.67 & 4.35 & 5.77 \\ .37 & 1.29 & 1.79 & 4.64 \\ .51 & .42 & .35 & .14 \end{bmatrix}$

Section 6.4 (Page 343)

1. 2×2; 2×2 **3.** 3×3; 5×5 **5.** AB does not exist; 3×2

7. columns; rows

9. $\begin{bmatrix} 5 \\ 9 \end{bmatrix}$ **11.** $\begin{bmatrix} -2 & 12 \\ 0 & 8 \end{bmatrix}$ **13.** $\begin{bmatrix} -4 & 1 \\ 2 & -3 \end{bmatrix}$

15. $\begin{bmatrix} 3 & -5 & 7 \\ -2 & 1 & 6 \\ 0 & -3 & 4 \end{bmatrix}$ **17.** $\begin{bmatrix} 16 & 14 \\ 37 & 38 \\ 58 & 62 \end{bmatrix}$

19. $AB = \begin{bmatrix} -30 & -45 \\ 20 & 30 \end{bmatrix}$, but $BA = \begin{bmatrix} 0 & 0 \\ 0 & 0 \end{bmatrix}$.

21. $(A+B)(A-B) = \begin{bmatrix} -7 & -24 \\ -28 & -33 \end{bmatrix}$, but $A^2 - B^2 = \begin{bmatrix} -37 & -69 \\ -8 & -3 \end{bmatrix}$.

23. $P(XT) =$

$\begin{bmatrix} (mx+nz)r+(my+nw)t & (mx+nz)s+(my+nw)u \\ (px+qz)r+(py+qw)t & (px+qz)s+(py+qw)u \end{bmatrix}$

$P(XT)$ is the same, so $(PX)T = P(XT)$.

25. $k(X+T) = k\begin{bmatrix} x+r & y+s \\ z+t & w+u \end{bmatrix}$

$\qquad = \begin{bmatrix} k(x+r) & k(y+s) \\ k(z+t) & k(w+u) \end{bmatrix}$

$\qquad = \begin{bmatrix} kx+kr & ky+ks \\ kz+kt & kw+ku \end{bmatrix}$

$\qquad = \begin{bmatrix} kx & ky \\ kz & kw \end{bmatrix} + \begin{bmatrix} kr & ks \\ kt & ku \end{bmatrix} = kX + kT$

27. no   **29.** yes   **31.** yes

33. $\begin{bmatrix} 2 & 3 \\ 1 & 2 \end{bmatrix}$ **35.** no inverse   **37.** $\begin{bmatrix} 2 & -3 \\ -\frac{1}{2} & 1 \end{bmatrix}$

39. $\begin{bmatrix} -4 & -2 & 3 \\ -5 & -2 & 3 \\ 2 & 1 & -1 \end{bmatrix}$ **41.** $\begin{bmatrix} 2 & 1 & -1 \\ 8 & 2 & -5 \\ -11 & -3 & 7 \end{bmatrix}$

43. no inverse   **45.** $\begin{bmatrix} 3 & -5 & 8 \\ -\frac{1}{2} & 1 & -1 \\ -\frac{1}{2} & 1 & -2 \end{bmatrix}$

47. $\begin{bmatrix} \frac{1}{2} & -1 & -\frac{1}{2} & \frac{1}{2} \\ \frac{1}{2} & 4 & \frac{5}{2} & -\frac{1}{2} \\ -\frac{1}{4} & -\frac{1}{2} & -\frac{1}{4} & \frac{1}{4} \\ \frac{1}{2} & -2 & -\frac{3}{2} & \frac{1}{2} \end{bmatrix}$ **49. (a)** $R = \begin{bmatrix} .024 & .008 \\ .025 & .007 \\ .015 & .009 \\ .011 & .011 \end{bmatrix}$

(b) $P = \begin{bmatrix} 1996 & 286 & 226 & 460 \\ 2440 & 365 & 252 & 484 \\ 2906 & 455 & 277 & 499 \\ 3683 & 519 & 310 & 729 \\ 4723 & 697 & 364 & 702 \end{bmatrix}$ **(c)** $PR = \begin{bmatrix} 63.504 & 25.064 \\ 76.789 & 29.667 \\ 90.763 & 34.415 \\ 114.036 & 43.906 \\ 143.959 & 53.661 \end{bmatrix}$

(d) Rows represent the years 1970, 1980, 1990, 2000, 2025. Column one gives the total births in those years, column two the total deaths.   **(e)** 114,036,000; 53,661,000

51. (a) $\begin{pmatrix} 278.1 & 31.6 & 37.4 & 126.8 \\ 300.1 & 34.3 & 41.1 & 127.3 \end{pmatrix}$

(b) $\begin{pmatrix} .01425 & .00865 \\ .01145 & .00775 \\ .0175 & .00755 \\ .00945 & .00925 \end{pmatrix}$ **(c)** $\begin{pmatrix} 6.177505 & 4.105735 \\ 6.591395 & 4.349520 \end{pmatrix}$

(d) The rows correspond to years. The entries in each column give the total number of births and deaths, respectively, in the four countries, taken together.   **(e)** 6,177,505

53. (a) $\begin{bmatrix} 128.4 & 73.0 \\ 140.8 & 73.5 \\ 158.7 & 73.4 \\ 182.1 & 73.2 \end{bmatrix}$ **(b)** $\begin{bmatrix} 47.37 & 48.4 & 49.91 & 50.64 \\ 35.33 & 36.47 & 37.64 & 38.79 \end{bmatrix}$

(c) $\begin{bmatrix} 8661.40 & 8876.87 & 9156.16 & 9333.85 \\ 9266.45 & 9495.27 & 9793.87 & 9981.18 \\ 10,110.84 & 10,357.98 & 10,683.49 & 10,883.75 \\ 11,212.23 & 11,483.24 & 11,843.86 & 12,060.97 \end{bmatrix}$

(d) $\begin{bmatrix} 30,039.29 & 14,385.65 \\ 22,708.48 & 10,861.84 \end{bmatrix}$

(e) All dollar figures are in millions. In AB, row 1, column 1, is the total combined monthly cost of both cell phones and basic cable for all subscribers in 2001; row 2, column 2, is the same total for 2002; row 3, column 3, is the same total for 2003; row 4, column 4, is the same total for 2004. All other entries in AB are meaningless in this context. In BA, row 1, column 1, is the total monthly cost for all cell phone subscribers over the four-year period; row 2, column 2, is the total monthly cost for all basic cable subscribers over the four-year period. The other two entries in BA are meaningless in this context.

55. (a)
	A	B
Dept. 1	57	70
Dept. 2	41	54
Dept. 3	27	40
Dept. 4	39	40

(b) supplier A: $164; supplier B: $204

Section 6.5 (Page 355)

1. $\begin{bmatrix} -2 \\ 2 \end{bmatrix}$ **3.** $\begin{bmatrix} 1 & 2 \\ 0 & -1 \end{bmatrix}$ **5.** $\begin{bmatrix} -24 \\ 50 \\ -13 \end{bmatrix}$ **7.** $(-63, 50, -9)$

9. $(-31, -131, 181)$ **11.** $(-60, 10, 14)$

13. $(2, 2, -1, -2)$ **15.** $\begin{bmatrix} -6 \\ -14 \end{bmatrix}$

17. either 10 buffets, 5 chairs, no tables, or 11 buffets, 1 chair, 1 table   **19.** 2340 of the first species, 10,128 of the second species, 224 of the third species   **21.** jeans $34.50; jacket $72; sweater $44; shirt $21.75

23. $\begin{bmatrix} \frac{32}{3} \\ \frac{25}{3} \end{bmatrix}$ **25.** about 1073 metric tons of wheat, about 1431 metric

tons of oil   **27.** gas $98 million, electric $123 million

29. (a) $\frac{7}{4}$ bushels of yams, $\frac{15}{8} \approx 2$ pigs   **(b)** 167.5 bushels of yams, $153.75 \approx 154$ pigs

31. (a) .40 unit of agriculture, .12 unit of manufacturing, and 3.60 units of households.   **(b)** 848 units of agriculture, 516 units of manufacturing, and 2970 units of households   **(c)** about 813 units

33. (a) .017 unit of manufacturing and .216 unit of energy   **(b)** 123,725,000 pounds of agriculture, 14,792,000 pounds of manufacturing, 1,488,000 pounds of energy   **(c)** 195,492,000 pounds of agriculture, 25,933,000 pounds of manufacturing, 13,580,000 pounds of energy   **35.** $532 million of natural resources, $481 million of manufacturing, $805 million of trade and services, $1185 million of personal consumption

37. (a) $\begin{bmatrix} 1.67 & .56 & .56 \\ .19 & 1.17 & .06 \\ 3.15 & 3.27 & 4.38 \end{bmatrix}$

(b) These multipliers imply that, if the demand for one community's output increases by $1, then the output of the other community will increase by the amount in the row and column of that matrix. For example, if the demand for Hermitage's output increases by $1, then output from Sharon will increase by $.56, from Farrell by $.06, and from Hermitage by $4.38.

39. $\begin{bmatrix} 23 \\ 51 \end{bmatrix}, \begin{bmatrix} 13 \\ 30 \end{bmatrix}, \begin{bmatrix} 45 \\ 96 \end{bmatrix}, \begin{bmatrix} 69 \\ 156 \end{bmatrix}, \begin{bmatrix} 87 \\ 194 \end{bmatrix}, \begin{bmatrix} 23 \\ 51 \end{bmatrix}, \begin{bmatrix} 51 \\ 110 \end{bmatrix}, \begin{bmatrix} 45 \\ 102 \end{bmatrix}, \begin{bmatrix} 69 \\ 157 \end{bmatrix}$

41. (a) 3   **(b)** 3   **(c)** 5   **(d)** 3

43. (a) $B = \begin{bmatrix} 0 & 2 & 3 \\ 2 & 0 & 4 \\ 3 & 4 & 0 \end{bmatrix}$ **(b)** $B^2 = \begin{bmatrix} 13 & 12 & 8 \\ 12 & 20 & 6 \\ 8 & 6 & 25 \end{bmatrix}$ **(c)** 12   **(d)** 14

45. (a) $C = $
	Dogs	Rats	Cats	Mice
Dogs	0	1	1	1
Rats	0	0	0	1
Cats	0	1	0	1
Mice	0	0	0	0

(b) $\begin{bmatrix} 0 & 1 & 0 & 2 \\ 0 & 0 & 0 & 0 \\ 0 & 0 & 0 & 1 \\ 0 & 0 & 0 & 0 \end{bmatrix}$; C^2 gives the number of food sources once removed from the feeder.

Chapter 6 Review (Page 361)

1. $(9, -14)$ **3.** $(9, 6)$ **5.** inconsistent, no solution   **7.** $(-18, 69, 12)$ **9.** 8000 standard clips, 6000 extra-large clips   **11.** $7000 in the first fund, $11,000 in the second fund   **13.** 50 from source I; 150 from source II; 100 from source III   **15.** $(-14, 17, 11)$ **17.** no solution   **19.** $(-41, 51, -4)$ **21.** 2×2; square   **23.** 1×4; row   **25.** 2×3

27. $\begin{bmatrix} 8 & 8 & 8 \\ 10 & 5 & 9 \\ 7 & 10 & 7 \\ 8 & 9 & 7 \end{bmatrix}$

29. $\begin{bmatrix} -1 & -3 & 2 \\ -2 & -3 & 0 \\ 0 & -1 & -5 \end{bmatrix}$ **31.** $\begin{bmatrix} 2 & 30 \\ -4 & -15 \\ 10 & 13 \end{bmatrix}$ **33.** not defined

35.

Next day	Two-day total

$\begin{bmatrix} 2310 & -\frac{1}{4} \\ 1258 & -\frac{1}{4} \\ 5061 & \frac{1}{2} \\ 1812 & \frac{1}{2} \end{bmatrix}; \begin{bmatrix} 4842 & -\frac{1}{2} \\ 2722 & -\frac{1}{8} \\ 10{,}035 & -1 \\ 3566 & 1 \end{bmatrix}$

37. $\begin{bmatrix} 18 & 80 \\ -7 & -28 \\ 21 & 84 \end{bmatrix}$ **39.** $\begin{bmatrix} 38 & 39 \\ 47 & 44 \end{bmatrix}$

41. $\begin{bmatrix} 622 & 596 \\ -217 & -210 \\ 651 & 630 \end{bmatrix}$

43. (a) $\begin{bmatrix} \frac{1}{4} & \frac{1}{2} \\ \frac{1}{3} & \frac{1}{3} \end{bmatrix}$ **(b)** cutting 34 hours; shaping 46 hours

45. Many correct answers, including $\begin{bmatrix} 1 & 2 \\ 3 & 4 \end{bmatrix}$

47. $\begin{bmatrix} -\frac{1}{4} & \frac{1}{5} \\ 0 & \frac{1}{5} \end{bmatrix}$ **49.** no inverse

51. $\begin{bmatrix} \frac{1}{4} & \frac{1}{2} & \frac{1}{2} \\ \frac{1}{4} & -\frac{1}{2} & \frac{1}{2} \\ \frac{1}{12} & -\frac{1}{6} & -\frac{1}{6} \end{bmatrix}$ **53.** no inverse

55. $\begin{bmatrix} -\frac{2}{3} & -\frac{17}{3} & -\frac{14}{3} & -3 \\ \frac{1}{3} & \frac{1}{3} & \frac{1}{3} & 0 \\ -\frac{1}{3} & -\frac{10}{3} & -\frac{7}{3} & -2 \\ 0 & 2 & 1 & 1 \end{bmatrix}$ **57.** $\begin{bmatrix} -\frac{7}{23} & \frac{2}{23} \\ \frac{8}{23} & \frac{1}{23} \end{bmatrix}$

59. $\begin{bmatrix} -\frac{1}{18} & -\frac{1}{6} \\ \frac{7}{18} & \frac{1}{6} \end{bmatrix}$ **61.** $\begin{bmatrix} -\frac{15}{19} & \frac{17}{19} & -\frac{6}{19} \\ \frac{10}{19} & -\frac{5}{19} & \frac{4}{19} \\ -\frac{2}{19} & \frac{1}{19} & \frac{3}{19} \end{bmatrix}$ **63.** $\begin{bmatrix} 18 \\ -7 \end{bmatrix}$

65. $\begin{bmatrix} -22 \\ -18 \\ 15 \end{bmatrix}$ **67.** $(-2, 4)$ **69.** $(4, 2)$ **70.** $(3, -2)$

73. no inverse; no solution for the system   **75.** 16 liters of the 9%, 24 liters of the 14%   **77.** 30 liters of 40% solution, 10 liters of 60% solution   **79.** 80 bowls, 120 plates   **81.** $12,750 at 8%; $27,250 at 8.5%; $10,000 at 11%

83. (a) $\begin{bmatrix} 1 & -\frac{1}{4} \\ -\frac{1}{2} & 1 \end{bmatrix}$ **(b)** $\begin{bmatrix} \frac{8}{7} & \frac{2}{7} \\ \frac{4}{7} & \frac{8}{7} \end{bmatrix}$ **(c)** $\begin{bmatrix} 2800 \\ 2800 \end{bmatrix}$

85. agriculture $140,909; manufacturing $95,455   **87. (a)** .4 unit agriculture, .09 unit construction, .4 unit energy, .1 unit manufacturing, .9 unit transportation   **(b)** 2000 units of agriculture, 600 units of construction, 1700 units of energy, 3700 units of manufacturing, 2500 units of transportation   **(c)** 29,049 units of agriculture, 9869 units of construction, 52,362 units of energy, 61,520 units of manufacturing, 90,987 units of transportation

89. (a) $\begin{bmatrix} 54 \\ 32 \end{bmatrix}, \begin{bmatrix} 134 \\ 89 \end{bmatrix}, \begin{bmatrix} 172 \\ 113 \end{bmatrix}, \begin{bmatrix} 118 \\ 74 \end{bmatrix}, \begin{bmatrix} 208 \\ 131 \end{bmatrix}$ **(b)** $\begin{bmatrix} 2 & -3 \\ -\frac{1}{2} & 1 \end{bmatrix}$

Case 6 (Page 366)

1. Boston, Hyannis, Martha's Vineyard, Nantucket, New Bedford, and Providence may be reached by a two-flight sequence from New Bedford; all Cape Air cities may be reached by a three-flight sequence.
3. The connection between Provincetown and Providence and the connection between Provincetown and New Bedford each take three flights.
5. All Big Sky cities may be reached by a three-flight sequence from Helena. At least three flights must be used to get from Helena to Havre, Glendive, and Bismarck.

Chapter 7

Section 7.1 (Page 374)

1. F   **3.** A   **5.** E

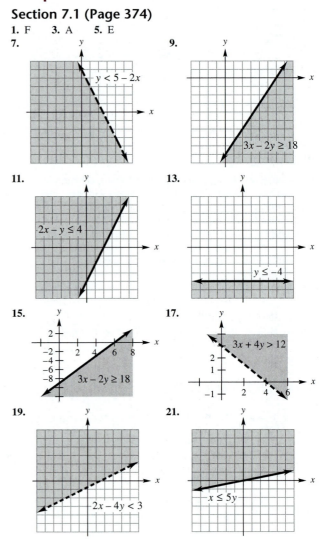

23.

25.

45.

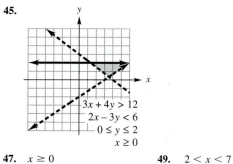

27. Answers vary.

29.

31.

47. $x \geq 0$ **49.** $2 < x < 7$
$0 \leq y \leq 4$ $-1 < y < 3$
$4x + 3y \leq 24$

51. (a)

Number		Hours Spinning	Hours Dyeing	Hours Weaving
Shawls	x	1	1	1
Afghans	y	2	1	4
Maximum Number of Hours Available		8	6	14

(b) $x + 2y \leq 8$; $x + y \leq 6$; $x + 4y \leq 14$; $x \geq 0$; $y \geq 0$

(c)

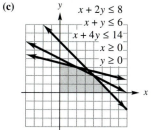

33.

35.

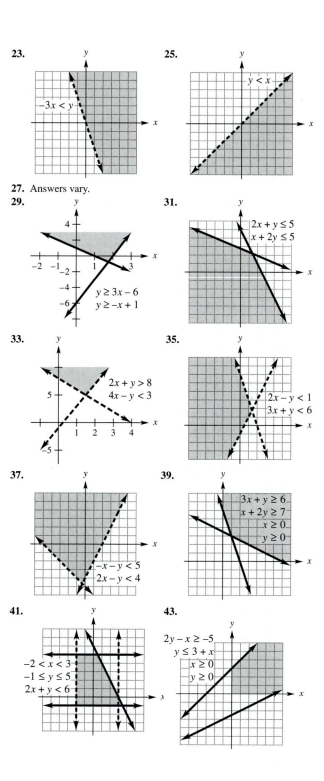

53. $x \geq 3000$; $y \geq 5000$; $x + y \leq 10{,}000$

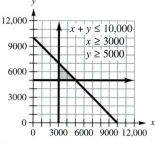

37.

39.

55. $x + y \geq 3.2$; $.16x + .20y \leq .8$; $.5x + .3y \leq 1.8$; $x \geq 0$; $y \geq 0$

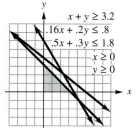

41.

43.

Section 7.2 (Page 382)

1. maximum of 65 at (5, 10); minimum of 8 at (1, 1)
3. maximum of 6.75 at (9, 0); minimum of 0 at (0, 0)
5. (a) no maximum; minimum of 12 at (12, 0)
(b) no maximum; minimum of 18 at (3, 4)
(c) no maximum; minimum of 21 at $\left(\dfrac{13}{2}, 2\right)$
(d) no maximum; minimum of 8 at (0, 8)
7. maximum of 10 at (0, 2)
9. minimum of 13 at (5, 3)
11. maximum of 100 at (0, 25)
13. no maximum; minimum 9
15. maximum 22; no minimum
17. (a) (18, 2) **(b)** $\left(\dfrac{12}{5}, \dfrac{39}{5}\right)$ **(c)** no maximum
19. Answers vary.

Section 7.3 (Page 388)

1. $8x + 5y \leq 110$, $x \geq 0$, $y \geq 0$ where x is the number of canoes and y is the number of rowboats **3.** $250x + 750y \leq 9500$, $x \geq 0$, $y \geq 0$ where x is the number of radio spots and y is the number of TV ads **5.** 12 chain saws; no chippers **7.** 10 lb deluxe; 40 lb regular **9. (a)** 6 units of policy A and 16 units of policy B, for a minimum cost of \$940 **(b)** 30 units of policy A and 0 units of policy B, for a minimum cost of \$750 **11.** $3\frac{3}{4}$ servings of A and $1\frac{7}{8}$ servings of B, for a minimum cost of \$1.69 **13.** 800 type 1 and 1600 type 2, for a maximum revenue of \$272 **15.** 0 plants and 18 animals, for a minimum of 270 hr **17.** From warehouse I, ship 60 boxes to San Jose and 250 boxes to Memphis; from warehouse II, ship 290 boxes to San Jose and none to Memphis, for a minimum cost of \$136.70 **19.** \$20 million in bonds and \$24 million in mutual funds, for maximum interest of \$2.24 million. **21.** 8 humanities, 12 science **23. (b)** **25. (c)**

Section 7.4 (Page 402)

1. (a) 3
(b) x_3, x_4, x_5
(c)
$$
\begin{aligned}
4x_1 + 2x_2 + x_3 &= 20 \\
5x_1 + x_2 + x_4 &= 50 \\
2x_1 + 3x_2 + x_5 &= 25
\end{aligned}
$$
3. (a) 3
(b) x_4, x_5, x_6
(c)
$$
\begin{aligned}
3x_1 - x_2 + 4x_3 + x_4 &= 95 \\
7x_1 + 6x_2 + 8x_3 + x_5 &= 118 \\
4x_1 + 5x_2 + 10x_3 + x_6 &= 220
\end{aligned}
$$
5.

	x_1	x_2	x_3	x_4	x_5	z	
	2	5	1	0	0	0	6
	4	1	0	1	0	0	6
	5	3	0	0	1	0	15
	-5	-1	0	0	0	1	0

7.

x_1	x_2	x_3	x_4	x_5	x_6	z	
1	2	3	1	0	0	0	10
2	1	1	0	1	0	0	8
3	0	4	0	0	1	0	6
-1	-5	-10	0	0	0	1	0

9. 4 in row 2, column 2
11. 6 in row 3, column 1
13.

x_1	x_2	x_3	x_4	x_5	z	
-1	0	3	1	-1	0	16
1	1	$\frac{1}{2}$	0	$\frac{1}{2}$	0	20
2	0	$-\frac{1}{2}$	0	$\frac{3}{2}$	1	60

15.

x_1	x_2	x_3	x_4	x_5	x_6	z	
$-\frac{1}{2}$	$\frac{1}{2}$	0	1	$-\frac{1}{2}$	0	0	10
$\frac{3}{2}$	$\frac{1}{2}$	1	0	$\frac{1}{2}$	0	0	50
$-\frac{7}{2}$	$\frac{1}{2}$	0	0	$-\frac{3}{2}$	1	0	50
2	0	0	0	1	0	1	100

17. (a) basic: x_3, x_5; nonbasic: x_1, x_2, x_4 **(b)** $x_1 = 0, x_2 = 0,$ $x_3 = 16, x_4 = 0, x_5 = 29, z = 11$ **(c)** not maximum
19. (a) basic: x_1, x_2, x_5; nonbasic: x_3, x_4, x_6 **(b)** $x_1 = 6, x_2 = 13,$ $x_3 = 0, x_4 = 0, x_5 = 21, x_6 = 0, z = 18$ **(c)** maximum
21. Maximum is 30 when $x_1 = 0, x_2 = 10, x_3 = 0, x_4 = 0, x_5 = 16.$
23. Maximum is 8 when $x_1 = 4, x_2 = 0, x_3 = 8, x_4 = 2, x_5 = 0.$
25. no maximum **27.** no maximum **29.** Maximum is 34 when $x_1 = 17, x_2 = 0, x_3 = 0, x_4 = 0, x_5 = 14$ or when $x_1 = 0, x_2 = 17, x_3 = 0, x_4 = 0, x_5 = 14.$ **31.** Maximum is 26,000 when $x_1 = 60, x_2 = 40, x_3 = 0, x_4 = 0, x_5 = 80, x_6 = 0.$
33. Maximum is 64 when $x_1 = 28, x_2 = 16, x_3 = 0, x_4 = 0,$ $x_5 = 28, x_6 = 0.$ **35.** Maximum is 250 when $x_1 = 0, x_2 = 0,$ $x_3 = 0, x_4 = 50, x_5 = 0, x_6 = 50.$ **37. (a)** Maximum is 24 when $x_1 = 12, x_2 = 0, x_3 = 0, x_4 = 0, x_5 = 6.$ **(b)** Maximum is 24 when $x_1 = 0, x_2 = 12, x_3 = 0, x_4 = 0, x_5 = 18.$ **(c)** The unique maximum value of z is 24, but this occurs at two different basic feasible solutions.

Section 7.5 (Page 409)

1.

x_1	x_2	x_3	x_4	x_5	
2	1	1	0	0	90
1	2	0	1	0	80
1	1	0	0	1	50
-12	-10	0	0	0	0

where x_1 is the number of Siamese cats and x_2 is the number of Persian cats

3.

x_1	x_2	x_3	x_4	x_5	x_6	x_7	
0	0	.375	.625	1	0	0	500
0	.75	.5	.375	0	1	0	600
1	.25	.125	0	0	0	1	300
-90	-70	-60	-50	0	0	0	0

where x_1 is the number of kilograms of P, x_2 is the number of kilograms of Q, x_3 is the number of kilograms of R, and x_4 is the number of kilograms of S **5. (a)** Make no 1-speed or 3-speed bicycles; make 2295 10-speed bicycles; maximum profit is \$55,080. **(b)** 5280 units of aluminum are unused; all the steel is used.   **7. (a)** 12 min to the senator, 9 minutes to the congresswoman, and 6 min to the governor, for a maximum of 1,320,000 viewers   **(b)** $x_4 = 0$ means that all of the 27 available minutes are used; $x_5 = 0$ means that the senator had *exactly* twice as much time as the governor; $x_6 = 0$ means that the senator and governor had a total time *exactly* twice the time of the congresswoman.   **9.** 4 radio ads, 6 TV ads, and no newspaper ads, for a maximum exposure of 64,800 people   **11. (a)** 22 fund-raising parties, no mailings, and 3 dinner parties, for a maximum of $6,200,000   **(b)** Answers vary.   **13. (a)** (3)   **(b)** (4)   **(c)** (3)   **15.** The breeder should raise 40 Siamese and 10 Persian cats, for a maximum gross income of $580   **17.** 163.6 kilograms of food P, none of Q, 1090.9 kilograms of R, 145.5 kilograms of S; maximum is 87,454.5.

Section 7.6 (Page 420)

1. $\begin{bmatrix} 3 & 1 & 0 \\ -4 & 10 & 3 \\ 5 & 7 & 6 \end{bmatrix}$ **3.** $\begin{bmatrix} 3 & 4 \\ 0 & 17 \\ 14 & 8 \\ -5 & -6 \\ 3 & 1 \end{bmatrix}$

5. Maximize $z = 4x_1 + 6x_2$
subject to $3x_1 - x_2 \le 3$
$x_1 + 2x_2 \le 5$
$x_1 \ge 0, x_2 \ge 0.$

7. Maximize $z = 18x_1 + 15x_2 + 20x_3$
subject to $x_1 + 4x_2 + 5x_3 \le 2$
$7x_1 + x_2 + 3x_3 \le 8$
$x_1 \ge 0, x_2 \ge 0, x_3 \ge 0.$

9. Maximize $z = 8x_1 + 12x_2$
subject to $3x_1 + x_2 \le 1$
$4x_1 + 5x_2 \le 2$
$6x_1 + 2x_2 \le 6$
$x_1 \ge 0, x_2 \ge 0.$

11. Maximize $z = 5x_1 + 4x_2 + 15x_3$
subject to $x_1 + x_2 + 2x_3 \le 8$
$x_1 + x_2 + x_3 \le 9$
$x_1 + 3x_3 \le 3$
$x_1 \ge 0, x_3 \ge 0, x_3 \ge 0.$

13. $y_1 = 0, y_2 = 100, y_3 = 0$; minimum is 100.   **15.** $y_1 = 0$, $y_2 = 12, y_3 = 0$; minimum is 12.   **17.** $y_1 = 0, y_2 = 0, y_3 = 2$; minimum is 4.   **19.** $y_1 = 10, y_2 = 10, y_3 = 0$; minimum is 320.
21. $y_1 = 0, y_2 = 0, y_3 = 4$; minimum is 4.   **23.** 4 servings of A and 2 servings of B, for a minimum cost of $1.76   **25.** 28 units of regular beer and 14 units of light beer, for a minimum cost of $1,680,000   **27. (a)**
29. (a) Minimize $w = 200y_1 + 600y_2 + 90y_3$
subject to $y_1 + 4y_2 \ge 1$
$2y_1 + 3y_2 + y_3 \ge 1.5$
$y_1 \ge 0, y_2 \ge 0, y_3 \ge 0.$
(b) $y_1 = .6, y_2 = .1, y_3 = 0, w = 180$ **(c)** $186   **(d)** $179

Section 7.7 (Page 432)

1. (a) Maximize $z = 5x_1 + 2x_2 - x_3$
subject to $2x_1 + 3x_2 + 5x_3 - x_4 = 8$
$4x_1 - x_2 + 3x_3 + x_5 = 7$
$x_1 \ge 0, x_2 \ge 0, x_3 \ge 0, x_4 \ge 0, x_5 \ge 0.$

(b)
$$\begin{array}{ccccc} x_1 & x_2 & x_3 & x_4 & x_5 \end{array}$$
$$\begin{bmatrix} 2 & 3 & 5 & -1 & 0 & 8 \\ 4 & -1 & 3 & 0 & 1 & 7 \\ \hline -5 & -2 & 1 & 0 & 0 & 0 \end{bmatrix}$$

3. (a) Maximize $z = 2x_1 - 3x_2 + 4x_3$
subject to $x_1 + x_2 + x_3 + x_4 = 100$
$x_1 + x_2 + x_3 - x_5 = 75$
$x_1 + x_2 - x_6 = 27$
$x_1 \ge 0, x_2 \ge 0, x_3 \ge 0, x_4 \ge 0, x_5 \ge 0, x_6 \ge 0.$

(b)
$$\begin{array}{cccccc} x_1 & x_2 & x_3 & x_4 & x_5 & x_6 \end{array}$$
$$\begin{bmatrix} 1 & 1 & 1 & 1 & 0 & 0 & 100 \\ 1 & 1 & 1 & 0 & -1 & 0 & 75 \\ 1 & 1 & 0 & 0 & 0 & -1 & 27 \\ \hline -2 & 3 & -4 & 0 & 0 & 0 & 0 \end{bmatrix}$$

5. Maximize $z = -2y_1 - 5y_2 + 3y_3$
subject to $y_1 + 2y_2 + 3y_3 \ge 115$
$2y_1 + y_2 + y_3 \le 200$
$y_1 + y_3 \ge 50$
$y_1 \ge 0, y_2 \ge 0, y_3 \ge 0.$

$$\begin{array}{cccccc} y_1 & y_2 & y_3 & y_4 & y_5 & y_6 \end{array}$$
$$\begin{bmatrix} 1 & 2 & 3 & -1 & 0 & 0 & 115 \\ 2 & 1 & 1 & 0 & 1 & 0 & 200 \\ 1 & 0 & 1 & 0 & 0 & -1 & 50 \\ \hline 2 & 5 & -3 & 0 & 0 & 0 & 0 \end{bmatrix}$$

7. Maximize $z = -y_1 + 4y_2 - 2y_3$
subject to $7y_1 - 6y_2 + 8y_3 \ge 18$
$4y_1 + 5y_2 + 10y_3 \ge 20$
$y_1 \ge 0, y_2 \ge 0, y_3 \ge 0.$

$$\begin{array}{ccccc} y_1 & y_2 & y_3 & y_4 & y_5 \end{array}$$
$$\begin{bmatrix} 7 & -6 & 8 & -1 & 0 & 18 \\ 4 & 5 & 10 & 0 & -1 & 20 \\ \hline 1 & -4 & 2 & 0 & 0 & 0 \end{bmatrix}$$

9. Maximum is 480 when $x_1 = 40, x_2 = 0$.   **11.** Maximum is 114 when $x_1 = 38, x_2 = 0, x_3 = 0$.   **13.** Maximum is 90 when $x_1 = 12, x_2 = 3$ or when $x_1 = 0, x_2 = 9$.   **15.** Minimum is 40 when $y_1 = 0, y_2 = 20$.   **17.** Maximum is $133\frac{1}{3}$ when $x_1 = 33\frac{1}{3}$, $x_2 = 16\frac{2}{3}$.   **19.** Minimum is 512 when $y_1 = 6, y_2 = 8$.
21.
$$\begin{array}{ccccccc} y_1 & y_2 & y_3 & y_4 & y_5 & y_6 & y_7 \end{array}$$
$$\begin{bmatrix} 1 & 1 & 1 & -1 & 0 & 0 & 0 & 10 \\ 1 & 1 & 1 & 0 & 1 & 0 & 0 & 15 \\ 1 & -\frac{1}{4} & 0 & 0 & 0 & -1 & 0 & 0 \\ -1 & 0 & 1 & 0 & 0 & 0 & -1 & 0 \\ \hline .30 & .09 & .27 & 0 & 0 & 0 & 0 & 0 \end{bmatrix}$$

23.

	y_1	y_2	y_3	y_4	y_5	y_6	y_7	y_8	y_9	y_{10}	
	1	1	0	0	−1	0	0	0	0	0	32
	1	1	0	0	0	1	0	0	0	0	32
	0	0	1	1	0	0	−1	0	0	0	20
	0	0	1	1	0	0	0	1	0	0	20
	1	0	1	0	0	0	0	0	1	0	25
	0	1	0	1	0	0	0	0	0	1	30
	14	12	22	10	0	0	0	0	0	0	0

25. Ship 200 barrels of oil from supplier S_1 to distributor D_1; ship 2800 barrels of oil from supplier S_2 to distributor D_1; ship 2800 barrels of oil from supplier S_1 to distributor D_2; ship 2200 barrels of oil from supplier S_2 to distributor D_2. Minimum cost is $180,400.   **27.** Use 1000 lb of bluegrass, 2400 lb of rye, and 1600 lb of Bermuda, for a minimum cost of $560.   **29.** Allot $3,000,000 in commercial loans and $22,000,000 in home loans, for a maximum return of $2,940,000. **31.** Make 32 units of regular beer and 10 units of light beer, for a minimum cost of $1,632,000.   **33.** $1\frac{2}{3}$ ounces of ingredient I, $6\frac{2}{3}$ ounces of ingredient II, and $1\frac{2}{3}$ ounces of ingredient III produce a minimum cost of $1.55 per barrel.   **35.** 22 from W_1 to D_1, 10 from W_2 to D_1, none from W_1 to D_2, and 20 from W_2 to D_2, for a minimum cost of $628.

Chapter 7 Review (Page 436)

1.

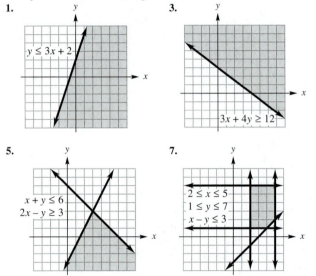

3.

5.

7.

9. Let x represent the number of batches of cakes, and let y represent the number of batches of cookies. Then

$$2x + \left(\frac{3}{2}\right)y \le 15$$
$$3x + \left(\frac{2}{3}\right)y \le 13$$
$$x \ge 0$$
$$y \ge 0.$$

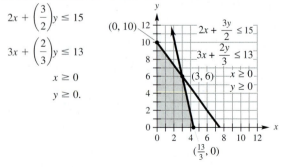

11. maximum of 38 at (7, 3); minimum of 7.5 at (1, 2.5)   **13.** maximum is 11 when $x = 3.5, y = 1$ **15.** minimum of 90 when $x = 30, y = 0$ **17.** Make 3 batches of cakes and 6 batches of cookies, for a maximum profit of $210.   **19. (a)** Let x_1 = number of item A, x_2 = number of item B, x_3 = number of item C.
(b) $z = 4x_1 + 3x_2 + 3x_3$
(c) $2x_1 + 3x_2 + 6x_3 \le 1200$
$\quad x_1 + 2x_2 + 2x_3 \le 800$
$\quad 2x_1 + 2x_2 + 4x_3 \le 500$
$\quad x_1 \ge 0, x_2 \ge 0, x_3 \ge 0$
21. (a) Let x_1 = number of gallons of Fruity wine and x_2 = number of gallons of Crystal wine.   **(b)** $z = 12x_1 + 15x_2$
(c) $2x_1 + \quad x_2 \le 110$
$\quad 2x_1 + 3x_2 \le 125$
$\quad 2x_1 + \quad x_2 \le 90$
$\quad x_1 \ge 0, x_2 \ge 0$
23. when there are more than 2 variables
25. any standard minimization problem
27. (a)
$$3x_1 + 5x_2 + x_3 \qquad\qquad = 47$$
$$x_1 + x_2 \qquad + x_4 \qquad = 25$$
$$5x_1 + 2x_2 \qquad\qquad + x_5 \qquad = 35$$
$$2x_1 + x_2 \qquad\qquad\quad + x_6 = 30$$

(b)

x_1	x_2	x_3	x_4	x_5	x_6	
3	5	1	0	0	0	47
1	1	0	1	0	0	25
5	2	0	0	1	0	35
2	1	0	0	0	1	30
−2	−9	0	0	0	0	0

29. (a)
$$x_1 + x_2 + x_3 + x_4 \qquad\qquad = 100$$
$$2x_1 + 3x_2 + \qquad\quad + x_5 \qquad = 500$$
$$x_1 \qquad + 2x_3 + \qquad\quad + x_6 = 350$$

(b)

x_1	x_2	x_3	x_4	x_5	x_6	
1	1	1	1	0	0	100
2	3	0	0	1	0	500
1	0	2	0	0	1	350
−5	−6	−3	0	0	0	0

31. Maximum is 80 when $x_1 = 16, x_2 = 0, x_3 = 0, x_4 = 12, x_5 = 0$.   **33.** Maximum is 35 when $x_1 = 5, x_2 = 0, x_3 = 5, x_4 = 35, x_5 = 0$, $x_6 = 0$.   **35.** Maximize $z = -18y_1 - 10y_2$ with the same constraints.   **37.** Maximize $z = -6y_1 + 3y_2 - 4y_3$ with the same constraints.   **39.** Minimum is 40 when $y_1 = 10$ and $y_2 = 0$.
41. minimum of 1957 at (47, 68, 0, 92, 35, 0, 0)   **43.** (9, 5, 8, 0, 0, 0); minimum is 62.   **45.** Get 250 of A, none of B or C, for a maximum profit of $1000.   **47.** Make 17.5 gal of Crystal and 36.25 gal of Fruity, for a maximum profit of $697.50.   **49.** Produce 660 cases of corn, no beans, and 340 cases of carrots, for a minimum cost of $15,100.   **51.** Use 1,060,000 kilograms for whole tomatoes and 80,000 kilograms for sauce, for a minimum cost of $4,500,000.

Case 7 (Page 440)

1. The answer in 100-gram units is 0.243037 unit of feta cheese, 2.35749 units of lettuce, 0.3125 unit of salad dressing, and 1.08698 units of tomato. Converting into kitchen units gives approximately $\frac{1}{6}$ cup feta cheese, $4\frac{1}{4}$ cups lettuce, $\frac{1}{8}$ cup salad dressing, and $\frac{7}{8}$ of a tomato.

Chapter 8

Section 8.1 (Page 450)

1. False   **3.** True   **5.** True   **7.** True   **9.** False
11. Answers vary.   **13.** $\subseteq$ **15.** $\nsubseteq$ **17.** $\subseteq$ **19.** $\subseteq$
21. 8   **23.** 128   **25.** $\{x \mid x$ is an integer ≤ 0 or $\geq 8\}$
27. Answers vary.   **29.** $\cap$ **31.** $\cap$ **33.** $\cup$ **35.** $\cup$
37. $\{3, 5\}$ **39.** $\{1, 7, 9, 11, 15\}$ **41.** $\{1, 11, 15\}$
43. $\{2, 3, 4, 5, 6\}$ **45.** All students not taking this course.
47. All students taking both accounting and philosophy.   **49.** C
and D, A and E, C and E, D and E **51.** $C' = \{$Allstate$\}$
53. $(A \cup B)' = \{$Goodyear$\}$ **55.** $M \cap E$ is the set of all male
employed applicants.   **57.** $M' \cup S'$ is the set of all female or married applicants.   **59.** $\{$Magazines, Internet$\}$ **61.** $\{$Comcast,
Time Warner$\}$ **63.** $\{$Comcast, Time Warner, Charter, Adelphia,
Cablevision$\}$ **65.** $\{$Comcast, Time Warner$\}$ **67.** $\{s, d, c\}$
69. $\{g\}$

Section 8.2 (Page 458)

1.
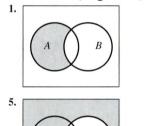

3.
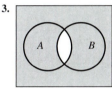

5.
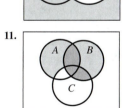

7. $\emptyset$ **9.** 8

11.
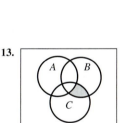

13.

15.
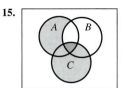

17. 22.9%   **19. (a)** 342   **(b)** 192   **(c)** 72   **(d)** 86
21. (a) 50   **(b)** 2   **(c)** 14   **23. (a)** 54   **(b)** 17   **(c)** 10   **(d)** 7
(e) 15   **(f)** 3   **(g)** 12   **(h)** 1   **25. (a)** 100   **(b)** 88   **(c)** 295
(d) 334   **27. (a)** 1,348   **(b)** 214,567   **(c)** 165,424   **(d)** 84,078
(e) 179,161   **(f)** 81,595   **29.** Answers vary.   **31.** 9
33. 27

35.
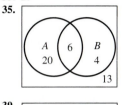

37.
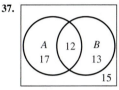

39.
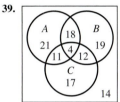

41.
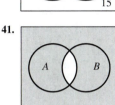

43.

45. The complement of A intersect B equals the union of the complement of A and the complement of B.   **47.** A union (B intersect C) equals (A union B) intersect (A union C).

Section 8.3 (Page 467)

1. Answers vary.   **3.** $\{$January, February, March, . . . , December$\}$
5. $\{0, 1, 2, . . . , 80\}$ **7.** $\{$go ahead, cancel$\}$ **9.** Answers vary.
11. $\{y_1, y_2, y_3, w_1, w_2, w_3, b_1, b_2, b_3, b_4, b_5, b_6, b_7, b_8\}$
(a) $\{y_1, y_2, y_3\}$ **(b)** $\{b_1, b_2, b_3, b_4, b_5, b_6, b_7, b_8\}$ **(c)** $\{w_1, w_2, w_3, w_4\}$
(d) $\emptyset$ **13. (a)** $\{$H, H, T$\}$, $\{$H, T, H$\}$, $\{$T, H, H$\}$ **(b)** $\{$T, T, T$\}$
(c) $\{$H, T, T$\}$, $\{$T, H, T$\}$, $\{$T, T, H$\}$ **15. (a)** $\{$2 inch, black$\}$, $\{$3 inch, black$\}$ **(b)** $\{$flats, light beige$\}$, $\{$flats, dark beige$\}$ **(c)** $\{$2 inch, light beige$\}$, $\{$2 inch, dark beige$\}$, $\{$3 inch, light beige$\}$, $\{$3 inch,

dark beige} **17.** $\frac{1}{6}$ **19.** $\frac{1}{3}$ **21.** $\frac{1}{3}$ **23.** $\frac{1}{13}$ **25.** $\frac{1}{26}$
27. $\frac{1}{52}$ **29.** $\frac{1}{13}$ **31. (a)** .065 **(b)** .139 **(c)** .165 **(d)** .75
(e) .242 **33. (a)** The person is not overweight. **(b)** The person has a family history of heart disease and is overweight. **(c)** The person smokes or is not overweight. **35. (a)** .118 **(b)** .168
(c) .122 **(d)** .132

Section 8.4 (Page 476)

1. Answers vary. **3.** No **5.** Yes **7.** No **9. (a)** $\frac{5}{36}$
(b) $\frac{1}{9}$ **(c)** $\frac{1}{12}$ **(d)** 0 **11. (a)** $\frac{5}{18}$ **(b)** $\frac{5}{12}$ **(c)** $\frac{1}{3}$ **13.** $\frac{1}{3}$
15. (a) $\frac{2}{13}$ **(b)** $\frac{4}{13}$ **(c)** $\frac{7}{13}$ **17. (a)** $\frac{3}{13}$ **(b)** $\frac{4}{13}$ **(c)** $\frac{7}{13}$
(d) $\frac{4}{13}$ **(e)** $\frac{8}{13}$ **19. (a)** $\frac{1}{2}$ **(b)** $\frac{3}{10}$ **(c)** $\frac{4}{5}$ **21.** $\frac{1}{5}$
23. (a) .42
(b) .88 **(c)** .72 **(d)** .28 **25.** 1:5 **27.** 2:1 **29. (a)** 1:4
(b) 8:7 **(c)** 4:11 **31.** 2:7 **33.** Answers vary. **35.** No
37. Yes **39.** Yes **41.** No **43.** Yes **45.** Possible
47. Not possible **49.** Not possible **51. (a)** .2778 **(b)** .4167
53. (a) .15625 **(b)** .3125 **55. (a)** .484 **(b)** .750 **(c)** .357
(d) .416 **57. (a)** .25 **(b)** .75 **(c)** .36 **59.** 13:37
61. (a) .961 **(b)** .491
(c) .509 **(d)** .505 **(e)** .004 **(f)** .544 **63. (a)** $\frac{1}{4}$ **(b)** $\frac{1}{2}$
(c) $\frac{1}{4}$ **65. (a)** .024 **(b)** .021 **(c)** .045 **(d)** .440
67. (a)

	A	B	C	D
O	.053	.058	.039	.005
E	.310	.273	.215	.047

(b) .310 **(c)** .044 **(d)** .694 **(e)** .889

Section 8.5 (Page 491)

1. $\frac{1}{3}$ **3.** 1 **5.** $\frac{1}{6}$ **7.** $\frac{4}{17}$ **9.** about .012 **11.** Answers
vary. **13.** Answers vary. **15.** No **17.** Yes **19.** No, Yes
21. $\frac{3}{7} = .43$ **23. (a)** .815 **(b)** .79 **25. (a)** .17 **(b)** .19
(c) .28 **(d)** .64 **27.** .499 **29.** 0 **31.** .30
33. Answers vary. **35.** .049 **37.** .534 **39.** .143
41. Dependent **43.** .05 **45.** .25 **47.** .08 **49.** .52
51. .26 **53.** .24 **55.** .0457 **57.** .999985
59. (a) 3 backups **(b)** Answers vary. **61.** Dependent

Section 8.6 (Page 498)

1. $\frac{1}{3}$ **3.** .0488 **5.** .5122 **7.** .4706 **9.** .2780
11. .7556 **13.** .5272 **15.** .30873 **17.** .2579 **19. (c)**
21. .481 **23.** .751 **25. (a)** .155 **(b)** .990 **(c)** about 59
27. (a) .4738 **(b)** .1665 **29.** .524 **31.** .283 **33.** .857
35. .059

Chapter 8 Review (Page 502)

1. False **3.** False **5.** True **7.** True **9.** {New Year's Day, Martin Luther King's Birthday, Presidents' Day, Memorial Day, Independence Day, Labor Day, Columbus Day, Veterans' Day, Thanksgiving, Christmas} **11.** {1, 2, 3, 4} **13.** {$B_1, B_2, B_3, B_6, B_{12}$}
15. {A, C, E} **17.** {$A, B_3, B_6, B_{12}, C, D, E$}, **19.** {All students majoring in business and with brown eyes} **21.** {All students majoring in business or younger than 25} **23.** {All students with GPA's $\geq$ 3.0 and who do not have brown eyes}
25. **27.**

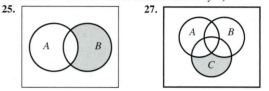

29. 5 **31.** {1, 2, 3, 4, 5, 6} **33.** {(red, 10), (red, 20) (red, 30), (blue, 10), (blue, 20), (blue, 30), (green, 10), (green, 20), (green, 30)}.
35. F = {(2, blue), (4, blue), (6, blue), (8, blue), (10, blue)}
37. No **39.** $E \cup F$ **41.** Answers vary. **43.** Answers vary.
45. $\frac{3}{13}$ **47.** $\frac{1}{2}$ **49.** 1 **51.** 1:25 **53.** .86
55.

	N_2	T_2
N_1	$N_1 N_2$	$N_1 T_2$
T_1	$T_1 N_2$	$T_1 T_2$

57. $\frac{1}{2}$ **59.** $\frac{5}{36} \approx .139$ **61.** $\frac{5}{18} \approx .278$ **63.** $\frac{1}{3}$
65. .79 **67.** .66 **69. (a)** .7 **(b)** $\frac{2}{15} \approx .1333$
71. Answers vary. **73.** $\frac{3}{22}$ **75.** $\frac{2}{3}$ **77. (a)** .271 **(b)** .379
(c) .625 **(d)** .342 **(e)** .690 **(f)** .514 **(g)** No. Answers vary.

Additional Probability Review (Page 505)

1. (a) .88 **(b)** .30 **(c)** .70 **3. (a)** .261 **(b)** .379 **(c)** .909
5. (a) 0 **(b)** $\frac{1}{3}$ **(c)** 1 **7.** .221 **9.** .068 **11.** No
13. $\frac{1}{6}$ **15.** 82.4% **17.** .644 **19.** .312 **21.** .286
23. .68 **25. (a)** $\frac{2}{5}$ **(b)** $\frac{3}{5}$ **(c)** $\frac{1}{10}$ **(d)** $\frac{7}{10}$

Case 8 (Page 507)

1. .076 **3.** .051

Chapter 9

Section 9.1 (Page 516)

1.

Number of boys	0	1	2	3	4
$P(x)$	.063	.25	.375	.25	.063

3.

Number of Queens	0	1	2	3
$P(x)$	.7826	.2042	.0130	.0002

5.

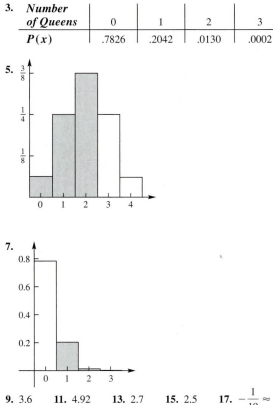

7.

9. 3.6   **11.** 4.92   **13.** 2.7   **15.** 2.5   **17.** $-\frac{1}{19} \approx -.05$

19. $-\$.50$ **21.** $-\$.97$ **23.** $-\$21.75$ **25.** 2.35

27. No, cannot have a probability <0.   **29.** No, cannot have a probability <0.   **31.** .15   **33.** .25   **35.** Answers vary; could be .15, .15.   **37.** \$4550   **39.** 1.15

41.

Account Number	Expected Value	Existing Volume + Expected Value of Potential	Class
3	2000	22,000	C
4	1000	51,000	B
5	25,000	30,000	C
6	60,000	60,000	A
7	16,000	46,000	B

43. (a) $68.51, $72.84 **(b)** Amoxicillin   **45. (a)** £472,500 **(b)** £350,000

Section 9.2 (Page 531)

1. 12   **3.** 56   **5.** 8   **7.** 24   **9.** 84   **11.** 156
13. 6,375, 600   **15.** 240,240   **17.** 8,568   **19.** 2,042,975
21. $\frac{24}{0}$ is undefined   **23. (a)** 8 **(b)** 64   **25.** 576

27. 1 billion; yes.   **29.** 1 billion   **31. (a)** 2,177,280
(b) 1,451,520   **33. (a)** 160; 8,000,000   **(b)** Some, such as 800, 900, etc., are reserved.   **35.** 1600   **37.** Answers vary.
39. 3,628,800   **41.** 288   **43.** 863,040   **45.** 272
47. (a) 210 **(b)** 210   **49. (a)** 0 **(b)** 792 **(c)** 658,008
(d) 370,832 **(e)** 22,308   **51.** Answers vary.   **53. (a)** 9
(b) 6 **(c)** 3, yes   **55. (a)** 120 **(b)** 20 **(c)** 60 **(d)** 100
57. (a) 441 **(b)** 9261   **59. (a)** 1,120,529,256
(b) 806,781,064,320   **61.** 81   **63.** Not possible, 4 initials
65. (a) 10 **(b)** 0 **(c)** 1 **(d)** 10 **(e)** 30 **(f)** 15 **(g)** 0
67. (a) 11,880 **(b)** 495   **69.** 5.524×10^{26} **71. (a)** 2520
(b) 840 **(c)** 5040   **73. (a)** 479,001,600 **(b)** 6 **(c)** 27,720

Section 9.3 (Page 540)

1. .424   **3.** $\frac{5}{8}$ **5.** $\frac{5}{28}$ **7.** .00005   **9.** .0163
11. .5630   **13.** 1326   **15.** .851   **17.** .765   **19.** .941
21. Answers vary.   **23.** .0067   **25. (a)** 8.9×10^{-10}
(b) 1.2×10^{-12} **27.** 6.9×10^{-17} **29. (a)** .011 **(b)** .330
(c) .670   **31. (a)** .003 **(b)** .005 **(c)** .057 **(d)** .252
33. $1 - \frac{{}_{365}P_{100}}{(365)^{100}} \approx 1$ **35.** .0031   **37.** .3083

39. We obtained the following answers; yours should be similar.
(a) .0399   **(b)** .5191   **(c)** .0226

Section 9.4 (Page 547)

1. .200   **3.** .000006   **5.** .999994   **7.** .075   **9.** .599
11. .401   **13.** $\frac{1}{32}$ **15.** $\frac{13}{16}$ **17.** Answers vary.   **19.** .0045
21. .9952   **23.** 6   **25.** .3006   **27.** .9735   **29.** .247
31. .193   **33.** .350   **35.** 3.85   **37.** .966   **39.** 330
41. (a) .9995 **(b)** Answers vary.   **43. (a)** 1 chance in 1024
(b) 1 chance in 1.1×10^{12} **(c)** 1 chance in 2.6×10^{6}
(d) Answers vary.

Section 9.5 (Page 557)

1. Yes   **3.** Yes   **5.** No   **7.** No   **9.** No   **11.** No
13. Not a transition diagram
15. $\begin{bmatrix} .6 & .20 & .20 \\ .9 & .02 & .08 \\ .4 & 0 & .6 \end{bmatrix}$ **17.** Yes   **19.** Yes   **21.** No

23. $\begin{bmatrix} \frac{6}{23}, \frac{17}{23} \end{bmatrix}$ **25.** $\begin{bmatrix} \frac{3}{11}, \frac{8}{11} \end{bmatrix}$

27. $[.4633, .1683, .3684]$ **29.** $[.4872, .2583, .2545]$

31. $A^2 = \begin{bmatrix} .23 & .21 & .24 & .17 & .15 \\ .26 & .18 & .26 & .16 & .14 \\ .23 & .18 & .24 & .19 & .16 \\ .19 & .19 & .27 & .18 & .17 \\ .17 & .2 & .26 & .19 & .18 \end{bmatrix}$

$$A^3 = \begin{bmatrix} .226 & .192 & .249 & .177 & .156 \\ .222 & .196 & .252 & .174 & .156 \\ .219 & .189 & .256 & .177 & .159 \\ .213 & .192 & .252 & .181 & .162 \\ .213 & .189 & .252 & .183 & .163 \end{bmatrix}$$

$$A^4 = \begin{bmatrix} .2205 & .1916 & .2523 & .1774 & .1582 \\ .2206 & .1922 & .2512 & .1778 & .1582 \\ .2182 & .1920 & .2525 & .1781 & .1592 \\ .2183 & .1909 & .2526 & .1787 & .1595 \\ .2176 & .1906 & .2533 & .1787 & .1598 \end{bmatrix}$$

$$A^5 = \begin{bmatrix} .21932 & .19167 & .25227 & .17795 & .15879 \\ .21956 & .19152 & .25226 & .17794 & .15872 \\ .21905 & .19152 & .25227 & .17818 & .15898 \\ .21880 & .19144 & .25251 & .17817 & .15908 \\ .21857 & .19148 & .25253 & .17824 & .15918 \end{bmatrix}; .17794$$

33. $[.389, .222, .167, .222]$ **35.** $[.802, .198]$

37. $\left[\frac{1}{4}, \frac{1}{2}, \frac{1}{4}\right]$ **39.** $[.666 \quad .333 \quad .001]$

41. (a) $[42{,}500 \quad 5000 \quad 2500]$ **(b)** $[37{,}125 \quad 8750 \quad 4125]$
(c) $[33{,}281 \quad 11{,}513 \quad 5206]$ **(d)** $[.475 \quad .373 \quad .152]$

43. (a) $\begin{bmatrix} .348 & .379 & .273 \\ .320 & .378 & .302 \\ .316 & .360 & .323 \end{bmatrix}$ **(b)** .348 **(c)** .320

45. (a) $\begin{array}{c} \\ 0 \\ 1 \\ 2 \end{array} \begin{array}{ccc} 0 & 1 & 2 \\ \begin{bmatrix} .4 & .3 & .3 \\ .4 & .3 & .3 \\ 0 & .5 & .5 \end{bmatrix} \end{array}$ **(b)** $\begin{array}{c} \\ 0 \\ 1 \\ 2 \end{array} \begin{array}{ccc} 0 & 1 & 2 \\ \begin{bmatrix} .28 & .36 & .36 \\ .28 & .36 & .36 \\ .2 & .4 & .4 \end{bmatrix} \end{array}$

(c) .36   **47.** $[0 \quad 0 \quad .102273 \quad .897727]$

Section 9.6 (Page 563)

1. (a) coast **(b)** highway **(c)** highway **(d)** coast   **3. (a)** Do not upgrade, $10,060 **(b)** Upgrade. **(c)** Do not upgrade.

5. (a)
$$\begin{array}{c} \\ \text{Overhaul} \\ \text{Don't Overhaul} \end{array} \begin{array}{cc} \text{Fails} & \text{Doesn't Fail} \\ \begin{bmatrix} -\$8600 & -\$2600 \\ -\$6000 & \$0 \end{bmatrix} \end{array}$$

(b) Don't overhaul the machine.
7. (a) No campaign **(b)** Campaign for all. **(c)** Campaign for all.
9. Environment, 15.3

Chapter 9 Review (Page 566)

1. (a) 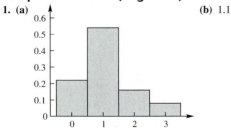 **(b)** 1.1

3. (a) 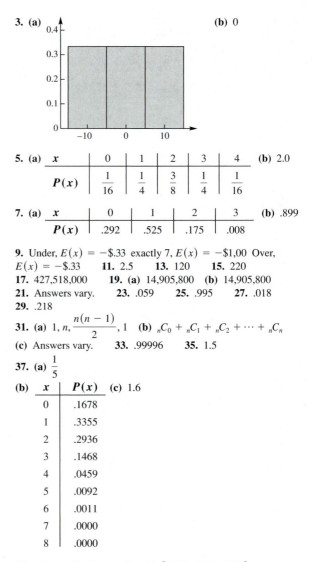 **(b)** 0

5. (a)

x	0	1	2	3	4
$P(x)$	$\frac{1}{16}$	$\frac{1}{4}$	$\frac{3}{8}$	$\frac{1}{4}$	$\frac{1}{16}$

(b) 2.0

7. (a)

x	0	1	2	3
$P(x)$	.292	.525	.175	.008

(b) .899

9. Under, $E(x) = -\$.33$ exactly 7, $E(x) = -\$1.00$ Over, $E(x) = -\$.33$ **11.** 2.5   **13.** 120   **15.** 220
17. 427,518,000   **19. (a)** 14,905,800 **(b)** 14,905,800
21. Answers vary.   **23.** .059   **25.** .995   **27.** .018
29. .218

31. (a) $1, n, \dfrac{n(n-1)}{2}, 1$ **(b)** ${}_nC_0 + {}_nC_1 + {}_nC_2 + \cdots + {}_nC_n$
(c) Answers vary.   **33.** .99996   **35.** 1.5

37. (a) $\dfrac{1}{5}$

(b)

x	$P(x)$
0	.1678
1	.3355
2	.2936
3	.1468
4	.0459
5	.0092
6	.0011
7	.0000
8	.0000

(c) 1.6

39. No   **41.** Yes   **43. (a)** $[.1515, .5635, .2850]$
(b) $[.1526, .5183, .3290]$ **(c)** $[.1509, .4697, .3795]$
45. (a) active learning **(b)** active learning **(c)** lecture, 17.5
(d) active learning, 48   **47. (c)**

Case 9 (Page 570)

1. (a) $69.01 **(b)** $79.64 **(c)** $58.96 **(d)** $82.54 **(e)** $62.88
(f) $64   **3.** Answers vary, but the answer needs to include the idea that 1, 2, and 3 are not the only events in the sample space.

Chapter 10

Section 10.1 (Page 582)

1. (a)–(b)

Interval	Frequency
15–17	4
18–20	7
21–23	8
24–26	10
27–29	7
30–32	7
33–35	3
36–38	3
39–41	1

(c)–(d)

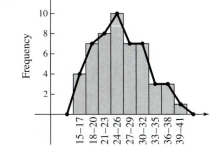

3. (a)–(b)

Interval	Frequency
0–9	2
10–19	12
20–29	14
30–39	11
40–49	7
50–59	3
60–69	0
70–79	1

(c)–(d)

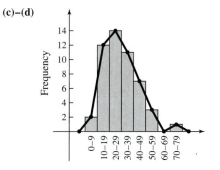

5.

Stem	Leaves
1	
1	6 6 6 7 8 8 8 9
2	0 0 0 1 1 2 2 2 3 3 3 4 4 4 4 4
2	5 5 6 6 6 7 8 8 9 9 9 9
3	0 0 1 1 1 2 2 3 4
3	5 7 7 7
4	1

Units: 4|1 = 41 years

7. Answers vary.   **9.** \$27,955   **11.** 4.8   **13.** 10.3
15. 21.2   **17.** 14.8   **19.** \$33,679   **21.** 98.5   **23.** 2
25. 65, 71   **27.** 5.8   **29.** Answers vary.   **31.** 29.1, 20–29
33. Answers vary.   **35. (a)** \$58,566,000   **(b)** \$49,347,000
(c) The high compensation for Reuben Mark   **(d)** \$60,023,000
37. 7.38, 7, there are three modes 7, 5, and 4   **39.** 55.5°F, 50.5°F
41. \$3.32, \$3.39   **43.** \$33,310   **45. (a)** about 14%
(b) about 7%   **(c)** 30–39 and 40–49

Section 10.2 (Page 592)

1. Answers vary.   **3.** 15, 6.83   **5.** 15, 6.83   **7.** 41, 13.89
9. 2240, 716.46   **11.** 45.2   **13.** $\bar{x} = 5.0876$, $s = .1087$

15. $\dfrac{3}{4}$ **17.** $\dfrac{5}{9}$ **19.** 93.75%   **21.** 11.1%

23. (a) Males: $\bar{x} = 175.25$, $s = 5.75$; Females: $\bar{x} = 165.25$, $s = 5.68$
(b) females   **(c)** males, no   **25. (a)** $\bar{x} = 6.9286$, 2001
(b) $s = 1.21$ **(c)** 4   **(d)** 7   **27. (a)** $\bar{x} = 14.8$, $s = 3.8$
(b) 10   **29. (a)** $\bar{x} = 127.71$ days, $s = 30.16$ days   **(b)** All
(c) Answers vary.

31. (a) $\dfrac{1}{3}, 2, -\dfrac{1}{3}, 0, \dfrac{5}{3}, \dfrac{7}{3}, 1, \dfrac{4}{3}, \dfrac{7}{3}, \dfrac{2}{3}$ **(b)** 2.1, 2.6, 1.5, 2.6, 2.5, .6, 1,
2.1, .6, 1.2   **(c)** 1.13   **(d)** 1.68   **(e)** −2.15, 4.41, the process is out
of control   **33.** $\bar{x} = 80.17$, $s = 12.2$ **35.** Answers vary.

Section 10.3 (Page 608)

1. The mean   **3.** Answers vary.   **5.** 45.99%   **7.** 16.64%
9. 7.7%   **11.** 47.35%   **13.** 91.20%   **15.** 1.64 or 1.65
17. −1.04   **19.** .5; .5   **21.** .889; .997   **23.** .3821
25. 5762   **27.** .2776   **29.** .0025   **31.** 5000   **33.** 642
35. 9,918   **37.** between 372 and 628 hours   **39.** 45.2 mph
41. 24.17%   **43.** Answers vary.   **45.** 665 units
47. 190 units   **49.** 15.87%   **51.** .0618   **53.** .9278
55. min = 2.3, $Q_1 = 3.8$, $Q_2 = 4.05$, $Q_3 = 6.0$, max = 8.3

57.

2003
Billions

59. min = 5.1, Q_1 = 5.4, Q_2 = 7.05, Q_3 = 8.0, max = 9.2

61.

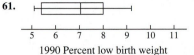

1990 Percent low birth weight

63. Yes. Explanations vary.

Section 10.4 (Page 615)
1. The number of trials and the probability of success on each trial
3. (a) .1964 **(b)** .1974 **5. (a)** .0021 **(b)** .0030 **7.** .0240
9. .9463 **11.** .0956 **13.** .7291 **15.** .1841 **17.** .0222
19. .0594 **21.** .9821 **23. (a)** .0005 **(b)** .0005 **(c)** .0000
25. (a) .0488 **(b)** .0425 **(c)** .9575 **27. (a)** .0764 **(b)** .121
(c) Yes, probability of such a result is .0049 **29.** .8643
31. (a) 1.2139×10^{-7} **(b)** essentially 0 **(c)** .7939

Chapter 10 Review (Page 618)
1. Answers vary.
3. **(a)**

Interval	Frequency
450–474	5
475–499	6
500–524	5
525–549	2
550–574	2

(b)–(c)

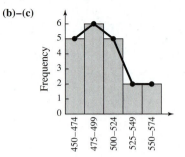

5. $\bar{x}$ = 498.9 **7.** 75.6 cm **9.** Answers vary. **11.** 72, 72
13. 2000–2999 **15.** Answers vary. **17.** range = 37,
s = 12.10 **19.** 1,117.7 calories **21.** Answers vary.
23. .6480 **25.** .0606 **27.** Answers vary. **29. (a)** .1295
(b) μ = .72, s = .796 **31. (a)** Diet A **(b)** Diet B
33. (a) $\bar{x}$ = 73.51, s = 4.46 **(b)** 1.42, −1.40 **35.** 1990:
min = 150, Q_1 = 265, Q_2 = 438, Q_3 = 1824, max = 2514 2003:
min = 285, Q_1 = 401, Q_2 = 559, Q_3 = 2028, max = 4553
37. (a) 2.28% **(b)** 15.87% **(c)** 68.26% **39. (a)** .0305
(b) .9938 **(c)** 1.000 **(d)** Normal approximation = .1056
Binomial calculation = .1326

Case 10 (Page 621)
1. z = −29.1 **3. (a)** 174 **(b)** 6.03 **(c)** −12.3 **(d)** < .004

Index of Applications

American Century Global Growth
   Fund, 505
Amortization, 283–284, 285, 287,
   289–290
Annuities due, 275, 277
Bankruptcy petitions, 576–577
Borrowing money, 323
CD rates, 266
Certificate of deposit, 267
Commercial and home loans, 434
Compound interest, 213, 216,
   256–257, 259, 261, 262–263, 264,
   265, 266, 267, 271, 272–273, 275,
   276, 277, 278, 289, 290
Consumer price indexes, 149, 290
Cost of filling a gas tank, 55
Credit cards, 219, 256, 268–269, 331
Debt settlement, 252
Deposits, 255, 273, 277, 278,
   291–292
Discount, 253, 254, 255, 289
Doubling money, 231
Effective rate, 261, 262, 266, 290
Federal income tax, 114
Finance charge, 54–55
Fund-raising, 410
Future value of an annuity, 251,
   270–271, 276–277, 289
Gross federal debt, 54
Income tax, 255
Inflation, 264–265, 290
Inheritance, 266, 267, 286
Insurance policy, 519
Interest paid, 250, 251, 266, 285
Interest rate, 68, 145–146, 252, 253,
   255, 258–259, 265, 285
Investment, 216, 255, 258–259, 266,
   267, 276, 290, 293–294, 310, 312,
   323, 324, 434, 468, 567
Laffer curve model, 187, 198
Lottery jackpot, 278, 286
Maturity value, 251
Maximum allowable IRA
   contribution, 148
Median income, 310
Money losing value, 213
Money market, 255, 267
Monthly payment, 278, 280, 281,
   282–283, 285, 286, 291, 292
Mortgage defaults, 500

National debt, 222
Paying off a loan, 255, 266,
   281–282, 286, 287, 291
Pension Benefit Guaranty
   Corporation, 64
Pension experts, 291
Periodic payments, 274–275
Pledging to a charity, 291
Present value, 252, 255, 264, 266,
   280, 285, 289
Profit-sharing, 290
Real estate firm investment, 68
Retirement, 47, 277, 278, 286, 290,
   292
Savings account, 277
Scholarship fund, 290–291
Simple interest, 250, 254–255, 256,
   257–258, 289
Sinking funds, 273, 274–275, 277,
   278, 289
Spending on education, 505
State income tax, 135, 137, 148
Stock prices, 449, 480–482
Stock returns, 255, 619
Student loans, 286
Student tuition, 255
Thirty-year mortgage, 81, 145–146
Treasury bonds, 390
Types of bills, 361
Value of property, 51
Yield to maturity, 293

Geometry

Architectural arches, 203–204
Area of a rectangular region,
   61–62
Area of a square, 138
Dimensions of a box-shaped object,
   56
Dimensions of a rectangular object,
   56, 62
Length of the side of a triangle, 56
Maximum area, 179, 201
Strip of floor around a rug, 64
Surface area of a right circular cone,
   82
Surface area of a right square
   pyramid, 82
Volume of the frustum of a square
   pyramid, 19

Health

AIDS, 64, 500
Alzheimer's disease, 106, 122, 324
Anticlot drug, 151–152
Blood alcohol level, 114
Blood antigens, 459
Blood types, 594
Body Mass Index (BMI), 10
Breakdown of foods, 323–324
Breast cancer, 494, 499–500
Cholesterol levels, 604–605, 608
Cigarette consumption, 215–216
Computer Aided Tomography
   (CAT), 312–313
Consumer price index for medical
   care, 149
Cooking, 439–440
Cost of employee's health insurance,
   32
Deaths from heart disease, 104–105,
   137, 221, 310, 332
Diet, 376, 389, 421
Drug effectiveness, 240, 329–330
Drugs effective in treating disease,
   533
Flu virus, 231
HIV/AIDS infection, 491–492
Head and neck injuries among
   hockey players, 362
Health care expenditures, 54, 107,
   122, 178, 201, 214, 219
Health insurance, 499
Heights, 593
Hepatitis, 499
Hospitals in the United States,
   214–215
Infant birth weight, 478, 610
Kidney transplants, 44, 242
Life expectancy, 18, 106, 114,
   126–127, 201
Live births and infant mortality, 345
Lung cancer and smoking, 494
Measuring cardiac output, 187
Medical costs, 68
Medical diagnosis, 507
Medical experiment, 469
Medicare costs, 124, 178
Number of deaths in the U.S., 82–83
Nursing home care, 356

Transportation

Index

COMPOUND AMOUNT

If P dollars is deposited for n time periods at a compound interest rate i per period, the **compound amount** A is

$$A = P(1 + i)^n.$$

The **future value** S of an ordinary annuity of n payments of R dollars each at the end of consecutive interest periods with interest compounded at a rate i per period is

$$S = R\left[\frac{(1 + i)^n - 1}{i}\right] \quad \text{or} \quad S = R \cdot s_{\overline{n}|i}.$$

The **matrix** version of the elimination method uses the following **matrix row operations** to obtain the augmented matrix of an equivalent system. They correspond to using elementary row operations on a system of equations.

1. Interchange any two rows.
2. Multiply each element of a row by a nonzero constant.
3. Replace a row by the sum of itself and a constant multiple of another row in the matrix.

To obtain the **inverse matrix** A^{-1} for any $n \times n$ matrix A for which A^{-1} exists, follow these steps.

1. Form the augmented matrix $[A \mid I]$, where I is the $n \times n$ identity matrix.
2. Perform row operations on $[A \mid I]$ to get a matrix of the form $[I \mid B]$.
3. Matrix B is A^{-1}.

UNION RULE

For any events E and F from a sample space S,

$$P(E \cup F) = P(E) + P(F) - P(E \cap F).$$

PROPERTIES OF PROBABILITY

Let S be a sample space consisting of n distinct outcomes $s_1, s_2, \ldots, s_n$. An acceptable probability assignment consists of assigning to each outcome s_i a number p_i (the probability of s_i) according to these rules.

1. The probability of each outcome is a number between 0 and 1.

$$0 \le p_1 \le 1, \quad 0 \le p_2 \le 1, \quad \ldots, \quad 0 \le p_n \le 1.$$

2. The sum of the probabilities of all possible outcomes is 1.

$$p_1 + p_2 + p_3 + \cdots + p_n = 1.$$

BAYES' FORMULA

For any events E and $F_1, F_2, \ldots, F_n$, from a sample space S, where $F_1 \cup F_2 \ldots \cup F_n = S$,

$$P(F_i \mid E) = \frac{P(F_i) \cdot P(E \mid F_i)}{P(F_1) \cdot P(E \mid F_1) + \cdots + P(F_n) \cdot P(E \mid F_n)}.$$

PERMUTATIONS

The number of permutations of n elements taken r at a time, where $r \leq n$, is

$$_nP_r = \frac{n!}{(n-r)!}.$$

COMBINATIONS

The number of combinations of n elements taken r at a time, where $r \leq n$, is

$$_nC_r = \binom{n}{r} = \frac{n!}{(n-r)!\,r!}.$$

BINOMIAL PROBABILITY

If p is the probability of success in a single trial of a binomial experiment, the probability of x successes and $n - x$ failures in n independent repeated trials of the experiment is

$$_nC_r p^x (1-p)^{n-x}.$$

MEAN

The mean of the n numbers, $x_1, x_2, x_3, \ldots, x_n$, is

$$\bar{x} = \frac{x_1 + x_2 + \cdots + x_n}{n} = \frac{\Sigma(x)}{n}.$$

SAMPLE STANDARD DEVIATION

The standard deviation of a sample of n numbers, $x_1, x_2, x_3, \ldots, x_n$, with mean $\bar{x}$, is

$$s = \sqrt{\frac{\Sigma(x - \bar{x})^2}{n-1}}.$$

BINOMIAL DISTRIBUTION

Suppose an experiment is a series of n independent repeated trials, where the probability of a success in a single trial is always p. Let x be the number of successes in the n trials. Then the probability that exactly x successes will occur in n trials is given by

$$_nC_r p^x (1-p)^{n-x}.$$

The mean μ and variance σ^2 of a binomial distribution are, respectively,

$$\mu = np \quad \text{and} \quad \sigma^2 = np(1-p).$$

The standard deviation is

$$\sigma = \sqrt{np(1-p)}.$$

The **derivative** of the function f is the function denoted f' whose value at the number x is defined to be the number

$$f'(x) = \lim_{h \to 0} \frac{f(x + h) - f(x)}{h},$$

provided this limit exists.

RULES FOR DERIVATIVES

Assume all indicated derivatives exist.

Constant Function   If $f(x) = k$, where k is any real number, then

$$f'(x) = 0.$$

Power Rule   If $f(x) = x^n$, for any real number n, then

$$f'(x) = n \cdot x^{n-1}.$$

Constant Times a Function   Let k be a real number. Then the derivative of $y = k \cdot f(x)$ is

$$y' = k \cdot f'(x).$$

Sum or Difference Rule   If $y = f(x) \pm g(x)$, then

$$y' = f'(x) \pm g'(x).$$

Product Rule   If $f(x) = g(x) \cdot k(x)$, then

$$f'(x) = g(x) \cdot k'(x) + k(x) \cdot g'(x).$$

Quotient Rule   If $f(x) = \dfrac{g(x)}{k(x)}$, and $k(x) \neq 0$, then

$$f'(x) = \frac{k(x) \cdot g'(x) - g(x) \cdot k'(x)}{[k(x)]^2}.$$

Chain Rule   Let $y = f[g(x)]$. Then

$$y' = f'[g(x)] \cdot g'(x).$$

Chain Rule (Alternative Form)   If y is a function of u, say $y = f(u)$, and if u is a function of x, say $u = g(x)$, then $y = f[g(x)]$, and

$$\frac{dy}{dx} = \frac{dy}{du} \cdot \frac{du}{dx}.$$

Generalized Power Rule   Let u be a function of x, and let $y = u^n$ for any real number n. Then

$$y' = n \cdot u^{n-1} \cdot u'.$$

Exponential Function   If $y = e^{g(x)}$, then

$$y' = g'(x) \cdot e^{g(x)}.$$

Natural Logarithmic Function   If $y = \ln |g(x)|$, then

$$y' = \frac{g'(x)}{g(x)}.$$